Advances in Intelligent and Soft Computing

168

Editor-in-Chief

Prof. Janusz Kacprzyk
Systems Research Institute
Polish Academy of Sciences
ul. Newelska 6
01-447 Warsaw
Poland
E-mail: kacprzyk@ibspan.waw.pl

For further volumes:
http://www.springer.com/series/4240

David Jin and Sally Lin (Eds.)

Advances in Computer Science and Information Engineering

Volume 1

Editors
David Jin
Wuhan Section of ISER Association
Wuhan
China

Sally Lin
Wuhan Section of ISER Association
Wuhan
China

ISSN 1867-5662
e-ISSN 1867-5670
ISBN 978-3-642-30125-4
e-ISBN 978-3-642-30126-1
DOI 10.1007/978-3-642-30126-1
Springer Heidelberg New York Dordrecht London

Library of Congress Control Number: 2012937219

© Springer-Verlag Berlin Heidelberg 2012
This work is subject to copyright. All rights are reserved by the Publisher, whether the whole or part of the material is concerned, specifically the rights of translation, reprinting, reuse of illustrations, recitation, broadcasting, reproduction on microfilms or in any other physical way, and transmission or information storage and retrieval, electronic adaptation, computer software, or by similar or dissimilar methodology now known or hereafter developed. Exempted from this legal reservation are brief excerpts in connection with reviews or scholarly analysis or material supplied specifically for the purpose of being entered and executed on a computer system, for exclusive use by the purchaser of the work. Duplication of this publication or parts thereof is permitted only under the provisions of the Copyright Law of the Publisher's location, in its current version, and permission for use must always be obtained from Springer. Permissions for use may be obtained through RightsLink at the Copyright Clearance Center. Violations are liable to prosecution under the respective Copyright Law.
The use of general descriptive names, registered names, trademarks, service marks, etc. in this publication does not imply, even in the absence of a specific statement, that such names are exempt from the relevant protective laws and regulations and therefore free for general use.
While the advice and information in this book are believed to be true and accurate at the date of publication, neither the authors nor the editors nor the publisher can accept any legal responsibility for any errors or omissions that may be made. The publisher makes no warranty, express or implied, with respect to the material contained herein.

Printed on acid-free paper

Springer is part of Springer Science+Business Media (www.springer.com)

Preface

In the proceeding of CSIE2012, you can learn much more knowledge about Computer Science and Information Engineering all around the world. The main role of the proceeding is to be used as an exchange pillar for researchers who are working in the mentioned field. In order to meet high standard of Springer, the organization committee has made their efforts to do the following things. Firstly, poor quality paper has been refused after reviewing course by anonymous referee experts. Secondly, periodically review meetings have been held around the reviewers about five times for exchanging reviewing suggestions. Finally, the conference organization had several preliminary sessions before the conference. Through efforts of different people and departments, the conference will be successful and fruitful.

During the organization course, we have got help from different people, different departments, different institutions. Here, we would like to show our first sincere thanks to publishers of Springer, AISC series for their kind and enthusiastic help and best support for our conference.

In a word, it is the different team efforts that they make our conference be successful on May 19–20, Zhengzhou, China. We hope that all of participants can give us good suggestions to improve our working efficiency and service in the future. And we also hope to get your supporting all the way. Next year, In 2013, we look forward to seeing all of you at CSIE2013.

March 2012 CSIE2012 Committee

Preface

Committee

Honor Chairs

Prof. Chen Bin	Beijing Normal University, China
Prof. Hu Chen	Peking University, China
Chunhua Tan	Beijing Normal University, China
Helen Zhang	University of Munich, China

Program Committee Chairs

Xiong Huang	International Science & Education Researcher Association, China
Li Ding	International Science & Education Researcher Association, China
Zhihua Xu	International Science & Education Researcher Association, China

Organizing Chair

ZongMing Tu	Beijing Gireida Education Co. Ltd, China
Jijun Wang	Beijing Spon Technology Research Institution, China
Quanxiang	Beijing Prophet Science and Education Research Center, China

Publication Chair

Song Lin	International Science & Education Researcher Association, China
Xiong Huang	International Science & Education Researcher Association, China

International Committees

Sally Wang	Beijing Normal University, China
LiLi	Dongguan University of Technology, China
BingXiao	Anhui University, China
Z.L. Wang	Wuhan University, China
Moon Seho	Hoseo University, Korea
Kongel Arearak	Suranaree University of Technology, Thailand
Zhihua Xu	International Science & Education Researcher Association, China

Co-sponsored by

International Science & Education Researcher Association, China
VIP Information Conference Center, China
Beijing Gireda Research Center, China

Reviewers of CSIE2012

Z.P. Lv	Huazhong University of Science and Technology
Q. Huang	Huazhong University of Science and Technology
Helen Li	Yangtze University
Sara He	Wuhan Textile University
Jack Ma	Wuhan Textile University
George Liu	Huaxia College Wuhan Polytechnic University
Hanley Wang	Wuchang University of Technology
Diana Yu	Huazhong University of Science and Technology
Anna Tian	Wuchang University of Technology
Fitch Chen	Zhongshan University
David Bai	Nanjing University of Technology
Y. Li	South China Normal University
Harry Song	Guangzhou Univeristy
Lida Cai	Jinan University
Kelly Huang	Jinan University
Zelle Guo	Guangzhou Medical College
Gelen Huang	Guangzhou University
David Miao	Tongji University
Charles Wei	Nanjing University of Technology
Carl Wu	Jiangsu University of Science and Technology
Senon Gao	Jiangsu University of Science and Technology
X.H. Zhan	Nanjing University of Aeronautics
Tab Li	Dalian University of Technology (City College)
J.G. Cao	Beijing University of Science and Technology
Gabriel Liu	Southwest University
Garry Li	Zhengzhou University
Aaron Ma	North China Electric Power University
Torry Yu	Shenyang Polytechnic University
Navy Hu	Qingdao University of Science and Technology
Jacob Shen	Hebei University of Engineering

Contents

Design on Architecture of Internet of Things

GuiPing Dai[1] and Yong Wang[2]

[1] College of Electronic Information and Control Engineering
Beijing University of Technology
Beijing, China
[2] College of Computer Science and Engineering
Beijing University of Technology
Beijing, China
{daigping,wangy}@bjut.edu.cn

Abstract. Internet of Things (IoT) gains more and more attentions in a global wide as a new information technology. But the architecture of IoT is still opaque now. In this paper, we do some design works on the architecture of IoT, including layered architecture model and system structure of IoT. We wish that our works can be helpful to the clarifications of concepts and architecture of IoT.

Keywords: Internet of Things, Architecture, Layered Model, System Structure.

1 Introduction

Comparing to Internet which is used to link computing devices, Internet of Things (IoT) can be used to link any thing and is more ubiquitous. As a new information technology, IoT has gained many attentions in many countries and districts, such as European communities[1]. In China, IoT has also gained more interests and became a country strategy[2-4].

Though IoT maybe origin from some related technologies, such as ubiquitous networking, RFID things identification technology and sensor networks, in 2005 ITU clearly put forward the concept of things, analyzes the characteristics of things, challenges and opportunities in a research reports of IoT[5]. More accepted that IoT is an integrated network, that is, IoT is unlikely to become another peer and global network with the Internet, but to use the Internet to connect a variety of things.

As we all know, the layered structure of computer network architecture model, such as the ISO OSI model and the Internet TCP/IP model, played an important role to the occurrence and development of computer networks. Similarly, the IoT architecture model will contribute to the development of things focused on the vital role to play.

By analyzing research and development results of the existing architecture of IoT to find their shortcomings, layered architecture model and system architecture for IoT are designed. The architecture of the results has a clear hierarchy of features and gives some key technologies to appropriate positions, and points out that the future development of IoT needs to focus on solving some new key technologies, and also clarifies the relationship between IoT and the Internet.

D. Jin and S. Lin (Eds.): Advances in CSIE, Vol. 1, AISC 168, pp. 1–7.
springerlink.com © Springer-Verlag Berlin Heidelberg 2012

2 State of the Art

In view of guiding role of architecture on IoT, there are many of the studies and standardization on the architecture of IoT[6-14]. Three of the more influential is the ITU Y.2002 high-level model for architecture of ubiquitous networking[12], RFID-based EPC architecture[13], and the common WSN architecture[14], a brief introduction to these three systems structure is following.

2.1 High-Level Model for Architecture of Ubiquitous Networking

ITU Y.2002 high-level model for architecture of ubiquitous networking[12] is as Fig.1 showing.

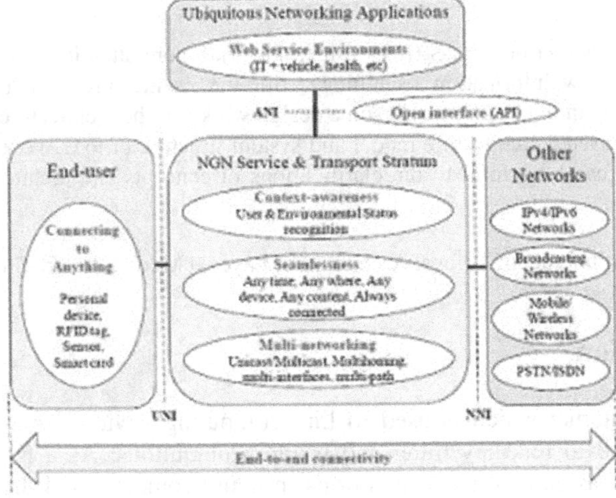

Fig. 1. High-Level Model for Architecture of Ubiquitous Networking.

The model illustrates:

- Personal device, RFID tag, sensor, smart card and other things connect to the NGN (Next Generation Network) through UNI (User-Network-Interface).
- IT + vehicle, health and other applications connect to the NGN through ANI (Application-Network-Interface).
- IPv4/IPv6 networks, broadcasting networks, mobile/wireless networks, PSTN/ISDN and other networks connect to the NGN through NNI (Network-Network-Interface).

2.2 Architecture of EPC Networks

The architecture of RFID-based EPC networks is as Fig.2 showing.

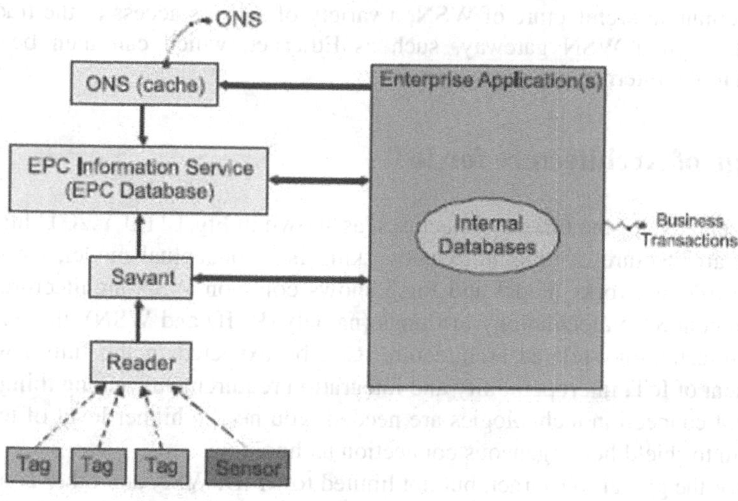

Fig. 2. Architecture of EPC Networks.

RFID tag reader reads the tags of things. And the tag information is stored into EPCIS (EPC Information Service) or is delivered to enterprise applications through EPC middleware, such as Savant[15], by using a "pull" mode or a "push" mode. The information for things can be searched through an ONS (Object Name Service).

2.3 Common Architecture of WSN

The common architecture of WSN is as Fig.3 showing.

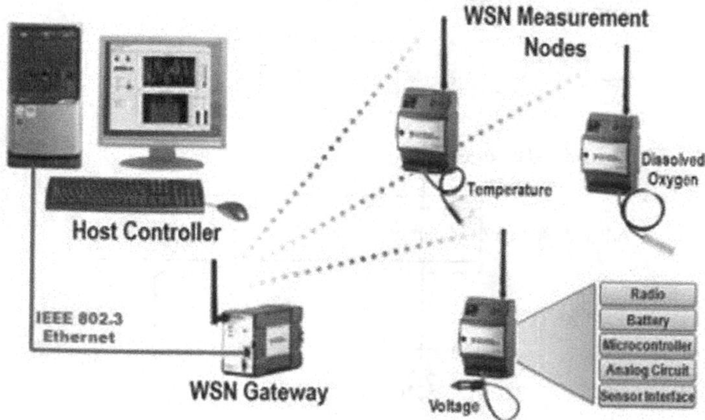

Fig. 3. Common Architecture of WSN

In this common architecture of WSN, a variety of sensors access to the traditional network through a WSN gateway, such as Ethernet, which can then be further connected to the Internet.

3 Design of Architecture for IoT

The representative of the three architectures, as shown in Fig.1 ITU Y.2002 high-level model for architecture of ubiquitous networking is a conceptual model, while Fig.2 shows the EPC networks model and Fig.3 shows common WSN architecture are all for special connection technology of things (namely, RFID and WSN), that is, are all connection-technology-related architecture. Can be expected in the future with the development of IoT, interoperability and integration requirements among things based on different connection technologies are needed, you need a higher level of technical mechanism to shield heterogeneous connection technologies.

Items for the global IoT target, but not limited to RFID, WSN and other connection technologies, in particular, to clarify the relationship between the IoT and the Internet, we give layered architecture model and system architecture for future IoT.

3.1 Layered Architecture Model of IoT

The development of IoT should be a great success leveraging the Internet, through Internet connectivity up to a variety of things and through the Internet to provide a variety of IoT applications.

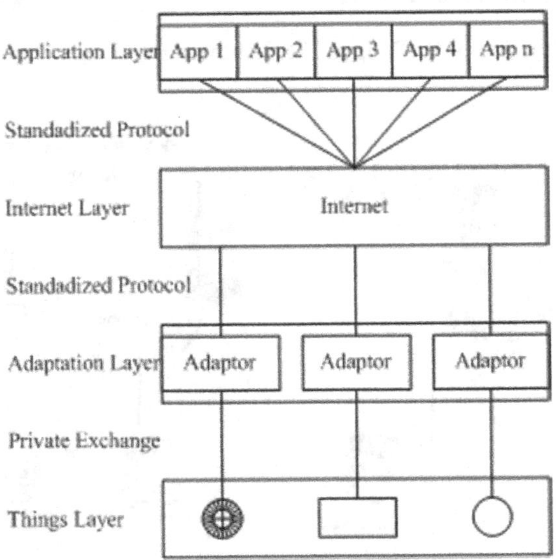

Fig. 4. Layered Architecture Model of IoT.

We believe that the general sense of IoT in the concept exists of four levels: things layer, adaptation layer, Internet layer and application layer, as shown in Fig.4.

Things layer locates at the bottom of the layered architecture in Fig.4, including a variety of things with different physical shapes and sizes, and different technical mechanisms, showing different physical, mechanical, electrical characteristics. These characteristics of things are the origination of information for things. Higher-level application exchanges information to achieve the ultimate purpose of the information integration needing a unified, standardized information representation processing. Note that to connect these things into IoT, their original shapes and technical mechanisms should not be changed or should be changed as little as possible.

Adaptation layer is located above the things layer, to offer a variety of adapters for things to connect to the IoT. Because of the different types of things, even things of the same species from different manufacturers, adaptor may be different. Although do not rule out several things for the same adapter developed, such as RFID and WSN and other sensing devices, in general the adapter depends on the things to adapt.

Between the adapter and things are private things interactive technology to be limited by a variety of things. The so-called private exchange technology is relative to the development of Internet protocols and switching technology, and does not preclude the use of standardized exchange of technology, such as standardized WSN.

Internet layer is located above the adaptation layer and uses the main body of the existing Internet, including its physical network, system software and open communication protocols and sophisticated technical mechanisms. Internet layer provided in exchange for the information of things as a transmission carrier.

The link to adapter layer and the Internet layer should be a standard communication protocol, known as things management protocol, used to exchange things information between things and things, things and IoT applications and even IoT applications and IoT applications. From the perspective of the Internet, things management protocol is an Internet application layer protocol. On the one hand, adapters exchange information with things through proprietary interactive technology, and on the other hand, they need to be able to understand the things management protocol, to be able to adopt the protocol to exchange information with IoT applications or other things (adaptors). To achieve the purpose exchange, the definition of information of things (including definition methods, information structure and definitions of concrete information) to be exchanged needs to understand by both sides, or even needs to be standardized, which requires intervention of standardization organizations, government departments or industry associations.

Application layer is located at the top of layered architecture, providing a variety of IoT applications such as monitoring the status of things, operational control, public inquiries and other value-added services to use, interconnection and collaboration service (which can be a focal point for the interconnection and collaboration between things, and even Ad Hoc collaboration) between things and things. In application layer, applications also need to understand things management protocol. In the application layer, depending on a variety of requirements for things, application middleware may occur.

3.2 System Structure of IoT

Based on the layered architecture model shown in Fig.4, we design the system architecture of IoT shown in Fig.5.

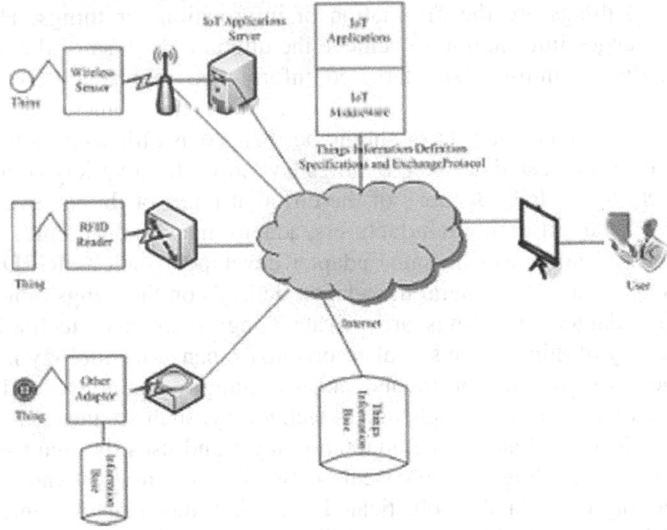

Fig. 5. System Structure of IoT.

The system structure of IoT shown in Fig.5, the things access to the Internet through various access technologies such as wireless sensor networks, RFID technology. Things information providers can maintain their own private repository of things, or can also build larger information centers for things; information providers can provide their own IoT applications, or can also be integrated into an application center. Things pass information on a uniform definition, and even standardized things management protocol that can understand both sides. The definition of things information and things management protocol are all connection-technology-independent protocols.

4 Conclusion

In the early stage of development of IoT, industry and researchers dedicated to connection technologies and the architecture of IoT is not yet clear. With the development of IoT, especially integration of information for things based on different connection technologies, higher-level representation of the information for things and things management protocol are needed to shield the heterogeneity of the different connection technologies. And then the architecture of IoT will gradually mature.

In this paper, based on the existing architecture of IoT and aimed on the basic needs of interconnection of a variety of things, to clarify the relationship between the IoT and the Internet, the layered architecture model and system architecture of IoT are designed.

Acknowledgments. This research was partly supported by Beijing Municipal Education Commission Projects grants JC007011201004, Beijing Municipal Education Colleges and Universities to Deepen Talents Scheme, and CSC Projects in China.

References

1. Commission of the European communities: Internet of Things-an Action Plan for Europe (2009), http://ec.europa.eu/information-society/policy/rfid/documents/commiot2009.pdf
2. Wen, J.B.: 2010 Report on the Work of the Government (2010), http://www.gov.cn/2010lh/content-1555767.htm
3. 2010 China Conference for IoT (2010), http://www.iotconference.com
4. The State Council of the People's Republic of China: State Council on Accelerating the Development of New Industries and Development of a Strategic Decision (2010), http://www.gov.cn/zwgk/2010-10/18/content-1724848.htm
5. International Telecommunication Union: The Internet of Things. ITU, Geneva (2005)
6. Shen, S.B., Fan, Q.L., Zong, P.: Study on the Architecture and Associated Technologies for Internet of Things. Journal of Nanjing University of Posts and Telecommunications 29(6), 1–11 (2009)
7. Sun, Q.B., Liu, J., Li, S.: Internet of Things: Summarize on Concepts, Architecture, and Key Technology Problem. Journal of Beijing University of Posts and Telecommunications 33(3), 1–9 (2010)
8. Huang, Y.H., Li, G.Y.: Descriptive Models for Internet of Things. In: Proceedings of the 2010 International Conference on Intelligent Control and Information Processing, pp. 483–486. IEEE, Dalian (2010)
9. Wu, M., Lu, T.J., Ling, F.Y.: Research on the Architecture of Internet of Things. In: Proceedings of the 3rd International Conference on Advanced Computer Theory and Engineering, pp. 484–487. IEEE, Chengdu (2010)
10. Miao, Y., Bu, Y.X.: Research on the Architecture and Key Technology of Internet of Things (IoT) Applied on Smart Grid. In: Proceedings of the 2010 International Conference on Advances in Energy Engineering, pp. 69–72. IEEE, Beijing (2010)
11. Castellani, A.P., Bui, N., Casari, P.: Architecture and Protocols for the Internet of Things: a Case Study. In: Proceedings of the 8th IEEE International Conference on Pervasive Computing and Communications Workshops, pp. 678–683. IEEE, Mannheim (2010)
12. ITU Y.2002: Overview of Ubiquitous Networking and of Its Support in NGN, ITU-T Recommendation (2009)
13. EPC: Architecture of EPC Networks (2011), http://www.gs1.org/epcglobal/
14. National Instruments: Common Architecture of WSN (2011), http://zone.ni.com/devzone/cda/tut/p/id/8707
15. Savant: Savant Middleware (2011), http://www.savantav.com/

XML-Based Structural Representing Method for Information of Things in Internet of Things

GuiPing Dai[1] and Yong Wang[2]

[1] College of Electronic Information and Control Engineering
Beijing University of Technology
Beijing, China
[2] College of Computer Science and Engineering
Beijing University of Technology
Beijing, China
{daigping,wangy}@bjut.edu.cn

Abstract. The development of Internet of Things has gained more attentions in many countries, but Internet of Things is still focusing on connection-technology-related developments. With the development of Internet of Things, especially integration of things' information based on different connection technologies is more required, higher level connection-technology-independent protocols will emerge to shield different connection technologies. Unified representation of things' information is one core technology of them. A representation method of things' information based XML is designed, including definition of structure for things' information and concrete definition of things' information. Finally, the implementation structure of this method is given.

Keywords: Internet of Things, Representation for Things Information, XML, Architecture.

1 Introduction

Internet of Things (IoT) brought a new wave of information technology wave, including many countries and regions concerned, such as EU[1]. China government come to realize the importance of IoT[2-4]. in October 2010, the China State Council issued the "State Council on accelerating the development of new industries and the development of strategic decision"[4], IoT was included in the new generation of information technology industry directories, IoT has officially become a national strategy in China.

In 2005, the ITU report of IoT[5], the concept of things is clearly stated, and analyze its characteristics, challenges and development opportunities. Generally believed that the IoT is an integrated concept that IoT ia an integrated network that a variety of things connection technologies, such as RFID, two-dimensional code and other identification technologies, sensor technologies, such as wireless sensor networks, are integrated with the current Internet as well as with the next generation Internet. And IoT are unlikely to become another independent, global network. It will make full use of the Internet as a vehicle to transport information of things.

D. Jin and S. Lin (Eds.): Advances in CSIE, Vol. 1, AISC 168, pp. 9–15.
springerlink.com © Springer-Verlag Berlin Heidelberg 2012

Early in the development of things, the existing research is still concentrated in the things connection technologies, and top level connection-technology-related R & D, such as RFID-based logistics system has a number of countries in the world with a more mature application. But with the development of IoT, especially things based on different connection technologies, information integration needs of an increasingly urgent need to connection-technology-independent mechanism to shield the lower technical heterogeneous connection technologies, and representation of things information that belongs a key technology to them.

In this paper, eXtensible Markup Language (XML)[10-11] is used as a representation tool for things information. The structure of things information and concrete things information defined by specific things of soft drinks that has been demonstrated, and gives the Java-based toolkit for the design and implementation of the things information representation method.

2 Architecture of IoT and Information Representation for Things

Existing research and development of IoT although includes both the architecture and key technology research mechanisms, work of standardization organization and the application of a more mature industry, but the architecture of IoT is not yet clear. Mostly existing technologies are connection-technology-related, and even global goal of interoperability of IoT is far from achieved. Such as the ITU high-level architecture model of ubiquitous networking[6] is a tight conceptual model, and RFID-based EPC network model[7] and the common WSN architecture model[8] are connection-technology-related.

Focusing on the future of IoT to achieve interoperability and integration among all the things based on heterogeneous connection technologies, we believe that the general sense of the concept of IoT in four layers: things layer, adaptation layer, the Internet layer and application layer[9], shown in Fig. 1.

Things layer locates at the bottom of the layered architecture in Fig.1, including a variety of things with different physical shapes and sizes, and different technical mechanisms, showing different physical, mechanical, electrical characteristics. Adaptation layer is located above the things layer, to offer a variety of adapters for things to connect to the IoT. Because of the different types of things, even things of the same species from different manufacturers, adaptor may be different. Between the adapter and things are private things interactive technology to be limited by a variety of things. The so-called private exchange technology is relative to the development of Internet protocols and switching technology, and does not preclude the use of standardized exchange of technology, such as standardized WSN. Internet layer is located above the adaptation layer and uses the main body of the existing Internet, including its physical network, system software and open communication protocols and sophisticated technical mechanisms. The link to adapter layer and the Internet layer should be a standard communication protocol, known as things management protocol, used to exchange things information between things and things, things and IoT applications and even IoT applications IoT applications. Application layer is located at the top of layered architecture, providing a variety of IoT applications such as monitoring the status of things, operational control, public inquiries and other

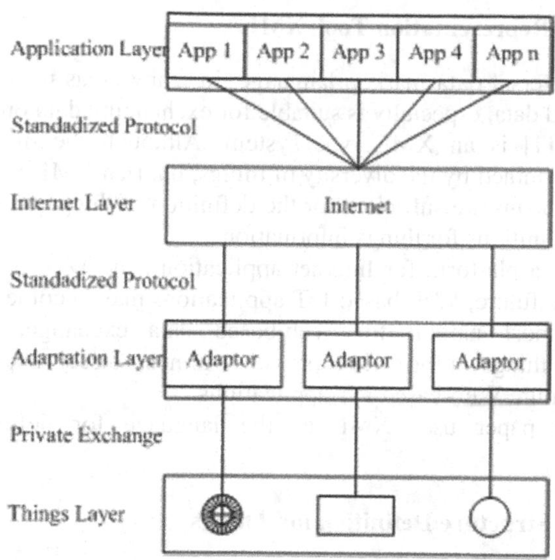

Fig. 1. Layered Architecture Model of IoT.

value-added services to use, interconnection and collaboration service (which can be a focal point for the interconnection and collaboration between things, and even Ad Hoc collaboration) between things and things.

Accessing to a variety of things to achieve interoperability between things and things, between applications and things, and between applications and applications to exchange things information is the goal of IoT. The exchanging of things information is a problem of information representation. From a technical perspective, the representation of information belongs to the representation layer in the computer network layered model.

To achieve interoperability between the sender and the receiver, the two parties must exchange information with a unified understanding of things information, including defined methods and tools for information, information structure, the definition of specific things information and also information encoding and decoding rules. To achieve a wide range of information integration and application, and even every things information needs to be standardized which requires government departments, standardization organizations or trade associations to be involved.

3 XML-Based Structural Representation Method for Thing Information

Representation of information includes the selection of representation methods, definitions of the structure of things information, and definitions of concrete things information.

3.1 Selection for Representation Tool–XML

XML[10] as a universal data markup language, in many areas to be widely used to represent structured data, especially is suitable for exchanging data on the Web.

XML Schema[11] is an XML type system. Although the diversity of things information is determined by the diversity of things, but rich XML Schema data types and custom mechanisms are sufficient for the definition of things information and are fully capable of definitions for things information.

In particular, as a platform for Internet applications, the Web is one of the most widely used. In the future, Web-based IoT applications may become the mainstream. The XML is almost natural for Web-based data exchange, so XML-based representation of things information provides almost direct application without modification for future Web-based IoT applications.

Therefore, this paper uses XML as the language for definition of things information.

3.2 Information Structure Definition for Things

In determining the XML as a tool for representation of things information, the data structures of things information are required to determine. Also relationship information between the various things data elements, various categories of things information defined in the data structure, and extension mechanisms need to be provided.

Structure definition of things information is related to taxonomy of things, such as the functional classification of things. The same kind of things have information of common interest, while the same kind of things from different manufacturers can also provide their own specific kind of information, which requires standardization of things information and also providing adequate extension mechanisms.

We envision a grand blueprint for the future, all of the things are to achieve uniform classification. This classification can be done by the national government departments, or even completed by the international standardization organization. Completed classification of different types of things information can be standardized by the functional departments, industry associations or standardization organization. We select things information as a tree data structure, such as information structure of China's things may be shown in Fig.2.

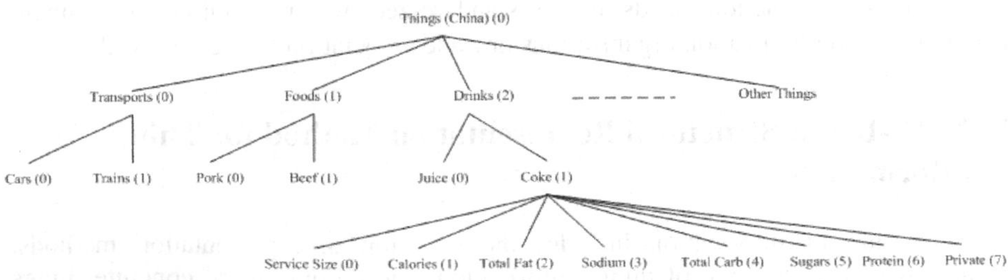

Fig. 2. Structure for Information of Things.

Object-oriented technology is adopted to represent things information. In the tree structure in Fig.2, all the things information (tree nodes) are treated as objects. Information object has a name, but also has a corresponding number as identification. For example: Things (China).Drinks.Coke.Sugars is the name of the sugar information of cola, but it also has a digital ID: 0.2.1.5, this identity is called an object identifier.

Below Things (China).Drinks.Coke.Private object, the manufacturer of various cola can define their own private information. The definition for the things information provides an extended way.

3.3 Definitions for Things Information

Each things information is an information object, as Fig.2 shows a node in the tree. Before definition of each information object, we first abstract information object metadata items.

Items of information object's metadata are shown in Table 1.

Table 1. Meta Data for Information Object of Thing

Name	Description
NAME	name of thing information object
TYPE	data type of thing information object, it should be a date type of XML
ACCESS	access of thing information object, its value may be read-only, read-write and write-only
DESCRIPTION	description text of thing information object
DEFAULTVALUE	default value of thing information object

Items shown in Fig.2 define a unified information structure based on things information objects and can be defined by XML Schema, shown inFig.3, is stored as a file *thinginformationobject.xsd.*

```
<?xml version="1.0" encoding="ISO-8859-1" ?>
    <xs:schema xmlns:xs=http://www.w3.org/2001/XMLSchema >
    <xs:element name="ThingInformationObject">
      <xs:complexType>
        <xs:sequence>
          <xs:element name="NAME" type="xs:string"/>
          <xs:element name="TYPE" type="xs:string"/>
          <xs:element name="ACESS" type="xs:string"/>
          <xs:element name="DESCRIPTION" type="xs:string"/>
          <xs:element name="DEFAULTVALUE" type="xs:string"/>
          <xs:element name="OID" type="xs:string"/>
          <xs:element name="VALUE" type="xs:string"/>
        </xs:sequence>
      </xs:complexType>
    </xs:element>
</xs:schema>
```

Fig. 3. XML Schema Definition for Information of Coke.

Things (China).Drinks.Coke. Sugars information object definition is instanced by use of XML Scehma shown in Fig.4.

```
<?xml version="1.0" encoding="ISO-8859-1"?>
 <Sugars type="ThingInformationObject"
  xs:schema xmlns:xs=http://www.w3.org/2001/XMLSchema
  xmlns:xsi="http://www.w3.org/2001/XMLSchema-instance"
    xsi:noNamespaceSchemaLocation="thinginformationobject.xsd">
    <NAME>Things (China).Drinks.Coke.Sugars</NAME>
    <TYPE>xs:float</TYPE>
    <ACCESS>read-only</ACCESS>
    <DESCRIPTION>The sugars contained in the Coke product</DESCRIPTION>
    <DEFAULTVALUE>46</DEFAULTVALUE>
    <OID>0.2.1.5</OID>
    <VALUE>46<VALUE>
 </Sugars>
```

Fig. 4. XML Instance for Information of Coke.

4 Implementation for XML-Based Representation Method of Things Information

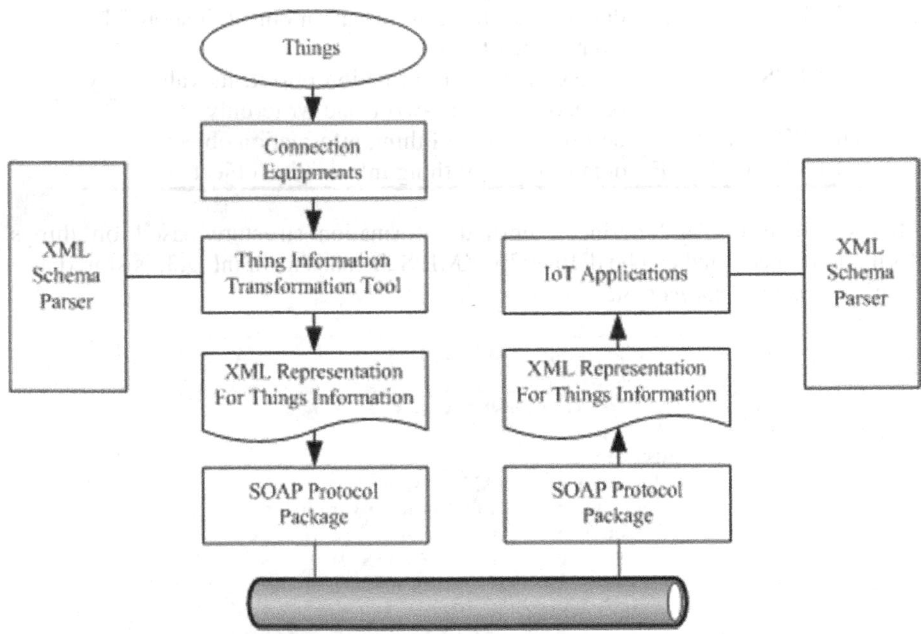

Fig. 5. Implementation Structure for XML-Based Representation Method of Things Information.

The implementation structure for Java-based representation method of things information are shown in Fig.5.

In this implementation structure, a typical thing information object acquisition process is as follows:

1) Access devices, such as wireless sensors or RFID tag readers, etc., read the thing information, and the information are passed to the Things Information Transformation Tool.
2) The Things Information Transformation Tool transforms the information into an XML document, which requires the use of an XML parser.
3) The XML document is encapsulated into a SOAP envelope, then they are transmitted through SOAP/HTTP to the receiver by the SOAP package.
4) The recipient receives the data through network equipment to SOAP package.
5) SOAP package reduces the XML document by parsing the SOAP envelope.
6) XML parser parses XML document to produce things information and these information are delivered to IoT applications.

5 Conclusion

Unified representation and standardization for things information are inevitable results with the further development of IoT and integration of things information. The representation for things information problem is discussed and XML-based representation method is proposed. We hope that these works can help practitioners of IoT.

Acknowledgments. This research was partly supported by Beijing Municipal Education Commission Projects grants JC007011201004, Beijing Municipal Education Colleges and Universities to Deepen Talents Scheme, and CSC Projects in China.

References

1. Commission of the European communities: Internet of Things-an Action Plan for Europe (2009), http://ec.europa.eu/information-society/policy/rfid/documents/commiot2009.pdf
2. Wen, J.B.: 2010 Report on the Work of the Government (2010), http://www.gov.cn/2010lh/content-1555767.htm
3. 2010 China Conference for IoT (2010), http://www.iotconference.com
4. The State Council of the People's Republic of China: State Council on Accelerating the Development of New Industries and Development of a Strategic Decision (2010), http://www.gov.cn/zwgk/2010-10/18/content-1724848.htm
5. International Telecommunication Union: The Internet of Things. ITU, Geneva (2005)
6. ITU Y.2002: Overview of Ubiquitous Networking and of Its Support in NGN, ITU-T Recommendation (2009)
7. EPC: Architecture of EPC Networks (2011), http://www.gs1.org/epcglobal/
8. National Instruments: Common Architecture of WSN (2011), http://zone.ni.com/devzone/cda/tut/p/id/8707
9. Dai, G.P.: Design on Architecture of Internet of Things. Journal of Beijing University of Technology (2011) (submitted)
10. W3C: XML (2011), http://www.w3.org/XML/
11. W3C: XML Schema (2011), http://www.w3.org/XML/Schema

A QoS-Aware Web Service Orchestration Engine Based on Actors

GuiPing Dai[1] and Yong Wang[2]

[1] College of Electronic Information and Control Engineering
Beijing University of Technology
Beijing, China
[2] College of Computer Science and Engineering
Beijing University of Technology
Beijing, China
{daigping,wangy}@bjut.edu.cn

Abstract. QoS-aware Web Service orchestration can satisfy not only functional requirements of the customers, but also QoS requirements. QoS-aware Web Service orchestration engine provides runtime support to Web Service Orchestration. Based on requirements of QoS-aware Web Service orchestration and Web Service orchestration engine, we design a typical QoS-aware Web Service orchestration engine based on actor system theory called Ab-QWSOE.

Keywords: Web Service, Web Service Orchestration, Web Service Composition, Actor Systems.

1 Introduction

Web Service (WS) is a new distributed component emerged about ten years ago, which uses WSDL[1] as its interface description language, SOAP[3] as its communication protocol and UDDI[2] as its directory service. WS uses the Web as its runtime platform, so it is suitable to be used to develop cross-organizational business applications.

A cross-organizational business process is a usual form in e-commerce that organizes some business activities into a process. WS Orchestration (WSO) provides a solution of cross-organizational business process base on the so-called Service-Oriented Architecture. A WSO represents a cross-organizational business process where business activities are captured as its component WSes.

In technical viewpoint, WSO provides a workflow-like pattern to orchestrate existing WSes to be encapsulated as a new composite WS. Specially, We use the term -- WSO but not another term -- WS Composition, because there are also other WS composition patterns, such as WS Choreography (WSC)[9]. About WSC and the relationship of WSO and WSC[8], we do not explain more, the readers please to refer some other materials[8-9].

In this paper, we focus on QoS-aware WSO engine (runtime of WSO). A QoS-aware WSO enables the customers to be satisfied with not only their functional requirements, but also their QoS requirements, such as performance requirements,

D. Jin and S. Lin (Eds.): Advances in CSIE, Vol. 1, AISC 168, pp. 17–22.
springerlink.com © Springer-Verlag Berlin Heidelberg 2012

reliability requirements, etc. An QoS-aware WSO engine provides runtime support for WSOs with assurance of QoS implementations.

The main efforts on WSO of the industry are trying to establish a uniform WSO description language specification, such as the early WSFL[5], XLANG[6], and lately converged WS-BPEL[7]. Such WSO description languages based on different mathematic models have constructs to model invocation of WSes, manipulate information transferred between WSes, control execution flows of these activities and inner transaction processing mechanisms. WSO description language can be used to define various WSOs under different requirements and acts as a so-called meta language. WSOs described by such meta languages actually are pure texts and must be enabled by the meta language interpreter called WSO engine, such as ActiveBPEL[4].

QoS-aware WSO engine can process not only the functional requirements modeled by the WSO description language, but also QoS requirements of the customers, for examples, a WSO must complete within three hours and the cost running some WSO must be under twenty dollars. A QoS-aware WSO translates the QoS requirements of the customers into QoS requirements of component WSes. Such a translation is accomplished by the implementation of the so-called QoS-aware Service Selection Algorithm (QSSA). There are many kinds such service selection algorithms, such as [10], which uses the so-called local service selection approach and global allocation approach based on integer programming, and another one in [11], which models the service selection problem (SSP) in two ways: one defines SSP as a multi-dimensional multi-choice 0-1 knapsack problem (MMKP) based on combinatorial model and the other captures SSP as a multi-constraint optimal path problem (MCOP) based on graph model, and gives heuristic algorithms of the two ways. We do not research SSP itself, but use the implementation of a WSSA above as a component of our WSO engine QoS-WSOE.

In this paper, we design and implement a QoS-aware WSO engine based on actors[12-13] called Ab-QWSOE. This paper is organized as follows. In section 2, we simply introduce actor systems theory. The requirements of a QoS-aware WSO engine are analyzed in section 3. In section 4, we give the architecture of Ab-QWSOE and discuss some issues on Ab-QWSOE. Finally, in section 5, we conclude the paper and point out the works in future.

2 Actors

An actor[12-13] is a basic concurrent computation unit which encapsulates a set of local states, a control thread and a set of local computations. It has a unique mail address and maintains a mail box to receive messages sent by other actors. Through processing the messages stored in the main box sequentially, an actor computes locally and blocks when its mail box is empty.

During processing a message from its mail box, an actor may perform three candidate actions: (1)(**send**)sending messages asynchronously to other actors; (2)(**create**) creating new actors with new behaviors; (3)(**ready**) becoming ready again to process the next message from the mail box or block if the mail box is empty. The actor model as shows in Fig.1 which is first illustrated in [13].

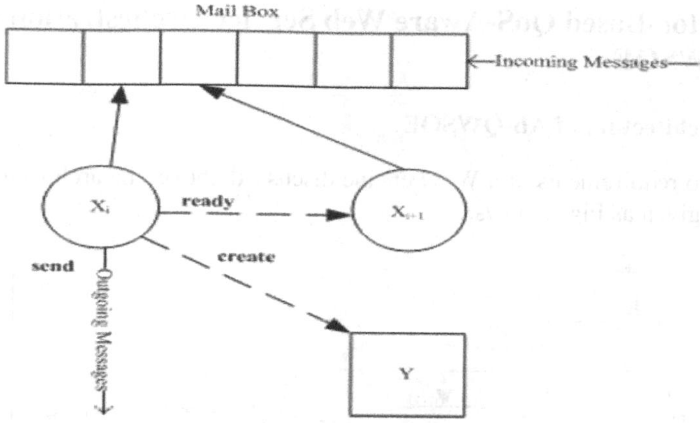

Fig. 1. Model of an Actor.

3 Requirements of A QoS-Aware Web Service Orchestration Engine

A WSO described by WS-BPEL is a program with WSes as its basic function units and must be enabled by a WSO engine. An execution of a WSO is called an instance of that WSO. The WSO engine can create a new WSO instance according to information included in a request of a customer. Once a WSO is created, it has a thread of control to execute independently according to its definition described by a kind of description language, such as WS-BPEL. During its execution, it may create activities to interact with WSes outside and also may do inner processings, such as local variable assignments. When it ends execution, it replies to the customer with its execution outcomes.

In order to provide the adaptability of a WSO, the bindings between its activities and WSes outside are not direct and static. That is, WSes are classified according to ontologies of specific domains and the WSes belonging to the same ontology have same functions and interfaces and different access points and different QoS. To make this possible, in a system viewpoint, a name and directory service -- UDDI is necessary. All WSes with access information and QoS information are registered into a UDDI which classifies WSes by their ontologies to be discovered and invoked in future. UDDI should provide multi interfaces to search WSes registered in for its users, for example, a user can get information of specific set of WSes by providing a service ontology and specific QoS requirements via an interface of the UDDI.

Above mechanisms make QoS-aware service selection possible. In a QoS-aware WSO engine, after a new WSO instance is created, the new WSO instance firstly selects its component WSes according to the QoS requirements provided by the customer and ontologies of component WSes defined in the description file of the WSO.

4 An Actor-Based QoS-Aware Web Service Orchestration Engine, Ab-QWSOE

4.1 An Architecture of Ab-QWSOE

According to requirements of a WSO engine discussed above, the architecture of Ab-QWSOE is given as Fig.2 shows.

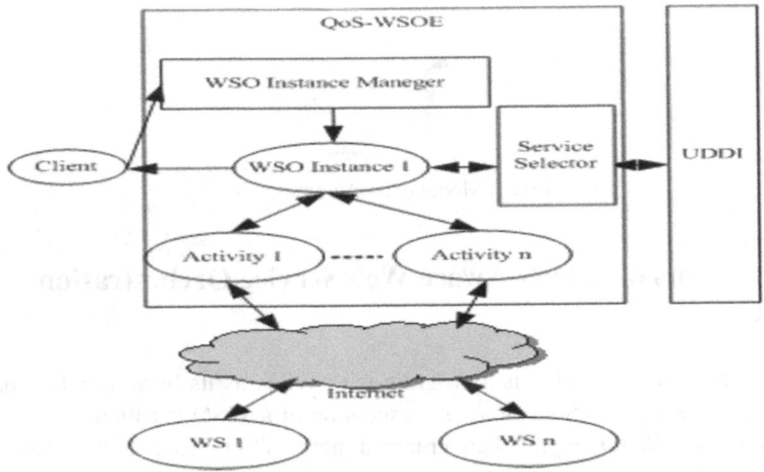

Fig. 2. Architecture of Ab-QWSOE.

In the architecture of Ab-QWSOE, there are external components, such as Client, UDDI and component WSes, and inner components, including WSO Instance Manager, WSO Instances, Activities, and Service Selector. Among them, UDDI, WSO Instance Manager and Service Selector are permanent components and Client, component WSes, WSO Instances, Activities are transient components. Component WSes are transient components since they are determined after a service selection process is executed by Service Selector.

Through a typical requirement process, we illustrate the functions and relationships of these components.

1) A Client submits its requirements including the WSO ontology, input parameters and QoS requirements to the WSO Instance Manager.
2) The WSO Instance Manager creates a new WSO Instance including its Activities and transmits the input parameters and QoS requirements to the new instance.
3) The instance transmits ontologies of its component WSes and QoS requirements to the Service Selector to perform a service selection process via interactions with a UDDI. If the QoS requirements can not be satisfied, the instance replies to the Client to deny this time service.
4) If the QoS requirements can be satisfied, each activity in the WSO Instance is bound to a WS outside.

5) The WSO Instance transmits input parameters to each activity for an invocation to its binding WS.
6) After the WSO Instance ends its execution, that is, every invocation to its component WSes by activities in the WSO Instance is returned, the WSO Instance returns the execution outcomes to the Client.

Components in Fig.2 are all implemented as actors. They are Client Actor (CA), WSO Instance Manager Actor (WSOIM), WSO Instance Actor (WSOI), Activity Actor (AA), Service Selector Actor (SS), component WS Actor (WS).

Actors in Ab-QWSOE are all customizations of common actors and are implemented based on a common actor runtime called Actor Foundry. Actor Foundry is developed by Open Systems Laboratory (OSL) of University of Illinois at Urbana-Champaign as an open source software.

Actor Foundry is not only a common actor runtime, but also an actor development framework that allows further developments and customizations.

Because an CA, WSOIM, WSOI, AA, SS are all common actors, and they are implemented as an actor class inheriting from the abstract class *osl.manager.ActorImpl*.

A WS is also a basic actor, but it adopts SOAP as its transport protocol. This requires a SOAP transport package *osl.transport.soap* must be implemented firstly. Then a WS is implemented by inheriting *osl.manager.basic.BasicActorImpl* class with a transport binding to the SOAP package.

4.2 Some Issues on Ab-QWSOE

Orchestration means that all AAs are orchestrated by control flows among them. That is, the executing of AAs should be in some orders, such as in sequence, in parallel, in choice, or in loop, etc. Sequence, parallel, choice and loop are four basic constructs for an orchestration.

Firstly, we discuss about sequence construct. Ab-QWSOE is a concurrent computing language since actor is a basic concurrent computation model. Actors are basic components in Ab-WSCL, that is, AA, WSO, WS, WSC are all actors. As Agha[13] pointed out, actor computing is intrinsic concurrent and sequential computing can only be implemented by causalities of messages sending among actors.

Parallel is intrinsical to AAs just because AAs are all concurrent actors.

We model an actor as a distributed object[14] and messages are transfered among invocations of methods of actors. Methods of actors (AA, WSO, WS, WSC) to receive messages from other actors have the form of *method-name (args) if condition*. There is a *condition* consisted of local states of actors to control whether or not to receive a message for an actor. Using this *condition* can implement the choice construct.

Similarly, loop construct can also be implemented through the *condition* of a method.

5 Conclusion

In this paper, we analyze the requirements of QoS-aware Web Service orchestration and design a QoS-aware Web Service orchestration engine called Ab-QWSOE based on actors.

In future, we will utilize actor systems theory to give Ab-QWSOE a firm theoretic foundation.

Acknowledgments. This research was partly supported by Beijing Municipal Education Commission Projects grants JC007011201004, Beijing Municipal Education Colleges and Universities to Deepen Talents Scheme, and CSC Projects in China.

References

1. Chinnici, R., Moreau, J.J., Ryman, A.: Web Services Description Language (WSDL) Version 2.0 Part 1: Core Language, W3C Recommendation (2007)
2. Clement, L., Hately, A., Riegen, C.: UDDI Version 3.0.2, OASIS Draft (2004)
3. Mitra, N., Lafon, Y.: SOAP Version 1.2 Part 0: Primer, 2nd edn., W3C Recommendation (2007)
4. Active Endpoints (2011), `http://www.activevos.com/`
5. Leymann, F.: Web Service Flow Language (WSFL) 1.0, IBM Tech Report (2001)
6. Thatte, S.: XLANG: Web Services for Business Process Design, Microsoft Tech. Report (2001)
7. Jordan, D., Evdemon, J.: Web Services Business Process Execution Language Version 2.0, OASIS Standard (2007)
8. Pelz, C.: Web Services Orchestration and Choreography. IEEE Computer 36(8), 46–52 (2003)
9. Kavantzas, N., Burdett, D., Ritzinger, G.: Web Services Choreography Description Language Version 1.0, W3C Candidate Recommendation (2005)
10. Zeng, L., Benatallah, B.: QoS-Aware Middleware for Web Services Composition. IEEE Tran. on Software Engineering 30(5), 311–327 (2004)
11. Yu, T., Zhang, Y., Lin, K.: Efficient Algorithms for Web Services Selection with End-to-End QoS. ACM Tran. on the Web 1(1), 1–26 (2007)
12. Hewitt, C.: View Control Structures as Patterns of Passing Messages. Artificial Intelligence 8(3), 323–346 (1977)
13. Agha, G.: Actors: A Model of Concurrent Computation in Distributed Systems. Ph.D. dissertation, MIT Lab. for Computer Science (1986)
14. Agha, G.: Concurrent Object-Oriented Programming. Communications of ACM 33(9), 125–141 (1990)

Emergency Management Application Based on Spatial Information Cloud Service

Ken Chen[1,2,4,*], Fang Miao[1,2,3,4], and WenHui Yang[1,2,4]

[1] Key Lab of Earth Exploration & Information Techniques of Ministry of Education
Chengdu University of Technology, 610059 Chengdu, China
ck@cdut.edu.cn

[2] State Key Laboratory of Geohazard Prevention and Geoenvironment Protection
Chengdu University of Technology, 610059 Chengdu, China

[3] Policy and Strategy Center, Institute of Digital Earth
Peking University, Beijing, 100871 Beijing, China

[4] Spatial Information Technology Institute
Chengdu University of Technology, 610059 Chengdu, China

Abstract. The research of emergency management is a frontier area. According to the imperfection of the theory and application, it has to build an effective architecture of emergency management. Problems with emergency management such as single user, time lag of response and cannot effectively handle huge data, etc. From the perspective of the users' experience, we take into diverse needs, characteristics and laws of the time of emergency management. To the concept of Data-Scatter, Information-Collection, Service-Aggregation, we propose G/S model based on EMML (Emergency Management Markup Language). By emergency management platform that based on spatial information cloud service, we can study typical cases and try to resolve the key scientific issues in emergency management system of our country in recent years.

Keywords: cloud service, spatial information cloud, emergency management, G/S model, EMML.

1 Introduction

Emergency managements have been designed from their own characteristics and various angles by users from different departments, industries or fields. And they give specific plan [1]. But with the increasing complex of social systems and advanced of information systems, an emergency management application platform with usability, stability, security and real-time need to be developed. Emergency management system can be regarded as an open and complex system with multi-agent, multi-media, multi-incentives, multi-scale, variability and other characteristics, which including rich, profound and complex of scientific problems. The existing emergency

* Ken Chen (1986-), male, from Leshan Sichuan, Han, Ph.D. student, Spatial Information Technology Institute in Chengdu University of Technology, research area: spatial information technology and application. E-mail: ck@cdut.edu.cn

D. Jin and S. Lin (Eds.): Advances in CSIE, Vol. 1, AISC 168, pp. 23–28.
springerlink.com © Springer-Verlag Berlin Heidelberg 2012

management system is aimed at specific target system. It couldn't resolve dates diversity and complex of technology and other issues because of its highly targeted.

To solve the above problems, combined with the characteristics and development trends of the present spatial information technology, Internet/Internet of Things technology, computer hardware technology/software technology, cloud computing technology and new computing models, to collect, manipulate, manage and transmit of temporal and spatial information on the net, it makes our demands for spatial information has shifted from data to services [2].

Sum up above all, emergency management applications based cloud service in the spatial information system not only can effectively ensure high capacity, high storage and high analytical ability, but also can ensure stability, security and real-time. It can solve the problems and effectively meet the various needs of users.

2 Emergency Management Cloud Service Based on G/S

2.1 G/S Model

According to the C/S model in the Client side and Server side usually require specific software support, and lacks of uniform standards, do not have the cloud service. While B/S model has a low response rate, it is difficult to achieve personalized functional requirements. And it does not meet mass storage requirements of the spatial data. According to some bottlenecks[3], we analyze the characteristics and structure of information flow of spatial information service based on Internet, using geographic markup language HGML (Hyper Geographic Markup Language) which can exchange vast amounts of heterogeneous data as the core, storage, organize, exchange, scheduling and display the various types and formats of data storage in the network. Upon request - aggregation - service "client aggregation service" working mechanism, achieve the various functions of geographic information browser/space information service cloud (Geographic Information Browser/Spatial Information Service Cloud) model (G/S model) through multi-client collaborative emergency management.

G/S model architecture use HGML as the core. Not only to support a variety of customer types, but also complete application logic by the browser. The data processing logic is on the spatial data server side, not only improve the functionality and efficiency of the G side, but also optimize data management, and reduce network. On the base of "loosely and coupled" distributed data, access and update data from a variety of application servers to bear. Not only to enhance the efficiency and convenience of data maintenance, but also to protect the data in real-time, timeliness and accuracy.

2.2 Emergency Cloud Service of Spatial Information Based on G/S

We can acquire, share, process rapidly, updates, display, assess spatial information data which could be huge, virtual, and high available in real-time .This characteristic coincides with the need and objective of emergency management.

As we all know, spatial information data is an important factor in the establishment. Traditional hardware-based servers will be difficult to meet management and processing requirements of the massive data. And, because the industries have differences in the cognitive perspective of space phenomena, needs, data collection methods, data modeling methods, leading to a variety of heterogeneous multi-dimensional spatial data, spatial data models, database schema, space operations structure and different data file formats. [4] Redundant data will cause a huge waste of resources. Relying on the cloud processing technology, we can build an information processing system to provide open information services on the Internet. With the use of data integration and integration technology of the cloud processing, we may achieve the information to be found and shared timely.

With features of G/S model architecture, according to the management features of massive heterogeneous data in the emergency managements refer to HGML, made EMML (Emergency Management Markup Language) as a data exchange system for emergency. It contains the standard emergency management terminology, data dictionary, metadata sets, and data program exchange parameters. Using EMML to unify emergency management data exchange standards in G/S model , aims to organize, classify, manage and specify the emergency management data in the entire NoSQL database through" Request - Aggregation - Service".

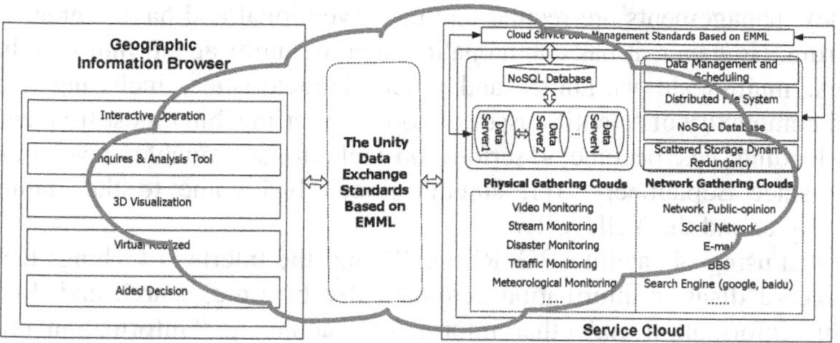

Fig. 1. Framework of G/S model

For example, scatter data in server-side. Through the spatial data cloud structure, increase capacity of the terminal services by data block framing, redundant storage, synchronization, multi download, load balance, automatic adaptation that can eliminate bottlenecks in accessing network. Gathering information and service on the client, and achieve demand for data, aggregation, processing, aggregation services, information display and decision support through EMML language. To achieve high computing power, high storage capacity, high analytical skills and time required for data description of the emergency data, multi-dimensional interaction description, parallel storage of the emergency data description, analysis and computing in a cluster.

As the G/S model is on the basis of stability, security, real-time, so it's easy to achieve functions through the emergency management service cloud. Such as quickly sense of spot information, remote transmission of spot scene and live video, collaborating of multi-client, searching and displaying emergency data cross the platform, inquiring plans and knowledge data base. On account of unconventional emergency, it can achieve to supply early warning before disaster, handling in disaster, assessment of reconstruction after disaster in order to protect social security and harmonious development.

3 Emergency Management Based on Spatial Information Cloud

Humans are not fully rational, externality of environment, and the rapid development of science and technology. It makes a variety of emergencies far beyond human ability to anticipate and respond. After the SARS, in the face of southern snow disaster, 5.12 earthquake, Tibet 3.14 events and a series of unusual emergencies, still due to "extreme environmental conditions and deficient resource, lack or surplus of information, a huge psychological pressure, conflict of the target of interest, complex changes in social structure [5]" and other factors, emergency management system is difficult to get accurate understanding on the likelihood for the emergence and extent of the damage. So it couldn't make a scientific and effective decision.

3.1 Emergency Management Cloud System

Emergency managements are regular and unconventional and have a great variety to choose from. However, it has commonality such as: multi-agent, multi-media, multi-incentives, multi-scale variability and other characteristics, including a rich and profound complexity of scientific issues. Floods occur in China most frequently in the world. Flooding has become a serious obstacle to sustainable development. For example, since September 2011, autumn floods had come to the Jialing River, Hanjiang River and the Yellow River.

Integrated using of satellite, 3G, RFID, 3S, and the Internet of Things technology, we can record disaster information automatically by time, search and display spot geographic information and other information, achieving "information on-demand access, service on-demand aggregation" cross the platform and multi-client under the cloud service architecture.

First, it can record situation of flood areas of automatically in time. Such as inundated area of arable land, water height, accurately recording data is first-hand information for analysis and learn.

Secondly, it can supply searching and displaying geographic information to help users to grasp the situation and then make right orders.

Third, it can search and display other information. Flood will be sure to bring economic losses to countries and people. It can collect local economic development situation, housing damage situation, casualties, property damage and analysis after the flood.

Fourth, through transformation, value-added process of the relevant data "data - information - knowledge", drawing on the various existing geographic information network service model architecture, it establish cloud service architecture system which has "request - Aggregation - Service" features.

Finally, this application can assess damage effectively after the flood. Figure2 shows emergency management platform interface based on G/S model structure. This interface has clear function blocks. It's easy to use and great to meet the needs of different levels of requirements.

3.2 The Superiority of Emergency Management

For example, we are faced with ever-changing environment, complex process and urgency situation of the event during using various types of flood emergency management. The advantages of flood emergency management based on G/S model are:

(1) This system is stability, security and real-time update. Through collaboration, it can achieve scattered storage and balancing load of the physical collection clouds and the internet collection clouds. And it can eliminate duplication and redundancy of data, in order to reduce power consumption of system and platform, and ultimately make the platform rapid and stable. And through the prevention of data interception, middle attacks, replay attacks, data organization and management of enhanced security protection.

(2) It has the characteristics of Identifying real-time information quickly. Classify arrival time of peak and number of water level and other massive heterogeneous data, then analyze methods of data acquisition, and establish mechanisms to achieve the required spatial, with time coding ,transmission and integration of data source, in order to percept real-time information quickly.

(3) The feature is simultaneous transmission of graphic information and live video. With the freedom of 3S technology and data exchange, once rescue teams exposed to the disaster site, they can send such as text, photo or video of the seriousness of the situation ,trapped casualties and the environmental conditions to the system in the form of information collection cloud, to achieve simultaneous transmission of information and graphic.

(4) The feature is personalized integrating client. It means built an implementation system to aggregate information services on the base of cloud service architecture. It can transmit the command from all levels of command authority, the scene commander or the command headquarters to the scene of the rescue team timely to reach a unified coordination and concert efforts. This is not only inheritance of the mode of traditional management systems, but also greatly reducing the complexity of space.

(5) It provides searching and displaying features of contingency plans, expert database and knowledge base. The scale and degree of harm of every flood are different. At this point we need to provide expert database for the query as a rescue plan for reference. Therefore, it uses the keyword to search the relevant contingency plans, and to be combine with expert database in order to support the emergency command center to make decisions.

4 Conclusion

Spatial information cloud service system architecture on the basis of G/S model is according to the cloud service in emergency management. To integrate, share, exchange, manage, display, analysis, service and decision of massive heterogeneous data, it will be more widely used in emergency management platform:

(1) This system has characteristics as multidimensional processing model, integration of 3S technology, liberalization of data exchange, inheritance of the traditional model. Through cloud storage dynamic management technology, it builds a real-time data using platform.

(2) Case confirmed that the emergency management system is able to break through the bottleneck in the use of GIS, truly realizing popular apply. Platform is easy to understand. We can easily get without professional. Through scattering the data, gathering information and services, it can achieve security, stability, and timely emergency management services.

(3) Unified data standard technology based on EMML, can standard the various types of spatial data and temporal data. From the perspective of safety and real-time, it studies organization and management issues of structured, unstructured, spatial and nonspatial data. According to characteristics of the emergency management cloud services, it can unify exchange, management and service of data.

Acknowledgments. The authors would like to acknowledge the Spatial Information Technology Institute the members of, the State Key Laboratory of Geohazard Prevention and Geoenvironment Protection as well as Key Lab of Earth Exploration & Information Techniques of Ministry of Education, owing to their help to produce these examples and ideas presented here and the financial support (Grant No. NSFC 61071121) provided by the National Nature Science Foundation of China.

References

1. Ji, L., Chi, H., Chen, A., et al.: Emergency management. The high education press (March 2006)
2. Miao, F., Guo, X., Ye, C.: Space information service polymerization and GIS popularization application. Geography and Geographic Information Science (2010)
3. Zhang, W., Tang, J., et al.: Cloud computing profound change the future. Science press, Beijing (2009)
4. Feng, J., Xiong, Y.: Cloud computing environment of the digital city construction. Digital Communication (2011)
5. Liu, X., Yan, X., Liu, S.: Unconventional emergency nature and characteristics of paper. Journal of Beijing University of Aeronautics (social science edition) (2011)
6. Chen, K., Miao, F., Yang, W., Guo, X.: Research on Meteorological Data Display Based on GS Model. In: The 2nd International Multimedia Technology Conference, ICMT 2011 (2011)
7. Chen, K., Miao, F., Yang, W., Guo, X., Leng, X., Tan, L., Liu, B.: Research on Multiple Spatial Data Gathering Mechanism Based on Geo-information Browser. In: 7th International Symposium on Digital Earth Secretariat, ISDE 2011 (2011)

Research on Complex Networks Evolution Model

JinJun Xie*

Polytechnic School of Shenyang Ligong University, China

Abstract. This paper is aim to introduce the research background and the significance of the complex network ,it make a explanation of the basic concepts involved in complex networks, then Discusses the complex network model research in detail.

Keywords: complex networks, small-world networks, scale-free networks.

1 Introduction

Research of complex network is the complexity of scientific research is an important part of. nineteen sixties by the famous mathematician Erdos and Renyi proposed ER random graph model. Until 1998, Watts and Strogatz published papers [1] in the "Nature" magazine, introduced the small world network model (Small world),and described the huge network of most nodes have short connection (path) of nature. Barabasi and Albert published article in the "Science" in 1999 [2], and pointed out that many real-world complex networks degree distribution have power index. As a result of the power-law distribution without obvious characteristic length, the network called scale-free network (Scale-free). Research shows that, from the organism in the brain structure into the new supersedes the old. Network, from Internet to WWW, from the large power network to the global transportation network, from the scientific collaboration network to a variety of political, economic, and social relationship network, the real world in many of the network are small world phenomenon or scale-free characteristic of complex network [3].

2 Complex Network Characteristics

With the studying of complex networks, people put forward many concepts and measurement methods for representing the complicated network structure characteristics.

(1)Degree distribution, degree distribution is an important statistical characteristics of networks. Here the degree is also known as connectivity. Degree of node refers to the number of edges between nodes connected. The meaning of degree is different in different networks, in social networks, degree can be expressed in terms of individual influence and importance, the greater the degree of the individual, their influence is

* JinJun Xie, male, born in 1965 in Shenyang, Polytecnic School of Shenyang Ligong University, associate professor, research direction: computer software and theory.

D. Jin and S. Lin (Eds.): Advances in CSIE, Vol. 1, AISC 168, pp. 29–33.
springerlink.com © Springer-Verlag Berlin Heidelberg 2012

bigger, the role in the organization is bigger also. Degree distribution presents node degree probability distribution function P(k), it refers to the probability that node has K edges connected. In the present, One of the two common degree distributions is index of degree distribution, namely P(k) as K increases exponentially, another distribution is a power law distribution, namely $P(k) \sim k^{-r}$, where r is known as the index. Kinetic properties are also different in different networks. In addition, the degree distribution and other forms, such as the star form network degree distribution are two points distribution, the rule networks degree distribution are single point distribution.

(2)Clustering coefficient, clustering coefficient is a measure of network group of degree, it is another important parameter of network.. To the social network, a group of shape is an important feature, group network in circle of friends or acquaintance, group members are often familiar with each other, scientists have proposed the concept of cluster coefficient as a measure of the cluster phenomenon. Node i cluster coefficient C_i is the network and the nodes directly connected to the connection between nodes, and the nodes directly adjacent nodes existing edge number accounts for the largest possible number of edges in the proportion, the expression for C is

$$C_i = 2e_i / k_i (k_i-1)$$

k_i says the degree of node i, e_i says the actual number of edges of adjacency point. between node i .

(3)The average path length (APL),The average path length in the network is another important feature. It refers to all of the nodes on the shortest distance between the average nodes. The distance between nodes (Distance) means from one node to another node need go through the side of the minimum number, in which all nodes on the maximum distance between the known as the network diameter (Diameter). The average path length and diameter is a measure of the network transmission performance and efficiency. The average path length of the formula is

$$APL = \frac{1}{N(N-1)} \sum_{i,j \in V} d_{ij}$$

where d_{ij} is the shortest distance of i node to j node.

(4) Betweenness, Betweenness have two kinds, one is nodes betweenness, another is edges betweenness. It reflects the node or edge influence. If a node has N different shortest path, which has n path through the node i in N. The node i to all of the nodes on the contribution of accumulated divided by the node to node i number is n/N.

3 Complex Networks Evolution Model

At present, the research on complex networks are mainly concentrated in three areas: one is network formation mechanism and evolution model, it make model through the formation mechanism, simulate the real network behavior; the other is a complex network stability, research limitation on network geometric characteristic influence, such as the complex network to withstand accidental fault and malicious attacks ability; another is the complex network dynamics, which is one of the complex network 's ultimate goal, which is beyond the topological structure of the network, to establish the network system working mode and mechanism, understanding of the dynamics in complex system.

(1) Rule network

Each node only and its surrounding neighbor nodes are connected, with periodic boundary conditions in the vicinity of coupled networks including N surrounded by a ring of points, where each node is connected to its neighbor point about K / 2 (K is even.)

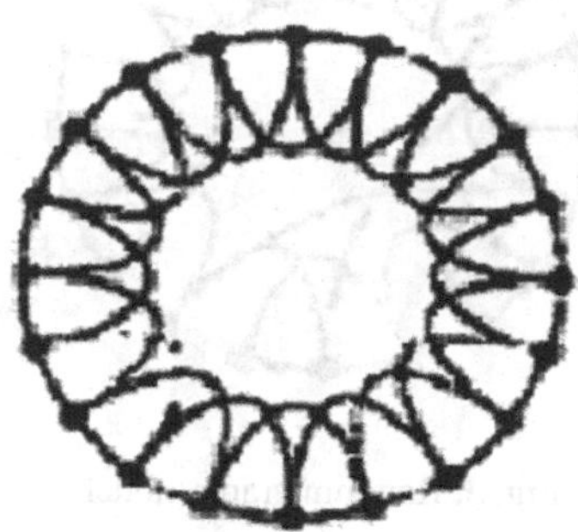

Fig. 1. Rule network

(2) Random graph model

Stochastic network theory is presented by the Hungarian mathematician Erdos and Renyi, their model is known as the classical ER model. The ER model is defined as : in the N vertices, C_n^2 = n (n-1)/2 edges in the graph constructed, randomly connected g edges to form a random network, denoted by GN, G, by the way of N nodes, G edges formed network of common species, constitute a probability space, each network is equal to the probability.

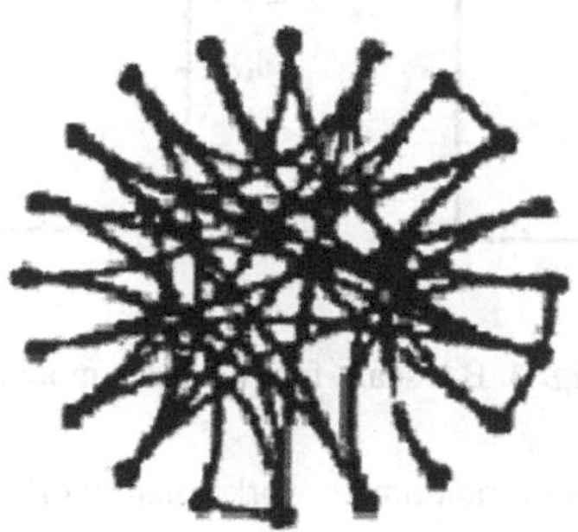

Fig. 2. Random graph model

(3) Small world network

The earliest model based on small world network is presented by Watts and Strogatz in 1998 (WS model), the model consists of a N node ring, ring on each node and on each side of the m edges connected to each side, then with probability p random reconnection (except of self connection and the heavy side). These heavy even side called "long-range connections", long-range connections greatly reduces network average path length.

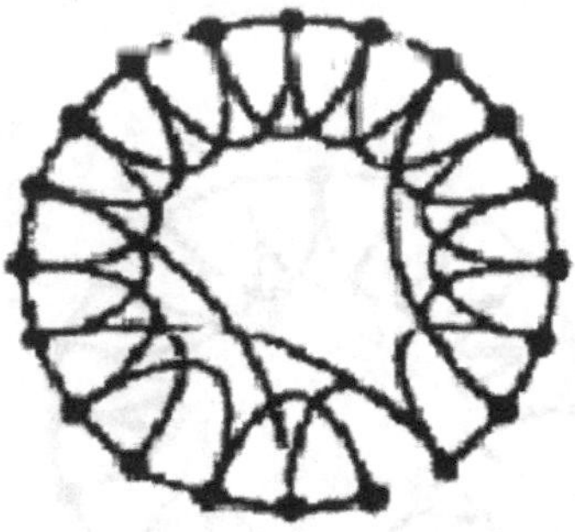

Fig. 3. Random graph model

(4) BA scale-free network model

In 1999, Barbasi and Albert found that many complex networks have large-scale highly self organization characteristic by tracing the evolution of the world wide web, the most complex network follows a power-law distribution of node degree, and power-law degree distribution network is called scale-free network.

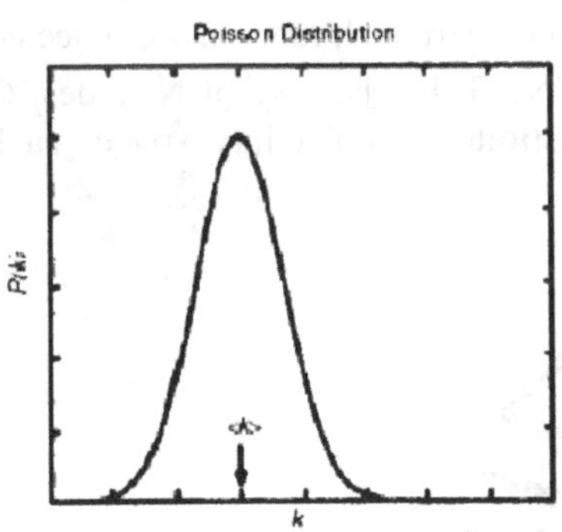

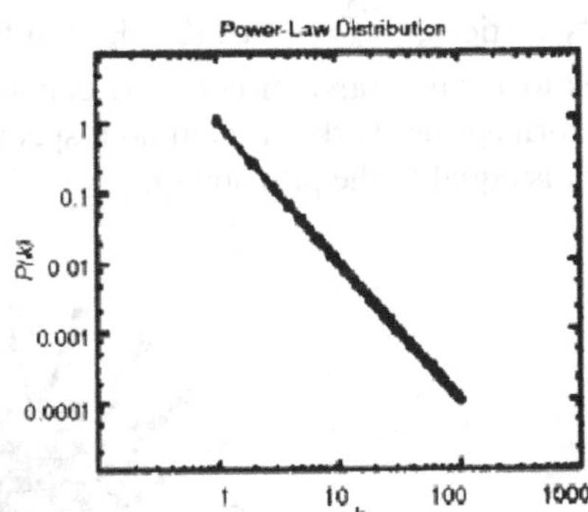

Fig. 4. BA scale-free network model

Table 1. Regular network, random network, small-world network comparison

Characteristic parameters	Regular network	random network	small-world network
Network diameter	large	small	small
clustering coefficient	large	small	large

4 The Problems of Researching on Complex Networks Evolution Model

(1) The relation of the model and the real network is not close enough

Despite the complex network has a large number of empirical studies, but empirical model evolution disjunction has not yet been resolved, network model setting of generation and evolution mechanism, also cannot reflect the real characteristics of network. According to the actual network to set the specific mechanism of the model is also compared with less.

(2) Considering the dynamics and other factors in the model research is less

At present, the BA model still have a lot of problems witch to be resolved, for example, it did not get better reflect and express the distributed nodes change in the real world of scale-free networks. In addition, many real networks have shown module characteristics. The network would be effected by module interactions, it is also a complex challenging problems of network.

(3) It still need further research on the specific reality network power-law form mechanism

At present, building a model applicable for all networks is the starting point. The general research is meaningful, but it ignores the specific network personality. The real world is different network may have a different formation mechanisms. We only focus on the commonness and neglect personality, we are difficult to interpret a single specific network generation mechanism.

(4) The number of generation mechanism network model that it not only be relatively simple, but also be in accordance with the reality is lacked.

The BA model is called the classical model, in addition to the groundbreaking solution to the scale-free property problem, but also is the model for the generation of a relatively simple mechanism. It grasped most essential feature of the real network.

(5)Research on the self Similarity of complex network is not too much

Model more just using the box counting dimension method to prove the fractal, fractal theory and no more content and complex network into consideration.

5 Conclusion

This paper introduces the basic characteristics of complex networks, it analysis and comparison on several kinds of complex network model, and it presents the advantages and disadvantages of each model, finally discusses the evolving model problems of complex network. It has a certain reference value of researching on the complex network model.

References

1. Newman, M.E.J.: Scientific collaboration networks. II. Shortestpaths, weighted networks, and centrality. Physical Review E 64(1), 016131 (2001)
2. Newman, M.E.J.: Scientific collaboration networks. II. Shortestpaths, weighted networks, and centrality. Physical Review E 64(1), 016132 (2001)
3. Liu, J., Lu, J.: A small scientific collaboration complex network and its analysis. Complex Systems and Complexity Science (3), 56–61 (2004)

Using ASP.NET to Develop Intelligent Online Ordering System in E-commerce

Jing Li*

Modern Education Technology Center, Huanghe Science and Technology College, Zhengzhou, 450063, Henan, China
lilijingjing2011@sina.com

Abstract. Online ordering system is a typical e-commerce site. The seemingly mysterious e-commerce is not a new concept, nor is it only in recent years the rise of new economic activity. ASP.NET is a unified Web development model, which includes the use of as little as possible of the code generation enterprise applications WEB necessary services. This paper proposes the online ordering sales system based on ASP.net by end users browse the goods online information, and launched online shopping cart.

Keywords: online ordering system, ASP.NET, e-commerce.

1 Introduction

Online ordering system is a typical e-commerce site. The seemingly mysterious e-commerce is not a new concept, nor is it only in recent years the rise of new economic activity. Exact, back in the 1970s, companies have started different types of electronic transactions, such as inter-bank funds transfer, etc. In a certain sense, the data interchange (EDI) is a prototype and predecessor of e-commerce, with Internet technology, so that e-commerce environment has greatly improved and to expand the concept of e-commerce [1]. From different angles can be divided into several types of e-commerce, the use of more, are more well-known in accordance with the participating subjects to be divided, so that can be divided into business to business e-commerce transaction model (B to B), business to consumer electronic business transaction model (B to C), consumer to consumer e-commerce transaction model (C to C), online ordering system belongs to B to C model. Online orders will be more attractive for large users, Internet users view online shopping will certainly never learned to understand, from denial to acceptance, from suspicion to trust, from negative to positive, the concept of online shopping will more deeply. "Online ordering system," this new inter-company transaction to achieve a paperless document exchange and capital, changing the product customization, distribution and exchange means. For customers, find and buy products as well as greatly improve the way services.

---

* Author Introduce: Jing Li(1975-), Female, lecturer, Master, Huanghe science and technology college, Research area: computer application, e-commerce, ASP.ENT.

D. Jin and S. Lin (Eds.): Advances in CSIE, Vol. 1, AISC 168, pp. 35–39.
springerlink.com © Springer-Verlag Berlin Heidelberg 2012

Order is an interactive feature of the business information system. It provides users with two types of static and dynamic information resources. The so-called static information is constantly changing or updating those resources, such as company profile, management practices and corporate system; dynamic information is subject to change information, such as commodity price, meeting arrangements, and training information [2]. Online ordering system with a powerful interactive feature enables businesses and users to easily transfer information to complete e-commerce or EDI transactions. For businesses and enterprises can take full advantage of multi-network infrastructure to provide, payment platforms, security platforms, management platform to share resources in order to effectively carry out their own low-cost business, so you want to create "online ordering system".

Online ordering system is based on Windows XP under Microsoft Visual Studio 2010 and Microsoft SQL Server as a development environment for e-commerce-related knowledge as a theoretical basis, using the Browser / Server (B / S) structure, to achieve the consumer's online shopping, online transactions between merchants and online e-payment business model. This paper proposes the online ordering sales system by end users browse the goods online information, and launched online shopping cart. Submitting orders online and other functions. Managing goods and goods can be added to the classification system, user management, order management, sales and other statistics. Compared with traditional sales behavior, online goods sales system provides users and businesses to have the highest efficiency and most convenient means.

Along with the constant development of e-commerce, enterprise networks have become a trend, and how to use the limited funds it is particularly important to build their own e-commerce website. In this paper, the goods sales a ASP.NET -based web site, by introducing a three-tier structure of the website JSP model fully embodies the advantages of ASP.NET technology, and through the development process in detail. Construction of a given SME e-commerce websites use ASP.NET specific method.

2 The Introduction of Development Environment

The system based on ASP.NET an online shopping system, in which the database using the popular, easy installation and use of Microsoft SQL Server database, while the production of web pages using a visual interface with Microsoft Visual Studio 2005 software and Internet Information Services (IIS), the following key for these types of technologies and software overview[3].

(1) ASP (Active Server Pages), said active server pages. ASP.NET is a unified Web development model, which includes the use of as little as possible of the code generation enterprise applications WEB necessary services. ASP.NET as. NET Framework provides a part of. With the further development of Web application technology, Microsoft introduced the ASP.NET 2.0, ASP.NET 2.0 allows users to build Web applications increasingly about easy. ASP.NET 2.0 high efficiency, flexibility and scalability.

(2) SQL Server is Microsoft's launch of a comprehensive, integrated, end to end database. It provides enterprise users with a safe, reliable and efficient platform for enterprise data management and business intelligence applications. In the. NET

Framework, called a data source (database) applications for the hosting provider, and specifically for SQL Server database provides a hosting provider, SQL Server managed provider, which is used to link and a SQL Server database. By using the SQL Server managed provider, developers can easily send SQL statements to virtually any database. Will. NET and SQL Server managed provider combination will only need to rewrite the application programmer can make it run on any platform.

(3) Internet Information Services (IIS, Internet Information Server is the acronym), is a Web service, including the WWW service, FTP server, etc. Through IIS, you can easily Internet (LAN) and Internet (Internet) on the distribution of information. Microsoft IIS is one of the main push of the Web server. Windows 2000 Advanced Server and Windows XP operating system used has been included in IIS 5.0; Windows Server 2003 operating system already includes IIS 6.0, so users can take advantage of Windows NT Server and NTFS (NT Filc System, NT file system) Built-in security features, build a strong, flexible and secure Internet and Intranet sites.

3 Outline Design and Detailed Design of Online Ordering System

System uses three-tier structure, the simple, the user interface layer through a unified interface to send requests to the service layer, business logic layer according to their rules after the request processing database operations, then the database returns the data package into a class form returned to the user interface layer. This user interface layer can even do not know the database structure, maintaining it as long as the interface between the business layer can be. In this way a certain extent, increase the security of the database, but also reduces the level of the user interface developers request because it does not require any database operations. Three-tier architecture works as shown in Figure 1.

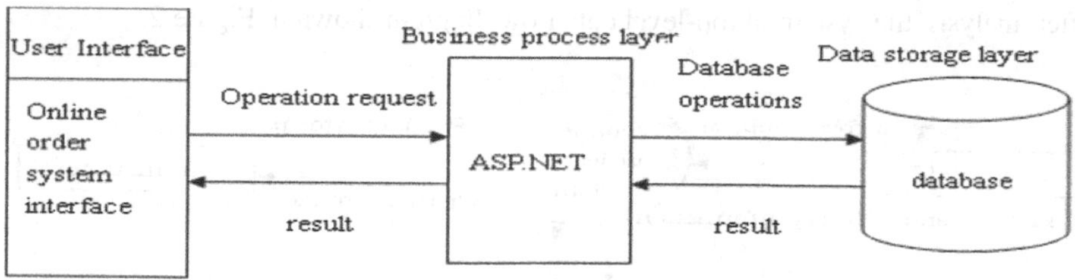

Fig. 1. Schematic three-tier working

The main feature of this system by the customer interface and management interface is divided into two parts. Compared with the traditional physical stores and online orders with the characteristics of the integrated system functional analysis, the system's data including user data, the role of data, types of goods data, commodity data, commodity image data, order data, message data, product reviews data, Notice of data, news data, etc.

ASP.NET form by returning the object to return data, within the class can specify which data can be accessed, what data is read-only, so by encapsulating the data to improve data security purposes. This module contains the main business online work order system, the core of the equivalent of a physical store salesperson.

Online order system for order management and traditional physical store sales work slightly different. Average salesperson only needs to process user to his orders, that is, online order system has been successfully submitted orders. But in fact, before the user submits the order, there will be more choices as the options selected, master list of these products will help us further understand the buying trends and user interest, so as to implement a more targeted and more effective marketing strategies. Quite simply, if we are able to provide users with more information on alternative products, may be able to contribute to the completion of purchase. So the master list of user options for the shopping cart, the next step has a very significant impact on sales. Front module of the system for users to browse, purchase products provide a complete system, it loads the user's role different functional modules.

4 The Implementing of Intelligent Online Ordering System Based on ASP.NET

Functional analysis based on the system, first use the new Asp.net Web application, the application can be logically divided into four layers: page presentation layer, business logic, data access layer, the database. Page indicates the specific application layer functions, usually by a Web page, the system controls custom controls, components. Business logic and application business logic associated with the data access layer uses the services provided, but also provide services for the upper layer. Business logic layer data access layer through the function to access the database, while the upper interface to provide access to databases and other services or functions. According to "top-down, layer by layer decomposition" thinking, the use of decomposition and abstraction are two means to control the system's complexity. After analysis, the system of top-level data flow diagram shown in Figure 2.

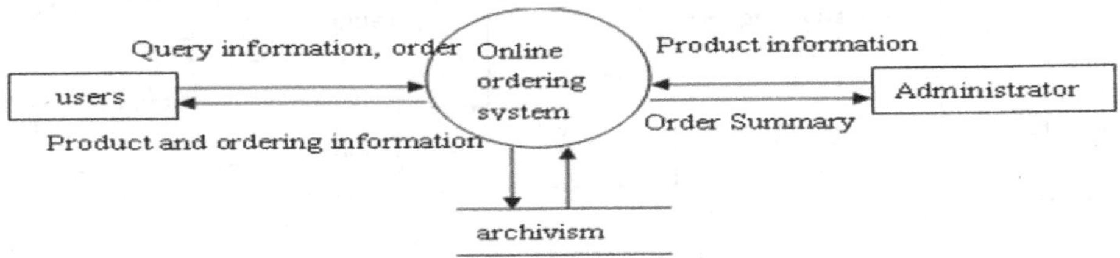

Fig. 2. System top-level data flow diagram.

Access the database data access layer encapsulates the various operations, such as connecting to the database, database operations, data conversion. The system's data access layer from the document SQLHelper.cs achieve, which defines the class SQLHelper designed to handle a variety of operations to access the database. Business logic and application business logic associated with the data access layer uses the services provided, but also provide services for the upper layer. Business logic layer data access layer through the function to access the database, while the upper interface to provide access to databases and other services or functions. When editing a good ASP.NET file, must be achieved after the release of its features, ASP.NET file itself does not run. We use Windows own tools IIS (windows2000 XP) to publish an ASP.NET file.

The system designs a model of online merchandising website. This site uses the popular three-tier architecture, by the logic of the entire system is divided into different modules, the system greatly reduces application development and maintenance costs and improve the reusability of system modules, while the scalability of the system greatly to improve. As the. Net benefits of its own, this website for a lower system requirements, compatibility, and good for promoting the use of the site provides a convenient and may.

5 Summary

This paper presents the development of online ordering system by end users browse the goods online information, and launched online shopping cart. Submitting orders online and other functions. Managing goods and goods can be added to the classification system, user management, order management, sales and other statistics.

References

1. Wei, M.: A Research on Statistical Information Applied to Tourist Traffic and Transport System Design Based on ASP. NET. JCIT 6(1), 147–156 (2011)
2. Li, Y., Shu, C., Xiong, L.: Information Services Platform of International Trade Based on E-commerce. AISS 3(1), 78–86 (2011)
3. Liu, D.: E-commerce System Security Assessment Based on Grey Relational Analysis Comprehensive Evaluation. JDCTA 5(10), 279–284 (2011)

Passive Energy-Dissipation Control for Aqueduct Structural under Earthquake Based on SIMULINK

Liang Huang[1], Yujie Hou[2], and Bo Wang[3,*]

[1] School of Civil Engineering, Zhengzhou University, 450001, Zhengzhou, China
[2] Florida International College, Zhengzhou University, 450001, Zhengzhou, China
[3] College of Water Conservancy and Environmental Engineering, Zhengzhou University, 450001, Zhengzhou, China

Abstract. The passive control for aqueduct under earthquake exaction is a complex process, especially when the control devices have multi-variety optimization problems, it is difficult to solve complex model by using conventional optimal methods. According to such problem, based on the MATLAB/SIMULINK platform, the finite element model of isolated aqueduct is established, a dynamic program is designed, and the lead rubber bearing (LRB) is used to control the seismic response. The result indicates the solution method based on the SIMULINK block is simple, precise, and facile, with intuitional and believable results. Setting LRB can efficiently reduce the input energy of earthquake. The results from the present study may serve as a reference base for seismic analysis of large-scale aqueducts.

Keywords: Aqueduct, Earthquake, Control, Simulink.

1 Introduction

As a building of transporting water spanning rivers, large-scale aqueduct plays an important role in South-to-North water diversion. The seismic safety problem is very important because the middle route of South-to-North water diversion project crossing the high intensity seismic region of China [1], the seismic intensity at the several region are 8 degree. During Wen-chuan earthquake, because a lot of aqueducts were damaged, local farmland irrigation water conservancy projects suffered big losses. Therefore, control of aqueduct structures for earthquake hazard mitigation is very essential.

As a building of transporting water spanning rivers, large-scale aqueduct plays an important role in South-to-North water diversion. The seismic safety problem is very important because the middle route of South-to-North water diversion project crossing the high intensity seismic region of China [1], the seismic intensity at the several region are 8 degree. During Wen-chuan earthquake, because a lot of aqueducts were damaged, local farmland irrigation water conservancy projects suffered big losses. Therefore, control of aqueduct structures for earthquake hazard mitigation is very essential.

* Corresponding author.

D. Jin and S. Lin (Eds.): Advances in CSIE, Vol. 1, AISC 168, pp. 41–46.
springerlink.com © Springer-Verlag Berlin Heidelberg 2012

With the further study of passive control theory, the isolated technical based on LRB has been widely used in engineering, the development of computer science also gives powerful support to it, however, some problems are difficult to solve because of their individuation during practical projects. SIMULINK is an environment for multi-domain simulation and Model-Based Design for dynamic and embedded systems. It provides an interactive graphical environment and a customizable set of block libraries that let you design, simulate, implement, and test a variety of time-varying systems, including communications, controls, signal processing, video processing, and image processing[2,3] Numerical simulations of the seismic responses of the large-scale isolated aqueduct are performed within the MATLAB environment through a SIMULINK block. The results showed that, compared with traditional analysis method, the solution method based on the SIMULINK block is simple, precise, and facile, with intuitional and believable results. It is also convenient for control calculation.

2 Analysis Model of Isolated Aqueduct

The aqueduct belongs to the thin-wall beam construction, according to this characteristics, the body of aqueduct is divided into several beam segment elements (BSE) [4] along its longitudinal direction. There are two nodes in each BSE; each one has 7 degree-of-freedoms (DOFs. The DOFs of the ith node for any BSE are denoted by *ui, ui', vi, vi', wi, Φi, Φi'*, the bending displacement in X-direction, the angle of distortion about X-axis, the bending displacement in Y-direction, the angle of distortion about Y-axis, the displacement in Z-direction, angle of distortion about X-axis, the angle of free torsion, and the angle of constrained torsion, respectively. The displacements within the BSE can be expressed by

$$\begin{cases} u=[N(z)]\{u\}^e \\ v=[N(z)]\{v\}^e \\ w=[N_1(z)]\{w\}^e \\ \phi[N(z)]=\{\phi\}^e \end{cases} \tag{1}$$

Where, $\{u\}^e=[\, ui\ ui'\ uj\ uj']^T$, $\{v\}^e=[\, vi\ vi'\ vj\ vj']^T$, $\{w\}^e=[\, wi\ \ wj]^T$, $\{\Phi\}^e=[\, \Phi i\ \Phi i'\ \Phi j\ \Phi j']^T$. Because the constrained torsion of the cross section is considered, the nodal displacements should include the change of the torsion angle along the BSE. Thus the shape function *N(z)* for *u, v,Φ,* are taken to be third-order Hermit polynomials, while the shape function N1(z) for *w* is taken to be a linear polynomial.

3 Mechanical Model of LRB

LRB is simulated by WEN model [5], the restoring force is given by

$$Q(x,\dot{x}) = \alpha \frac{F_y}{Y} x + (1-\alpha) F_y Z \tag{2}$$

Where

$$Y\dot{Z} = -\gamma |\dot{x}| Z |Z|^{\eta-1} - \beta \dot{x} |Z|^{\eta} + A\dot{x} \tag{3}$$

Here α is the ratio of yield stiffness to initial stiffness, Y and Fy are yield displacement and force of isolation bearing, x and are horizontal displacement and velocity of isolation bearing, η is the characteristic coefficient, A, γ and β are the parameters of hysteretic curve, respectively. The study shows that different parameters have different hysteretic curve [6], as Figure 1 shows. The SIMULINK model of LRB is shown in Figure 2.

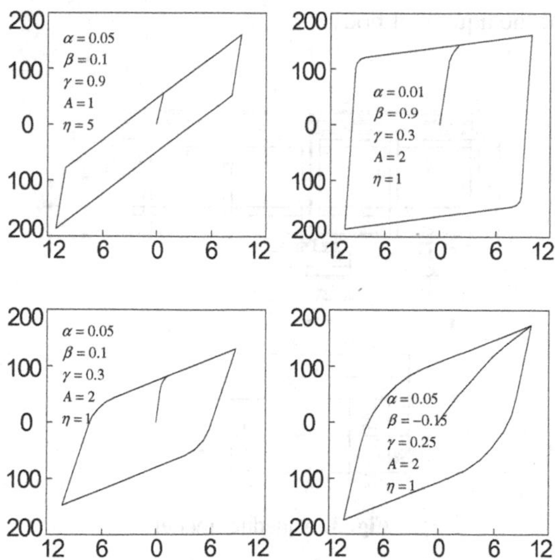

Fig. 1. The dynamic performance of LBR

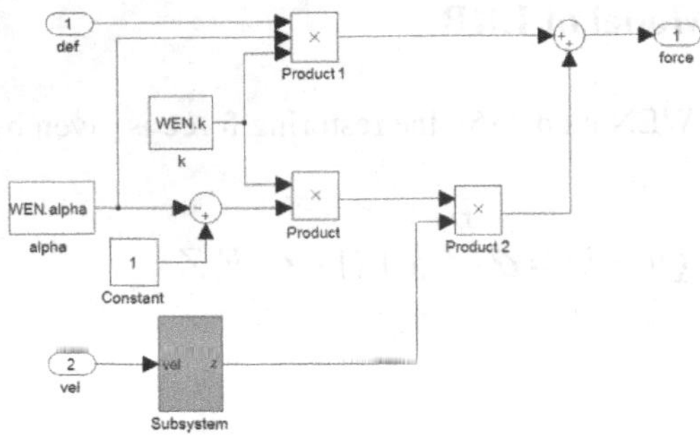

Fig. 2. The SIMULINK block of WEN model

4 Numerical Example

A large aqueduct in the South-to-North Water Convey Project is analyzed by seismic time-history analysis using the BSE model. The structure of the aqueduct is shown in Figure 3; the aqueduct has three spans with the length of 28.0m for each span. The aqueduct body is supported with a frame via a basin-shaped rubber support.

The frames of the aqueduct are in H-shape with the height of 11.2m and fixed bottoms. Both ends of the aqueduct are simply supported. Sixteen beam elements with seventeen nodes were used to model the frame; twenty-one BSEs with each length of 4.0m were used for the aqueduct body.

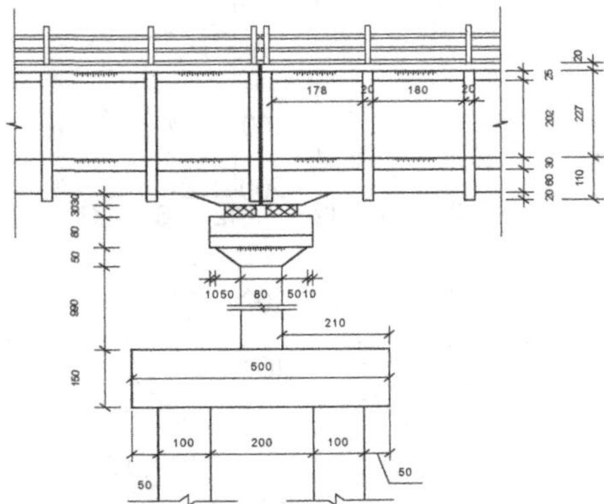

Fig. 3. Aqueduct model

The model of the aqueduct is subjected to the NS component of the 1940 El Centro earthquake, in simulation, the PGA is adjusted to 0.4 Gal.

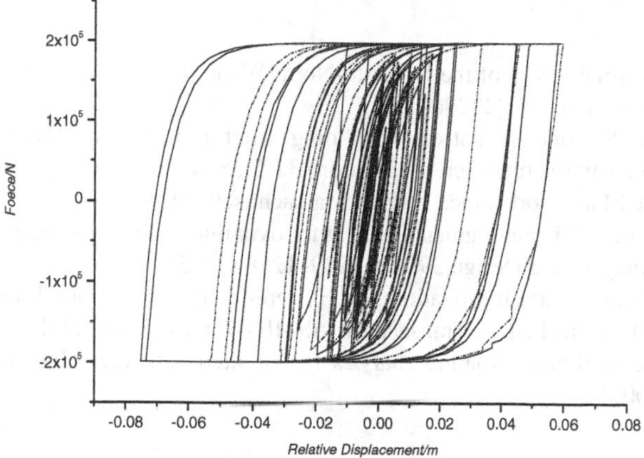

Fig. 4. Restoring force curve of LBR

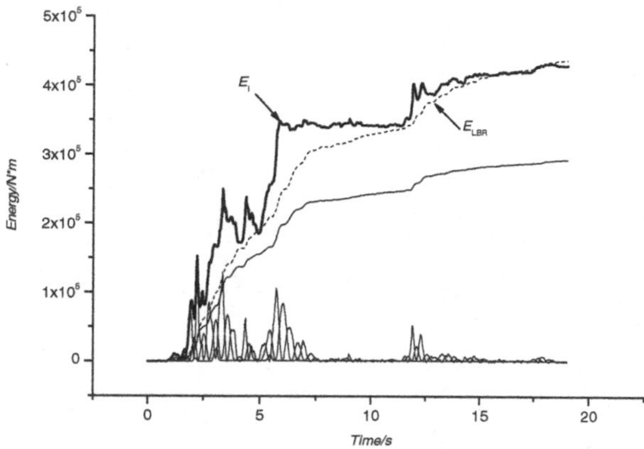

Fig. 5. Energy history of aqueduct under earthquake

The results show that the SIMULINK model can express the dynamic performance of LRB preferable (Figure 4). The LRB can dissipate most input energy of earthquake, as shown in Figure 5, it is a useful and economic method to protect the aqueduct structural during earthquake happens.

5 Summary

Based on the MATLAB/SIMULINK platform, the finite element model of isolated aqueduct is established, the LRB block is built, a dynamic program is designed, and the lead rubber bearing is used to control the seismic response. The result indicates the solution method based on the SIMULINK block is simple, precise, and facile, with intuitional and believable results. Setting LRB can efficiently reduce the aqueduct responses.

References

1. Chen, H.Q.: Seismic safety of the South-to-North Water Transfer Project. Journal of China Institute of Water 1(1), 17–22 (2003)
2. Mao, L., Li, A.: Seismic simulation of a sliding structure based on SIMULINK. Journal of Southeast University (Natural Science Edition) 32(5), 804–808 (2002)
3. MATLAB, The Math Works, Inc., Natick, Massachusetts (1994)
4. Wang, B., Li, Q.B.: A beam segment element for dynamic analysis of large aqueducts. Finite Elements in Analysis and Design 39(13), 1249–1258 (2003)
5. Wen, W.K.: Equivalent Linearization for Hysteretic Systems under Random Excitation. Journal of Applied Mechanics, Transactions ASME 47(1), 417–432 (1980)
6. Liang, C.: The nonlinear dynamic analysis for isolated aqueduct. Zhengzhou University, Zhengzhou (2005)

Research on Psychological Characteristics of Undergraduates and Construction of Study Style for Electronic-Information-Type Specialties

Liqun Huang, Fulai Liu, Hongfei Mo, and Xiang Cheng

Department of Electronic Information, Northeastern University at Qinhuangdao, Qinhuangdao, China
persistent_hlq@sina.com

Abstract. In this paper, we analyze psychological and behavioral characteristics of the post-90s college students, and summarize the facing problems of the construction of study style. On this basis, some effective measures are adopted to boost study-style construction. Good teaching style and rigorous examination discipline play an important role in promoting study-style construction; then, we emphasize the outstanding roles of college counselors, extracurricular scientific innovation activities, and career guidance. By establishing the standardized management system of study style, optimizing the guidance mode of student scientific innovation activities, and reinforcing the guidance of postgraduate admission examination and employment, students have clear learning objectives in each stage, which lays a solid foundation for forming a good style of study.

Keywords: study-style construction, scientific innovation activity, career guidance, teaching quality.

1 Introduction

Style of study is the stable spiritual outlook of learning objectives, attitude and behavior, which is formed in the long-term learning process [1]. Creating a good style of study is one of the important prerequisites for improving the quality of talents training, which is not only an important observation target for evaluating the education quality, but also an important means for cultivating good learning habits and self-management ability.

At present, owing to the expansion of college enrollment, the rapid change of social environment, and the diversification of value conception of the post-90s college students, some alarming problems of study style have emerged. Therefore, the construction of study style must become the most important task of universities; how to create a good style of study has become a hot topic that many teachers focus on specially [1-3]. Firstly, the current facing problems of study style are analyzed in this paper, and then the systematic and effective methods are presented to promote the improvement of study style.

D. Jin and S. Lin (Eds.): Advances in CSIE, Vol. 1, AISC 168, pp. 47–52.
springerlink.com © Springer-Verlag Berlin Heidelberg 2012

2 The Facing Problems of Study-Style Construction

As the reform and the open policy continue to deepen in china, our economy has rapidly developed, and the Internet is growing fast in popularity, which leads to the diversification of value conception. Thereinto, various utilitarianism and undesirable conduct have come forth in large numbers; rapid and drastic social changes will inevitably have a great effect on contemporary undergraduates.

The post-90s college students have some outstanding personality features, such as self-confidence, openness, and desire for self-expression. However, their attention paid to tasks and targets is not enough and continuous. They often blindly choose an unrealistic goal, and do a job carelessly and hastily. Thus, they usually have fine start and bad finish without enough patience. In addition, everyone faces keen competition in our fast-paced modern society. The post-90s undergraduates just contact this society, and their psychological bearing ability is relatively weak. Hence, they easily have the feeling of frustration, and then lose learning motivation and goals [4].

Currently, our overall situation of study style is good. However, due to the influence of above factors, some alarming problems of study style have emerged. These outstanding problems mainly includes that: the lack of learning objectives and motivation, bad learning habits and attitudes, poor self-control ability, and serious phenomenon of cheating in examinations [3, 5].

3 Practical Measures for Promoting Study-Style Construction

The construction of study style require the close attention of leaders at all levels, and the full participation of the whole teaching staff. The attention of leaders is of prime importance, and the idea should be built that educating people depends on everyone, environment and practice. On the basis of a perfect management system, we need to emphasize specially the important role of three teams, including college counselors, teachers, and student cadres [1].

3.1 Outstanding Emphasis on the Role of College Counselors

College counselors play a key role in the construction of study style. They are responsible for the daily management of student life and learning, keep the most close contact with students, grasp students' psychological, living and learning situation in time, and have a responsibility to select and train student cadres team. In the process of study-style construction, college counselors should be concentrating on the following three tasks:

(1) Establishing a standard management system of study style

Through classroom attendance checking, bedroom visits, individual conversations and other forms, college counselor can grasp the psychological state of students in time, and find out their deficiencies and undesirable tendencies of study style. In order to do well the management of study style, we must pay more attention to some important data, such as classroom attendance rate, the number of students studying by themselves after class, examination pass rate, and the number of students with psychological barriers in learning [5].

On this basis, we should provide efficient psychological guidance for these students with psychological barriers. The students violating the discipline will be criticized and educated, and even subject to disciplinary sanction. On the contrary, good students will be commended and awarded, which is beneficial to promoting the construction of study style effectively. In addition, through regular class meeting, student cadres are required to summarize the successful experience of learning, find the deficiencies existing in learning, and propose the methods of improving the style of study.

(2) Selecting and training student cadres carefully, and creating excellent class culture

Student cadres play a leading role in creating excellent class culture. We must do well in the construction of student cadres team, and give play to the subject function of students, so that we can achieve better results with less efforts. Through the establishment of class information members system, leaders at all levels and college counselor can keep abreast of the teaching quality of teachers, know the situation of student learning and living in time, and do well in prediction and prevention of the unexpected incidents.

The selection of student cadres must adhere to the principle of "being beneficial to the construction of study style", and the relevant standards should be established, such as candidate's learning achievement, working attitude, team spirit, and organizational skills. After selecting student cadres team, they will be responsible for creating excellent class culture. In this process, we should focus on training student cadres, know their work achievements and their facing problems in time, give them effective guidance, and help them overcome difficulties. Creating excellent dormitory and class cultures is the most effective approach, lays a solid foundation for constructing good study style, and thereinto, student cadres team and class information members system play a key role.

(3) Organizing various academic meetings, and carrying out rich and colorful extracurricular activities

The post-90s students have strong self-consciousness and vigorous energy; they are full of passion for scientific knowledge, practical skills, and sports activities. However, most of them have not very clear learning targets, and their self-control ability also isn't enough strong. They often do a job carelessly and hastily without enough patience, moreover, a small number of students are addicted to online games, and cut class, fail or cheat in examinations. Even as final-year students, they still haven't found the direction of learning, feel so confused, and do not know what to do.

Considering above psychological and behavioral characteristics of the post-90s college students, college counselors should organize various activities, help students quickly find learning targets and learning methods, and enrich students with culture, sports, and sci-tech activities. Thus, the construction of study style will be effectively promoted through cultivating the campus culture. The activities which need to be elaborately organized should include: the exchange meeting of learning experiences, the experience exchange meetings of postgraduate admission examination and employment, the conversation meeting between teachers and students, special subject academic seminars, extracurricular scientific competitions, and colorful recreational and sports activities.

3.2 Outstanding Emphasis on the Role of Extracurricular Scientific Activities

Promoting study style through good teaching style is the most important approach for improving the teaching quality, which need to be firmly grasped. Good teaching style and rigorous examination discipline play an important role in promoting the construction of study style.

On the other hand, The post-90s college students have vigorous energy; they are full of passion for scientific knowledge and practical skills. In the process of constructing study style, we must highlight the important leading role of extracurricular scientific innovation activities. Through extracurricular scientific competitions, students' learning interest and potential are fully stimulated, which further enhance students' engineering practical skills, and cultivate students' team spirit and innovation ability.

For extracurricular scientific innovation activities, management systems, incentive mechanism, and service guarantee mechanism must be scientifically established. On this basis, we give full play to the important role of backbone instructors, teaching team, and scientific research team, and provide relative equipments and dedicated laboratories. Students' research projects in innovation activities include students' scientific contest projects, teachers' scientific research sub-projects, and some projects of interest to students [6, 7]. We should focus on researching the effective guidance mode of scientific innovation activities. After summarizing successful experiences of the interest group of electronic technology, we further explore better organizational forms and guidance modes, and these guidance modes will be widely promoted in other scientific research groups..

The department of electronic information in our school has established students' dedicated innovation lab. students team up annually for the application of scientific innovation projects, and backbone teachers with abundant practical experiences are responsible for guiding them. The department academic committee evaluates these application projects periodically, and some excellent innovative projects will be recommended to our university. Then, these approved projects are funded, and a certain amount appropriations are allocated for purchasing the necessary electronic components and devices. Under the guidance of instructors, students carry out the work according to the progress plan of projects. Meanwhile, the department will organize the mid-term project inspection. After final appraisal and acceptance of innovative projects, our school will give outstanding projects the university-level awards.

The winning project team continue to improve their works, and then participate in various provincial or national technology innovation competitions, such as "Challenge Cup" national undergraduate extracurricular academic science and technology works competition, national undergraduate electronic design contest, "Freescale Cup" national undergraduate smart car competition, China undergraduate mathematical contest in modeling, national software professional talent design and development competition, national computer game competitions, and so on. In recent years, through the above exercise in extracurricular science and technology activities, our students have achieved good results in various national innovation races; the winning rate and the award level have raised significantly. Furthermore, students' practical ability has been enhanced, which promote the construction of study style in our school.

3.3 Outstanding Emphasis on the Role of Conversation Meetings and Career Guidance

In each semester, our school will hold the conversation meetings between teachers and students during mid-term teaching inspection, student representatives reflect the existing problems in teaching and learning, and experienced teachers record and answer these problems. Finally, the formal written records of the conversation meetings must be sent to the department heads.

Teaching research sections should take full advantage of the conversation meetings, find out the teaching problems in time, do well in guiding students, answer their questions patiently, and further improve the teaching quality. Frequently asked questions need to be clearly explained, such as the arrangements of course hours in each semester, the features of professional experiment courses, how to choose optional courses, learning methods and the main points of learning at different learning stages. Professional teachers should guide students to increase learning initiative, and enhance the communication between teachers and students, and the communication among students from same or different grades.

In addition, professional teachers must reinforce the guidance of postgraduate admission examination and employment[3], and take full advantage of the exchange opportunity with students during the break. For representative problems, we may spend a few minutes to give a brief answer in classroom. For example, how to choose between taking postgraduate admission exam and finding a job? which university is more appropriate for postgraduate admission application? how about the employment situation of recent graduates? how to prepare the postgraduate admission exam? how to prepare a job interview? how to select the correct direction of job hunting for our specialty? how to draw up the career planning? what's the difference for the professional knowledge required by different jobs? Teachers should have a full understanding of the above questions, and be able to answer the similar questions asked by students at any time.

Through the exchange during the break and the conversation meetings during mid-term teaching inspection, teachers do well in guidance of postgraduate admission examination and employment, so that students have clear learning targets at each stage. In this way, students' interest, and initiative in learning are mobilized and inspired, thus, the shortcomings of the post-90s undergraduates, such as the lack of learning motivation, bad learning attitudes, and poor self-control ability, can be surmounted to a certain extent, which is beneficial to forming a good study style.

By using the above methods, Study style has been significantly improved in our department. The rate of postgraduate admission, the employment rate and the employment quality have been heightened every year, which fully validate the effectiveness of these measures. Specific data are given in Table 1.

Table 1. The rate of postgraduate admission and the employment rate in our Department

Specialized Subject	The rate of postgraduate admission			The employment rate		
	2009	2010	2011	2009	2010	2011
Computer Science and Technology	11.34%	18.32%	16.49%	70.10%	82.20%	85.64%
Communication Engineering	7.05%	16.77%	19.23%	82.69%	77.64%	88.46%
Electronic and Information Engineering	12.84%	19.64%	24.22%	79.05%	87.34%	89.44%

4 Conclusion

Teaching activities are in close connection with teaching and learning, students are the subjects of learning, and a good study style is one of important prerequisites for improving the teaching quality. Teaching research sections and college counselors jointly study the psychological characteristics of current undergraduates, keep abreast of the new trends of thought changes, and find out new problems and new contradictions. Consequently, we can better guide students in fostering correct views on life and values, and stimulate their sense of responsibility and learning interest. The construction of study style should be regarded as one of the most important tasks for college counselors.

References

1. Yu, X.: Practice and exploration of reinforcing the construction of undergraduates' study style. China off-Campus Education (4), 11–12 (2010)
2. Li, Z.: Survey analysis and countermeasures research on the study-style construction of universities in Nanjing region. Business Culture (academic version) (10), 204–205 (2010)
3. Kuang, Y., Xu, J.: Discussion on the role of undergraduates' career planning education in constructing the study style. Entrepreneurs World (academic version) (2), 142–143 (2010)
4. Chen, L.: Strengthening the study-style construction of the post-90s undergraduates. Read and Write Periodical 7(6), 81–82 (2010)
5. Liu, L., Huang, J., Liu, X.: Research on countermeasures and the role of college counselors in constructing the study style. Ideological and Political Education Research 26(4), 126–128 (2010)
6. Hao, Y.: Research on the guidance mode of student scientific innovation activities in vocational colleges. Communication of Vocational Education (3), 90–93 (2010)
7. Song, B.: Scrolling-support guidance mode of undergraduate scientific innovation activities. Heilongjiang Researches on Higher Education 119(3), 146–147 (2004)

Research on Self-Organization Mechanism of Discipline Cluster System

Changhui Yang* and Fengjuan Cao

School of Business, Zhengzhou University, P.R. China, 450001
yang0825@zzu.edu.cn

Abstract. We have dicsussed the concept of discipline cluster and discipline cluster system has been put forward, and then analyses their basic connotation and function of discipline cluster system. This paper discusses the dissipation structure character of discipline cluster system, puts forward the disturbing factors of discipline cluster system evolution, at last, analyses the self-organization connection of discipline cluster system, favorable discipline cluster construction will benefit from these in a certain extent.

Keywords: Discipline Cluster, Dissipation Structure, Self-Organization.

1 Discipline Cluster and Discipline Cluster System

1.1 Discipline Cluster

The concept of "Cluster" roots in ecology, the original intention of cluster is the biology colony that perch in the same habitat with symbiotic relation. And now is widely used in industry economy, that is industry cluster, that means the interconnected companies and associated institutions in geographically close, they are the same or related industries in a specific field, which connect together by the common nature and complementarities.

Applying the concept of cluster into the field education, that is discipline cluster, it is the aggregation of different and closed discipline as well as their relations that taking one certain or consanguineous relation discipline or based on one basic theory as the core, relying on related disciplines, and supported by basic disciplines (such as mathematics and English etc basic disciplines). It can be divided into nuclear and non- nuclear discipline cluster (Or called Collaborative Discipline cluster). The nuclear discipline cluster is take one certain discipline as the core or based on the together basic theory; the non- nuclear discipline cluster (Or called Collaborative Discipline cluster) is to take the related discipline as the core. Discipline Cluster is a opening and complex system and a dissipative framework. It requires constantly exchange with the outside world for material, personnel, information and absorb the negative entropy flow -- enough material, Energy and information, and then by the processing in the system, then export new information and release energy to the

* This paper was supportted by DIA090238.

D. Jin and S. Lin (Eds.): Advances in CSIE, Vol. 1, AISC 168, pp. 53–58.
springerlink.com © Springer-Verlag Berlin Heidelberg 2012

external environment. Meanwhile, the discipline is a cluster of non-equilibrium system of very complexity; there does exist complicatedly nonlinear interaction and feedback loop among composed factors. The basic configuration of discipline is as figure 1 showing.

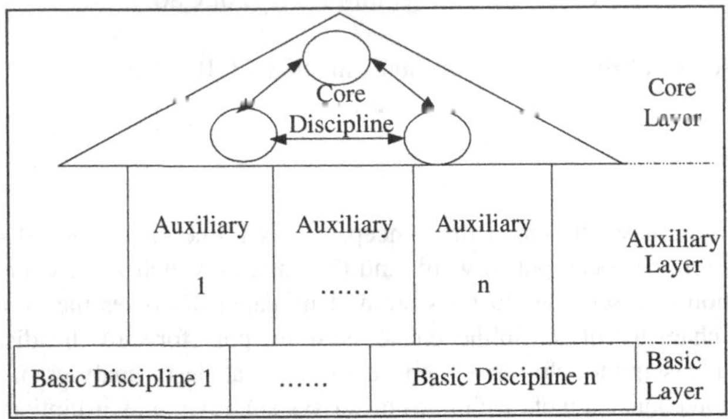

Fig. 1. Figure of Discipline Cluster basic Construction

The core layer indicates the certain discipline, or multi-discipline that have high relation or based on the same theory. The auxiliary layer indicates the related auxiliary discipline that derived from the core discipline or the same theory. The basic layer indicates the basic disciplines, such as such as mathematics, foreign languages.

From the point of system, university is composed by discipline cluster sub-system and other assistant subsystems. Because of the economy, the subsystem that consists of basic discipline lies in university, and so we believe that the main research disciplines inspection group is the relationship between core disciplines and supplementary disciplines.

1.2 Discipline Cluster System

Discipline cluster system consists of discipline cluster, academic team, the material and technical foundation, cultural atmosphere and the system (or mechanism). From the previous analysis, we know that discipline is the base that university performs its functions and the base that specialty and scientific establish. The deploying profit of people, finance and material in university is realized by discipline, and the character, advantage and science report is embodied by discipline. And so, discipline system is most important part in university, the material and technical foundation which is the inputting factors, in a certain cultural atmosphere, relying on some systems (mechanism), acts on discipline by academic team, thereby completes university functions. From that perspective, the basic function of discipline cluster system is to cultivate people, academic research, social services and the conversion of academic results.

2 Analysis of the Dissipative Structure Character of Discipline Cluster System

From the point of system, discipline cluster system is a dissipative structure. Discipline cluster system is opening systems, which need exchange material, personnel, and information with the outside world. Meanwhile, it is a very complex and non-equilibrium systems. There is existence the complex nonlinear reciprocity relation and feedback loop among the consisted factors Therefore, discipline cluster system possesses of the conditions of forming dissipative structures.

Except as the purpose, wholeness, open, and so the nature of information feedback of ecological system, physical system, engineering system and other natural system, discipline cluster system has the particularity, just as following:

1. The function of people dominates the operation of discipline cluster system. People are the most active and most fundamental factor. People designs, dominates and changes discipline cluster system, and enjoys the benefits of discipline cluster system.
2. The communication between discipline cluster systems is enslaved to vary factors, and cannot spontaneously progress. The communication of people, technology and information will be enslaved to the restriction of many factors including government, social and natural conditions, and the operating mechanism. If the system that has dissipative structure character will be created, there will need the favorable external environment and internal conditions.
3. Uncertainty and randomness, there are many random factors and uncertainties in disciplines cluster system. Because of the influence of random factors and uncertainty factors of uncertainty, it is very difficulty to hold the law of discipline cluster system than any other natural systems.
4. Multi-target and multi-function, discipline cluster system is a syntheses with multi-target and multi-function, it requires to achieve the overall optimization of objective function under the limited conditions. The very importantly all-around indicator of discipline system is to achieve more academic fruit. To achieve the greatest academic achievements, we must ensure the best quality, lowest cost, optimal output and more reliability.
5. Anti-jamming character, discipline cluster system has a strong anti-jamming capability. External and internal mutations cause of the larger impact may break through the restrictions on the return, dissipative structure will collapse, and it is not going to disappear, but under the right conditions in a stable and orderly formation of a new and evolving dissipative structure, structural change to a higher dissipative structure.
6. The development of discipline cluster depended on advancing promotion force of system. Because discipline cluster system constantly exchange material, people and information with outside environment, there will have the negative entropy flow, and will form the force that promote inter-act between subsystem in discipline cluster system. The more frequent, the bigger the negative entropy flow, and the bigger of the promoting and cooperating force.

3 Analyses of Disturbing Factors of Discipline Cluster System

Prigogine had put the concept of dissipative structure forward in the early years. It means a stable micro system framework that formed by means of exchanging energy and material with outside world when system is at the time of non-equilibrium state of thermodynamics. it is self-organization of macro system under the non-equilibrium state [2]. The equilibrium phase is a quiet phase that isolation system formed after keeping long time. It is not changing with time and has no relation with the outside world; the entire system has a single uniform, silence and disorganized state [4].

Discipline cluster system is an opening system far from equilibrium phase. It is a dissipative structure that takes academic research as center, and including personnel cultivation, academic achievements change, society service. The most discipline cluster system evolves from the single direction to personnel cultivation, academic research, academic achievement transition and society service. Clearly, discipline cluster system with dissipative structures character need open to the outside world and obtain resource of material, energy and information that sustainable develop. Moreover, due to the existence of extremely complex nonlinear system, the change of any factor may cause the change of all other factors, and give feedback to the initial changing factors. The entire self-organization of discipline cluster system development is controlled by the feedback mechanism of nonlinear relation. The self-organization phase of discipline cluster system is disturbed by different factors, and the disturbance has dynamic character and produces positive effect or negative effect, that will spur the structure and function of the original system to intense phase change.

Under the condition of guarantying the exchange material, energy and information with the outside, the phase transformation of discipline cluster system is enslaved to the following factors: development of university, academic team, development strategy of discipline cluster system, foundation of material and technology, culture and university environment (micro-environment). Affected by the above-mentioned factors, discipline cluster system continued to expand in all directions, such as the rich fruits of academic research, personnel cultivation, academic achievements conversion increased, the number of scientific research personnel, external financing and material and technical basis for the large-scale introduction of the improvement. The expansion will result in the development of discipline cluster system, the level of local economic and social progress.

4 Analyses of Self-Organization of Discipline Cluster System

The self-organization of discipline system is to spontaneously form the structure full of organizational character from chaos phase under the condition of guarantying the exchange material, energy and information with the outside. Leading by the order parameter, nonlinear interaction and restraint of the internal and external factors and other subsystems will form the dissipative structure, just as figure 2 showing.

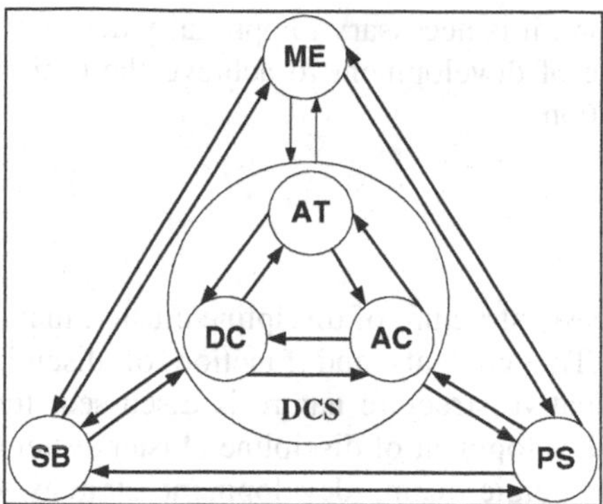

Fig. 2. Figure of self-organization of discipline cluster system

DCS indicates discipline cluster system, DC indicates discipline cluster, AT indicates academic team, AC indicates cultural atmosphere, PS indicates development status of university, SB indicates material and technical foundation, ME indicates university environment. In Figure 2, the change of the various factors of DC, AT, AC, PS, SB and ME changes is decided by their self0development condition and the nature of various factors. Discipline cluster system is enslaved to the influence and restraint of subsystem, and also a have feedback, change the relationship between the subsystems, the whole system nature depends on the impact of the interaction between subsystems.

Due to the unstable variable exceed the threshold value, structure of discipline self-organization system will mutate. Research level of academic team in discipline cluster system is an un-damping and unstable variable, research level of academic team always developed to the more senior level. The original level of scientific research is an impetus and condition for further development. The development and change of scientific research level decides the status of discipline cluster system. When it exceeds a certain threshold, discipline cluster system will change, which will be replaced by the new structure.

When the old structure has not completely disappeared, the new structure is far from being established, discipline cluster system is the chaotic system. At this time, the self-organization process in discipline cluster system is very complex. The structure of discipline cluster system is to realize self-organization by diffusing and competing. At this juncture, the change scope of variables in the system is within the threshold, the effect that the variables to system structure is to be control.

When the linkage between the subsystems in discipline cluster system can restrict the phase of subsystem, and the system is to show as the certain structure, this is called order. It is the fruit that interacted and diffused by every elements and subsystems.

The process of realizing orderly development of system by self-organization can show that the evolution of discipline cluster system structure is a self-organization process. It is not according to the optimal model that had been predetermined, but is gradually evolving, and forms the macro-structure at last. Therefore, in the development of discipline cluster system, it is not necessary to assume and optimally design the macro-structure; it is necessary to optimally design the realizing process of every stage. Every stage of development to achieve the optimum design is the real significance of optimization.

5 Conclusions

The paper presents the basic structure of discipline cluster, that is core layer, auxiliary layer and basic layer. The contents and function of discipline cluster system is discussed, and the dissipative structure nature is discussed, too. And the disturbing factors of affecting the development of discipline cluster system include development status of university, academic team, development strategy of discipline cluster, material and technological foundation, cultural atmosphere and environment of schools (micro-environment). At last, the self-organization relationship of discipline cluster system is analyzed.

References

1. Porter, M.E.: Clusters and the new economics of competition. Harvard Business Review, 77–90 (November/December 1998)
2. Nicolis, G., Prigogine, I.: Self-organization in Non-equilibrium Systems. John Wiley & Sons (March 1997)
3. Kan, K., Shen, X.: Prigogine and theory of dissipative structure. Shaanxi Science and Technology Publishing House, Xi'an (1982)
4. Ebeling, W., Ulbricht, H.: Self-organization by Nonlinear Irreversible Processes. Springer (1996)

Study on Clustering Algorithms of Wireless Self-Organized Network*

Haitao Wang[1], Lihua Song[2], and Hui Chen[1]

[1] Institute of Communication Engineering, PLA Univ. of Sci. & Tech., Nanjing, China
[2] Institute of Command Automation, PLA Univ. of Sci. & Tech., Nanjing, 210007, China

Abstract. Clustering algorithm is one of key technologies for constructing hierarchical network structure and its quality directly affects performance of wireless self-organized network (WSON). In this paper study backgrounds of clustering algorithms in WSON are introduced firstly and then principles and objectives of clustering algorithms are expounded. Afterwards, existent clustering algorithms in WSON are categorized, compared and analyzed in detail, including clustering based on node ID, degree, mobility and weights, whether selecting cluster-heads, single-hop cluster or multi-hop cluster and geographical information based clustering algorithms, etc. Finally, conclusions are given.

Keywords: Wireless Self-organized Network, Clustering Algorithms, QoS, VBN, MDS.

1 Introduction

Wireless self-organizing network (WSON) is the product of integration and development of computer network and mobile radio communications network, which is characterized by using wireless communicated, no network center, self-organization and multi-hop relay. In general, Wireless ad hoc networks (MANET), wireless sensor network (WSN) and wireless mesh network (WMN) all can belong to WSON [1-3]. In order to improve the wireless ad hoc network scalability and QoS support hierarchical network structure is often used. Clustering algorithms organize numerous

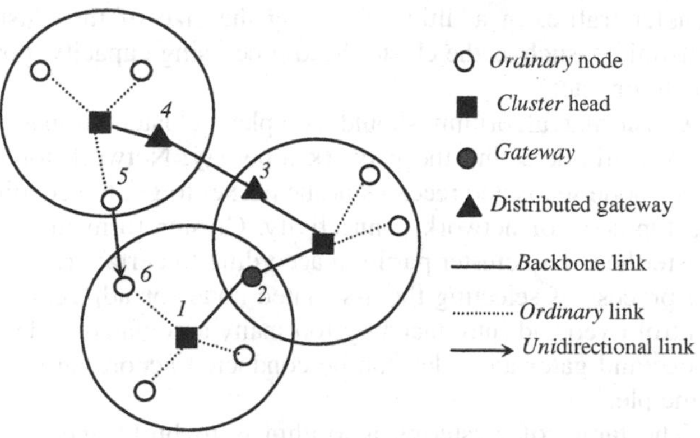

Fig. 1. Typical clustering structure of wireless self-organizing network

* This paper is supported by NSFC (61072043).

D. Jin and S. Lin (Eds.): Advances in CSIE, Vol. 1, AISC 168, pp. 59–65.
springerlink.com © Springer-Verlag Berlin Heidelberg 2012

nodes into manageable clusters according to system requirements, as the key technology to construct hierarchical network structure; its quality directly affects various indexes of wireless self-organizing network performance [4].

Cluster based hierarchical network structure is shown in figure 1. Network is divided into a number of clusters, and each cluster consists of a cluster head and a plurality of ordinary nodes. Cluster heads and gateways/distributed gateways form a virtual backbone network (VBN). In cellular network, the allocation of resources can be achieved more easily, because mobile terminal can learn bandwidth requirements directly or with the aid of a base station. Through dividing the network into clusters, this method can be extended to WSON. In clusters, the cluster head can control nodes' service access request and allocate the bandwidth properly. Based on cluster structure, hybrid routing algorithm can be adopted, such as using prior routing in cluster and reactive routing between clusters to reduce routing cost. In addition, the QoS and security of network can be improved with the help of virtual backbone network. Therefore, dividing network into clusters by clustering algorithm can largely improve performance of WSON.

2 Principle and Target of Clustering Algorithm

According to system requirements clustering algorithms divide network into multiple clusters which can communicate with each other and covers all nodes. When network environment changes, cluster structure is updated to maintain normal network function. Size of cluster should be decided based on the transmission power of each node and the cluster's own characteristics. If the cluster is too large, the cluster head's burden is heavy, and the distance between ordinary nodes to the cluster head is so far that excessive energy will be consumed. By contrast, if the cluster is too small, channel spatial reuse rate can be increased, and the nodes' power consumption for transmission can be reduced. However, small size of the cluster will lead to a large number of clusters in the network. As a result, the hops on the route between source node and destination node are more, which will increase the packet delivery delay and transfer traffic. In addition, selecting the size of the cluster should consider many constraints, such as the cluster head processing capacity, power loss and geographical condition, etc.

Clustering algorithm should complete cluster formation and cluster-connecting work while detecting the network topology. Network topology detection refers that every node sends and receives probe packet to get the condition of adjacent nodes and the topology of network connectivity. Cluster formation is the process of selecting cluster head and cluster partition according to certain rules. Cluster connection means the process of selecting the associated node for adjacent clusters. In order to reduce control overhead introduced by too many associated nodes, the selection of gateway nodes and gateway nodes can be conducted according to the minimum node degree principle.

The target of clustering algorithm is to build and maintain a connected cluster ensemble which can better support resource management and routing protocol with full network coverage at the cost of less computation and communication overhead. Clustering algorithm should be simple and efficient, and the original structure should be maintained when only a few nodes move and the topology changes slowly, thereby reducing the cost for reforming clusters and improving overall network performance. Ideally the entire network are covered with least cluster heads, namely the set of cluster heads form a minimum dominating set (MDS). The problem of cluster heads selection optimization problem that meeting the MDS is NP complete problem [5], so heuristic clustering algorithms are often used. A good clustering mechanism should try to keep the network topology stable, reduce the time of re-clustering, optimize the connection within cluster and among clusters, and also consider the energy level of the node, network load balance and used routing protocols, etc.

3 Classification and Comparison of Clustering Algorithms

To date, a lot of clustering algorithms are put forward to build and maintain a cluster networks structure in WSON. The choice of clustering algorithm depends on the requirements of application, network environment and node features. Different clustering algorithms differ in optimization objectives, such as reducing cluster computing and maintenance cost, minimizing cluster heads, maximizing cluster stability and survival time and so on.

3.1 Clustering Algorithm Based on Node's ID

Minimal node ID (LowID) clustering algorithm is proposed by C.R. Lin firstly which is a typical representative of clustering algorithms that based on ID [6]. In LowID algorithm the node having minimum ID is selected as cluster head among adjacent nodes, and nodes one hop away from cluster heads become members of clusters and no longer involved in the cluster head election process. The process is repeated until all nodes belong to clusters. LowID clustering algorithm is simple and easy to realize. In mobile environment, the frequency of updating cluster heads in the LowID algorithm is rather slow and the cost required for maintaining clusters is small. Its drawback is that this algorithm tends to select the node with smaller ID as cluster head, which make these nodes consuming more energy, moreover it does not consider network load balance and other factors.

3.2 Clustering Algorithm Based on Maximum Node Degree

Maximum node degree clustering algorithm is also called maximum connectivity algorithm, and its goal is to improve the network's ability of controlling and reduce the number of clusters [7]. The idea of this algorithm is simple: the node having the maximum node degree among adjacent nodes is selected as the cluster head. Node degree refers to the number of adjacent nodes who are one hop away. When node degrees are the same, the node with less ID is selected as cluster head. The remaining process is the same as LowID algorithm. This algorithm has advantages that the cluster number of the network is less, and the average number of hops between source node and destination node is less, thereby reducing the packet delivery delay. However, less clusters number also reduced channel's spatial reuse rate. In addition, when the mobility of the node is high, the frequency of cluster head update will increase sharply and the cost of maintaining cluster structure becomes intolerant.

Maximum node degree algorithm is often applied to occasions where the mobility of nodes is rather low and the density of nodes is rather sparse.

3.3 Clustering Algorithm Based on Nodes' Mobility

In order to adapt to the node mobility and increase the stability of cluster structure, nodes can be assigned weights according to the mobility, and based on weights the cluster head is elected. In the lowest node mobility clustering algorithm the higher is the node mobility, the lower is its assigned weight, nodes having the highest weight among neighborhood are selected as cluster heads [8]. In this clustering algorithm, a mechanism is required to quantify the node mobility. A simple method supposes that the mobility for any pair of nodes is the average on time of absolute value of their relative speed, but this method requires GPS to obtain the positions of nodes. For this, a kind of collective local mobility index is proposed in which nodes compare he strength of two successive signals received from a neighbor node to estimate their relative mobility. When the mobility is high, the lowest mobility clustering algorithm can significantly reduce the cluster head updating frequency, its drawback is that computing overhead of cluster heads is large since node weights are changing frequently, and not considering the load balance of the system and the node energy consumption.

3.4 Clustering Algorithm Based on Combination of Weights

The election of cluster heads is crucial for the performance of the network, a variety of factors needs to be considered, and a reasonable compromise should be taken based on the network environment and application needs. Based on above considerations, combined weight clustering algorithm (WCA) can be used [9]. In WCA algorithm, each node is assigned a value to indicate the level of the node is suitable to act as cluster head. Node weights can be expressed by a general formula which considers multiple factors: Weight=a×mobility+b×degree+c×power+d×energy, in which a, b, c and d are determined by the application and the network environment. In above formula, mobility represents the node velocity or the mobility relative to neighbor nodes, degree represents the node degree, and power stands for node transmission power, energy stands for node residual energy. In addition, and more variables may also be added based on requests.

3.5 Clustering Algorithm without Cluster Heads

In the realization of clustering algorithms, there are two kinds of views: selecting cluster heads or not. The latter considers that the cluster head may become the bottleneck of network because of heavy burden. When cluster heads fail, the performance of network will be seriously affected. On the other hand, if not selecting cluster heads, each node need maintain the routing information within clusters and between clusters, the cluster maintenance overhead is too high, and the benefits of cluster structure can not be utilized fully, such as convenient network management and resource allocation. When cluster heads exist, they act as a coordinators and managers and can form a virtual backbone network with associated nodes. As a result, cluster structure with heads can be conveniently used to realize hierarchical routing, mobile management and resource allocation, and the cost required for maintaining cluster structure is reduced, especially when the scale of network is large. In the condition that the network size is moderate, node density is high and the node processing capacity is low, in order to prevent cluster heads become bottleneck, no cluster head clustering algorithm can be considered. Partition based hierarchical link state routing protocol (ZHLS) takes advantage of the clustering structure without cluster head to improve the performance of routing protocols, in which each node in cluster are treated equally, and thus traffic bottleneck can avoid to enhance the robustness of the system [10].

3.6 Clustering Algorithm Considering Cluster Size

Clusters generated by most clustering algorithms is one-hop clusters with heads or two-hop clusters without heads, this kind of cluster structure can be easy to achieve, but not considering the number of nodes in clusters which will affect the performance of cluster-based network protocol in some conditions. For example, when nodes in one cluster is excessive, the burden of the cluster head will be heavy; on the other hand if nodes in cluster are too small cluster number will be excessive, making cluster maintaining cost and packet delivery delay become large. Accordingly, some mechanism is needed to control the size of cluster and the cluster number. A feasible solution is to determine the maximum distance D between the cluster head and cluster nodes based on the node's density and number. Max-Min D hops clustering algorithm is multi-hop clustering algorithm which selects cluster heads based on node ID, the algorithm can effectively reduce the number of cluster heads, and the temporal complexity of the algorithm is O(d) [11]. In addition, the size of the cluster can also be adjusted by limiting the number of cluster nodes. For example, the MMWN system architecture limits the number of cluster nodes through cluster's separation and merge.

3.7 Clustering Algorithm Based on Geographic Location

Clustering algorithms above does not consider the node location information, if GPS can be used or the node location coordinates can be estimated, clustering algorithm based on geographic location can be taken. A simple clustering method based on nodes' location coordinates is depicted in reference [12] (see Figure 2): assuming that each cluster is square, not mutually overlapped and equal in size, there is no cluster head within the cluster, cluster nodes can communicate directly, i.e., the nodes' communication range is equal to the length of the diagonal of a square cluster. Then the node can determine their situated clusters according the following formula: x' = floor($\sqrt{2}$ x/r+0.5); y' = floor($\sqrt{2}$ y/r+0.5). In above formula, x and y are physical coordinates of the

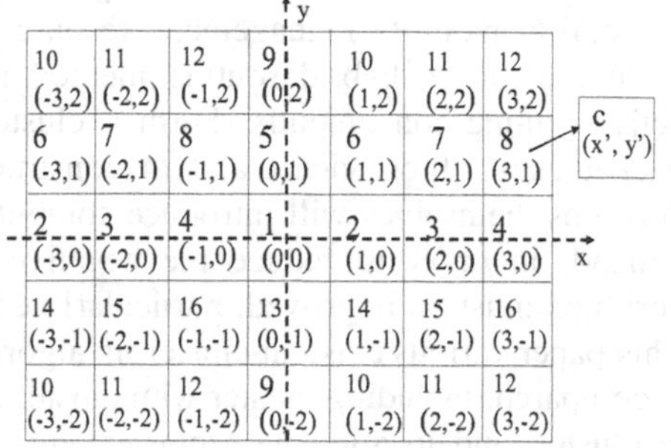

Fig. 2. Clustering and channel distribution based on geographical information

node, r is the communication range, (x', y') represents the cluster that the node situates. Considering channel spatial reuse, nodes 2 hops away can reuse channel without interference. In addition, after dividing nodes into clusters in terms of location information, the size of the cluster can be adjusted dynamically according to density, mobility and transmitting power of nodes to optimize network performance. This location-based clustering algorithm can manage nodes effectively, but the message overhead is rather large.

3.8 Comparison and Analysis

To sum up, different clustering algorithms have distinct optimization objectives and application scenarios, which should be chosen reasonably according to network environment and system requirements. For example, LowID clustering algorithm is simple and convenient, but did not consider node energy consumption and load balance. Maximum node degree clustering algorithm can reduce the number of cluster heads, but will further aggravate the burden of cluster heads. Minimum mobility clustering algorithm can better adapt to node mobility, but need accurate estimation of the node mobility. Combined weighted clustering algorithm considers a variety of factors affecting the performance of the cluster network structure, thus having higher flexibility and adaptability. The clustering algorithm based on regulation of cluster size can be used to optimize the performance of the cluster structure, but the realization is rather complex. The geographic location based clustering algorithm emphasizes geographic distribution of nodes and can better manage the mobile nodes, but requests that nodes can learn their own position coordinates, which leads message overhead is rather large. In addition, No cluster head clustering algorithm is intend to avoid cluster head becoming the bottleneck node, but weakens the cluster structure efficiency.

4 Conclusions

Cluster network structure based on clustering algorithm is very important to improve the performance of wireless self-organizing network. Cluster structure is rather beneficial to mobility management, resource allocation and channel access, and can be suitable to use hybrid routing method integrating advantages of proactive and reactive routing conveniently. Further, cluster structure can also be used to improve the efficiency of broadcast and to construct a Bluetooth network. But clustering algorithms themselves will introduce computation, communications and maintenance overhead. In order to reduce the negative impact, the performance of clustering algorithms must be improved, particularly communication and maintenance expense. In this paper various existent clustering algorithms in WSON are introduced, analyzed and compared, including cluster with cluster head or not, single hop cluster and multi-hop cluster, and location base cluster, etc. After all, each cluster algorithm has its merits and defects, must be selected reasonably based on users' demand and applicable environment.

Acknowledgment. This paper is supported by NSFC (NO: 61072043).

References

1. Zheng, S., Wang, H., Zhao, Z., et al.: Ad hoc network technology. Posts and Telecom Press (January 2005)
2. Wang, X.: Wireless sensor network measuring system. Machinery Industry Press, Beijing (2008)
3. Lee, M.: Emerging Standards for Wireless Mesh Technology. IEEE Wireless Communication 13(2), 56–63 (2006)
4. Wang, H., Zheng, S.: The clustering algorithm in mobile Ad hoc network. Journal of PLA University of Science and Technology 5(3), 28–32 (2004)
5. Abbasi, A., Younis, M.: A survey on clustering algorithms for wireless sensor networks. Computer Communications 30(6), 2826–2841 (2007)
6. Lin, C.R., Gerla, M.: Adaptive Clustering for Mobile Wireless Networks. IEEE Journal on Selected Areas in Communications 15(7), 1265–1275 (1997)
7. Gerla, M., Tsai, J.T.C.: Multicluster, Mobile, Multimedia Radio Network. Wireless Networks 1(3), 255–265 (1995)
8. Basagni, S.: Distributed Clustering for Ad Hoc Networks. In: International Symposiun on Parallel Architectures, Algorithms and Networks, Perth, pp. 310–315 (June 1999)
9. Chatterjee, M., Das, S., Turgut, D.: WCA: A Weighted Clustering Algorithm for Mobile Ad Hoc Networks. Journal of Cluster Computing 5(2), 193–204 (2002)
10. Ng, M.J.: Routing Protocol And Medium Access Protocol For Mobile Ad Hoc Networks. Phd Thesis, Polytechnic University (January 1999)
11. Amis, A.D., Prakash, R., Vuong, T.H.P., Huynh, D.T.: Max-Min D-Cluster Formation in Wireless Ad Hoc Networks. In: Proceedings of IEEE INFOCOM 2000. Tel Aviv (March 2000)
12. Guo, H., Liu, L.: Tactical Ad hoc network clustering algorithm based on group mobility. Computer Application 29(7), 1871–1874 (2009)

Study on Speech Control of Turning Movements of the Multifunctional Nursing Bed

Yingchun Zhong and Xing Hua

Faculty of Automation, Guangdong University of Technology, Guangzhou City, Guangdong Province, China
gzzhw@126.com, huaxing06_@126.com

Abstract. As it takes long time to train while Hidden Markov Model(HMM) is employed to speech recognition, an improved method is proposed and successfully applied to speech control of turning movements of the multifunctional nursing bed. The experimental results show that the improved algorithm reduces training time greatly and the accuracy of speech recognition remains unchanged, which results in the multifunctional nursing bed being easier to use.

Keywords: Nursing Bed, Turning Movements, Speech Recognition, Improved HMM.

1 Introduction

The paralysis, disabled and senseless patients are increasing daily with the growth of aging population in China. Many paralyzed patients need to be turned over the body by nurses constantly to avoid decubitus. However, artificial turning position of the patients is very troublesome and labour-intensive. Additionally, the nurses can not ofen turn position of the patients in time because of negligence, which is easy to result in the patients suffering decubitus [1]. So some nursing beds with automatic turning function appeared on the market. But most of these nursing beds only have the manual control function and can't meet convenience and efficiency requirements of patients [2].

Speech recognition control technique is an advanced control method. When speech recognition is applied to controling the nursing beds, it will facilitate the users greatly, especially for those who have lost self-care ability. It can help them boost their confidence, reduce dependence of others and increase their courage in life [3]. So it's necessary to study the speech recognition system of the multifunctional nursing bed, which will provide a better way to control the multifunctional nursing bed conveniently and effectively. The dynamic time warping (DTW) , hidden Markov models (HMM) and artificial neural networks (ANN) are often employed to speech recognition [4]. Because the speech recognition based on HMM has the better recognizing adaptability, it is employed to nursing bed control in this paper. First the training algorithm of the speech recoginition based on HMM is improved. Then the improved algorithm is applied to the multifunctional nursing bed. Finally, the experiments have been done to testify the effectiveness of improvement.

D. Jin and S. Lin (Eds.): Advances in CSIE, Vol. 1, AISC 168, pp. 67–72.
springerlink.com © Springer-Verlag Berlin Heidelberg 2012

2 HMM Model and Its Improvement

Speech recognition includes three parts: speech signal pre-processing, speech feature extraction and speech recognition training [5].

HMM model is usually expressed by $\lambda = (A, B, \pi)$ where A indicates the state transition probability matrix, B represents the output probability density function and π represents the initial state probability distribution [6].

For the HMM, there are three basic problems to be solved [7]:

The problem of calculating the output probability: given observation sequence $O = (o_1, o_2, \cdots, o_T)$ and the HMM model $\lambda = (A, B, \pi)$, how to calculate the output probability of the HMM observation sequence $P(O \mid \lambda)$. The solution of the problem can be used by the forward and backward algorithms.

The problem of the state sequence decoding: given observation sequence $O = (o_1, o_2, \cdots, o_T)$ and the HMM model $\lambda = (A, B, \pi)$, how to determine an optimal state transition sequence. It can be solved by using the Viterbi algorithm.

The problem of estimating the model parameters: how to adjust the parameters $\lambda = (A, B, \pi)$ to make $P(O \mid \lambda)$ maximum. It can be solved by using Baum-Welch training algorithm [8-10].

During the training of HMM parameters, a large number of observation sequences need to be trained. Therefore, it causes a shortage of large computation and long training time [11-14].

When training HMM parameters, the output probability of all the observation sequences needs to be calculated and accumulated to get the sum of the output probability. If the sum of the output probability changes relatively small and reaches a certain value (threshold D), the iteration will be ended. In addition, set a constant as the maximum number of iterations. If the number of iterations exceeds the constant, the iteration will be ended too. When the iteration is ended, HMM parameters will be achieved.

The improvement steps of training HMM parameters are as follows:

Step 1: Initialize the number of iterations and the threshold, train and recognize the speech, note the training time and recognition results.

Step 2: Maintain the initial number of iterations and increase the threshold D, then train and recognize the speech, note the training time and recognition results.

Step 3: Compare the training time and recognition results which are got from Step 1 and Step 2. If the training time is reduced and recognition accuracy changes little, continue to increase the threshold.

Step 4: Repeat step 2 and 3 to achieve suitable threshold.

Step 5: Maintain the threshold and reduce the number of iterations until to get the appropriate number of iterations.

3 Application of Improved HMM Algorithm to Speech Control of Turning Movements of the Mutlitfunctional Nursing Bed

In the paper, the improved HMM algorithm is applied to speech control of turning movements of the multifunctional nursing bed. And 14 voice commands are used to the speech control of the turning movements. The turning movements and vocie commands are shown in Table 1.

Table 1. Turning movements of the multifunctional nursing bed

Function	Turning movements	Voice commands
Left upward and downward adjustments	The bed turns left upward	Left up
	The bed turns left downward	Left down
Right upward and downward adjustments	The bed turns right upward	Right up
	The bed turns right downward	Right down
Overall upward and downward adjustments	The bed-body rises up	Bed up
	The bed-body falls down	Bed down
Legs upward and downward adjustments	Legs turn upward	Leg up
	Legs turn downward	Leg down
Bed-back upward and downward adjustments	Bed-back turns upward	Back up
	Bed-back turns downward	Back down
Bed-head forward and backward adjustments	The bed-head turns forward	Forward
	The bed-head turns backward	Backward
Actions of the bed stop	The actions stop	Stop
Actions of the bed reset	The actions reset	Reset

The steps of applying the improved HMM algorithm to the control of turning movements are as follows:

Step 1: Train voice commands. Record 14 voice commands of turning movements. Then train the voice commands with improved HMM algorithm to establish the library of HMM models.

Step 2: Recognize the voice commands. Set the multifunctional nursing bed as speech control mode, and speak one voice command of turning movements by the microphone. Then the system will obtain a recognition result with the speech recognition algorithm.

Step 3: Execute the voice commands. The control system of the multifunctional nursing bed will turn the recognition result into a control signal to drive the motors mounted on the bed to work, so as to complete appropriate turning movements.

4 Experiments and Analysis

In the paper, PC is used as an experimental platform. Its operating system is Windows XP Professional, CPU is AMD 4200+, dominant frequency is 2.19 GHz, internal memory is 1G. And MATLAB 7.11 is used as simulation software. The experimental steps are as follows:

Step 1: HMM training and recognition algorithm are programmed at the platform MatLab2010b.

Step 2: Record 4 groups of voice commands with the recording format of 8 kHz sampling frequency, 16 bits sampling precision and single channel PCM. Each group of voice commands are 14 commands about turning movements of the nursing bed.

Step 3: Initialize the number of iterations and the output probability threshold. Then train 3 groups of voice commands for the HMM parameters. After training, recognize the voice commands with the rest group of voice commands and note the result.

Step 4: Maintain the initial number of iterations, change the output probability threshold. Then train and recognize the voice commands, note the result.

Step 5: Maintain the output probability threshold, change the number of iterations. Then train and recognize the voice commands, note the result.

The results are shown in Table 2 and Table 3. In Table 2, the number of iterations equals 40. And in Table 3, the output probability threshold equals $5e-3$.

Table 2. Effect caused by threshold change on training time

Threshold	Training time[s]	Recognition accuracy
D= $5e-6$	1347	0.87
D= $5e-5$	942	0.87
D= $5e-4$	413	0.85
D= $5e-3$	213	0.85
D= $5e-2$	160	0.79
D= $5e-1$	178	0.60

Table 3. Effect caused by iterations change on training time

Iterations	Training time[s]	Recognition accuracy
N=40	213	0.85
N=35	174	0.85
N=30	197	0.85
N=25	172	0.83
N=20	166	0.79
N=15	160	0.64

Table 2 and Table 3 show that the training time is decreasing when the output probability threshold increases, and the training time will be reduced too when the number of iterations is reduced. When the threshold increases from $5e-6$ to $5e-3$, which means the threshold increases to 1000 times, the training time reduces $(1347-213)/1347=84\%$, while recognition accurate decreases from 87% to 85%. Correspondingly, when the number of iterations changes from 40 to

20, which means the number of iterations decreases by 50%, the training time reduces $(213-166)/213=22\%$, while recognition accurate decreases from 85% to 79%. So it can be concluded that increasing the threshold can reduce training time greatly in the condition of maintaining recognition accurate basically unchanged.

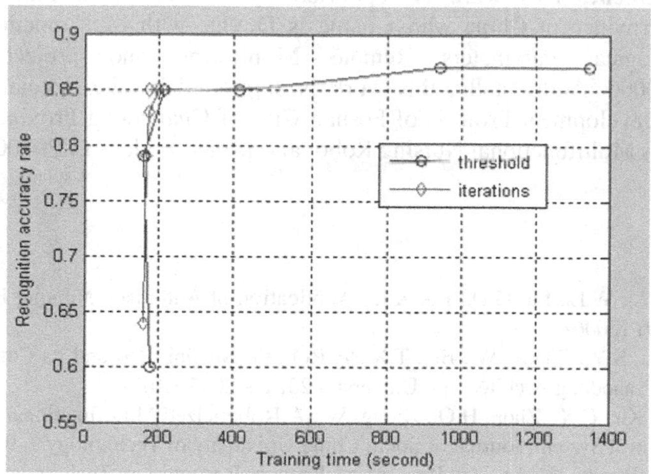

Fig. 1. Recognition accuracy rate and training time curve diagram

The effect on training time which is caused by change of the output probability and the number of iterations is showed in Fig.1. Apparently, the change of the output probability threshold has great effection on training time while the change of the number of iterations has quite small effection on training time. According to Fig.1, there is a intersection point of the red curve and the blue curve. At the point, the output probability threshold equals $5e-3$ and the number of iterations equals 35. When the output probability threshold equals $5e-3$ and the number of iterations equals 35, the training time is 174 seconds, and the recognition accuracy rate is 0.85. It has achieved relatively ideal training time and recognition accuracy.

5 Conclusion

As it takes long time to train when the HMM is applied to speech recognition, an improved algorithm is proposed in this paper, in which the output probability threshold and the number of iterations in the parameters training algorithm of HMM are studied. Then the improved algorithm is employed to multifunctional nursing bed and the experiments have been done. The results show that the change of the output probability threshold has great effection on training time while the change of the number of iterations has quite small effection on training time, and increasing the threshold can reduce training time greatly in the condition of maintaining recognition accuracy basically unchanged. When the output probability threshold equals $5e-3$ and the number of iterations equals 35, the training time is reduced greatly and the recognition accuracy maintains basically unchanged. It will reduce the training time when patients use the speech control of the multifunctional nursing bed, so as to improve the practicality of the multifunctional nursing bed.

Acknowledgement. This work is supported by the Technical Plan Project of Guangdong Province of China whose name is Device with Multifunctional Caring and Physiological Parameters Remote Monitoring and project code is 2010A030500006. Additionally, this work is supported by the Special Funds for Technology Development Projects of Foshan City of Guangdong Province of China whose name is Multifunctional Nursing Robot and project code is FZ2010013.

References

1. Liu, S.N., Li, W.L., Lu, G.D., Lu, X.L.: Application of Automatic Nursing Bed. Medical Forum 3, 91 (2006)
2. Tan, L., Lu, S.Y., Zhang, W., Bei, T.X.: A Robotic Nursing Bed and its Control System. Journal of Shandong Architectural University 25, 19–20 (2010)
3. Zhang, T., Xie, C.X., Zhou, H.Q., Xiong, W.: A Roboticized Multifunctional Nursing Bed and its Control System. Journal of South China University of Technology 3, 95–96 (2006)
4. Ma, Z.X., Wang, H., Li, X.: Review of Speech Recognition Technology. Journal of Changji College 3, 95–96 (2006)
5. Zhao, L.S., Han, Z.Y.: Speech Recognition System Based on Integrating Feature and HMM. In: International Conference on Measuring Technology and Mechatronics Automation, pp. 449–452 (2010)
6. Hu, L., Lu, L.X., Huang, T.: Application of an Improved HMM to Speech Recognition. Information and Control 36, 716–719 (2007)
7. Yang, Y.M., Wang, C.L., Sun, Y.: Speech Recognition Method Based on Weighed Autoregressive HMM. National Science Foundation, 946–947 (2010)
8. Zhou, D.X., Wang, X.R.: The Improvement of HMM Algorithm Using Wavelet Denoising in Speech Recognition. In: 3rd International Conference on Advanced Computer Theory and Engineering, pp. 439–440 (2010)
9. Zhang, X., Wang, Y., Zhao, Z.: A Hybrid Speech Recognition Training Method for HMM Based on Genetic Algorithm and Baum Welch Algorithm, pp. 2882–2886. IEEE (2007)
10. Mozes, Weimann, Ziv-Ukelson: Speeding up HMM Decoding and Training by Exploiting Sequence Repetitions. In: 18th Annual Symposium on Conbinatorial Pattern Matching, pp. 4–15 (2007)
11. Li, X., Wang, D.K.: Test Platform of Isolated-word Speech Recognition Based on MATLAB. Journal of SiChuan University of Science & Engineering 19, 97–100 (2006)
12. Shi, X.F., Zhang, X.Z., Zhang, F.: Design of Speech Recognition System Based on HTK. Computer Technology and Development 16, 37–38 (2006)
13. Xi, X.J., Lin, K.H., Zhou, C.L., Cai, J.: Key Technology Research for Speech Recognition. Computer Engineering and Applications 42, 66–69 (2006)
14. Seward, A.: A Fast HMM Match Algorithm for Very Large Vocabulary Speech Recognition. Speech Communication 42, 191–206 (2004)

Robust Control for a Mixed Leg Mechanism Four-Legged Walking Robot

ChuangFeng Huai, XueYan Jia, and PingAn Liu

School of Mechanical and Electronic Engineering, East China Jiaotong University, Jiangxi Nanchang 330013
hcf811225@163.com

Abstract. This paper designs a mixed leg mechanism four legged robot with excellent mobility and performance. Robot's virtual model is constructed and a control algorithm is proposed by applying virtual components at some strategic locations. For adjusting the planned control law continuously in response to the current movement of the feet of the robot, a control scheme has been presented to make the four-legged robot control systems robust to parametric and unstructured uncertainties.

Keywords: Robust Control, Four-legged robot, Virtual Leg, Contact Force, Friction Pyramid.

1 Introduction

One of the main reasons to develop a four-legged walking robot is to overcome the lack of mobility of wheeled vehicles on irregular terrains. The ability to traverse uneven or varying terrain at high speeds, turn sharply, and start or stop suddenly are all ordinary aspects of four-legged locomotion for a variety of cursorial mammals. Researches on four-legged robots have been widely carried out[1–3].

A quadruped robot with Multi-degrees of freedoms is a high nonlinear system. Robust control maybe a suitable method to solve the problems, but the literatures that use robust control to control the quadruped robot is sparse.

E.H. Mamdani applied fuzzy control to practical robot initially. Then, fuzzy system has been widely used and studied in robot modeling[4], flexible arm control[5], position control[6], fuzzy compensation control[7] and tracking design of robot system[8]. In addition, variable structure control system had developed some improved algorithm to weak chattering, such as dynamic adjust sliding mode parameter[9] and line estimate sliding mode parameter[10].

In this paper, the method of multi-body dynamics is employed to construct the equations of motion for the complicated biped robot system and designed a robust controller for stable dynamic biped robot system. We address how to generate a robust controller for a quadruped walking robot, and how to convert this plan into an appropriate joint-space walking pattern. What sets our approach apart from the large number of previous projects on similar topics is that we wish to adjust the planned control law continuously in response to the current movement of the feet of the robot, and that we wish to accomplish this adjustment in a very smooth way. Under the

D. Jin and S. Lin (Eds.): Advances in CSIE, Vol. 1, AISC 168, pp. 73–79.
springerlink.com © Springer-Verlag Berlin Heidelberg 2012

assumption of the bounded estimation errors on the unknown parameters, the proposed controller provides a successful way to achieve the stability and can provide a powerful safeguard for the robot system.

2 Robot System

2.1 Walking Machine and Foot Contact Model

The quadruped model described in this section and the nominal parameters are shown in Fig. 1. The total mass of the system is about 13.6 kg with a nominal leg length of 0.74 m. Unlike some other studies, the model has mass in each of the leg links and a symmetric distribution of mass in the trunk, and the body's center of mass (COM) is not located at its geometric center. Articulated knee joints are used instead of prismatic joints to better model biological legs.

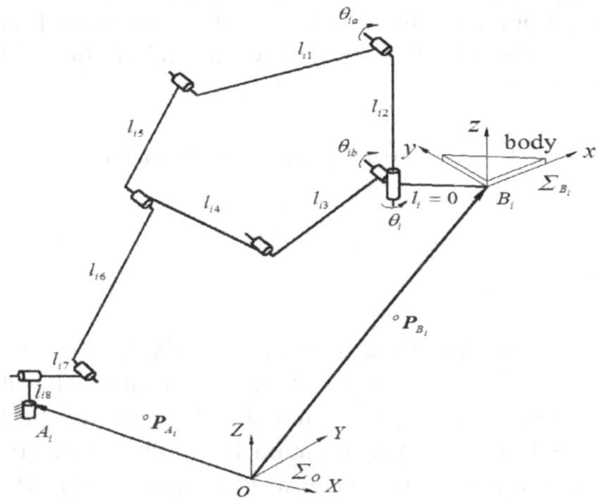

Fig. 1. Structural symmetrical quadruped model

Ideal actuators are modelled at each of the abductor/ adductor, joints so that the results of this study are independent of assumptions regarding specific actuator models. Principal moment of inertia is perpendicular to the axial direction of the thigh or shank link. Inertia along the axial direction is negligible.

2.2 Dynamics Formulation

Let $q \in R^n$ denote the joint coordinate of the four-legged robot, where, n is the rigid links of robot. When the robot links is rigidly in contact with an uncertain surface, the environmental constraint is expressed as an algebraic equation of the coordinate q and time t, namely, $\Phi(q,t)=0$, where $\Phi: R^n \times R \mapsto R^m$. Denote $A(q,t)$ as the

Jacobian matrix of $\Phi(q,t)$ with respect to q, i.e., $A=\frac{\partial \Phi(q,t)}{\partial q}$. Without loss of generality, the uncertain constraint is decomposed into a nominal part $\Phi_*(q)$ and a constraint modeling error part $\delta\Phi_*(q,t)$. This implies that the Jacobian matrix $A(q,t)=A_*(q)+\Delta A(q,t)$ with nominal $A_*(q)$ and uncertain $\Delta A(q,t)$. Meanwhile, several assumptions on the system are made as follows.

The dynamic equation of the constrained robot is written as

$$M(q)\ddot{q}+C(q,\dot{q})\dot{q}+F(q,\dot{q},t)=\tau-\tau_d=Y_1(q,\dot{q},\ddot{q})\eta \tag{1}$$

3 LNCS Online

Assumed that there exists a known bound ρ on parametric uncertainty such that

$$\|\tilde{\eta}\|=\|\eta-\eta_0\|\le\rho \tag{2}$$

where, η_0 represents the fixed parameters in dynamic model. Also, for unstructured uncertainty it is assumed that the norm of $\tau_d(t)$ is bounded as

$$\|\tau_d\|<w_1+w_2\|q\|+w_3\|\dot{q}\| \tag{3}$$

where, w_1, w_2 and w_3 are known positive constant. Equation (3) can be written as

$$\|\tau_d\|<W^{\mathrm{T}}Q \tag{4}$$

where, $W=[w_1,w_2,w_3]^{\mathrm{T}}$ and $Q=[1,\|q\|,\|\dot{q}\|]^{\mathrm{T}}$.

The robust controller design problem is as follows: given the desired joint trajectory $q_d(t)$, derive a control law for joint torques such that the manipulator joint position $q(t)$ accurately tracks $q_d(t)$ in the presence of parametric and unstructured uncertainties.

Joint position and velocity errors are defined as $e=q_d-q$ and $\dot{e}=\dot{q}_d-\dot{q}$, respectively. Define

$$r=\dot{e}+\Lambda e;\ \ v=\dot{q}_d+\Lambda e \tag{5}$$

where, Λ is a constant diagonal positive definite (pd) matrix.

Assume that q_d is a given twice continuously differentiable reference trajectory. As mentioned before, the robot dynamics is linear in parameter, thus we can write

$$M(q)\dot{v}+C(q,\dot{q})v+F(q,\dot{q},t)=Y(q,\dot{q},v,\dot{v})\eta \tag{6}$$

Substituting equations (2) and (3) into equation (6), yields

$$M(q)\dot{r}+C(q,\dot{q})r=Y(q,\dot{q},v,\dot{v})\eta-\tau-\tau_d \tag{7}$$

Now, we consider three compensation parts for the control law. In the first part, we define a nominal control law as

$$\tau_0=Y(q,\dot{q},v,\dot{v})\eta_0+Kr \tag{8}$$

where, η_0 represents the fixed parameters in dynamic model and Kr is the vector of PD control effort. The second and the third parts are considered to compensate for $\tilde{\eta}$ and τ_d, and are denoted by u_ρ and u_d, respectively. Thus the control law is

$$\tau=Y(q,\dot{q},v,\dot{v})\eta_0+Kr+u_\rho+u_d \tag{9}$$

We define u_ρ and u_d as

$$u_\rho=\begin{cases}\dfrac{\rho}{\|Y^{\mathrm{T}}r\|}YY^{\mathrm{T}}r & Y^{\mathrm{T}}r\neq 0\\ 0 & Y^{\mathrm{T}}r=0\end{cases}\qquad u_d=\begin{cases}\dfrac{1}{\|r\|}W^{\mathrm{T}}Qr & r\neq 0\\ 0 & r=0\end{cases} \tag{10}$$

To analyse the stability of the whole system the following Lyapunov function candidate is defined.

$$V(x,t)=\frac{1}{2}r^{\mathrm{T}}Mr+e^{\mathrm{T}}\Lambda Ke \tag{11}$$

where, K is a constant diagonal pd matrix.

Computing the time derivative of V along the trajectory and making some simplification by considering equations (1) and (5), yield

$$\dot{V}=2e^{\mathrm{T}}\Lambda K\dot{e}+r^{\mathrm{T}}(M\ddot{q}+C\dot{q}+F+\tau_d-\tau+M\Lambda\dot{e}+\frac{1}{2}\dot{M}r) \tag{12}$$

Based on (5), $\dot{q}=v-r$ and $\dot{v}=\ddot{q}_d+\Lambda e$. Also, due to skew symmetric property $r^{\mathrm{T}}[\dot{M}(q)-2C(q,\dot{q})]r=0\ \ \forall r\in R^n$. Therefore,

$$\dot{V}=2e^{\mathrm{T}}\Lambda K\dot{e}+r^{\mathrm{T}}(M\dot{v}+Cv+F+\tau_d-\tau) \tag{13}$$

which based on equation (6) is arranged as

$$\dot{V}=2e^{\mathrm{T}}\Lambda K\dot{e}+r^{\mathrm{T}}(Y\eta+\tau_d-\tau) \tag{14}$$

Substituting equation (7) into equation (14) and simplifying the result, we have

$$\dot{V}=-V_1+r^{\mathrm{T}}(Y\tilde{\eta}-u_\rho-u_d+\tau_d) \tag{15}$$

where, $V_1 = e^{\mathrm{T}}\Lambda K \Lambda e + \dot{e}^{\mathrm{T}} K \dot{e}$. Equation (15) can be written as the inequality

$$\dot{V} \le -V_1 + \left\| r^{\mathrm{T}} Y \right\| \left\| \tilde{\eta} \right\| - r^{\mathrm{T}} u_\rho + \left\| r^{\mathrm{T}} \right\| \left\| \tau_d \right\| - r^{\mathrm{T}} u_d \tag{16}$$

It follows from equations (1), (2) and (16) that,

$$\dot{V} \le -V_1 + z \tag{17}$$

where, $z = \left\| Y^{\mathrm{T}} r \right\| \rho - r^{\mathrm{T}} u_\rho + \left\| r^{\mathrm{T}} \right\| W^{\mathrm{T}} Q - r^{\mathrm{T}} u_d$.

Substituting equations (10) and gives $z = 0$, thus according to (17) we have

$$\dot{V}(x,t) \le -V_1(x) \tag{18}$$

V_1 is a pd function and based on equation (15), $\dot{V}$ becomes zero at the origin, thus, $\dot{V}$ is negative definite. We can rewrite V as $V = x^{\mathrm{T}} B x$ where

$$B = \begin{bmatrix} \Lambda K & 0 \\ 0 & \frac{1}{2} M(q) \end{bmatrix} \tag{19}$$

$M(q)$ is a pd matrix, as a result, $B(q)$ is a pd matrix. If $\lambda(B(q))$ represents the eigenvalues of $B(q)$, and we define $\underline{B} = \min_q \lambda(B(q)) I$, $\overline{B} = \max_q \lambda(B(q)) I$, where I is the identity matrix, then

$$\min_q \lambda(B(q)) \left\| x \right\|^2 \le V \le \max_q \lambda(B(q)) \left\| x \right\|^2 \tag{20}$$

We can define a continuous control law by defining u_ρ and u_d as

$$u_\rho = \begin{cases} \dfrac{\rho}{\left\| Y^{\mathrm{T}} r \right\|} Y Y^{\mathrm{T}} r & \left\| Y^{\mathrm{T}} r \right\| > \varepsilon \\ \dfrac{\rho}{\varepsilon} Y Y^{\mathrm{T}} r & \left\| Y^{\mathrm{T}} r \right\| \le \varepsilon \end{cases} \qquad u_d = \begin{cases} \dfrac{1}{\left\| r \right\|} W^{\mathrm{T}} Q r & \left\| r \right\| > \delta \\ \dfrac{1}{\delta} W^{\mathrm{T}} Q r & \left\| r \right\| \le \delta \end{cases} \tag{21}$$

where ε and δ are positive constants. It is obvious that the control law is continuous for any $\varepsilon > 0$ and $\delta > 0$. If $\left\| Y^{\mathrm{T}} r \right\| > \varepsilon$ and $\left\| r \right\| > \delta$ then based on equations (17), (21), we achieve again equation (18), thus, $\dot{V} < 0$. If $\left\| Y^{\mathrm{T}} r \right\| > \varepsilon$ and $\left\| r \right\| \le \delta$ then based on equations (17), (21), we have $\dot{V} \le -V_1 + \frac{1}{4} W^{\mathrm{T}} Q$. If

$\|Y^{\mathrm{T}} r\| \le \varepsilon$ and $\|r\| > \delta$ then similar to the second case, we can achieve $\dot{V} \le -V_1 + \frac{\rho\varepsilon}{4}$. If $\|Y^{\mathrm{T}} r\| \le \varepsilon$ and $\|r\| \le \delta$ then with an approach similar to previous cases, we have

$$\dot{V} \le -V_1 + \frac{1}{4} W^{\mathrm{T}} Q + \frac{\rho\varepsilon}{4} \tag{22}$$

The fourth case is the worst case and the inequality equation (22) is satisfied for all cases. On the other hand, if we define $\hat{x} = [e^{\mathrm{T}} \quad \dot{e}^{\mathrm{T}}]^{\mathrm{T}}$, then $V_1 = \hat{x}^{\mathrm{T}} P \hat{x}$ where

$$P = \begin{bmatrix} \Lambda K \Lambda & 0 \\ 0 & \Lambda \end{bmatrix} \tag{23}$$

It is clear that P is a pd matrix. V_1 is bounded with a class-K function as

$$\dot{V} \le \min \lambda(P) \|\hat{x}\|^2 \tag{24}$$

Substituting equation (24) into equation (22) and doing some straightforward calculation, we can show that if

$$\|\hat{x}\| > [\frac{1}{\min \lambda(P)} (\frac{\delta}{4} W^{\mathrm{T}} Q + \frac{\rho\varepsilon}{4})]^{0.5} \tag{25}$$

then $\dot{V} < 0$.

4 Conclusions

In this paper, a control scheme has been presented to make the quadruped robot control systems robust to parametric and unstructured uncertainties. This controller is designed based on a priori knowledge of uncertainties bounds. A constant bound is considered for parametric uncertainty and a position-velocity dependent bound is considered for unstructured uncertainty. The stability properties of the system with two different control laws were studied and stated via two theorems. For the first control law, it is guaranteed that the whole system is globally uniformly asymptotically stable. The second control law is continuous, and uniform ultimate boundedness of the tracking error is established for it.

Acknowledgments. This paper is supported by DEF of Jiangxi(Gjj11102) and NSF of Jiangxi (20114BAB216003).

References

1. Hirose, S., Fukuda, Y., Kikuchi, H.: The gait control system of a quadruped walking vehicle. Advanced Robotics 1(4), 289–323 (1986)
2. Wong, D.P., Orin, D.E.: Control of a quadruped standing jump over irregular terrain obstacles. Autonomous Robots 1, 111–129 (1995)
3. Jindrich, D.L., Full, R.J.: Many-legged maneuver-ability: Dynamics of turning in hexapods. The Journal of Experimental Biology 202, 1603–1623 (2005)
4. Miller, W.T.: Real-time application of neural networks for sensor-based control of robots with vision. IEEE Trans. System, Man, and Cybernctics, 825–831 (1989)
5. Ananthraman, S., Garg, D.P.: Training backpropagation and CMAC neural networks for control of a SCARA robot. Engng Applied Artif. Intell. 6(2), 105–115 (2003)
6. Stepanenko, Y., Su, C.Y.: Varible structure control of robot manipulators with nonlinear sliding manifolds. Int. J. Control 58(2), 285–300 (1993)
7. Fei, M., Chen, B.: Intelligent control method intercross synthesis and application. Control Theory and Applications 13(3), 273–281 (1996)
8. Wang, F.Y., Lever, P.L.A.: Rule generation and modification for intelligent control using fuzzy logic and neral networks, work report. The University of Arizona, Tucson (1995)
9. King, H., Litz, L.: Inconsistency detection-a powerful means for the design of MIMO fuzzy controllers. In: Int. Conf. Fuzzy Systems, pp. 1191–1197 (2006)
10. Takagi, T., Sugeno, M.: Fuzzy identification of systems and its applications to modeling and control. IEEE Trans. System, Man, and Cybernetics 11(1), 116–132 (1985)

Pedometer Algorithm Research Based-Matlab

Bin Huang and Xinhui Wu

Liuzhou Railway Vocational Technology College
hb1083@sina.com

Abstract. Pedometer can help people grasp the exercise in real time, through the detection of human walking and running to calculate the distance. In order to improve the accuracy of pedometer, the paper,with the help of Matlab simulation tools, take advantage of the acceleration sensor (ADXL330) axis output signals and the signal energy-based adaptive threshold to detect the number of peak for acceleration signal, in order to accurately calculate the number of steps of human walking.The experimental results show that compared with traditional method, based on signal energy adaptive threshold detection method has better performance, can effectively improve the accuracy of a pedometer.

Keywords: Pedometer, Adaptive threshold, Simulation tools.

1 Introduction

As society develops, people increasingly focused on their health, running as a convenient and effective form of exercise. Pedometer can help people grasp the exercise in real time, its main function is testing step, stride length can be calculated by walking away.

At present, The pedometer have mechanical pedometer and electronic pedometer for the composition.Pedometer for mechanical vibration caused by human walking pedometer or elastic small balls inside the reed vibrations to generate electrical pulses, the internal processor achieve the total-step function by determining the electrical impulses.The pedometer is the use of electronic acceleration sensor (ADXL330) [1], when the acceleration by measuring changes in human walking, to achieve the total-step function.

As everyone walking posture, stride are different, so in addition to steps number data other data is not accurate. But for the gait of people showed symptoms of the steps may not even accurate. In view of this, we use a pedometer counting algorithm matlab on a simulation study, with some reference value.

2 Model of Human Walking

To achieve the detection steps, we could have some understanding for walk on the attitude the first. There are many parameters to describe the behavior of human walking,so as distance, velocity, acceleration.This paper uses this acceleration parameter to simulate the human walking [2]. Walking, the foot arm, leg, waist, they

D. Jin and S. Lin (Eds.): Advances in CSIE, Vol. 1, AISC 168, pp. 81–86.
springerlink.com © Springer-Verlag Berlin Heidelberg 2012

will have a corresponding movement of the acceleration, so it will have a peak at some point. Acceleration from the feet to detect the number of steps is the most accurate, and the most obvious displacement is the waist down , and easy to carry, so we use the waist to detect the movement of steps. Walking the waist has the vertical movement up and down, and each step will have a larger start acceleration, we can get the walking steps by detection the peak. Figure 1 shown three different directions acceleration: vertical, forward and lateral.

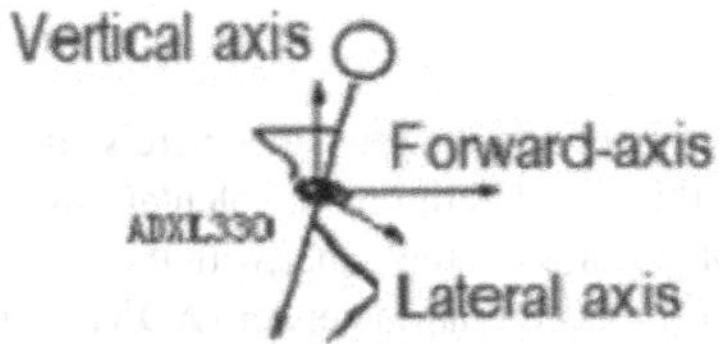

Fig. 1. Model of human walking

ADXL330 is a three-axis (X axis, Y axis and Z axis) analog output accelerometer, just as vertical, forward and lateral direction of the three sensors. ADXL330 the X, Y, Z axis represents the output data of human walking acceleration in three directions.

3 Adaptive Threshold and Algorithm

Different people have different sized acceleration of walking, but also by the impact of vibration and noise vibration, so the signal acceleration is changed and the impact of interference is unknown and time varying. If you use a fixed threshold of detection methods, they can not get a good detection performance, which requires we use adaptive threshold detection techniques.

Adaptive threshold is determined by the received signal of average power, that is used by estimating the signal power before the matched filter to construct the adaptive threshold. The construction of adaptive threshold need estimate the received signal power. Signal power P is the N sampling points of the average power, the signal average power P is

$$P = (\sum_{i=1}^{N} |r_i(t)|^2)/N \tag{1}$$

The $r_i(t)$ is the signal of i point sampling.Signal power-based detection methods are adaptive CFAR detection [3]. By the constant C can make the system maintain a constant false alarm probability, when the system is constant false alarm probability P, the coefficient is taken as

$$C = 1-(P_a)^{1/(N-1)} \tag{2}$$

The adaptive threshold based on signal power detection algorithm is

$$V_T = CP = \frac{1}{N}\left[1-(P_a)^{1/(N-1)}\right]\sum_{i=1}^{N}|r_i(t)|^2 \tag{3}$$

Generally, the human body to walk 0.5 to 2 steps per second, up to no more than 5 steps. Therefore, a reasonable pedometer output is 0.5 ~ 5Hz. We use the FIR cutoff frequency of 5Hz low-pass filter to filter high frequency noise.Figure 2 is a low-pass filter frequency response.

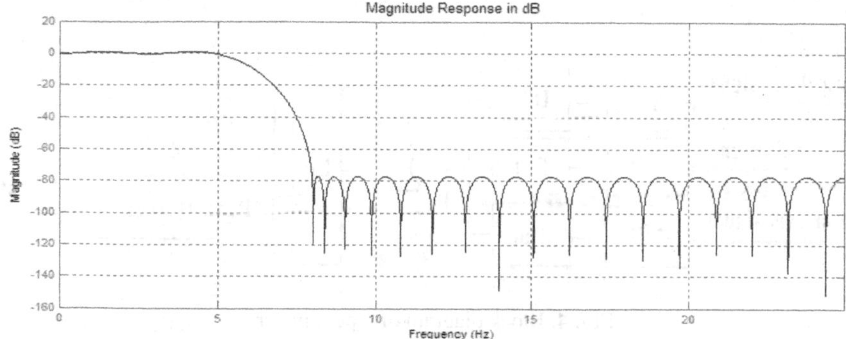

Fig. 2. Low-pass filter

Figure 3 is collected at the waist worn pedometer to the vertical axis acceleration waveforms, From the graph it is clear that there are eight peaks, representing eight walking steps, indicating that is feasible use of the number of steps the acceleration of the waist to detect.

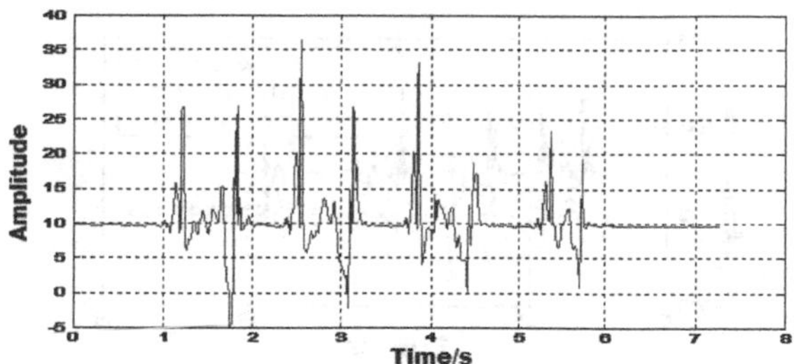

Fig. 3. The vertical axis signal waveform

According to statistics, the frequency of people walking normally at 110 steps / min (1.8Hz) [4], when the frequency of running no more than 5Hz, 100Hz sampling rate can be selected more accurately reflect the acceleration of change.

Different people have different acceleration of the size of walking, the traditional method is to use a maximum acceleration of the output shaft as a valid output, acceleration sensor peak - which is used to determine the output shaft output is valid [5]. However, this method easily lost count of points. So take full advantage of this three-axis acceleration sensor output signal, to overcome this shortcoming. Three-axis acceleration sensor output signal, after sampling, low pass filter, and a series of differential treatment, the peak detection, signal energy-based adaptive threshold algorithm, accurately detect the number of steps of human walking.

4 Experiment and Simulation

In this paper, ADXL330 acceleration sensor detects acceleration changes in human movement. Block diagram shown in Figure 4, ADXL330 accelerometer output three signals, each 0.01S a data collection, both the sampling rate of 100Hz.

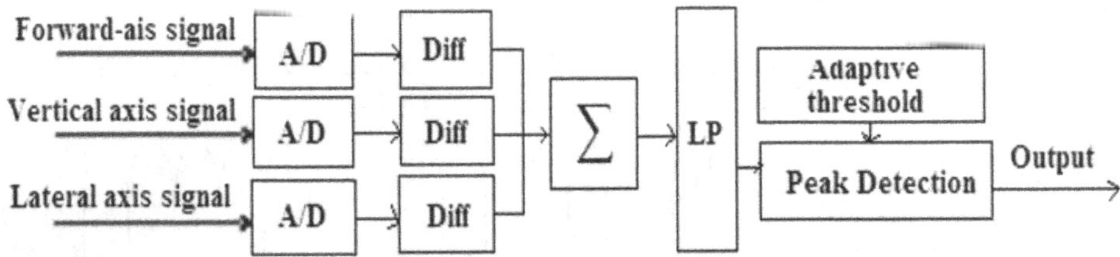

Fig. 4. Block diagram of a pedometer

From the figure we can see that the output of the ADXL330 accelerometer acceleration signals were treated with three-way sampling and differential treatment, the modulus of the three signals are summed, and then filtered through a low pass filter signal frequencies above 5Hz, and through Adaptive threshold for peak detection, and thus the number of human motion detection. Figure 5 shows the three-way acceleration sensor output signal.

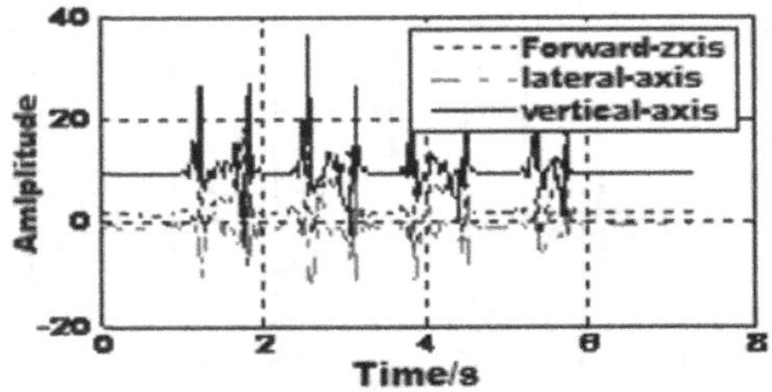

Fig. 5. Three-axis acceleration sensor signal

The three signals by sampling,respectively differential treatment,differential treatment is intended to interfere with flattened segments,the signal segment is more acute.

The three-way differential signal after summation and the modulus, through the low-pass filter, filter out the clutter of more than 5Hz, the waveform shown in Figure 6, after low-filtered signal smoother.

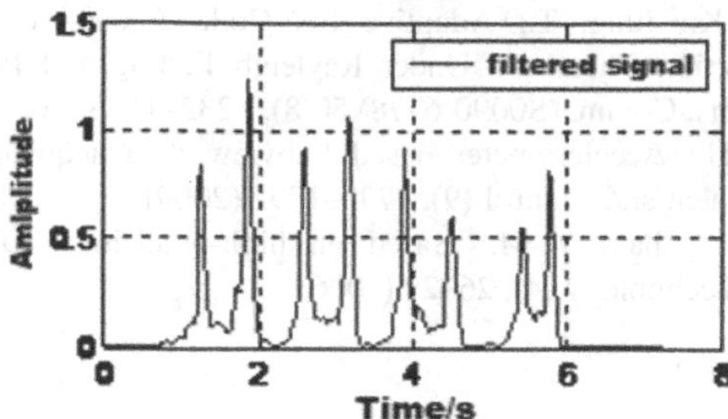

Fig. 6. Filtered Signal

Adaptive threshold based on signal energy detection, N=16,pa=0.00015,Figure 7 for the adaptive threshold test conditions can be seen from the figure, that with the adaptive threshold signal changes. When the signal power is unknown and constantly changing, the fixed threshold can not adapt to signal fading, interference and other dynamic changes, and automatically adjust the threshold can effectively solve these problems.

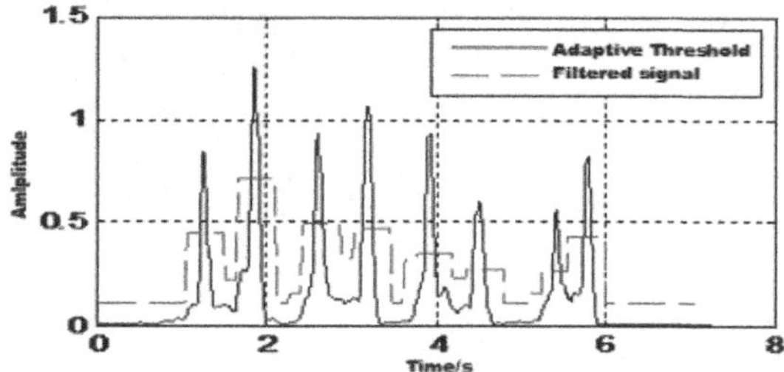

Fig. 7. Adaptive Threshold Detection

Pedometer accuracy by calculating the peak to calculate the number of human movement. The large number of experiments show that the proposed adaptive threshold based on signal energy, more accurate testing methods.

5 Summary

In this paper, we process three-axis accelerometer ADXL330 output signal using Matlab, use of adaptive threshold to detect human walking steps, and can improve the detection probability than the traditional fixed threshold detector.The method can effectively eliminate the detection of small signal low probability phenomenon.

References

1. ADXL330 Datasheet (EB/OL). Analog Devices Inc., Norwood, MA, `http://www.analog.com`
2. Chen, financial, Xing, moving autumn: Matlab-based acceleration sensor vibration signal processing method. Microcontroller and Embedded Systems (9), 16–19 (2006)

3. Choi, K., Cheun, K., Jung, T.: Adaptive PN Code Acquisition Using Instantaneous Power-Scaled Detection Threshold Under Rayleigh Fading and Pulsed Gaussian Noise Jamming. IEEE Trans. Comm. (S0090-6778) 50(8), 1232–1235 (2002)
4. Guo, Z., Han, X.-H.: Accelerometer based Labview data acquisition platform motion. Computer Measurement and Control (9), 1790–1792 (2009)
5. Song, H., Liao, S., Zhao, Y.-M.: Based on high-precision ADXL330 accelerometer pedometer. Sensor Technology (4), 26–29 (2006)

Multi-user Detection Based on EM Gradient Algorithm

Jinlong Xian and Zhi Liu

College of Information Science and Engineering, Henan University of Technology,
Lianhua Street, Zhengzhou, 450001, Henan, P.R. China
13991339876@vip.sina.com

Abstract. The EM algorithm is commonly used for the maximizing likelihood of missing data in recent years. Even if E-step is easy to be implemented, the calculation of M-step is also a great difficulty.In order to reduce the computational complexity of the M-step, we proposed the EM gradient algorithm.In MUD system within Gaussian noise, simulation numerical is carried out. The results show that the EM gradient algorithm has a good BER performance curve .It also has the same convergence speed as standard EM algorithm.

Keywords: Multi-user detection, EM algorithm, EM gradient algorithm.

1 Introduction

In 1977, Dempster, Laird and Rubin (DLR) firstly proposed the EM algorithm model in their research paper [1].H. Poor apply the EM algorithm to Multi-User Detection (MUD) in 1996. EM algorithm is so popular because it could simply constantly tend to global optimal solution by iterative, but the application was limited by the shortcomings of itself. In some cases, even if it is easy to find the full-likelihood expectation, it is difficulty to achieve the maximization of the full-likelihood expectation [2]. What's more, the convergence speed is a weakness. How to improve these problems? The EM gradient algorithm is chosen. It reduces the computational complexity of the M-step. In MUD system with in Gaussian noise, the simulation results show that EM gradient algorithm has a good BER performance curve; it has the same convergence speed as the EM algorithm.

2 EM Gradient Algorithm

Suppose θ^i is the parameter i-th iterations value, and the EM algorithm steps are:

E-step: Compute expectation of full likelihood:

$$Q(\theta,\theta^i) = E[\log f(\theta \mid y,z) \mid y,\theta^i] = \int \log f(\theta \mid y,z) k(z \mid y,\theta^i) dz \tag{1}$$

M-step: Maximum $Q(\theta^{i+1},\theta^i) = \max Q(\theta,\theta^i)$, $\theta^{i+1} = \arg\max Q(\theta,\theta^i)$ (2)

D. Jin and S. Lin (Eds.): Advances in CSIE, Vol. 1, AISC 168, pp. 87–90.
springerlink.com © Springer-Verlag Berlin Heidelberg 2012

Repeat E-step and M-step iterations, and don't stop the loop until $\left\|\theta^{i+1}-\theta^{i}\right\|$ small enough. If the maximum method can't be resolved, we can use iterative optimization approach for each M-step. It will generate a nested loop. In the EM algorithm, each M-step iterative inserted two maximum steps, which will also cause the nested iterative. In order to avoid the computational burden of nested loops, EM Gradient Algorithm was proposed by Lange which was about how to use single-step Newton method instead M-step [4].Thus we can approximate the maximum instead of exact solution.

The M-step is equivalent to

$$\theta^{(i+1)}=\theta^{(i)}-Q''(\theta\mid\theta^{(i)})^{-1}\mid_{\theta=\theta^{(i)}} Q'(\theta\mid\theta^{(i)})\mid_{\theta=\theta^{(i)}} \tag{3}$$

We can get the approximate equation

$$\theta^{(i+1)}=\theta^{(i)}-Q''(\theta\mid\theta^{(i)})^{-1}\mid_{\theta=\theta^{(i)}} l'(\theta^{(i)}\mid x) \tag{4}$$

$l'(\theta^{(i)}\mid x)$ is the valuation of the current scoring function.

3 EM Gradient Algorithm Apply in Multi-user Detection

The k-th receive signal is:

$$r(t)=\sum_{k=1}^{K} A_k(t)g_k(t)b_k(t)+n(t)\,. \tag{5}$$

In the formula, A_k means the amplitude of the k-th signal; g_k means spread spectrum waveform of the k-th signal, and the value is ±1 ; b_k means the *k*-th user data, the value is ±1 ; $n(t)$ is background noise that is selected by different type of the noise. Suppose y_k is the output of the k-th matched filter and its expression is:

$$y_k=\int_0^T r(t)g_k(t)dt\,,1\le k\le K\,. \tag{6}$$

$$y_k=\int_0^T\left(\sum_{k=1}^{K} A_k(t)g_k(t)b_k(t)+n(t)\right)g_k(t)dt=A_kb_k+MAI_k+n_k \tag{7}$$

In Synchronous DS-CDMA system, the EM multi-user detection is where adding the expectation maximum function in the classic detector. Select here decor-relation detection, and get the initial estimate $\hat{b}$, then apply EM gradient algorithm, suppose the k-th user information bit estimate is b_k , b_k^i means the i-th iterative value of EM[5,6]. As in ref. [3] the expectation complete likelihood function is;

$$Q(b_k\mid b_k^i)=\frac{A_k^2}{2\sigma^2}(-(b_k)^2+2b_k\frac{1}{A_k}(y_k-\sum_{j\neq k}R_{kj}A_jb_j)) \tag{8}$$

Maximize expectation: $b_k^{i+1} = \arg\max Q(b_k \mid b_k^i)$

We obtain the simple formula $b_k^{i+1} = (y_k - \sum_{j \neq k} R_{kj} A_j b_j)$.The update of equation the Gradient algorithm about θ maximizes $l(\theta \mid x)$, refer to formula Eq.(4).Because of $l'(\theta^{(t)} \mid x) = Q'(\theta \mid \theta^{(t)})|_{\theta=\theta^{(t)}}$, we will find an alternative of $l'(\theta^{(t)} \mid x)$ in the EM framework. Expand Q' and $l'(\theta^{(t)} \mid x)$ in $\theta^{(t)}$.In the formula Eq.(9), $\hat{i}_Y$ is complete signal. Because $\theta_{EM}^{(t+1)}$ maximizes $Q(\theta \mid \theta^{(t)})$ about $Q(\theta \mid \theta^{(t)})$, Get $Q'(\theta \mid \theta^{(t)})|_{\theta=\theta^{(t)}} \approx \hat{i}_Y(\theta^{(t)})(\theta_{EM}^{(t+1)} - \theta^{(t)})$.

$$l'(\theta^{(t)} \mid x) = l'(\theta \mid x)|_{\theta=\theta^{(t)}} \approx l''(\theta \mid x)|_{\theta=\theta^{(t)}} (\theta_{EM}^{(t+1)} - \theta^{(t)}) \tag{9}$$

Just as introduced in the above article, the iterative equation can be written as:

$$\theta^{(t+1)} = \theta^{(t)} + i_{obs} \cdot i_{com}{}^{-1} \cdot (\theta_{EM}^{(t+1)} - \theta^{(t)}) \tag{10}$$

In the iterative equation, θ_{EM}^{t+1} is the $t+1$ time estimate of M-step of EM Gradient algorithm based on θ_{EM}^{t}, $\theta^{(t)}$ is the i-th iterative estimate of EM Gradient algorithm, i_{obs} is observational data, and i_{com} is complete information. EM Gradient algorithm iteration steps:

Step1 : Initialization estimates $\theta_{EM}^{0} = 0$;

Step2 : Get $\theta_{EM}^{(i+1)}$ by calculating $\theta^{(t)}$ through EM gradient algorithm;

Step3 : Substitute $\theta^{(t)}$, $\theta_{EM}^{(i+1)}$ into the iterative formula Eq.(10),get $\theta^{(t+1)}$;

Step4 : Judge $\| \theta^{(t+1)} - \theta^{(t)} \| < \sigma$, repeat the above steps until the required convergence.

4 Simulation Results and Conclusions

In the DS-CDMA system, applying EM gradient algorithm, and we select 5 users, 1000 information bits, 31-bit gold spread-spectrum code, and the user power partial value is 12. Simulation result shows: In Fig.1, we get that the EM gradient algorithm has a good BER performance curve; From Fig.2, we know that the EM gradient algorithm has the same convergence speed as standard EM algorithm. In the future, the focus research is to modify algorithm, and to improve convergence speed combine with some accelerate algorithms.

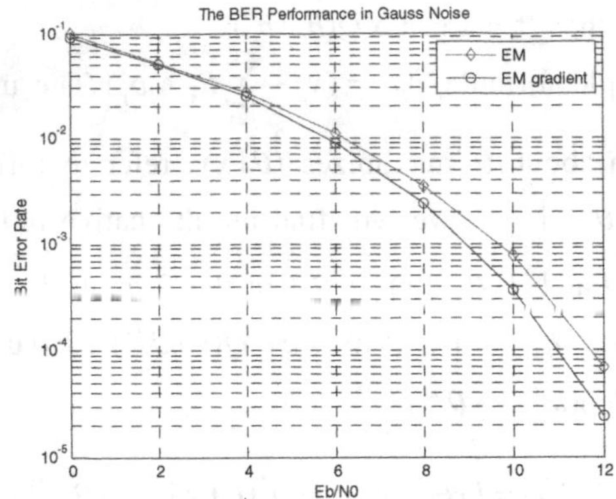

Fig. 1. The BER curve of EM gradient and EM algorithm in Gaussian noise

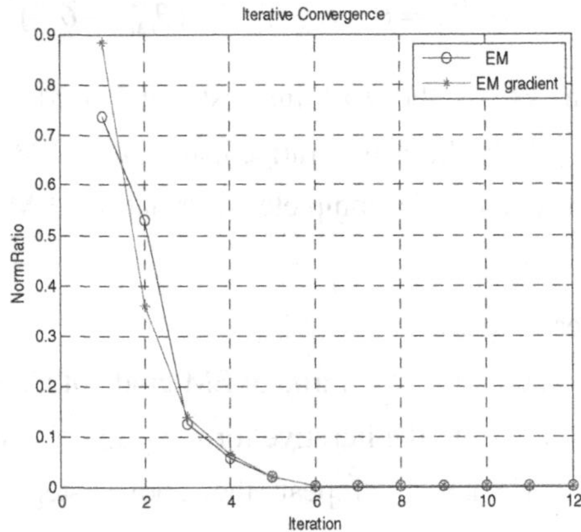

Fig. 2. The convergence of EM gradient and EM algorithm in Gaussian noise

References

1. McLachlan, G.J., Krishnan, T.: The EM Algorithm and Extensions, 2nd edn., pp. 149–155. Wiley-Interscience (2008)
2. Kay, S.M. (ed.): Statistics based on signal processing: estimation and detection theory. Electronic Industry Press, Beijing (2006)
3. Kay, S.M.(ed.), Luo, P.-F.(trans.): Statistics based on signal processing: estimation and detection theory. Electronic Industry Press, Beijing (2006)
4. Lang, K.: A gradient algorithm locally equivalent to the EM algorithm. Journal of the Royal Statistical Society, Series B 57, 425–437 (1995)
5. Kocian, A., Fleury, B.H.: EM-Based Joint Data Detection and Channel Estimation of DS-CDMA Signals. IEEE Transactions on Communications (October 2003)
6. Laurie, B., Nelson, H., Poor, V.: Iterative Multi-user Receivers for CDMA Channels: An EM-Based Approach. IEEE Transactions on Communications 44(12), 1700–1710 (1996)

Multi-user Detection Based on the ECM Iterative Algorithm in Gaussian Noise

Jinlong Xian and Zhi Liu

College of Information Science and Engineering, Henan University of Technology, Lianhua Street, Zhengzhou, 450001, Henan, P.R. China
13991339876@vip.sina.com

Abstract. Generally, it is easier to compute the derivation and maximization of the full-likelihood expectation than the calculations of incompletely data maximizing likelihood function. In some cases, even if it is easy to find the full-likelihood expectation, it is difficult to achieve the maximization of the full-likelihood expectation. So a novel approach for multi-user detection based on the ECM iterative algorithm is proposed. Compared with the EM algorithm, the ECM algorithm reduces the computational complexity of the M-step. The results show that the proposed algorithm has well performance and Convergence in Gaussian noise.

Keywords: Multi-user detection, EM algorithm, ECM iterative algorithm.

1 Introduction

The principle of EM Algorithm: the observe data is Y, complete data is $X = (Y \mid Z)$, Z is missing data, θ is model parameter [1]. The MAP $P(\theta \mid Y)$ is very complex and difficult to calculate different statistical signal while not added missing data, but if added, it is easy to get the $P(\theta \mid y, z)$. $h: x \to y$ is a many-to-one mapping from X to Y , $g(\theta \mid y)$ means a posteriori distribution density function that base on observe data y ,called the observation posterior distribution; $f(\theta \mid y, z)$ means about the posteriori distribution function of θ while adding missing data; $k(\theta \mid y, z)$ means conditional distribution function of Z while θ and Y are given[2]. The EM algorithm is commonly used for the missing data of iterative algorithm[3].Compared with EM algorithm, background noise chose the Gaussian noise, and compare the error performance of different algorithm, the results show that the ECM algorithm has good performance, but the convergence of variations is well.

2 ECM Algorithm

Suppose θ^i is the parameter i-th iterations value, and the EM algorithm steps are:

D. Jin and S. Lin (Eds.): Advances in CSIE, Vol. 1, AISC 168, pp. 91–94.
springerlink.com © Springer-Verlag Berlin Heidelberg 2012

E-step: Compute expectation of full likelihood:

$$Q(\theta,\theta^i) = E[\log f(\theta \mid y,z) \mid y,\theta^i] = \int \log f(\theta \mid y,z) k(z \mid y,\theta^i) dz \quad (1)$$

$$\text{M-step: Maximum } Q(\theta^{i+1},\theta^i) = \max Q(\theta,\theta^i)\,,\ \theta^{i+1} = \arg\max Q(\theta,\theta^i) \quad (2)$$

Repeat E-step and M-step iterations, and don't stop the loop until $\left\|\theta^{i+1} - \theta^i\right\|$ small enough. The ECM algorithm is a natural extension of the EM algorithm in situations where the maximization process on the M-step is relatively simple when conditional on some function of the parameters under estimation. The ECM algorithm therefore replaces the M-step of the EM algorithm by a number of computationally simpler conditional maximization (CM) steps.[4]. $G = \{\vartheta_s(\theta) : 1, 2, .., S\}$ is defined as the pre-selected constraint functions of muster which is used to estimate the parameter θ .The function is defined by $\vartheta_s(\theta) = \vartheta_s(\theta^{i+(s-1)/S})$. $\theta^{i+(s-1)/S}$ is the obtained maximum point in the CM-step of the (*s-1*)th in the current cycle. At each iteration, let $\theta^{i+1} = \theta^{i+S/S}$ and calculated the next iteration of E-step when finished all the CM-step cycle.S is the number of CM-step in every CM cycle. It is simple to show that

$$\vartheta_s(\theta) = (\theta_1, ..., \theta_{s-1}, \theta_{s+1}, ..., \theta_S,), s = 1, ..., S. \quad (3)$$

For this case, the next iteration of the s-th CM-step with the definition

$$(\theta_1, ..., \theta_{s-1}, \theta_{s+1}, ..., \theta_S,) = (\theta_1^{(t)}, ..., \theta_{s-1}^{(t)}, \theta_{s+1}^{(t)}, ..., \theta_S^{(t)}). \quad (4)$$

3 ECM Algorithm Apply in Multi-user Detection

3.1 MUD

The k-th receive signal is:

$$r(t) = \sum_{k=1}^{K} A_k(t) g_k(t) b_k(t) + n(t). \quad (5)$$

In the formula, A_k means the amplitude of the k-th signal; g_k means spread spectrum waveform of the k-th signal, and the value is ± 1 ; b_k means the *k*-th user data, the value is ± 1 ; *n(t)* is background noise that is selected by different type of the noise. Suppose y_k is the output of the k-th matched filter and its expression is:

$$y_k = \int_0^T r(t) g_k(t) dt, 1 \le k \le K. \quad (6)$$

This result can be obtained by expansion:

$$y_k = \int_0^T \left(\sum_{k=1}^{K} A_k(t) g_k(t) b_k(t) + n(t) \right) g_k(t) dt = A_k b_k + MAI_k + n_k \quad (7)$$

It can be known clearly that the first item is the data that the *k*-th user wants to receive; the second item is multiple access interference (MAI) which is generated by other users; and the third is noise [6].

3.2 Multi-user ECM Detection

In Synchronous DS-CDMA system, the EM multi-user detection is where adding the expectation maximum function in the classic detector. Select here decor-relation detection, and get the initial estimate $\hat{b}$, then apply EM iterative algorithm, suppose the k-th user information bit estimate is b_k , b_k^i means the i-th iterative value of EM[5].As in ref. [3] the expectation complete likelihood function is;

$$Q(b_k \mid b_k^i) = \frac{A_k^2}{2\sigma^2}(-(b_k)^2 + 2b_k \frac{1}{A_k}(y_k - \sum_{j\neq k} R_{kj} A_j b_j)) \tag{8}$$

Maximize expectation: $b_k^{i+1} = \arg\max Q(b_k \mid b_k^i)$. (9)

We obtain the simple formula $b_k^{i+1} = (y_k - \sum_{j\neq k} R_{kj} A_j b_j)$. (10)

At *i*-th iteration, suppose the conditions of $b_2 = b_2^i ..., b_S = b_S^i$ are met, calculated the value of b_1 when $Q(\theta \mid \theta^i)$ reaches the maximum.The value of b_1 is defined as b_1^{i+1}.Under conditions of $b_1 = \mathrm{b}_j^{i+1},...,b_j = b_j^i,...,b_S = b_j^i, j = 2,3,...,S$, b_j could be obtained when $Q(\theta \mid \theta^i)$ reaches the maximum. Based on that outcome, the value of b_j is treated as $b_j^{i+1}, j = 2,3,...,S$ .The final result is expressed as $b^{i+1} = (b_1^{i+1}, b_2^{i+1}..., b_S^{i+1})$ .We will define normal deviation ratio is $Norm = \frac{\| b_k^{i+1} - b_k^i \|}{b_k^i}$. ECM algorithm concrete steps:

Step1 : Obtain the initial estimate value $b_k^0 = \hat{b}$.

Step2 : Compute b_j and b_k^{i+1}.

Step3 : Judge $\| b_k^{i+1} - b_k^i \| < 0.01$, the permanent can be set by requirement.

Step4 : If "no", $i++$ and $b_k^i = b_k^{i+1}$, return to Step2.If "yes", $b_{opt} = b_k^{i+1}$, and b_{opt} is the final optimum value of iteration.

4 Simulation Results and Conclusions

In the DS-CDMA system, applying EM algorithm, and we select 5 users, 1000 information bits, 31-bit gold spread-spectrum code, and the user power partial value is 12. Simulation result shows: The performance and Convergence of ECM algorithm is almost as well as the EM algorithm in Gaussian noise in Fig.1; In Fig.2,we know that the convergent speed of ECM algorithm is faster than the convergent speed of EM algorithm enough, in the future, the focus research is to modify algorithm,and to improve convergence speed combine with some of accelerate algorithm.

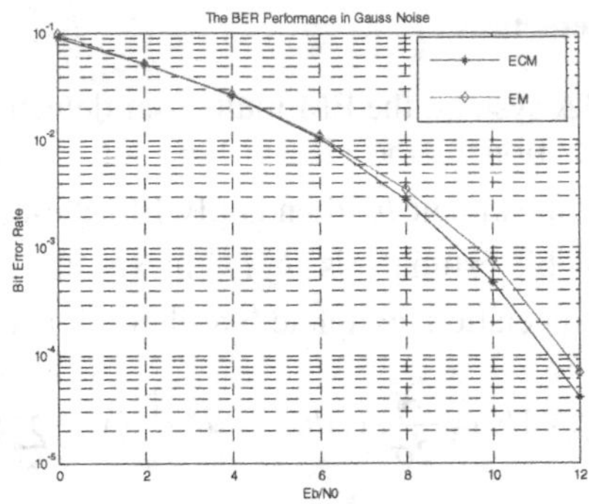

Fig. 1. The BER curve of ECM and EM algorithm in Gaussian noise

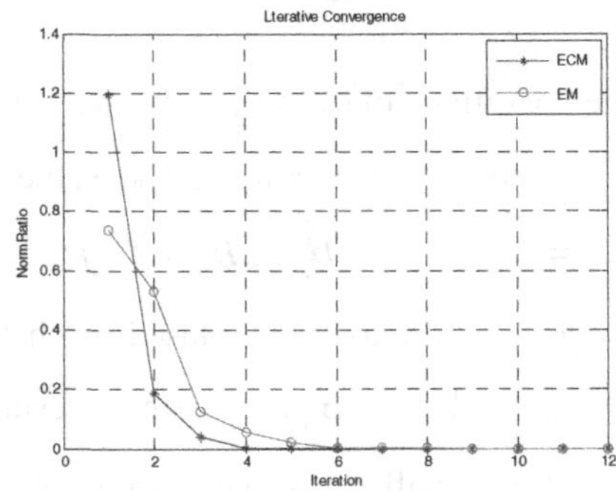

Fig. 2. The Iterative convergence of ECM and EM algorithm in Gaussian noise

References

1. Feder, M., Weinstein, E.: Parameter estimation of superimposed signals using the EM algorithm. IEEE Trans. Acoust., Speech, Signal Processing 36, 477–489
2. Borran, M.J., Nasiri-Kenari, M.: An Efficient Detection Technique for Synchronous CDMA Communication Systems Based on the Expectation Maximization Algorithm. IEEE Transactions on Vehicular Technology, 1663–1668 (September 2000)
3. Kay, S.M.(ed.), Luo, P.-F.(trans.): Statistics based on signal processing: estimation and detection theory. Electronic Industry Press, Beijing (2006)
4. Meng, X.-L., Rubin, D.B.: Maximum likelihood estimation via the ECM algorithm: ageneral framework. Biometrika 80, 267–278 (1993)
5. Kocian, A., Fleury, B.H.: EM-Based Joint Data Detection and Channel Estimation of DS-CDMA Signals. IEEE Transactions on Communications (October 2003)
6. Nelson, L.B., Poor, V.: Iterative Multi-user Receivers for CDMA Channels: An EM-Based Approach. IEEE Transactions on Communications 44(12), 1700–1710 (1996)

Research on Metamorphic Testing for Oracle Problem of Integer Bugs

Yi Yao[1,2], Song Huang[1], and Mengyu Ji[1]

[1] Institute of Command Automation, PLA University of Science and Technology, Nanjing, China 210007
[2] Mobile Post Doctoral Station, PLA University of Science and Technology, Nanjing, China, 210007
yaoyi226@yahoo.com.cn, hs0317@sohu.com, txmxr@163.com

Abstract. The integer bugs play a vital role in functionality and security of software. For the Oracle problem, integer bugs are always ignored unless program throws an exception obviously. In this paper, a metamorphic relation is provided which is based approach to detect invisible integer bugs without oracle. It is shown in our case study that this method can detect some invisible errors which are difficult to be found in conventional approach and improve the efficiency of integer bugs detection.

Keywords: metamorphic relation, integer bugs, integer overflows, test oracle.

1 Introduction

The basic process of software testing is in certain conditions, to provide test input, and record output, then to compare actual output with expected output to determine whether the test passes [1]. Software testing theory assumes that there must be a clear expected output, as a criterion to determine whether the test passes, however. But the reality is: software is not clear or difficult to obtain expected output, and thus difficult to determine the correctness of the actual output. Therefore, there is a 'non-test program' which is determined by character of the test itself, and not because of human factors (such as software development process lacks the necessary testing documentation, etc.). This question is so-called 'Test Oracle' problem in the software testing [2].

University of Hong Kong Professor Tsong Yueh Chen [3] proposed that Test Oracle problem can be resolved to some extent by making use of the metamorphic relationship and source test case to build follow-up test case. This testing technique is called Metamorphic Testing (MT) technology. It was originally used to determine the correctness of software by validating some independent multi versions of a program. In the actual test, because of the lack of multiple versions of the program, of this technology is evolved. It becomes that as long as the program design obeys a certain character, we can determine the output from different inputs whether to obey this character, and thus it can be judged that whether program obey this character. Accordingly, we can determine the correctness of the program to some extent.

D. Jin and S. Lin (Eds.): Advances in CSIE, Vol. 1, AISC 168, pp. 95–100.
springerlink.com © Springer-Verlag Berlin Heidelberg 2012

With effect of software testing in software quality assurance increasingly, 'Test Oracle' has become the most critical problem in software testing development [4]. This problem in the software security testing is very prominent. Because the requirement of software security testing is more difficult to obtain than that of other types of tests, and it is difficult to predict what kind of the behavior of software is safe, and it is also difficult to give the guidelines of security testing pass. This software security test oracle seriously hampered the development of software security testing technology. Therefore, it has an important research value to solve the problem that software security testing can be carried out in the lack of determinant guidelines. In aerospace, weapons, process control, nuclear energy, transportation and health care and other mission-critical areas, the correctness of the results the software is the part of security assurances.

The integer bug is one of the main reasons that cause software calculation error. In computer program, the integer variables are expressed by fixed-bit-wide vector. When the value got by instruction operation is more than the value of storage capacity, integer overflow take places. When Europe launches rocket of Ariane5 firstly, because of an integer overflow in the procedure of a 64-bit floating point number into a 16-bit signed integer, the rocket control system instructing incorrectly resulted in disastrous consequences that the rocket vacated. In addition, because it can't afford to test every result of calculation, integer overflow in commercial software has not been detected by and large. For example, if an integer is disposed to be an unexpected value by a program and this unexpected value is then used for the array indexes or loop variable, it will produce software security vulnerabilities in the program.

This paper presents a method to detect integer bugs by metamorphic relationship, and validates this method by case study.

2 Traditional Method for Integer Bugs Detection

Robert C.Seacord [5] concludes three categories of integer bugs detection, which consist of precondition, error detection and post condition.

Precondition: Check the program whether integer bugs occur before the execution. Static analysis is often used in precondition checking where the checking rules are formalized before static analysis starts.

Error Detection: Detect errors during the runtime. The operating system and the compiler usually have the mechanism to report and deal with the errors. Nowadays many binary-based runtime checking technique belongs to this category.

Post Condition: After execution, check the result value and make sure the value is valid. Fanping Zeng [6] summarize the mutant operators for the integer bugs and propose a mutation testing based method to measure the adequacy of the test data. But mutation based testing can't help tester to generate the test data.

From Table 1 we conclude that post condition-based category is more dependable than precondition one and easier to be operated than error detection one. So if our goal is to find the implementation failure caused by the integer bugs only, rather than vulnerability analysis, post condition-based category is more effective.

Table 1. Comparison the Performance of Three Categories.

Categories	False Positive	False Negative	Difficulty of the detection	Difficulty of source code -based fault localization
Precondition	High	High	Low	Low
Error detection	Low	Low	High	High
Post condition	Low	Medium	Medium	Medium

3 Methods of Integer Bugs Detection Based on Metamorphic Testing

For the 'Test Oracle' problem in the method of the integer bugs detection, method of the integer bugs detection based on metamorphic test is proposed, which is essentially a detection method based on validation of correctness. If an integer error occurs, the program must calculate to get an inaccurate value, and then if this inaccurate value is used in the next step calculation, the program will eventually either collapse, or calculate to get an unexpected output. That is, when the inputs of software meet some certain characters, the corresponding output of software will also meet the corresponding character.

Before the specific description of this method, formal definitions of the concepts which are required in the method are given.

Definition 1: It is assumed that program P is an implementation of the function f. $x_1,x_2,\ldots,x_n$, (n>1) are n-group variables for function f, and $f(x_1),f(x_2),\ldots,f(x_n)$ are corresponding outputs for function f. If x_1, x_2,..., x_n satisfy the relation r among themselves, and $f(x_1)$, $f(x_2)$,..., $f(x_n)$ satisfy the relation rf:, (r, rf) is recognized as the metamorphic relationship of program P .

Definition 2: For the same program P, Metamorphic relationships (r, rf) which need to verify or to extract always are not only one. It is shown that $R_i=(r_i,r_{fi})$ denotes the i-th metamorphic relationship of the program P. and that $S(R)=\{R_1,R_2,\ldots\}$ denotes the set of metamorphic relationships of the program P.

Method of the integer bugs detection based on metamorphic relationship includes three steps:

Step1: select the source test cases. For program P, I_1, I_2, …, I_n are selected as inputs corresponding to x_1, x_2, ..., x_n in program P. That is, source test case (I_1, I_2, …, I_n) are gained.

Step2: select the correct metamorphic relationship to generate follow-up test cases. We choose the appropriate metamorphic relation R = (r, rf) of program P. It is assumed that the program P is correct, and then $r(I_1,I_2,\ldots,I_n)\Rightarrow r_f(P(I_1),P(I_2),\ldots,P(I_n))$ is generated from Definition 1. It means that follow-up test cases to be derived from the test case. If there are a variety of metamorphic relationships, can also choose a number of metamorphic relations, some metamorphic relationships R_1, R_2, ..., R_n can be selected to build many follow-up test cases in order to enhance the veracity of test.

Step3: compared results from source test cases with results from follow-up test cases in order to judge whether metamorphic relationship is obeyed. If program P is correct, P obeys $r(I_1, I_2, ..., I_n) \Rightarrow r_f(\mathrm{P(I_1),P(I_2),...,P(I_n)})$ in which $\mathrm{P(I_1),P(I_2),...,P(I_n)}$ are the corresponding output. So if the test case which is running does not meet the formula above, the assumption does not correct, and it means the program has errors.

Fig. 1 shows process of integer bugs detection based on metamorphic testing.

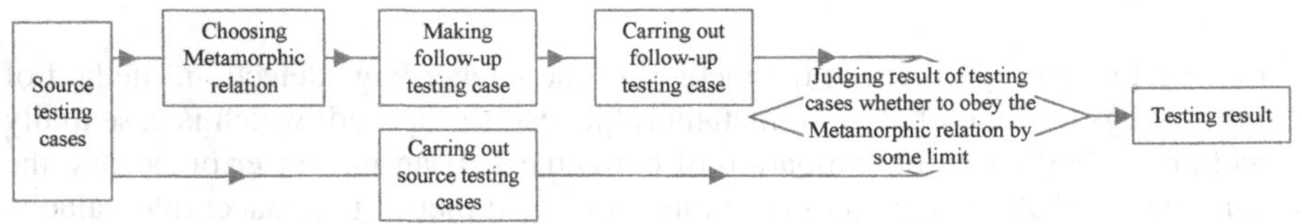

Fig. 1. Process of integer bugs detection based on metamorphic testing

4 A Simple Case Study

For there is few algebra operation in tcas.c as below, the most possible situation of integer error is that programmer doesn't estimate the range of the inputs by defining a wrong type that represents a shorter bit vector. For the rang of integer type is from -2147483648 to +2147483647 and char type is from-128 to+127 in GCC standard, we replace type of parameter own_Tracked_Alt from "int" to "char" to model the integer bug.

The modified program is called version *mutant* and the original one is called version *original.*

A case of computer program, named tcas.c from Traffic Collision Avoidance System (TCAS), which can be downloaded from Software-artifact Infrastructure Repository [7].

```
Program
  ...
  Char Own_Tracked_Alt; // correct version:
                        // int  own_Tracked_Alt;
  int Own_Tracked_Alt_Rate;
  int Other_Tracked_Alt
...
```

Safety properties of TCAS which were proposed by Livadas and Coen-Porisini are formalized by Arnaud Gotlieb in [8]. We propose a metamorphic relation based on both black-box and white box information.

MR(Metamorphic Relation): Program tcas.c is P, $I_1, I_2, \ldots, I_{12}$ are inputs of P, I_4=Own_Tracked_Alt=a, I_6=Other_Tracked_Alt=b,

$P(I_1, I_2, I_3, a, I_5, b, I_7, I_8) = P(I_1, I_2, I_3, a-\gamma, I_5, b-\gamma, I_7, I_8)$ let γ =500 in this experiment.

Definition 3 [9]: Failure detection ratio(FD), a microscopical measurement to analyze to which extent a testing method can detect a fault in program, is the percentage of test cases that could detect certain mutant m, that is

$$FD(m,T) = \frac{N_f}{N_t - N_e} \quad (1)$$

where N_f is the number of times program fails, N_t the number of tests, and N_e the number of infeasible tests.

From Table 2, we conclude that metamorphic testing method is more effective than formal safety property method for the mutant in this experiment. FD of MR is 11.7% for version mutant which is much higher than 3.3% which is FD of formal safety property method.

In the experiment the number of actual mutants result because of mutant injection is 15. Neither of the methods has false positive. Metamorphic testing method can detect about 50% actual mutants and formal safety property method only 13.3%. For the mutant in this experiment, the false negative ratio of metamorphic testing method is much lower than formal safety property method.

However, safety property method can detect a design failure which is not detected by MR, because both outputs of the two versions are equal and this is a design bug that programmer can't take into account in the implementation or design process.

Table 2. Execution Result.

Categories	Failure Tests/Total Tests		Failure Tests/Actual Mutants	
	Formal Safety Property	Metamorphic Relation	Formal Safety Property	Metamorphic Relation
original	1/60	0/60	1/15	0/15
mutant	2/60	7/60	2/15	7/15

5 Conclusion

In this paper, method of integer bugs detection based on metamorphic relationship is proposed. It is proved by case studies that this method of metamorphic relationship can detect the hidden unexpected failure which traditional testing techniques can't detect.

Because of certain blindness of choosing source test input and metamorphic relationship, optimization algorithm of test case generation and selection of metamorphic relationship are the research direction for the future. At present, research method by use of the evolutionary algorithms combined with metamorphic relationship are proposed, and error detection efficiency of test cases is improved [10], but whether this method is effective for the integer bugs detection need to be further studied.

Acknowledgments. This work is supported by National High Technology Research and Development Program of China (No: 2009AA01Z402) and China Postdoctoral Science Foundation (No: 20110491843). Resources of the PLA Software Test and Evaluation Centre for Military Training are used in this research.

References

1. Heitmeyer, C.: Applying *Practical* Formal Methods to the Specification and Analysis of Security Properties. In: Gorodetski, V.I., Skormin, V.A., Popyack, L.J. (eds.) MMM-ACNS 2001. LNCS, vol. 2052, pp. 84–89. Springer, Heidelberg (2001)
2. Weyuker, E.J.: On testing non-testable programs. Computer Journal 25, 465–470 (1982)
3. Chen, T.Y., Cheung, S.C., Yiu, S.M.: Metamorphic testing: a new approach for generating next test cases. Technical Report HKUST-CS98-01 (1998)
4. Manolache, L.I., Kourie, D.G.: Software testing using model programs. Software: Practice and Experience 31, 1211–1236 (2001)
5. Robert, C.: Secure Coding in C and C++. Person Education (2010)
6. Zeng, F., Mao, L., Chen, Z., Cao, Q.: Mutation-based Testing of Integer Overflow Vulnerabilities. In: The 5th International Conference on Wireless Communications, Networking and Mobile Computing, WiCOM 2009, September 24-26. IEEE Press, Beijing (2009)
7. Chen, T.Y., Huang, D.H., Tse, T.H., et al.: Case studies on the selection of useful relations in metamorphic testing. In: Proceeding of the 4th Ibero-American Symposium on Software Engineering and Knowledge Engineering, JIISIC 2004, Polytechnic University of Madrid, Madrid Spain, pp. 569–583 (2004)
8. Gotlieb, A.: TCAS software verification using Constraint Programming. The Knowledge Engineering Review, vol. 00:0, pp. 1–15. Cambridge University Press (2009)
9. Wu, P., Shi, X.C., Tang, J.J., Lin, H.M., Chen, T.Y.: Metamorphic testing and special case testing: A case study. Journal of Software 16(7), 1210–1220 (2005)
10. Dong, G.W., Wu, S.Z., Wang, G.S., Guo, T., Huang, Y.G.: Security Assurance with Metamorphic Testing and Genetic Algorithm. In: IEEE/WIC/ACM International Conference on Web Intelligence and Intelligent Agent Technology, pp. 368–373 (2010)

Application of Wireless Distributed Sensor Networks in Subway Energy Saving Systems*

Jun Fan[1], Bing Xu[2,3], and Peng Sun[3]

[1] Liaoning Geology Engineering Vocational College, Dandong, China
[2] Logistics Engineering College, Shanghai Maritime University, Shanghai, China
[3] School of Engineering Innovation, Shanghai Institute of Technology, Shanghai, China

Abstract. This research in order to achieve the energy saving effect in the subway station ventilation and air conditioning systems[1], Wireless sensors based on embedded systems, distributed artificial intelligence by the control method to achieve the subway platform of distributed temperature and humidity measurements.

Keywords: Distributed-Intelligence, Wireless Sensor, Embedded System.

1 Introduction

Using wireless sensor for data collection. Measuring terminals use embedded system to complete the signal acquisition, processing and wireless transmission. The signal temperature and humidity of various points collected by the receiver and sent to the host computer.Obtain the overall measurement of comfort timely by distributed measurement. Establishing mathematical models of Subway platforms' air temperature and humidity measurement and control. To change the traditional air-conditioning systems' control methods which is rigid model and parameters of curing.Full account of both the waiting crowds' requirements of air-conditioned comfort , but also to meet the energy saving needs of subway air-conditioning system.To achieve subway air-conditioning control systems' flexible regulation and energy efficient control and to minimize subway air-conditioning systems' energy consumption.

2 Distributed Artificial Intelligence

Distributed artificial intelligence [2] the rise of new disciplines in recent years, and it is the product of artificial intelligence, knowledge engineering, distributed computing, parallel processing, computer networks and communication technology cross-development. Distributed artificial intelligence use artificial intelligence technology. To Study a group of geographically dispersed, loosely coupled intelligence agencies how to coordinate and organizations. Their knowledge, skills, goals and

* Supported by 2009 Shanghai STCSM Project, Project Number: 09220502500.

 © Springer-Verlag Berlin Heidelberg 2012

planning for effective united solution. The research includes a parallel artificial intelligence, distributed knowledge systems two parts.

Distributed Computer Automated Measurement and Control System, DCAMCS is the distributed applications which refers to a collection of independent computer systems,and through the network communications to develop, deploy, manage and maintain, resource sharing and collaboration as the main target application. It has a strong real-time and space constraint and other characteristics, field bus control system (FCS) is the distributed control system [3] which has the most typical structure and widely used in industrial applications.

3 Research of Artificial Intelligence Control Strategy in the Subway Air Conditioning Systems for Energy Savings

In 2010, we took part in the Shanghai Science and Technology Commission of longitudinal research projects "research of intelligent information processing technology, energy-saving control of rail traffic"[4] and technology development.

The initial results have been achieved (in a retrieval system for scientific papers published):

1) Xu Bing, Qian Ping.Multiobjective Evolutionary Algorithms Applied to Compressor Stations Network Optimization Scheduling Control System. Procedia Engineering (ISSN: 1877-7058)

2) Xu Bing, Qian Ping. Network Sniffer Component Program Design based on Promiscuous Pattern. ICISE2010.

3) The construction of "artificial intelligence laboratory"

4 Wireless Sensors Based on Embedded Systems

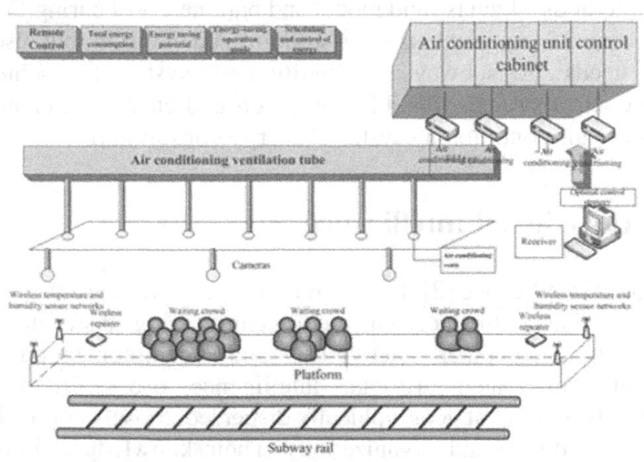

Fig. 1. Subway Station Platform Control System

5 The Research and Application of Subway Platform Distributed Temperature and Humidity Measurement Technology

According to the Ministry of Construction, "Ventilation and Air Conditioning Engineering Construction Quality Acceptance Construction" (standard [2002] No. 60) B6.3 item states:

Table 1. Temperature and humidity measuring points

Fluctuation range	Room area$\leqslant$50m^2	Each additiona20~50m^2
$\Delta t=\pm 0.5\sim\pm 2$℃	5	Increase 3~5
$\Delta RH=\pm 5\%\sim\pm 10\%$		
$\Delta t\leqslant\pm 0.5$℃	Spacing not greater than 2m, should not be less than 5	
$\Delta RH\leqslant\pm 5\%$		

This topic can be applied to existing domestic "point" temperature and humidity measurement technology. Consider issues such as building wiring pilot phase .Using wireless sensor for data collection. Measuring terminals use embedded system to complete the signal acquisition, processing and wireless transmission. The signal temperature and humidity of various points collected by the receiver and sent to the host computer.Obtain the overall measurement of comfort timely by distributed measurement.Establishing mathematical models of Subway platforms' air temperature and humidity measurement and control.

To take into account future extensions, such as air quality testing, detection of combustible gases and toxic gas testing and other needs.Field measurement equipment use the CPU chip and peripheral circuit of the measuring end. The Signal communication protocol uses the **Zig-Bee** protocol which has anti-interference ability. At the same time transmission using hardware relay mode to protect the normal transmission of measurement signals.

References

1. Science and Technology Commission of Shanghai vertical issues "intelligent information processing based on energy-saving control of rail transport" (vertical item number: 09220502500)
2. Blake, M.B.: Rule-driven coordination agents: a self-configurable agent architecture for distributed control. In: Proceedings of the 5th Inter. Symposium on Autonomous Decentralized Systems, ISADS 2001, pp. 271–277. IEEE Press (2001)
3. Yang, X.: The technology optimization of Subway station ventilation and air conditioning energy saving. Railway Engineering (12) (2009)
4. Tang, M.: Based on load forecast of the subway ventilation and air conditioning system energy optimization. City Rapid Rail Transport (4) (2008)
5. Mu, G.Y., Yin, L., Marvin: The energy analysis and energy-saving measures of subway station power system. Shanghai Electrical (4) (2010)
6. Technology Advice Herald. Evolutionary Algorithms Preliminary Study 25 (2007)

The Model of Face Recognition in Video Surveillance Based on Cloud Computing

Shijun Yi, Xiaoping Jing, Jing Zhu, Jinshu Zhu, and Heping Cheng

Chengdu Electro-mechanical College, Chengdu, China
voicent@163.com

Abstract. As more and morc using of the video surveillance systems, how to find criminal information through these vast amounts of video data in a large-scale cross-platform has been a difficult problem. Based on discussion about the high performance computing and mass storage capabilities of cloud computing, this paper puts forward a new idea that dealing with face recognition in a wide range of video surveillance system through the cloud platform. And provides the model of face recognition in video surveillance based on cloud computing.

Keywords: Cloud Computing, Video surveillance, Face recognition, Model.

1 Introduction

Now, with the rapid development of information technology and social stability needs, many important sites are equipped with video surveillance equipment, which for people in dealing with security incidents provides the most direct evidence of the scene. People can use these data to achieve real-time video tracking, or based on the video data in real time or after the events of all kinds to make the most direct and fair ruling. However, the current monitoring systems used by many units are independent and can only find in a certain range or tracking. Due to the lack of large-scale multi-camera (such as the whole city, down to the a unit or a residential area) complexes, on tracking and distinguishing target, the coordination of many units to find a person will waste a lot of manpower and resources, make the system processing delay, and even lead to failure.

Therefore, in these cases, the local unit of the camera can not play well the role of identification and tracking, we only need to coordinate and integrate these cameras, which can be achieved by collaboration and video surveillance tracking purposes. However, in multi-camera work, it will produce large amounts of video information, and collaborate on a very enormous computation, so that the whole system performance of less than real-time requirements.

Cloud computing in high performance computing and mass storage has unparalleled advantages[1], so the use of cloud computing can solve such problems. Currently, cloud-based applications more and more research; cloud computing as a new method of shared infrastructure, built on top of a large-scale clusters of cheap servers, through infrastructure and applications together to build the upper to maximum efficient use of hardware resources. The huge combined system pool provides a variety IT services, which is of the high scalability and high reliability.

D. Jin and S. Lin (Eds.): Advances in CSIE, Vol. 1, AISC 168, pp. 105–111.
springerlink.com © Springer-Verlag Berlin Heidelberg 2012

This instruction file for Word users (there is a separate instruction file for LaTeX users) may be used as a template. Kindly send the final and checked Word and PDF files of your paper to the Contact Volume Editor. This is usually one of the organizers of the conference. You should make sure that the Word and the PDF files are identical and correct and that only one version of your paper is sent. It is not possible to update files at a later stage. Please note that we do not need the printed paper.

2 Cloud Computing for Face Recognition of Video Surveillance

Video surveillance technology is toward high-definition, digital, networked and intelligent direction. Video surveillance must constantly improve the picture quality and extend the system for special needs; all these factors lead to doubling the storage capacity of the video surveillance system, but also significantly increase the system's processing requirements. Especially the identification process of huge video data, is far from a local server or a cluster of processing power can bear.

II. In this article, the high-performance processing power of cloud computing and low-cost mass storage will be introduced into the video surveillance system to achieve a large range of face recognition and tracking, this may supply a new way of thinking such application.

2.1 What Is Cloud Computing

What is Cloud ComputingAfter grid computing and service computing, cloud computing is a new computing model, which is gradually being recognized and accepted by the industry. In fact, cloud computing is a super-computing model based on internet. In the remote data center, the computer cloud combined thousands of computers and servers can provide massive storage capabilities and powerful computing capability, and is of highly scalable. According to their needs and operations, users may access to the cloud computing by computers, mobile phones and other intelligent terminal, etc[2].

Cloud computing includes two aspects: one describes the infrastructure, which is used to construct the upper application basis; the second describes cloud computing applications built on the infrastructures. Cloud computing is inherited and evolved in grid computing, cluster technology, and super-computing, but it has more flexibility, and is of enlarged power in computing[3]. It may provide dynamic resource pools, virtualization and high availability of the following computing platforms.

The users of cloud computing platform or third-party application developers do not have to care about the realization of the under-layer of cloud platform[4], and just call the provided interface of cloud platform to complete their work. They will feel that cloud platform provides an unlimited resource.

2.2 Application of Cloud Computing in Face Recognition

In this model, the video data captured by the front-end system with camera and video server, can be directly sent to the cloud computing system, and also be handled by the video server. Under normal circumstances, only the varied image data is sent to the cloud platform for the purpose of identification through the cloud platform, but ordinary video data is stored locally[5].

The remote terminal can request the real-time data from the video surveillance system. Those data are provided by the cloud platform, and also can be offered by the specific video server requested by the cloud platforms, and then are transmitted to the monitoring terminal. On dealing with the requested identification, cloud computing platform process video data of an investigated region by virtual computing. Users of system only contact with the cloud computing platform, including video surveillance, recognition of monitored site-specific target and track, the data of video surveillance is locally stored in this model, the varied data is sent to the cloud platform. This method ensures the redundancy of local and remote terminal, even if disconnecting the network does not affect local use.

3 The Model Design Face Recognition Video Surveillance Based on Cloud Computing

On tracking or investigating somebody, the huge video surveillance system needs massive computing and storage capabilities, a single common server is clearly no longer appropriate for dealing with videos data; cloud computing platform provides a newer idea for it's characters, such as low-cost, high performance computing and mass storage capacity.

3.1 System Architecture of Face Recognition in Video Surveillance Based on Cloud Computing

With the development of technology in video surveillance, video surveillance system has been experienced a rapid evolution from the initial analog cameras to today's network camera. Resolution ratio of analog camera is lower, the amount of storage required is also small, but the HD pixels network camera are generally required large storage space, but also need high-performance computing power, if you want to track the overall treatment linkage, small storage and a single basic server can not be achieved, so the need to use this latest technology - "cloud computing" to deal with.

As the complexity of video surveillance system, we make the following assumptions:

Assumption 1: the model of any single monitoring system with motion detection capability; once the image take the initiative to change, it will be initiatively sent to the cloud platform.

Assumption 2: In the model, each video services (including DVR, video services with a single camera, and other video services) have been active and passive features to send video streaming and mobile detection history of a certain period of time video stream, but also be able to send live video stream data to the cloud platform, but most of the video remains in the local, to make up for insufficient network bandwidth, or disconnected from the network defects.

Assumption 3: each device connected with cloud platforms has enough network bandwidth.

Based on the assumption 1-3, we give the model shown in Figure 1.

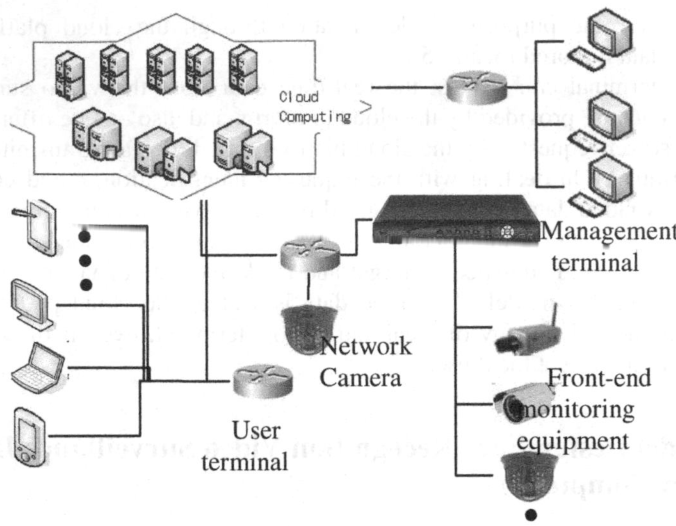

Fig. 1. Model structure

To realize the model of Figure1, we need to make some definition for the data of camera, the source face image, the target data etc.

Definition 1: The data structure of source face image matching

We set Sp as the source image data, it's structure is defined as follows:

$$Sp=\{T,P,Ts,Te,S,A\} \qquad ......(1)$$

T=Ticket, unique serial number assigned by system when source image arrives cloud platform.

P=Picture, source image data including the information of image data and image format

Ts: the time of that the source image arrives cloud platform.

Te: the time of that the system successfully complete the recognition.

S: a sign to mark the success(True) or failure(False) of system.

A: the address of last and successful recognition.

Definition 2: the data structure of video camera

Video service, the provider of video data, includes data from one or more video camera. This requests encoding the camera and server to distinguish the enormous cameras in the system.

Set Cb as the camera code, Cb={ XXXX(international area code)XXXX(state or province area code)XXXX(county or district)XXXX(reserved code)XXXXXXXX}
......(2)

The encoding must ensure the unique code, 0 is ready to replenish the insufficient digit. For example, the code of some or other camera in Chenghua district(0001;Chengdu, 0028; China, 0086) is expressed as Cb={00860028000100000000001}.

Cs={Cb, Ip, Name, Time, Work, Demo}

Set Cs as the data structure of camrea, Cs={Cb, Ip, Name, Time, Work, Demo}, Cb is as above.

Ip is the IP address of video server which the camera is subordinate to, a general public IP address. if IP is private, the camera must be connected and found in time, in some case, one IP is correspond with several cameras.

Time: the installation time of camera.

Name: designation of camera named as it's application or location passably.

Work: mark the operative mode, off-working-state includes maintenance, dismantlement etc.

Demo: the user of camera or other remarks.

3.2 Algorithm Design of Face Recognition

Face recognition is to find the proximal face image for recognizing in awaiting face images according to the basis of the facial features[6]. Current facial features in using are generally the shape characteristic, gray-scale distribution and frequency characteristic. Shape characteristic include shape of major parts of the face and their relative position; gray-scale distribution is the calculated distance between different gray patterns, when the face image is regarded as a one-dimensional or two-dimensional gray-scale mode; when face image is transformed into frequency spectrum, the characteristic is the frequency characteristic.

In the specific character of the video evidence, it may only get one image, that is to say we may use a single sample for face recognition. Face recognition methods can be divided into two categories, which are based on geometry and template matching. In the face recognition method based on single samples, there are sample-expansion method, characteristic subspace-expansion method, common learning framework method, image enhancement, neural network method and three-dimensional identification methods.

The focus of these methods is different. Sample-expansion method composites multiple virtual images from the original sample using various technology, expand the number of training samples per class, then transform a single training sample face recognition into a general face recognition. This method is more effective for dealing with the varied expression of face, the disadvantage is that virtual face images, not truly independent image of the face, are highly relevant to original face image. Neural network approach has the advantage of avoiding the complex feature extraction, and can significantly improve the speed by parallel processing of information. Three-dimensional face recognition method to solve single sample face recognition is more effective, but the three-dimensional image data and computation are very enormous.

Using powerful computing capabilities of cloud computing platform, and combining with variety of methods to avoid the low rate of identification as far as possible effected by causes, such as gesture, light and other factors. To improve the accuracy of face recognition, the two-step detection and identification methods is applied in this paper:

a) Pre-identification: using the above method for detection and identification. When the identifying similarity is above a certain threshold value K1 (the K1 can artificially adjust it's value), then the accurate detection function starts by operating and adjusting the PTZ control module to a more appropriate location for more clear objectives such as human faces.

b) Precise identification: a more clear face image may be gained through a series of operations after the pre-identification. Recognition is again based on the gained image, In this process, if you find similarity with the target to be identified is less than the threshold K2 (K2 can also be artificially pre-set), the goal is not the specific tracking target, otherwise, it is. Immediately you can start tracking module and collaborative tracking module.

Based on this idea, the process of identification is shown as Figure 2:

1). The image and range of identification: in this case of multiple images submitted for recognizing by the user, these images are sent to the image processing module by queue-processing with FIFO method. The range of identification is a required particular area in which the face images recognition will be carried out, it's expression refers to the definition Cb(definition 2), such as: China (0086), Chengdu (0028) Chenghua District(0001),the range of identification is expressed as follows: {008600280001}. If more ranges are required, they can be assembled as a set expressed as {Range 1, Rang 2, ...}, expression of each range is encoded as the former example.

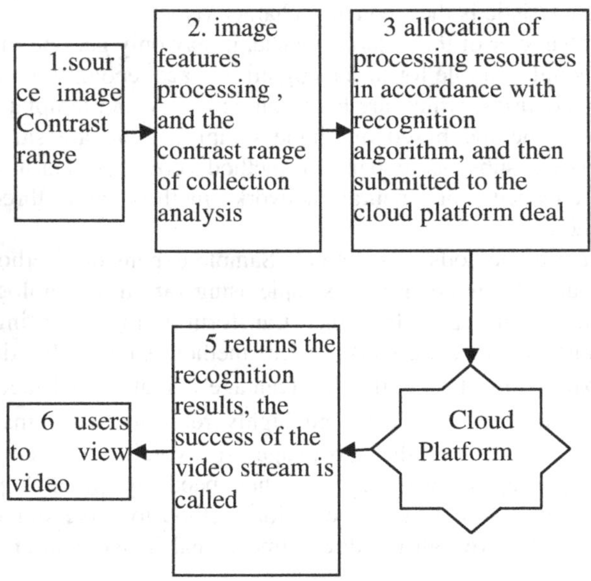

Fig. 2. Identification Process

Face recognition is to find the proximal face image for recognizing in awaiting face images according to the basis of the facial features[6]. Current facial features in using are generally the shape characteristic, gray-scale distribution and frequency characteristic. Shape characteristic include shape of major parts of the face and their relative position; gray-scale distribution is the calculated distance between different gray patterns, when the face image is regarded as a one-dimensional or two-dimensional gray-scale mode; when face image is transformed into frequency spectrum, the characteristic is the frequency characteristic.

2). The processing of image features: on acquired the source image and the system range of identification, the system process the image features, and factorize the range of identification into cameras recognized by the system, then get the code set and submit to the next processes.

3) Resource allocation: allocating resources into n small tasks, which will respectively be submitted to the cloud platform by multi-threading.

4) Cloud platform: completing the image contrast and the call of video surveillance images, and then generating results recognition.

5) Returns the result: the result is a collection combined with zero or a low and time.

4 Conclusion

Cloud computing is a hot spot in recent years for its research and application prospects. This paper presents a novel face recognition video surveillance system solutions using virtual mass storage capacity and powerful virtual computing power of cloud computing. Target data for identifying is from the existing installed video cameras and real-time data, video data is transmitted over the network to the cloud computing system, the cloud computing platform offers video data storage and computing functions. Collaborative communication of camera and computing is also provided by the cloud computing platform. The remote client only need to submit the source face images and range of identification, and then may get data and results.

References

1. China cloud computing network. Definition and characteristics of cloud computing, February 15 (2009), http://www.chinacloud.en/how.aspxTid=741&eid=17
2. Chen, K., Zheng, W.: Cloud computing: system instance and Research. Journal of Software 20(5), 1337–1348 (2009)
3. Yin, G., Wei, H.: Cloud computing: a method to realize conceptual computing. Journal of Southeast University 33(4), 502 (2003)
4. Jianxun, Gu, Z.-M., Zheng, C.: Cloud computing research overview. Application Research of Computers 27(2), 429–433 (2010)
5. Peng, Z., Zhou, Y., Wen, C.-K.: Monitoring system of digital video forensics based on cloud computing platform. Application Research of Computers 28(8), 2975–2976 (2011)
6. Jiang, J.-G., Sun, H.-Y., Qi, M.-B.: Real-time algorithm for face tracking based on mean-shift. Application Research of Computers 25(7), 2226–2227 (2008)

A Research on the Difference, Equilibrium and Evolution of the Wage in Inter-province: An Analysis Based on the Equilibrium Theory

Junbo Xue[1] and Zheng Wang[1,2]

[1] Institute of Policy and Management, Chinese Academy of Sciences, Beijing, P.R.C.
[2] Key Laboratory in Geography Information Science of the Ministry of Education, East China Normal University, Shanghai, P.R.C.
jbxue@casipm.ac.cn, wangzheng@casipm.ac.cn

Abstract. Based on the equity theory of Adams and the bureaucrats utility model of Niskanen, together with the modified human capital model of Mulligan and Sala-I-Martin, the author calculated the equilibrium wage of each province and its evolution. The results show the difference between the real wage and equilibrium of the West of China is bigger than that of the East of China. The gap between the real wage and the equilibrium level are decreasing during the recent years after the adjustment of the officials' wage. The results provide a conference for the regulation of officials' wage.

Keywords: equity theory, equilibrium wage, officials, human capital.

1 Introduction

Official wages play an important role in the public economics and labor economics. As far as this issue is concerned, scholars in domestic an oversea have done many researched from the points of their views, such as the comparison between the office and other sectors, whether the official wage is too high or not, whether the official wage should be increased and whether increasing official wage can promote the probity or not[7][15].

As we all know, the official wage are belong to the theory of wage. Therefore, the research on the wage theory is an important base of it. The history of the research on the wage theory is long, During the early period, the economists pay much attention to the decisive factor of the wage, such as Smith, Richado, Muler, etc. After that, Clark set up the theory of marginal productive. Marshall set up a theory of equilibrium between the supply and demand. With the enforce of the labor union, the theory based on the collective negotiation entered into the scope of the economists, such as Pigou and Hicks[16].

Most of the mentioned theory above are put much weight on the microeconomic agent. After the research of Marshall, many economists developed the wage theory during the research on the macro-economics [2][3][11][17] [18] [19].

D. Jin and S. Lin (Eds.): Advances in CSIE, Vol. 1, AISC 168, pp. 113–118.
springerlink.com © Springer-Verlag Berlin Heidelberg 2012

However, all of the wage theory above don't consider the social action and psychological factors. Adams, a scientist in behavior, did some research on the equality theory which can compliment the theory. Adams(1956) figured out that a man will consider not only the absolute value of what he get but also consider the relative value. After that, many scholars did the deeper research [4][5][6][8][9] .And one of the important contribution of the official wage theory is maximum budget model of bureaucracy and bureaucrats utility model which is made by Niskanen in 1971 and 1975 respectively[13][14]. Niskanen(1975) figured out that the utility of the bureaucrats is composed of two main parts, one is the current value of his incomings, *Y*, the other is the additional non-monetary incomings that come from his position, *P*. Therefore, the utility function can be written as follows:

$$U=\alpha Y^{\beta}P^{\gamma} \tag{1}$$

Where β and γ is the elasticity coefficient.

2 Model and Data Analysis

According to Niskanen's theory (1975), there are mainly two parts of the officials' wages on considering of their effectiveness. One part is the earnings due to his position (present value) and the other part is the additional revenue in forms of non-monetary. Among these two contents, the former part is much higher than the later one. Miger, Belanger believed that (1974) the officials would like to budget the surplus as "discretionary budget" in terms of utility function. Currently, the administrative cost is the main part which can be controlled by the officials. Hence, in the utility function, except for the wages, there should be administrative cost. However, administrative cost varies a lot in different provinces as the number of officials, population, and economic levels are different. That is why we add average administrative cost as it is not precise to only consider the total amount of the administrative cost.

In addition, there are differences of the price levels in different provinces (autonomous regions and municipalities), we can not only take account into the nominal wage and make some adjustment of the consumer price index for wages. The real wages equal to the nominal wage divided by the consumer price index.

Generally, the main body of personal income is the wage, which can be used discretionarily. However, the administrative cost cannot be regarded as discretionary spending as it is one part of the work. Hence, it plays smaller roles in the utility function. Niskaned argued (1975) that in the utility function a significant feature is $\beta>\gamma$. In Gong and Zou's paper (2000), the elastic coefficient in their utility function are $\beta:\gamma:\lambda\approx 0.75:0.15:0.1$ respectively. According to the survey results of actual research, in the officials' opinion, the ratio of these three parts is around 0.68:0.22:0.1. Consequently, the coefficients in our research are 0.7, 0.2, and 0.1, respectively. The utility function was listed in equation 2:

$$U_i=(W_i/P_i^{index})^{0.7}(M_i/L_i)^{0.2}(M_i)^{0.1} \tag{2}$$

Where i stands for the order of ith province, U_i is the effectiveness of the officials in the ith province, W_i is the wages of the ith province, M_i is the administrative cost of the ith province, P_i^{index} is the consumer price index of the ith province, L_i is the number of employees in the society of the ith province .

In this paper, Shanghai was selected as the standard, and other provinces(autonomous regions and municipalities) were used for comparative analysis. According to equity theory, the effectiveness of other provinces should be the same as the one of Shanghai. Thus, equation (3) was obtained:

$$U_i=(W_i/P_i^{index})^{0.7}(M_i/L_i)^{0.2}(M_i)^{0.1}=U_{shh}=(W_{shh}/P_{shh}^{index})^{0.7}(M_{shh}/L_{shh})^{0.2}(M_{shh})^{0.1} \quad (3)$$

According to equation (3), the equilibrium of wage level EW_i in the ith province is

$$EW_i=((M_{shh}/L_{shh})/((M_i/L_i))^{0.2587}(M_{shh}/M_i)^{0.1429}(P_i^{index}/P_{shh}^{index})W^p_{shh} \quad (4)$$

On considering of the effectiveness of equilibrium, the officials' wages in different provinces should be of equilibrium. In equation (4), there are relations between the equilibrium wage levels, average administrative cost per capita, total administrative cost, the price index and the levels of wages in Shanghai. The higher the average administrative cost per capita and total capita administrative cost is, the lower the equilibrium wage levels are. However, the higher the price index is, the higher the equilibrium wage levels are.

According to the equation (4), the level of wages of each province (in one year) can be calculated under equilibrium state. Compared with the actual level, $\eta_{i,t}$ can be calculated, which stands for in the year t, in the ith province, the ratio of equilibrium wage $EW_{i,t}$ and real wage $W_{i,t}$ (1996-2006 samples).

$$\eta_{i,t}= EW_{i,t}/W_{i,t} \quad t=1996\text{-}2006 \quad (5)$$

In addition, it is noted that, it is not enough to only consider equitable factors, as the payment of working and economic development of each province (autonomous regions and municipalities). The wage level of a provincial (autonomous regions and municipalities) should be related with the output of their own provinces. In that way, the human capital model developed by Mulligan, Sala-I-Martin (1997) and Jeong (2002) is a good example that put output, human capital input and wage together well. Wu, Wang(2004) revised the model of Mulligan, Sala-I-Martin (1997) and Jeong (2002) and calculated the human capital of all of the provinces. The core equation is as follows(see equation 6):

$$H_s/H_u=(y_s/y_u)(w(u,\tilde{r})/w(s,\dot{s})) \quad (6)$$

Where H_s and H_u is the human capital input of the two regions respectively, y_s and y_u is the per capita GDP of the two regions respective, $w(u,\tilde{r})$, $w(s,\dot{s})$ is the wage of he two regions respectively.

Then we can get the human capital input of all of the provinces, see table 1

Table 1. The human capital input of all of the provinces

provinces	*h*	provinces	*h*	provinces	*h*
Beijing	0.81	Anhui	0.63	Sichuan	0.65
Tianjin	0.81	Fujian	0.86	Guizhou	0.65
Hebei	0.66	Jiangxi	0.63	Yunnan	0.7
Shanxi	0.55	Shandong	0.83	Tibet	1.58
Inner Mongolia	0.64	Henan	0.6	Shaanxi	0.57
Liaoning	0.83	Hubei	0.67	Gansu	0.61
Jilin	0.63	Hunan	0.58	Qinghai	0.81
Heilongjiang	0.77	Guangdong	0.70	Ningxia	0.61
Shanghai	1.00	Guangxi	0.58	Xinjiang	0.62
Jiangsu	0.83	Hainan	0.64		
Zhejiang	0.87	Chongqing	0.59		

We can revise equation (4) with the result of table1, see equation (7), and we can get the ratio of the equilibrium wage to the real wage[1].

$$\eta_i = \eta'_i \, h_i \tag{7}$$

In order to see the evolution trend of the wage, we can see ratio of the equilibrium wage to the real wage among the east coast region and the middle-west region. See figure1 and figure 2.

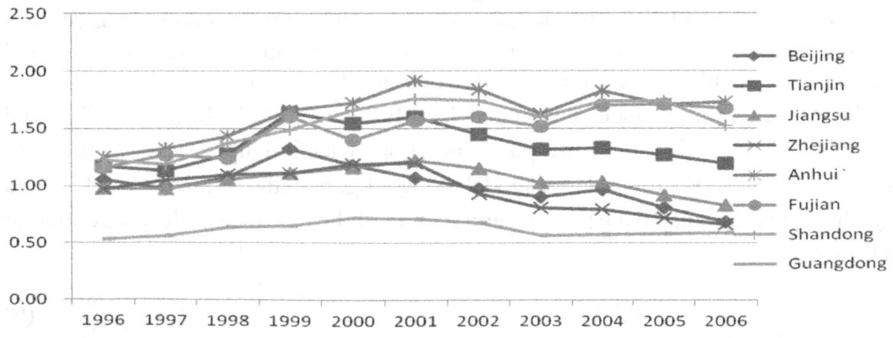

Fig. 1. The evolution trend of the ratio of the equilibrium wage to the real wage among the east coast region

[1] Because of the length limitation of the paper, we don't list all of the results here.

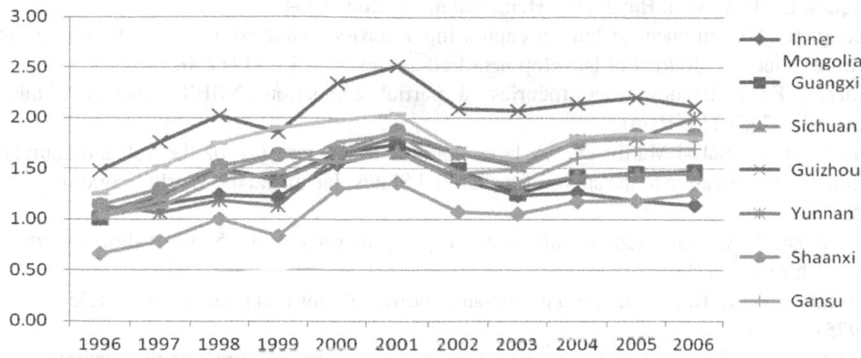

Fig. 2. The evolution trend of the ratio of the equilibrium wage to the real wage among the middle-west region

3 Conclusion

In consideration of the human capital and equal, we calculated the equilibrium wage, the ratio of equilibrium wage to real wage and the evolution trends in inter-provinces . From the results we can see that the gap between the equilibrium wage and real wage in the west and west-south are higher than that of the east coast region. Therefore, the central government should carry out some favorable policy.

Because of the data limitation, the analysis dimension is inter-provinces. What is more, the allowance is another important part of the incomings, especially in the west and north-west part of china.

Acknowledgement. This paper is supported by the NSFC(70933002,40701068).

References

1. Adams, J.S.: Inequity in Social Exchange. In: Berkowitz (ed.) Advances in Experimental Social Psychology, pp. 267–299. Academic Press, New York (1965)
2. Akerlof, G.A., Katz, L.F.: Workers' trust funs and the logic of wage profiles. Quarterly Journal of Economics 104, 525–536 (1989)
3. Akerlof, G.A., Yellen, J.L.: The fair wage-effort hypothesis and unemployment. Quarterly Journal of Economics 105, 255–283 (1990)
4. Atchison, T.J., Belcher, D.W.: Equity, Rewards, and Compensation Administration. Personnel Administration, 32–36 (March-April 1971)
5. Belcher, D.W., Atchison, T.J.: Equity Theory and Compensation Policy. Personnel Administration, 22–33 (July-August 1970)
6. Belcher, D.W., Atchison, T.J.: Compensation for Work. In: Dubin, R. (ed.) Handbook of Work, Organization, and Society, pp. 567–611. Rand McNally, Chicago (1976)
7. Bygren, M.: Pay reference standards and pay satisfaction: what do workers evaluate their pay against? Social Science Research 33, 206–224 (2004)
8. Jaques, E.: Equitable Payment. John Wiley, New York (1961)

9. Jaques, E.: Time Span Handbook. Heinemann, London (1964)
10. Jeong, B.: Measurement of human capital input across countries: a method based on the laborer's income. Journal of Development Economics 67, 333–349 (2002)
11. Katz, L.F.: Efficiency wage theories: a partial evaluation. NBER Macroeconomics Annual 1, 235–276 (1986)
12. Mulligan, C., Sala-i-Martin, X.: A labor–income-based measure of the value of human capital: an application to the states of the United States. Japan and the World Economy 9(2), 159–191 (1997)
13. Niskanen, W.A.: Bureaucracy and representative government, p. 15, 38. Aldine-Atherton. Inc., Chicago (1971)
14. Niskanen, W.A.: Bureaucrats and politicians. Journal of Law and Economics, 18 (December 1975)
15. Rahman, A.T.R.: Legal and administrative measures against bureaucratic corruption in Asia. In: Carino (ed.) Bureaucratic Corruption in Asia: Causes, Consequences, and Controls, pp. 109–162. NMC Press, Philippines, Quezon City (1986)
16. Romer, D.: Advanced Macroeconomisc, 2nd edn., pp. 412–432. Shanghai University of Finance & Economics Press, Shanghai (2001)
17. Shapiro, C., Stiglitz, J.E.: Equilibrium unemployment as a worker discipline device. American Economic Review 75, 433–444 (1984); Reprinted in Mankiw and Romer (1991)
18. Weitzman, M.L.: The Share Economy. Harvard University Press, Cambridge (1984)
19. Yellen, J.L.: Efficiency wage models of unemployment. American Economic Review 74, 433–444 (1984); Reprinted in Mankiw and Romer (1991)
20. Wu, B., Wang, Z.: The Human capital calculation of the China. Science and Research Management 4, 60–65 (2004)

Study on the Deployment Process of Low Attitude Interception Net

Xiaofei Zhou[1], Yi Jiang[1], and Yongyuan Li[2]

[1] School of Aerospace Engineering, Beijing Institute of Technology, Beijing, China
3120100052@bit.edu.cn
[2] R&D Centre, China Academy of Launch Vehicle Technology, Beijing, 100076, China

Abstract. In this paper, Dynamic analysis for the process of opening net is based on a new and low attitude towards interception device. Depending on the problem that dynamics modeling of fully compliant interception net during the process, a new equivalent method is proposed. This method regards the interception net as a variable drag coefficient rigid body, and regards four traction heads as a suppositional traction head, so the process of opening net can be transformed to the problem for movement of two rigid bodies. Furthermore, the dynamic equations for the process of opening net were derived from this method. These equations are used to simulating the process of opening net; the simulation results are compared with test results to verify the practicability of the method.

Keywords: Dynamics; projecting deployment, Interception net, Opening net.

1 Introduction

With the development of missile technology, the defense ability of high altitude, middle altitude and low altitude has been strengthened. At the same time as development of stealth technology, detection technology and electronic jamming, low altitude and very low altitude penetration means more and more are widely used. Enemy or terrorists may use model airplane or delta wing carrying small explosion source or other disturbance weapons to infringe important place, to create chaos and terror effects. The target-spacecraft of this attack pattern has characteristics such as small radar cross-section, low speed, low altitude, small lethality and simple control and so on. Thus, even the antiaircraft gun which entry into the range of fire attack cannot do anything for this kind target-spacecrafts. The very low attitude interception device this paper studies can search the omnidirectional target, track the target and intercept target softly, it shoots an interception net after the target is locked, the interception net covers the target then can fully guarantee the security of the region.

The deployment of interception net in the air is a very complex dynamic process, and is similar to the deployment process of space net technology which is extensive concerned at space On-Orbit Capture. The concept of space net is first used by Nakasuka and other authors at the Furoshiki satellite missions[1~3], then the following researchers made further research from both simulations and experiments,

D. Jin and S. Lin (Eds.): Advances in CSIE, Vol. 1, AISC 168, pp. 119–127.
springerlink.com © Springer-Verlag Berlin Heidelberg 2012

and have achieved some results[4~5]. In domestic, Zhai and other authors made research on attitude dynamics with time-varying inertia for space net capture robot system[6], Chen and other authors made research on the expansion of space net from both numerical simulation and ground test[7~8], Yu and other authors established three dimensional finite element model for the deployment of space net by using the absolute nodal coordinate formulation, and made great deal of calculation and simulation[9]. But for the low attitude projecting deployment process of interception net which is fully compliant, there are no mature mathematical model or test results to be referenced, so this paper focuses on the establishment of projecting deployment model for interception net, and tests the model by experiment.

2 Working Principle of Interception Device and Simplified Model of Projectile Body

The Interception device carries the interception net in the head, and the parachute in the tail which is mainly used to recovery the remaining load of interception device after the interception device was launched. Its working principle is as follows: The target data which is obtained from the detection device is passed to fire control computer; the fire control computer starts to estimate the track of target and to bind the launch elements, then, sends instruction to servo; the servo accepts instruction from the fire control computer and turns around the specified location to prepare for launching.

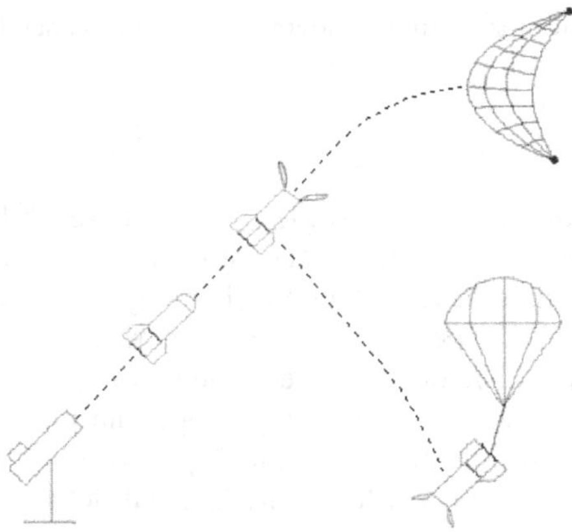

Fig. 1. The work process schematic diagram of interception device

Interception device starts the interception net according to the launch elements; the net is deploying by drive of the traction heads, to wind the target and make it lose power and crash. After the deployment of interception net, the interception device starts the parachute which carries the remaining load by slow landing, to complete the recovery of interception device.

The arrangement for projecting body is in the head of interception device, the head shape is hemispherical that in order to increasing capacity utilization. The projecting body consists of interception net, traction heads, ignition, propellant, steel tubes and other structure. The angles between the every tube and axis of interception device are certain. Traction heads finish acceleration in the tube and pull the interception net to flight after the ignition is ignited, the interception net to be deployed during the flight course.

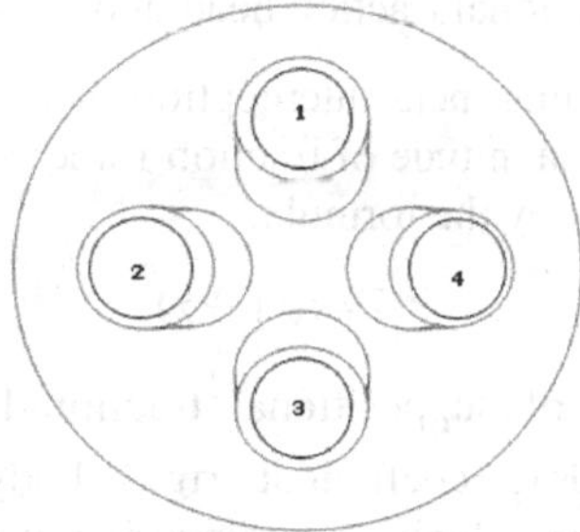

Fig. 2. Schematic diagram of layout for traction heads in the opening net device

Interception net is circular reticulations which are woven with lightweight ropes; layout of traction heads in the opening net device is shown in Fig. 2.

3 Projecting Deployment Dynamics Model of Interception Net

There are continuum dynamics model and lumped parameter model two categories modeling method for the interception net which is fully compliant structure[10-14]. Continuum dynamics model can meticulously depict the propagation process of stress wave for the whole rope segment, can obtain the shape and deformation condition for rope segment reliably, but its equations are complex second order partial differential equations, integral calculation work is heavy. Furthermore, it would be a very complex boundary condition if take into account the braided structure of interception net. Lumped parameter model predigests spatial configuration and stress distribution condition of rope segment, merely obtain the approximate configuration and stress distribution for the fully compliant rope in the system.

This paper proposes a method that using variable drag coefficient rigid body to instead interception net which is fully compliant as a whole; the dynamic equations for the process of interception net deployment are derived from this method. Due to the time that interception net was pulled out of intercept device is very short, in order to convenient for calculation, the hypotheses are as follows:

a) Trajectory inclination angle of intercept device holds on line during the moment opening net;

b) Scalar of initial velocity for four traction heads is equal, and the interception net is equably deploying during the process;

c) The four traction heads are replaced with a suppositional traction head, the velocity of suppositional traction head is equal to that four traction heads velocity projects at the axis of interception device, the mass of suppositional traction head is equal to sum of four traction heads;

d) The interception net is replaced with a variable drag coefficient rigid body, its drag coefficient changes with the radius of net during deployment process;

e) The friction force that the interception net is pulled from the intercept device is ignored;

f) The lift force for suppositional traction head and interception net are ignored.

During the process of opening net, interception net is moving by drive of the suppositional traction head, magnitude of traction force is mainly relate to the relative velocity, it can be represented by the formula:

$$T=k(V_1-V_2) \tag{1}$$

Where, V_1 is the velocity of suppositional traction head, V_2 is mass centroid velocity of the variable drag coefficient rigid body that interception net is equivalent, k is a const that can obtain by engineering experience. It can be seen from the above formula, when V_1 and V_2 are equal, then traction force that traction head acts on interception net be zero, interception net is completely deploying. In the following process, due to drag coefficient of interception net is much larger than the traction head's, traction head starts to decelerate under reaction force of interception net, and pulls the interception net to fall into the ground under gravity action.

Besides the traction force and gravity, the Interception net is also affected by air resistance D_2. The magnitude of air resistance is mainly related to mass centroid velocity of the variable drag coefficient rigid body, and the deployment area of interception net. Due to the time deployment of interception net is a very short, this paper describes the radius of deploying during the process as follow formula:

$$\begin{cases} r(t)=ae^{bt}+c & 0\le t\le t0 \\ r(t)=ae^{bt0}+c & t>t0 \end{cases} \tag{2}$$

$$D_2 = 0.5\rho V_2^2 f(r(t)) \tag{3}$$

Here, t_0 is the time that interception net completely deployed. $f(r(t))$ is function of characteristic area and drag coefficient that is mainly related to the deployment radius.

$$\frac{dV_1}{dt}=\frac{-k(V_1-V_2)-0.5\rho V_2^2 C_d S_{ref}-m_1 g\sin\theta}{m_1} \tag{4}$$

$$\frac{d\theta}{dt}=-\frac{g\sin\theta}{V_1} \tag{5}$$

$$\frac{dV_2}{dt} = \frac{k(V_1 - V_2) - 0.5\rho V_2^2 f(r(t)) - m_2 g \sin\theta}{m_2} \tag{6}$$

$$\begin{cases} r(t) = ae^{bt} + c & 0 \le t \le t0 \\ r(t) = ae^{bt0} + c & t > t0 \end{cases} \tag{7}$$

$$\frac{dr}{dt} = \begin{cases} abe^{bt} & 0 \le t \le t0 \\ 0+ & t > t0 \end{cases} \tag{8}$$

$$\frac{dx_1}{dt} = V_1 \cos\theta \qquad \frac{dy_1}{dt} = V_1 \sin\theta \tag{9}$$

$$\frac{dx_2}{dt} = V_2 \cos\theta \qquad \frac{dy_2}{dt} = V_2 \sin\theta \tag{10}$$

$$\frac{dx_{10}}{dt} = V_{10} \cos\theta \cos\sigma \qquad \frac{dy_1}{dt} = V_{10} \sin\theta \sin\sigma \tag{11}$$

Through the force analysis, we can establish the motion equations for suppositional traction head and equivalent rigid. In the above equations, parameter ρ is air density, parameter S_{ref} is reference area of suppositional traction head, parameter C_d is the drag coefficient of suppositional head, parameter x_1, y_1 are horizontal displacement and height of the suppositional traction head respectively, parameter x_2, y_2 , are horizontal displacement and height of the mass centroid of equivalent rigid body for interception net respectively, parameter , x_{10}, y_{10} are horizontal displacement and height of No.1 traction head.

4 Comparison of the Test and Numerical Simulation

This segment studies the projecting deployment process through test and numerical simulation.

4.1 Test

Test weather as follows: ground temperature is 8.2 Celsius; the average of ground wind speed is 2.6m/s. Test condition includes static test and dynamic test two kinds; the interception device is fixed at a certain angle during the static test; the interception device is launched from launch canister during the dynamic test, then, the intercept device starts the interception net. According to the location that is pre-judged, the test instruments such as high-speed camera, laser rangefinder and Doppler radar are preparing to work.

The test of projecting deployment for interception net needs that every operating preparation is strictly done according to the operating rules, such as propellant charging, interception net folding, traction heads placing etc. Traction heads are launched by the drive of propellant in the sloping tubes, to draw out the interception net in sequence from edge to bottom. With flight of traction heads, the interception net starts forming in the air.

Multiple test results show that the projecting deployment process of interception net is smooth and steady, and there is no severe vibration; although there is some overlap about interception net, it does not form a wing, ultimately, it completely deploys to the braided shape about 300 milliseconds after traction heads are launched from the interception device. Through tests we can find that regular folding and placing of the interception net can lead to the net be deployed effectively by the drive of traction heads.

Fig.3 is measured velocity comparison of No.1 traction head (see Fig. 2) in dynamic test and static test.

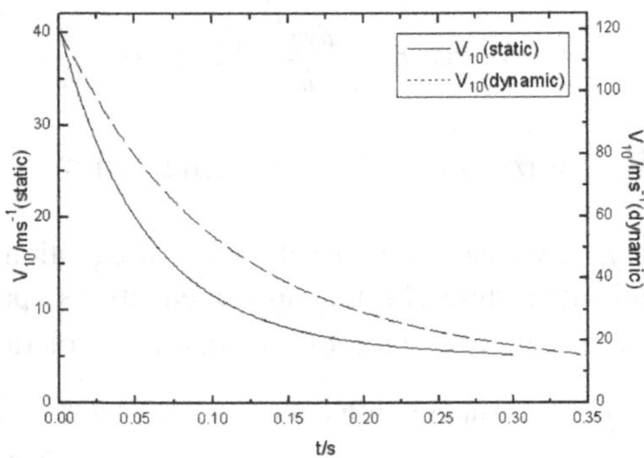

Fig. 3. Dynamic test and static test for measured velocity comparison

4.2 Numerical Simulation

The differential equations which are established by this paper to describe the deployment process of interception net are used to simulate, to analysis the deployment process of interception net. This paper simulates the process of static and dynamic opening net. The initial conditions of static opening net as follows: $\theta_0=25.7^o$, the initial velocity of No.1 traction head is given according to the test result, herein, $V_{10}=40$ m/s, the initial velocity of interception net $V_{20}=0$ m/s,. The initial conditions of dynamic opening net as follows: $\theta_0=30^o$, $V_{10}=120$, $V_{20}=80$.

The simulation results are shown in Fig.4, Fig.5 and Fig.6.

Figure 4 shows the velocity change during the process of static opening net, the velocity of No.1 traction head rapidly decreases rapidly from 40m/s after it was launched from the oblique tube, on the contrary, the velocity of interception net gradually increasing speed from 0m/s by the drive of traction head. Finally, the two velocities are equal that is t=0.299s which means the interception net is completely deploying.

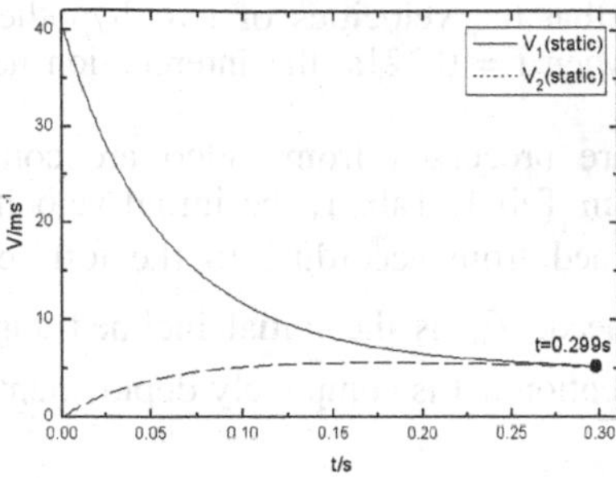

Fig. 4. The velocity of hypothesis traction head and velocity of the mass centroid for hypothesis rigid body (static test)

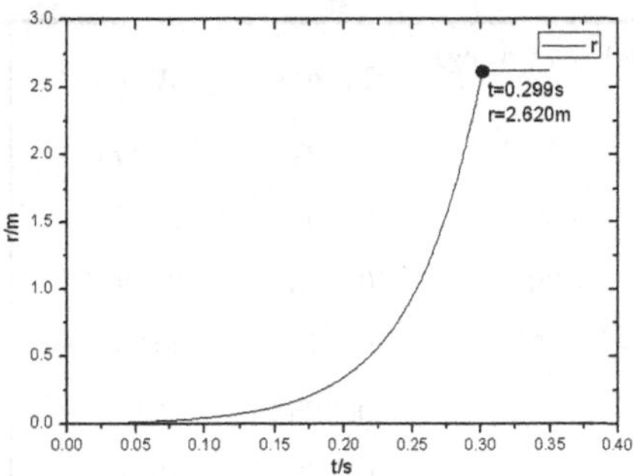

Fig. 5. The radius of interception net deployment change with time (static)

It can be seen from Fig. 5, the radius of interception net deployment reach 2.620m at t=0.299s, then the radius no longer increases, that also means the interception net has been completely deployed.

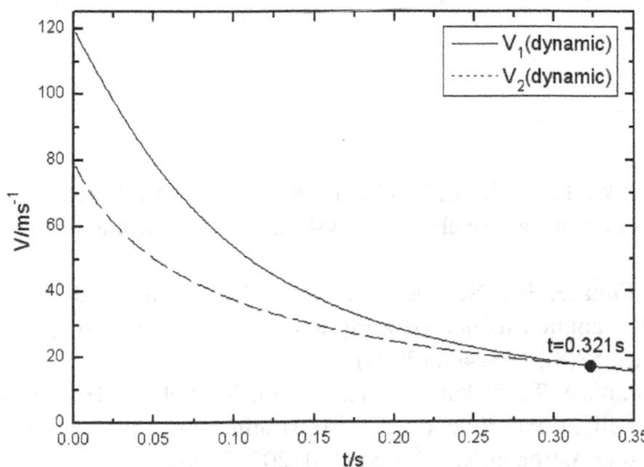

Fig. 6. The velocity of hypothesis traction head and velocity of the mass centroid for hypothesis rigid body (dynamic)

Fig. 6 shows the time that the velocities of two hypothesis rigid body achieve balance need more time, when t = 0.321s, the interception net has been completely deployed.

In addition the data are processed from video are consistent with the result simulation data, as shown in Tab 1. Tab. 1, the initial velocity and initial trajectory inclination angle are obtained from according to the test result. V_{10} is the initial velocity of No 1 traction head, θ_0 is the initial inclined angle of intercept device, Time is time that the interception net is completely deployment and x is the horizontal displacement.

Table 1. Comparison of simulation and ground test

	Test				Simulation	
Test condition	V_{10} (m/s)	θ_0 (deg)	Time(s)	x (m)	Time(s)	x (m)
static	39. 87.	30	0. 287	2. 90	0. 299	3. 088
static	38. 75	30	0. 305	2. 95	0. 297	2. 977
static	40. 53	30	0. 295	3. 18	0. 300	3. 137
dynamic	119. 50	25. 7	0. 311	11. 90	0. 320	12. 844
dynamic	120. 23	25. 7	0. 325	12. 80	0. 322	12. 883
dynamic	122. 25	25. 7	0. 332	13. 70	0. 323	13. 005

5 Conclusions

By comparing test and simulation can be seen, the simulation results agree well with the experimental results, to illustrate the mathematical model can describe the projecting deployment process. At present, this method has been applied to the model developed.

References

[1] Nakasuka, S., Aoki, T., Ikeda, I., Tsuda, Y., Kawakatsu, Y.: Furoshiki satellite—a large membrane structure as a novel space system. Acta Astronautica 48(5-12), 461–468 (2001)

[2] Nakasuka, S., Funase, R., Nakada, K., Kaya, N., Mankins, J.C.: Large membrane furoshiki satellite applied to phased array antenna and its sounding rocket experiment. Acta Astronautica 58(8), 395–400 (2006)

[3] Nakasuka, S., Funane, T., Nakamura, Y., Nojira, Y., Sahara, H., Sasaki, F., Kaya, N.: Sounding rocket flight experiment for demonstrating furoshiki satellite for large phased array antenna. Acta Astronautica 59(1-5), 200–205 (2006)

[4] Gardsback, M., Tibert, G.: Deployment control of spinning space webs. Journal of Guidance, Control, and Dynamics 32(1) (2009)

[5] The Grapple, Retrieve and Secure Payload (GRASP)Experiment, http://www.tethers.com/GRASP.html
[6] Zhai, G., Qiu, Y., Liang, B., Li, C.: Research of Attitude Dynamics with Time-Varying Inertia for Space Net Capture Robot System. Journal of Astronautics 29(4), 1131–1136 (2008)
[7] Chen, Q., Yang, L.-P.: Research on Casting Dynamics of Orbital Net Systems. Journal of Astronautics 30(5), 1829–1833 (2009)
[8] Chen, Q., Yang, L.-P., Zhang, Q.-B.: Dynamic Model and Simulation of Orbital Net Casting and Ground Test 31(3), 16–19 (2009)
[9] Yu, Y., Baoyin, H.-X., Li, J.-F.: Modeling and Simulation of Projecting Deployment Dynamics of Space Webs. Journal of Astronautics 31(5), 1289–1296 (2010)
[10] Mankala, K.K., Agrawal, S.K.: Dynamic Modeling and Simulation of Impact in Tether NetPGripper Systems. Multibody System Dynamics 11(3), 235–250 (2004)
[11] Kim, E., Vadali, S.R.: Modeling Issues Related to Retrieval of Flexible Tethered Satellite Systems. AIAA -92-4661 (1992)
[12] Cosmo, M.L., Lorenzini, E.C.: Tethers in Space Handbook, 3rd edn. Smithsonian Astrophysical Observatory (1997)
[13] Jin, D.-P., Wen, H., Hu, H.-Y.: Modeling, dynamics and control of cable systems. Advance in Mechanics 34(3), 304–313 (2004)
[14] Buckham, B., Nahon, M.: Dynamics simulation of low tension tethers. In: Riding the Crest Into the 21st Century, Ocean 1999, pp. 757–766. MTS/ IEEE (1999)

Construction and Study of Computer Agongqiang Corpus

Guofu Yin

Weinan Normal University, Shaanxi China
yinguofu@126.com

Abstract. construction of computer corpus is a complicated engineering. The basic concept of corpus is introduced and six processes of the corpus construction is analyzed in this paper: demand analysis and planning stage, corpus design stage, language materialcollection & computer pre-phase design stage, corpus system realization stage, corpus label and computer system development & input stage and usage and maintenance of corpus stage; the corpus system construction principle based on content theme is expounded the design and optimization of Agongqiang corpus is discussed by combining the reality, thus playing a positive role in promoting conservation of non-material culture.

Keywords: Corpus, Computational linguistics, Agongqiang, label & coding.

1 Introduction

The corpus in modern sense was born in 1960s. The landmark work is the construction and use of American brown corpus, which only has one million words. Though a small corpus from the today's prospective, it is the world's first corpus can be read by machine. Over the past five decades, great progresses have already been witnessed in corpus and corpus methods at home and abroad. The scale of corpus is not only larger and larger and the processing depth is deeper and deeper and the corpus technology is increasingly applied.

2 Corpus and Its Study in Linguistics

Corpus is the storage and set of language materials, it indicates the set of passages collected purposefully in accordance with the definite design standard.

Corpus can be divided into the three categories according to the content:

1) Heterogeneous corpus: it collects various texts without selection principle and specific standards determined in advance, containing various language materials and widely collected stored in the original form.

2) Homogeneous corpus: it collects the data of the same category. That is, the contained language materials have the same attribute according to the content;

3) Special type corpus, for certain specific purpose, the language materials are collected and corpus established.

D. Jin and S. Lin (Eds.): Advances in CSIE, Vol. 1, AISC 168, pp. 129–133.
springerlink.com © Springer-Verlag Berlin Heidelberg 2012

Corpus is widely applied into the linguistics research field. It is the modern-based significant foundation for language study, the corpus is utilized to study language, thus the problems in the traditional linguistics study that can be overcome: (1) weak objectivity of language materials; (2) small proportion of language materials; (3) heavy workload and small efficiency; (4) insufficient sharing of language materials.

3 Design of Computer Corpus System

Corpus system is the core of corpus, including a complete system of computer hardware, software, corpus user, corpus collection and processing rule, corpus management and application programme. A corpus system contains technique system and social system, in which the former contains linguistics knowledge and computer-related techniques while the latter contains whether the work scheduling, work take-over and quality can coordinate the talents in each aspect and various problems related to copyright can be solved effectively.

Construction of computer corpus system is a complicated “corpus project”. From the angle of computer software engineering, we can see the construction of corpus system. The process of the construction of computer corpus in this paper can be divided into the following stages: demand analysis and planning stage, corpus design stage, language material collection and computer pre-phase design stage, corpus system realization stage, corpus labeling and computer system development & input stage as well as corpus application and maintenance stage.

The process of the overall construction is as shown by the Fig. 1.

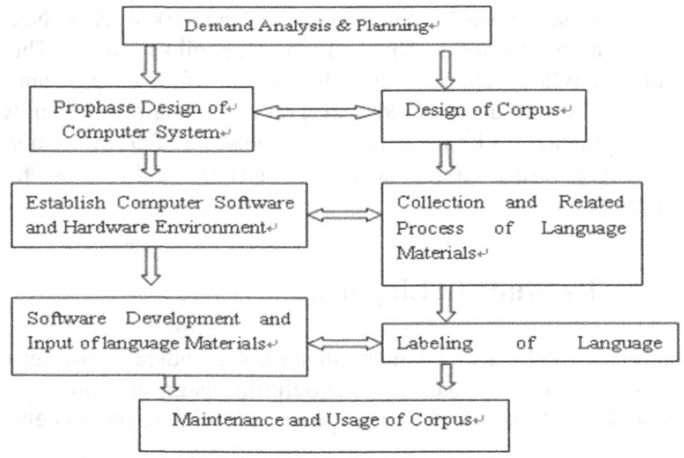

Fig. 1. The process of the overall construction

(1) Demand analysis and planning stage. The demand analysis mainly indicates determining the meaning and purpose of corpus system and the applications which should be completed in this system in the future; planning mainly indicates planning the category, scale and development cost and involved techniques of this corpus and

planning the job scheduling for language collector, disposer and computer system developer.

(2) Computer system pre-phase design and corpus design is started synchronically. The design of corpus generally involves three aspects: corpus balance structure design, language material sampling principle design and logical structure design of corpus. The design of these three aspects shall be started synchronically with the design of computer system pre-phase. That is, how to select computer hardware environment, software technology (XML technology, relation database system technique and programming technique, etc.), how use and release computer corpus and what technique should be adopted to develop the corpus software to better use corpus, etc.

(3) Computer hardware environment shall be established with the collection and related disposal of language materials. Language material collection stage, computer system should establish the environment of the experiment property, the collected language material samples shall be analyzed to determine the storage technique or data structure in the computer system and the experiment environment shall be evaluated to prepare for the genuine input in the future and make pre-phase preparation for the software development at the next step;

(4) Software development, input and labeling of language materials. This part is a core content of the corpus construction: inputting the collected language materials into computer system according to the corpus inputting method determined before, starting software development & testing according to the input language materials and goal at the earlier stage; as for language material, it is necessary to start part-of-speech, participle and semantic label, etc. to satisfy the demands of the software development.

(5) Corpus maintenance and application stage. To establish good corpus, it is necessary to: revise balance proportion of corpus, upgrade corpus, revise the logical structure of language materail, etc. It is also necessary to test and revise the computer hardware environment and developed software and repeatedly test the application of corpus to fulfill the anticipated application purpose. The whole process of computer corpus system construction is the process for project manager, linguistics worker and computer technical personnel to mutually communicate coordinate and cooperate with each other and also the process for mutually learning from each other.

4 Discussion on Computer Agongqiang Corpus Construction

With the development of era, the non-material cultural heritage-Agongqiang is on the verge of extinction. Due to the entertainment diversification, fast food, modernization and utility, the Agongqiang is gradually marginalized under the impact of the stormy wave of the market-based economy. It is facing the most critical life and death crisis. Therefore, it is an imperative to have in-depth study, rescue and pass on it. The script language is mainly taken as the starting point to tap the artistic charm of the local opera from the angle of the folk-custom culture. It should be conserved and passed on through establishing corpus. As for the design and construction of Agongqiang corpus, it shall be conducted according to that aforementioned but that of Agongqiang corpus should be certainly of specialty.

4.1 Problems to Be Solved for Design and Construction of Agongqiang Corpus

(1) Agongqiang, as a graceful and wonderful local type of drama with song lyrics and aria having bright local characteristic, contains a lot of local dialects, proverbs and common sayings, etc. However, the study on its opera history is still not made. Additionally, Agongqiang did not have script in the past. The Agongqiang actor expresses according to the story plot orally narrated by certain person. For example, the famous Agongqiang actor Duan Tianhuan ever orally narrated more than 30 scripts. Afterwards, some hand-copied scripts were generated but all were sporadic.

(2) Retrieval and network transmission of Agongqiang character: nowadays, there is still no standard character set for Agongqiang character. The computer can not retrieve, display and transmit Agongqiang character on the webpage. It is necessary to dispose international standard information based on Unicode to solve these problems.

4.2 Selection of Agongqiang Corpus

(1) Principle for selection of Agongqiang sample: as for the selection of the Agongqiang samples, it is necessary to select some original samples, especially the song periodicals popular in Fuping civilian society except selecting some classic scripts put on stage under the guidance of drama expert, like Female Inspector in *Tang Dynasty, Romance of the Western Chamber, Story about Jinlin and Wang Kui is thankless*, etc. In this case, the samples are of authenticity. Simultaneously, it is also necessary to select the Agongqiang samples widely applied in many areas and related to the realistic life, thus the selection coverage of the Agongqiang samples is wide.

(2) Classification index of Agongqiang corpus: as for the selection of corpus, it is necessary to design classification index in advance and scientifically determine the proportion of each type of corpus in the corpus to make it reach balanced. Classification index indicates the balance factor of corpus and the key representative characteristic influencing the corpus. The theme, literary, age and school system can be selected as the balance factors in accordance with the characteristics of Agongqiang.

4.3 Design of Agongqiang Corpus Mainly Includes

(1) Agongqiang character dictionary base: contains the Unicode coding, classification coding, character pattern and Chinese meaning of the character symbols, etc.

(2) Agongqiang text: consisting of the texts of Agongqiang passage grade. There is such information as Agongqiang title, Agongqiang script, Chinese translation, theme, literary type, age and application place, etc. The retrieval can be performed according to various demands by establishing corresponding retrieval catalogs.

(3) Agongqiang label corpus: label the passages, sentences, words and characters of corpus by using some specific symbols representing certain meaning through the labeled Agongqiang corpus text base is a key content for establishing Agongqiang.

5 Conclusion

Study on computation linguistics helps establish corpus for Agongqiang script language. Through field investigation, network and video media collection and ancient books and records collection, etc. the Agongqiang scripts are collected and sorted and the Agongqiang raw corpus database and related bases are established. Based on the word list established before and related bases and by combining the characteristics of Agongqiang, it shall be divided preliminarily with computer. As for the result of the divided Agongqiang, it shall be calibrated and labeled artificially to establish Agongqiang annotated corpus. Simultaneously, the word list and related knowledge bases shall be further perfected. On one hand, it is for convenience of retrieval and reference; on the other hand, the language fact of the script can be observed and grasped to analyze and study the rule of language system.

Acknowledgments. This work was supported by the Qin East History and Culture Research Center research projects ,the number is QDYB1119.

References

1. Yu, S.: Introduction to Computational Linguistics. Commercial Press, Beijing (2003)
2. Huang, C., Li, J.: Corpus Linguistics. Commercial Press, Beijing (2003)
3. Wang, J.: Construction & Application of Computer Corpus. Tsinghua University Press, Beijing (2005)
4. Tokuda, K., Kobayashi, T., Imai, S.: Speech parameter generation from HMM using dynamic features. In: Proc. IEEE International Conference on Acoustics Speech and Signal Processing, pp. 660–663 (1995)
5. Kawahara, H.: Straight, exploration of the other aspect of vocoder: Perceptually isomorphic decomposition of speech sounds. Acoustic Science and Technology 27(6), 349–353 (2006)
6. Olohan, M.: Introducing Corpora in Translation Studies, pp. 94–96. Rougtledge, London and New York (2004)
7. Mc Cathy: Issues in Applied Linguistics. Cambridge University Press, Cambridge (2001)

A Semantic Context-Based Access Control Model for Pervasive Computing Environments

HaiBo Shen

School of Computer, Hubei University of Technology, Wuhan 430068, China
jkxshb@163.com

Abstract. In pervasive computing environment, computing devices and applications can easily interoperate each other and interact with human activities by context-awareness. Therefore, it is necessary to have a formal representation that represents semantics of the contexts, reflects the change of the situation, and can be shared and understood by system security policies. Traditional security policy and subject-based access control mechanisms are inadequate in such an environment. So this paper presents a semantic context-based access control model (SCBAC) which adopts OWL-based ontologies to express context/policy and SWRL rules to express context-aware access control policies. The proposed model adopts a context-centric policy method, and grants permissions to users according to current context information and allows high-level description and reasoning about contexts and policies.

Keywords: Pervasive Computing Environment, Context-Based Access Control, Semantic Web Technologies, SWRL.

1 Introduction

Pervasive computing and all kinds of mobile devices allow users to access services or resources anytime and anywhere even when they are on the move. Because of the high unpredictability, heterogeneity and dynamicity of new pervasive scenario, traditional subject-based access control methods, such as identity-based access control (IBAC) and role-based access control (RBAC), break down in such an environment. In contrast with this, user's contexts and environment context may play a more important role. Granting a user access without taking the user's current context into account can compromise security as the user's access privileges not only depend on "who the user is" but also on "where the user is" and "what is the user's state and the state of the user's environment".

In pervasive computing environments, it is crucial to define context-based policy and have a policy system that understands and interprets semantics of the context correctly. Therefore, a new trend of research in computer security formed towards designing semantic context-aware security infrastructures and access control models. It is to say that context-based solutions need to adopt semantic web technologies, especially ontology technologies, as key building blocks for supporting expressive context/policy modeling and reasoning [1].

This paper proposes a semantic context-based access control model (called SCBAC) for pervasive computing environments by combining ontology technologies

D. Jin and S. Lin (Eds.): Advances in CSIE, Vol. 1, AISC 168, pp. 135–140.
springerlink.com © Springer-Verlag Berlin Heidelberg 2012

with context-based access control mechanism. In SCBAC, this paper introduces context-based access control policies in controlling access based on the context of the situation; Furthermore, this paper specifies access control policies as rules over ontologies representing the concepts, and uses semantic web rule language (SWRL [2]) to form policy rule and infer those rules by JESS inference engine.

The remainder of this paper is organized as follows: Section 2 presents SCBAC model for pervasive computing environments and its authorization architecture. In the last section, the conclusion is given.

2 Semantic Context-Based Access Control Model

2.1 Context Defining and Representation

Unlike traditional subject-based access control, the SCBAC model adopts context-centric access control solutions, which defines permissions based on just about any security relevant contexts. For access control purposes, we are concerned with five types of contexts:

1. Subject Contexts (SC). A subject is an entity that takes action on an object or resource. Subject Contexts define the specific subject-related contexts that must be held or exercised by the subject in order to obtain rights to an object or resource. In SCBAC model, subject contexts are used to determine access rights for an entity requesting access privileges. Subject-related contexts include: the subject's role, identity, credentials, name, organization, activity, location, task; services, devices, network, and platform provided to subject, social situation, and so on.

2. Object Contexts (OC). An object is an entity that is acted upon by a subject. As with subjects, objects have contexts that can be leveraged to make access control decisions. Object contexts are any object-related information that can be used for characterizing the situation in which the protected object was created and its current status, which is relevant for making access control decisions.

3. Transaction Contexts (TC). In pervasive computing environment, transactions involve the user, the mobile platform, the specific resource or service being accessed, and the physical environment of both the user and platform. A transaction specifies a particular action to be performed in the system. For example, a transaction may be initiated by a user from a specific location, to access a resource that is currently in a state, at a particular time of day. Transaction contexts capture significant information about the transactions that are occurring or have occurred in the past.

4. Environment Contexts (EC). Environment Contexts describe the operational, technical, and even situational environment at the time a transaction takes place. Environment contexts, such as current date and time, or other contextual information that is relevant to access control, are not associated with a particular subject or a resource, but may nonetheless be relevant in applying an access control policy. The state of the environmental conditions can be captured via sensors.

5. Social Contexts (SRC). Social Contexts are any information relevant to the characterization of a situation that influences the interactions of one user with one or more other users, such as social relationships (e.g., nearby persons and nearby friends), proximity of other people, task-related activities, and so on. In SCBAC model, SRC is mainly related to social relationships.

The SCBAC model is a context-centric access control solutions, context is the first-class principle that explicitly guides both policy specification and enforcement process. But context-centric access control solutions need to adopt ontology technologies as key building blocks for supporting expressive context modeling and reasoning. Context ontology examples can be found in [3,4].

2.2 SCBAC Model

The basic SCBAC model has the following components:

1. S, O, E, T and SRC are subjects, objects, environments, transactions, and social relationships, respectively;

2. SC_i $(1 \leq i \leq K)$, OC_j $(1 \leq j \leq M)$, EC_k $(1 \leq k \leq N)$,TC_n $(1 \leq n \leq J)$ and SRC_m $(1 \leq m \leq I)$ are the contexts for subjects, objects, environments, transactions, and social relationships, respectively;

3. CONT(s), CONT(o), CONT(e), CONT(t) and CONT(sr) are context assignment relations for subject s, object o, environment e, transaction t, and social relationship sr respectively:

$CONT(s) \subseteq SC_1 \times SC_2 \times \ldots \times SC_K$; $CONT(r) \subseteq OC_1 \times OC_2 \times \ldots \times OC_M$;
$CONT(e) \subseteq EC_1 \times EC_2 \times \ldots \times EC_N$; $CONT(t) \subseteq TC_1 \times TC_2 \times \ldots \times TC_J$;
$CONT(sr) \subseteq SRC_1 \times SRC_2 \times \ldots \times SRC_I$;

4. Action (Act): an action is an event that a subject seeks to perform. An action can be given a list of parameters defining how the action must be performed.

5. Permission Assignments (PA): An permission grants the right to a subject to perform an action on an object or resource. Permission assignments (PA) capture the privileged actions that a subject is authorized to hold or exercise on an object. The authorization is determined based on all type of contexts. One significant advantage of the SCBAC model is that rights can be assigned to contexts only.

The following function captures the rights that are assigned to a user when a given set of contexts are active and she is attempting to access an object:

$$(< Act, SC, OC, EC, TC, SRC >, Perm) \in PA, \text{ where } Perm = \{Allow, Deny\}$$

As indicated above, the permission assignment (PA) not only associates an permission with the user context(s), but makes it conditional on a set of active environment contexts. Clearly, rights may change for the same user accessing a resource if the object contexts, environment contexts, or even user contexts vary between requests. In this system, a request will be granted access rights if and only if:

(1) The policy rule assigning a specified action (Act) to an access request exists with the specified subject contexts (SC), object contexts (OC), environment contexts (EC), transaction contexts (TC), and Social contexts (SRC) that match those specified in the set of permission assignments (PA)

(2) The subject contexts (SC) are active for the user making the current request

(3) The object contexts (OC) are active for the object being accessed by the user

(4) The environment contexts that are made active by the current environmental conditions are contained in the set EC.

6. In the most general form, a Policy Rule that decides on whether a subject s can access an object o in a particular environment e and within a transaction t and with a social relationship sr, is a Boolean function of s, o, e , t, and sr's contexts:

Rule: can_access(s,o,e,t,sr)←*f*(CONT(s),CONT(o),CONT(e),CONT(t), CONT(sr))

Given all the context assignments of s, o, e, t, and sr, if the function's evaluation is true, then the access to the resource is granted; otherwise the access is denied.

The SCBAC model is illustrated in Figure 1. SCBAC model consists of seven fundamental components: Subject Contexts (SC), Object Contexts (OC), Transaction Contexts (TC), Environment Contexts (EC), Social Contexts (SRC), Action (Act) and Permission Assignments (PA)

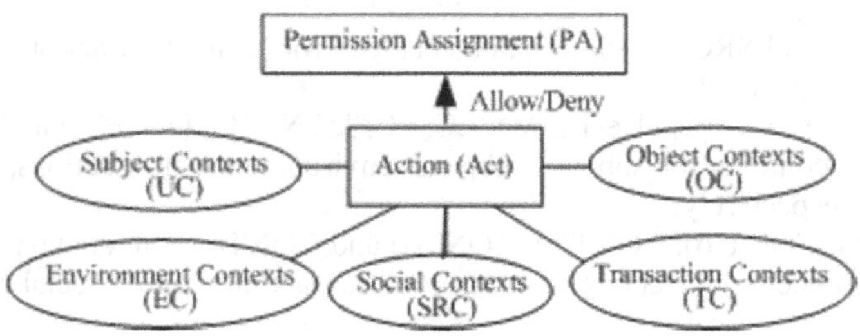

Fig. 1. SCBAC model components

2.3 SCBAC Authorization Architecture

Authorizations are functional predicates that must be evaluated in order to return whether the subject is allowed to perform the requested right (action) on the object. Authorizations evaluate all contexts and the requested rights. Semantic context-based authorization architecture is illustrated in Figure 2 below. The diagram reflects the following logical actors involved in CASBAC model: (1) The Knowledge Base (KB) is a data repository of domain ontology. KB is composed of the set of security policy rules in SPR, subsumption relations between concepts (SUB) and context information (CI). Inference of implicit authorization rules is based on the facts and rules in the KB. (2) The Ontology Manager is responsible for gathering and updating ontologies in domains of subjects, objects, actions, and policies and also reducing the semantic relations to the subsumption relation. We can use the Protégé-OWL ontology editor in the Protégé-OWL ontology development toolkit [5] to create all ontologies. (3) The Context Handler is responsible for getting the contextual information from external environment, and performing assertion to the KB database according to the model of the domain knowledge as well as maintaining the consistency of the KB since the KB has to accurately reflect the dynamic changes of the environment. (4) The Policy Enforcement Point (PEP) is responsible for requesting authorization decisions and enforcing them. In essence, it is the point of presence for access control and must be able to intercept service requests between service requester and providers. (5) The Policy Decision Point (PDP) is responsible for evaluating the applicable policies and making the authorization decision (allow or deny) by making use of an inference engine, based on facts, contexts and rules in the KB. The PDP is in essence a policy execution engine. When a policy references a subject contexts, object contexts, or context that is not present in the request, Jess Rule Engine [6] in PDP contacts the KB to retrieve the contextual information. (6) The Policy Administration Point (PAP) is responsible for creating a policy or policy set.

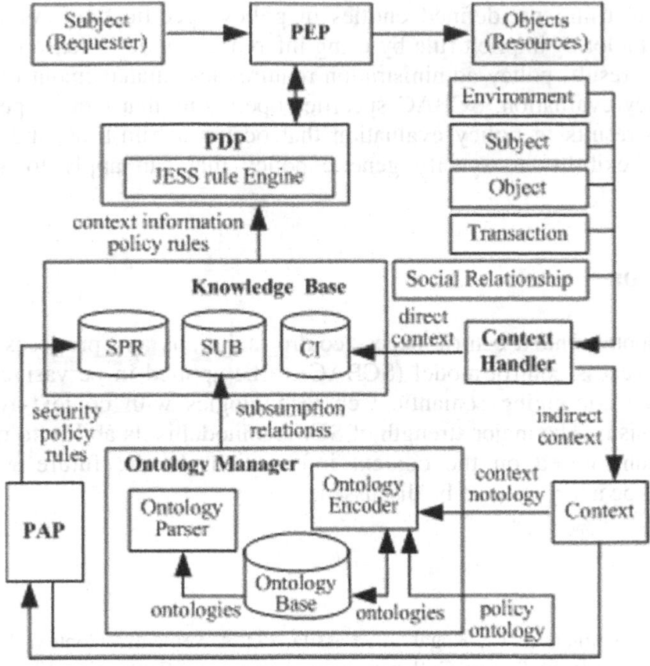

Fig. 2. SCBAC authorization architecture

2.4 Advantages of SCBAC Model

SCBAC model provides several advantages over existing authorization models. (1) SCBAC may encompass IBAC, RBAC and ABAC (attribute-based access control). By treating identity, role and attributes as contexts of a principal, SCBAC fully encompasses the functionality of IBAC, RBAC and ABAC approaches. Therefore, we believe that the SCBAC is the natural convergence of existing access control models and surpasses their functionality. (2) SCBAC can realize more fine-grained access control. Policy/context representation is semantically richer and more expressive, and can be more fine-grained within SCBAC because it can be based on any combination of subject, resource, and environment contexts. Semantically-rich policy/context representations permit description of policies/contexts at different levels of abstraction and supporting reasoning about both the structure and properties of the elements that constitute a pervasive system, thus enabling policy analysis, conflict detection, and harmonization. (3) Better suited for dynamic context-aware environments. Current authorization models utilize static concepts - such as unique user identities, object names, and subject roles-to specify policy. In dynamic context-aware environments, however, subjects and objects may not be known ahead of time due to the dynamic nature of interactions. SCBAC is a context-centric solution, which supports authorization in these dynamic environments by using contextual information instead of static concepts in policy specification. (4) Less administrative overhead for policy specification. In contrast to traditional access control models, SCBAC does not require the definition of static policies based on users, objects or roles. Instead, SCBAC uses contextual information that do not limit the policy administrator to using pre-defined entities in policy specification. As well as, the system can get a lot of implicit rule by using inference engine to reason about policy ontology. As a result, policy administration requires less management overhead. (5) Run-time policy evaluation. SCBAC specifies operations that can be performed on contexts. This results in policy evaluation that occurs at run-time, thus giving the administrator flexibility to specify generic policy that can apply to a variety of situations.

3 Conclusion

This paper adopts context-centric access control solutions and proposes a semantic context-based access control model (SCBAC) to be applied in pervasive computing environment by combining semantic web technologies with context-aware access control mechanism. The major strength of SCBAC model is its ability to make access control decisions based on the context information. In the future work, policy adaptation will be a key issue to be discussed.

References

1. Toninelli, A., Montanari, R., Kagal, L., Lassila, O.: A Semantic Context-Aware Access Control Framework for Secure Collaborations in Pervasive Computing Environments. In: Cruz, I., Decker, S., Allemang, D., Preist, C., Schwabe, D., Mika, P., Uschold, M., Aroyo, L.M. (eds.) ISWC 2006. LNCS, vol. 4273, pp. 473–486. Springer, Heidelberg (2006)
2. McGuinness, D.L., van Harmelen, F.: OWL web ontology language semantics and abstract syntax (2004), http://www.w3.org/TR/owl-semantics/
3. Chen, H., Finin, T., Joshi, A.: An ontology for context-aware pervasive computing environments. Special Issue on Ontologies for Distributed Systems, Knowledge Engineering Review 18(3), 197–207 (2004)
4. Moussa, A.E., Morteza, A., Rasool, J.: Handling context in a semantic-based access control framework. In: Proceedings of the 2009 International Conference on Advanced Information Networking and Applications Workshops, pp. 103–108. IEEE Computer Society (2009)
5. Protégé Editor and API , http://protege.stanford.edu/plugins/owl
6. JESS : The Rule Engine for Java Platform, http://herzberg.ca.sandia.gov/jess

On the Mean Value of an Arithmetical Function*

MingShun Yang and YaJuan Tu

College of Mathematics and Information Science, Weinan Teachers University, Weinan, Shaanxi, 714000, China
yangms64@sohu.com

Abstract. Let p be a prime , $e_p(n)$ denote the largest exponent of power p which divides n .In this paper ,we use elementary and analytic methods to study the asymptotic properties of $\sum_{n\le x} e_p(n)\varphi(n)$, and give an interesting asymptotic formula for it.

Keywords: Asymptotic formula, Largest exponent, Perron formula.

1 Introduction

Let p be a prime , $e_p(n)$ denote the largest exponent of power p which divides n .In problem 68 of [1], Professor F.Smarandache asked us to study the properties of the sequence $e_p(n)$. About this problem, it seems that none had studied it, at least we have not seen related papers before. In this paper, we use elementary and analytic methods to study the asymptotic properties of the mean value $\sum_{n\le x} e_p(n)\varphi(n)$ ($\varphi(n)$ is the Euler totient function), and give an interesting asymptotic formula for it. That is, we will prove the following:

Theorem. Let p be a prime , $\varphi(n)$ is the Euler totient function. Then for any real number $x \ge 1$, we have the asymptotic formula

$$\sum_{n\le x} e_p(n)\varphi(n)= \frac{3p}{(p^2-1)\pi^2}x^2+O\,(x^{\frac{3}{2}+e}).$$

2 Some Lemmas

To complete the proof of the theorem, we need the following:

* **Foundation project:** Scientific Research Program Funded by Shaanxi Provincial Education Department (11JK0485) and he students of innovation projects of Weinan Teachers University (11XK052) Education Reform Project of Weinan Teachers University (JG201132).

D. Jin and S. Lin (Eds.): Advances in CSIE, Vol. 1, AISC 168, pp. 141–145.
springerlink.com © Springer-Verlag Berlin Heidelberg 2012

Lemma 1. Let p be a given prime. Then for any real number $x \ge 1$, we have the asymptotic formula

$$\sum_{\substack{n\le x\\(n,p)=1}} \phi(n) = \frac{3p}{(p+1)\pi^2} x^2 + o\,(x^{\frac{3}{2}+e})$$

Proof. Let F (s)= $\sum_{\substack{n=1\\(u,p)=1}}^{\infty} \frac{\phi(n)}{n^s}$ Re(s)>1. Then from the Euler product formula [3] and the multiplicative property of φ(n), we have

$$\sum_{\substack{n=1\\(n,p)=1}}^{\infty} \frac{\phi(n)}{n^s} = \prod_{q\ne p}\sum_{m=0}^{\infty}\frac{\phi(q^m)}{q^{ms}}$$

$$=\prod_{q\ne p}(1+\frac{q-1}{q^s}+\frac{q^2-q}{q^{2s}}+\frac{q^3-q^2}{q^{3s}}+\cdots)$$

$$=\prod_{q\ne p}(1+\frac{1-\frac{1}{q}}{q^{s-1}}(1+\frac{1}{q^{s-1}}+\frac{1}{q^{2(s-1)}}+\cdots))$$

$$=\prod_{q\ne p}(1+\frac{1-\frac{1}{q}}{q^{s-1}}\frac{q^{s-1}}{q^{s-1}-1})$$

$$=\frac{\zeta(s-1)}{\zeta(s)}\frac{p^s-p}{p^s-1}$$

Where $\zeta(s)$ is the Riemann zeta-function. By Perron formula [2] with $s_0=0$, T= x and b = $\frac{5}{2}$, we have

$$\sum_{\substack{n\le x\\(n,p)=1}} \frac{\varphi(n)}{n^s} = \frac{1}{2\pi i}\int_{\frac{5}{2}-iT}^{\frac{5}{2}+iT} \frac{\varsigma(s-1)}{\varsigma(s)}\frac{p^s-p}{p^s-1}\frac{x^s}{s}ds + O(\frac{x^{\frac{5}{2}}}{T}).$$

To estimate the main term

$$\frac{1}{2\pi i}\int_{\frac{5}{2}-iT}^{\frac{5}{2}+iT} \frac{\zeta(s-1)}{\zeta(s)}\frac{p^s-p}{p^s-1}\frac{x^s}{s}ds,$$

We move the integral line from $s=\frac{5}{2}\pm iT$ to $s=\frac{3}{2}\pm iT$. This time, the function

F (s) = $\frac{\zeta(s-1)}{\zeta(s)}\frac{p^s-p}{p^s-1}\frac{x^s}{s}$ Has a simple pole point at s = 2, and the residue is

$\frac{3px^2}{(p+1)\pi^2}$. So we have

$$\frac{1}{2i\pi}\left(\int_{\frac{5}{2}-iT}^{\frac{5}{2}+iT}+\int_{\frac{5}{2}+iT}^{\frac{3}{2}+iT}+\int_{\frac{3}{2}+iT}^{\frac{3}{2}-iT}+\int_{\frac{3}{2}-iT}^{\frac{5}{2}-iT}\right)\frac{\zeta(s-1)}{\zeta(s)}\frac{p^s-p}{p^s-1}\frac{x^s}{s}=\frac{3px^2}{(p+1)\pi^2}.$$

$$\frac{1}{2i\pi}\left(\int_{\frac{5}{2}+iT}^{\frac{3}{2}+iT}+\int_{\frac{3}{2}+iT}^{\frac{3}{2}-iT}+\int_{\frac{3}{2}-iT}^{\frac{5}{2}-iT}\right)\frac{\zeta(s-1)}{\zeta(s)}\frac{p^s-p}{p^s-1}\frac{x^s}{s}\ll x^{\frac{3}{2}+\varepsilon}.$$

From above we may immediately get the asymptotic formula:

$$\sum_{\substack{n\le x\\(n,p)=1}}\phi(n)=\frac{3p}{(p+1)\pi^2}x^2+O(x^{\frac{3}{2}+\varepsilon})$$

This completes the proof of the Lemma 1.

Lemma 2. Let a is an any fixed integer, and p is a prime. Then for any real number $x\ge 1$,We have the asymptotic formula

$$\sum_{\alpha\le\frac{\log x}{\log p}}\frac{\alpha}{p^\alpha}=\frac{p}{(p-1)^2}+O(x^{-1}\log x);$$

$$\sum_{\alpha\le\frac{\log x}{\log p}}\frac{\alpha}{p^{\frac{\alpha}{2}}}=\frac{p^{\frac{1}{2}}}{(p^{\frac{1}{2}}-1)^2}+O(x^{-\frac{1}{2}}\log x);$$

Proof. From the properties of geometrical series, we have

$$\sum_{\alpha\le\frac{\log x}{\log p}}\frac{\alpha}{p^\alpha}=\sum_{t=1}^{\infty}\frac{t}{p^t}-\sum_{\alpha>\frac{\log x}{\log p}}\frac{\alpha}{p^\alpha}$$

$$=\sum_{t=1}^{\infty}\frac{t}{p^t}-\frac{1}{p^{[\frac{\log x}{\log p}]}}\sum_{t=1}^{\infty}\frac{[\frac{\log x}{\log p}]+t}{p^t}$$

$$=\sum_{t=1}^{\infty}\frac{t}{p^t}+O(x^{-1}(\frac{[\frac{\log x}{\log p}]}{p-1}+\sum_{t=1}^{\infty}\frac{t}{p^t}))$$

$$=\frac{p}{(p-1)^2}+O(x^{-1}\log x),$$

And

$$\sum_{\alpha\le\frac{\log x}{\log p}}\frac{\alpha}{p^{\frac{\alpha}{2}}}=\sum_{t=1}^{\infty}\frac{t}{p^{\frac{t}{2}}}-\sum_{\alpha>\frac{\log x}{\log p}}\frac{\alpha}{p^{\frac{\alpha}{2}}}$$

$$=\sum_{t=1}^{\infty}\frac{t}{p^{\frac{t}{2}}}-\frac{1}{p^{\frac{1}{2}[\frac{\log x}{\log p}]}}\sum_{t=1}^{\infty}\frac{[\frac{\log x}{\log p}]+t}{p^{\frac{t}{2}}}$$

$$=\sum_{t=1}^{\infty}\frac{t}{p^{\frac{1}{2}}}+O(x^{-\frac{1}{2}}(\frac{[\frac{\log x}{\log p}]+t}{p^{\frac{1}{2}}-1}+\sum_{t=1}^{\infty}\frac{t}{p^{\frac{t}{2}}}))$$

$$=\frac{p^{\frac{1}{2}}}{(p^{\frac{1}{2}}-1)^2}+O(x^{-\frac{1}{2}}\log x).$$

This completes the proof of the Lemma 2.

3 Proof the Theorem

In this section, we complete the proof of the theorem.

$$\sum_{n\le x}e_p(n)\varphi(n)$$

$$=\sum_{p^\alpha\le x}\sum_{\substack{p^\alpha u\le x\\(u,p)=1}}\alpha\varphi(p^\alpha u)=\sum_{p^\alpha\le x}\alpha\varphi(p^\alpha)\sum_{\substack{u\le\frac{x}{p^\alpha}\\(u,p)=1}}\varphi(u)$$

$$=\frac{p-1}{p}\sum_{\alpha\le\frac{\log x}{\log p}}\alpha p^\alpha(\frac{3p}{(p+1)\pi^2}(\frac{x}{p^\alpha})^2+O((\frac{x}{p^\alpha})^{\frac{3}{2}+e}))$$

$$=\frac{3(p-1)}{(p+1)\pi^2}x^2\sum_{\alpha\le\frac{\log x}{\log p}}\frac{\alpha}{p^\alpha}+O(x^{\frac{3}{2}+e}\sum_{\alpha\le\frac{\log x}{\log p}}\frac{\alpha}{p^{\frac{\alpha}{2}}})$$

$$=\frac{3(p-1)}{(p+1)\pi^2}x^2\left(\frac{p}{(p-1)^2}+O(x^{-1}\log x)\right)+$$

$$O\left(x^{\frac{3}{2}+e}\left(\frac{p^{\frac{1}{2}}}{(p^{\frac{1}{2}}-1)^2}+O(x^{-\frac{1}{2}}\log x)\right)\right)$$

$$=\frac{3p}{(p^2-1)\pi^2}x^2+O(x^{\frac{3}{2}+\varepsilon}).$$

This complete the proof of the theorem.

References

1. Smarandache, F.: Only problems, not Solutions. Xiquan Publ. House, Chicago (1993)
2. Pan, C., Pan, C.: Elements of the Analytic Number Theory. Science Press, Beijing (1991)
3. Apostol, T.M.: Introduction to Analytic Number Theory. Springer, Heidelberg (1976)
4. Zhang, W., Liu, D.: Primitive numbers of power *P* and its asymptotic property. Smaramche Notions Journal 13, 173–175 (2008)
5. Yang, H.: Yan' an University master's degree dissertion, pp. 19–23 (2006)
6. Pan, C., Pan, C.: The Elementary Number Theory. Beijing University press, Beijing (2003)

A 1-D Image Random Scrambling Algorithm Based on Wavelet Transform

Qiudong Sun, Yunfeng Xue, Xueyi Chang, and Yongping Qiu

School of Electronic and Electrical Engineering, Shanghai Second Polytechnic University, Shanghai 201209, China
{qdsun,yfxue,xychang,ypqiu}@ee.sspu.cn

Abstract. Although the general random scrambling based on pixel can achieve a good chaotic effect, but it can not change the histogram of a digital image. We introduce the random scrambling into the domain of wavelet transform of image and scramble the coefficients of wavelet transform to improve the performance of scrambling. Firstly, we decomposed the original image by 2-D discrete wavelet transform into four sub-images. Secondly, we scrambled these sub-images separately. Thirdly, we reconstructed the scrambled image from all chaotic sub-images by 2-D inverse discrete wavelet transform. Finally, to further improve the scrambling degree of the reconstructed result, we scrambled it again in space domain to gain the encryption image. Experiments show that this algorithm is effective at visual evaluation and is more stable in scrambling degree than Arnold transformation.

Keywords: wavelet transform, 1-D random scrambling, dual scrambling, image encryption.

1 Introduction

Due to media data (such as digital image, video etc.) are very easy to be intercepted illegally when they are transmitted in the network. Therefore, people pay more attention to the security of media data and have developed four types of information hiding techniques, which are scrambling, watermarking, share storing and hiding [1].

This paper will discuss the scrambling technique for digital image encryption. Many typical image encryption methods scramble the pixels in an image directly [1-6]. These methods restrict the scrambling degree to a certain extent [7]. Some algorithms, also, do scrambling in transformation domain [7, 8]. But they, in general, are too complicated to be realized.

This paper will introduce a simple 1-D random scrambling algorithm [6] to the image in its wavelet transform domain. We will do first scrambling separately to the four sub-images of 1-level wavelet transform of image. After reconstructing the shuffle sub-images to a chaotic image, we will also do second scrambling to this image in space domain to further improve its scrambling degree.

D. Jin and S. Lin (Eds.): Advances in CSIE, Vol. 1, AISC 168, pp. 147–152.
springerlink.com © Springer-Verlag Berlin Heidelberg 2012

2 Wavelet Decomposition and Reconstruction of Image

The process of wavelet transform to an image is just a multi-resolution decomposition from large scale to small scale for it. The 2-level wavelet decomposition of image is illustrated as in Fig. 1. The wavelet decomposing result of image is a set of several sub-images. In 1-level wavelet decomposing, the original image can be decomposed into a low frequency sub-image LL_1 (corresponding to the approximate image of the original image in the low resolution) and three high frequency sub images HL_1、LH_1 and HH_1 (corresponding to the horizontal, vertical and diagonal detail image of the original image in the low resolution). In 2-level wavelet decomposing, the low frequency sub-image LL_1 can be decomposed continuously into a low frequency sub-image LL_2 and three high frequency sub-images HL_2、LH_2 and HH_2 in more lower resolution.

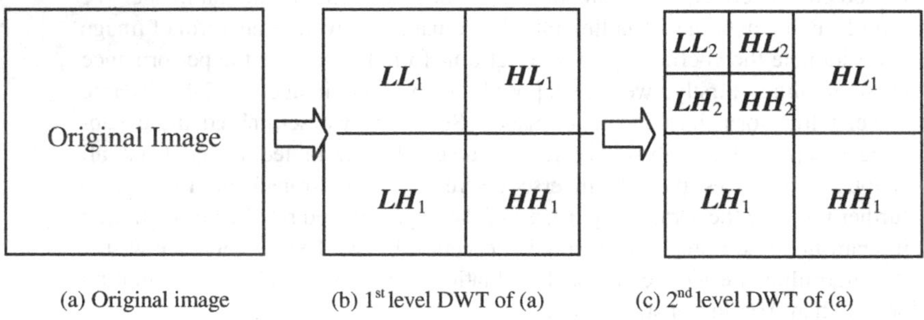

(a) Original image (b) 1st level DWT of (a) (c) 2nd level DWT of (a)

Fig. 1. 2-level wavelet decomposition of image

The decomposing process described above can be expressed by the iterative formulas follows.

$$LL_i \Rightarrow \begin{bmatrix} LL_{i+1} & HL_{i+1} \\ LH_{i+1} & HH_{i+1} \end{bmatrix}. \tag{1}$$

where the symbol "$\Rightarrow$" represents the wavelet decomposition. This formula means that the low frequency sub-image LL_i in i-level decomposing can be decomposed into four sub-images LL_{i+1}, HL_{i+1}, LH_{i+1} and HH_{i+1} in more lower resolution. In this way, we can obtain more and more sub-images in lower resolution. The process of wavelet reconstruction is just the reverse action of Eq. 1 or Fig. 1.

3 One-Dimensional Random Scrambling

3.1 Random Scrambling of Image

It is the random scrambling of digital image that choosing two pixels randomly from an image and interchanging them, after interchanging a certain couple of pixels to achieve the goal of chaotic result. In order to facilitate the implementation of 2D scrambling, we firstly scan the 2D matrix X by column direction, which denotes an

$M \times N$ digital image, and transform it into a 1-D vector $\boldsymbol{V}$. Thus, the position of 2-D image can be represented by the relevant subscript of 1-D vector. The interchanging of two pixels also becomes the interchanging of two elements in the 1D vector $\boldsymbol{V}$. In order to complete the random scrambling of 1-D vector $\boldsymbol{V}$, we need two random sequences $\boldsymbol{R}_S$ and $\boldsymbol{R}_D$, whichever has a length same as that of $\boldsymbol{V}$. $\boldsymbol{R}_S$ and $\boldsymbol{R}_D$ can be generated by a random natural number generator with the property of uniform distribution using two different seeds respectively. Which two elements in $\boldsymbol{V}$ to be exchanged are determined by the values of relative positions of $\boldsymbol{R}_S$ and $\boldsymbol{R}_D$ [6].

Because both $\boldsymbol{R}_S$ and $\boldsymbol{R}_D$ are of property of uniform distribution and generated by different seeds, they are almost not correlative. The scrambled 1-D vector $\boldsymbol{V}$, therefore, must be of properties of random and uniform distribution. Consequentially, it will have real scrambling effect.

3.2 Random Anti-scrambling of Image

Due to the interchanging rule is reversible, hence, when two random sequences $\boldsymbol{R}_S$ and $\boldsymbol{R}_D$ in anti-scrambling are same as that of in scrambling, the anti-scrambling can also be completed by doing the random scrambling again [6].

4 Dual Scrambling

4.1 First Scrambling

After the image transformed by DWT, we choose different couple of seeds and do scrambling to all sub-images separately in DWT domain. That is the first scrambling.

4.2 Second Scrambling

To improve the efficiency of the algorithm, we just do scrambling one time in DWT domain. In the result, the gray distribution of reconstructed image may not be well-proportioned. Thus, it is necessary to scramble the reconstructed image again in space domain. After that, the gray distribution of scrambled image looks like very uniform at once. That is the second scrambling.

4.3 Scrambling Security

Because of the fact that there are two seeds requiring in random number generating both in scrambling and anti-scrambling, the proposed algorithm must be an encryption method based on the key. The key space composed of several couples of seeds, theoretically, is infinite. It is obvious that the proposed algorithm is very safe for image encryption.

5 Experiment Result Analysis and Comparison

In the experiment, we choose N=1 and do the dual random scrambling to the gray image Lena as shown in Fig. 2(a), which is of size 256×256. The result is shown as in

Fig. 2(e). The four sub-images in Fig. 2(b) are the results of 1-level wavelet decomposing from Fig. 2(a). The scrambled sub-images in Fig. 2(c) are the results of first scrambling to Fig. 2(b), and the first scrambled result is reconstructed from the scrambled sub-images as shown in Fig. 2(d). Fig. 2(e) is the result of doing second scrambling to Fig. 2(d), and it is the final scrambled image. The anti-scrambling result is shown in Fig. 2(f), which is completely same as the original image.

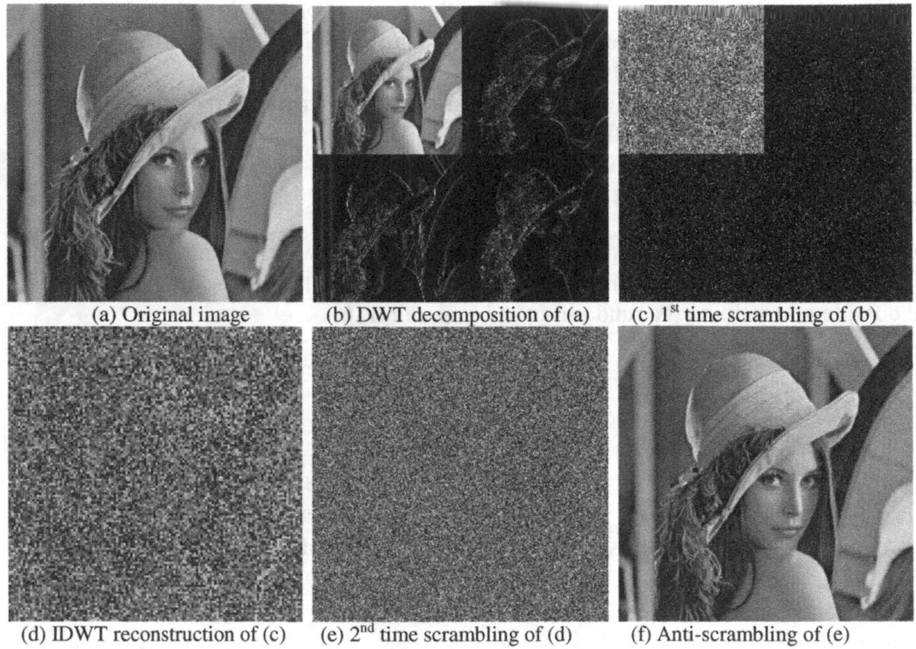

(a) Original image (b) DWT decomposition of (a) (c) 1st time scrambling of (b)

(d) IDWT reconstruction of (c) (e) 2nd time scrambling of (d) (f) Anti-scrambling of (e)

Fig. 2. Scrambling results of Lena

From the experiment results, we know that the scrambled result of image Lena is very like the white noise although it just goes though one time dual random scrambling. The scrambling effect of proposed method, apparently, is very good and better than that of those algorithms need iteration, such as the classical method - Arnold transform.

The grade of image scrambling can be evaluated quantificationally by scrambling degree [9, 10]. According to the scrambling degree evaluation method in reference [9], after being transformed into another image ***B*** by a transformation *T*, a digital image ***A*** has scrambling degree as follows:

$$S(\boldsymbol{A}) = \frac{\sum_{k=1}^{K}\left(\mu(k)\times\frac{H(k)}{H}\right)}{\sum_{k=1}^{K}\mu(k)}. \tag{2}$$

where H and $H(k)$ are the entropy of whole image $\boldsymbol{B}$ and the entropy of kth block of $\boldsymbol{B}$, $\mu(k)$ is the average of differences between all two connective pixels in block k of $\boldsymbol{B}$.

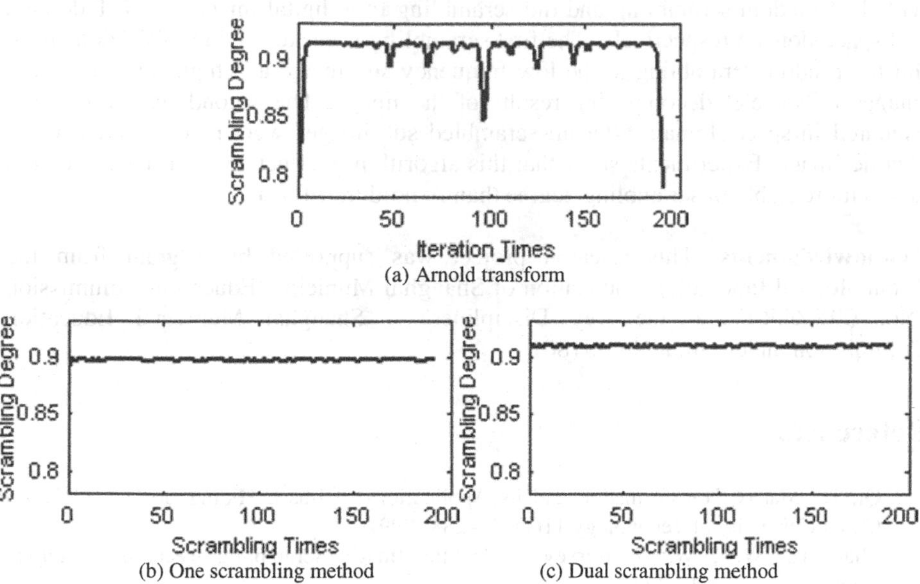

(a) Arnold transform

(b) One scrambling method

(c) Dual scrambling method

Fig. 3. Scrambling degree curves and comparing with Arnold transform

We now use Eq. 1 to evaluate stability of a scrambling algorithm, which can be represented by the relationship between scrambling degree and iteration times or scrambling times. For Arnold transform, our proposed one scrambling method (just do scrambling in DWT domain) and our proposed dual scrambling method, the results are shown as Fig. 3(a), (b) and (c) respectively. Comparing with unstable Arnold transform, the scrambling degree of our method, apparently, is very stable. We also know that the scrambling degree of one scrambling method is lower than that of dual scrambling method. The former is about 0.9, while the later is close to 0.92.

When being iterated 192 times by Arnold transformation, the scrambling degree of the image will change substantially from 0.78 to 0.93. In foremost, midterm and aftermost iterations, the scrambling degrees are quite bad. After 192 times iterations, the scrambled image recovers to the original image exactly. It is shown that the Arnold transformation has the property of periodicity and its security is worse. If being scrambled 192 times by our method, the scrambling degree of the image will be almost kept over 0.91 and close to 0.92, namely the scrambling degree is almost invariableness regardless scrambling times. That is to say, our method can reach the goal of being scrambled rapidly, and is quite stable, also has no security problem of periodically recovering.

6 Conclusion

This paper presented a dual scrambling algorithm, which combined wavelet transform with 1-D random scrambling and did scrambling to a digital image in DWT domain and space domain respectively. The first scrambling was executed in DWT domain. It did the random scrambling to the low frequency sub-image and high frequency sub-images of wavelet decomposing results of the image. The second scrambling was executed in space domain after all scrambled sub-images were reconstructed into a chaotic image. Experiments show that this algorithm is effective at visual evaluation and is more stable in scrambling degree than Arnold transform.

Acknowledgments. This research project was supported by a grant from the Technological Innovation Foundation of Shanghai Municipal Education Commission (No. 09YZ456) and the Key Disciplines of Shanghai Municipal Education Commission under Grant No. J51801.

References

1. Qi, D.: Matrix Transformation and its Applications to Image Hiding. Journal of North China University of Technology 11(1), 24–28 (1999)
2. Zhao, H., Wen, W.: A Survey of Digital Image Scrambling Techniques. Fujian Compute (12), 10, 12 (2007)
3. Tan, Y., Liang, X., Zhang, J., Liu, K.: The Study of Scrambling Technology Based on Arnold Transformation and Sampling Theoretics. Control and Automation 22(36), 74–76 (2006)
4. Li, G.: The Scrambling Method of Digital Image Based on Latin Square Transformation. Journal of North China University of Technology 13(1), 14–16 (2001)
5. Zhang, S., Chen, J.: Digital Image Scrambling Technology Based on Matrix Transformation. Journal of Fujian Normal University (Natural Science Edition) 22(4), 1–5 (2004)
6. Sun, Q., Ma, W., Yan, W., Dai, H.: A Random Scrambling Method for Digital Image Encryption: Comparison with the Technique Based on Arnold Transform. Journal of Shanghai Second Polytechnic University 25(3), 159–163 (2008)
7. Hou, Q., Yang, X., Wang, Y., Huang, X.: An Image Scrambling Algorithm Based on Wavelet Transform and Knight's Tour. Journal of Computer Research and Development 41(2), 369–375 (2004)
8. Yin, D., Tang, Y., Li, B.: Image Scrambling and Gray-level Diffusion Algorithm Based on Wavelet Transform. Journal of Sichuan University (Natural Science Edition) 42(4), 706–709 (2005)
9. Zhang, H., Lv, H., Weng, X.: Image Scrambling Degree Evaluation Method Based on Information Entropy. Journal of Circuits and Systems 12(6), 95–98 (2007)
10. Shang, Y., Zheng, Z., Wang, Z.: Digital Image Scrambling Technology and Analysis of Scrambling Degree. Journal of Tangshan Teachers College 28(2), 80–82 (2006)

A Delay-Guaranteed Two-Level Polling Model

Zheng Guan and Dongfeng Zhao

School of Information Science & Technology, Yunnan University
650091 Kunming, China
gz_627@sina.com

Abstract. We present a discrete time single-server two-level mixed service polling systems with two queue types, one center queue and N normal queues. The center queue will be successive served under the exhaustive scheme after each normal queue with a parallel 1-limited scheme. The proposed model is zero-switchover time when the buffers are not empty, and then conserve the cycle time. We propose an imbedded Markov chain framework to drive the closed-form expressions for the mean cycle time, mean queue length and mean waiting time. Numerical examples demonstrate that theoretical and simulation results are identical the new system efficiently differentiates priorities.

Keywords: Polling model, parallel 1-limited service, two-level, mean waiting time.

1 Introduction

A classical polling model is a single-server system with N queues $Q_1,Q_2,\ldots,Q_N$ in cyclic order. Customers arrive at each queue are served by a discipline. It finds a wealth of applications in the area of computer-communication system and the system performance evaluation [1,2]. Furthermore, polling models are applicable in traffic and transportation systems, and production systems.

Consequently, we study the cycle times, queue length, waiting time distributions and mean value of these characterizes, and stable conditions of different polling models. Resing [3] discussed the stability condition of the model, in which each queue has the so-called Bernoulli-type service discipline (exhaustive and gated service). Mean queue length and mean waiting time of the symmetric polling model with exhaustive, gated and 1-limited service have been analyzed in [4, 5, 6]. Model with limited-k service has been studied in [7], in which achieve the system mean waiting time by using iterative algorithm.

Boon, Adan and Boxma have studied a polling model with two queues which contains customers of priority classes in [8]. In these models, priority levels are distinguished by the order in which customers are served within a certain queue not among different queues. L. Qiang et al. have introduced the principle of two-level queueing priority station polling system with N+1 queues[9], one key queue and N normal queues, are polled according to the polling table as $1 \rightarrow H \rightarrow 2 \rightarrow \cdots \rightarrow N \rightarrow H$. Center queue and normal queues are provided with different service disciplines in [10, 11, 12]. We explore these models in this paper, a

D. Jin and S. Lin (Eds.): Advances in CSIE, Vol. 1, AISC 168, pp. 153–158.
springerlink.com © Springer-Verlag Berlin Heidelberg 2012

polling model with N+1 queues and two priority levels called DGTP (delay-guaranteed two-level polling) model is proposed. In the proposed model, parallel 1-limited service discipline [13] is assigned for normal queues. Proposed polling model is discussed to achieve the exact closed-form expression of mean queue length and mean waiting time are achieved. It is compared with two-level polling model with exhaustive, gated and 1-limited service. The simulation experiments show the superiority of DGTP on services distinguishing and transmission delay.

The rest of the paper is organized as follow: Section 2 gives new results on the DGTP model. We designed and successively propose the generation functions of the model. Mean cycle time, mean queue length and mean waiting time are discussed in Section 3. In Section 4, numerical results obtained with the proposed analytical models are shown and their very good agreement with realistic simulation results is discussed. Finally, concluding remarks are given in the end.

2 Model Description

2.1 Working Conditions

The mathematic model is proposed under the assumption as follow:

(1) Customers arrive at normal queues according to independent Poisson processes with rate λ_i, and their generating function is $A_i(z_i)$. The arrival of customers waiting for service meets the independent Poisson distribution with arrival rate λ_h, and the generating function is $A_h(z_h)$ in node h.
(2) The service time of a customer at each queue is independent of each other. The mean value is β_i in node i and β_h in node h, and their generating function are $B_i(z_i)$ and $B_h(z_h)$.
(3) The switch-over times is a random variable with mean value γ_i and generating function $R_i(z_i)$.
(4) Further assume that each queue has enough storage so that no customer is lost under the first-come-first-serve rule.

2.2 Delay-Guaranteed Two-Level Polling Model Description

Based on the proposed model [9], Define a random variable $\xi_i(n)$ (i=1, 2,…, N) as number of customers in storage at queue i at time t_n. The sever polls Q_i at time t_n switches to poll Q_h at time t_n^* and then polls Q_{i+1} at time $t_{n+1}(t_n < t_n^* < t_{n+1})$Then the status of the entire polling model can be represented as$\{\xi_1(n), \xi_2(n),\ldots, \xi_N(n), \xi_h(n)\}$, $\{\xi_1(n+1), \xi_2(n+1),\ldots, \xi_N(n+1), \xi_h(n+1)\}$and $\{\xi_1(n^*), \xi_2(n^*),\ldots, \xi_N(n^*), \xi_h(n^*)\}$.

For $\sum_{i=1}^{N}\rho_i+\rho_h<1$ ($\rho_i=\lambda_i\beta_i$, $\rho_h=\lambda_h\beta_h$), we will always assume the queues are stable. Then, the probability distribution is defined as

$$\lim_{n\to\infty}P\left[\xi_j(n)=x_j; j=1,2,\cdots,N,h\right]=\pi_i\left(x_1,\cdots,x_i,\cdots,x_N,x_h\right)$$

By using the embedded Markov chain theory and the probability generating function method to characterize the proposed system model, the generating function of the common queue is the following:

$$G_i(z_1,\ldots,z_i,\ldots,z_N,z_h)=\sum_{x_1=0}^{\infty}\cdots\sum_{x_i=0}^{\infty}\cdots\sum_{x_N=0}^{\infty}\sum_{x_h=0}^{\infty} z_1^{x_1}\ldots z_i^{x_i}\ldots z_N^{x_n} z_h^{x_h}\pi_i(x_1,\ldots,x_i,\ldots,x_N,x_h)$$

$$i=1,2,\ldots,N \qquad (1)$$

According to the proposed mechanism, the generating function of the key station Q_h at the time t_n is the following.

$$\begin{aligned}G_{ih}(z_1,\cdots,z_i,\cdots,z_N,z_h)&=\lim_{n\to\infty}E\left[\prod_{j=1}^{N} z_j^{\xi_j(n^*)} z_h^{\xi_h(n^*)}\right]\\&=\frac{1}{z_i}B_i\left(A_h(z_h)\prod_{j=1}^{N}A_j(z_j)\right)\left[G_i(z_1,\cdots,z_i,\cdots,z_N,z_h)-G_i^0\right]+R_i\left(A_h(z_h)\prod_{j=1}^{N}A_j(z_j)\right)G_i^0\end{aligned} \qquad (2)$$

Where $G_i^0=G_i(z_1,\ldots,z_i,\ldots,z_N,z_h)|_{z_i}=0$, $G_i(z_1,\ldots,z_i,\ldots,z_N,z_h)$ is the generation function for the number of data packets present at polling instants of Q_i.

And the generating function for the number of customers present at polling instants t_{n+1}.is

$$G_{i+1}(z_1,z_2,\cdots,z_N,z_h)=G_{ih}\left(z_1,z_2,\cdots z_N,B_h\left(\prod_{j=1}^{N}A_j(z_j)F_h\left(\prod_{j=1}^{N}A_j(z_j)\right)\right)\right) \qquad (3)$$

3 Performance Evaluation

Let the average number of message packets at Q_j at t_n be defined as $g_i(j)$ when Q_i is polled and at t_n^* as $g_{ih}(j)$ when Q_h is polled. Then $g_i(j)$ and $g_{ih}(j)$ can be given as the following :

$$g_i(j)=\lim_{z_1,\cdots,z_i,\cdots,z_N,z_h\to 1}\frac{\partial G_i(z_1,\cdots,z_j,\cdots,z_N,z_h)}{\partial z_j},\quad i=1,2,\ldots,N\,;j=1,2,\ldots,N,h \qquad (4)$$

$$g_{ih}(j)=\lim_{z_1,\cdots,z_i,\cdots,z_N,z_h\to 1}\frac{\partial G_i(z_1,\cdots,z_j,\cdots,z_N,z_h)}{\partial z_j},\quad i=1,2,\ldots,N\,;j=1,2,\ldots,N,h \qquad (5)$$

And let

$$g_i(j,k)=\lim_{z_1,\cdots,z_i,\cdots,z_N,z_h\to 1}\frac{\partial^2 G_i(z_1,z_2,\cdots,z_j,\cdots,z_k,\cdots,z_N,z_h)}{\partial z_j\partial z_k}, \qquad (6)$$

$$g_{ih}(j,k)=\lim_{z_1,\cdots,z_i,\cdots,z_N,z_h\to 1}\frac{\partial^2 G_{ih}(z_1,z_2,\cdots,z_j,\cdots,z_k,\cdots,z_N,z_h)}{\partial z_j\partial z_k},\quad h;k=1,2,\ldots,N,h \qquad (7)$$

3.1 Mean Cycle Time

The mean cyclic period is the mean value of the time between two successive visit beginnings to Q_i; it consists of service time and switchover time. In the proposed mechanism, it is zero-switchover time system when the buffer is not empty, and then decrease the cycle time. Considering the characteristics of the generating function, the related expressions can be given as the following:

$$1-G_i\left(z_1,\cdots,z_i,\cdots,z_{N,}z_h\right)\Big|_{z_i=0}=\lambda_i\overline{\theta} \qquad j=1,2,\ldots,N,h; \tag{8}$$

Take Eq.(2),and Eq.(3) into Eq.(4) and Eq.(5), Simplify these using Eq.(8)we have:

$$\overline{\theta}=\frac{N\gamma}{1-\rho_h-N\rho+N\lambda\gamma} \tag{9}$$

3.2 Mean Queue Length

According to the mechanism of exhaustive service and parallel limited-1 service, the mean queue length of the center queue can be derived from substituting Equations (2) and (3) into Equations (4) and (5), respectively, and $g_{ih}(h)$ can be expressed as the following.

$$g_{ih}(h)=\lambda_h\left[\gamma+\lambda(\beta-\gamma)\overline{\theta}\right] \tag{10}$$

The mean queue length of the common station can be derived from substituting Equations (2) and (3) into Equations (6) and (7), considering the special characteristic of the symmetric polling system as well as the set of discrete-time equations, then $g_i(i)$ can be expressed as the following.

The mean queue length for the center queue within interval service can be derived from substituting Equation (3) into Eq. (7), then $g_{ih}(h,h)$ can be expressed as the following.

$$g_{ih}(h,h)=\left[A_h''(1)\beta+\lambda_h^2B''(1)-\lambda_h^2R''(1)-A_h''(1)\gamma\right]\lambda\overline{\theta}+\lambda_h^2R''(1)+A_h''(1)\gamma \tag{11}$$

$$\begin{aligned}g_i(i)=&\frac{1-\rho_h}{2(1-N\rho-\rho_h)}\Big\{N\left(\gamma A''(1)+\lambda^2R''(1)\right)+\frac{2N\rho_h}{1-\rho_h}\left(\lambda^2R''(1)+\lambda^2r\right)\\&+\frac{N\rho_h^2}{(1-\rho_h)^2}\left(\gamma A_i''(1)+\lambda^2R''(1)\right)+\frac{N\lambda}{1-N\rho-\rho_h+N\lambda\gamma}\Big[N\gamma(\beta-\gamma)A''(1)-N\lambda^2\gamma R''(1)\\&+\frac{1}{1-\rho_h}\left(2N\lambda^2\gamma\rho_h(\beta-\gamma)-2N\lambda^2\gamma(1+\rho_h)R''(1)+2\gamma(1-\rho-\rho_h)+\lambda\gamma\rho_h(1+\rho_h)\right.\\&\left.+2(N-1)\lambda\gamma(\gamma-\beta)\right)-(N-1)\lambda\gamma^2\\&+\frac{\lambda}{(1-\rho_h)^2}\left(N\lambda\gamma B''(1)-N\lambda\gamma\rho_h^2R''(1)+\gamma\beta_h^2\left(\rho_h+N\lambda(\beta-\gamma)\right)A_h''(1)+\lambda_h\gamma B_h''(1)\right)\Big]\Big\}\end{aligned} \tag{12}$$

3.3 Equations Mean Waiting Time

The waiting time of customers w_j , denotes the time from when a customer enters into the queue at Q_j to when it is served, and $E[w_i]$ as well as $E[w_h]$ denotes the mean waiting time of customers for the common station and the key station, respectively.

The normal node Q_i is served in the parallel 1-limited scheme, under the theory in [12], we could get the expression of the mean waiting time as follow:

$$E[w_i] = \frac{1}{\lambda C} g_i(i) - \frac{1}{\lambda} - \frac{A''(1)}{2\lambda^2} \tag{13}$$

Where $C = 1 - G_i(z_1, \cdots, z_i, \cdots, z_N, z_h)\big|_{z_i=0}$ and i=1,2,..., N.

Take with Eq. (12), it is easy to obtain the closed form expression of mean waiting time.

The high-priority node Q_h is served in the exhaustive scheme, under the theory in [6], we could get the expression of the mean waiting time as follow:

$$E[w_h] = \frac{g_{ih}(h,h)}{2\lambda_h g_{ih}(h)} - \frac{A_h''(1)}{2\lambda_h^2(1+\rho_h)} + \frac{\lambda_h B_h''(1)}{2(1-\rho_h)}$$

4 Numerical Analysis and Comparison

We performed numerical experiments to test the efficiency of the DGTP for different values of the workload of the system. We make a comparison between the DGTP and [10,11, 12] which work with exhaustive, gated and 1-limited service policy.

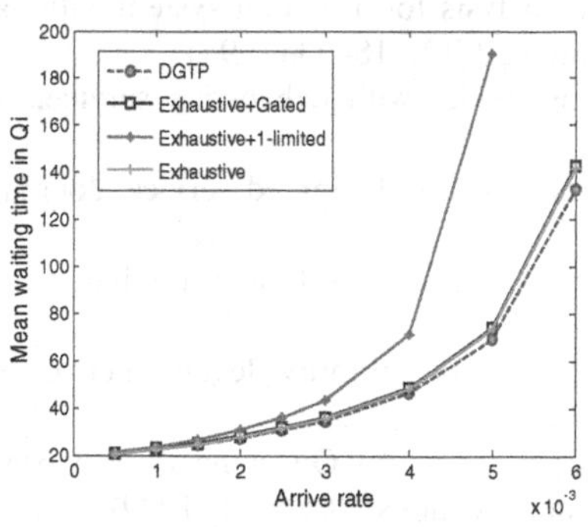

Fig. 1. (a) Mean waiting time of Q_i

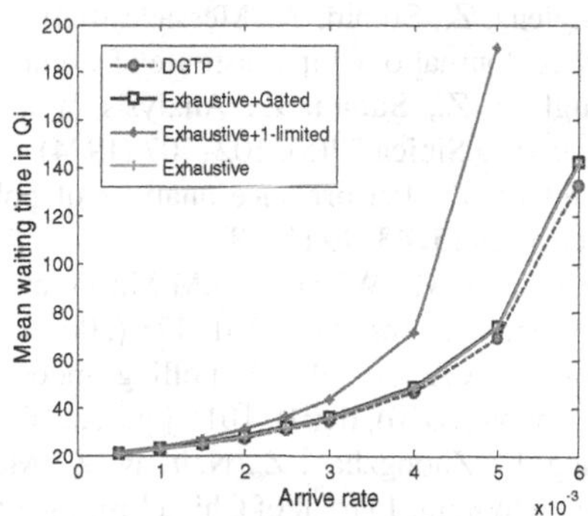

Fig. 1. (b) Mean waiting time of Q_h

Fig. 1. Comparison with the two-level polling model ($\beta = \beta_h = 10$, $\gamma = 5$, M=10)

In this example we illustrate the accuracy of theoretical analysis and the model performance with superiors in the high priority queue. Consider a polling system with ten queues--one high priority queue and nine normal queues.

In Figure 1, we compared the present DGTP with the other two-level models with gated, exhaustive, and 1-limited scheme in Q_i and Q_h.. Comparing with the forgoing, cause of parallel process between packets and token transmission, DGTP achieves a better performance in delay guarantee and system stability in both. For lower load, in most of the case, there is no customer in the buffers, thus a switch over time is necessary when the server switch between Q_i and Q_h. While for higher work load, the DGTP performs obviously better. That drive from when the buffers are not empty, parallel dispatching of service and switch over time improves the service.

5 Conclusion

In this paper, we proposed a delay-guaranteed two-level polling (DGTP) model, in which queues are provided a priority distinguished service, parallel dispatching of service and switch over time improves the service efficiency. We got the closed from expressions of mean queue length and average waiting time, which are exactly coincide with the computer simulation results. As we have show, high level queue has a superior performance in queue length and delay, as well as low level queues still worked stable.

References

1. Levy, H., Sidi, M.: Polling systems: applications, modeling and optimization. IEEE Transactions on Communications 38, 1750–1760 (1990)
2. Takagi, H.: Queueing analysis of polling models: an update. In: Takagi, H. (ed.) Stochastic Analysis of Computer and Communication Systems, pp. 267–318. North-Holland, Amsterdam (1990)
3. Resing, J.A.C.: Polling systems and multitype branching processes. Queueing Systems 13, 409–426 (1993)
4. Dongfeng, Z., Sumin, Z.: Message waiting time analysis for a polling system with gated service. Journal of China Institute of Communications 15(2), 18–23 (1994)
5. Dongfeng, Z., Sumin, Z.: Analysis of a polling model with exhaustive service. Acta Electronica Sinica 22(5), 102–107 (1994)
6. Dongfeng, Z.: Performance analysis of polling systems with limited service. Journal of Electronics 15, 43–49 (1998)
7. van Vuuren, M., Winands, E.M.M.: Iterative approximation of k-limited polling systems. Queueing Systems 55(3), 161–178 (2007)
8. Boon, M.A.A., et al.: A polling model with multiple priority levels. Performance Evaluation (2010), doi:10.1016/j.peva.2010.01.002
9. Qiang, L., Zhongzhao, Z., Naitong, Z.: Mean Cyclic Time of Queueing Priority Station Polling System. Journal of China Institute of Communications 20, 86–91 (1999)
10. Zhijun, Y., Dongfeng, Z., Hongwei, D., et al.: Research on two class priority based polling system. Acta Electronica Sinica 37(7), 1452–1456 (2009) (in Chinese)
11. Zhuguan, L.: Research on discrete-time two-level polling system with exhaustive service. Ph.D. Thesis, Yunnan University (2010)
12. Qianlin, L., Dongfeng, Z.: Analysis of two-level-polling system with mixed access policy. In: IEEE International Conference on Intelligent Computation Technology and Automation, vol. 4, pp. 357–360 (2009)
13. Chunhua, L., Dongfeng, Z., Hongwei, D.: Study of parallel schedules for polling systems with limited service. Journal of Ynnan University 25(5), 401–404 (2003)

E-commerce Personalized Recommendation System Design Based on Multi-agent*

Ya Luo

Guizhou University of Finance and Economics, Guiyang, Guizhou, China
141756@qq.com

Abstract. In order to make functional modules in traditional personalized recommendation systems componentized, this paper introduces the method of using multi-agent technology with combination of Web log mining to optimize the framework model of an e-commerce personalized recommendation system as a whole, as well as to design the system workflow in the model as well as the functions and structure of each Agent.

Keywords: Agent, Mining, Personalized Recommendation System.

1 System Requirement Analysis

The requirements for an e-commerce personalized recommendation system are as following:

(1) To establish the commodity association model related to a customer's interests, preference and access modes, etc., by effectively analyze the customer's existing purchase behaviors, and eventually recommend commodities that are most appropriate for the customer.

(2) To actively track and analyze how a user browses by the personalized e-commerce recommendation technology to effectively provide specific personalized recommendation service for the user.

(3) To use different recommendation strategies based on different users' requirements and how users are interested by commodities so as to implement the personalized recommendation service better.

(4) Due to the reason that the variety and amount of commodities as well as users' interests and browsing behaviors are under constant and dynamic change, the problem of the system personalized recommendation is required to be dynamic and have quite strong self-adaptation in the whole process of problem solving.

(5) To adjust recommendation methods and recommended commodities in real time based on changes of a user's interests by tracking the purchase process of the user who browses, so as to generate recommendations in real time.

* Guizhou S&T Foundation Funded Project Contract No.: QKHJ.Zi.[2009]2121# Research on Customized Recommendation Based E-shop System.

D. Jin and S. Lin (Eds.): Advances in CSIE, Vol. 1, AISC 168, pp. 159–164.
springerlink.com © Springer-Verlag Berlin Heidelberg 2012

2 Overall System Structure Design

By blending the communication and cooperation of multi-Agent into e-commerce personalized recommendation systems, the advantages of Agent and multi-Agent systems can be fully implemented to optimize personalized recommendation systems for e-commerce, shown as following,

(1) To provide the recommendation service with the best quality for commercial systems in different application areas. When a commercial system needs to use the recommendation service, it is possible to call the unified web service interface provided by the Agent, and reach a relatively stable status after several times of weight adjustment. After that, it is ready to provide the recommendation service with high quality for commercial systems.

(2) To support all kinds of existing recommendation algorithms to adapt to recommendation requirements of different types of products. All recommendation algorithms are centralized and stored in a unified recommendation algorithm base so that the recommendation engine Agent can use based on different products and recommendation requirements. Eventually the commendation engine Agent can make different recommendation strategies for different products with strong flexibility.

(3) It is very easy to add the recommendation engine Agent into e-commerce recommendation systems. When recommendation algorithms have gone through optimization or new algorithms are discovered, these algorithms can be used for new recommendations by adding into the recommendation engine Agent only if the recommendation engine Agent provides a standard web service interface. Then the Agent can automatically identify the new algorithms and use the recommendation service provided by them.

3 Design of Recommendation Engine Agent

In e-commerce personalized recommendation systems, the core recommendation generation based on multi-Agent is realized by the recommendation engine Agent. However, the main task of the recommendation engine Agent is to search for products and user information, evaluation of recommendation, recommendation generation, and optimization, etc. The design of the recommendation engine Agent is detailed as below:

3.1 Analysis for Essential Factors of the Recommendation Engine Agent

The recommendation engine Agent consists of business logic layer and data service layer.

(1) Business logic layer

Business logic layer is composed by search Agent, recommendation generation Agent and evaluation Agent.

(2) Data service layer

The data service layer includes recommendation case base, preference knowledge base, rule base, recommendation record base, recommendation algorithm base and evaluation method base.

3.2 Recommendation Procedure

When the system receives a user's purchase info and request for recommendation, one work is to search the product information base to generate the product set, and another one is to search the user information. Then, the system can search for the commodity cases user may be interested from the recommendation case base based on the obtained user information. If there is any existing case, the recommendation can be generated based on the case and the system also keeps tracking user's browsing and purchase behavior so as to evaluate and improve the recommendation list based on changes of the user's purchase behavior and interests. If there is no case existing, then the system searches the customer's preference based on the user information. The recommendation engine Agent would generate its recommendation by using appropriate recommendation algorithms based on the customer being searched and the selection of product information, so as to generate the list for recommended commodities. What needs to be noted is that all the recommendation generation lists and the recommendation algorithms used should be recorded into the recommendation record base. In the meantime, the recommendation lists should be evaluated and put in order by the evaluation model chosen by the evaluation Agent, and also judged to check if they are proper. If they are proper, the lists should be saved into the recommendation case base. Otherwise, they should also be saved into the recommendation log base and another recommendation algorithm should be selected again to generate recommendations until the recommendation is evaluated to be proper. After that, the successful recommendation lists should be saved into the recommendation case base. By then, the recommendation is completed.

4 Agent Structural Design in System

In different e-commerce systems, the functions of the Agent are different depending on the role and implementation functions of the Agent. The Agent in the systems can be classified into three types, management Agent, function Agent and auxiliary Agent.

(1) Management Agent

The management Agent has the functions of managing, coordinating and controlling other Agents based on a system's actual situation and function requirement. For example, the functions of the recommendation engine Agent are to control and coordinate the search Agent, the recommendation generation Agent, and the evaluation Agent, so as to realize its recommendation function and goal of self-study by working together.

(2) Function Agent

The function Agent plays the role of an abstract entity of a functional sub-module, being responsible for the implementer of specific functions and the unified design oriented to specific function requirements. A function Agent is commonly used exclusively to implement the functions of a functional sub-module. Additionally, the function Agent cooperates and works together with the management Agent to complete a task. The main functions of a system is composed and completed by the

Agent, mainly including web log Agent, preference analysis Agent, data conversion Agent, data mining Agent, and recommendation engine Agent.

(3) Auxiliary Agent

The auxiliary Agent is an undertaker of some simple tasks. For example, interface Agent, monitor Agent. Its main purpose is to provide the demanded services for the functional Agent. Normally it undertakes no specific application task.

Based on the above analysis, the organization structure of the Agent in a system is combination of hierarchical structure and alliance structure, as below.

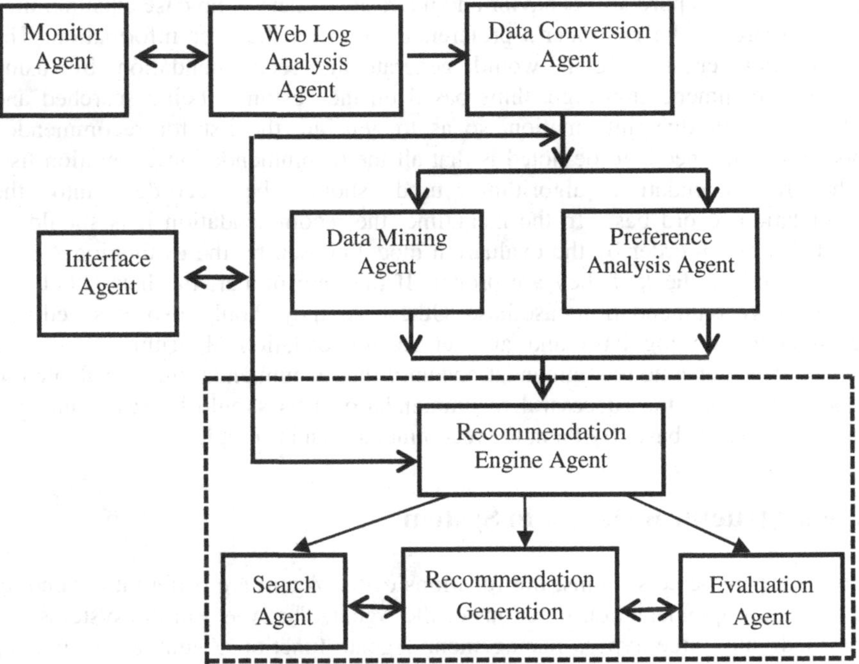

Among the figure, the arrow represents the logic order; the solid line for the flow relation; the dotted line for the organization relation of the management Agent.

5 Agent Design in System

The personalized recommendation is completed and implemented by the interconnected functional modules. These modules refer to the Agents communicating with each other and working together. Due to the reason that the Agents implement their functions respectively, there is no need to change other functional modules in an Agent when new functions are added into the functional modules so as to improve the functionality of the system as a whole. The Agents act as the main representation of abstraction and conceptualization of entities in systems. Each Agent can represent a different role and be used for implementing specific functions. The Agents can be classified as below:

(1) Interface Agent

The main functions of the interface Agent are to provide an interface for human-machine interaction for experts and sales management personnel. The Agent can set up corresponding parameters for data conversion Agent, data mining Agent, recommendation engine Agent, and preference analysis Agent.

(2) Monitor Agent

The functions of the Agent are to monitor user behaviors of logon, browsing and purchase. Its main purpose is to provide the related information for the recommendation engine Agent on time, and offer the recommendation results generated by the recommendation engine Agent for user in real time. The monitor Agent can also act in an active or passive way.

(3) Data conversion Agent

The data conversion Agent refers to extracting, decomposing and merging the related data from web log base, customer information base, product information base, and sales record information base for format conversion so as to establish and maintain data sets, and then interact with the preference analysis Agent, the data mining Agent and the recommendation engine Agent so as to generate the data required by the personalized recommendation.

(4) Web log analysis Agent

The functions of the Web log analysis Agent are to automatically record all the customer's browsing behavior on a website in his/her purchase process, and to manage and maintain the Web log base. It can also collect and process the content of the log file in the Web server periodically, and update the Web log base, aiming to provide for the data conversion Agent for processing in next step.

(5) Data mining Agent

The tasks of the data mining Agent are to receive the requests from the recommendation engine Agent and analyze a customer's purchase or access behaviors by the ways of regression algorithm, association rules, decision tree, neural network and Web log mining, etc., to generate the modes of the association rules for the product accessed by the customer in order, customer classification and similar customer groups and save them into the rule base in order to provide the basis for the recommendation tasks for the recommendation engine Agent.

(6) Preference analysis Agent

The functions of the preference analysis Agent are to acquire, update and maintain the information based on customers' preference. The processes of the preference analysis Agent acquiring user preference can be classified into two approaches – passive acquisition and active acquisition. The passive acquisition is to ask a user to fill in a complete preference information table, and then the system can normalize and save it into the preference knowledge base. And the active acquisition is that the Agent actively acquires the user's behavior trends so as to revise the existing preference knowledge in the preference knowledge base based on the user purchase behavior and click streams saved in the data mart. The preference acquisition is mainly by the active approach facilitated by the passive approach.

(7) Recommendation engine Agent

The functions of the recommendation engine Agent are to use certain recommendation strategies and proper recommendation algorithms to generate the recommendation list based on customers preference provided by the preference analysis Agent and the product association rules generated by the data mining Agent, aiming to provide the personalized products for customers.

The work principles of the recommendation engine Agent: To receive the requests for recommendation from the monitor Agent, inform the search Agent to search for commodities and generate the set of the candidate commodities, and then obtain certain user preference data from the user preference knowledge base or the preference analysis Agent. Finally, the recommendation engine Agent can create the initial recommendation commodity list based on recommendation algorithms. Now the evaluation Agent can make certain assessment for the candidate items of the candidate commodities in the initial list. The candidate set is ordered from high to low based on the assessment result. The candidate item having the high assessment should be kept as the set of the recommendation commodities. With the search Agent transferring new data constantly, the evaluation Agent would unite the recommendation item set and the new candidate item set and starts a new run of evaluation screening. The set of the recommendation commodities are constantly updated so that the set of the recommendation commodities can be finalized. In the mean time, the recommendation engine Agent would save the recommendation rules used and results in the recommendation process into the recommendation case base for usage and study in next time.

6 Conclusion

By the analysis above, it is not hard to find out that the method of combining multi-Agent with Web log mining can make the functional modules in the traditional personalization recommendation systems componentized so as to increase the adaptability and extensibility of e-commerce personalization systems.

References

1. Yi, M.: Personalized Information Recommendation based on Web Mining, May 1. Science Press (2010)
2. Li, S., Han, Z.: Personalized Search Engine Principles and Technologies, June 1. Science Press (2008)
3. Yu, L.: E-commerce Personalization – Theories, Methods and Applications, January 1. Tsinghua University Press (2007)
4. Qing, H.: A Study on the Core Technologies of Recommender Systems for E-commerce. Beijing University of Technology (2009)

Codebook Perturbation Based Limited Feedback Precoding MU-MIMO System

Hongliang Fu, Beibei Zhang, Huawei Tao, Yong Tao, and Zheng Luo

Henan University of Technology, Institude of Information Science and Engineering, Zhengzhou, China
jackfu_zz@163.com

Abstract. In this paper, considering the problem of performance loss due to limited feedback in multiuser multiple-input multiple-output (MU-MIMO) downlink systems, limited feedback precoding based on codebook perturbation is proposed. The maximum signal-to-interference ratio (SIR) criterion is used for selecting optimal codeword from the Grassmannian codebook and perturbation codeword at the receiver. And the Grassmannian precoding codeword index and perturbation codeword index are fed back to the transmitter. Then the array preprocessing is used at the transmitter to get optimal capacity and compensate for the performance loss due to the limited feedback. Simulation results show that the proposed method ensures the cost of the feedback link and improves the system throughput and bit error rate (BER) performance effectively.

Keywords: MU-MIMO, Precoding, Limited Feedback, Codebook Perturbation.

1 Introduction

In recent years, MU-MIMO limited feedback precoding based on codebook design not only becomes the focus of wireless communications research but also be included as one of the 3GPP LTE standards [1]. Because of the dispersion of the multi-user in MIMO downlink, the received signals can't be co-processed among the receivers. That leads to serious deficiencies in the system error performance.

In this paper, we propose limited feedback precoding based on codebook perturbation to solve the problem of receiving signal which can't be co-processed in the downlink of MU-MIMO with limited feedback. Simulation results show that system throughput and error performance are actively improved using the method proposed in this paper.

2 Construction and Selection Criteria of Perturbation Codebook

Because of a perturbation matrix is constructed using the rotation matrix and the Householder transformation [2], [3], the system is composed of a main codebook and a perturbation codebook. Each codeword F_i of the main codebook which is the traditional Grassmannian space belongs to space $G(M,M)$. Linear transformation

D. Jin and S. Lin (Eds.): Advances in CSIE, Vol. 1, AISC 168, pp. 165–169.
springerlink.com © Springer-Verlag Berlin Heidelberg 2012

triggered by the perturbation codebook does not change the distribution characteristic of the main codebook. To achieve optimal performance of the system, perturbation codebook in which each codeword U_i belongs to space $\Psi(M,M)$ is used to assist the main codebook. In the precoding, the receiver respectively selects the best codeword F_{op} in the main codebook $G(M,M)$ and the best codeword U_{opt} in the perturbation codebook $\Psi(M,M)$. Then the final precoding matrix is $W_{opt} = F_{opt}U_{opt}$.

Perturbation matrix U_{opt} which is based on the optimal channel capacity is constituted by the eigenvectors of the matrix $F^H H^H HF$. From the mathematical point of view, U_{opt} has the role of diagonalizing the matrix $F^H H^H HF$. According to the matrix theory, the rotation matrix can diagonalize any symmetric matrix. Therefore, we use rotation matrix to construct the perturbation sub-codebook. The sub-codebook is given by

$$U_{opt} = \begin{bmatrix} \cos\theta_{opt} & -\sin\theta_{opt} e^{j\varphi_{opt}} \\ \sin\theta_{opt} e^{-j\varphi_{opt}} & \cos\theta_{opt} \end{bmatrix} \tag{1}$$

where $\theta_{opt} = -\frac{\pi}{4} + \frac{(g-1)}{2(G-1)}, \varphi_{opt} = -\frac{\pi}{2} + \frac{(g-1)\pi}{(G-1)}$.

According to the matrix theory, unitary matrix can be constructed by the Householder transformation of a vector. The Householder transformation of a vector can be defined as follow:

$$H = I - 2\frac{VV^H}{V^H V} \tag{2}$$

If a complex conjugate unitary matrix meets the conditions as a perturbation matrix, then we can construct codebook using the Householder transformation of the vector. The channel capacity gain obtains the maximum value if and only if

$$U^H F^H H^H HFU = (I - 2\frac{VV^H}{V^H V})F^H H^H HF(I - 2\frac{VV^H}{V^H V}) = \begin{bmatrix} a_{11} & a_{12} \\ a_{12}^* & a_{22} \end{bmatrix} \tag{3}$$

where $a_{12} = a_{12}^* = 0$.

3 Capacity Analysis

Currently, MU-MIMO downlink capacity schemes include Per-User Unitary and Rate Control (PU2RC) and Zero Forcing Space Division Multiple Access (ZF-SDMA) [4]. Compared with PU2RC, ZF-SDMA is not only using the beamforming to pre-code to reduce interferences between users but not strict with the user group that any two users can be paired. Therefore, the system still has a higher multi-user diversity gain while the number of users is small.

We assume that the receiver has got the full channel state information (CSI). Using the Grassmannian codebook, the system capacity of MU-MIMO precoding with limited feedback can be expressed as

$$C_{MU-MIMO} = \sum_{k=1}^{K} \log_2(1 + SIR_k) \tag{4}$$

were SIR_k is the SIR of the k-th user. For the linear receiver, the SIR of the k-th user can be written as

$$SIR_k = \frac{|\boldsymbol{G}_k \boldsymbol{H}_k \boldsymbol{F}_k \boldsymbol{U}_k \boldsymbol{w}_k|^2}{\sum_{i=1, i \neq k}^{K} |\boldsymbol{G}_k \boldsymbol{H}_k \boldsymbol{F}_k \boldsymbol{U}_k \boldsymbol{w}_k|^2 + |\boldsymbol{G}_k|^2 N_0} \tag{5}$$

So the maximum capacity criterion is

$$\mathrm{Fw} = \arg\max_{\{Fw\}} (C_{MU-MIMO}) \tag{6}$$

4 Simulation

The communication channels over which the signals are typically transmitted can be modeled as Rayleigh flat-fading channels. There are 4 transmitting antennas and 2 receiving antennas in the MIMO system. With increasing number of users, for different SNRs, Fig. 1 shows the throughput curves of the MU-MIMO system with the codebook perturbation based limited feedback and the conventional limited feedback. The number of feedback bits is 4. As shown in Fig. 1, with increasing number of users, throughput of the system has increased. But when the system reaches a certain number of users, the throughput of the system tends to a limit value for a given SNR. At high SNR, the system throughput has improved significantly. With increasing SNRs, for different number of users, When the number of users is the same as well as the SNR is the same, codebook perturbation based limited feedback can always bring more system throughput than that of the traditional limited feedback.

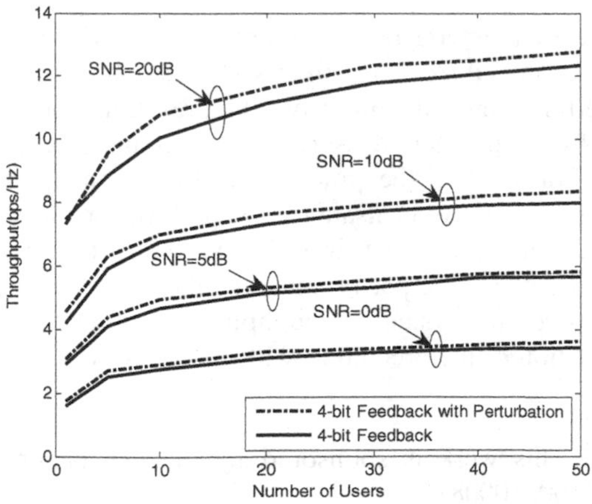

Fig. 1. The throughput curves of the MU-MIMO system

Fig. 2 show the BER curves of the MU-MIMO system with codebook perturbation based limited feedback and the BER curves of the same system with conventional limited feedback according to ZF criterion. Simulation results show that the error rate performance is better in the case when the number of the feedback bits is 6 than in the case when the number of the feedback bits is 4. When the number of the feedback bits is same, the proposed perturbation code book based limited feedback precoding scheme can achieve better BER performance than that of the conventional limited feedback precoding scheme.

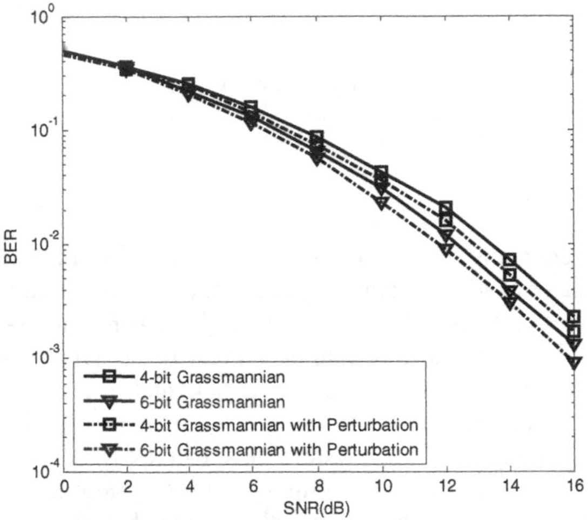

Fig. 2. The BER curves of the MU-MIMO system

5 Conclusion

In this paper, we present a codebook perturbation based MU-MIMO downlink precoding scheme with limited feedback. In this scheme, the receiver selects the main codebook and the perturbation codebook from the presetting codebook set according to the maximum SNR of the desired user. Then their indexes are fed back to the transmitter. At the transmitter, the linear transformation (perturbation) of the precoding codeword is done to compensate for the performance loss due to the limited feedback. Simulation results show that, in the condition of ensuring the overhead of the feedback link and better BER performance of the system, the proposed method can effectively improve the system throughput and contribute to the practical application and promotion of the precoding with limited feedback in the MU-MIMO system.

Acknowledgements. This work is sponsored by the Outstanding Youth Fund of Henan Province (104100510008).

References

1. Fang, S., Li, L., Zhang, P.: Non-Uuitary Precoding Base on Codebook for Multi-user MIMO System with Limited Feedback. Journal of Electronics & Information Technology 30(10), 2419–2422 (2008)

2. Love, D.J., Heath, R.W.: Limited feedback unitary precoding for spatial multiplexing systems. IEEE Transactions on Information Theory 51(8), 2967–2976 (2005)
3. Huang, Y., Xu, D., Yang, L., Zhu, W.: A limited feedback precoding system with hierarchical codebook and linear receiver. IEEE Transactions on Wireless Communications 7(12), 4843–4848 (2008)
4. Zhou, S., Li, B.: BER criterion and codebook construction for finite-rate precoded spatial multiplexing with linear receivers. IEEE Transactions on Signal Processing 54(5), 1653–1665 (2006)

Compressed Sensing Image De-noising Method Based on Regularization of Wavelet Decomposition

Hongliang Fu, Huawei Tao, Beibei Zhang, and Jing Lu

College of Information Science and Engineering, Henan University of Technology, Zhengzhou, 450001, Henan, P.R. China.
ttkltao@163.com

Abstract. A method called compressed sensing image de-noising method based on wavelet decomposition was presented. In view of lack of sparsity in the above method, A new method called compressed sensing image de-noising method based on regularization of wavelet decomposition was presented, First, the image signal was decomposed by multi-scale wavelet, then high-frequency coefficients of each level was divided into two positive and negative by regularization; each level high-frequency coefficients was sampled by compressed sensing, and measured coefficients can be acquired; At last, measured coefficients was reconstructed, and then de-noising image can be acquired according to inverse wavelet transform. The simulation show that regularization can effectively increase the capacity of compressed sensing de-noising.

Keywords: Multi-scale wavelet decomposition, regularization, compressed sensing.

1 Introduction

Compressed sensing is a new sampling strategy that has been proposed jointly by Candes, Tao and Dohono[1,2,3]. Instead of performing the acquisition with a high sampling rate and then compressing the data in an orthogonal basis, compressed sensing will directly acquire signal in a compressive form and then reconstruct the original signal perfectly through a certain algorithm.

Since wavelet threshold de-noising was presented by Donoho and Johnstone in 1995, wavelet de-noising as a classical de-noising method has made a wide range of application. However, the general threshold given by Dohono sometimes "over kill" wavelet coefficients, so most of people has to study new threshold; [4] presents BayesShrink threshold based on wavelet coefficients obeyed the generalized Gaussian distribution, [5] presents mapshink threshold threshold based on wavelet coefficients obeyed the Laplace distribution.

Most of the transform domain small coefficients were turn into zero during reconstruction of CS, Meanwhile, signal noise were always small coefficients in transform domain, so CS can be used to de-noise. [6] present a new CS image de-noising based on GPSR and simulation shows well; In 1-D signal, [7] present a new wavelet de-noising based on sparse representation which change traditional wavelet

D. Jin and S. Lin (Eds.): Advances in CSIE, Vol. 1, AISC 168, pp. 171–176.
springerlink.com © Springer-Verlag Berlin Heidelberg 2012

de-noising idea and use sparse representation to de-noise. Taking into account the wavelet image de-noising, a wavelet image de-noising method based on sparse representation was presented in this paper. However, most of image wavelet high-coefficients were lack of sparsity, so a new method called compressed sensing image de-noising method based on regularization of wavelet decomposition was presented in this paper. As regularization can increase the sparsity of coefficients, simulation shows good result

2 Relevant Theories

2.1 Compressed Sensing

Suppose that x is a length –N signal. It is said to be K-sparse if x can be well approximated using $K \ll N$ coefficients under some linear transform. According to the CS theory, such a signal can be acquired through the following linear random projections:

$$y = \Phi x \quad (1)$$

Where y is the sampled vector with $M \ll N$ data points, Φ represents an $M \times N$ random projection matrix. Although the encoding process is simply linear projection, the reconstruction requires some non-linear algorithms to find the sparsest signal from the measurements. There exist some famous reconstruction algorithms, such as basis pursuit aims at the l_1 minimization (formula 2) using linear programming, iterative greedy pursuit and so on.

$$\hat{x} = \arg\min \|x\|_1 \quad s.t. \quad \|\Phi\Psi x - y\|_2 \le \varepsilon . \quad (2)$$

2.2 Wavelet Threshold De-noising

Suppose the noisy signal:

$$f(k) = x(k) + n(k) . \quad (3)$$

Where $x(k)$ is the noise-free signal and $n(k)$ is the Gaussian white noise. $w_{j,k}$ is the wavelet coefficients where $j = 1, \cdots, J$ is the scale factor and $k = 1, \cdots, N$ is the time. Donoho proposed two wavelet threshold de-noising methods called soft threshold and hard threshold. $\hat{w}_{j,k}$ is the de-noising coefficients according wavelet de-noising.

Soft threshold de-noising $$\hat{w}_{j,k} = \begin{cases} sign(w_{j,k}) \cdot (|w_{j,k}| - \lambda), & |w_{j,k}| \ge \lambda \\ 0, & |w_{j,k}| < \lambda \end{cases} . \quad (4)$$

Hard threshold de-noising $\hat{w}_{j,k} = \begin{cases} w_{j,k}, & |w_{j,k}| < \lambda \\ 0, & |w_{j,k}| < \lambda \end{cases}$. (5)

3 Compressed Sensing Image De-noising Method Based on Regularization of Wavelet Decomposition

Algorithm flow chart was shown in figure 1, and specific steps are as follows:

Initialization: Suppose that x is an $n \times n$ image, $x_0 = x + n$ where n is Gaussian white noise of variance σ^2, φ is scaling function, ψ is wavelet function, T is threshold, Φ is Gaussian random matrix, Y is measured signal.

1. Expand the object x_0 in the wavelet basis

$$x_0 = A_{j_0} + \sum_{j=j_0}^{j_1} \sum_k w_{jk} \psi_{jk} = \sum_k \beta_{j_0 k} \varphi_{j_0 k} + \sum_{j=j_0}^{j_1} \sum_k w_{jk} \psi_{jk} . \quad (5)$$

We can obtain the decomposed coefficients $[(\beta_{j_0,\cdot}), (w_{j_0,\cdot}), (w_{j_0+1,\cdot}) \cdots, (w_{j_1-1,\cdot})]$.

2. Each of high-coefficients was preformed by formula (6) where operator $(\cdot)_+$ says that the positive parts of the matrix is retained and negative parts were turn into zero. The coefficients change into $[(\beta_{j_0,\cdot}), (w^+_{j_0,\cdot}), (w^-_{j_0,\cdot}), (w^+_{j_0+1,\cdot}), \cdots, (w^+_{j_1-1,\cdot}), (w^-_{j_1-1,\cdot})]$.

$$s = (w)_+ - (-w)_+ . \quad (6)$$

3. We apply Nyquist to sample coefficients $(\beta_{j_0,k})$ and separately apply CS scale-by-scale, sampling data y_j^+ about the coefficients $(w^+_{j_0,k})$ at level j using a $n_j \times 2^j$ CS matrix Φ_j, and $Y = [y_{j_0}, \cdots, y^+_{j_1-1}, y^-_{j_1-1}]$ to obtain a reconstruction, we then solve the sequence of problems

$$\arg\min \left\| \bar{w} \right\|_1 \quad s.t. \quad \left\| \Phi_j \hat{w} - y_j \right\|_2 \le \varepsilon \quad j = j_0, \cdots, j_1 - 1 . \quad (7)$$

We can obtain the decomposed coefficients $[(\beta_{j_0,\cdot}), (\hat{w}_{j_0,\cdot}), (\hat{w}_{j_0+1,\cdot}) \cdots, (\hat{w}_{j_1-1,\cdot})]$ according to (8)

$$\hat{w} = \bar{w}^+ - \bar{w}^- . \quad (8)$$

4. We can reconstruct the signal according to formula (9)

$$\hat{x}_0 = \sum_k \beta_{j_0,k}\varphi_{j_0,k} + \sum_{j=j_0}^{j_1-1}\sum_k \hat{w}_{j,k}\psi_{j,k} \,. \tag{9}$$

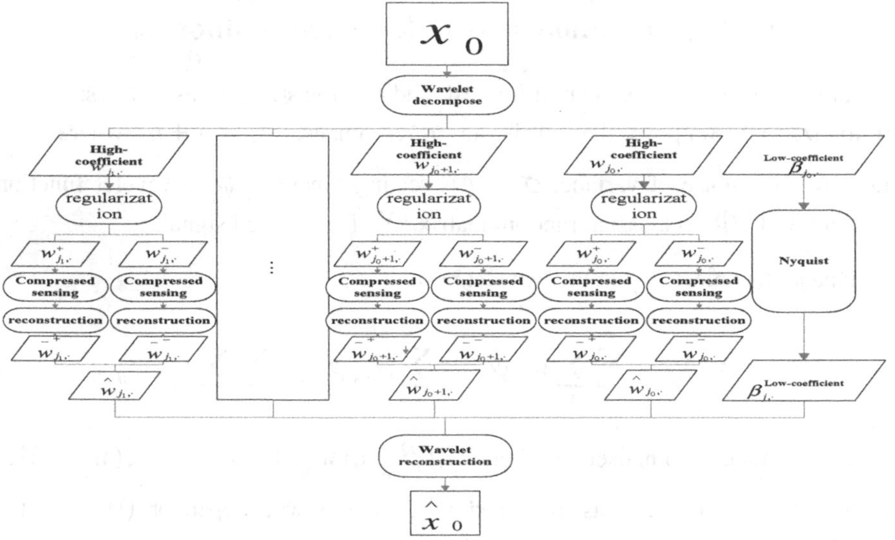

Fig. 1. Compressed sensing image de-noising method based on regularization of wavelet decomposition

4 Experimental Simulation

In order to verify the validity of the algorithm, we use matlab7.9 wavelet toolbox "sym4" to decompose the image with decomposition level of 3 layers and use "randn" function to add Gaussian white noise; compressed sensing observation matrix is Gaussian matrix, and reconstruction algorithm is OMP; Objective evaluation criteria is SNR.

Figure 2 show the four different algorithms de-noising effect under Gaussian white noise, which σ is form 1 to 29 (db). Due to the sparsity, CS de-noising based on wavelet decomposition was greatly affected with the lowest de-noising capability; due to increasing sparsity, improved regularization CS wavelet de-noising show well result in the figure, and the de-noising capability improve a lot. According to the figure, the capability of regularization CS wavelet de-noising is much stronger than the CS de-noising based on wavelet decomposition and soft threshold de-noising method, but lower than hard threshold de-noising.

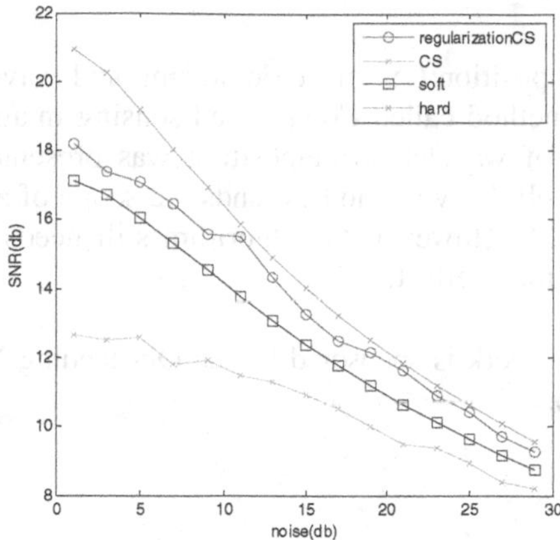

Fig. 2. Comparative result based on different db noise.

Figure 3 show the de-nosing result under different de-noising methods. From the visual point of view, CS de-noising based on wavelet decomposition is worst. The soft threshold de-noising is stronger than the former, but there is visual blurred. Compare with the former, the hard threshold de-noising and the regularization CS wavelet de-noising look well.

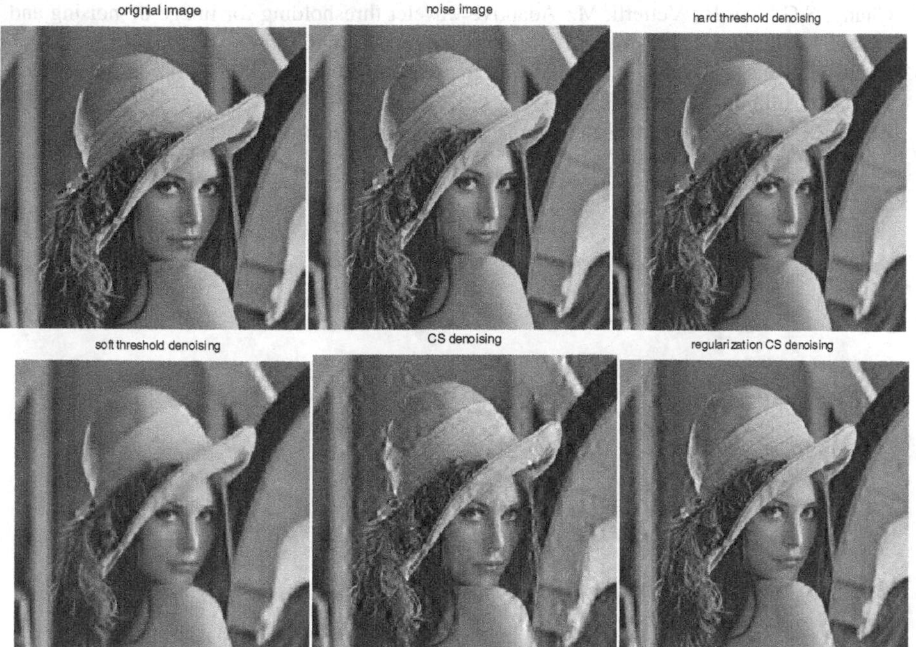

Fig. 3. De-noising result under different method

5 Conclusion

Based on wavelet decomposition CS image de-noising and wavelet high-coefficients lack of sparsity, a new method called Compressed sensing image de-noising method based on regularization of wavelet decomposition was presented in this paper, the simulation result show well. New method expands the scope of application of CS and contributed to improve CS. However, the algorithm still needs to improve in some areas, so we still need to make effort.

Acknowledgements. This work is sponsored by the Outstanding Youth Fund of Henan Province (104100510008).

References

1. Donoho, D.L.: Compressed sensing. IEEE Transactions on Information Theory 52(4), 1289–1306 (2006)
2. Candès, E.: Compressive sampling. In: Proceedings of International Congress of Mathematicians, pp. 1433–1452. European Mathematical Society Publishing House, Zürich (2006)
3. Tsaig, Y., Donoho, D.L.: Extensions of Compressed Sensing. Signal Processing 86(3), 533–548 (2006)
4. Xu, Y., Weaver, J.B., Jetal, H.M.: Wavelet transform domain filters: A spatially selective noise filtration technique. IEEE Trans. Image Processing 3(6), 743–758 (1994)
5. Chang, S.G., Yu, B., Vetterli, M.: Adaptive wavelet thresholding for image de-noising and compression. IEEE Trans. Image Processing 9(9), 1532–1546 (2000)
6. Chen, M., Guo, S., Wang, Y.: Finger vein image de-noising based on compressive sensing. Journal of Jilin university 41(2), 559–562 (2011)
7. Zhen, R., Liu, X., et al.: Wavelet denoising based on sparse representation, vol. 40(1), pp. 33–40. Science China Press (2010)

The Optimal Antenna Array Group for a Novel QOSTBC Decode Algorithm

Yong Wang and Hui Li

State Key Lab. of ISN, Xidian Univ., Xian, 710071, China
xd.wangyong@gmail.com, lihui@mail.xidian.edu.cn

Abstract. The quasi-orthogonal matrix is used in the STBC for complex. Signal constellation, the interference signal components will occur in the decoding process. In this paper, a novel decoding algorithm is proposed for STBC with the full code rate and quasi-orthogonal design. The proposed algorithm divides the received signals into two groups according to the encoded matrix. It can eliminate the interference from the neighboring signals dependent on different orthogonal design encoding scheme. The simulation results demonstrate that the proposed algorithm can provide the improved performance.

Keywords: MIMO system, QOSTBC, antenna array groups, beamforming.

1 Introduction

It has been shown that multiple-input multiple-output (MIMO) systems substantially increase the capacity of wireless channels [1]. The two main approaches for exploiting this spatial diversity have been proposed [2],[3]. The first one [2] is spatial multiplexing which aims at improving the spectral efficiency and the second one [3] consists in designing space-time codes to optimize diversity and coding gain. These two approaches can be combined to maintain a high spectral efficiency together with full antenna diversity gain.

The effort for combining of beamforming and transmit diversity at the downlink has been researched [4]. However, diversity antenna spacing is usually required to be large enough, say ten or more times wavelength of the carrier for a uniform linear array in small angular spread environments, to obtain low correlation or independent fading channels [5], while the beamforming antenna spacing needs to small enough, for example, half wavelength for a uniform linear array [6], to achieve spatial directivity and signal correlated or coherent. It is necessary to make an optimum configuration of antenna array for getting the maximum performance and capacity in the combined beamforming with transmit diversity.

2 Conventional Combined Beamforming with STBC Decoder

Let us consider square n by n QOSTBC matrix which was designed for getting the full rate transmission at the expense of a loss in the full transmit diversity. The STBC send to receiver by different antenna array groups. A block of n bits information is

D. Jin and S. Lin (Eds.): Advances in CSIE, Vol. 1, AISC 168, pp. 177–183.
springerlink.com © Springer-Verlag Berlin Heidelberg 2012

denoted by $\boldsymbol{S}=(s_1, s_2,\cdots s_n)$. For each symbol, the encoded signals through STBC transmission matrix can be given by

$$\boldsymbol{S}_T=\begin{bmatrix}\boldsymbol{S}_1 & \cdots & \boldsymbol{S}_n\end{bmatrix} \tag{1}$$

Depending on the encoding matrix, the channel coefficient matrix can be expressed in the generalized form $\boldsymbol{H}=\{h_{ij}\}$. Here, each of the matrix elements $h_{ij}=\alpha_{ij}e^{j\beta_{ij}}$ can have positive or negative sign and the conjugated transform of the fading coefficient which is corresponding to signal s_{ij} transmission.

Suppose that we have the best estimated channel information and perfect knowledge of the channel at the receiver. In this case, the beamforming weighted vector $\boldsymbol{w}_i$ may be set using the steering vector $c\boldsymbol{a}(\theta_i)$, where c is a constant which can be assumed as [7]

$$k=\boldsymbol{w}_i^{\mathrm{H}}\cdot\boldsymbol{a}(\theta_i)=\boldsymbol{w}_i^{\mathrm{T}}\cdot\boldsymbol{a}^*(\theta_i) \tag{2}$$

where $\boldsymbol{a}(\theta_i)$ is the downlink steering vector at θ_i. The received signals can be get

$$\begin{bmatrix} r_1 \\ \vdots \\ r_n \end{bmatrix} = k\boldsymbol{H}\begin{bmatrix} s_1 \\ \vdots \\ s_n \end{bmatrix}+\begin{bmatrix} n_1 \\ \vdots \\ n_n \end{bmatrix}=k\begin{bmatrix} h_{11} & \cdots & h_{1n} \\ \vdots & \ddots & \vdots \\ h_{n1} & \cdots & h_{nn} \end{bmatrix}\begin{bmatrix} s_1 \\ \vdots \\ s_n \end{bmatrix}+\begin{bmatrix} n_1 \\ \vdots \\ n_n \end{bmatrix} \tag{3}$$

Where r_i can be conjugate transform of the received signal at ith time slot, and n_i can be conjugate transform of additive white Gaussian noise at ith time slot.

Therefore the estimation of signals can detect the jth signal of each symbol as

$$\tilde{s}_j=k\sum_{i=1}^{n}h_{ij}^*(h_{i1}s_1+h_{i2}s_2+\cdots+h_{in}s_n)+\sum_{i=1}^{n}h_{ij}^*n_i,(j=1,2,\cdots n) \tag{4}$$

It is worth to mention that some terms can be zero depending on different STBC encoding matrix. In (4), there are totally n^2 terms including the information of signals. However, just n terms, which can be expressed as $h_{ij}^*h_{ij}s_j$, are related with signal s_j. Therefore there are n^2-n-p interference terms from neighboring signals. Here, p is the number of terms cancelled out by orthogonal relation between the signals. The actual value of the result will be dependent on the encoding scheme.

3 The Proposed Decode Algorithm

In the proposed scheme with the aid of beamforming, we devide the received signals into two groups according to the encoded matrix. Fig. 1 show the scheme model. According to the encoded matrix, the signals are divided equally into two groups of *Tx* antennas which

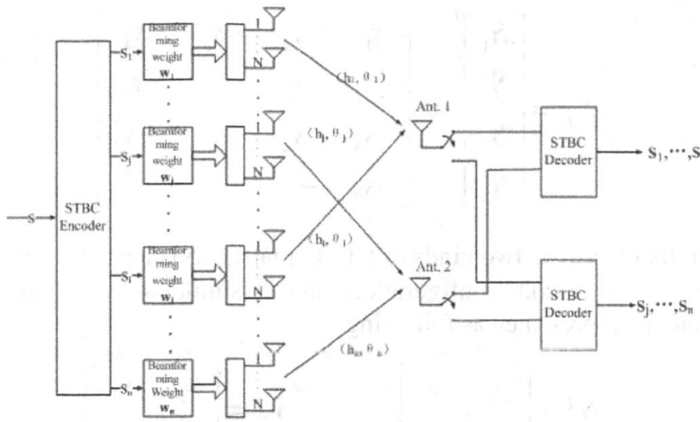

Fig. 1. The proposed decode scheme model

make a beamforming toward two different receiving antennas. At the receiver, the signals in two groups are decoded separately in different STBC decoders.

Considering the received signal at receiving antenna 1 in one symbol, it can be expressed as

$$\begin{bmatrix} r_1 \\ \vdots \\ r_{n/2} \end{bmatrix} = k\boldsymbol{H}' \begin{bmatrix} s_1 \\ \vdots \\ s_{n/2} \end{bmatrix} + \begin{bmatrix} n_1 \\ \vdots \\ n_{n/2} \end{bmatrix} \tag{5}$$

where $\boldsymbol{H}' = \{h_{ij}\}, i = 1,\ldots,n,\ j = 1,..n/2$. The detection of the *j*th signal at the proposed receiver structure can be rewritten as

$$\tilde{s}_j = k\sum_{i=1}^{n} h_{ij}^*(h_{i1}s_1 + h_{i2}s_2 + \cdots + h_{in/2}s_{n/2}) + \sum_{i=1}^{n} h_{ij}^* n_i, (j = 1, 2, \cdots n/2) \tag{6}$$

There are totally $n^2/2$ terms including the information of signals in (6). However, just n terms, which can be expressed as $h_{ij}^* h_{ij} s_j$, are related with signal s_j. Therefore there are $n^2/2 - n - q$ interference terms from neighboring signals. Here, q is the number of terms cancelled out by orthogonal relation between the signals. The actual value of the result will be dependent on the encoding scheme.

Comparing (6) with (4), we can easily find that our proposed scheme can reduce at least half interference terms from neighboring signals. If the non-orthogonal signals are separated perfectly, there will be no interference terms from neighboring signals such as the example with four transmitters.Let us consider QOSTBC with four transmit antennas, its coding matrix show as [7]

$$S_T = \begin{bmatrix} S_1 \\ S_2 \\ S_3 \\ S_4 \end{bmatrix}^T = \begin{bmatrix} s_1 & s_2 & s_3 & s_4 \\ -s_2 & s_1 & -s_4 & s_3 \\ -s_3^* & s_4^* & s_1^* & -s_2^* \\ -s_4^* & -s_3^* & s_2^* & s_1^* \end{bmatrix} \quad (7)$$

It can be found that there are two kinds of STBC matrix $S1$ and $S2$ obtained, both of which have orthogonal code configuration same or similar as Alamouti code. These two matrixes can be represented as following

$$S1 = \begin{bmatrix} s_1 & s_3 \\ -s_3^* & s_1^* \end{bmatrix} \qquad S2 = \begin{bmatrix} s_2 & s_4 \\ s_4^* & -s_2^* \end{bmatrix} \quad (8)$$

Those signals in the first signal set of $S1$ are transmitted from first antenna group and third antenna group at odd time slot and those signals in the second signal set of $S1$ are transmitted from second and fourth antenna group at even time slot. While the two signal sets of $S2$ will transmit in the opposite way from $S1$, the first decoder for $S1$ will be connected to the upper *Rx* antenna 1 at odd time slot, but switched to the lower *Rx* antenna 2 at even time slot. Meanwhile, the signals belonging to $S2$ will be decoded at the second decoder, whose connection to the *Rx* antenna is the reverse of the first decoder.

These signals are transmitted separately towards two different receiving antennas. The first and third *Tx* antennas will make a beamforming towards receiving antenna 1, and the second and fourth *Tx* antennas towards receiving antenna 2. In the case, the received signals at receiving antenna 1 correspond to S_{TB}^1 and S_{TB}^3, whereas the received signals at receiving antenna 2 correspond to S_{TB}^2 and S_{TB}^4:

$$\boldsymbol{r}^1 = \begin{bmatrix} S_{TB}^1 \\ S_{TB}^3 \end{bmatrix}^T \begin{bmatrix} h_1 \cdot \boldsymbol{a}(\theta_1) \\ h_3 \cdot \boldsymbol{a}(\theta_3) \end{bmatrix} + \boldsymbol{n}_1 \qquad \boldsymbol{r}^2 = \begin{bmatrix} S_{TB}^2 \\ S_{TB}^4 \end{bmatrix}^T \begin{bmatrix} h_2 \cdot \boldsymbol{a}(\theta_2) \\ h_4 \cdot \boldsymbol{a}(\theta_4) \end{bmatrix} + \boldsymbol{n}_2 \quad (9)$$

Where $\boldsymbol{r}^i$ is the received signal from receiving antenna I, and $\boldsymbol{n}_i$ the additive white Gaussian noise at the *i*th receiver according to different time slot, S_{TB}^i represents the signals which are transmitted from *i*th antenna group.

The detection of $S1$ at the proposed receiver structure under perfect channel estimation can be expressed as

$$\tilde{s}_1 = k(\alpha_1^2 + \alpha_2^2 + \alpha_3^2 + \alpha_4^2)s_1 + n_1 h_1^* + n_2 h_2^* + n_3^* h_3 + n_4^* h_4 \quad (10)$$

It could be showed that, after detecting the signal, our new scheme has no interference terms, which come from neighboring signals in a same symbol.

4 Simulation and Discussion

In this section, we use the QPSK system model with four array groups and receiver, which was introduced in section 3 as those examples for comparing the performances of the conventional receiver scheme. In all simulation, we assume four uniform linear array antennas. The total emitted power from all the transmit antennas is kept identical. We use a quasi-static and flat Rayleigh fading channel, and set the angle spread (AS) to 10° and 50°.

Firstly, the numerical example of the performance equation is presented according to the number of divided groups. We assume the total transmit power was fixed and total number of transmit antenna M equal to 100.

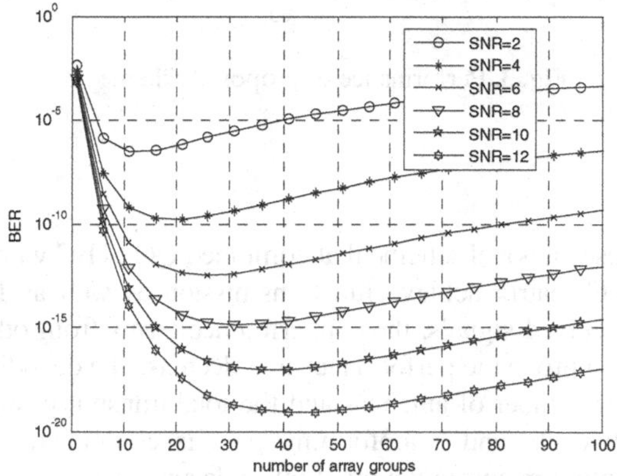

Fig. 2. The optimal number with different SNR

Fig. 2 illustrates the performance according to the number of groups with different SNR. As expected, the optimal number of groups can be derived differently under the different SNR. The simulation results show that using the analysis result, the fluctuant channel condition under different SNR can also be managed. We can also combine the transmit diversity and beamforming, by finding the optimal configuration of multiple antennas, in order to achieve the best performance.

Secondly, we use the above simulation results and decide the optimal array groups' configuration. Then, we compare the bit error rate (BER) performance under different SNR. Because the proposed scheme used two *Rx* antennas for decode (called scheme 2), the performance can be better than the conventional scheme (called scheme 1) with one *Rx* antenna.

Fig. 3 demonstrates that the performance of the proposed scheme can achieve performance enormously, when compared with the conventional scheme. It could be showed that our new scheme has no interference terms after detecting the signal, which come from neighboring signals in same symbol.

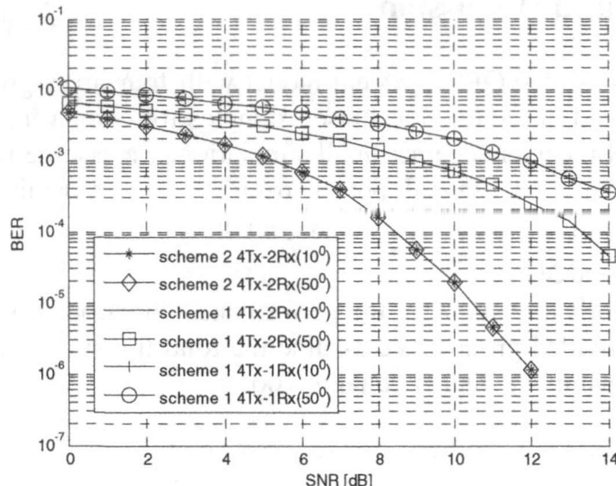

Fig. 3. Performance of proposed scheme

5 Conclusions

In this paper, we proposed a novel scheme that combined QOSTBC with beamforming for making complex STBC matrix achieve full transmission as well as full diversity. The scheme separate the received signals, those interference terms from other signals can be minimized, which can improve the performance and decrease the decoding complexity. At the same time, the total number of antennas and the total transmit power is fixed and the requirement for *Tx* diversity and beamforming gain is contradictive. We present the method to obtain the optimum configuration of multiple antennas.

Acknowledgement. The Project Supported in part by National Natural Science Foundation of China (No. 61101147), The National Basic Research Program of China (973 Program, No.2012CB316100), Specialized Research Fund for the Doctoral Program of Higher Education (No. 20110203120004), Natural Science Basic Research Plan in Shaanxi Province of China (Program No. 2011JQ8033), Science and Technology Projects in Xi'an of China (Program No. CXY1117(2)), The Fundamental of Research Funds for the Central Universities(No. K50510010028), The 111 Project (No. B08038).

References

1. Telatar, I.: Capacity of multi-antenna Gaussian channels. Eur. Trans. Telecommun. 10(6), 585–595 (2004)
2. Foschini, G.: Layered space-time architecture for wireless communication in fading environments when using multiple antennas. Bell Labs. Tech. J. 2 (2006)
3. Tarokh, V., Jafarkhani, H., Calderbank, A.: Space-time block codes from orthogonal designs. IEEE Trans. Inf. Theory 45, 1456–1467 (2005)
4. Lei, Z.D., F.P.S., C., Liang, Y.-C.: Combined beamforming with space-time block coding for wireless downlink transmission. In: IEEE 56th, Conf. on Vehicular Technology Proceedings, vol. 4, pp. 24–28 (2009)

5. Lee, W.C.Y.: Antenna Spacing Requirement for a Mobile Radio Base-Station Diversity. Bell System Tech. Journal 50, 1859–1876 (2001)
6. Ertel, R.B., Cardieri, P., Sowerby, K., Rappaort, T.S., Reed, J.: Overview of spatial channel models for antenna array communication systems. IEEE Personal Communications Magazine 5, 10–22 (2004)
7. Hou, J., Lee, M.H., Park, J.Y.: Matrices analysis of quasi-orthogonal space-time block codes. IEEE Commun. Lett. 7(87), 385–387 (2009)

The Application of Semantic Vocabulary in Building Intelligent Search System

HongSheng Xu* and GuoFang Kuang

College of Information Technology, Luoyang Normal University, Luoyang, 471022, China
xhs_ls@sina.com

Abstract. Nowadays, there are still many problems in present searching systems. There are some defects between keyword search and category search in many aspects. This paper presents the application of semantic vocabulary in intelligent search system to makes up these defects. The system builds categories by analyzing the relationships and meaning between words. Users can efficiently retrieve contents from a variety and plenty of information through the classification of documents, and reduce the cost of information management.

Keywords: Information Classification, Taxonomy, Vocabulary Analysis, Information Search.

1 Introduction

With the development of information science and the Internet, the information resources of network has become more and more abundant, On-line information assumes the explosively to grow[1] .users want to find the useful information in the vast online world, as the main application access to modern information technology——search system is essential. However, using the current search systems often face the following problems: First, the main problem of keyword search is that the search result is obscure and great quantity. It can not really understand the user's intention by the key words , semantic gap and search results far exceeds the reference range may be happened, causing the search results of fuzziness and huge, resulting in search results obscure and great quantity; Second, the problem of categories search is that information management consume large human resources and slowly updates. Such systems constructed index of classification and file of classification networks by the artificial way, although reach to the purpose of accurately classification, but it takes too much human resources, and different person do the classciftion, results may be different.

For those reasons, this paper proposes the research of intelligent search system model based on semantic vocabulary. System model's aim is to collect the network information automatically to achieve the data gathering timeliness, builds categories by analyzing file information automatically, and increase the flexibility of categories.

* Author Introduce: HongSheng Xu(1979-), Male, lecturer, Master, College of Information Technology, Luoyang Normal University, Research area: computer application, Search.

D. Jin and S. Lin (Eds.): Advances in CSIE, Vol. 1, AISC 168, pp. 185–190.
springerlink.com © Springer-Verlag Berlin Heidelberg 2012

By using position that the file in the classification struction after classified automatically to understand the semantic content of the file, provides reference of search information, and help users find the information more efficiently and quickly.

2 Search System Model

To be able to builds categories, classify of web documents automatically, and present the semantic categories, this paper presents an intelligent search system based on semantic vocabulary.

The whole model is divided into file collection, file information extraction, analysis, file categories and search present five parts.

(1) File Information Extraction

This model is divided into three parts: the file structure analysis, broken-word processing, and file feature extraction.

File structure analysis, also known as Parser, is dedicated to analysis the file data structure of web page, the current web documents HTML, XML structure, and divided web file into different parts by signs.

The main function of file structure analysis has two aspects: on the one hand it may give different weight of characteristic according to the data in the document's different position when the document characteristic extraction in the future. This is mainly divided into the key words in the title or paragraphs appearance; on the other hand, the file structure analysis need to give the file a unique Process Identifier (PID), to distinguish between different files; For example: title, paragraph, author, and then have the same Process Identifier.

(2) Broken-word processing

Broken-word processing is to break down the sentences into words. In the field of computational linguistics is called participle. And the word is the smallest meaningful units of language that can be used freely. Any language processing system must analyze the word in the text first and then do the further processing, such as machine translation, linguistic analysis, and information extraction.

In the study of broken-word or participle, the broken-word accuracy has been the focus; there are kinds of methods to improve the accuracy of broken-word now. The methods of dealing with Chinese broken-word are divided into the following ways [2]: dictionary-based method、statistical method and mixed method .Hybrid method is the integration of dictionary-based method and statistical method. Broken out the senses by using dictionary into different combinations of vocabulary words, and then use corpus statistics to find the best combination of broken words, lower down the chances of meaningless words. The disadvantage is that it needs to maintain the dictionary and provide a large statistical information from corpus.

(3) File Characteristic extraction

File characteristic extraction is to extract the key words which represent file concepts and semantics in order to classify the documents. This part is mainly counts the position, the frequency and the lexical category of the key words in the document as the basis when extracting the file feature. The position of the key words has the intense relations with the whole concept and the semantics. For example, a word in the title of the document is usually the subject, so the vocabulary is more important.

However, the vocabularies in paragraphs are less importance. In addition, the frequency of the key words should also take into account. If a word has a large frequency in a chapter, it means that this article focuses on the conception of this word, so using these words to represent one of the file features.

Given different weight according to appearance of the key words in title or paragraphs, compile the whole weight score of the words that appear in the file as its eigenvalue. The weight score of the words is calculated as follows: weight score of title words = title weight×the number title appeared +paragraph weight×paragraph. Therefore, the higher eigenvalues of the word, the better it represent the characteristics of the file.

3 Glossary Sorting Algorithm

For base of the correlation between vocabulary: the more two words appear in the same article the closer semantic relations they have. And this relationship may be based on relationship of subordinate, the relationship between similarities of meaning, or any other relationship. For example, if the "diet" and "health" appear in high frequency in "Life" article together, it means "diet" and "healthy" have a great relationship because "diet" is one of essential conditions to obtain "health which is a very common relationship. However, this relationship seems so abstract for computer to deal with, so we propose correlation coefficient to represent the correlation between the degrees of correlation terms, using the mathematical model to concrete such a concept.

Because the relationship between words is two-way, we define the correlation coefficient between words as follows. Suppose that there are two terms A, B, we define the correlation between A and B as :the multiplication of the probability of B appears in one document under the conduction that A appears in the same document and the probability of A appears in one document under the conduction that B appears in the same document ,as is shown in formula 1, γ presents correlation coefficient between A and B , No presents the total number of documents, N(A) presents the number of document in which A appears, N(B) presents the of documents in which B appears;

N ($A \cap B$) presents the number of document in which A and B appear simultaneously. γhas a maximum value 1 when NA = NB = N ($A \cap B$),as is shown by equation1.

$$\gamma \equiv P(A \mid B) \times P(B \mid A) = \frac{P(AB)}{P(B)} \times \frac{P(AB)}{P(A)} = \frac{\frac{N(A \cap B)}{N_0}}{\frac{N(B)}{N_0}} \times \frac{\frac{N(A \cap B)}{N_0}}{\frac{N(A)}{N_0}} = \frac{[N(A \cap B)]^2}{N(A) \times N(B)} \leq 1 \qquad (1)$$

This formula is similar with the one using to calculate the similarly of glossary mentioned by Salton, G. and McGill. However, this formula has a low complexity, so we use this formula to calculate the correlation between glossaries and reduce the amount of computing. Considerations are as following.

First, the main ideas of using tree structure to express Glossary classification is to browse conveniently and its biggest characteristic is the succession, the glossary relations also have such characteristic. So the paper is using tree structure to show the complex relationship between correlated glossaries [3].

Second, remove low related links by providing relation threshold; show the glossary whose relation threshold is greater than the critical value to reduce the complexity of the classification structure.

In constructing the classification tree structure, the two words which have a certain correlation with each other(correlation coefficient is greater than the critical value), the word which has a higher weight score will be placed in the top of classification structure while the lower one be placed in the bottom of classification structure. Because the words which have a high weight score are generally the common terms or popular vocabulary in certain areas, in addition, the word which has a higher weight score usually distributed in more documents and is a larger concept. Vice reverse, the smaller concept. This classification structure is in keeping with the most people's cognitive models of the classification structure, and when using the class to research, it can find necessary details slowly by a large concept.

For example, Matrix is used to show the correlation between glossaries such as table 1. We assume that the weighting score in the computer field as: computer >software > system software >application software.

Suppose the correlation coefficient γ= 0.5 and use it as critical value to build classification struction of tree, the results is shown in figure1.

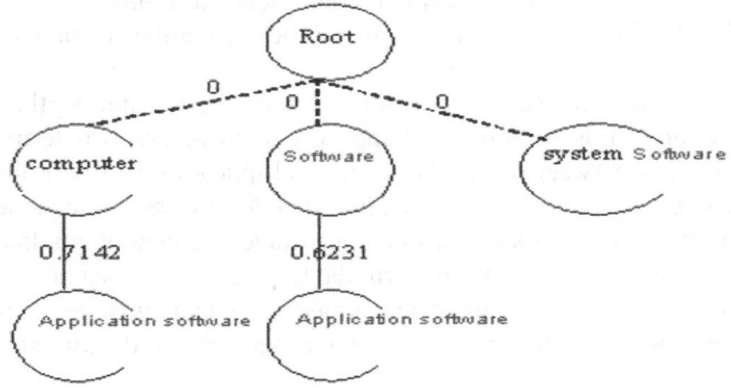

Fig. 1. Classification structure tree diagram example

4 Document Classification and Search Show

Responsible of the documents classification is to classify document to the appropriate category according to their characteristics. It is purpose--classifying the collected documents to the classification structure that has been built up.

The basic idea is: in the file of K, the key word A is the representative characteristics vocabulary of the file K if the keyword A weighted highest score, so it will refer A when classified K. This node will be define as the master node of classifying file K, and the crossing nodes from the roots of a tree to the master node A is the classifieds path represent locates of the file in the whole classification structure and semantic information as well as concepts.

We take electronic term as an example to illustrate the works of lexical analysis system. Statically list of all the key words in file K is Figure 2 for classification structure, the number between nodes and node in the chart means the correlation coefficient. The words "machine" weighted highest value in the file K so be viewed as the master node thus the file path may be classified as " electron→ Computer →machine" or " electron→Software→ machine". It means the paper is mainly about something that " computer" for " machine " in the area of " electron " if the document is classified as " electron→ Computer →machine".

In addition to decide the master node, it should also calculate the similarity of K in the file path. The document is classified in the highest path. In the above example, the file K classification in the classification structure in Figure 2. And the node called "machine" is candidate master node of K, then K will have the file path " electron→ Computer →machine" or " electron→Software→ machine" two classifications path, the path known as the classification of candidate . Similarity calculation can be as the base of classification for Hierarchical search. In this paper vector space model is used to represent the file and the candidate path in the similarity calculation.

Therefore, whether the file itself or the category classification path, we view as the vector that amount of features with different characteristics is a combination of terms. And different words to each other for the vector space orthogonal to the base, so the path similarity is the cosine of the angle between two vectors of values.

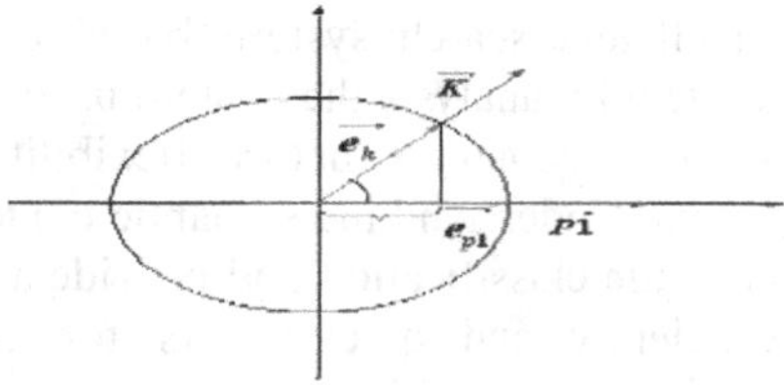

Fig. 2. For classified documents, said its candidate path vector map.

In Figure 2, the vector space can be expressed as $(\vec{e_1}, \vec{e_2}, \vec{e_3}, \vec{e_4}, \vec{e_5})$ = (application software, computers, software, electronics, software), $\vec{K}$ as the vector for the classification documents and $\vec{P_1}$ as one of file path for Path1: " electron→ Computer →machine", $\vec{e_k}$ 、 $\vec{e_{p1}}$ represent the unit vector of $\vec{K}$ 、 $\vec{P_1}$, θ for the angle between two vectors. Therefore, the similarity of $\vec{K}$ and $\vec{P_1}$ namely the projection length $\vec{e_k}$ and $\vec{e_{p1}}$ and its value is $\cos\theta$.

The system designs a model of online merchandising website. This site uses the popular three-tier architecture, by the logic of the entire system is divided into different modules, the system greatly reduces application development and maintenance costs and improve the reusability of system modules, while the scalability of the system greatly to improve. As the. Net benefits of its own, this website for a lower system requirements, compatibility, and good for promoting the use of the site provides a convenient and may.

In the hierarchical directory-style search, using the classification structure to provide hierarchical index and directory to allow users use the correlation between the levels, starting from the root of the tree layer goes down by layer, from the abstract concept to concrete concept to retrieve the required details. When the user selects the type of classified directory, the system will pick up the document matches required from the database, and according to the similarity of classification path, do descending order on selected documents, making the high similarity file is first. The search method is very similar with Yahoo, Yam, etc.

In the keyword search, by providing classified documents that match your search path users can understand the semantic and position in the classification of the file then to filter required information. Classification structure in Figure 4, for example, when you type "machine", there may be classification path "electron→ Computer →machine" and " electron→Software→ machine" then sort those documents according to the keyword score rate. The document is ordered by its key word's weight score in descending. Present document and the classification of the file path together to provide more information about the file.

5 Summary

This paper presents an intelligent search system based on semantic vocabulary, indicating the role of the system by analysis the system used in the field of computer terms: Increasing category management structure flexibility in the structure that analysis information by system. Understand the semantic content through the position in the Taxonomy after Automatic classification and provide a search data to help find the information more efficiently and quickly. As for classifying construction optimization, although this article provides the critical value of the correlation coefficient building as the reference of building the classification struction, but how to adjust critical value to achieve a better effects need further consideration and exploration. In addition, how to assess the effect of classification of documents is also worth considering.

References

1. Chi, C.H., Ding, C., Lim, A.: Word segmentation and recognition for web document framework. In: ACM CIKM 1999, vol. (11), pp. 458–465 (2004)
2. Salton, G., Wong, A., Yang, C.S.: A Vector Space Model for Automatic Indexing. Communication of the ACM 18(11), 613–620 (2005)
3. Tsoi, A.C.: Structure of the Internet. In: IEEE Conference on Intelligent Multimedia Video and Speech Processing, pp. 449–452 (2001)

Different Characteristics of MCP and PAB in Electricity Market by Evolutionary Game Theory

Xian Huang and Jialei Tan

School of Control and Computer Engineering, North China Electric Power University, Beijing, 102206, China
hx@ncepu.edu.cn

Abstract. There is a spirited debate on which auction mechanism is better, PAB or MCP in electricity generating-side market. This paper applies evolutionary theory game to study this problem under assumption of full competition. Our simulation results show that MCP is much better than PAB whenever the bidders with complete market information or not, represented by different kinds of payoff matrixes, with regards to social welfare and market stability.

Keywords: Evolutionary game theory, Electricity Market, MCP, PAB.

1 Introduction

According to the statistic reports of China Electricity Council [1]: 1) Both China's installed and generated capacities are ranking the second worldwide; 2) The averaged unit annual utility hour is around 2300 hrs these years; 3) In the first half of 2011, 182.052 Billion kWh have been totally transacted by China National Electricity Market. It tells the truth that the scale of China's electric industry has developed up to world top level thanks to decades of rapid development and now is in the eve of further marketization. In the 1990s, there were six zones set up for trial test of generation-side electricity market. Most zones selected PAB (pay-as-bid) as their auction mechanism while the rest selected MCP (market clearing price).

Which auction mechanism is better, PAB or MCP is still a quite spirited debate. Several approaches have been applied to study and answer this question, such as game theory, multi-agent, experimental economics and so on [2],[3],[4],[5].

In this paper, we use evolutionary game theory to model and simulate the long term market behaviors under MCP and PAB. By now, marketization has been introduced into the generating side only in China while the consumers buying prices are controlled firmly by the government, so unlike other studies based on game theory, it does not use the concepts such as demand curve and Cournot-Nash equilibrium. As China's electrical power generating industry consists of five huge independent groups, the competition in the generating market is rather tense and supposed to be of none coalition. In this context, our study conclusion is that: MCP is better than PAB both in social welfare and market stability.

D. Jin and S. Lin (Eds.): Advances in CSIE, Vol. 1, AISC 168, pp. 191–196.
springerlink.com © Springer-Verlag Berlin Heidelberg 2012

2 The Electrical Generating Market

The market demand is supposed to be N % (N=0~100) of the capacity of all the generating units concerned, which implies a rather competitive market. That is to say all generating units are competing for the demand which is equivalent to 0~100% of their rated capacity. N<1.0 represents a market status of "supply exceeds demand" while N=1.0 represents equilibrium of supply and demand. Only the situation of N<1.0 is discussed in this paper.

For simplicity, all generating units are supposed to be of the same capacity and cost function. Each of them has three bidding strategies such as H (high price), M (medium price) and L (low price), which stands for a net profit of S_h=162534.4, S_m=117351.7 and S_l=66814.3 units if its electric power is sold at the corresponding price. And the percentages bidding at L and M are denoted by x and y respectively.

2.1 Under MCP Auction Mechanism

There are totally three scenarios which are classified by different kinds of market competition. The detailed scenarios and their corresponding payoff matrixes are as follows.

(1) $\begin{cases} x < N \\ (x+y) < N \end{cases}$. The MCP is high price. Table 1 is its payoff matrix.

Table 1. Payoff matrix when MCP is high price

	$L(x)$	$M(y)$	$H(1\text{-}x\text{-}y)$
$L(x)$	162534.4,162534.4	162534.4,162534.4	162534.4,162534.4×(N-x-y)/(1-x-y)
$M(y)$	162534.4,162534.4	162534.4,162534.4	162534.4,162534.4×(N-x-y)/(1-x-y)
$H(1\text{-}x\text{-}y)$	162534.4×(N-x-y)/(1-x-y),162534.4	162534.4×(N-x-y)/(1-x-y),162534.4	162534.4×(N-x-y)/(1-x-y),162534.4×(N-x-y)/(1-x-y)

(2) $\begin{cases} x < N \\ (x+y) \geq N \end{cases}$. The MCP is medium price. Table 2 is its payoff matrix.

Table 2. Payoff matrix when MCP is medium price

	$L(x)$	$M(y)$	$H(1\text{-}x\text{-}y)$
$L(x)$	117351.7,117351.7	117351.7,117351.7×(N-x)/y	117351.7,0
$M(y)$	117351.7×(N-x)/y,117351.7	117351.7×(N-x)/y ,117351.7×(N-x)/y	117351.7×(N-x)/y,0
$H(1\text{-}x\text{-}y)$	0,117351.7	0,117351.7×(N-x)/y	0,0

(3) $x \geq N$. The MCP is low price. Table 3 is its payoff matrix.

Table 3. Payoff matrix when MCP is low price

	L(*x*)	*M*(*y*)	*H*(1-*x*-*y*)
L(*x*)	66814.3×*N*/*x*,66814.3×*N*/*x*	66814.3×*N*/*x*,0	66814.3×*N*/*x*,0
M(*y*)	0,66814.3×*N*/*x*	0,0	0,0
H(1-*x*-*y*)	0,0	0,0	0,0

2.2 Under PAB Auction Mechanism

There are also three scenarios totally in accordance with different kinds of market competition. The detailed scenarios and their corresponding payoff matrixes are as following.

(1) $\begin{cases} x < N \\ (x+y) < N \end{cases}$. It is the least competitive status. Its payoff matrix is Table 4.

(2) $\begin{cases} x < N \\ (x+y) \geq N \end{cases}$. It is a moderate competitive status. The payoff matrix is Table 5.

(3) $x \geq N$. It is the most competitive status. Its payoff matrix is the same as Table 3.

Table 4. Payoff matrix at the least competition under PAB

	L(*x*)	*M*(*y*)	*H*(1-*x*-*y*)
L(*x*)	66814.3,66814.3	66814.3,117351.7	66814.3,162534.4×(*N*-*x*-*y*)/(1-*x*-*y*)
M(*y*)	117351.7,66814.3	117351.7,117351.7	117351.7,162534.4×(*N*-*x*-*y*)/(1-*x*-*y*)
H(1-*x*-*y*)	162534.4×(*N*-*x*-*y*)/(1-*x*-*y*),66814.3	162534.4×(*N*-*x*-*y*)/(1-*x*-*y*),117351.7	162534.4×(*N*-*x*-*y*)/(1-*x*-*y*),162534.4×(*N*-*x*-*y*)/(1-*x*-*y*)

Table 5. Payoff matrix at moderate competition under PAB

	L(*x*)	*M*(*y*)	*H*(1-*x*-*y*)
L(*x*)	66814.3,66814.3	66814.3,117351.7×(*N*-*x*)/*y*	66814.3,0
M(*y*)	117351.7×(*N*-*x*)/*y*,66814.3	117351.7×(*N*-*x*)/*y* ,117351.7×(*N*-*x*)/*y*	117351.7×(*N*-*x*)/*y*,0
H(1-*x*-*y*)	0,66814.3	0,117351.7×(*N*-*x*)/*y*	0,0

3 Simulation of Market Behavior

3.1 With Awareness of Market Demand and Supply Information

Firstly, the market behavior simulation can be carried out based on the above payoff matrixes if all generating units are informed of the relationship about demand and supply. Fig.1 and Fig.2 are such examples which are under MCP at N=0.7 and PAB at N=0.8 respectively. Fig.1 converges at (1,0), implying all generating units bidding at low price; while Fig.2 may converges at (1,0), (0,1) and (0,0) with different initial conditions which implying that all generating units may bidding at low, medium and high prices.

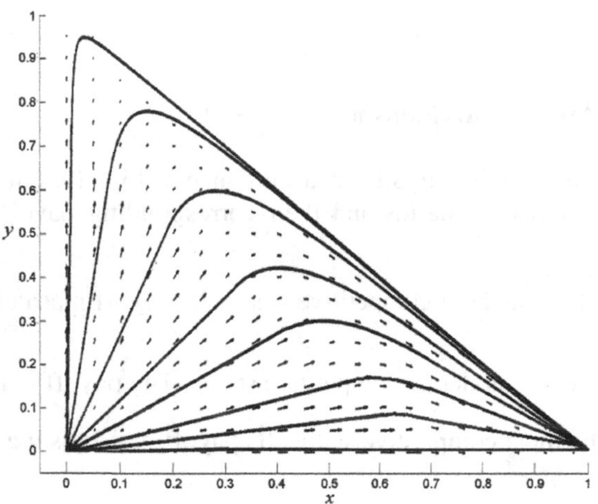

Fig. 1. N=0.7 under MCP With awareness of market demand and supply information

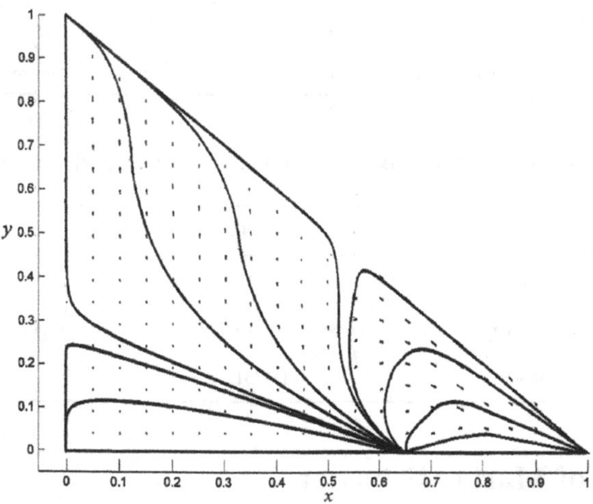

Fig. 2. N=0.8 under PAB With awareness of market demand and supply information

3.2 With No Awareness of Market Demand and Supply Information

In such a case, the payoff matrixes have to been modified with the concept of expectation utility according to Fig.3. Isoscele striangle Δ011 can be divided into triangle ΔNM1 and trapezia "0NM1", of which area is represented by p_1 and p_2 respectively. And trapezia "0NM1" can be divided into trapezium "NM1N" and isosceles triangle Δ0NN, of which area is represented by q_1 and q_2 respectively. So, the comprehensive payoff matrix of MCP and PAB can be described as Table 6 and 7.

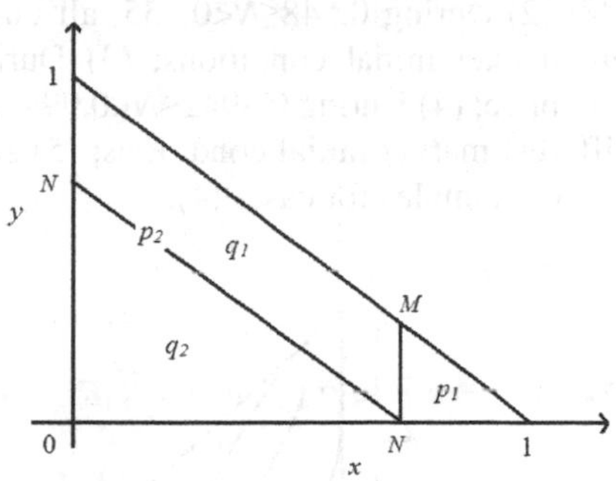

Fig. 3. Schematic map for expected utility calculation

Table 6. Comprehensive Payoff matrix under MCP

	$L(x)$	$M(y)$	$H(1\text{-}x\text{-}y)$
$L(x)$	$p_1S_lA+p_2q_1S_m+p_2q_2S_h$, $p_1S_lA+p_2q_1S_m+p_2q_2S_h$	$p_1S_lA+p_2q_1S_m+p_2q_2S_h$, $p_2q_2S_h+ p_2q_1S_mB$	$p_1S_lA+p_2q_1S_m+p_2q_2S_h$, $p_2q_2S_hC$
$M(y)$	$p_2q_2S_h+ p_2q_1S_mB$, $p_1S_lA+p_2q_1S_m+p_2q_2S_h$	$p_2q_2S_h+ p_2q_1S_mB$, $p_2q_2S_h+ p_2q_1S_mB$	$p_2q_2S_h+ p_2q_1S_mB$, $p_2q_2S_hC$
$H(1\text{-}x\text{-}y)$	$p_2q_2S_hC$, $p_1S_lA+p_2q_1S_m+p_2q_2S_h$	$p_2q_2S_hC$, $p_2q_2S_h+ p_2q_1S_mB$	$p_2q_2S_hC$, $p_2q_2S_hC$

Table 7. Comprehensive Payoff matrix under PAB

	$L(x)$	$M(y)$	$H(1\text{-}x\text{-}y)$
$L(x)$	$p_1S_lA+p_2S_l$, $p_1S_lA+p_2S_l$	$p_1S_lA+p_2S_l$, $p_2q_2S_m+ p_2q_1S_mB$	$p_1S_lA+p_2S_l$, $p_2q_2S_hC$
$M(y)$	$p_2q_2S_m+ p_2q_1S_mB$, $p_1S_lA+p_2S_l$	$p_2q_2S_m+ p_2q_1S_mB$, $p_2q_2S_m+ p_2q_1S_mB$	$p_2q_2S_m+ p_2q_1S_m B$, $p_2q_2S_hC$
$H(1\text{-}x\text{-}y)$	$p_2q_2S_hC$, $p_1S_lA +p_2S_l$	$p_2q_2S_hC$, $p_2q_2S_m+ p_2q_1S_m B$	$p_2q_2S_hC$, $p_2q_2S_hC$

The variables in Table 7 and 8 are as follows: $p_1=(1\text{-}N)^2$, $p_2=2N\text{-}N^2$, $q_1=(2\text{-}2N)/(2\text{-}N)$, $q_2=N/(2\text{-}N)$, $A=\begin{cases} N/x & x>N \\ 1 & x\le N \end{cases}$, $B=\begin{cases} 0 & x>N \\ (N-x)/y & x\le N\ \&\ (x+y)>N \\ 1 & x\le N\ \&\ (x+y)\le N \end{cases}$,

$$C=\begin{cases} 0 & x>N \\ 0 & x\le N\ \&\ (x+y)>N \\ (N-x-y)/(1-x-y) & x\le N\ \&\ (x+y)\le N \end{cases}.$$

The simulation shows that all units will bid low price when $N<1.0$ under MCP. However, it is quite different under PAB: (1) All units will bid low price when $N<0.548$; (2) During $0.548 \leq N<0.755$, all units may bid low or moderate price with different market initial conditions; (3) During $0.755 \leq N<0.8942$, all units will bid moderate price; (4) During $0.8942 \leq N<0.9999$, all units may bid high or moderate price with different market initial conditions; (5) All units will bid high price when $N=1.0$. Figs.4 is an examples for cases (4)

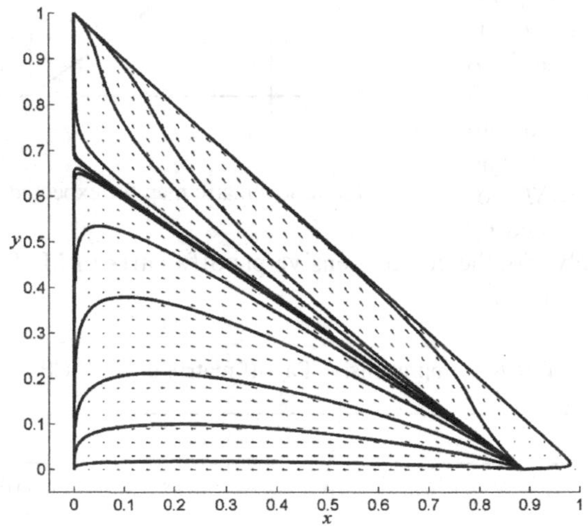

Fig. 4. N=0.94 under PAB with comprehensive payoff matrix

4 Summary

There is a spirited debate on which auction mechanism is better, PAB or MCP in electricity generating-side market. This paper applies evolutionary theory game to study this problem at $N<1.0$ which means of a full competition market. The simulation results show that MCP is better than PAB whenever the bidders with complete market information or not, represented by different kinds of payoff matrixes, with regards to social welfare and market stability.

References

1. Information on, `http://www.cec.org.cn`
2. Chen, H.: Experimental analysis of uniform price and PAB auctions in electricity markets. In: The 8th International Power Engineering Conference, pp. 24–29 (2007)
3. Wang, J., Shahidehpour, M., Li, Z., Botterud, A.: Strategic Generation Capacity Expansion. IEEE Transactions on Power Systems 24, 1002–1010 (2009)
4. Hobbs, B.F., Metzler, C.B., Pang, J.-S.: Strategic Gaming Analysis for Electric Power Systems: An MPEC Approach, vol. 15, pp. 638–645 (2000)
5. Liu, Z., Zhang, X., Lieu, J.: Design of the incentive mechanism in electricity auction market based on the signaling game theory. Energy 35, 1813–1819 (2010)

Study on the Stable Conditions of 3×3 Symmetrical Evolutionary Games

Xian Huang

School of Control and Computer Engineering, North China Electric Power University, Beijing, 102206, China
hx@ncepu.edu.cn

Abstract. Firstly all its seven rest points of a general 3×3 evolutionary game are worked out; then their stable necessary and sufficient conditions are deduced and specified in details; finally their relationships of coexistence and mutual repulsion are summarized. The results of this paper provide a guideline applicable for the study and design of 3×3 games.

Keywords: Evolutionary game theory, Replicator equations, Population dynamics.

1 Replicator Dynamic Equation

A symmetric 3×3 evolutionary game can be described in a normal form as

	A	*B*	*C*
A	*a*	*b*	*c*
B	*d*	*e*	*f*
C	*g*	*h*	*i*

Supposing the percentages of people who take strategy *A*, *B* and *C* are *x*, *y* and (1-*x*-*y*) respectively, their average expected payoffs, when the population is well-mixed and infinite large, will be correspondingly

$$\begin{cases} f_A = ax + by + c(1-x-y) \\ f_B = dx + ey + f(1-x-y). \\ f_C = gx + hy + i(1-x-y) \end{cases} \tag{1}$$

And the population average payoff is

$$\varphi = f_A x + f_B y + f_C\,(1-x-y). \tag{2}$$

Herein, the population evolutionary dynamics are governed by the following replicator equation

D. Jin and S. Lin (Eds.): Advances in CSIE, Vol. 1, AISC 168, pp. 197–202.
springerlink.com © Springer-Verlag Berlin Heidelberg 2012

$$\begin{cases} x' = x\left[\left(f_C - f_A\right)(x-1) + \left(f_C - f_B\right)y\right] \\ y' = y\left[\left(f_C - f_B\right)(y-1) + \left(f_C - f_A\right)x\right] \end{cases}. \tag{3}$$

2 The Rest Points of Replicator Equation

The rest points of (3) should be firstly worked out in order to study its dynamics. Here, we regard its right-sides are the product of M, N and P, Q respectively

$$\underset{M}{x}\,\underbrace{[(f_C - f_A)(x-1) + (f_C - f_B)y]}_{N},\quad \underset{P}{y}\,\underbrace{[(f_C - f_B)(y-1) + (f_C - f_A)x]}_{Q}.$$

Thus, it is quite obvious that $\begin{cases} x'=0 \\ y'=0 \end{cases}$ can be satisfied by one of the combinations of (M,P), (M,Q), (N,P) and (N,Q) of which both elements are 0 simultaneously. In such a way, all the rest points of (3) can be worked out. The four combinations are discussed as follows:

(i) (M, P) combination: $M=P=0$. We get the first rest point $(0,0)$.

(ii) (M,Q) combination: $M=Q=0$. We need to discuss two scenarios here: (*a*) if $f_C \neq f_B$, then there must be $y=1$. So, we get the second rest point $(0,1)$; (*b*) if $f_C = f_B$, we get the third rest point $\left(0, \frac{i-f}{e-h-f+i}\right)$.

(iii) (N,P) combination: $N=P=0$. Here we also need to discuss two scenarios: (*a*) if $f_C \neq f_A$, then there must be $x=1$. Thus, we get the forth rest point $(1,0)$; (*b*) if $f_C = f_A$, we get the fifth rest point $\left(\frac{i-c}{a-g-c+i}, 0\right)$.

(iv) (N,Q) combination: $N=Q=0$. Here we need to discuss two scenarios: (*a*) if $\begin{cases} f_C \neq f_A \\ f_C \neq f_B \end{cases}$, then there must be $x+y=1$. We get the sixth rest point $\left(\frac{e-b}{a-b-d+e}, \frac{a-d}{a-b-d+e}\right)$. (*b*) if $f_C = f_A$ or $f_C = f_B$, then we can further have the seventh rest point, which can be denoted as $\left(\frac{U}{U+V+W}, \frac{V}{U+V+W}\right)$, where

$$\begin{cases} U = bi - bf + ce - ch + fh - ei \\ V = af - ai + cg - fg + di - cd \\ W = -ae - bg + eg + bd + ah - dh \end{cases}. \tag{4}$$

3 Stability of the Rest Points

For the seven rest points worked out in last section, what concerned here are their necessary and sufficient conditions to be stable or fixed points, each of which stands for an evolutionarily stable strategies profile. They can be classified into three types: (*a*) two strategies are dominated by the third one. This kind of rest point includes $(0,0)$, $(0,1)$ and $(1,0)$; (*b*) one strategy is dominated by the other two. This kind of rest point includes $\left(0,\frac{i-f}{e-h-f+i}\right)$, $\left(\frac{i-c}{a-g-c+i},0\right)$ and $\left(\frac{e-b}{a-b-d+e},\frac{a-d}{a-b-d+e}\right)$; (*c*) three strategies compete so intensely that there is none dominating or dominated. This kind of rest point is $\left(\frac{U}{U+V+W},\frac{V}{U+V+W}\right)$ lonely.

The complexity of the stability situation of above three kinds of rest point increases according to their ordering. The simplest is the three rest points of type (*a*), while the most complicated is type (*c*).

Based on the books of [1] and [2] for the study of the stability of rest points, the ordinary nonlinear equation (3) can be expanded in the form of $\begin{cases} x'=\Phi(x,y) \\ y'=\Psi(x,y) \end{cases}$.

To simplify the analysis procedure, it is reasonable to assume all entries of the payoff matrix *a*~*i* are nonnegative. The stability discussion for the seven rest points is summarized as follows:

$(0,0)$. Its characteristic equation is

$$\lambda^2+[(c-i)+(f-i)]\lambda+(c-i)(f-i)=0. \tag{5}$$

Its stable necessary and sufficient condition is $\begin{cases}(c-i)+(f-i)<0\\(c-i)(f-i)>0\end{cases}$, which is equivalent to

$$\begin{cases} i>f \\ i>c \end{cases}. \tag{6}$$

$(0,1)$. Its characteristic equation is

$$\lambda^2+[(b-e)+(h-e)]\lambda+(b-e)(h-e)=0. \tag{7}$$

Its stable necessary and sufficient condition is $\begin{cases}(b-e)+(h-e)<0\\(b-e)(h-e)>0\end{cases}$, which is equivalent to

$$\begin{cases} e>b \\ e>h \end{cases}. \tag{8}$$

$(1,0)$. Its characteristic equation is

$$\lambda^2 + \left[(g-a)+(d-a)\right]\lambda + (g-a)(d-a) = 0. \tag{9}$$

Its stable necessary and sufficient condition is $\begin{cases}(g-a)+(d-a)<0\\(g-a)(d-a)>0\end{cases}$, which is equivalent to

$$\begin{cases}a > d\\ a > g\end{cases}. \tag{10}$$

$\left(0, \dfrac{i-f}{e-h-f+i}\right)$. Its characteristic equation is

$$\begin{aligned}&\lambda^2 + (-e+h+f-i)(ef-ce+hi-2hf+hc+bf-bi)\lambda + \\ &(hf+bi-bf+ce-hc-ie)(i-f)(e-h)(e-h-f+i)^2 = 0.\end{aligned} \tag{11}$$

The corresponding stable necessary and sufficient condition is

$$\begin{cases}h > e\\ f > i\\ hf > b(f-i)+c(h-e)+ei\end{cases}. \tag{12}$$

$\left(\dfrac{i-c}{a-g-c+i}, 0\right)$. Its characteristic equation is

$$\begin{aligned}&\lambda^2 + (a-g-c+i)(af-ac-fg+id+2cg-cd-gi)\lambda - \\ &(ai-id-cg+cd-af+fg)(i-c)(a-g)(a-g-c+i)^2 = 0.\end{aligned} \tag{13}$$

The corresponding stable necessary and sufficient condition is

$$\begin{cases}a < g\\ i < c\\ gc > d(c-i)+f(g-a)+ai\end{cases}. \tag{14}$$

$\left(\dfrac{e-b}{a-b-d+e}, \dfrac{a-d}{a-b-d+e}\right)$. Its characteristic equation is

$$\lambda^2 + (-d+e+a-b)(ge-de-bg-ab-dh+ah+2bd)\lambda - (-ge+ae-ah-bd+dh+bg)(a-d)(e-b)(-d+e+a-b)^2 = 0. \tag{15}$$

The corresponding stable necessary and sufficient condition is

$$\begin{cases} d > a \\ b > e \\ bd > h(d-a) + g(b-e) + ae \end{cases}. \tag{16}$$

$\left(\frac{U}{U+V+W}, \frac{V}{U+V+W}\right)$. Its characteristic equation is

$$\lambda^2 + K\lambda + L = 0. \tag{17}$$

Where,

$$\begin{cases} K = \theta \cdot \vartheta \\ \theta = (f-i)(b-e)(a-g) + (c-i)(h-e)(a-d), \\ \vartheta = \pi_1 + \pi_2 + \pi_3 \end{cases} \tag{18}$$

$$\begin{cases} L = \pi_1 \cdot \pi_2 \cdot \pi_3 \cdot \pi_4 \\ \pi_1 = -U = c(h-e) + f(b-h) + i(e-b) \\ \pi_2 = -V = i(a-d) + c(d-g) + f(g-a) \\ \pi_3 = -W = e(a-g) + b(g-d) + h(d-a) \\ \pi_4 = (-ah - hf + ch + hd + ea + ie + ai - af - eg + fg - ib + bf + gb - ce - cg - di - bd + cd)^2 \end{cases}. \tag{19}$$

The corresponding stable necessary and sufficient condition is

$$\begin{cases} d > a, g > a, b > e, h > e, c > i, f > i \\ hf < b(f-i) + c(h-e) + ei \\ gc < d(c-i) + f(g-a) + ai \\ bd < h(d-a) + g(b-e) + ae \end{cases}. \tag{20}$$

4 Coexistence and Mutual Repulsion

Mutual repulsion is a phenomenon that some rest points are exclusively non-stable when the payoff matrix satisfies the stable condition of one else rest point. On the contrary, coexistence is a phenomenon that some rest points can be stable simultaneously under the same payoff matrix. However, coexistence does not mean at all the evolution has multiple solutions such as the situation of multiple Nash equilibriums in classical Game Theory. Whenever the initial condition is specified, the population evolution result is one and only. For the sake of simple expression, the seven rest points are denoted by I~VII by their order in last Section. Their coexistence and mutual repulsion relationships are summarized in Table 1.

Table 1. Coexistence and mutual repulsion relationships

Rest point	Mutual repulsion	Coexistence
I	IV、V、VII	II、III、VI
II	IV VI VII	I、III、V
III	V、VI、VII	I、II、IV
IV	I、II、VII	III、V、VI
V	I、III、VII	II、IV、VI
VI	II、III、VII	I、IV、V
VII	I、II、III、IV、V、VI	N/A

5 Summary

By far there are literatures reporting detailed analytical results about 2×2 symmetrical evolutionary games [3,4], but there lacks of such results for the 3×3 games which is of more representativeness. In this paper, firstly all its seven rest points of a general evolutionary game are worked out; then their stable necessary and sufficient conditions are deduced and specified in details; finally their relationships of coexistence and mutual repulsion are summarized. The results of this paper provide a guideline applicable for the study and design of 3×3 games.

References

1. Lu, Q.S., Peng, L.P., Yang, Z.Q.: Ordinary Differential Equations and Dynamic Systems. Beihang University Press, Beijing (2010)
2. Perko, L.: Differential Equations and Dynamical Systems. Springer, NY (1993)
3. Hofbauer, J., Sigmund, K.: Evolutionary Games and Population Dynamics. University Press, Cambridge (1998)
4. Zeeman, E.C.: Population dynamics from game theory. In: Global Theory of Dynamical Systems. Springer Lecture Notes in Mathematics, vol. 819, pp. 471–497 (1980)

The Parameter Matching Study of Hybrid Time-Share Drive Electric Vehicle on a High Mobility Vehicle

Yan Sun[1,2], Shishun Zhu[1], Sujun Luo[3], Jianwen Li[1], and Peng Ye[4]

[1] Military Transportation University Department of Automobile Engineering, Tianjin, China
[2] 66417 Troops , Xuanhua, China
[3] Military Transportation University Department of Military Logistics, Tianjin, China
[4] Military Transportation University Department of Scientific Research

Abstract. A parameters matching scheme for HEV powertrain is proposed with concurrent consideration of driving conditions, power flow control strategy and power performance.The scheme is then applied to the parameters matching design of the powertrain of hybrid time-share drive electric vehicle on mengshi. In the investigation, the power and the torque requirements for the powertrain were calculated, and parameters for powertrain components were designed preliminarily. Finally, the hybrid electric time-share drive vehicle was modeled in advisor 2002. The simulation results show that the parameter design of each powertrain component is reasonable.

Keywords: hybrid electric vehicle, advisor 2002, parameter matching, simulation.

1 Introduction

The optimization of the parameters matching of Hybrid system key parts can improve the performance of the vehicle and bring all the superiority of the hybrid cars into play . Hybrid system parameters optimization matching is more than a goal and it is a multivariate optimization problem. Optimization goal are one or several portfolio of the vehicle dynamic performance, fuel economy and emission performance. Variables include engine parameters (power, quality), motor parameters (power, rated speed, the highest speed, quality and volume), the battery parameters (the battery unit number, capacity, quality, volume, unitage power and unitage energy) and transmission system parameters (speed). In addition, the running condition of and control strategy of the vehicle also has very big effect on parameters optimization matching of complete vehicle dynamic system. In a hybrid car, there are very much literature about the hybrid system parameters matching, but most of the documents act in accordance with only hybrid car performance index of the hybrid system parameters matching, then with the simulation analysis. The optimum matching method of the hybrid power system parameters ignores the driving road conditions and control strategy for the comprehensive influence on the performance of the vehicles. This paper will fully consider the influence of running condition and control strategies on hybrid system matching parameters.

D. Jin and S. Lin (Eds.): Advances in CSIE, Vol. 1, AISC 168, pp. 203–208.
springerlink.com © Springer-Verlag Berlin Heidelberg 2012

2 The Simplified Model of Driving Power

The actual running condition of vehicles is relatively complex. For convenience of analysis, in this paper the running mode will be simplified to four typical working conditions as shown in figure 1 below. And typical working conditions can be divided into two categories: steady-state condition pi transient condition pj. Therefore, according to the vehicle dynamics, the vehicle driving drive demand for power is following:

$$p_r = \begin{cases} p_i & \text{steady-state condition} \\ p_j & \text{transient condition} \end{cases} \tag{1}$$

In Eq. (1),Pr is drive power demand, Pi is steady-state power demand and Pj is transient power demand, kW.

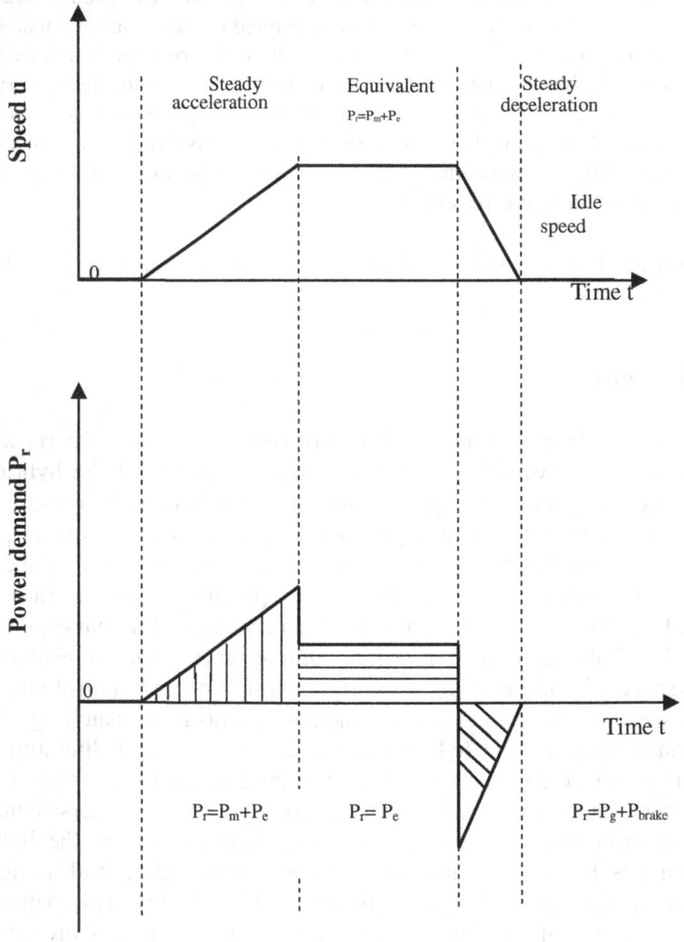

Fig. 1. Drive power allocation of typical working conditions

3 MengShi Hybrid Suvs Parameters Matching

The parameters of hybrid transmission system match preliminary on the basis of the vehicle dynamic performance requirements. The total required power and the maximum torque are calculated according as the specific hybrid transmission structure, machinery property requirements of engine, transmission, and motor and voltage and capacity requirements of the battery pack. Hybrid tactics to vehicle power transmission system first should meet dynamic performance requirements of the vehicle. The vehicle dynamic performance is comprehensively evaluated as the following several major indexes:

1)The highest speed of vehicle;
2) the biggest PaPoDu of vehicle;
3) the acceleration time of vehicle.

3.1 The Parameters of the Engine

The engine is the main power source in military hybrid suv and drive motor and ISG motor are auxiliary power sources. The power matching principle of the three parameters is that the power of engine just meets the daily power demand. So the most power of the engine should be provided by Eq. (1) and Eq. (2).

$$p_1 = \frac{1}{\eta_T}\left(\frac{Gf}{3600}u_{\max} + \frac{C_D A}{76140}u_{\max}{}^3\right) \tag{1}$$

$$p_2 = \frac{1}{\eta_T}\left(\frac{Gf\cos\alpha}{3600}u + \frac{G\sin\alpha}{3600}u + \frac{C_D A}{76140}u^3\right) \tag{2}$$

In Eq.(1)and Eq.(2) , η_T is transmission efficiency of vehicle ; f is rolling resistance coefficient ; C_D is air resistance coefficient ; A is vehicle windward acreage ; α t is PaPoDu for vehicles ; G is tactical vehicle running weight.

According to the technical requirements, on the flat road the maximum speed of the vehile reachs 105 km/h, the corresponding resistance power of engine is about 63.24 kW. Under minimum speed stability the vehile climbs 60% slope, the corresponding resistance power of engine is 23.87 kW. Therefore,reserved 8 kW to charge the generator, engine power needs for 71.24 kW corresponding top speed 105 km/h.

3.2 The Parameters of Drive Motor

The matching of the motor power must meet peak driving power demand of vehicles in certain reference condition. The motor can provide drive power in pure electric conditions, otherwise the vehicles will fails to keep up with demand condition speed and the acceleration time of the vehicle doesn't reach the drivers expect in pure electric mode. Drive motor should appropriately adjust drive motor power demand according to different control strategies and pure electric power requirements and energy can maximum be reclaimed in the braking.

The parameters of drive motor include motor power rating p_e, the most high power p_m, the rated speed N_e, the maximum speed N_m, the rated torque T_e, the maximum torque T_m and assurance λ.

The motor power of hybrid tactics vehicle is mainly used for the silence run of the vehicle in the depth region of the battlefield in addition to power auxiliary for vehicle acceleration and climbing. On the basis of the biggest range of conventional weapons, infrared equipment investigation distance and raider vehicle attack deep, the hybrid suvs selected driving speed of deep silence 60 km/h. sure.

By the calculation, drive motor parameters as is shown in table 1.

Table 1. Motor parameters of mengshi hybrid suv

name	parameters	Value
motor	Rating power	20kw
	maximum power	30kw
	rating revolution	750 r/min
	maximum revolution	3000 r/min
	rating torque	150N.m
	maximum torque	200N.m

3.3 Power Battery Parameters

For calculating battery energy, the trip mileage must be clear. Hybrid tactics vehicles should have certain pure electric drive capacity, namely,the vehicle has been in stealth operation in the front line depth attack and its pure electric trip range isn't smaller than the depth of the attack. On the basis of vehicle the biggest range of conventional weapons, infrared equipment investigation distance and the depth of the launch attacks, pure electric trip mileage should be no less than 30 km.In this article , we choose 30 km to calculate.

1) The battery number

According to the voltage demand, the battery number n_s is confirmed as following

$$n_s = \frac{U}{U_0} \tag{3}$$

In Eq.(3), U is the voltage of the batteries ; U_0 is the voltage of single battery.

2) The battery power

In some cases, hybrid tactics vehicles need have great discharge battery power in very short time. the power output is following

$$p_{\max} = U_c I_c \tag{4}$$

In Eq.(4), U_c is the discharge voltage of the batteries ; I_c is the discharge current.

3) The battery capacity

Based on the power balance equation hybrid military vehicle need power by Eq. (5) to compote in pure electric road.

$$p=\frac{1}{\eta_T}(\frac{Gf}{3600}u+\frac{C_DA}{76140}u^3) \tag{5}$$

To The military hybrid suvs driving system the total pure electric power needed as following:

$$W\geq\frac{p\cdot T}{\eta_t}=\frac{p}{\eta_t}\cdot\frac{s}{u} \tag{6}$$

By the calculation, power battery parameters are shown in table 2.

Table 2. Battery parameters

assembly	parameters	Value
battery	type	lithium battery
	voltage	320v
	capability	55A.h
	energy	12.5kw.h

The online version of the volume will be available in LNCS Online. Members of institutes subscribing to the Lecture Notes in Computer Science series have access to all the pdfs of all the online publications. Non-subscribers can only read as far as the abstracts. If they try to go beyond this point, they are automatically asked, whether they would like to order the pdf, and are given instructions as to how to do so.

Please note that, if your email address is given in your paper, it will also be included in the meta data of the online version.

4 Vehicle Performance Simulation

4.1 Control Strategy

Hybrid cars control strategy is relatively complex and will affect the simulation results for the different control strategy.In this paper charge of battery needed to keep in a interval range.This belongs to charge keep type and the final energy comes from the engine.so battery control strategy is auxiliary method in this paper .The basic idea is as follows:

(1) the engine work alone when the power demand is in optimization interval of the engine;

(2) When the vehicle demand greater than the engine maximum power , drive motor provide power;

(3) When need to pure electric run, motor work alone to drive vehicle and the engine shut;

(4) when SOC below the lowest setting, the engine will provide energy to battery;
(5) when braking, drive motor is in power generation state, recycling energy to the battery.

4.2 The Simulation Analysis

By ADVISOR2002 application software and the development control strategy, the MengShi hybrid suvs power performance and the simulation of economic behavior are completed.the results satisfy the design index such as table 3.

Table 3. Performance parameters of MengShi hybrid suvs

Performance parameters	hybrid driving mode
fuel consume	9.8L. (100km) -1
Maximum speed	112 km.h-1
maximum climbing capability	61%
accelerate time (0-80km/h)	21.8s
accelerate time (60-90 km/h)	16.4s

5 Summary

This paper puts forward parameters matching method of double-axis parallel hybrid vehicle power system considering running condition, control strategy and and vehicle dynamic performance. This method guidance completed mengshi hybrid time-share drive tactics vehicle design.In MATLAB/Simulink environment, using ADVISOR simulation software mode of auxiliary control strategy simulation model is established based on batteries work and the simulation results show that the matching parameters of dynamic assembly parts can satisfy the requirements of vehicle performance. This mothod provide the basis for product development.

References

1. Montazeri-Gh, M., Poursamad, A., Ghalichi, B.: Application of genetic algorithm for optimization of control strategy in parallel hybrid electric vehicles. Journal of the Franklin Institute 343, 420–435 (2006)
2. Langari, R., Won, J.-S.: A Driving Situation Awareness-Based Energy Management Strategy for Parallel Hybrid Vehicles, SAE PAPER 2003-01-2311
3. Salmasi, F.R.: Designing control strategies for hybrid electric vehicles. In: Tutorial Presentation in EuroPes 2005, Benalmadena, Spain, June 15-17 (2005)

An Improved Multiple Attributes Decision Making Based on Ideal Solution and Project

HouXing Tang[1,*] and WeiFeng Lu[2]

[1] Business Administration College, Nanchang Institute of Technology, Nanchang, China
tanghouxing@yahoo.com.cn

[2] Academic Affairs Division, Nanchang Institute of Technology, Nanchang, China

Abstract. The standard TOPSIS has many defects on the definition of ideal solution and negative-ideal solution, setting of weight and the calculation of relative approachable degree, which leads to rank reversal. Therefore, in this paper, the relative ideal solution and negative-ideal solution will be substituted with absolute ideal solution and negative-ideal solution. And then, the formula of relative approachable degree is improved based on project method. Through an example, we show that the improved method has better stability and consistency.

Keywords: multiple-attribute decision making, ideal solution, project, TOPSIS.

1 Introduction

At present, the multiple-attribute decision making is the hot topic in decision-making science and system engineering. The TOPSIS is a common used method. But standard TOPSIS has many defects. For instance, Yonghong Hu(2002) points out that it will cause illogical result if there are particular samples. The author solves this problem by introduce a virtual negative-ideal solution to improve the formula of relative approachable degree[1]. Gensheng Qiu, Shuimu Zou and Rihua Liu (2005) think that the standard TOPSIS will lead to rank error because of the defect of relative approachable degree[2]. Wei Chen (2005) analyses the cause of rank reversal existing in TOPSIS method and presents a modified method which not only can eliminate rank reversal but can reflect the relative importance of different criteria on the final results[3]. Qiaofeng Fu (2008) introduces an improved method which give a new method of decision making matrix ranking and getting weight. This method simplifies standardization of decision making matrix and calculation of ideal solution and put forward a relative approachable degree equal to old one. At the same time, it calculated the index' weight by optimization model so as to weaken subjective factors[4].

As seen in above literatures, the improvements are focused on three ways: the first is the redefinition of ideal solution and negative-ideal solution; the second is the setting of attributes' weight; the third is redefinition of relative approachable degree. No doubt, we will follow it. The structure of this paper is as follow: The defects of

[*] Corresponding author.

D. Jin and S. Lin (Eds.): Advances in CSIE, Vol. 1, AISC 168, pp. 209–214.
springerlink.com © Springer-Verlag Berlin Heidelberg 2012

standard TOPSIS are exposed and improvements are given in section 2; The effect of improved method is showed by an example in section 3; the conclusion is given in section 4.

2 Some Problems and Improvements of Standard TOPSIS

2.1 Rank Reversal Caused by Unreasonable Selection of Negative-Ideal Solution and Its Improvement

In Fig.1, the x^* is ideal solution and x^0. The x_1 and x_2 are two of candidate alternatives. Especially, we suppose that $|x_1 - x_0| = |x_2 - x_0|$ and $|x^* - x_1| = |x^* - x_2|$. So, we draw a conclusion that the alternative x_1 equals to x_2. However, if we add a new alternative A, the negative-ideal solution will changed from x^0 to A according to the formula (1) and (2).

$$x_j^* = \begin{cases} \max\limits_i x_{ij} & for \quad revenue \\ \min\limits_i x_{ij} & for \quad \cos t \end{cases} \qquad j = 1, \cdots, n \tag{1}$$

$$x_j^0 = \begin{cases} \max\limits_i x_{ij} & for \quad revenue \\ \min\limits_i x_{ij} & for \quad \cos t \end{cases} \qquad j = 1, \cdots, n \tag{2}$$

Obviously, we get $x_2 \succ x_1$. On the contrary, if we add a new alternative B, the negative-ideal solution will change from x^0 to B. Also, we get $x_2 \prec x_1$. So we find the result is inconsistent and rank reversal arises. The reason is that the ideal solution and negative-ideal solution is relative rather than fixed. The rank reversal can be eliminated by fixing the ideal solution and negative-ideal solution. Well, we normalize the decision matrix as follow formulates.

$$r_{ij} = \frac{a_{ij}}{\max\limits_i (a_{ij})} \quad i = 1, \cdots, m; j = 1, \cdots, n \quad \text{for revenue attribute}$$

$$r_{ij} = \frac{\min\limits_i (a_{ij})}{a_{ij}} \quad i = 1, \cdots, m; j = 1, \cdots, n \quad \text{for cost attribute}$$

The attribute value will be between 0 and 1, that is $r_{ij} \in [0,1]$ and the more the better. So, the absolute ideal solution is $x^* = [1,1,\cdots,1]_n^T$ and negative-ideal solution is $x^0 = (0,0,\cdots,0)_n^T$. No doubt, however the alternatives changes, the relative approachable degree of any alternative is fixed. Therefore, it keep the consistency and stability of evaluation.

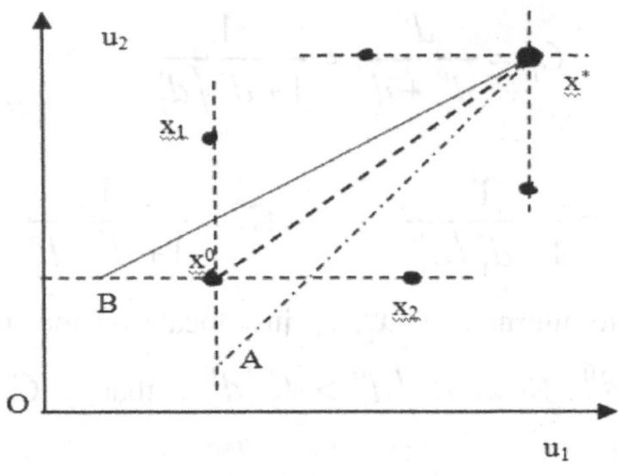

Fig. 1.

2.2 Rank Reversal Caused by Unreasonable Measure and Its Improvement

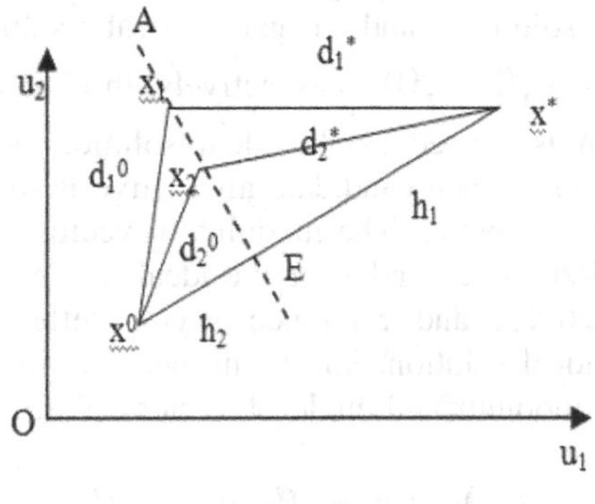

Fig. 2.

In Fig.2, x^0 is the negative-ideal solution and x^* is the ideal solution, the x_1, x_2 are two of candidate alternatives. The distances from x_1 and x_2 to ideal and negative-ideal solution are d_1^*, d_1^0 and d_2^*, d_2^0 respectively. The line AE is vertical with line of x^0 and x^*. The distances from E to ideal solution and negative-ideal solution are h_1, h_2 respectively. According to the TOPSIS ideology, whatever the line AE moves along with line of x^0x^* (the relative position of x_1, x_2 keep unchanged), we will get the same result $x_1 \prec x_2$ because of $d_1^* > d_2^*$. However, the author Hu Yonghong(2002) had proved that according to formula (3), the C_1^*, C_2^* is determined by the ratio of h_1, h_2. That is to say, if $h_2 < h_1$, then $x_1 \succ x_2$; if $h_2 > h_1$, then $x_1 \prec x_2$; if $h_2 = h_1$, then $x_1 \sim x_2$. This indicates there is rank reversal. And the author introduced a virtual negative-ideal solution to solve this problem. According to the formula (3), we get

$$C_i^* = \frac{d_i^0}{d_i^0 + d_i^*} = \frac{1}{1 + d_i^* / d_i^0}$$

then

$$C_1^* = \frac{1}{1 + d_1^* / d_1^0} \qquad C_2^* = \frac{1}{1 + d_2^* / d_2^0}$$

We find if the candidate alternatives x_1, x_2 just locate on the line of $x^0 x^*$, and if $d_1^* > d_2^*$, then $d_1^0 < d_2^0$. Now, $d_1^* / d_1^0 > d_2^* / d_2^0$, that is $C_1^* < C_2^*$, so we get $x_1 \prec x_2$ and vice versa.

In another words, only the relative poison of x_1, x_2 keeps unchanged, the result will keep consistent. Under this case, the results are the same according to formula (3) and figure illustration. So, we can draw a conclusion that the formula (3) in standard TOPSIS has limitation. In this paper, a project method is introduced to improve it.

Because the ideal solution and negative-ideal solution are fixed as $x^* = [1, 1, \cdots, 1]_n^T$, $x^0 = (0, 0, \cdots, 0)_n^T$ respectively, the line of from ideal solution to negative-ideal solution is treated as the ideal solution vector which is called reference vector in this paper. Each candidate alternative is treated as a vector. The closer to reference vector, the better. The modulus of vectors measures the distance between the candidate alternatives and negative-ideal. As well, the angel between vector (candidate alternatives) and reference vector reflects the distance from candidate alternatives to ideal solution. So, it can measure the relative approachable degree by combining the modulus and angle of vectors. Well, the project method is introduced as follows.

Let $\alpha = (\alpha_1, \alpha_2; \cdots, \alpha_n)$ and $\beta = (\beta_1, \beta_2, \cdots, \beta_n)$ be two vectors, and the definition of cosine of angel between vector α and β as follow:

$$\cos(\alpha, \beta) = \frac{\sum_{j=1}^{n} \alpha_j \beta_j}{\sqrt{\sum_{j=1}^{n} \alpha_j^2} \bullet \sqrt{\sum_{j=1}^{n} \beta_j^2}}$$

And the definition of project from α to β is as follow:

$$\Pr \mathrm{j}_\beta(\alpha) = \frac{\sum_{j=1}^{n} \alpha_j \beta_j}{\sqrt{\sum_{j=1}^{n} \alpha_j^2} \bullet \sqrt{\sum_{j=1}^{n} \beta_j^2}} \sqrt{\sum_{j=1}^{n} \alpha_j^2} = \frac{\sum_{j=1}^{n} \alpha_j \beta_j}{\sqrt{\sum_{j=1}^{n} \beta_j^2}}$$

Generally speaking, the more $\Pr \mathrm{j}_\beta(\alpha)$, the nearer α and β.

Let

$$\Pr \mathrm{j}_{x^*}(x_i) = \frac{\sum_{j=1}^{n} x_j^* x_{ij}}{\sqrt{\sum_{j=1}^{n} (x_j^*)^2}} \quad i = 1, 2, \cdots, m$$

And, the more $\Pr \mathrm{j}_\beta(\alpha)$, the better the alternativc x_i.

3 An Example

Now, we construct an example to test the improved method. Suppose that alternative set $X=\{X_1,X_2,X_3,X_4,X_5,X_6,X_7\}$, the attribute set is $U=\{U_1,U_2,U_3,U_4\}$ and the U_1,U_2,U_3 are revenue attributes and U_4 is cost attribute. The attribute values are listed in Tab.1. X_8 means that we add a new alternative.

Table 1. A sample decison matrix

	U_1	U_2	U_3	U_4
X_1	92	30	331	47
X_2	96	26	332	41
X_3	95	24	329	50
X_4	97	28	356	67
X_5	98	28	378	40
X_6	94	28	352	41
X_7	96	26	352	41
*X_8	98	30	378	300

Now, we calculate the relative approachable degree by standard TOPSIS and improved method respectively. The results are as follows:

For standard TOPSIS:

D_1=[0.6938,0.7344,0.5155,0.2324,0.8919,0.8323,0.7697];

The rank of alternatives is $x_5 \succ x_6 \succ x_1 \succ x_7 \succ x_2 \succ x_3 \succ x_4$

D_2=[0.9326,0.9197,0.8879,0.8890,0.9690,0.9527,0.9320,0.1053];

The rank of alternatives is : $x_5 \succ x_6 \succ x_1 \succ x_7 \succ x_2 \succ x_4 \succ x_3 \succ x_8$

Obviously, the $x_3 \succ x_4$ becomes the $x_4 \succ x_3$. It indicates that there is rank reversal.

For improved method:

D_1=[0.4582,0.4625,0.4300,0.4327,0.4917,0.4749,0.4691];

The rank of alternatives is : $x_5 \succ x_6 \succ x_7 \succ x_2 \succ x_1 \succ x_4 \succ x_3$

D_2=[0.4582,0.4625,0.4300,0.4327,0.4917,0.4749,0.4691,0.3917];

The rank of alternatives is : $x_5 \succ x_6 \succ x_7 \succ x_2 \succ x_1 \succ x_4 \succ x_3 \succ x_8$

There is no rank reversal. This results show that the improved method be much more consistent and stability than standard TOPSIS.

4 Conclusions

For the defects of standard TOPSIS, we improve it from the perspectives of normalization of decision matrix, redefinition of ideal solution and negative-ideal solution, setting of weights and redefinition of relative approachable degree. The result shows that the improved method has much better consistent and stability.

Acknowledgments. The authors thank the financial support of Soft Science Research Project of Jiangxi Provincial Department of Science and Technology, grant 2009DR05100 and The Youth Fund Project of College Humanities and Social Sciences of Ministry of education, grant 11YJC630193.

References

[1] Hu, Y.: The Improved Method for TOPSIS in Comprehensive Evaluation. Mathematics in Practice and Theory 32(4), 572–575 (2002) (in Chinese)

[2] Qiu, G., Zou, S., Liu, R.: Improvement of the TOPSIS for multiple criteria decision making. Journal of Nanchang Institute of Aeronautical Technology (Natural Science) 19(3), 1–4 (2005) (in Chinese)

[3] Chen, W.: On the Problem and Elimination of Rank Reversal in the Application of TOPSIS Method. Operation Research and Management Science 14(3), 39–43 (2005) (in Chinese)

[4] Fu, Q.: Study on TOPSIS. Journal of Xi'An University of Science and Technology 28(1), 190–193 (2008) (in Chinese)

Unequal Error Protection for Image Transmission Based on Power Water-Filling Policy over MIMO Channels

Wu-ping Zhang[1], Xiao-rong Jing[1,2], Jun Li[1], Zu-fan Zhang[1,2], and Qian-bin Chen[1,2]

[1] School of Communications and Information Engineering,
Chongqing University of Posts and Telecommunications
[2] Chongqing Key Laboratory of Mobile Communication (CQUPT)
Chongqing, China 400065
zhangwuping315@163.com

Abstract. In this paper, a power water-filling scheme for unequal error protection (UEP) based on set partitioning in hierarchical trees (SPIHT) coded image transmission over multiple input multiple output(MIMO) channels is proposed. SPIHT coded image information is divided into different sub-streams according to their different importance, which is allocated to different channels. Then unequal error protection (UEP) of image with singular value decomposition and power water-filling policy is achieved. The simulation results show that the quality of reconstructed image of using UEP is better than equal error protection (EEP).

Keywords: SPIHT, MIMO, singular value decomposition, power water-filling.

1 Introduction

The design of efficient communication systems for progressive transmission[1],[7] of multimedia over wireless channels has recently attracted lots of interests because of the increasing demands for wireless applications[4]. The output of the image compression based on SPIHT coding is decomposed into different sub-streams with different importance in terms of image progressive performance[1]. The first part can be used as the base sub-stream including the most important information to reconstruct the image, the second part can be used as enhance sub-stream including the less important information to improve the image resolution and so on. The MIMO technology, which makes use of multiple antennas at both the transmitter and receiver ends, has been shown to be capable of achieving extraordinary throughput without additional power or bandwidth consumption and providing high bit rate for wireless communication systems[3][4]. Some literatures have proposed the application of the image transmission over MIMO channels, such as[1],[2],[5]. In[1], a low-complexity technique, which preserves the progressive transmission property, and is simple to implement in practice, is proposed . Following that significant work[2], an alternative exposition of subset partitioning technique is developed. The optimal design of a joint source-channel coder for a given embedded source bit-streams is considered in [5].

In this paper, we propose a power water-filling scheme for different levels of error protection to different sub-streams. The power water-filling scheme can improve the

D. Jin and S. Lin (Eds.): Advances in CSIE, Vol. 1, AISC 168, pp. 215–221.
springerlink.com © Springer-Verlag Berlin Heidelberg 2012

quality of the reconstructed image by combining the SPIHT-based source coding [2] with Turbo channel coding. The proposal carries out the UEP strategy by allocating different transmitted power to different sub-streams with different importance. This UEP solution[5],[6] improves the quality of the reconstructed image effectively.

2 SPIHT Source Encoding

In this section, the SPIHT encoding can be described briefly.

Before the SPIHT encoding, three lists are introduced to store the important information, which named list of insignificant pixels(LIP), list of significant pixels(LSP), list of insignificant sets(LIS). Also three sets of coordinates are defined. O(i,j) denotes the set coordinates of all offspring of node (i,j); D(i,j) denotes the set of all descendants of the node(i,j); and L(i,j) denotes the set excluding its immediate child nodes, L(i,j)= D(i,j)- O(i,j).

The coding processes of SPIHT algorithm can be illuminated as follows,

Initialization. Initialize the output $n = [\log 2(\max\{|c_{i,j}|\})]$ and the ordered list. Empty the LSP and add starting root coordinates to LIP and LIS;

Sorting Pass. The purpose of the sorting pass is to code significant coefficient of current bits;

Refinement Pass. For each entry (i, j) in the LSP, output the nth most significant bit of $|c_{i,j}|$ excepting those included in the last sorting pass;

Quantization-step Update. Decrement n by 1 and go to the step of sorting pass.

The algorithm stops at the desired rate or distortion. Normally, good quality of images can be recovered after a relatively small fraction of the pixel coordinates are transmitted.

3 System Model

The system model with N_t transmit antennas and N_r receive antennas is considered in this paper. The block diagram of the image transmission system is illustrated in Fig.1.

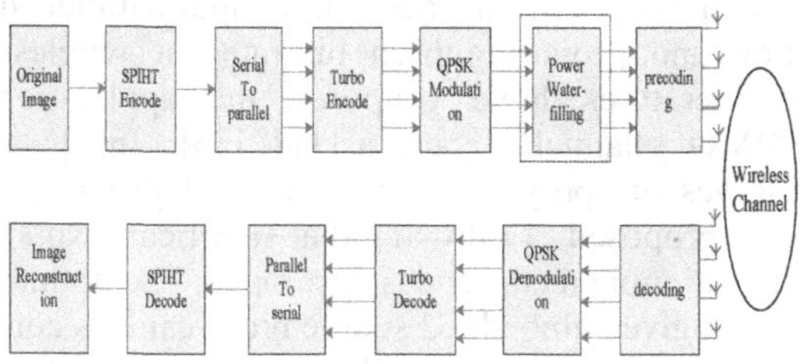

Fig. 1. Block diagram of the system

We assume that the channel is Rayleigh flat fading channel. **H** is the complex channel matrix, where $h_{i,j}$ is the complex gain coefficient of the channel from the jth transmit antenna to the ith receive antenna.

The input-output relationship of this system is given by

$$\mathbf{y} = \mathbf{Hx} + \mathbf{w} \tag{1}$$

where **x** is the $N_t \times 1$ transmitted signal vector, **y** is the $N_r \times 1$ received signal vector, **w** is the received noise vector, which is additive white Gaussian noise with zero mean and variance σ^2.

At the transmitting end, the original image is coded with SPIHT algorithm [2], then the SPIHT encoded bit-stream is divided into several parts of equal length. The first sub-stream includes lower frequency information, which is the most important part of the image. The remainder sub-streams contain higher frequency information, which is less important part of the image. Each of the sub-streams is coded with Turbo code and modulated with QPSK. Furthermore, the outputs of the modulator are precoded and finally transmitted through transmitting antennas.

At the receiving end, it should perform such process as the inverse transformation of precoding, demodulation, turbo decoding, SPIHT decoding and then reconstruct the original image.

4 UEP Based on Power Water-Filling Policy

In our work, we carry out the UEP by the power water-filling policy. The process can be summarized as follows.

Taking the singular value decomposition (SVD) of the channel matrix,

$$\mathbf{H} = \mathbf{USV}^H \tag{2}$$

where **U** and **V** are the unitary matrix, and **S** can be represented as

$$\mathbf{S} = diag(\sqrt{\lambda_1}, \sqrt{\lambda_2}, \cdots \sqrt{\lambda_r}) \tag{3}$$

The elements satisfy the following relationship

$$\sqrt{\lambda_1} \geq \sqrt{\lambda_2} \geq \cdots \geq \sqrt{\lambda_r} \tag{4}$$

After SVD, by multiplying the (1) with $\mathbf{U}^{\mathrm{H}}$ and considering the preprocessing of the precoding matrix **V**, then we have,

$$\begin{aligned}\tilde{\mathbf{y}} &= \mathbf{U}^H \mathbf{HVx} + \mathbf{U}^H \mathbf{w} \\ &= \mathbf{Sx} + \tilde{\mathbf{w}}\end{aligned} \tag{5}$$

The formula above can be further expressed as,

$$\tilde{\mathbf{y}}_i = \sqrt{\lambda_i}\mathbf{x}_i + \tilde{\mathbf{w}}_i \quad i = 1, 2, \ldots, r \tag{6}$$

From formula (6), we can find that the MIMO transmission channel is divided into r parallel sub-channels, which are uncorrelated with each other. Here r is the rank of the channel matrix. Thereby the different level information can be transmitted through these sub-channels respectively. Although the image reconstruction quality through such transmission mode is acceptable at the region of the high SNR, the performance of the recovering quality of the image can be further improved by UEP strategy with power water-filling solution.

With the power water-filling policy, more transmission power is allocated to the better channels to maximize the achievable data rate and capacity of the communication system.

The power allocated to each sub-channel is:

$$\gamma_i = \max\{(\mu - \frac{1}{\lambda_i}), 0\} \tag{7}$$

$$\sum_{i=1}^{r} \gamma_i = P_t \tag{8}$$

where P_t is the total transmission power.

The more detailed process of power water-filling algorithm is,

Step1. Let p=1 calculate

$$\mu = \frac{1}{r-p+1}[P_t + \sum_{1}^{r-p+1} \frac{1}{\lambda_i}] \tag{9}$$

Step2. Calculate the sub-channels power,

$$\gamma_i = \mu - \frac{1}{\lambda_i} \tag{10}$$

where $i = 1, 2, \ldots, r$;

Step3. If the channel energy with minimum gain is negative, assume $\gamma_{r-p+1} = 0$, then $p = p+1$, and return to Step1, Until any γ_i is non-negative and $\sum_{i=1}^{r} \gamma_i = P_t$, the process is finished.

5 Experimental Results

In our experiment, we use 512×512 gray-scale 'Lena' image as the test image with 9-level wavelet decomposition. Four transmitting and receiving antennas are employed in the system. The result will be evaluated by bit error rate (BER) and peak-signal-to-noise-ratio (PSNR) of the reconstructed image.

The PSNR of an image can be calculated by the following formula:

$$PSNR = 10\log_{10}(K^2 / MSE)(dB) \tag{11}$$

where MSE is the mean square error between the original image and the reconstructed image, while K is the maximal gray value of the image.

Experiment1. BER comparison between UEP and EEP

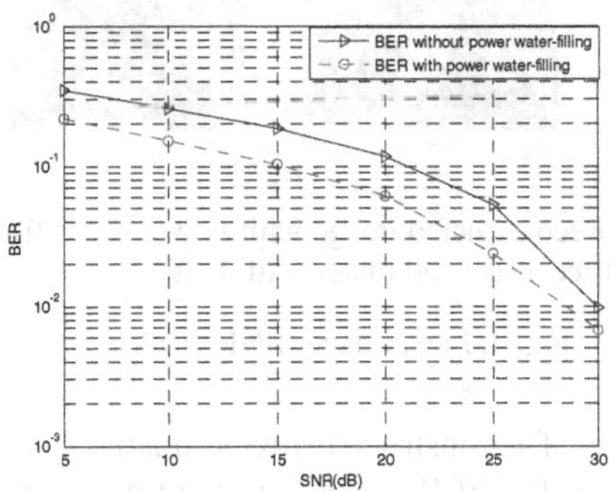

Fig. 2. BER comparison between UEP and EEP

Fig.2 shows the BER performance of the proposed UEP scheme with power water-filling algorithm compared versus the EEP scheme without power water-filling algorithm. It is clearly seen that the BER of UEP is lower than the EEP scheme at the same channel SNR. That is, the performance of BER is better than the latter.

Experiment2. PSNR comparison between UEP and EEP;

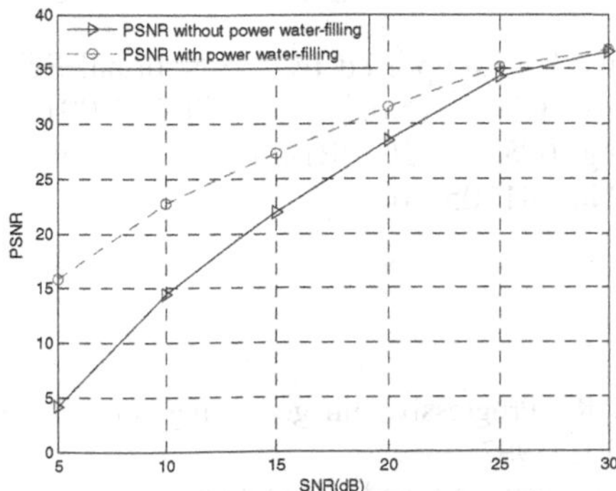

Fig. 3. PSNR comparison between UEP and EEP

From Fig.3, the PSNR improvement is achieved by the proposed UEP scheme with power water-filling against the EEP scheme without power water-filling. Especially when the channel condition is not so good, the gap between the two schemes is more obvious. With the increase of channel SNR, the PSNR performance of power water-filling algorithm will gradually get close to the EEP scheme.

Experiment3. The performance of reconstructed image;

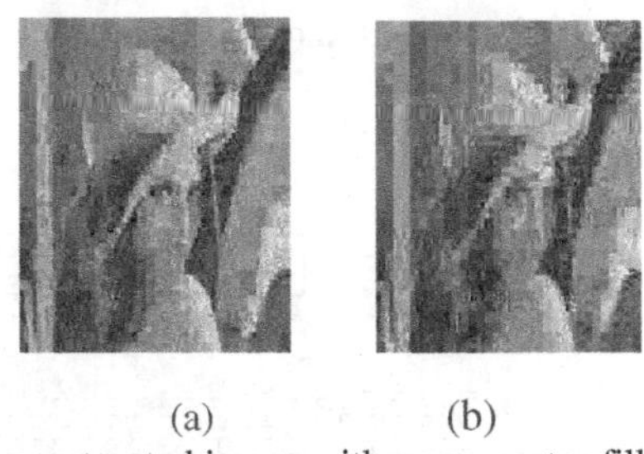

(a) (b)

(a)Reconstructed image with power water-filling
(b)Reconstructed image with EEP

Fig. 4. Lena image (SNR=10dB)

Fig.4 gives the comparison of reconstructed image quality between the UEP scheme with power water-filling and the EEP scheme without power water-filling. When the channel SNR=10dB, the PSNR of reconstructed image of the UEP scheme is much higher than that of EEP scheme, thereby the reconstructed image is more clear than that of EEP scheme.

6 Conclusions

In this paper, we proposed a UEP of SPIHT-Coded Images by using power water-filling scheme. From the experiment results, the new UEP scheme can decrease BER effectively and improve the recovering quality of the original image significantly.

Acknowledgment. The work is supported by the National Science and Technology Major Special Project of China (No. 2011ZX03003-001-01), Natural Science Foundation of Chongqing (CSTC, 2010BB2417) and Foundation of Chongqing Educational Committee (No.KJ110526).

References

1. Sherwood, P.G., Zeger, K.: Progressive image coding for noisy channels. IEEE Signal Processing Lett. 7, 188–191 (1997)
2. Said, A., Pearlman, W.A.: A new, fast, and efficient image codec based on set partitioning in hierarchical trees. IEEE Trans. Circuits Syst. Video Techol. 6(3), 243–250 (1996)
3. Teletar, E.: Capacity of multi-antenna Gaussian Channels, AT&T-Bell Labs, Internal TechMemo (1995)
4. Foschini, G.J., Gans, M.J.: On limits of wireless communication in a fading environment when using multiple antennas. Wireless Pers. Commun. 6(3), 311–335 (1998)

5. Chande, V., Farvardin, N.: Progressive Transmission of Images over Memoryless Noisy Channels. IEEE J. Selected Areas in Comm. 18, 850–860 (2000)
6. Hamzaoui, R., Stankovic, V., Xiong, Z.: Fast Algorithm for Distortion-Based Error Protection of Embedded Image Codes. IEEE Trans. Image Processing 14(10), 1417–1421 (2005)
7. Said, A., Pearlman, W.A.: Image Compression using the Spatial-Orientation Tree. In: Proc. IEEE Int. Symp. Circuits and Syst., pp. 279–282 (May 1993)

The Delay of State Detection in Wireless Monitoring System Based on ZigBee

Renshu Wang, Yan Bai, and Jianjun Zhao

School of Control and Computer Engineering
North China Electric Power University, 102206, Beijing, P.R. China
{wangrenshu_1,by_ncepu,rainaider}@126.com

Abstract. This paper introduces the application of wireless sensor networks (WSNs) based on ZigBee to monitor the states of device in the workshop of chemical water treatment in power plant. For the sleep mechanism is adopted, the delay is introduced at the same time. To limit the delay in the permit range, the analysis of the delay is carried out with data collected from the practical system. And an appropriate work mode of wireless nodes is designed to satisfy the requirement of the process production and save the energy consumption.

Keywords: wireless networks, ZigBee, delay, energy consumption.

1 Introduction

In recent years, with the rapid development of microprocessor and wireless technology, the application of wireless sensor networks (WSNs) in industrial field will inevitably become a direction of research. And ZigBee is considered as one of the most suitable technology for industrial application. There are many merits of ZigBee such as low energy consumption, ease of deployment, scalability, reliable data transmission and low cost [1].

In our research, the network of ZigBee nodes is applied to monitor the state of the device in the sequence control system of chemical water treatment in power plant. The architecture of wireless monitoring system with wireless networks is shown in Fig. 1. The states of the valves can be detected through the switches attached to the Pneumatic valve and the states of the motors are detected by the relays in the motor control cabinets. All the detected states are transmitted to base station by the ZigBee nodes deployed in the field. And the base station changes the format of the data in wireless network to the format of profibus. Then, the data is forwarded to controller. And at last, the states of the devices in field are displayed in the operator station.

The rest of the paper is organized as follows. In Section 2, the problem caused by delay is described in detail while introducing the wireless technology into the monitoring system. Section 3 presents the delay analysis in theory of the state detection. And in Section 4, according to the collected data from the field, the solution is derived. At last, the conclusion is presented in Section 5.

D. Jin and S. Lin (Eds.): Advances in CSIE, Vol. 1, AISC 168, pp. 223–228.
springerlink.com © Springer-Verlag Berlin Heidelberg 2012

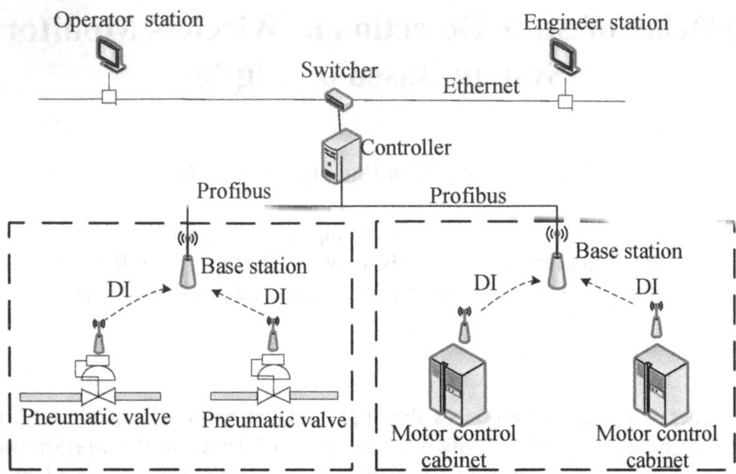

Fig. 1. The architecture of wireless monitoring system with wireless networks

2 Problem

In traditional cable system, because the power of whole system is supplied by cables and there is no necessary to consider the power consumption, the detection of the states can be carried out in small period with little delay introduced. However, in wireless monitoring system, the nodes attached to the device in field, is powered by batteries. Besides the inherent problems of wireless networks [2], for the ZigBee nodes adopt sleep mechanism to extend the life of the nodes, the sleep mode is adopted while there is no business to tackle with and nodes are woken by a timer [3]. And there is delay for the nodes to wake up to detect the states of the devices which will degrade the performance of the monitoring system greatly. So there should be a settlement for this problem.

The work process of the nodes is described in Fig. 2. Work duration of t_w and sleep duration of t_s are included in a period T. Once the change happens in the work duration, the nodes can detect it. However, when the state of device changes in the sleeping period, such as E_2, it has to wait until the next working period to detect the state. Thus, the delay is introduced.

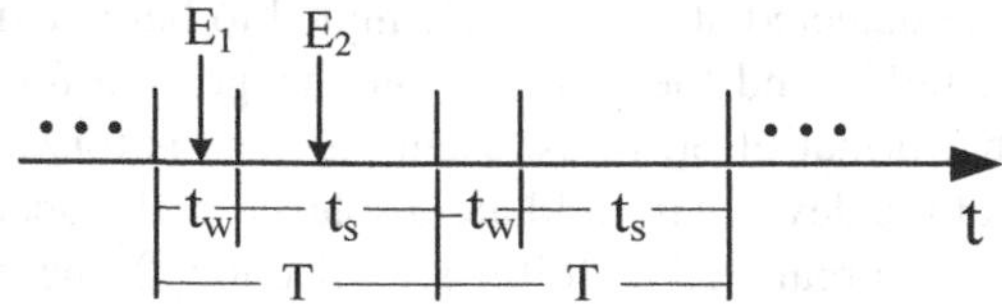

Fig. 2. The work process of wireless nodes.

In Fig. 3, it shows the states of two conditions that the designed state coordinates with the requirement of the production process and the state sampled with delay varies the designed one. And we can see $\Delta_1=t_{n+1}-t_n$, $\Delta_2=t_{p+1}-t_p$. Once Δ_2 is greater than Δ_1 too much, there will be an accident. In [4], it also refers the problem caused by the delay of wireless network in the binary control system. So, what we focus on is how to save the energy of the wireless node and detect the state in permit delay which meets the requirement of the production process at the same time.

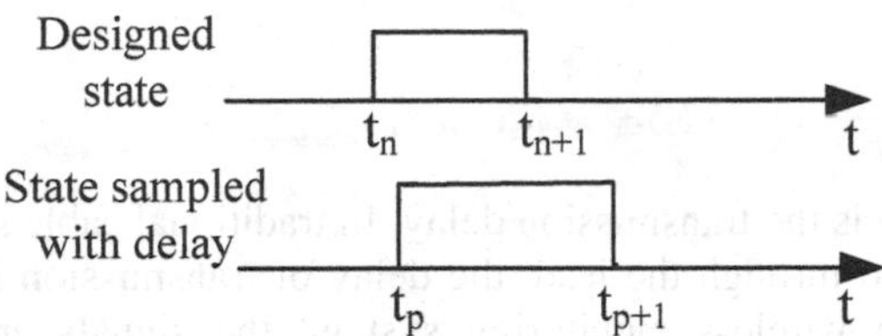

Fig. 3. The states of two conditions

3 Delay Analysis

Suppose that probability density of the state changes in every period is f(t) and f(t) obeys normal distribution, expressed as [5]:

$$f(t)=\frac{1}{\sqrt{2\pi}}e^{-\frac{(t-nT-\mu)^2}{2\delta^2}} \qquad nT<t<(n+1)T \tag{1}$$

Where n means the nth period of work process, μis the mean value and δ is the variance. The expectation of the delay caused by sleep can be derived:

$$E(t)=\int_{t_w}^{T} f(x)(T-x)dx = T\int_{t_w}^{T} f(x)dx - \int_{t_w}^{T} xf(x)dx \tag{2}$$

For t_w is greatly less than T, the approximation can be made:

$$T\int_{t_w}^{T} f(x)dx \approx T\int_{0}^{T} f(x)dx = T \tag{3}$$

$$\int_{t_w}^{T} xf(x)dx \approx \int_{0}^{T} xf(x)dx = \mu \tag{4}$$

So we have:

$$E(t) = T - \mu \tag{5}$$

And the variance is:

$$D=E(t^2)-[E(t)]^2$$

$$=\int_{t_w}^{T} f(x)(T-x)^2 dx-[\int_{t_w}^{T} f(x)(T-x)dx]^2$$

$$=T^2\int_{t_w}^{T} f(x)dx+\int_{t_w}^{T} x^2 f(x)dx-2T\int_{t_w}^{T} xf(x)dx-[\int_{t_w}^{T} Tf(x)dx-\int_{t_w}^{T} xt(x)dx]^2 \quad (6)$$

$$\approx \delta$$

Another kind of delay is the transmission delay. In traditional cable system, the binary signals are transmitted through the lead, the delay of transmission is so small that it can be neglected. In wireless monitoring system, the signals are transmitted by wireless channel with delay introduced. Because ZigBee adopts CSMA/CA in the MAC protocol, the delay of each node accesses wireless channel is stochastic [6]. And the end-to-end delay is nearly 15ms. Compared to the delay caused by sleep which is in seconds, the transmission delay is so small that it can be neglected.

So the problem is how to select the value t_s to derive the E(t), meeting the requirement of the production process and energy consumption as possible. To settle this problem, the parameters in f(t) should be estimated first.

4 Analysis with Collected Data

In this project, we build three wireless networks in different channel bands. And the nodes in each network have the fixed work duration to finish the designated tasks, but different sleep duration. The parameters of each network are shown in Table 1.

Table 1. Parameters of different wireless networks

Networks	work duration (tw)	sleep duration (ts)
NO.1 Network	500ms	18s
NO.2 Network	500ms	10s
NO.3 Network	500ms	5s

The delay of these networks is presented in Fig. 4, Fig. 5, Fig. 6 respectively. With the collected data, we can derive the parameter of all networks that E_1=16.05, δ_1 = 22.10, E_2= 8.76, δ_2 =6.21, E_3= 3.25, δ_3 =0.35.

According to Eq. 5, we have $\mu = \text{T-E}$. So μ_1=2.45, μ_2=1.74, μ_3=2.25. The mean value can be derived:

$$\mu=(\mu_1+\mu_2+\mu_3)/3=2.14$$

$$\delta=(\delta_1+\delta_2+\delta_3)/3=8.57$$

Suppose that the practical duration should be less than $T_r(1+d\%)$, and T_r is the designed duration according to the requirement of the production. For the sake of saving the energy, let the expected delay $E=T_r \times d\%$ to maximize the sleep duration. Then we can set $T=E+\mu=2.14+ T_r \times d\%$ to ensure that the delay can't affect the performance of the monitoring system. And set $ts=T-t_w=1.64+T_r \times d\%$ to save energy as possible.

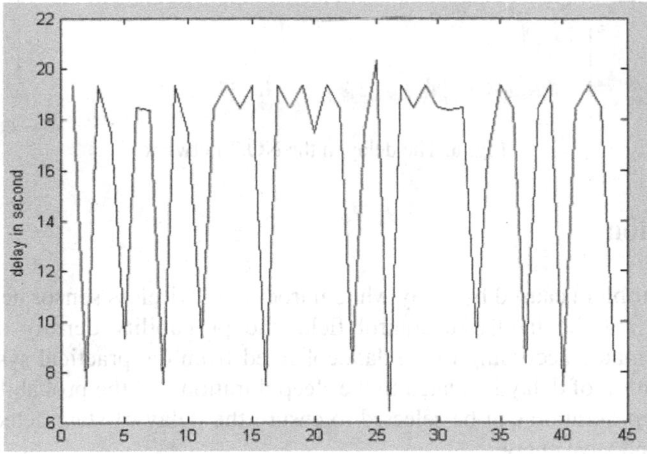

Fig. 4. The delay in the NO.1 network

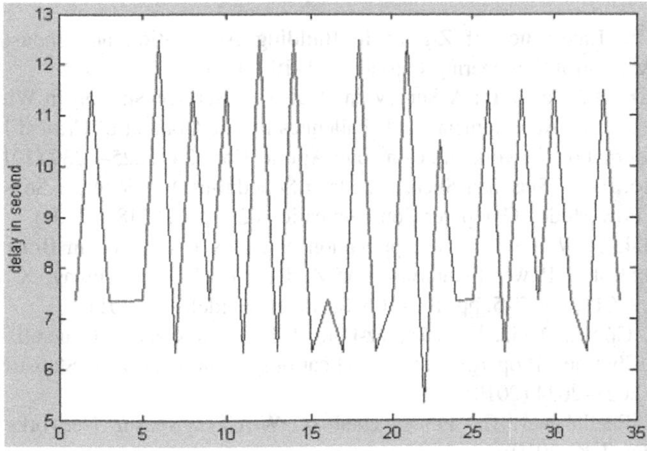

Fig. 5. The delay in the NO.2 network

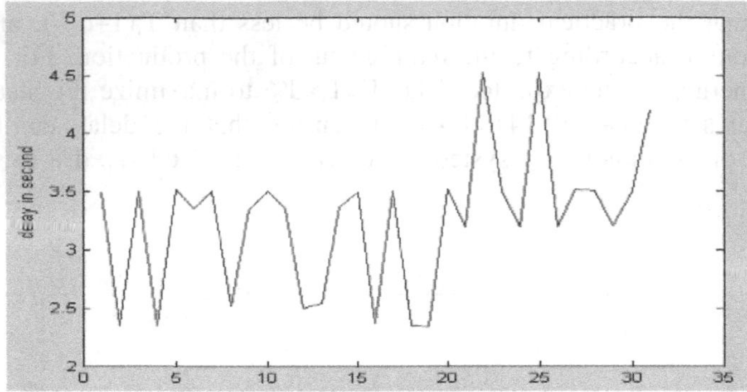

Fig. 6. The delay in the NO.3 network

5 Conclusion

To settle the problem caused by delay while introducing wireless sensor networks into the application of the industrial control field, the probability density of the state changes is estimated according to the data collected from the practical system. Then, for the expectation of delay is related to the sleep duration and the probability density, the proper sleep duration can be selected to ensure the delay of state detection in the permit range and save energy.

References

1. Egan, D.: The Emergence of ZigBee in Building Automation and Industrial Control. Computing & Control Engineering Journal 2, 14–19 (2005)
2. Chen, J., Díaz, M., Llopis, L.: A Survey on Quality of Service Support in Wireless Sensor and Actor Networks: Requirements and challenges in The Context of Critical Infrastructure Protection. Journal of Network and Computer Applications 34, 1225–1239 (2011)
3. Ha, R.W., Ho, P.-H., Sherman Shen, X.: Sleep Scheduling for Wireless Sensor Networks via Network Fow Model. Computer Communications 29, 2469–2481 (2006)
4. Wang, R., Bai, Y., Wang, F.: The Application of Delay Estimation in Binary Control of Sequence Control in Power Plant Based on ZigBee. In: Lin, S., Huang, X. (eds.) CSEE 2011, Part II. CCIS, vol. 215, pp. 102–106. Springer, Heidelberg (2011)
5. García, V.J., Gómez-Déniz, E., Vázquez-Polo, F.J.: A New Skew Generalization of The Normal Distribution. Properties and Applications. Computational Statistics and Data Analysis 54, 2021–2034 (2010)
6. Dargie, W., Poellabauer, C.: Fundamental of Wireless Sensor Networks:Theory and Practice. Wiley, UK (2010)

A Rough Set Approach to Online Customer's Review Mining

Wei Wang, Huan Yuan, Ye Chen, Ping Yang, Leilei Ni, and Yu Li

College of Economics and Management, Nanjing University of Aeronautics and Astronautics, Nanjing, Jiangsu 210016, China
tongxin002@yeah.net

Abstract. With rapid development of information technologies, vast data and information have been generated and accumulated which requires new methods to support decisions. In this paper under the research topic of multiple criteria decision analysis, a novel rough set-based analytical framework is designed to investigate customers' reviews and offer support for online buying decision. The proposed method is applied to a case study of online phone buying decision to demonstrate the designed process.

Keywords: Customer reviews, data mining, rough set approach, multiple criteria decision analysis.

1 Introduction

Nowadays, in customer's review mining area, we need powerful tool to discriminate and integrate all the information to make informed decisions. Various websites provide the search rank of products according to single index. Yet it could be more useful through aggregation of multiple copies of information, such as price, reviews and technical tests. This paper intends to establish a decision supporting system under multiple criteria decision analysis (MCDA) area and to design a rough set analysis (RSA) method based on internet information's mining.

2 Literature Reviews of Customer Review Mining

Zhu (2006) has selected k pairs of positive and negative benchmark words and used Equation (1) to calculate the orientation of a word. The similarity can be calculated by software HowNet (www.keenage.com). If Orientation (w) >0, it means that the word is positive, if not, then negative.

$$O(w)=\sum_{i=1}^{k}Sim(i-positive,w)-\sum_{i=1}^{k}Sim(i-negative,w) \tag{1}$$

D. Jin and S. Lin (Eds.): Advances in CSIE, Vol. 1, AISC 168, pp. 229–234.
springerlink.com © Springer-Verlag Berlin Heidelberg 2012

Yang and Wu (2009) have improved the selection of the basic word as well as the calculation formula of similarity, which obtained higher accuracy.

3 RSA-Based Model of Customer Reviews Mining Method

3.1 Information Collection of Customer Reviews

The basic means of mining comments is: grab page code through the URL, and then analyze the information environment of the page code. Furthermore, get the selection rule and use regular expressions convey it. Then, extract reviews.

For convenient, we extracted the reviews from Jingdong by December 30, 2011 according to the law of statistics. The principle of extracting samples is: praise, medium and bad review each accounts 40, according to user's level, diamond & double diamond, gold, silver & bronze each accounts 50%, 30% and 20%.

The Semantic information process is as follows:

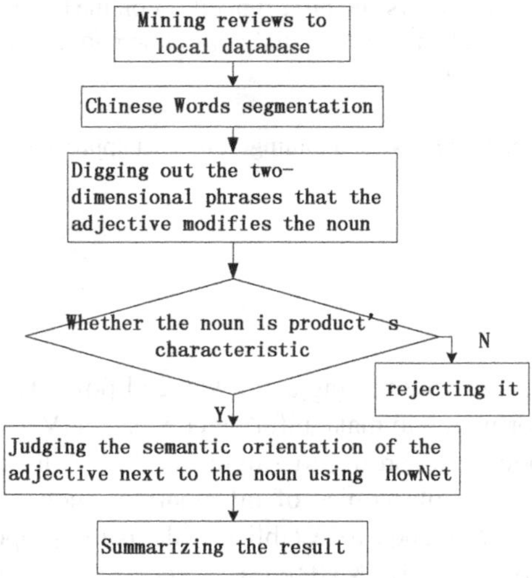

Flag. 1. The follow chart of semantic information process.

First suppose that there are 5 kinds of mobile phones, recorded as A、B、C、D、E, according to the above algorithm, taking A as an example, we can get Table 1:

Table 1. Scoring sheet

W_i	
adj	**Semantic orientation**
a_{wi1}	O_{wi1}
a_{wi2}	O_{wi2}
…	…
Positive side's number	P_{wi}
Negative side's number	N_{wi}
Score of high praise A_{wi}	$\frac{P_{wi}}{Q_{wi}}\times 100$

wn expresses the nth product's characteristic, a_{wi2} is one of the adjectives that modify the characteristic W_i and Owni is a_{wni}'s orientation value. Besides, Pwi plus Nwi is Qwi. So the total score among all the characteristics of A can be written as:

$$A_w = \sum_i A_{wi} \times \frac{Q_{wi}}{Q_A}, Q_A = \sum_i Q_{wi} \tag{2}$$

J_A is the average review score of product A, because the full mark is 5, so we need to transform it into hundred mark system to calculate the decision value:

$$D_A = \frac{1}{2}(20J_A + A_w) \tag{3}$$

3.2 The Sales and Price Information Collection

What consumers want most is the one with larger sales volume and relative cheaper price. But the search ranking index in current shopping website is sales volume or price. So we consider creating an index which is the ration of quantity and price:

$$K_i = \frac{quantity}{price}, \text{ by normalized, } Y_i = \frac{99(R - K_i)}{R - S} + 1 \tag{4}$$

In equation (4), R、S is the optimal value and worst value.

3.3 Decision Analysis Model

We already have 2 major categories of data, commentary and quantity&price module. Use the full index decision data in chapter 3.1, add every product's Y_i, and finally get the ultimate multi-objective decision table:

Table 2. The multi-objective decision table

	w_1	w_2	w_3		w_m	Decision values
A	A_{w1}	A_{w2}	A_{w3}		A_{wm}	D_A'
B	B_{w1}	B_{w2}	B_{w3}		B_{wm}	D_B'
C	C_{w1}	C_{w2}	C_{w3}	…	C_{wm}	D_C'
D	D_{w1}	D_{w2}	D_{w3}		D_{wm}	D_D'
E	E_{w1}	E_{w2}	E_{w3}		E_{wm}	D_E'

Taking product A as an example, in which A's new decision value is:

$$D_A' = \alpha_1 \times Y_i + \alpha_2 \times D_A, \alpha_1 + \alpha_2 = 1 \tag{5}$$

In equation (5), α_1 and α_2 are the weight of the two data modules, which can be decided by users or regarded as both 0.5 by the system.

Arranging the new decision values from largest to smallest, then the order is the multi-objective recommended order of the candidate products. We can also calculate the form using RSA to get the reduction and generate decision rules.

4 Data Analysis and Results

Type the following search criteria in Jingdong --brand: Sony Ericsson, price: 2000-2999, system: Android and candy bar phone. There are 12 candidates.

Step1: Combine the similar characteristics of phones. After combination, we get 11 criteria: W1,Sensor type; W2, Business function; W3, Appearance; W4, Entertainment; W5, Screen; W6, Operating system; W7, Battery; W8, Hardware; W9, Calling; W10, Accessories; W11, Keyboard.

Step2: Use the proposed method in chapter 3.1 to mine the data of the 12 products and calculate with chapter 3.2 and 3.3's method. From Table 3, we can see the final recommendation order is D>C>E>L>I>K>H>F>A>G>J>B.

Step3: Use 4eMka2 (ICS,2008) to get the criteria reduction and generate decision rules. Divide the 12 phones into three kinds according to their decision values, that is $High=\{D,C,E,L\}, Mid=\{I,H,K,A\}, Low=\{G,F,J,B\}$. The results are as Flag.3 and Flag.4 showed:

Reducts:

	Cardinality	Attributes Set
Core:	3	W6, W7, W8
Reduc...		
1.	4	W3, W6, W7, W8
2.	5	W4, W5, W6, W7, W8

Flag. 3. The reduction of criteria.

Filter Options:
Relative Strength: - Support: - Rule Type: All Length: -

Generated Rules: 9 **Displayed Rules: 9**

Number	Condition	Decision	Support	Relative Strength [%]
1.	(W6 <= 66.67)	Class at most L	3	75.00
2.	(W5 <= 75) & (W7 <= 25)	Class at most L	1	25.00
3.	(W4 <= 0)	Class at most M	1	12.50
4.	(W7 <= 0) & (W8 <= 0)	Class at most M	4	50.00
5.	(W9 >= 100)	Class at least H	2	50.00
6.	(W8 >= 100) & (W6 >= 100)	Class at least H	1	25.00
7.	(W7 >= 50) & (W4 >= 85.714)	Class at least H	1	25.00
8.	(W7 >= 50)	Class at least M	2	25.00
9.	(W6 >= 93.33) & (W3 >= 95.45)	Class at least M	4	50.00

Flag. 4. The rules.

From Flag.3 and 4, we know the preferences of people who bought such products and their decision rules, which can help us find what we most want. In the future, we will compare customers of different levels on their purchasing habits with RSA.

5 Conclusions

A definite trend is evident that we need such a tool help us generate and accumulate vast data to support our decisions. To fulfill this purpose, specifically in this paper a novel rough set-based analytic framework is designed to offer decision support to online buying under the research topic of multiple criteria decision analysis. It is hoped that the study will stimulate further study in this field.

Table 3. Final decision list

	w1	w2	w3	w4	w5	w6	w7	w8	w9	w10	w11	Decision values
A	0	0	100	75	100	50	0	0	0	0	0	45.62
B	0	0	100	33	87	50	40	100	0	100	0	39.42
C	0	0	77	69	91	83	38	50	100	0	0	64.42
D	0	0	91	86	7	73	50	0	0	0	0	87.77
E	0	0	93	80	100	100	33	0	100	0	0	52.86
F	0	0	80	0	100	90	67	0	0	0	0	45.91
G	0	0	95	100	75	100	25	0	0	50	0	45.37
H	0	0	95	50	88	93	0	0	0	0	0	46.72
I	0	0	100	75	83	100	0	0	0	50	0	48.50
J	0	0	94	100	71	67	44	0	0	0	0	42.90
K	0	0	100	100	100	100	0	0	0	0	0	47.72
L	0	0	95	100	100	100	0	100	0	0	0	52.69

References

1. Zhu, Y.-L., Min, J., Zhou, Y.-Q., Huang, X.-J., Wu, L.-D.: Semantic Orientation Computing Based on HowNet. Journal of Chinese Information Processing, 14–20 (2006)
2. Dong, Z., Dong, Q.: HowNet (EB/OL) (2012), http://www.keenage.com/zhiwang/e-zhiwang.html
3. Yang, Y.-B., Wu, X.-W.: Improved lexical semantic tendentiousness recognition computing. Computer Engineering and Applications 45(21), 91–93 (2009)
4. ICS, 2008. 4eMka2 software. Institute of Computing Science (ICS), Poznan University of Technology, http://idss.cs.put.poznan.pl/site/4emka.html

Cluster Center Initialization Using Hierarchical Two-Division of a Data Set along Each Dimension

Guang Hui Chen

Department of Mathematics and Computational Science,
Guang Dong University of Business Studies, Guangzhou, China, 510320
chenguanghui@gdcc.edu.cn

Abstract. This paper proposes a hierarchical two-division method that divides each mother subset of a data set at the same layer into two subsets along a dimension, and hierarchically divides the data set into a series of leaf subsets when the two-division process passes through each dimension of the data set. Then the initial cluster centers are picked out from the series of leaf subsets according to the rule that optimizes the dissimilarities among the initial cluster centers. Thus a new cluster center initialization method is developed. Experiments on real data sets show that the proposed cluster center initialization method is desirable.

Keywords: cluster center initialization, k-Means algorithm, two-division method.

1 Introduction

Clustering is an unsupervised pattern classification method that divides a set of given data into clusters, such that data in the same group are more similar to each other than to data from different clusters. Clustering is one of the most important tasks in data analysis. It has been used for decades in image processing and pattern recognition.

Many clustering algorithms were proposed to perform a partition on the given data set. The goal of any clustering algorithm is to reveal the natural structure of the data set. Crisp k-Means algorithm [1] is a partitioning clustering method that divides data into k mutually excessive groups. It is simple and fast to run, and most popular with researchers and practitioners. However, Crisp k-means algorithm is very sensitive to the initial cluster centers that have a direct impact on the formation of final clusters. Crisp k-Means algorithm may be strapped in a local optimum, or generates an empty cluster for the given data set, because of inappropriate initial cluster centers. Therefore, it is quite important to provide k-means algorithm with good initial cluster centers.

Several methods were proposed to solve the cluster initialization for k-means algorithm [2-4]. Each of them has its strong and weak points, and no one performs better than others in any cases, as shown by literatures [5, 6]. This paper also proposes a cluster center initialization method for k-means algorithm. The proposed method divides a data set into a set of subsets that are small enough by hierarchically dividing

D. Jin and S. Lin (Eds.): Advances in CSIE, Vol. 1, AISC 168, pp. 235–241.
springerlink.com © Springer-Verlag Berlin Heidelberg 2012

big subsets into two subsets along each dimension until all subsets are small enough or each dimension has been used once. Then, a minimum spanning tree is constructed on each subset and the data point with biggest degree is selected as its representative. Finally, initial cluster centers are picked out from the set of representatives according to the principle that maximize the dissimilarities among the initial cluster centers.

2 A Method for Hierarchical Two-Division of a Data Set along Each Dimension

Supposing that $X=\{x_1,x_2,\cdots,x_n\}$ is a data set, where $x_j=(x_{1j},x_{2j},\cdots,x_{mj})^T\in\Re^m$. If a subset of ***X*** contains more than $\sqrt{n}$ data points of ***X***, it is called a mother subset of ***X***, denoted by *Msubset*. If $A\subset B\subset X$ and $|A|>\sqrt{n},|B|>\sqrt{n}$, ***B*** is called a mother set of ***A***. A set of mother subsets of ***X*** resulting from the division of their mother sets along the *i*-th dimension is denoted by $Msets_i$. The number of mother subsets in $Msets_k$ is denoted by n_Msets_k. If a subset of ***X*** contains less than $\sqrt{n}$ data points of ***X***, it is called a leaf subset of ***X***, denoted by *Lsubset*. The set of leaf subsets of X is denoted by *Lsets*.

Let $X_i=(x_{i1},x_{i2},\cdots,x_{in})$ denote the *i*th dimension of ***X***. Sort X_i in ascending order to yield the sorted *i*th dimension $\hat{X}_i=(x_{ij_1},x_{ij_2},\cdots,x_{ij_n})$, where $j_1,j_2,\cdots,j_n$ is a permutation of $1,2,\cdots,n$. The data set ***X*** is rearranged to be $X=\{x_{j_1},x_{j_2},\cdots,x_{j_n}\}$ correspondingly. Calculate the difference along the sorted *i*th dimension as follows.

$$diff_k=x_{ij_{k+1}}-x_{ij_k},k=1,2,\cdots,n-1 \tag{1}$$

The position $p=\arg\max_{1\le k\le n-1}\{diff_k\}$ is called a division point along the *i*th dimension of ***X***. Dividing ***X*** along the *i*-th dimension at the division point *p* into two subsets, $A_1=\{x_{j_1},x_{j_2},\cdots,x_{j_p}\}$ and $A_2=\{x_{j_p+1},x_{j_p+2},\cdots,x_{j_n}\}$. If a subset of ***X*** becomes a leaf subset or all of data points in the subset are identical along the *i*th dimension, it is indivisible along the *i*th dimension. When all subsets of ***X*** are indivisible along each dimension, or the data set *X* has been divided once along each dimension, the division of ***X*** is terminated.

Calculate the standard deviations along each dimension of ***X***, denoted by $\sigma_1,\sigma_2,\cdots,\sigma_m$. Sort them in descending order to yield a sorted standard deviations, denoted by $\sigma_{i_1},\sigma_{i_2},\cdots,\sigma_{i_m}$, where $i_1,i_2,\cdots,i_m$ is a permutation of 1, 2, …,*m*. The initial set $Msets_0$ of mother subsets of ***X*** is initialized by $Msets_0=\{X\}$. At the division of ***X*** along the sorted *k*-th dimension in turn, *k*=1, 2, …, *m*, each mother subset *Msubset* in $Msets_{k-1}$ is divided at the dividing point *p* into two subsets,

$Msubset_1 = \{x_j \in Msubset | j = 1,2,\cdots, p\}$ and $Msubset_2 = Msubset - Msubset_1$. Put each of $Msubset_1$ and $Msubset_2$ in $Msets_k$, if it is a mother subset of $\boldsymbol{X}$, or *Lsets*, if it is a leaf subset of $\boldsymbol{X}$. When the division process along the last dimension of $\boldsymbol{X}$ is finished, output *Lsets* and $Msets_m$ if it is not empty. The union set of *Lsets* and $Msets_m$ forms the final division of $\boldsymbol{X}$. We call this division process hierarchical two-division of $\boldsymbol{X}$ along each dimension.

The main idea of hierarchical two-division of $\boldsymbol{X}$ along each dimension is hierarchically dividing each mother subset at the dividing point p from top to bottom until the division termination term is satisfied. The pseudo of hierarchical two-division of $\boldsymbol{X}$ along each dimension is given in Algorithm 1.

Algorithm 1. Hierarchical two-division of $\boldsymbol{X}$ along each dimension

---

1 calculate the standard deviation along each dimension of $\boldsymbol{X}$, denoted by $\sigma_1, \sigma_2, \cdots, \sigma_m$;

2 sorts the standard deviations $\sigma_1, \sigma_2, \cdots, \sigma_m$ in ascending order to be $\sigma_{i_1}, \sigma_{i_2}, \cdots, \sigma_{i_m}$, where $i_1, i_2, \cdots, i_m$ is a permutation of 1, 2, ...,*m*, and rearranges the dimensions of $\boldsymbol{X}$ in the order of the sorted standard deviation;

3 hierarchically divides $\boldsymbol{X}$ in the following way;

3.1 initialize $Msets_0$ and n_Msets_0 by $Msets_0=\{X\}$ and $n\_Msets_0=1$, respectively;

3.2 for k=1:m

if $n\_Msets_{k-1}>0$ and $\sigma_{i_k} > 0$

for j=1:n_Msets

calculate the standard deviation along the i_k-th dimension of the j-th mother subset $Msets_{k-1}(j)$ in $Msets_{k-1}$, denoted by $\sigma_{i_k}(Msets_{k-1}(j))$;

if $\sigma_{i_k}(Msets_{k-1}(j)) == 0$

Put $Msets_{k-1}(j)$ in $Msets_k$;

else

sort the values of the i_k-th dimension of data points in $Msets_{k-1}(j)$ in ascending order, and rearrange the data points in $Msets_{k-1}(j)$ in the order of the sorted values of the i_k-th dimension correspondingly;

---

calculate the differences of the sorted values along the i_k-th dimensionof

$Msets_{k\text{-}1}(j)$by $diff_{i_k}^{(j)} = \{x_{i_k,2}^{(j)} - x_{i_k,1}^{(j)}, x_{i_k,3}^{(j)} - x_{i_k,2}^{(j)}, \cdots, x_{i_k,n_j}^{(j)} - x_{i_k,n_j-1}^{(j)}\}$,

where $x_{i_k,p}^{(j)}$ is the value of the i_k-th dimension of the pth data point in

$Msets_{k\text{-}1}(j)$;

find the dividing point p_k by $p_k = \arg\max\{diff_{i_k}^{(j)}\}$;

divide $Msets_{k\text{-}1}(j)$ into two subsets A_1 and A_2,

where $A_1 = \{x_l \in Msets_{k-1}(j), 1 \le l \le p_k\}$ and

A_2=$Msets_{k\text{-}1}(j)$-A_2;

put each subset of $Msets_{k\text{-}1}(j)$ in $Msets_k$ if it is a mother subset of ***X***,

otherwise put it in *Lsets*;

end if

end for

else

break the for-loopse;

end if

end for

4 if *n*_ $Msets_m$>0

the result of hierarchical division of ***X***, denoted by HD(***X***),

is $\mathrm{HD}(\mathrm{X}) = \mathrm{Lsets} \cup \mathrm{Msets}$;

else

HD(***X***)=Lsets;

End if

5 output the result of hierarchical division of ***X*****,** HD(***X***).

3 Cluster Center Initialization Based on Hierarchical Two-Division of a Data Set along Each Dimension

When the result of hierarchical division of ***X***, denoted by HD(***X***), is output, the initial cluster center candidates will be selected from the subsets of ***X*** in HD(***X***) in the following way. Over the subset of ***X*** in HD(***X***) that contains no less than three data points, a minimum spanning tree ***T*** is constructed. The data point in ***T*** that has biggest degree is selected as an initial cluster center candidate. When all the initial cluster center candidates are picked out, *k* (the number of clusters) cluster centers will be selected from the candidates in the following way. The candidate whose degree is biggest is selected as the first initial cluster center. From the second initial cluster center to the last one, the candidate that is farthest away from the existed initial cluster centers is selected as the initial cluster center. Combining hierarchical two-division of X with the initial cluster center selection process, a new cluster center initialization method is developed, which is denoted by HTDD. The pseudo of HTDD is given in Alogirthm 2.

Alogirthm 2. The cluster center initialization method using hierarchical two-division of ***X*** along each dimension

---

1 produce a division HD(***X***) of ***X*** using Algirthm I;

2 for j=1 to | HD(***X***)| (the cardinality of the set HD(***X***) of subsets of ***X***)

If the *j*-th subset in HD(***X***) has no less than three data points, construct a minimum spanning tree ***T*** over it;

Calculate the degrees of data points in ***T*** and select the point with biggest degree as its representative;

end for

3 select initial cluster centers from the set of candidates;

3.1 select the candidate with biggest degree as the first initial cluster center;

3.2 for *j*=2 to *k* (*k* is the number of clusters)

Compute the distance between each existed initial cluster center and each candidate;

For each candidate, search the existed initial cluster center nearest to it;

Select the candidate whose nearest existed initial cluster center is farthest of all as the j-th initial cluster center;

end for

4 output *k* initial cluster centers;

---

4 Experiments

Three real data sets are employed to test the proposed cluster center initialization method HTDD. Their brief descriptions are given in Table 1 that shows they are of different sizes and different number of clusters.

Table 1. Brief description of data sets

Data set	Number of attributes	Number of data	Number of clusters
Optdigits (test) [7]	64	1797	10
Lung cancer[7]	56	32	3
ionosphere[8]	34	351	2

To demonstrate the effectiveness of HTDD, it was compared with two existed cluster center initialization methods, denoted by kd-tree and CCIA respectively, and two indexes, pattern recognition rate of hard c-Means [2] whose initial cluster centers are from each cluster center initialization method and CPU time of each cluster center initialization method, are employed to evaluate them. The comparative results are listed in Table 2 that shows HTDD improves on pattern recognition rate of hard c-Means HCM [2] relative to *kd*-tree [3] and CCIA [2]. In terms of CPU time, HTDD is faster than CCIA , but slower than *kd*-tree.

Table 2. Comparative results on real data sets

Data set	Pattern recognition rate (%)			CPU time (second)		
	HTDD	kd-tree	CCIA	HTDD	kd-tree	CCIA
Optdigits	**66.22**	57.31	62.10	0.3541	0.2283	73.8824
Lung cancer	**62.50**	56.25	43.75	0.0049	0.0042	0.1150
ionosphere	**71.22**	70.94	71.22	1.4665	0.4674	0.7163

5 Conclusions and Discussions

The differences among data points along each dimension may be an indicator of cluster. We hierarchically divides each mother subset of a data set into two subsets along each dimension until all subsets of the data sets are leaf subsets or the two-division process passes through all dimensions of the data sets. Therefore, a series of leaf subsets are produced, of which initial cluster centers are picked out according to the rule that optimizes the differences among intial cluster centers. Experiments demonstrate that the proposed cluster center initialization method is desirable relative to two existed cluster center initialization methods. However, it needs further investigation and tests from applications.

References

[1] MacQueen, J.B.: Some methods for classification and analysis of multivariate observation. In: Le Cam, L.M. (ed.) Berkeley Symposium on Mathematical Statistics and Probability, pp. 281–297. University of California Press (1967)

[2] Khan, S.S., Ahmad, A.: Cluster center initialization algorithm for k-means clustering. Pattern Recognition Lett. 25(11), 1293–1302 (2004)

[3] Redmond, S.J., Heneghan, C.: A method for initialising the K-means clustering algorithm using kd-trees. Patt. Recog. Letters 28(8), 965–973 (2007)

[4] Erisoglu, M., Calis, N., Sakallioglu, S.: A new algorithm for initial cluster centers in k-means algorithm. Patt. Recog. Letters 32, 1701–1705 (2011)

[5] Pena, J.M., Lozano, J.A.: An empirical comparison of four initialization methods for the k-means algorithm. Patt. Recog. Lett. 20(10), 1027–1040 (1999)

[6] Steinley, D., Brusco, M.J.: Initializing K-means Batch Clustering: A Critical Evaluation of Several Techniques. J. of Classification 24(1), 99–121 (2007)

[7] http://www.ics.uci.edu/~mlearn/MLRepository.html

[8] Chang, C.-C., Lin, C.-J.: LIBSVM: a library for support vector machines (2001), http://www.csie.ntu.edu.tw/~cjlin/libsvm

The Design and Implementation of a Driving Simulation System Based on Virtual Reality Technology

Hongyu Lai[1], Wei Huang[2], Peng Wei[3], and Jingchang Pan[1]

[1] Computer Department, Shandong University at Weihai, 264209, China
[2] School of Electronic Information, Wuhan University, 430072
[3] Key Laboratory of Optical Astronomy, NAOC, Chinese Academy of Sciences, Beijing, 100012, China
reachk@126.com, hw1293381721@163.com, lywei1987@sina.com, pjc@sdu.edu.cn

Abstract. Virtual Reality (VR) refers to the technology which allows a user to immerse into and interact with a computer-simulated environment using a variety of special equipments. This paper designs and implements a Driving Simulation System based on this technology, which includes simulation seats, simulation steering wheel and simulation program. The program implements rendering, roaming and real-time control of the three-dimensional scene through OpenGL, DirectX, etc. It's of great help for drive training and leisure.

Keywords: Virtual Reality, OpenGL, DirectX, Collision Detection, Virtual Driving.

1 Introduction

Virtual Reality[1] (VR), also known as virtuality, is an advanced human-machine interface characterized by immersion, interactivity and imagination, which is a combination of computer graphics, simulation technology, multimedia technology, artificial intelligence, computer networking, parallel processing and multi-sensor technology. It can simulate the functions of human's sensory organs such as vision, hearing and tactus, enabling human to immerse a computer-simulated virtual world, interact with the machine by languages and gestures in the real time. VR creates a comfortable humanizing multi-dimensional information space, thus it has a wide application prospects [2-6]. This paper develops a simulated driving system based on VR to create a virtual 3D model giving the person a vivid immersed sense.

This paper uses OpenGL [7, 8] to implement the rendering and roaming of the scene which are main parts of the virtual reality system. OpenGL is a graphics software interface with strong functions and cross-platform characteristics, which is irrelative to the hardware[9]. The three functions in this paper are the rendering of three-dimensional scene, collision detection, and information display and roaming. Microsoft DirectX [10-12] is a series of lower application programming interfaces which are prepared for developing computer games and other high-performance multimedia applications. The controller in the system is designed by DirecInput and its various sound effects are simulated by DirectSound.

D. Jin and S. Lin (Eds.): Advances in CSIE, Vol. 1, AISC 168, pp. 243–247.
springerlink.com © Springer-Verlag Berlin Heidelberg 2012

2 System Structure and Operation Mechanism

The system is composed of five subsystems whose functions are relatively independent and are invisible to each other. Five subsystems are integrated into an engine which call the interfaces of different subsystem and update and maintain data of the system. At the same time the engine exposes its interface to the application, which allows the interaction between engine and application. The application can not directly call the functions of the sub-systems. The block diagram of the system is shown in Figure 1:

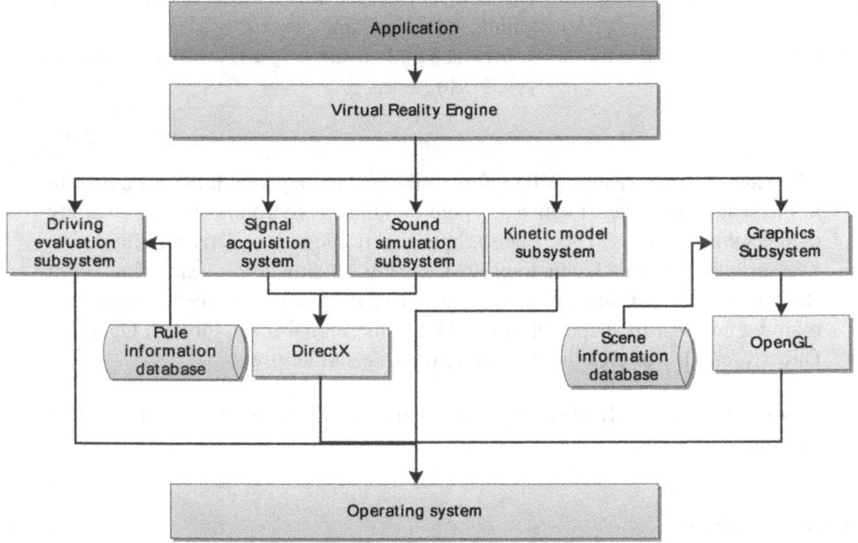

Fig. 1. Structure of system function

System engine functions are: to call the interface functions of various subsystems, to update the data of system and call the functions of subsystem; maintain and update the shared data; to interact with applications. Graphics subsystem renders the 3D scenes by reading the data edited by Map editor subsystem. Signal acquisition system detects and process the peripherally set state which can be returned to and be further processed by the system engine. Kinetic model subsystem is responsible for maintaining the kinetic data and respond to the user inputs. Sound simulation subsystem simulates various sounds in the process of driving. Driving evaluation subsystem enables users to learn the correct operation methods and driving technology and thus they can follow the traffic regulations. Scene information database and Rule information database is applied to store scene information and regulations which are imported into the system in the initialization of system engine which distributes the scene information to the rendering model to generate the virtual scene, and makes regulations organized in the form of queue which can be used by system engine.

3 System Implementation

This paper designs and realizes a simulated driving system based on virtual reality technology, whose operation interface are as shown in figure 2.

Fig. 2. Rendering of the program

Fig. 3. Road mapping

3.1 Rendering of 3D Scene

This paper constructs the sky and clouds to add the sense of reality and make users experience the real immersed sense. A cuboid box with texture is drawn, which can achieve a good visual effect as long as the texture mapping is good enough. Then some facilities are be set in the roads with green texture, and straight roads and arc roads are distinguish by the road texture in the divided rectangles. The principle of drawing these two kinds of roads is the same, first they are divided into small pieces of rectangles with the same size and then different textures are added in the surface, for those roads with higher surface, walls are erected between two sides of the road and the surface, which can make the experience more realistic. As shown in Fig. 3.

3.2 Steering Control Simulation

DirectInput is a component interface of the DirectX, which provide lots of interface functions to deal with the inputs by the keyboards, mouse, joysticks, force feedback and other game equipments. DirectInput deals directly with hardware drivers and is independent to hardware virtual input system. First, Preset the DirectInput and then set every input device using the same process. The whole processes are packaged to Class CVD_INPUT. The main functions include initialization and setting, releasing equipment, reading device data, judgment of the key state and displacements of directions, etc.

3.3 Collision Detection

After the scene rendering in OpenGL, video camera starts the roaming according to signals and data input. Video camera stops the roaming when moving to the side of the road, which is called as collision detection. Instead of adopting virtual height, this paper uses 128*128 points to store the height and obtain all the other heights through interpolation. In this way collision detection is achieved by fewer points. We judge the road range by current point and the x and z coordinate of the next point. The detection method is the same as the initialization of the height. In this way, we can detect whether the next point is in the range of the road, thus we can control the roaming of the camera. The flowchart is as shown in Fig3:

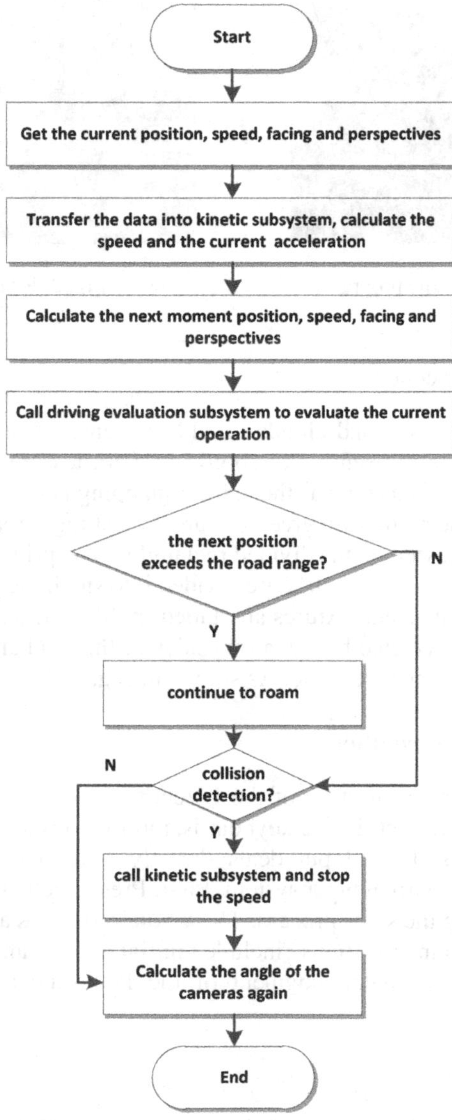

Fig. 3. The flowchart of rendering

4 Summaries and Prospect

This paper mainly studies the design and implement of driving simulation system based on the virtual reality technology. It first introduces the structure of the system engine and the functions and design principles of the subsystem, and then discuss various technologies used in the system construction. At present, the system still need improvements: further improvements of the scene authenticity, to add various effect using rendering function in OpenGL, such as multiple texture map, to use stricter collision detection to generate better collision effect, to add support of force feedback, to improve kinetic model of the car, to gain better simulation of car functions, to add more testing items, to achieve a better application.

References

1. Burdea, G., Coiffet, P.: Virtual reality technology. Presence: Teleoperators & Virtual Environments 12, 663–664 (2003)
2. Seymour, N.E., Gallagher, A.G., Roman, S.A., O'Brien, M.K., Bansal, V.K., Andersen, D.K., Satava, R.M.: Virtual reality training improves operating room performance: results of a randomized, double-blinded study. Annals of Surgery 236, 458 (2002)
3. Guttentag, D.A.: Virtual reality: Applications and implications for tourism. Tourism Management 31, 637–651 (2010)
4. Hsu, K.S.: Virtual reality: Application of a Virtual Reality Entertainment System with Human-Machine Sensor Device. Journal of Applied Sciences 11, 2145–2153 (2011)
5. Cha, S.J., Kim, G., Jang, C.H., Jo, G.S.: An Effective Learning Method in Art Using Virtual Reality and Ontology. In: International Conference on IT Convergence and Security 2011, pp. 425–434. Springer (2012)
6. Wu, X., Jiang, Y.: Industrial product design theory and method based on the advanced virtual design concept. In: Electric Information and Control Engineering, ICEICE, pp. 470–473. IEEE Press (2011)
7. Shreiner, D.: The OpenGL Reference Manual-The Bluebook (2010)
8. NeHe Productions - Everything OpenGL, http://nehe.gamedev.net/
9. Guha, S.: Computer Graphics Through OpenGL: From Theory to Experiments (2010)
10. Luna, F.D.: Introduction to 3D game programming with DirectX 9.0 c: a shader approach. Jones & Bartlett Learning (2006)
11. Snook, G.: Real-time 3D terrain engines using C++ and directx 9. Charles River Media, Inc. (2003)
12. Parberry, I.: Learn Computer Game Programming with DirectX 7.0. Wordware Publishing Inc. (2000)

A Novel PCA-GRNN Flow Pattern Identification Algorithm for Electrical Resistance Tomography System

Yanjun Zhang[1] and Yu Chen[2,*]

[1] Heilongjiang University Harbin, China, 150080
[2] Northeast Forestry University Harbin, China, 150040
lg_chenyu@yahoo.com.cn, zyj716@hotmail.com

Abstract. The two-phase flow measurement plays an increasingly important role in the real-time, on-line control of industrial processes including fault detection and system malfunction. Many experimental and theoretical researches have done in the field of tomography image reconstruction. However, the reconstruction process cost quite long time so that there are number of challenges in the real applications. An alternative approach to monitor two-phase flow inside a pipe/vessel is to take advantage of identification of flow regimes. This paper proposes a new identification method for common two phase flow using PCA feature extraction and GRNN classification based on electrical resistance system measurement. Simulation was carried out for typical flow regimes using the method. The results show its feasibility, and the results indicate that this method is fast in speed and can identify these flow regimes correctly.

Keywords: Electrical Resistance Tomography, Flow regime identification, Principal component analysis, GRNN.

1 Introduction

Two-phase flow is a mixed-flow pattern widely found in nature, especially in the chemical, petroleum, electricity, nuclear power and metallurgical industries[1]. Two-phase flow regime identification plays an increasingly important role in the automation process of energy industry. It can provide valuable information for a rapid and dynamic response which facilitates the real-time, on-line control of processes including fault detection and system malfunction, many experimental and theoretical researches have done in the field of image reconstruction. However, the visualization process cost quite long time so that there are number of challenges in the real applications. An alternative approach to monitor two-phase flow inside a pipe/vessel is to take advantage of identification of flow regimes instead of image reconstruction, especially for those stable and simple Two-phase flows. Two-phase flow regimes not only affect the two-phase flow characteristics and mass transfer, heat transfer performance, but also affect the system operation efficiency and reliability, while the other parameters of the two-phase flow measurements have a great impact. Therefore,

* Corresponding author.

D. Jin and S. Lin (Eds.): Advances in CSIE, Vol. 1, AISC 168, pp. 249–254.
springerlink.com © Springer-Verlag Berlin Heidelberg 2012

two-phase flow regimes on-line identification for oil mixed transportation systems such as two-phase process of analysis, testing and operation are important[2,3,4].

In the horizontal pipe and vertical pipe flow pattern is different, mainly because of the horizontal pipe by gravity due to the role of media in the pipeline at the bottom of a sedimentation trend. For different media, the classification of flow patterns are also differences in the actual should be divided according to specific circumstances. Present a common flow patterns are: the level of flow, core flow, circulation and trickle and so on[5].

In this paper, a principal component analysis of GRNN algorithm for flow pattern identification [6,7], algorithm to meet the convergence conditions and to simplify the complex pre-processing steps, greatly reducing the computational Experimental results show that the algorithm is effective, complexity, improve the speed of the identification. Experimental results show that the algorithm can obtain a higher recognition rate, and provide a new and effective method for flow pattern identification algorithm in electrical resistant tomography.

2 The Basic Principles of Principal Component Analysis

Principal component analysis (PCA) is a statistical analysis of data in an effective way[8-9]. With the aim of space in the data as much as possible to find a set of data variance explained by a special matrix, the original projection of high dimensional data to lower dimensional data space, and retains the main information data in order to deal with data information easily.

Principal component analysis is a feature selection and feature extraction process, its main goal is to enter a large search space characteristics of a suitable vector, and the characteristics of all the main features extracted. Characteristics of the selection process is to achieve the characteristics of the input space from the space map, the key to this process is to select feature vectors and input at all the features on the projector, making these projectors feature extraction can meet both the requirements of the smallest error variance. For a given M-dimensional random vector $X = [x_1, x_2, \dots x_m]^T$, For its mean E[X]=0, the covariance C_x can be expressed as follows:

$$C_X = E[(X - E[X])(X - E[X])^T] \tag{1}$$

Because of E[X]=0, covariance matrix is therefore auto-correlation matrix as follow:

$$C_x = E[XX^T] \tag{2}$$

Calculate eigen values of C_x $\lambda_1, \lambda_2, \dots, \lambda_m$ and the corresponding normalized eigenvector $\omega_1, \omega_2, \dots \omega_m$ as the following equation:

$$C_x \omega_j = \lambda_i \omega_j \qquad i = 1,2,\dots m \tag{3}$$

express the matrix as follows:

$$Y = \omega^T X \tag{4}$$

With a linear combination of eigenvectors can be reconfigurable X , The following formula:

$$X = \omega Y = \sum_{i=1}^{m} \omega_i y_i \tag{5}$$

Characteristics obtained through the selection of all the principal components, and in the feature extraction process, then select the main features to achieve the purpose of dimensionality reduction.The mean of the vector analysis can be written as:

$$E[Y] = E[\omega^T X] = \omega^T E[X] = 0 \tag{6}$$

The covariance matrix C_y is the autocorrelation matrix of Y , it can be written as:

$$C_Y = E[YY^T] = E[\omega^T XX^T \omega] = \omega^T E[XX^T \omega] \tag{7}$$

In the truncated Y, it is necessary to ensure the cut-off in the sense of mean square deviation is the optimal. $\lambda_1, \lambda_2, \ldots, \lambda_m$ can only consider the first L largest eigenvalue, with these characteristics for reconstruction of X , the estimated value of reconstruction is as follows:

$$\hat{X} = \sum_{i=1}^{L} \omega_i y_i \tag{8}$$

Its variances are met as follows:

$$e_L = E[(X - \hat{X})^2] = \sum_{i=L+1}^{m} \lambda_i \tag{9}$$

According to the formula (9), the current characteristic value L is larger, the minimum mean square error can be achieved. Also the formula as follows:

$$\sum_{i=1}^{m} \lambda_i = \sum_{i=1}^{m} q_{ii} \tag{10}$$

Where q_{ii} is the diagonal matrix element of C_X , the contribution rate of variance as follows:

$$\varphi(L) = \frac{\sum_{i=1}^{L} \lambda_i}{\sum_{i=1}^{m} \lambda_i} = \frac{\sum_{i=1}^{L} \lambda_i}{\sum_{i=1}^{m} q_{ii}} \tag{11}$$

When $\varphi(L)$ is large enough, you can pre-L constitute a feature vector space $\omega_1, \omega_2, \ldots \omega_L$ as a low-dimensional projection space, thus completing the deal with dimensionality reduction.

3 Flow Pattern Identification Based on GRNN

GRNN was first proposed by the specht is a branch of RBF neural network is a nonlinear regression based on the theory of feed-forward neural network model[10-12]. GRNN structure constitutes a general from the input layer, hidden layer and output layer. Input layer put variables into the hidden layer of the sample does not participate in a real operation. The number of hidden layer neurons is equal to the number of training samples, the weight of the layer functions as a Euclidean distance function. The transfer function of hidden layer is radial basis function, Gaussian function is usually used as the network's transfer function. The third layer of the network output layer is linear, the weight function is normalized dot product weight function.

Let the training sample input matrix T and output matrix P as follow:

$$p = \begin{bmatrix} p_{11} & p_{12} & \cdots & p_{1Q} \\ P_{21} & P_{22} & \cdots & P_{2Q} \\ \vdots & \vdots & \vdots & \vdots \\ P_{R1} & P_{R2} & \cdots & P_{RQ} \end{bmatrix}, \qquad T = \begin{bmatrix} t_{11} & t_{12} & t_{13} & t_{1Q} \\ t_{21} & t_{22} & \cdots & t_{2Q} \\ \vdots & \vdots & \vdots & \vdots \\ t_{s1} & t_{s2} & \cdots & t_{sQ} \end{bmatrix} \tag{12}$$

P is the training sample of input variables, T the training samples of the output variable, R is the dimension of input variables, S is the dimension of the output variable, Q is the training set samples.Q hidden layer neurons corresponding to the threshold as follows:

$$b_1 = [b_{11}, b_{12}, \cdots, b_{1Q}]' \tag{13}$$

Where $b_{11} = b_{12} = \cdots b_{1Q} = \frac{0.8326}{spread}$, spread is the pace of expansion of the radial basis function. The output of hidden layer neurons have the following formula:

$$a^i = \exp(-\|c - p_i\|^2 b_1, i = 1,2,\cdots,Q \tag{14}$$

Where $p_i = [p_{i1}, p_{i2}, \cdots p_{i3}]', a^i = [a_1^i, a_2^i, \cdots a_Q^i]$.

The hidden layer and output layer connection weights W is taken as the output matrix for training set in GRNN, When hidden layer and output layer neuron c onnection weights determined, we can calculate the output of output layer neuron s, with the following formula:

$$n^i = \frac{LW_{2,1}a^i}{\sum_{j=1}^{Q} a_j^i}, i = 1,2,\cdots,Q \tag{15}$$

$$y^i = purelin(n^i) = n^i, i = 1,2,\cdots,Q \tag{16}$$

4 Experiments Results

This paper implemented experiments for several common flow regimes: bubbly flow, core flow, annular flow, laminar flow, empty pipe and full pipe. The simulations were implemented by using MATLAB on a computer running at the AMD Phenom Triple-Core 2.31Ghz with 2G of RAM.

The approach based on PCA and GRNN identifier is tested on laminar flow, droplet flow, air traffic control, full pipe, the core flow and circulation flow. 40 samples were collected for each flow regimes in the experiments are.

After repeated experimental verification, set the threshold to improve recognition accuracy. Experiments results are as table 1.

The identification accuracy rate are computed to verify the confidence of the system. They are listed in Table 1. The average identification accuracy is 95.833%. This proved that PCA feature extraction method combined with GRNN classifier has good classification ability for flow regime identification. It can be observed that the simpler flow regime, the higher accuracy rate. The accuracy rate of identification of trickle flow is significantly lower than other flow regimes. It can be explained by the complexity of trickle flow.

Table 1. The Identification Accuracy Rate

Flow regime	samples	Accuracy Rate
Empty pipe	40	100%
Full pipe	40	100%
Core	40	90.0%
Circulation	40	97.5%
laminar	40	92.5%
Trickle	40	95%

5 Conclusions

This paper presents a principal component analysis feature extraction of the symmetric subspace flow pattern identification algorithm and GRNN classifier in ERT system. This method simplifies the complex pre-processing steps, greatly reduces the computational complexity and improves the recognition speed. The experimental results show that the algorithm is effective, the approach presented in this paper to obtain a higher recognition rate in the ERT. Future research should focus on improving the recognition rate while ensuring at the same time, as much as possible to reduce noise or changes in media distribution factors, the impact of flow pattern identification. In practical industry process, however, the flow regimes will be more complex and volatile. To improve the identification accuracy, more flow regimes should be collected and trained.

Acknowledgement. The Corresponding author is Yu Chen for this paper ,and this work is partially funded by the Science and Technology Foundation of Education Department of Heilongjiang Province (11551344, 12513016), Natural Science Foundation of Heilongjiang Province (F201019), Postdoctoral Fund of Heilongjiang Province.

References

1. Yang, X.D., Ning, X.B., Yin, Y.L.: Fingerprint image preprocessing technique and feature extraction algorithm. Journal of Nanjing University Natural Sciences 4(42), 351–361 (2006)
2. Areekul, V., Watchareeruetai, U.: Separable gabor filter realizationfor fast fingerprint enhancement. In: IEEE International Conference, vol. 3(1), pp. 5 10 (2005)
3. Karhumen, J., Joutsensalo, J.: Represention and Separation of Signals Using Nonlinear PCA Type Learning. Helsinki University of Technology 8, 1083–1091 (2001)
4. Ham, F.M., Kim, I.: Extension of the Generalized Hebbian Algorithm for Principal Component Extraction. Appliction and Science of Neural Networks 3455, 85–274 (1998)
5. da Rocha Gesualdi, A., Manoel de Seixas, J.: Characterrecognition in car license plates based on principal components and neural processing. Neural Networks 14, 206–211 (2002)
6. Yang, Z., Shi, Z.: Fast approach for optimal brain surgeon. Journal of System Simulation 17(1), 162–165 (2005)
7. Martinez, A.M., Kak, A.C.: PCA versus LDA. IEEE Trans., Pattern Anal., Machine Intelligent 23(2), 228–233 (2008)
8. Wang, L., Wang, X., Feng, J.: In: On image matrix based feature extraction algorithms. IEEE Transactions on Systems, Man, and Cybernetics-Part B: Cybernetics: Accepted for Future Publication 99, 194–197 (2005)
9. Tao, J., Jiang, W.: Improved two-dimensional principal component analysis based on the palmprint identification. Optical Technology 33(2), 283–286 (2005)
10. Xing, X., Luo, W.: State estimation in power system based on GRNN algorithm. Heilongjiang Electric Power 33(1), 50–54 (2011)
11. Liu, Z., Zhou, T.: Application of GRNN Model Based on SSA in Financial Time Series 28(2), 29–33 (2011)
12. Wang, H., Lin, J.: Sales Predictio n Based on Improved GRNN 32(1), 153–157 (2010)

A Temperature Difference Determination Method of Infrared Diagnostic for Petrochemical Equipment

Hua Lai[1,*], Yongwang Tang[2], Na Liu[3], and Jianlin Mao[1]

[1] Faculty of Information Engineering and Automation, Kunming University of Science and Technology 650500 Kunming, China
[2] Faculty of Information Engineering, The PLA Information Engineering University 450002 Zhenzhou, China
[3] Department of Automation, Beijing Institute of Petrochemical Technology 102617 Beijing, China
kmlh1966@sina.com

Abstract. A relative temperature difference discriminance of infrared diagnosis is presented, which is for petrochemical industry equipments. Based on the principle of infrared thermography, we analyze three factors of affecting relative temperature difference discriminance, i.e., the emissivity on the measured object surface, the environment temperature and the working waveband. Then we give a judgment method of equipment fault grades, including the general defects, the major defects and the emergency defects. Our conclusion can be helpful for making a proper judgment on the equipment status.

Keywords: Relative Temperature Difference, Working Waveband, Environment Temperature, Emissivity.

1 Introduction

Infrared diagnosis technology is used to realize and master the state of measured object in operation process, and confirm whether it works normally, and discover fault and adopt correspond countermeasure contraposing the concrete circs. It divides into three basic elements: detection, analysis and diagnosis, treatment and precaution[1]. The fault diagnosis methods of petrochemical industry equipment are as follows: surface temperature discriminance, relative temperature difference discriminance, homogeneous compare method, thermal image analysis method, archive analysis method.

There are multiple factors of influencing infrared diagnosis results, for example: the object emissivity, the wind power, the atmosphere, the ray radiation of sun, the distance coefficients, the adjacent equipments heat radiation and the device loads difference etc. They are all likely to influence the diagnosis results of petrochemical industry equipments. In practical application, surface temperature discriminance and homogeneous compare method is commonly used to judge if petrochemical industry equipments existing limitations, but these distinguishes are all based on equipments have high temperature reflects.

* The corresponding author.

D. Jin and S. Lin (Eds.): Advances in CSIE, Vol. 1, AISC 168, pp. 255–259.
springerlink.com © Springer-Verlag Berlin Heidelberg 2012

As we all known, when the environment temperature is low or equipment load is light, the equipment temperature certainly lower than that of high-temperature environment and heavy load. But when environment temperature is low or equipment load is light, the measured temperature cannot show that there is no fault in petrochemical industry equipments. Generally speaking, it will cause equipment accident after work load augments or environment temperature rises. Thus the relative temperature difference discriminance is brought forward for solving the influence[2].

2 Relative Temperature Difference Discriminance Theory

According to the Reference [1], relative temperature difference means the percent of the ratio between the two corresponding measured points temperature difference and with the obvious hotspot rise, which two equipments state are same or similar, such as the equipment model, the surface status and the load. The math expression of the relative temperature difference is as follows:

$$\delta_t = \frac{\tau_1 - \tau_2}{\tau_1} \times 100\% = \frac{T_1 - T_2}{T_1 - T_0} \times 100\% \tag{1}$$

In formula (1), τ_1 and T_1 are the temperature rise and temperature of hotspots respectively; τ_2 and T_2 are the temperature rise and temperature of normal relative points respectively; T_0 is the temperature of environment reference system.

However, the real temperature calculation equation [3] of practice measured petrochemical industry equipments is as follows:

$$T_0 = \left\{ \frac{1}{\varepsilon} \left[\frac{1}{\tau_\alpha} \bullet T_r^n - \left(\frac{1}{\tau_\alpha} - 1 \right) T_\alpha^n - (1 - \alpha) T_u^n \right] \right\}^{\frac{1}{n}} \tag{2}$$

In formula (2), T_0 means surface temperature of measured objects; ε means measured objects surface emissivity; τ_α means atmosphere transmission rate; α means surface absorptivity; T_r means indicative radiation temperature of thermal image instruments; T_α means atmosphere temperature; T_u means environment temperature; the value of n will change with the work wave band of thermal image instruments. In short wave band 2~5μm, n=9.2554; in long wave band 8~13μm, n=3.9889.

Due to formula (1) and (2), we can get calculation equation of relative temperature difference, as follows:

$$\delta_t = \frac{\tau_1 - \tau_2}{\tau_1} \times 100\% = \frac{T_1 - T_2}{T_1 - T_0} \times 100\%$$

$$= \frac{\left\{ \frac{1}{\varepsilon} \left[\frac{1}{\tau_\alpha} \bullet T_{r1}^n - \left(\frac{1}{\tau_\alpha} - 1 \right) T_\alpha^n - (1 - \alpha) T_u^n \right] \right\}^{\frac{1}{n}} - \left\{ \frac{1}{\varepsilon} \left[\frac{1}{\tau_\alpha} \bullet T_{r2}^n - \left(\frac{1}{\tau_\alpha} - 1 \right) T_\alpha^n - (1 - \alpha) T_u^n \right] \right\}^{\frac{1}{n}}}{\left\{ \frac{1}{\varepsilon} \left[\frac{1}{\tau_\alpha} \bullet T_{r1}^n - \left(\frac{1}{\tau_\alpha} - 1 \right) T_\alpha^n - (1 - \alpha) T_u^n \right] \right\}^{\frac{1}{n}} - \left\{ \frac{1}{\varepsilon} \left[\frac{1}{\tau_\alpha} \bullet T_{r0}^n - \left(\frac{1}{\tau_\alpha} - 1 \right) T_\alpha^n - (1 - \alpha) T_u^n \right] \right\}^{\frac{1}{n}}} \times 100\% \tag{3}$$

From formula (3), we can know that relative temperature difference is related to the parameters, i.e., ε, τ_α, α, T_α, T_u and the infrared thermal image instrument wave bands.

When measured surface satisfies greybody approximation value, namely satisfies $\alpha = \varepsilon$, formula (2) changes into:

$$T_0 = \left\{ \frac{1}{\varepsilon} \left[\frac{1}{\tau_\alpha} \bullet T_r^n - \left(\frac{1}{\tau_\alpha} - 1 \right) T_\alpha^n - (1-\varepsilon) T_u^n \right] \right\}^{\frac{1}{n}} \tag{4}$$

Formula (4) can be seen as calculation equation of calculating greybody surface temperature.

When temperature is measured in short distance, τ_α=1, then the formula (4) becomes:

$$T_0 = \left\{ \frac{1}{\varepsilon} \left[\frac{1}{\tau_\alpha} \bullet T_r^n - (1-\varepsilon) T_u^n \right] \right\}^{\frac{1}{n}} \tag{5}$$

When measured surface satisfies greybody approximation value and the temperature is measured in short distance, the calculation equation of relative temperature difference is as follows:

$$\delta_t = \frac{\tau_1 - \tau_2}{\tau_1} \times 100\% = \frac{T_1 - T_2}{T_1 - T_0} \times 100\%$$
$$= \frac{\left\{ \frac{1}{\varepsilon} \left[\frac{1}{\tau_\alpha} \bullet T_{r1}^n - (1-\varepsilon) T_u^n \right] \right\}^{\frac{1}{n}} - \left\{ \frac{1}{\varepsilon} \left[\frac{1}{\tau_\alpha} \bullet T_{r2}^n - (1-\varepsilon) T_u^n \right] \right\}^{\frac{1}{n}}}{\left\{ \frac{1}{\varepsilon} \left[\frac{1}{\tau_\alpha} \bullet T_{r1}^n - (1-\varepsilon) T_u^n \right] \right\}^{\frac{1}{n}} - \left\{ \frac{1}{\varepsilon} \left[\frac{1}{\tau_\alpha} \bullet T_{r0}^n - (1-\varepsilon) T_u^n \right] \right\}^{\frac{1}{n}}} \times 100\% \tag{6}$$

Here, relative temperature difference is related to ε, T_u and the infrared thermal image instrument wave bands.

3 Analysis of Influence Factors

3.1 Influence of the Measured Objects Surface Emissivity

Measured objects surface emissivity is the main factors causing relative temperature difference, the simulation and experiments show that the objects surface emissivity is the most sensitive to wavelength, the second is the measured objects surface and temperature. (see Table 1).

Table 1. The measuring temperature difference following ε

Error changes (%)	Emissivity					
	0.5	0.6	0.7	0.8	0.9	1
2～5μm	2.83	2.29	1.89	1.64	1.38	1.21
8～13μm	6.83	5.56	4.72	3.99	3.49	3.11

From Table 1 we can see, along with the enlargement of emissivity ε, the percent of causing error change will reduce, consequently it will cause the reduce of relative temperature difference percent; in addition that the error will bigger if choosing long wave than choosing shot wave to measure[4].

However, mostly infrared diagnosis is decided via compare method, usually it only needs finding relative changes of temperature, does not need to pay much attention to the precision of the error changes. But when do accurately measure of thermodynamic temperature, we should know the measured objects emissivity. Otherwise, there will be large error between measured temperature value and practical value, the maximum error can be achieved in 19%. The products data of infrared measuring temperature instrument usually have the corresponding range of the measured objects emissivity that can be used with this instrument. Due to there are many influence factors, so every kind of objects emissivity offers reference value, and only can be used in restrictive working waveband and temperature of the instrument. In addition, emissivity connects with the test direction, it is better to keep testing the test angle lower than 0°, and better not exceeding 45°. When the angle is not exceed 45°, a further correction of emissivity is needed.

3.2 Influence of Environment Temperature

The relative temperature difference changes (see Table 2) according to the environment temperature are simulated in laboratory (the temperature of detected equipment is 298K). From this we can draw the following conclusions: if the temperature difference between the detected objects and the environment is small, the caused relative temperature difference is large; if the temperature difference between the detected objects and the environment is large, the caused relative temperature difference is small. In addition, the different workload of equipment also can cause the changes of temperature difference.

Table 2. Relative temperature difference δ_t changes with environment temperature

δ_t (%)	Environmental temperature (K)					
	300	310	320	330	340	345
2～5μm	11.60	2.25	1.89	1.46	1.45	1.45
8～13μm	6.52	4.99	3.90	3.20	3.15	3.10

3.3 Influence of Working Waveband

According to the characteristics of infrared radiation, the higher the temperature the greater the radiation energy, the higher the temperature the shorter the wavelength.

The working wavelength of commonly used infrared thermometer usually has two kinds, short wave band 2~5μm and long wave band 8~13μm.

The temperature in 300~500K (27~227°C) corresponds to the peak wavelength radiation between 5.8~9.65μm, therefore the surface radiation energy is mainly in the range of 8~13μm long wave band of the spectra zone. Only when the aim of testing temperature is around 800 K (527°C), the Infrared equipment with short wave band of 2~5μm spectra zone should be chosen. So a suitable working waveband needs to be chosen before measuring temperature.

3.4 Discrimination Method

Above experiments indicate that when the relative temperature difference discrimination method is used in fault determination, the following conclusion can be referred to: the equipments have general defects under normal circumstances; the equipments have significant defects; the equipments are considered as to have emergency defects. According to different values, the fault-level of petrochemical equipment failures is made, and then the corresponding methods are taken.

4 Conclusion

From the above analysis, we know that relative temperature has relationship with ε, T_u and infrared imaging system, we can synthetically consider the influence of various factors in practical use, and then make a proper judgment on device status.

References

1. Zhao, Y.-L.: Infrared scene practical diagnostic technology. Machinery Industry Press, Beijing (2004)
2. Lang, C., Jin, G., Xu, D., Liu, T.: Analysis of influence factors on relative temperature difference judgement method of electrical equipments. Infrared and Laser Engineering (01) (2011)
3. Peng, C., Zhao, J., Miao, F.: Distributed temperature system applied in cable temperature measurement. High Voltage Engineering 32(8), 43–45 (2006)
4. Liu, J.-H., Chen, C.-K., Yang, Y.-T.: An inverse method for simultaneous estimation of the center and surface thermal behavior of a heated cylinder normal to a turbulent air stream. Journal of Heat Transfer, Transactions of the ASME 124(8), 601–608 (2002)
5. Hsieh, C.K., et al.: A general method for the solution of inverse heat geometries. Int. J. Heat Mass Transfer 29(1), 47–58 (1986)

Key Technologies in Digital Library Ontology

Qiong Su

Guangxi Economic and Management Cadre College, Nanning, China, 530007
1185561517@qq.com

Abstract. Several ontology-based digital library tasks such as, ontology building, ontology mapping. Scarlet1 is a technique for discovering relations between two given concepts by exploring ontologies available on the Semantic Web as a source of background knowledge. By relying on semantic web search engines such as Sraet automatically identifies and combines relevant information from multiple and heterogenicnexus online ontologism. This paper will be accompanied by a discuss on functionality available in key technologies in digital library ontology. The Sraet-based digital library ontology construction approach has already been used successfully to support a variety of tasks, but is also available as a stand alone component that can be reused in various other applications.

Keywords: Key Technologies, Digital library, Ontology.

1 Introduction

In recent years, China began working on library and information sector ontology in the field of digital libraries applications to digital library resources for the effective organization, use and sharing new breakthroughs. In this paper, Tongfang cnki database, VIP database and Wanfang database as a search tool to find out which collection of the five years 2005-2009, published in China's ontology-based digital library research literature. Through the analysis of the literature, attempts to describe the past five years, domestic scholars in the field of ontology research, analyze its characteristics and trends.

Semantic network of Digital Library of knowledge organization and changes in the emergence of new technology, traditional knowledge organization can only provide one-dimensional, linear information retrieval, can not provide semantic retrieval, semantic network environment can not meet the requirements under . Semantic Web Ontology as the core, can describe very complex concepts, attributes and relationships, the formation of semantic knowledge networks, semantic knowledge retrieval, it becomes the semantic network environment, the new digital library knowledge organization method.

Ontology (ontology) is a philosophy of the first category, and later with the development of artificial intelligence, artificial intelligence community has been given a new definition. However, the initial understanding of people's ontology is not perfect, these definitions are constantly evolving, the more representative definition.

D. Jin and S. Lin (Eds.): Advances in CSIE, Vol. 1, AISC 168, pp. 261–266.
springerlink.com © Springer-Verlag Berlin Heidelberg 2012

2 Related Work

The ontology is domain-specific knowledge of the objective world of abstract conceptual model. Conceptual model refers to the objective world in a number of phenomena related to the concept by identifying its abstract model obtained, the concept system. Concept system, including concept definition, concept property value and property value limitations, the concept of class hierarchy system (is_a, kind_of, part_of, etc.), the concept of the logical relationships between classes (caused_by, used_by, interact_with, supervised_by, etc.) and inference rules.

Niches's ontology defines the vocabulary related fields constitute the basic terms and relationships, and the use of these terms and provisions of these terms constitute the relationship between the extension of the rules. TRGruber pointed out: Ontology is a clear conceptual model specification. In Borst's paper: Ontology is a shared conceptual model of the formal specification.Studer (1998 years): Ontology is a shared conceptual model of a clear formal specification.F. Fonseca (2002 years): the ontology is a clear view of the glossary with detailed description of entities, concepts, features, and related functions of the theory. Zhang Xiaolin discussed that the so-called set of concepts (Ontology, also known as application of knowledge), refers to the specific application area recognized on the field of object (including the actual objects and logical objects) and object relations conceptualization of expression. Liu Wei claimed that Ontology is the specification of the abstract domain knowledge and description, is to express, share and reuse knowledge approach. Although people do not have a ontology of clear definition, but from the above definition of the statements we have been able to summarize the basic content of the ontology:

3 Key Technologies

Theoretical Study of Ontology. Domestic ontology of research begins in the theoretical study of the ontology, present, scholars have reached consensus on the concept of the ontology, are used Borst's notion that "ontology is a formal specification of shared concepts." The research papers, many scholars of the ontology and the classification, metadata, relationships and their impact on digital library research and other aspects.

Ontology and taxonomy, thesauri and metadata Ontology is the study of the relationship because the relationship between concept and the concept, and library and information science in the standardization of vocabulary has many similarities, thus causing the domestic Library information science researchers of great interest. About Ontolo a gy with the traditional classification and thesauri and metadata between the differences and study to become a common concern of many research topics.

Scholars say that: classification and thesauri are used specification language, with a certain standard style, structure, stability, and relatively conservative and difficult to modify, you can express simple semantic relations; and ontology concepts using natural language or semi-natural language to express, is an open integrated system, you can always revise, update and reuse, semantic relations described in more depth and extensive. Metadata and ontologies in common are: the use of standard coding language for formal treatment, which can provide the semantic basis for the resource can be used to organize resources and resource discovery. The difference in performance: meta-data is difficult for different knowledge systems, to describe the different size of resources, and provides a ontology of metadata between different mapping mechanism that can achieve interoperability between heterogeneous systems.

The bulk of the impact of digital libraries in the library and information field of the ontology, many aspects of applications, such as indexing the literature, knowledge management, knowledge base building, library information resources to build and so on. Especially in the digital library research and construction, play an important role in the ontology, the ontology which is being used in all aspects, such as information organization, information retrieval and interoperability of heterogeneous information systems, etc. Scholars believe that with the library and information science theory and development of information technology in various disciplines will be their unique identity, which will give the traditional use of information organization and information to bring a fundamental change.

Ontology Construction. The study of ontology should not be confined to the theoretical model, the more important is the actual operation should be - to build ontology, and its application to digital libraries. At present, domestic scholars to build the ontology, there are two main methods:

The use of automated or semi-automated approach to building the ontology from the dictionary or structured, semi-structured data or text extract or learning or discovery domain ontology. According to ontology learning different knowledge sources. For the use of automated or semi-automated approach to building domain ontology approach to Category: Ontology Learning from the dictionary, will build a ontology built on the existing machine-readable dictionary, based on the extract from the relevant concepts and relationships between concepts; learn from the knowledge base by learning from existing knowledge base to build the ontology; extract ontology from a relational database; data from semi-structured learning; learning from text. If Liubai Song presents a digital library for the automatic ontology building methods, including term selection, extraction ontology concepts, semantic relations extraction, classification system build, ontology building and ontology trim and evaluation. Automatic extraction of the basic process domain ontology can be summarized as follows: the source of knowledge extracted from a representative of the concept, the formation of domain ontology concepts; concept by analyzing the properties and clustering in the relationship between elements and other elements in the extraction of ontology the concept of relationship; ③ the relationship between concepts to solve the conflict.

Ontology-based Organization and Integration of Information Resources. Integration of information resources is to achieve the goal of building digital libraries a key, and the integration of information resources related to the rapid development of many new technologies and new methods, the ontology is that these new methods. Digital library information resources, including a number of levels, such as text information resources, multimedia information resources and knowledge management warehouse. Based on this, the current ontology-based digital library information resources to build a multi-layered system, from the macro consists of three hierarchies, namely ontology-based information resources to build, Web information resources to build the knowledge base and knowledge management in construction and so on. Such as Zhang Minqin discussed the number of ontology-based library information resources to build buildings operational methodology. Wang discussed the XML-based ontology language to describe digital library Web information resource integration function of the system and its implementation approach.

In addition, the ontology can use the dynamic organization of information resources, in a retrieval system, the literature identifies the user to effectively docking the question, that question based on the user model constructed questions, and retrieve results based on the model structure identification, the question model and identify the model in the field of semantic level through ontology mapping, in order to achieve.

Ontology Mapping. Now more and more application of the ontology, but because of the distributed nature of the Semantic Web, resulting in a large number of ontology heterogeneity of different sources of information that exists in the ontology between the heterogeneous phenomenon, a system of mutual understanding, exchange of information to achieve interoperability of the main obstacles. To complete the task of information exchange in the ontology must be erected between the semantic mapping of the bridge. Ontology mapping can solve the problem of heterogeneous ontologies. It is found that two of the same domain ontology between the concept of correlation (mapping) process, but also the ontology integration ontology mapping, ontology merging, ontology correction, ontology translation technology base. To solve the problem of heterogeneous ontology mapping, domestic researchers have made a lot of mapping and mapping technologies, such as Bi Qiang, Han Yiji in semantic grid, exploring ontology-based digital library metadata, interoperability between systems strategy, proposed the next generation of digital library systems interoperability framework - based on the DL ontology metadata interoperability framework. Liu Cheng-shan, Zhao did not give the peer holding the digital library environment, an ontology mapping algorithm, from grammar, vocabulary and context aspects of the concept of three matches and focuses on the context for the similarity computation, Simulation results show that the system through the construction and mapping algorithm is effective.

4 Approach

Domestic ontology-based digital library research and the lack of features

In the research process, I found that the domestic research on this topic reflect certain characteristics, there need to be perfected:

Ontology-based national study of the characteristics of the digital library

has grown rapidly, the initial formation of the core authors, such as the preceding analysis, although the domestic ontology-based digital library research every year to heat up, and the rapid growth of the number of research papers. At the same time there has been such a team led by DONG Hui and other follow-up study of the ontology, the authors have formed the core, the existence of the core authors and core authors to the amount of a certain extent, reflects the maturity of a research topic. Can be seen, the majority of scholars currently study on this research topic is heating up.

The study of specific applications and gradually move closer analysis of the characteristics from the current study, surrounding the ontology and the digital library research has gradually developed from a theoretical introduction to the ontology in a variety of practical applications of digital libraries. At first, the ontology often occur, such as the impact of digital library content of some of this literature, and as scholars study the deepening of the ontology, from the resource organization, information retrieval, personalized service, more detailed mapping technology research point of view to carry out research on this topic is becoming mainstream. Xie M developed an ontology cloud, which includes open liberary ontology data.

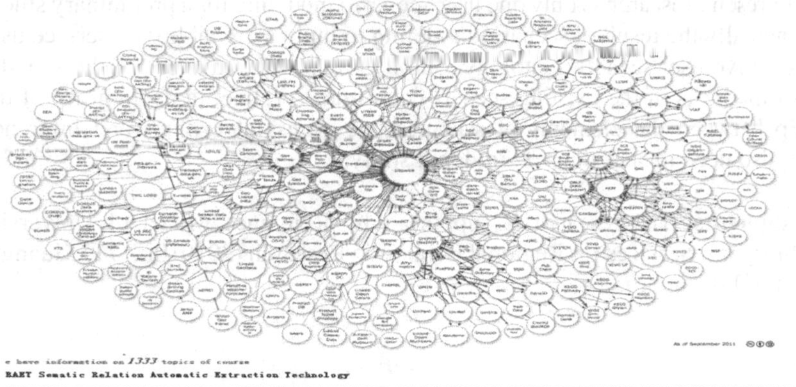

Fig. 1. Library Ontology Cloud

Although the ontology in artificial intelligence, knowledge representation has been widely discussed and practice, but in our intelligence community and the library in the field of digital libraries research study confined to the definition of the concept and preliminary theoretical research level, the ontology methodology, description languages, build tools, the lack of concrete to build the practice. From the research results, although the number of articles reflecting the domestic academia or industry ontology-based digital library research is more enthusiastic, but most of the articles to be discussed only in theory, its application to build ontology and little research carried out . There are many papers simply presented some model or ontology-based mechanism, the lack of deep understanding of the practical application of the ontology. Only a few of the several documents in practice to build a real ontology, and applied to the digital library, in fact, it is this research really help promote the construction of digital libraries in the introduction of better ontology, the application the ontology.

5 Summary

According to the above analysis of the domestic ontology-based digital library research shortcomings, combined with foreign ontology of recent research on the analysis of results, I believe that the future of the domestic ontology-based digital library research may have the following trends :

The use of ontology technology in-depth personalized service. Currently on the ontology-based personalized services focused on user modeling, personalized recommendation system, etc., point for future research are: How accurately reflect the user ontology user interest in attenuation, personalized recommendations to further improve the rules, how to integrate application of the recommended techniques personalized recommendations ontology learning resources strategy, that is content-based recommendation and collaborative filtering techniques to effectively integrate technology to achieve in different situations with different recommended techniques. User scenarios involving sensitive digital library services for the user scenarios, scenario modeling description, scenario reasoning, scene services and other aspects. At present, research is only one user scenario modeling for a preliminary study, the user is actually the result of scenario modeling to build a digital library service user scenario sensitive user scenario ontology, ontology construction needs of the specific elements of the basis of the existing scenario continue to refine, in-depth grasp of the relationship between the elements of scene, scene access algorithms and inference algorithms also need to start the next in-depth study.

Acknowledgments. The research is supported by Reference Model of University Digital Library in Services for Regional Economic Research (201010LX715 Guangxi Department of Education).

References

1. Yan, Q.: Ontology and its application in the field of Library and Information Summary. Heilongjiang Science and Technology Information (35), 18–182 (2008)
2. Xie, M., et al.: Open Rainbow Services-Oriented Testbed: Low Cost Cloud Computing Platform for Open Systems Research. In: Intelligent Systems and Applications, ISA (2010)
3. Su, Q.: A hidden Markov model of library users classification. In: 2010 Second International Conference on Computational Intelligence and Natural Computing Proceedings, CINC, pp. 117–120 (September 2010)
4. Xie, M., et al.: Intelligent Knowledge-Point Auto-Extracting Model in Web Learning Resources. Journal of Computational Information Systems, 2223–2330 (July 2010)
5. Xie, M., et al.: A New Intelligent Topic Extraction Model on Web. Journal of Computers 6(3), 466–473 (2011)
6. Xie, M.: Knowledge Topic Aware-based Modeling, Optimzation and on Web of Collaborative Logistics System. In: International Conference of China Communication Technology (2010)
7. Xie, M.: Semantic Knowledge Mining on Web. Journal of Computational Information Systems, 3787–3794 (November 2011)
8. Xie, M.: Intelligent Knowledge-Point Auto-Extracting Model in Web Learning Resources. Journal of Information and Computational Science (December 2012)
9. Qiao, Y.: The domestic field of Library and Information Science Research of Ontology. Modern Intelligence (9), 121–124 (2006)
10. Peng, C., Min, L., Yang, F.: Ontology-based Knowledge Retrieval Digital Library Research. Information Theory and Practice (5), 78–80 (2009)

Factor Analysis on M-learning Effect and Strategy Based on Information Technology

Lin Hu[1], Honghua Xu[2], Shuangzi Sun[2], and Chunfeng Xu[2]

[1] Jilin University of Finance and Economics, Changchun, Jilin, China
[2] Changchun University of Science and Technology, Changchun, Jilin, China
{huhu315,honghuax}@126.com
shuangzis@yahoo.com, chunfengx@hotmail.com

Abstract. Because of the limitation of handheld devices factors influencing mobile learning effect are different from traditional learning pattern. Theory shows that effective information processing strategies adopted by teachers according to the features of m-learning environment and students are key to success.

Keywords: m-learning, learning effect, influencing factor, processing strategy.

1 Introduction

M-learning is that learners can study anytime, anywhere with mobile devices (e.g. cell phone, PDA etc.) and wireless net to obtain learning materials and communicate and collaborate with others according to their personal needs in order to realize personal and social knowledge construction. Because mobile devices have the features of convenience, collaboration and flexibility etc, m-learning study has caused a hot study tendency in education circle within several years [1]. At present, the main study fields include the following: a) m-learning system structuring; b) m-learning material base structuring; c) m-learning relative theory; d) m-learning teaching pattern study etc. and all these studies have achieved abundant fruits. However, the author finds out that the study based on m-learning resources is mostly concerned with techniques related with resources system after analyzing relative studies of m-learning resources. Of course, effective technical support is the basis of learning method [2]. However, m-learning is a model of teaching and also is a teaching activity. So it is more important to obtain the best teaching effects under the support of an effective technical platform. M-learning, in essence, is a way of transmitting teaching information [3]. It will inevitably follow the basic rules of information transmission. But because handheld mobile devices own shortages of low storage, weak calculating ability and small screen, bad connection with internet and environmental influence, factors of influencing m-learning effects may be the same with those of influencing traditional e-learning effects. It bears its own features and rules [4]. Only when we are clear about the factors of influencing teaching under the circumstances of m-learning can we design a successful teaching model. Hence, this issue has cause a wide range of attention.

D. Jin and S. Lin (Eds.): Advances in CSIE, Vol. 1, AISC 168, pp. 267–272.
pringerlink.com © Springer-Verlag Berlin Heidelberg 2012

2 Factor Analysis on Influencing M-learning Effect

M-learning is a way of transmitting educational information. Hence, it must follow the theory of classic information transmission (as is shown in Fig. 1) and if we'd like to apply it to teaching procedure, we can see that teachers or courseware designers act as information source coding [5]. Teachers or courseware designers provide their code to the system which chooses certain and specific method to transmit information. The reception end receives signals and then restores information to its original multi-media elements. Students (recipient) receive these multi-media information elements and then digest them to change them into their own information [6]. After analyzing the whole procedure of teaching information transmission, we can see that the whole transmission system in fact involves in information source coding (teacher), information transmission system and information sink (student). So in order to obtain the success of teaching activity, three aspects are needed: a) a better information-coding ability; b) a reasonable information transmission channel; c) a students' effective reception.

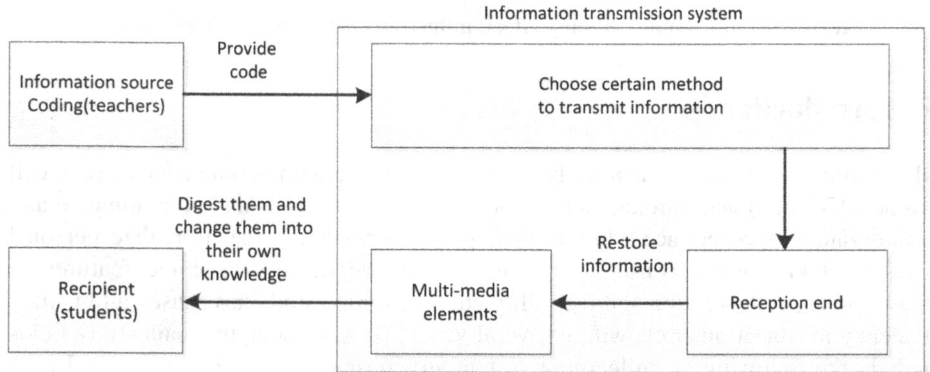

Fig. 1. The theory of information transmission

Any type of learning method is measured through students' learning effects [7]. However, the improvement of students' learning records depends on effective understanding to teaching information. Generally speaking, the clearer information students receive ,the better result they will obtain. To get this target, m-learning must satisfy the following conditions: a) teachers must produce small noise as possible as they can while coding teaching information, which includes the following two elements: students should accurately describe and demonstrate teaching information and teachers' coding must adapt to students' features, i.e, information flow provided by the system must match students' reception ability so that they can realize the best teaching information reception. If the provided information flow is too much, students cannot understand them in a short time, information loss and students' understanding obstacle may be resulted. If the provided information flow is not enough, students may get bored [8]. All these situations can cause students to get bored with learning. b) the information channel used to transmit m-learning information must receive small interference as small as possible. Generally speaking, m-learning information channels are all wireless channels which technically adopt microwave transmission. It is well known that microwave bears the feature of linear transmission. While transmitting, they are easily obstructed by obstacles such as high buildings, mountains etc. and weakened by signal blind areas. These situations may badly damage students' initiative to study. Besides, this kind of information channel has low efficiency of information transmission. If large amount of teaching information is needed to be transmitted, it is quite easy to fail. c) the complexity of environment should be overcome while students are doing m-learning. Teachers must consider all the environmental factors which may cause low learning efficiency to design strategies.

2.1 Strategy Analysis

At present, according to m-learning features, the following strategies should be followed:

Media Selection Strategy. From cognitive aspect, any knowledge acquisition (as is shown in Fig. 2)must experience procedures of perception, thinking, remembering and imagining etc. It means that knowledge acquisition should be realized on the basis of perception. So what kinds of perception knowledge we provide may provide great influence on learning effect. Generally speaking, through vision perception, people obtain the largest amount of information and can get the best result, whereas through hearing perception the result may be less favorable. Hence, while designing courseware, we should first design vision perception materials.

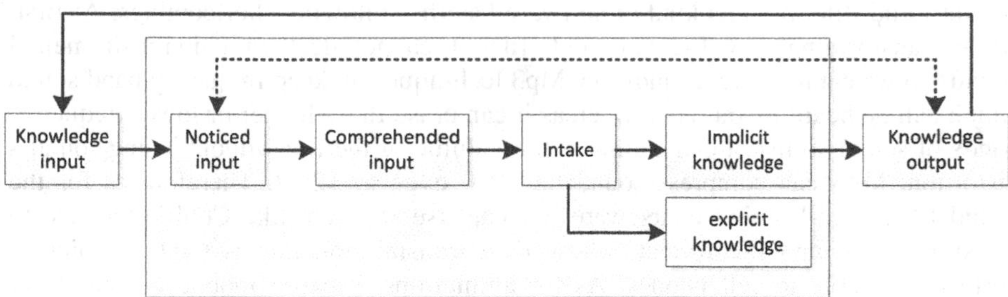

Fig. 2. Knowledge acquisition

Among perception elements, we shouldn't only use one element to demonstrate all the materials [9]. Because different vision elements have different demonstration features and there also exist different costs. To obtain the most favorable teaching effect and the most optimum performance cost ratio, the following aspects should be seriously considered while selecting media elements. Firstly, we should decide on whether or not we need vision images to demonstrate. If so, whether or not should we choose static images to meet the demands. If they can meet the demands, dynamic pictures or videos are chosen to demonstrate. Secondly, if teaching content doesn't need vision information, then we should analyze whether or not we should use texts and notes to demonstrate. If this cannot meet the teaching demands, we should adopt appropriate music as background information.

Technical Strategy. Though learning efficiency can be improved by stimulation of various senses, we cannot deal with various kinds of multi-media information as in traditional e-class because of the limitation of data transmission speed of wireless net [10]. So we need techniques to deal with this problem. The general rule of technique strategy is to make multi-media data smaller. The basic principle is to cope with multi-media elements which should be operated in m-learning devices as follows. As for graphs and images, because the demonstration of vector chart needs corresponding multi-media players, it is quite difficult to realize in handheld devices with limited resources. At present, cell phones only support the following bitmap files such as bmp, jpg, gif etc. Among these three types of files, because the compression ration of gif format is high and it needs small space to store and also has high speed, it bears greater advantage in dealing with pictures we need in m-learning courseware [11]. Generally speaking, as for video information, most video cameras take video saved as avi files which have the features of high resolution, better compatibility and they are supported by many media players but this kind of file has large video file so that handheld devices like cell phones cannot operate them. Generally, cell phones can identify and play video files like mp4 and 3gp. Mp4's biggest bit per second is 384kbps, whereas 3gp's biggest resolution is 176*144, 30fps and it also supports H.263 and MPEG4-SP encoders. Therefore, videos in courseware should firstly use corresponding software to change the format. As for sound information, cell phones mainly support midi file and mp3 file not avi file, whereas sounds in courseware are usually used to explain and create special effects. The explanation sounds and special effects sounds are usually saved as wav file. So before using them, we must change file format. Currently, we use software such as Amazing MIDI etc. to transform wav file into midi file, but this kind of software has the following shortcomings. At first, while transforming wav file into midi file, it cannot deal with non-instrumental sounds so we cannot save as midi file. Mp3 technique can keep frequency band sound which can be heard by our ears, whereas it can erase those higher or lower frequency bands of sound so that it can compress the volume of the file without losing sound's distortion. Mp3 can compress sound to 12:10 even to 12:12. Therefore as for the sound files in m-learning courseware, we can use software like CoolEditpro etc to transform into mp3 file format, whose data transmission rate is 64kbps which is perfectly suitable to cell phones. As for animations, because mobile devices have limited resources, we cannot install corresponding software into cell phones. So cell phones cannot play them. Generally, we adopt the following strategy: we transform flash animation into mp4/3gp file format and then play them in cell phones. Besides, we can issue animation as gif format but this format doesn't support sound.

2.2 Organizing Strategy of Teaching Information

Because m-learning has its own features, coping strategy different from traditional e-teaching strategy should be needed while organizing and dealing with m-teaching information [12]. To be specific, we should use small-step strategy. Because of the complexity of m-learning environment, the content for a whole class should be broken into smaller units to study. While designing this courseware, we should pay close attention to the efficient connection among knowledge points and adopt gradual strategy. Generally speaking, while studying with handheld devices such as cell phone, students may not have good interaction as in a traditional class. If they meet too many difficulties while studying, it will badly damage their study initiative. So when we organize and design teaching activities, the order of the logical system of the subject should be obeyed and the systematicness and coherence of knowledge should be strictly obeyed [13]. As for the way of demonstration, we should adopt the method from known knowledge to new knowledge, from easy knowledge points to difficult ones. Another strategy is the strategy of brief teaching information demonstration. Because handheld devices have the features of small screen, students don't have field of view width like using a CRT screen, not speaking of repeated appreciation. So the core content of the picture and the main body of the subject must be outstanding. Organization of the content should be brief and accurate so that students can grasp the content at a glance of it. As for the description of texts, we should adopt brief strategy. If necessary, explanation strategy should be adopted to make students be clear about them. The third strategy is the combination strategy of theory and practice. We should try to be good at connecting students' actual situation with reality. The content of courseware should base on students' knowledge level, living experience, reception ability and thoughts etc. to form combination between basic knowledge and students' direct experience [14]. Simultaneously, according to the subject feature and specific teaching content, we should put industrial production practice and modern scientific knowledge fruits into the strategy. The last strategy is the combination between intuition and abstract. While making courseware, various multi-media elements should be adopted to show teaching information, which may make students receive teaching information with various senses to realize improvement of teaching efficiency. So with the permission of technology, we should try our best to use intuitive pictures and sounds to express teaching content. At the same time, with intuitive images, texts are also needed to summarize the main idea of the teaching content so that students may grasp the knowledge system and cultivate abstract ability of thinking [15].

3 Summary

Because of the technological limitation of m-learning, traditional teaching and methods are not suitable while adopting m-learning to teach. We should sufficiently consider the technological features of mobile techniques as well as the particularity of students' learning environment to effectively re-organize teaching information, which will make teaching activities conform to the reality of students. If so, we can avoid the embarrassment that we experience m-teaching only through short message. Finally, we would like to realize the true value of m-teaching.

References

1. Chatham, M.A.: Knowledge lost in information. Report of the NSF Workshop on Research Directions for Digital Libraries (6), 15–17 (2003)
2. Cook, J., Bradley, C., Lance, J., Smith, C., Haynes, R.: Generating Learning Contexts with Mobile Devices. In: Pachler, N. (ed.) Mobile Learning: Towards a Research Agenda. Occassional Papers in Work-Based Learning, WLE Center for Excellence, London (2007)

3. Wang, X., Li, Y.: Mobile Learning Based on the Service of Cell Phone Short Messages. China Audio-Visual Education Journal (1), 114–117 (2007)
4. Flavell. Metacognition and cognitive monitoring: A new area of psychology inquiry. In: TON also ed. Metacognition: Core Readings, pp. 3–8. Allyn and Bacon, Boston (1979)
5. Brown, T.H.: Beyond constructivism: Exploring future learning paradigms. Education Today (2), 1–11 (2005)
6. Li, Y., Li, R.: Cultivating Meta-cognitive Ability to Teach Students How to Learn. China Education Journal (1), 32–35 (1999)
7. Wang, Y., Zhang, L.: How to Cultivate Students Meta-cognitive Ability. China Audio-Visual Education Journal (8), 21–23 (2000)
8. Li, D.: Problems on Teach Students How to Learn. Journal of Northwest Normal University (Social Sciences) (1), 3–9 (1994)
9. Dickinson, L.: Self-instruction in Language Learning. Pergamen, Oxford (1981)
10. Pelletier, C.M.: Successful Teaching Strategies: An Effective Instruction of Teaching Practice. China Light Industry Press, Beijing (2002)
11. Druck, P.F.: Knowledge Management. China Renmin University Press, Harvard Business School Press, Beijing (1999)
12. Vail III, E.F.: Knowledge Mapping: Getting Started with Knowledge Management. Information Systems Management 16(4), 16–23 (1999)
13. Barth, S.: Creating and Managing the Knowledge-Based Enterprise. Knowledge Management World Magazine 12(1) (2003)
14. Technologies for Personal and Peer-to-Peer (P2P) Knowledge Management, http://www.csc.com
15. Xu, C.: Two Tendencies of Personal Knowledge Management. Modern Educational Technology (4) (2009)

Design and Exploitation of Energy MIS Based on Data Integration

QingHua Pang and TianTian Lu

Business School, HoHai University, Changzhou, 213022, China
{pangqh, lutt}@hhuc.edu.cn

Abstract. Under a detailed system analysis, the paper designed and exploited the energy management information system (MIS) for one petrochemical company based on data integration with the principles of software engineering. The energy MIS had three functions: to collect data automatically from heterogeneous database, to process the complicated energy balance, and to optimize energy assignment with linear programming. The energy MIS can also offers decision-making support for the head office to comprehend the entire information of the company at appropriate time and manage the energy optimization.

Keywords: Energy Management, Data Warehouse, Decision Making Support.

1 Introduction

With the rapid development of the web, more and more information has been transferred from static web pages (that is Surface Web) into web databases (that is Deep Web) managed by web servers. Public information on the Deep Web is currently 400-500 times than that of Surface Web. Compared with Surface Web, Deep Web databases generally belongs to specific domains, which the subjects are simplex, therefore, the idea that organizing and integrating Deep Web information according to specific domains has been recognized by the most of researchers. In order to achieve the integration of Deep Web databases, an important step is to automatic classification to web databases. Artificial classification is a time-consuming work, so it is imperative to accelerate research on automatic classification of Deep Web databases.

The petrochemical company, discussed in this paper, built in September of 1983, is one of the super national petrochemical enterprises. It engages in the production, processing and sale of petroleum refining and hydrocarbon ramification mainly. At present, it has 5 big energy consumption units: chemical plant, aromatic hydrocarbon plant, alkene plant, refinery plant and plastic plant. Each of these plants has its production management system, but these systems are each independent and can't share data. In addition, there are two important energy consumption units: water plant and steam power plant. The two plants directly use EXCEL to carry on statistics and manage the energy consumption information. In each quarter, the head office needs to count and deal with the energy consumption of the whole company to make decision to distribute new energy. In this process, it needs to invest numerous financial

D. Jin and S. Lin (Eds.): Advances in CSIE, Vol. 1, AISC 168, pp. 273–276.
springerlink.com © Springer-Verlag Berlin Heidelberg 2012

resources of manpower and material. Especially, the time-validity of the whole course data is worse, so the work is time-consuming, laborious, and apt to make mistakes. Under the environment of implementing Enterprise Resource Planning (ERP) in the chemical trade, aiming to design and exploit a set of energy management system for the department of energy management, this paper launches the company to support its decision.

2 System Function Analysis

This system is mainly designed for the department of energy management of the company. It involves control center, chemical plant, aromatic hydrocarbon plant, alkene plant, refinery plant, plastic plant, water plant, and steam power plant. Water plant and steam power plant can't directly produce chemical production, but these two plants offer the whole company all water and electricity. They are important departments to offer and consume energy.

According to the principles of simplification, module, practicality and optimization, the paper sets up the framework of the system. It includes system set-up, energy management, system template, system maintenance, net service and help. The overall framework is like Fig.1.

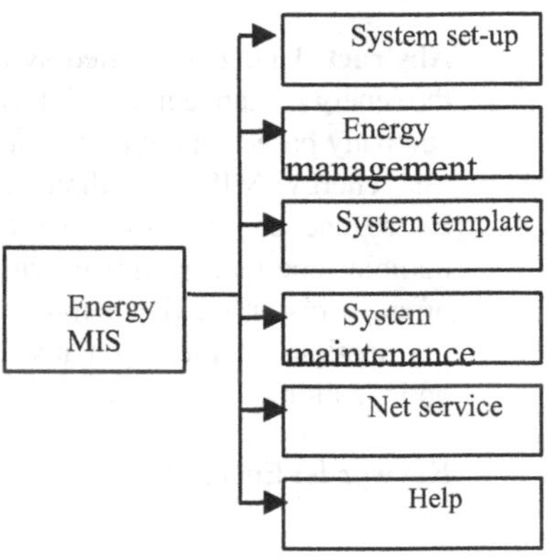

Fig. 1. System Framework

3 Design of Decision Optimization Model

The core of management is decision, and one of the final purposes of this system is to support decision making in enterprise. The system can unify the energy information distributed in every production management system to serve the head office. In this aspect, the system offers 3 modules to support decision: early warning (to control the energy consumption of every plant), equilibrium treatment (to balance the energy data), and optimized distribution (to optimize the energy assignment according to production plan). Here, we use optimized distribution model to explain how we design the decision optimization model.

The optimized distribution model in this system is a typical linear programming model. Its goal is to minimize the total energy consumption under the condition of guaranteeing production, and to optimally distribute energy according to the standard. Here, linear programming method is used in this system to construct optimized energy distribution model. Concrete to this system, the energy consumption is the goal function:

$$Z = W_2 = \sum_{i=1}^{14}\sum_{j=1}^{7} k_i (X_{ij} - Y_{ij}) \tag{1}$$

There are two kinds of constraints: Production constraints and equipment ability constraints (mainly means water and electricity

$$\sum_{j=1}^{7}\sum_{i=1}^{14} k_{ij} X_{ij} \geq b_j \tag{2}$$

$$\sum_{i=1}^{14}\sum_{j=1}^{7} X_{ij} \leq d_i \tag{3}$$

In formulas above: X_{ij} means the *i*th energy consumption of the *j*th plant, k_{ij} means consumption coefficient of the *i*th energy for products of the *j*th plant, b_j means the amount of production of the *j*th plant, d_i means the max that the *i*th energy can produce. Here, b_j is set by users, k_{ij} can be set through experience and not often changed. Only when equipments are worn out or upgraded, can user alter while need.

4 Database Design

The energy information of the company is not only numerous and complicated, but also distributed in the production management system of every plant. The production management system and database platform are not unified, so it needs to gather data from heterogeneous platforms and then transform them into uniform format. Therefore, this paper introduces the concept of data warehouse. It has solved the problem of data acquisition from heterogeneous platforms by using the characteristic of data warehouse and enables these data to support the decision better.

One of the main tasks of energy MIS is to calculate a large number of miscellaneous energy data through searching, and to offer information for decision by analysis. So, the system has to store and manage a large amount of data. Therefore a well-organized data structure and database can enable the whole system to transfer and manage the necessary data high-efficiently and accurately. The design of database is one of the important measuring indexes to weigh the work of energy MIS.

There are 8 databases in energy MIS, one of which is main database and the left are transferring databases. In addition, there are also 4 shared datasheets. The transferring databases are used to store the data gathered from 7 plants and the main database is used by the department of energy management. In the transferring databases, the database of chemical plant, alkene plant, aromatic hydrocarbon plant, refinery plant and plastic plant mainly has time-accumulator datasheets, statistical template and statistical bill of energy consumption. In database of steam power plant, consumption and statistic are separated, and each has its different template, bill and time-accumulator datasheets. The structure of the datasheets has its own mode except time-accumulator datasheets, which is different to the 5 plants mentioned above. The database of water plant has its own energy load template and bill besides time-accumulator datasheets, template and bill of energy consumption. The data of the main database comes from the 7 transferring databases. There are about 57 datasheets in the main database, including consumption template of purified water, desalted water, electricity, 103K, 95K, 41K, 14K, 8K, nitrogen and oxygen, bill, report forms, air separation, parameter of energy etc..

The database has solved the problem of data acquisition from heterogeneous platform by absorbing the concept of the data warehouse, and can offer help to carry on the complicated energy processing. By utilizing the network, the system can also carry on the long-range data transmission and processing function. In addition, it has reserved the interfaces, which makes the system have great expansibility. The company can expand and link the system when needed in more complicated environment.

5 Conclusion

Aiming at the actual demand of the company, by detailed analysis of energy information management, the authors design and exploit the energy MIS. The energy MIS has solved the problems of data gathered from heterogeneous platform automatically, balance processing automatically and optimized energy distribution.

According to the operation condition, the system can meet the demand of energy information management. It has improved the efficiency and accuracy of the energy management and production. It offers one effective means to reduce the energy consumption.

References

1. Zhao, L., Wang, H.: Primary Study on the Develop of the National Energy Resources Management Information System. Industrial Heating 12, 15–19 (2007)
2. Amundsen, A.: Joint Management of Engery and Environment. Journal of Cleaner Production 8, 483–494 (2009)
3. Green, J.F.: Mangagement Information Systems and Corporate Planning. Long Range Planning 4, 74–78 (1990)
4. Antje, J.: Use of Enterprise Resource Planning Systems for Life Cycle Assessment and Product Stewardship. Carnegie Mellon University (2001)
5. Li, S.F.: Informationization in Chemical Trade and ERP Selection. Chemical Industry Management 5, 32–33 (2008)
6. Pang, Q.H., Zhou, B., Pan, Y.: Design and Explotation of a Flexible Assistant Scheduling System: AJ-System. Computer Engineering and Design 12, 43–46 (2008)
7. Huang, T.S.: Study of Material Guarantee System Based on Minimum Cost and Its Application. Central South University (2006)
8. Zhou, X.Z., Li, Z.Y.: Supply-demand Models for Time and Space Resources of Urban Road and Their Application. Journal of Shanghai Maritime University 3, 16–22 (2008)
9. Wang, M.L., Guo, S.L.: Analysis of Balance Model of Monthly Water Quantity and Its Application. Yangtze River 6, 32–33 (2007)
10. Zhang, B.M.: Study of Data Reconciliation Model and Arithmetic in Flow Industry. Zhejiang University (2005)
11. Wang, X.Z., Zhang, S.W.: The Application of Data Warehouse in Modern Companies. Journal of Jingmen Vocational Technical College 6, 28–33 (2007)

An Improved Incremental Learning Approach Based on SVM Model for Network Data Stream

Na Sun and YanFeng Guo

School of Electronics & Information Engineering, Liaoning University of Technology, Jinzhou, China, 121001
yaodoctor@gmail.com

Abstract. Network attack detection is an important aspect in maintaining a stable network. The normal method is using the rule-based approach, such as firewall area. However, the rule-based method can't be changed with outside data dynamic environment under data stream. In this paper, we proposed an incremental learning-based model based on SVM and try to optimize the model in two aspects, time consuming and classification accuracy. Apart from this, we also compare our method with two other models, which are k-nearest neighbor algorithm (KNN) and decision tree. The experimental results show the robustness of the proposed model.

Keywords: incremental learning, support vector machine, data stream, network data classification.

1 Introduction

In network, many models have been proposed for attack detection. Most of them are employed in the static data environment usually. However, in network data stream situation, the tradition method now is facing some new challenges. One of the weakness is these methods cannot modified with data environment dynamically in real time.

In general, the rule-based method is the regular method for network attack detection [1, 2]. The advantage of rule-based method is conveniently disposing and efficiently running. In order to overcome the weakness of the rule-based method, one of the most well-known methods is SVM, which has employed the training dataset to obtain the hyper-plane [3, 4].

In order to build a model to fit data stream, a new incremental learning was proposed in 1999 [5] and the experimental results show the feasibility of SVM model incremental learning. Recently, many research study the incremental learning deeply in three directions, which are data itself [6, 7], single model modified [8-10] and multi-model structure [11-13].

In this paper, we propose a novel incremental learning strategy based on the SVM model and optimize the model parameters through the experiment. The final experimental results show that the computation time cost was reduced and the accuracy of classification was improved in some way.

D. Jin and S. Lin (Eds.): Advances in CSIE, Vol. 1, AISC 168, pp. 277–282.
springerlink.com © Springer-Verlag Berlin Heidelberg 2012

2 Methodology Formulation

2.1 SVM Model Introduction

The main idea of the SVM model is finding a hyper-plane that can separate a data set of positive examples from a set of negative examples with maximum margin.

Let the training data set $\{x_i, y_i\}$, $i=1,2,\ldots,l$, $y_i \in \{-1,+1\}$, $x_i \subset R^n$ where x_i denotes an input feature vector with the number of dimension n, and y_i is the corresponding class label. For the linear separable condition, the goal of the SVM model is to find a hyper-plane $w*x+b=0$. In order to find the best suitable hyper-plane, the hyper-plane should separate the two classes in the largest margin.

For the linearly separable condition, the maximum separating margin is equivalent to solving

$$\begin{aligned} &\min \frac{1}{2} w^T \cdot w \\ &\text{s.t. } y_i(w \cdot x_i + b) \geq 1 \end{aligned} \tag{1}$$

However, the data set is not always linear separable. The SVM model induces the kernel function to define a mapping from low dimension space to a higher dimension space. Thus, for the linear separable situation, the final decision function is

$$f(x) = \operatorname{sgn}(\sum_{i=1}^{l} \alpha_i y_i K(x, x_i) - b) \tag{2}$$

2.2 Data Initialization

Because of the different attributes having different meanings, which would affect the performance of SVM model. In this paper, three common normalization approaches are employed, which are Logarithm normalization, Inverse cotangent normalization and Mapminmax normalization.

2.3 Incremental Learning Method

In this section, we mainly introduce the process of the incremental learning, which contains two kernel contents.

The first part is the multi-classifier structure. The main idea of the model is to let a batch of incoming data of size m, and the number of SVM classifiers maintained in the memory being n, using 1, 2, …, n to represent the SVM model respectively. Thus, the total number of training data sets in first training is $w=m*n$ samples. Before the model begins running, the n SVM models should be trained with the data stream. The training process of the model is shown in Fig. 1.

Let all classifiers be trained at time t. The trained SVM model is represented by SVM_1^t, SVM_2^t, …, SVM_n^t respectively. When a new batch of data comes in, at time t+1, the model employs the SVM_n^t model to predict the class label of new batch data and then employs a new batch samples that combines with the class label predicted just now to train the new SVM_n^{t+1} model. The key point of the model is to discard SVM_n^t and remains SVM_1^t, SVM_2^t, …, SVM_{n-1}^t.

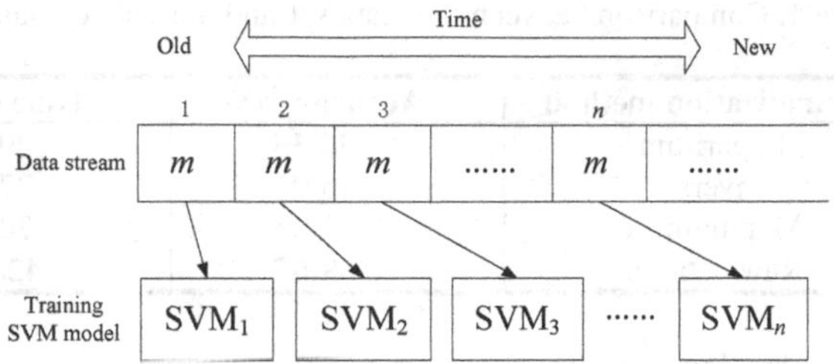

Fig. 1. SVM model training process

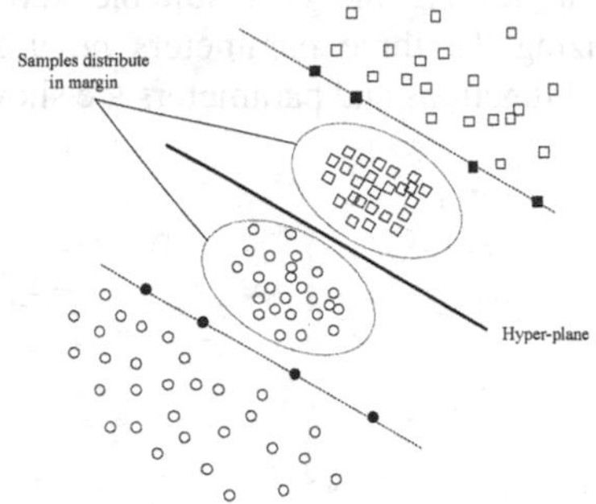

Fig. 2. The samples distribute in margin

The second part is the incremental learning strategy. Let parameter e denote the new batch samples distributed in the margin of the remaining SVM models. We believe that the number of new batch samples distributed in margin is in direct proportion to concept drifts. Thus, we compute the value of e before incremental training. We would optimize the parameter e through an experiment in the next section. The sample distributed in the margin is shown in Fig. 2.

3 Experimental Results

In this section, we use KDD 99 as the experimental data set, which contains 494,021 samples and 42 dimensions respectively.

3.1 Data Normalization

Three normalization methods are used to compare the classification accuracy between normalized data sets and raw data sets. Table 1 shows the experimental results.

According to the results given in Table 1, Logarithm normalization method has the best accuracy. For time cost, the fastest is Mapminmax. Overall, the Logarithm normalization method achieves the best performance than the others. The normalization method can significantly improve the classification accuracy and reduce the time cost. We would employ the Logarithm method in the experiment.

Table 1. Comparison between raw data set and normalized data set.

Normalization method	Accuracy (%)	Time cost (s)
Logarithm	**96.35**	29.31
Inverse	95.01	77.09
Mapminmax	91.83	**24.93**
Raw data set	**69.67**	**428.74**

3.2 Model Optimization

In this section, two optimal directions would be used. Firstly, two kernel functions employed in this experiment, finding the best suitable kernel function. The second function is involves optimizing the three parameters *m*, *n* and *e*. The classification accuracy with different kernel functions and parameters are shown in Fig. 3 and Fig. 4.

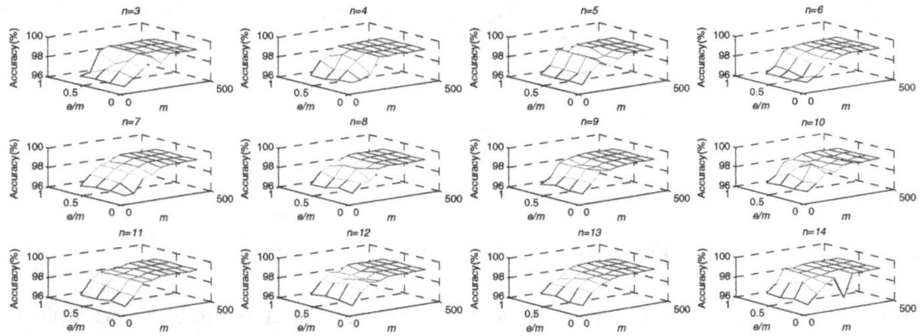

Fig. 3. Classification accuracy with RBF kernel

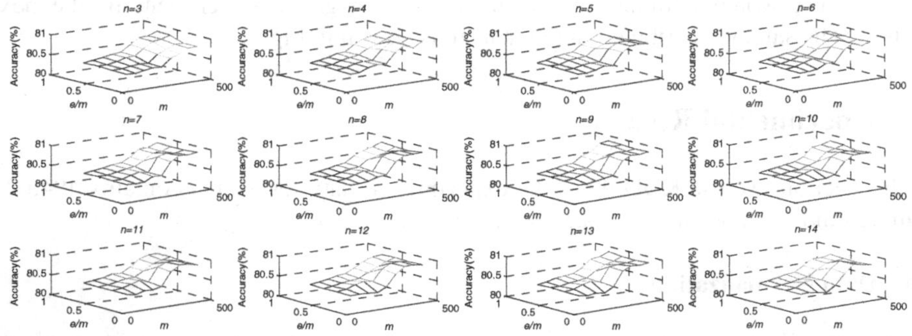

Fig. 4. Classification accuracy with Sigmoid kernel

From Fig. 3 and Fig. 4, the experimental result shows that the RBF has better classification performance than Sigmoid. Apart from this, the classification accuracy does not increase with the growth of parameter *n*. On the whole, the RBF kernel function is suitable for network data stream classification. Finally, we select RBF as the kernel function and also optimize the parameters *m*, *n* and *e*, which equal to 300, 6 and 0.6 respectively.

3.3 Comparison with Different Models

In this section, we compare the optimized model with two single models: KNN and decision tree. The experimental results are shown in Table 2.

Apart from the introduction just given, in order to clearly reveal the performance of each model, we employ three types of accuracy measurement.

$$\text{Type 1} = \frac{BB}{BB + BA} \tag{3}$$

$$\text{Type 2} = \frac{AA}{AA + AB} \tag{4}$$

$$\text{Type 3} = \frac{AA+BB}{AA + AB+BB+BA} \tag{5}$$

where letter B denotes class B, letter A denotes class A, BB denotes the number of vectors in which the original class as well as the output class is B, AA denotes the number of vectors in which the original class as well as the output class is A, AB denotes the number of vectors in which the original class is A and the output class is wrongly classified as B, and BA denotes the number of vectors in which the original class is B and the output class is wrongly classified as A.

Table 2. Comparing between optimized model and two other single models.

Model	Type 1(%)	Type 2(%)	Type 3(%)	Time cost(s)
KNN	99.67	**99.77**	**99.75**	**437.05**
Decision tree	**99.33**	99.63	99.24	**1.62**
Incremental learning model	**99.85**	**98.54**	**98.80**	33.08

From Table 2, it can be seen that the KNN model has the best classification accuracy, but the time consuming is out of the experiment requirement. So it is not suitable for the data stream. The decision tree model has the best performance with regard to accuracy and time cost, while it is not a learning-based method, but a rule-based method. The rule-based method cannot be changed by itself in a dynamic data environment, such as a data stream. The incremental learning method proposed is almost the same as the decision tree in classification accuracy, and its time-consuming nature is acceptable in the network data stream.

4 Conclusion

In this paper, we proposed an improved incremental learning model based on the SVM model and optimize the parameters. The model was more suitable for the network data stream environment. In addition, we proposed a new SVM incremental leaning strategy, which can reduce training frequency as well as time cost. We compare the performance of our model with two other models. Experimental results show the robustness of the proposed model.

References

1. Zhao, G., Xuan, K., Taniar, D., Srinivasan, B.: Incremental K-nearest-neighbor search on road networks. Journal of Interconnection Networks, 455–470 (2008)
2. Safar, M., Ebrahimi, D., Jane, M.M.F., Nadarajan, R.: Restricted continuous KNN queries on road networks with caching enhancement. Journal of Digital Information Management, 38–50 (2008)
3. Vapnik, V.: The nature of statistical learning theory. Springer (1995)
4. Cortes, C., Vapnik, V.: Support-vector networks. Machine learning 20, 273–297 (1995)
5. Syed, N.A., Liu, H., Sung, K.K.: Handling concept drifts in incremental learning with support vector machines. In: Proceedings of the Fifth ACM SIGKDD International Conference on Knowledge Discovery and Data Mining, pp. 317–321. ACM, San Diego (1999)
6. Getta, J.R., Vossough, E.: Optimization of data stream processing. SIGMOD Record 33, 34–39 (2004)
7. Kim, J.-H., Park, S.: Flexible Selection of Wavelet Coefficients for Continuous Data Stream Reduction. In: Kotagiri, R., Radha Krishna, P., Mohania, M., Nantajeewarawat, E. (eds.) DASFAA 2007. LNCS, vol. 4443, pp. 1054–1057. Springer, Heidelberg (2007)
8. Tang, Y., Zhang, Y.-Q., Chawla, N.V.: SVMs modeling for highly imbalanced classification. IEEE Transactions on Systems, Man, and Cybernetics, Part B: Cybernetics 39, 281–288 (2009)
9. Li, C., Kulkarni, P.R., Prabhakaran, B.: Segmentation and recognition of motion capture data stream by classification. Multimedia Tools and Applications 35, 55–70 (2007)
10. Khadam, I.M., Kaluarachchi, J.J.: Use of soft information to describe the relative uncertainty of calibration data in hydrologic models. Water Resources Research 40, W1150501–W1150515 (2004)
11. Domeniconi, C., Gunopulos, D.: Incremental support vector machine construction. In: 1st IEEE International Conference on Data Mining, ICDM 2001, November 29-December 2, pp. 589–592. Institute of Electrical and Electronics Engineers Inc., San Jose (2001)
12. Wang, H., Fan, W., Yu, P.S., Han, J.: Mining concept-drifting data streams using ensemble classifiers. In: 9th ACM SIGKDD International Conference on Knowledge Discovery and Data Mining, KDD 2003, August 24- August 27, pp. 226–235. Association for Computing Machinery, Washington, DC (2003)
13. Feng, L., Yao, Y., Jin, B.: Research on Credit Scoring Model with SVM for Network Management. Journal of Computational Information System 6, 3567–3574 (2010)

Improved Region Local Binary Patterns for Image Retrieval

Xiaosheng Wu[1] and Junding Sun[1,2]

[1] School of Computer Science and Technology, Henan Polytechnic University, Jiaozuo 454003, China
[2] Image Processing & Image Communication Lab, Nanjing University of Post & Telecommunication, Nanjing, 210003, China
{wuxs, sunjd}@hpu.edu.cn

Abstract. Two extended patterns, direction local binary pattern (D-LBP) and variance direction local binary pattern (vD-LBP) were introduced in the paper. More texture features are obtained in the new operators by introducing a variant threshold. The descriptors mentioned in the paper were firstly tested and evaluated on CUReT texture image database. Experimental results show that the two extended operators give better performance than D-LBP and vD-LBP respectively. Finally, the operators were used for trace fossils image retrieval and the results denote that the proposed operator, tvD-LBP is the most effective method for trace fossils image description in the four descriptors.

Keywords: center-symmetric local binary pattern (CS-LBP), D-LBP, vD-LBP, extended region local binary patterns.

1 Introduction

Recently, local features extracted from images have performed very well in many applications [1, 2]. Heikkilä, et al [3] introduced CS-LBP operator for interest region description based on the traditional local binary pattern (LBP) [4]. Different from LBP, it is defined by the gray-level difference of the pairs of opposite pixels and produces a rather short histogram. However, only the absolute gray-level differences are considered and the center pixel is ignored in CS-LBP. Consequently, it is difficult to denote the directionality, coarseness and contrast of a region in a full way. Two improved methods, D-LBP and vD-LBP were presented in our previous work [5, 6] to reduce such shortcomings. It has been proved that these two methods give better performance than CS-LBP.

In this paper, D-LBP and vD-LBP were further extended to contain much more information of a region by introducing a variant threshold. The new operators are called thresholded D-LBP (tD-LBP) and thresholded vD-LBP (tvD-LBP). The performance of the proposed approaches were fistly demonstrated on CUReT texture images. Finally, they were used for trace fossils image retrieval.

The rest of the paper is organized as follows. In Section 2, we briefly discussed the local patterns and the enhanced operators. Similarity measure is given in Section 3. The experimental results are described in Section 4. Finally, we conclude the study in Section 5.

D. Jin and S. Lin (Eds.): Advances in CSIE, Vol. 1, AISC 168, pp. 283–288.
springerlink.com © Springer-Verlag Berlin Heidelberg 2012

2 Region Texture Spectrum

2.1 Local Binary Patterns

In [3], Heikkilä proposed the center-symmetric local binary pattern (CS-LBP) descriptor. It is defined by the gray-level difference of the pairs of opposite pixels of a local region. A threshold is then set to obtain the robustness on flat image regions. For a 3×3 neighborhood (Fig. 1), the CS-LBP value is given as follows,

$$CS\_LBP_T(x,y)=\sum_{i=0}^{3} v(p_i-p_{i+4})\times 2^i, v(x)=\begin{cases}1, & x\geq T\\ 0, & \text{otherwise}\end{cases} \quad (1)$$

where T denotes the threshold.

p_5	p_6	p_7
p_4	p_c	p_0
p_3	p_2	p_1

Fig. 1. 3×3 neighborhood

It is various that the center pixel pc is ignored in the definition of CS-LBP, and it is hard to find a unchanged adaptable threshold T for all the local regions. An extended method, direction local binary pattern (D-LBP) was introduced in our previous work [5]. D-LBP is defined based on the relation of the center pixel and the center-symmetric pairs of pixels instead of the gray-level difference of the pairs of opposite pixels as CS-LBP. The D-LBP value is given as,

$$D\_LBP(x,y)=\sum_{i=0}^{3} w(p_i,p_c,p_{i+4})\times 2^i,\ w(p_i,p_c,p_{i+4})=s(p_i-p_c)\odot s(p_c-p_{i+4})$$
$$s(x)=\begin{cases}1, & x\geq 0\\ 0, & \text{otherwise}\end{cases} \quad (2)$$

In [6], D-LBP was further extended to variance D-LBP (vD-LBP). Both the gray-level intensity invariance property and the intensity inversion invariance property was fused in the new descriptor. The definition of vD-LBP is given as,

$$vD\_LBP(x,y)=\sum_{i=0}^{3}(s(p_i-p_c)\odot s(p_c-p_{i+4})\odot s((p_i-p_c)-(p_c-p_{i+4})))\times 2^i \quad (3)$$

2.2 Improved Local Binary Patterns

It has been proved that vD-LBP and D-LBP give better performance than CS-LBP [5, 6], and vD-LBP shows the best performance in the 3 descriptors. Based on CS-LBP, the two descriptors vD-LBP and D-LBP were further extended in the paper by introducing a variant threshold. The new proposed operators are named tvD-LBP and tD-LBP. For a 3x3 neighborhood, they are defined by the Eq. (4) and (5) respectively.

$$tD\_LBP(x,y)=\sum_{i=0}^{3} w(p_i,p_c,p_{i+4})\times 2^i,\ w(p_i,p_c,p_{i+4})=v(p_i-p_c)\odot v(p_c-p_{i+4}) \tag{4}$$

where $v(x)=\begin{cases}1, & x\geq t\times\mu\\ 0, & \text{otherwise}\end{cases}$, $\mu=\frac{1}{9}(p_c+\sum_{i=0}^{7}p_i)$ and t is the threshold. It is obvious tD-LBP is same as D-LBP when $t=0$. Experimental results also show that good performance can be achieved when $t\in[0.01,0.1]$.

By the same method, tvD-LBP is given as,

$$tvD\_LBP(x,y)=\sum_{i=0}^{3}(v(p_i-p_c)\odot v(p_{i+4}-p_c)\odot s(p_i+p_{i+4}-2p_c))\times 2^i \tag{5}$$

It is also clear that tvD-LBP is the same as vD-LBP when $t=0$. In the new definition of tvD-LBP and tD-LBP, $t\times\mu$ is more suitable than T of CS-LBP, because μ changes with different local regions, wheras T is unchanged.

3 Similarity Measure

Suppose $\boldsymbol{H}_1$ and $\boldsymbol{H}_2$ be the texture spectrum histograms of the query image and one image in the database, χ^2-distance is chosen as a measurement criterion.

$$\chi^2(\boldsymbol{H}_1,\boldsymbol{H}_2)=\sum_{i=1}^{K}\frac{(h_{1i}-h_{2i})^2}{h_{1i}+h_{2i}} \tag{6}$$

where K denotes the dimension of the operators.

4 Experimental Results

In order to evaluate the proposed methods, the retrieval performance is assessed in terms of the commonly used precision and recall. Precision is the ratio of the number of retrieved images that are relevant to the number of retrieved images. Recall is the ratio of the number of retrieved images that are relevant to the total number of relevant images. The precision and recall are defined as follows,

$$precision={n}/{L},\ recall={n}/{N} \tag{7}$$

where L is the number of retrieved images, n is the number of relevant image in the retrieved images and N is the number of all relevant images in the database. In the experiments, we set $t=0.05$.

4.1 The Performance on Texture Image Set

Texture images are chosen form CUReT database, which includes 45 classes and each class has 20 texture images. Fig. 2 shows the 45 example textures. This experiment employs each image in the first 20 classes as query images, and 400 retrieval results are obtained. Fig. 3 presents the comparision graph of the average results. The retrieval results show that the extended methods tvD-LBP and tD-LBP produces the better performance than vD-LBP and D-LBP respectively for texture images.

4.2 Trace Fossils Image Retrieval

We chose 521 trace fossils images as image set. Some examples are given in Fig. 4. Eighteen classes, each class has 16 similar images, are chosen as query images in this experiment, and 288 retrieval results were obtained. Fig. 5 presents the precision-recall graph of the average results for the four operators, tvD-LBP, tD-LBP, vD-LBP and D-LBP. The results show that D-LBP and vD-LBP are not effective for trace fossiles images description, though they can obtain good performance for the texture images. On the other hand, tvD-LBP operator gives better results than tD-LBP.

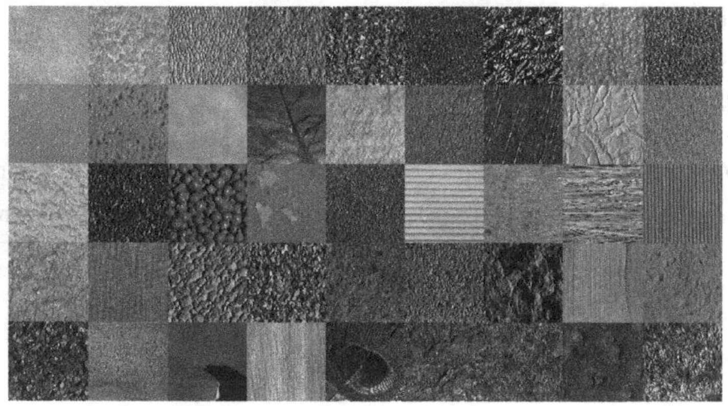

Fig. 2. Example texture images

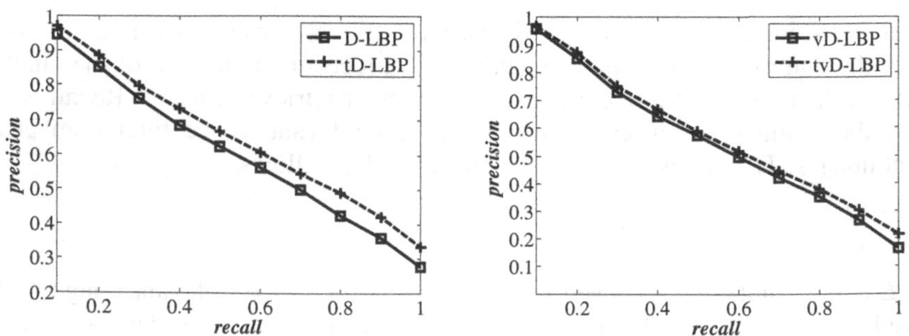

(a) Comparison of D-LBP and tD-LBP. (b) Comparison of tvD-LBP and vD-LBP.

Fig. 3. Average recall-precision graph of 400 query images.

Fig. 4. Example Trace Fossils images

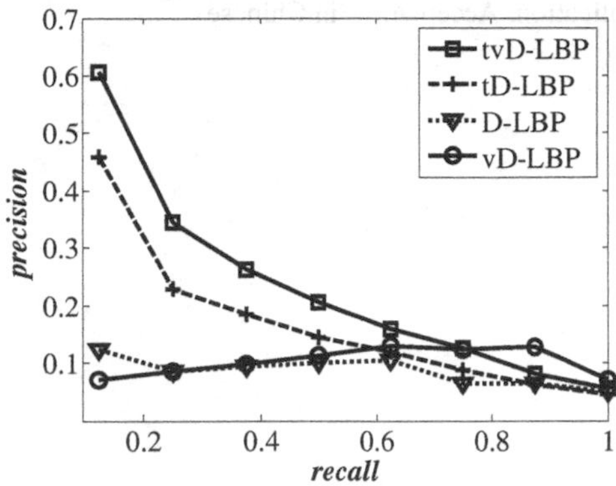

Fig. 5. Average recall-precision graph of 288 query images.

5 Conclusion

Based on the discussion of local binary patterns, center-symmetric local binary pattern, direction local binary pattern and variance direction local binary pattern, two extended descriptors, tD-LBP and tvD-LBP were introduced in the paper for region description. The methods mentioned in the paper are evaluated on CUReT texture image database firstly. Experimental results show that they are all effective for texture image description. After that, they were used for trace fossils image retrieval, and the experimental results prove that tvD-LBP and tD-LBP are adaptable, wheras vD-LBP and D-LBP are not. On the other hand, tvD-LBP gives the best performance in the four operators.

Acknowledgments. This work was supported by the Key Project of Chinese Ministry of Education (210128), The Backbone Teacher grant of Henan Province (2010GGJS-059), the Internal Cooperation Science Foundation of Henan province (104300510066), Image Processing & Image Communication Lab of Nanjing University of Post & Telecommunication (LBEK2011002).

References

1. Tuytelaars, T., Gool, L.V.: Matching widely separated views based on affine invariant regions. International Journal of Computer Vision 59(1), 61–85 (2004)
2. Lowe, D.G.: Distinctive image features from scale-invariant keypoints. International Journal of Computer Vision 60(2), 91–110 (2004)
3. Heikkilä, M., Pietikäinen, M., Schmid, C.: Description of interest regions with local binary patterns. Pattern Recognition 42(3), 425–436 (2009)
4. Ojala, T., Pietikäinen, M., Mäenpää, T.: Multiresolution gray-scale and rotation invariant texture classification with Local Binary Patterns. IEEE Transactions on Pattern Analysis and Machine Intelligence 24(7), 971–987 (2002)
5. Sun, J.D., Zhu, S.S., Wu, X.S.: Image Retrieval Based on an Improved CS-LBP Descriptor. In: The 2nd IEEE International Conference on Information Management and Engineering, pp. 115–117 (2010)
6. Wu, X., Sun, J.: Image Retieval Based on Region Direction Local Binary Pattern. Computer Engineering and Application, Acceptance (in Chinese)

A New WSN-Based Routing Optimization in Smart Distribution Grid

GangJun Gong, XingChuan Wang, and JuanXi Zhu

School of Electrical and Electronic Engineering,
North China Electric Power University, Beijing 102206, China
{948553964,421488273,546582715}@qq.com

Abstract. The method of routing optimization aimed at problems such as numerous-nodes, large-scale, low reliability of transmission link and inferior real-time in smart distribution grid based on wireless sensor network has been proposed in this paper. Theoretical analysis and simulations show that this routing strategy meets the requirements of electricity demand side and improves the quality of power supply and management efficiency.

Keywords: Smart Distribution Grid, WSN, Routing Optimization, Activation-by-Turns.

1 Introduction

Power distribution system is an important link in connecting power generation and transmission system with electricity customers. Wireless sensor networks (WSN) containing a great number of miniature sensor node matures gradually. WSN is a kind of brand-new information acquisition platform which can instantly monitor, collect and process environmental information. It is beneficial to performance enhancements applying WSN to distribution network with appropriate algorithm and creating a self-organization network [1]. Combining with the safety, reliability and economy requirements that the construction of power distribution network must comply with, this paper puts forward a technical solution in the light of smart distribution grid communication system that uses optical fiber communication to guarantee important nodes, wireless communication to achieve broad coverage and carrier communication to add access. The focus of research is routing strategy of WSN in smart distribution grid [2].

2 Distribution Network Topology Modeling

For structure of ring network design and the open loop running, it can be regarded as a tree structure in normal operation, and weak ring only in original network replace operation. According to the weak ring topology structure of distribution network, our subject structures topology model of information gathering node, as shown in Fig.1 [3].

D. Jin and S. Lin (Eds.): Advances in CSIE, Vol. 1, AISC 168, pp. 289–294.
springerlink.com © Springer-Verlag Berlin Heidelberg 2012

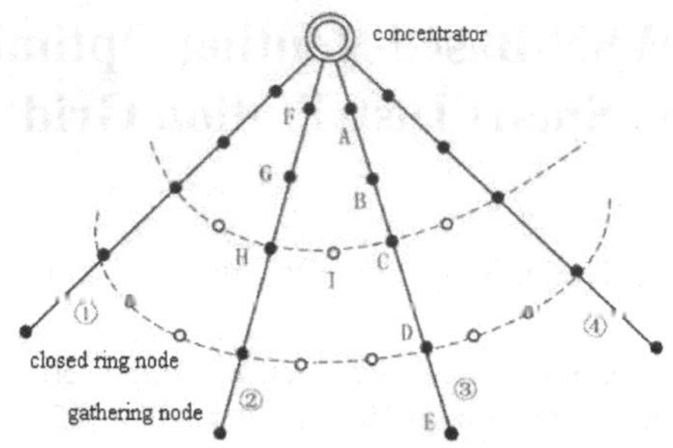

Fig. 1. Topology model of information gathering node

The following example illustrates weak ring communication system structure of distribution network. A to F are information gathering node on net rack ② or ③ of distribution network, and I is closed ring node. In an ordinary way, information transmission route in feeder ① is E→D→C→B→A. If node A occurs fault in a moment, node I starts to form weak ring network, so information will be transmitted automatically along B→I→H→G→F. Thus it can be seen that weak ring communication network can improve the reliability of the data transmission.

3 Routing Optimization Strategy

3.1 Analysis of Routing Requirements

Distribution communication system should be closely combined with application functions. Various electricity information and physical information of transmission network should be transmitted to the control center in time. Communication of distribution network consists of station to substation, station to terminal, substation to terminal and terminal-to-terminal mode. What information needs to be passed are about switch substation, converting station, distribution transformer, far meter reading information and image transmission [4].

3.2 Arrangement of Nodes

According to the topological model, sensor network has the following characteristics: (1) sensor nodes distribute densely, so data transmission can choose multiple routes; (2) single sensor node can directly communicate with nearby nodes, and data interacts with gateway through multi-hop routing way; (3) communication distance between nodes is moderate, so energy consumption is slowly since transmitting power will not be great; (4) nodes close to gateway load more, so analysis and consolidation of information are taken into account; (5) weak ring network topology can guarantee the reliability of communication [5].

3.3 Routing Protocol

Routing protocol is responsible for transferring data packet from source node to destination node through network. Information in routing table is the basis of transmission. In distribution network system, geographic location of sensor node is easy to confirm and fixed relatively, so there is no need for repetitive addressing. For this reason, we adopt static routing protocol to create route, to transfer data and to maintain route in our research. Hence, control message is simplified and routing overhead is decreased.

3.4 Activation the Routes by Turns

Sensor nodes have three statuses of activation, surveillance and dormancy. When transmitting and receiving data, node is in the status of activation. This status is the major cause of energy consumption. So we design a scheme of activation the routes by turns. This scheme not only decreases the working time of single node, but also relatively counterbalances energy consumption of various sensors.

In the actual environment, concrete structure of distribution power system varies considerably, so it is difficult to put forward a uniform scheme of activation. It is need to carry on the design according to the specific situation. In the model of Fig.2, five information flow routes are built, marked respectively by Ⅰ, Ⅱ, Ⅲ, Ⅳ and Ⅴ.

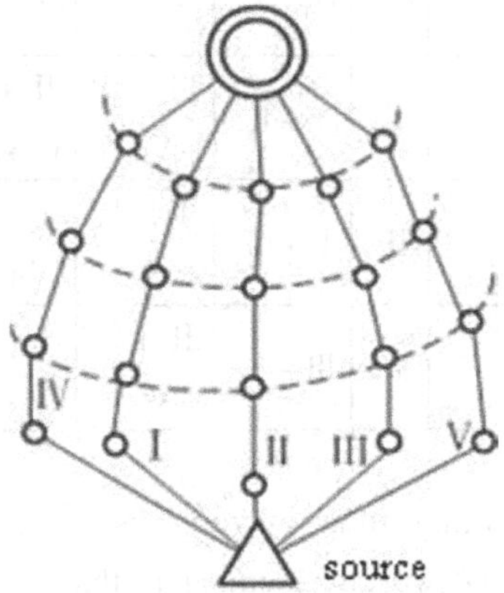

Fig. 2. Information flow routes

Within 24 hours, communication traffic changes periodicity and regularly over time, so activate different number of working routes according to different communication levels. The relation of working route number and communication level is shown in Table 1.

Table 1. Relation of working route number and communication level

communication level	A	B	C	D	E
route number	1	2	3	4	5

Supposing every adjacent routes of some route in working status are not activated, then activates sensor node in ring point next to that route. If the node next to the route is disabled, just activates node in open ring and use other routes to transfer data. This method can ensure the normal operation of communication network and buy time to workers' maintenance.

Supposing energy consumption is in proportion to working time, all nodes in route can work normally and listening state doesn't consume energy. Three days for cycle, we activate the routes by turns and probably balance the working time of nodes in routes. The scheme of activation by turns is shown in Table 2.

Table 2. Scheme of activation

communication level	A	B	C	D	E
first day	Ⅰ	Ⅰ+Ⅱ	Ⅲ + Ⅳ +Ⅴ	Ⅱ + Ⅲ + Ⅳ+Ⅴ	all
second day	Ⅱ	Ⅱ+Ⅲ	Ⅰ + Ⅳ +Ⅴ	Ⅰ + Ⅲ + Ⅳ+Ⅴ	all
third day	Ⅲ	Ⅲ+ Ⅰ	Ⅱ + Ⅳ +Ⅴ	Ⅰ + Ⅱ + Ⅳ+Ⅴ	all

Supposing the total working time of route is T_n , hour contained in every communication level is t_n and working number in that communication level in three days is N . In this way: $T_n = \sum t_n \times N$. If the scheme of activation by turns is not adopted, in three days, total working time of each route is $T_m = 24 \times 3 = 72h$, so saving energy of node is $P = T_m / T_n \times 100\%$. Simple calculations can g that the scheme of activation by turns can save energy by 45.8% to 51.4%.

4 Simulations and Analysis

In this section of simulations, we compare the result of turn-scheme with before the scheme at a. m (6:00-12:00). Before the scheme, all nodes are activated. Afterwards, two routes （Ⅲ+Ⅰ） are activated to transmit data. Supposing that simulation data are the same, then we select media access delay and end-to-end delay as contrastive performance parameters. Simulation results are shown in Fig.3.

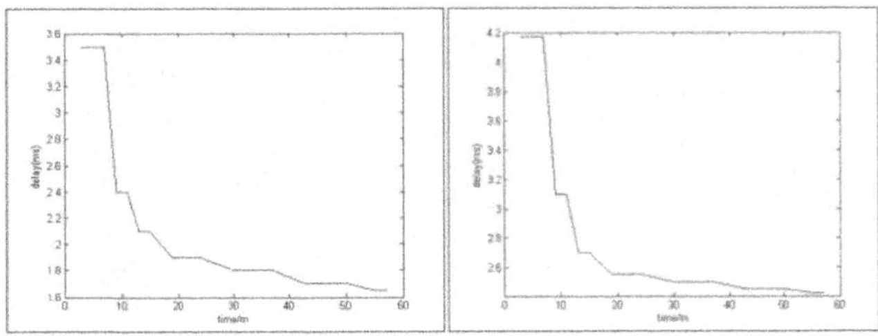

Fig. 3. Media access delay and end-to-end delay

Calculating delay time twenty minutes later, we know the sum of media access delay time of 2-route is 0.046449s and 0.114695s of 5-route. Hence, media access delay decreases by (0.114695-0.046449) /0.114695=59.5% after adopting activation-by-turns. In the process of simulation, the sum of end-to-end delay time of 2-route is 0.091461s and 0.160395s of 5-route. Hence, media access delay decreases by (0.160395-0.091461) /0.160395=43% after adopting activation-by-turns. Simulation result shows that transmission delay decreases after adopting activation-by-turns. In the period of time of low power consumption, only two route and making part of the node dormancy save the energy.

5 Summary

Wireless sensor network has advantages which traditional communication network doesn't have. WSN is one of development directions of smart distribution grid. The routing optimization strategy this subject research proposes can effectively prolong the life cycle of wireless sensor network, improve the throughput of communication network and enhance power distribution automation system in the real time and reliability. Our routing optimization strategy provides a reference for the construction of the smart distribution grid.

References

1. Lele: Network Topology Analysis of Distribution Automation. Zhejiang University, Hangzhou (2008)
2. Gu, S.: Distribution Automation. China Electric Power Press, Beijing (2008)

3. Huang, S.: Communications Business Demand Analysis and Technical Proposal of Smart Distribution Grid. Telecommunications for Electric Power System 31(212), 10–17 (2010)
4. Gao, C., Yang, M., Mao, D.: Studies Review of Routing Protocol of WSN. World Sci. Tech. R & D 27(4), 1–6 (2005)
5. Shu, J.: Research of Multichannel AODV Routing Protocol Mobile Ad Hoc. University of Electronic Science and Technology of China, Chengdu (2005)

Patent Evaluation Method of Automotive Parts Manufacturing Enterprise Based on Analytic Hierarchy Process and Fuzzy Matter-element Method

Jun Chen[1,2], He Huang[1], He Lu[1], Jian Hu[1], and Gangyan Li[1]

[1] School of Mechanical and Electronic Engineering, Wuhan University of Technology, 122, Luoshi Road, 430070, Wuhan, Hubei Province, P.R. China
{cjund,flying-0415}@163.com, birdbox0207@sina.com, {hujian,gangyanli}@whut.edu.cn

[2] Mechanical and Electronic Engineering College, Hubei Polytechnic University 16, Guilin North Road, 435003, Huangshi, Hubei Province, P.R. China

Abstract. Patent evaluation and analysis is the basis of patent information using in automotive parts manufacturing enterprises. Based on the analytic hierarchy process, the patent evaluation index system of automotive parts manufacturing enterprises is constructed in this paper according to several aspects, including research and development situation, technology situation, market situation and legal situation. The index weight is determined and the consistency is tested. Taking X company for example, the evaluation index ranges are divided and the measured value is obtained. Using fuzzy matter-element method, the patent rating is calculated and the enterprise patent situation is evaluated.

Keywords: Automotive Parts, Patent, Evaluation, Analytic Hierarchy Process, Fuzzy Matter-element Method.

1 Introduction

The development speed and scale of the car industry depends on the technology development of automotive parts, so it is very important to formulate some patent strategies correspond with the technology development of automotive parts manufacturing enterprise. Choosing appropriate evaluation method and constructing a scientific evaluation system, determining appropriate index weight, evaluating enterprise patent situations, analyzing enterprise patent in relative position are the basis of making an enterprise patent strategy. Xiao-li Wan, Shui-li Yang and W.D.Yu have put forward the index weight distribution method in patent index system and evaluation system, and make the fuzzy comprehensive evaluation [1]-[3]. Michael W.Brinn have given a detailed explanation on the standard custom program, the factors affect the value and evaluation methods in the patent evaluation [4]. Xiao-ping Lei established the patent application evaluation index system based on the SMART standards, and used intelligence analysis method and analytic hierarchy process to determine the index weight [5]. Song-xi Liu established the national defence

D. Jin and S. Lin (Eds.): Advances in CSIE, Vol. 1, AISC 168, pp. 295–302.
springerlink.com © Springer-Verlag Berlin Heidelberg 2012

industry's value evaluation model of science and technology achievement according to perspectives [6]. Minghua Zeng introduced the basic methods in the patent evaluation process and pointed out some existing problems of these methods, then advanced some correction method to solve these problems [7].

This paper mainly research on patent evaluation method of automotive parts manufacturing enterprise based on analytic hierarchy process and fuzzy matter-element method. Subsequent contents are as follow: In the second section, according to the automotive parts manufacturing enterprise patent assessment requirements, we choose the patent evaluation method. In part 3, based on hierarchy analysis method, we establish patent evaluation index system of the automotive parts enterprise. In part 4, based on fuzzy matter-element method, we evaluate the automotive parts manufacturing enterprise's patent situations. Finally, we summarize the full text work, and put forward some further research directions.

2 Patent Evaluation Method Selection

The common patent evaluation methods are Delphi method, regression analysis method, Analytic Hierarchy Process (AHP), principal component analysis method, grey correlation degree analysis method, and fuzzy comprehensive evaluation method, etc. The selection of patent evaluation method should be combined with the actual situation.

The patent evaluation method of automotive parts manufacturing enterprise should be appropriate, the evaluation process should be reasonable, the evaluation results should be scientific, the selection of evaluation methods should meet the following requirements:

(1) It can not only be used for quantitative analysis of qualitative analysis alone, but can also be combined for analysis.

(2) The index set must be flexible. When the index system changes, so index can also make some adjustment, and it has little effect on the evaluation process.

(3) The selection of the index is analytical, indexes are both relatively independent and certain integrity.

(4) The evaluation result is intuitive.

(5) Evaluation system can fully embody the relationship of the child objects and the objects which they are subordinated, and can reflect the relationship of different levels between the evaluation objects.

According to the patent situation and the requirements of the evaluation methods of the automotive parts manufacturing enterprise, in this paper we choose analytic hierarchy process and fuzzy matter-element method for patent evaluation..

3 The Automotive Parts Manufacturing Enterprise Patent Evaluation Index System Based on Analytic Hierarchy Process

3.1 The Process of AHP

The AHP application process is as follows:

(1) Analyze the objects' characteristics and classify levels, classify the evaluation factors in a level, and structure a hierarchical analysis evaluation model.

(2) Structure judgment matrix.

(3) Determine level weights and test consistency.

(4) Calculate the combination weight of the elements in the same level.

3.2 The Automotive Parts Manufacturing Enterprise's Patent Evaluation Indexes

Through analysis, the research and development situation, technology situation, market situation and legal situation, evaluation indexes are shown in Table 1.

Table 1. Automotive parts manufacturing enterprise's patent evaluation indexes

Target	Rule Layer	Index layer
A : Enterprise patent situations	B1 : the research and development situation	C1:the research and development team situation
		C2:research and development's software and hardware platform situation
		C3:capital investment in research and development
		C4:research and development's management system
		C5:research and development process conditions
	B2 : the technology situation	C6:technologies' maturity and related technical cooperation
		C7:technology application situation
		C8:technology advancement situation
		C9:technology innovation situation
		C10:alternative technology development situation
	B3 : the market situation	C11:market competitiveness
		C12:the market demand degrees
		C13:marketization ability
		C14:economic benefits
	B4 : the legal situation	C15:technology defensive ability
		C16:patent protection ability
		C17:patent right's get time and protection period

3.3 Indexes Weights of Automotive Parts Manufacturing Enterprise Patent Evaluation System Based on Analytic Hierarchy Process

Using the evaluation indexes, combining with the provisions of the quantitative scale in judgment matrix elements and considering experts' judgment, Each judgment matrix is obtained. With the biggest characteristic root method, we find out biggest root and the corresponding characteristic vectors, which are the corresponding weights of each index. Additional, we test the consistency, the formula (1) is as follows:

$$CR = \frac{CI}{RI} \tag{1}$$

And $CI = \frac{\lambda_{\max} - m}{m-1}$ ($\lambda_{\max}$ is the biggest root of the judgment matrix, m is the order number of judgment matrix) is the consistency index of judgment matrix, RI is the average random consistency index of judgment matrix.

Then, we get the calculation result and the test results of the second level single sort judgment matrix, it is shown in Table 2.

Table 2. A-B judgment matrix

	B1	B2	B3	B4
B1	1	1/2	5	7
B2	2	1	4	6
B3	1/5	1/4	1	4
B4	1/7	1/6	1/4	1

With the biggest characteristic root method, we get the biggest root: λ_b =4.2307, characteristic vector: W =(0.3647, 0.4661, 0.1198, 0.0494).

And the consistency ratio for the second level is $CR_{(2)}$=0.08755<0.1, so consistency inspection is qualified.

Similarly, we can find out the calculation result and the test results of the third level single sort judgment matrix, and the consistency inspection is qualified.

Next, Use formula (2) to find out the overall consistency of the first K level

$$CR_{(k)} = \frac{CI_{(k)}}{RI_{(k)}} \tag{2}$$

The whole consistency index is: $CI_{(3)}$=(0.0297,0.0991,0.0501,0)* W =0.0630

The whole average random consistency index is: RI ($_3$) =(1.12,1.12,0.9,0.58)* W =1.0669

The whole consistency ratio is: $CR_{(3)}=CI_{(3)}/RI_{(3)}$=0.0591<0.1, so the whole consistency is qualified.

Lastly, we can get the weight of each index as shown in Table 3.

Table 3. Index weights of the automotive parts manufacturing enterprise patent evaluation

Target layer	Rule layer	The index weight from rule layer to the target layer	Index layer	The index weight from index layer to target layer	The allover index weight from index layer to target layer
A	B1	0.3647	C1	0.4178	0.1524
			C2	0.2934	0.1070
			C3	0.1550	0.0565
			C4	0.0669	0.0255
			C5	0.0669	0.0355
	B2	0.4660	C6	0.3111	0.1450
			C7	0.4323	0.2015
			C8	0.1395	0.0650
			C9	0.0768	0.0358
			C10	0.0403	0.0188
	B3	0.1198	C11	0.5554	0.0665
			C12	0.2662	0.0319
			C13	0.1195	0.0143
			C14	0.0589	0.0071
	B4	0.0494	C15	0.4286	0.0212
			C16	0.4286	0.0212
			C17	0.1428	0.0071

4 Patent Evaluation of Automotive Parts Manufacturing Enterprise Based on Fuzzy Matter-element Method

4.1 Determine the Scopes of Evaluation Indexes and the Measured Values

In this paper, we divide the patent situation into five ranks: best, good, normal, bad, worst. Each rank is represented by I, II, III, IV and V respectively. Taking X company for example, we determine the scope of each index by company consultation and experts' scores, as shown in Table 4.

Table 4. Scopes of evaluation indexes and the measured values

Patent situation							
Evaluation object	**Evaluation index**	**Best (I)**	**Good (II)**	**Normal (III)**	**Bad (IV)**	**Worst (V)**	**Measure d value**
B1	C1	(80,100)	(60,80)	(40,60)	(20,40)	(0,20)	72
	C2	(80,100)	(60,80)	(40,60)	(20,40)	(0,20)	81
	C3	(80,100)	(60,80)	(40,60)	(20,40)	(0,20)	85
	C4	(80,100)	(60,80)	(40,60)	(20,40)	(0,20)	73
	C5	(80,100)	(60,80)	(40,60)	(20,40)	(0,20)	82

Table 4. *(continued)*

B2	C6	(80,100)	(60,80)	(40,60)	(20,40)	(0,20)	85
	C7	(80,100)	(60,80)	(40,60)	(20,40)	(0,20)	81
	C8	(80,100)	(60,80)	(40,60)	(20,40)	(0,20)	90
	C9	(80,100)	(60,80)	(40,60)	(20,40)	(0,20)	81
	C10	(80,100)	(60,80)	(40,60)	(20,40)	(0,20)	82
B3	C11	(80,100)	(60,80)	(40,60)	(20,40)	(0,20)	85
	C12	(80,100)	(60,80)	(40,60)	(20,40)	(0,20)	85
	C13	(80,100)	(60,80)	(40,60)	(20,40)	(0,20)	82
	C14	(80,100)	(60,80)	(40,60)	(20,40)	(0,20)	90
B4	C15	(80,100)	(60,80)	(40,60)	(20,40)	(0,20)	75
	C16	(80,100)	(60,80)	(40,60)	(20,40)	(0,20)	83
	C17	(80,100)	(60,80)	(40,60)	(20,40)	(0,20)	82

We analyze the situation of B1 firstly. Based on the matter-element model, the most advantage point of C1-C5 appears on the right side, so choose the formula for this situation and get the corresponding dependent relational function value is:

$$K_1 = \begin{bmatrix} -0.222 & 12 & -0.30 & -0.533 & -0.65 \\ 1 & -0.05 & -0.525 & -0.683 & -0.7625 \\ 5 & -0.25 & -0.625 & -0.75 & -0.8125 \\ -0.206 & 13 & -0.325 & -0.55 & -0.6625 \\ 2 & -0.1 & -0.55 & -0.7 & -0.775 \end{bmatrix}$$

We have gotten all the index weight, then use relational formula (3):

$$R_i = W_i K_i \qquad (3)$$

In this formula, R_i is the correlation of B_i, W_i is the weight vectors of the indexes of $B_{i,}$. K_i is the relational function value of B_i.

So, the answer is:

$$R_1 = [1.0958 \quad 5.8232 \quad -0.4348 \quad -0.6230 \quad -0.7175]$$

From this calculation results, maximum is 5.8283, and it suggests that the situation of research and development is II, so the team situation is good.

Similarly, we find out the relational function values of technology situation (B2), market situation(B3) and legal situation(B4):

$$R_2 = [3.5402 \quad -0.1770 \quad -0.5885 \quad -0.7255 \quad -0.7943]$$

$$R_3 = [4.9355 \quad -0.2468 \quad -0.6233 \quad -0.7488 \quad -0.8116]$$

$$R_4 = [1.5000 \quad 6.3504 \quad -0.4858 \quad -0.6572 \quad -0.7429]$$

4.2 Patent Comprehensive Evaluation in the Automotive Parts Manufacturing Enterprise

In this paper, we divide the patent situation into five ranks: best, good, normal, bad, worst. Each rank is represented by I, II, III, IV and V respectively. Take X company for example, we determine the scope of each index by company consultation and experts' scores, as shown in Table 4.

After evaluating each patent index, we begin patent comprehensive evaluation. From the results before, the allover relational function value of B1、B2、B3、B4 is:

$$K = \begin{bmatrix} R_1 \\ R_2 \\ R_3 \\ R_4 \end{bmatrix}$$

The weight of B1、B2、B3、B4 is: $W = (0.3647, 0.4661, 0.1198, 0.0494)$

So the correlation of patent situation is:

$$R = WK = [0.3647 \quad 0.4661 \quad 0.1198 \quad 0.0494] \begin{bmatrix} 1.0958 & 5.8232 & -0.4348 & -0.6230 & -0.7175 \\ 3.5402 & -0.1770 & -0.5885 & -0.7255 & -0.7943 \\ 4.9355 & -0.2468 & -0.6233 & -0.7488 & -0.8116 \\ 1.5000 & 6.3504 & -0.4858 & -0.6572 & -0.7429 \end{bmatrix}$$

$$= [2.7151 \quad 2.3254 \quad -0.5315 \quad -0.6875 \quad -0.7658]$$

According to this calculation results, the maximum is 2.7151 and its rank is I, so the patent situation of X company is best.

5 Conclusions

In this paper, the patent evaluation index system of automotive parts manufacturing enterprises is constructed. The indexes and their weight are determined. Based on fuzzy matter-element method, taking X company for example, we provided a intuitive evaluation result, which suggested the patent situation of this company, and it provided a reference for company to make further suitable patent strategy.

Acknowledgment. We would like to give our acknowledgments to Huangshi Institute of Technology, China because the research work reported here has been sponsored by "key project of Huangshi Institute of Technology" (10yjz05A).

References

1. Wan, X., Zhu, X.: The Patent Value Evaluation Index System and the Fuzzy Comprehensive Evaluation. Scientific Research Management 3, 187–191 (2008)
2. Yang, S.-L.: Research on the Project Approval Evaluation Index System of Science and Technology Industrialization Projects. In: International Conference on Information Management, Innovation Management and Industrial Engineering, pp. 15–19 (2009)

3. Yu, W.D., Lo, S.S.: Patent Analysis-based Fuzzy Inference System for Technological Strategy Planning. Automation in Construction 18, 770–776 (2008)
4. Brinn, M.W., Fleming, J.M., Hannaka, F.M., Thoms, C.B., Beling, P.A.: Investigation of Forward Citation Count as a Patent Analysis Method. In: System and Information Engineering Design Symposium (2003)
5. Lei, X., Zhu, D., Zhou, C.: Patent Appraisal Method of the Evaluation of Science and Technology Project Plan. Scientific Research 6, 573–577 (2008)
6. Liu, X., Li, Y., Yu, D.: Value Evaluation Research of National Defence Industry Science and Technology Achievement Based on Several Angles. Science and Technology's Policy and Management 10, 31–35 (2007)
7. Zeng, M., Pan, X., Liu, Q.: Algorithms on Discretizing Continuous Attributes Values and Its Application to Synthetical Test and Evaluation of Patent Strength. In: Sixth IEEE International Conference on Data Mining-Workshops (2006)

Application of Improved Genetic Algorithm in Logistics Transportation

Jixian Xiao and Bing Lu

College of Science, Hebei United University, Tangshan, Hebei, 063009, China
xiaojix@yahoo.com.cn, lubing311@126.com

Abstract. In recent years, modern logistics industry is a rapidly emerging industry, a new economic growth point, transportation is an important element of modern logistics, it can reduce transportation costs, and improve economic efficiency if the arrangements of transport vehicles are reasonable. Vehicle routing problem is a class of typical combinatorial optimization problems and is known for its physical transportation vehicle scheduling, which is proved to be a NP-Hard problem. in this paper, to improve the traditional genetic algorithm, and prevent the new individual of invalid solutions by crossing, it proposed a new method of cross-parents, making the solution process is always carried out in the effective solution set, eventually achieve the optimal route of transport vehicles.

Keywords: genetic algorithm, logistics transportation, combinatorial optimization, cross-parents.

1 Introduction

In China, modern logistics industry is an emerging industry, a new economic growth point, under the care of government at all levels, has the conditions of developing economic environment and market with a modern logistics industry. logistics transportation is an important component part of logistics system, cargo transportation logistics costs account for about 35%-60% of total operating costs, so the level of transportation costs directly related to the size of the logistics costs. Vehicle routing problem is the core and hot research in the field of logistics, which attracted many scholars and attention.

2 Modeling

Vehicle routing problem can be simply described as: a vehicle of loading goods, starting from the logistics center, do not repeat to finish the remaining N-receiving points, and return to the logistics center, find a shortest path length in all possible paths. This problem is a typical combinatorial optimization problem, which belong to NP-hard problems, in order to facilitate the solution, first establish the mathematical model of the vehicle routing problem.

D. Jin and S. Lin (Eds.): Advances in CSIE, Vol. 1, AISC 168, pp. 303–309.
springerlink.com © Springer-Verlag Berlin Heidelberg 2012

Assuming collection points for the receipt: $C = \{C_1, C_2, \cdots, C_N\}$, The distance between the two receiving points is $d(C_i, C_j)$,among $C_i, C_j \in C$ ($1 \le i, j \le N$), The mathematical model is,

$$\min\left(\sum_{i=0}^{N} d(C_{I(i)}, C_{I(i+1)})\right)$$

among $C_0 = C_{N+1}$ is logistics center, I(1), I(2),$\cdots$(N) is a full array of $1, 2, \cdots, N$.

3 Introduction to Genetic Algorithms

Genetic algorithm [1-2] is a random search algorithm based on natural selection and natural genetic mechanism, has a good parallelism and global optimization ability to adaptively adjust the search direction. This is a relatively simple algorithm, because it need not have a very complete knowledge to solve the problem, it can automatically generated solution of the problem similarly Organisms reproduce offspring, but the solution is often not the optimal solution, but relatively speaking, the second-best solution. It is because of the genetic algorithm has simple idea and powerful search capabilities, it has been quite widely used in the robot find its way, function optimization and combinatorial optimization and many other areas.

Genetic algorithm[3-5]is an iterative algorithm, it has a solution at each iteration, this set is randomly generated initial solution, it generates a new set of solutions by the simulation evolution and inheritance of genetic manipulation at each iteration, each solution has a given objective function evaluation, one iteration of a generation. algorithm steps are as follows:

For a given problem, given the variable encoding method, defined fitness function.

① initialized. make $t = 0$, given a positive integer T (the maximum number of iterations), the crossover probability p_c and mutation probability p_m, randomly generated initial population $p(0)$ formed by M individuals;

② individual evaluation. calculate the fitness of each individual from $p(t)$;

③ select: selecting from groups $p(t)$ to get middle groups;

④ crossover: the crossover operator acting on the intermediate group;

⑤ mutation: the mutation to be acting on the group after the crossover, then get the ($t+1$) generation group $p(t+1)$;

⑥ If $t < T$, then make $t := t+1$, turn ②;

If $t = T$, places the individual in the evolutionary process with the maximum fitness as the optimal solution, stop the operation.

4 Genetic Algorithm of the Vehicle Routing Problem

Definition: If I(1), I(2), ···, I(N) is a full array of 1,2,...,N in the chromosome, called the chromosome is an effective solutions of the vehicle routing problem. otherwise, call a non-effective solution.

Solving transport vehicles using genetic algorithms as follows:

① determine the variable encoding method, the problem with receiving point traversal order as arithmetic coding, that is $\{C_0, C_{I(1)}, C_{I(2)}, \cdots, C_{I(N)}, C_{N+1}\}$,

among I(1), I(2), ···, I(N) a full array of 1, 2, ..., N.

② Define the fitness function. the fitness function is defined as the inverse distance lines of the delivery vehicle , namely:

$$f = \frac{1}{\sum_{i=0}^{N} d(C_{I(i)}, C_{I(i+1)})}. \tag{1}$$

③ initialize groups. Determine population size, a randomly generated initial population (set of feasible solutions).

Population size indicates the number of chromosomes contained in the species, population is small, the genetic algorithm can improve the computational speed, but reduces the diversity of population, may not find the optimal solution; population is large, will increase the computing speed that reduce the efficiency of genetic algorithm. Generally the number of population in chromosome is about from 20 to 100.

④ Select operator. calculate the population fitness f_i of each chromosome as follow (1), and calculate the probability of each chromosome is selected values:

$$f_i^* = \frac{f_i}{\sum_{j=1}^{M} f_j}, \quad (i = 1, 2, \cdots, M)$$

Among M is the number of individuals in population. selection using roulette. :

⑤ crossover

At first generate two random integers ranging between 1 ~ M, so that ensure the number of chromosomes to participate in cross. then generates a random number determines the pairing of two chromosomes the gene locus cross position. Then, put the position as a dividing point, exchange part of the gene.

However, this method is generally used to solve the problem of transport vehicles is not applicable, for example: five receipt points for transportation vehicles, we use the symbols A, B, C, D, E represents the corresponding receiving point, O representative of logistics center. with the sequence of the six symbols that may be the solution (chromosome), then genetic manipulation.

Assume

$$S_1 = (O, A, C, B, E, D, O)$$

$$S_2 = (O, E, D, C, B, A, O)$$

Implementation of the conventional cross, such as exchange after the three,

$$S_1' = (O, A, C, B, B, A, O)$$

$$S_2' = (O, E, D, C, E, D, O)$$

It can be seen to be above the operating line S_1', S_2' are illegal. In order to avoid the above, it is proposed parents across every other gene, the specific operation as follows:

Starting from the second gene of the first Parent, diagonal cross to the third gene of the second parent, then, the fourth gene of the first parent, diagonal cross to the fifth gene of the second parent, a gene on the diagonal connection every so on, until the next line to connect to the last gene. every other gene on the diagonal connection, until the next line to connect to the last stop before a gene.

As follows

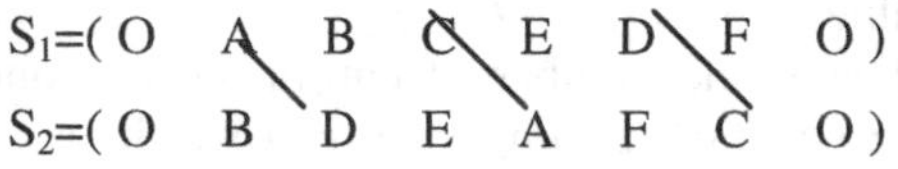

S_1=(O A B C E D F O)

S_2=(O B D E A F C O)

Interchange each of the two genes of crossover to connect, get two new chromosomes:

$$S_1' = (O, D, B, A, E, C, F, O)$$

$$S_2' = (O, B, A, E, C, F, D, O)$$

Through the cross, may have the following situations:

Case 1: the new chromosome is effective solution, as above S_1', S_2', then using S_1', S_2' instead of S_1, S_2, now cross operation was successful.

Case 2: the new chromosome is non-effective solution, it must be amended. the methods are: get rid of duplicate genes, in the front of the last gene, add the missing gene, making it an effective solution.

For example:

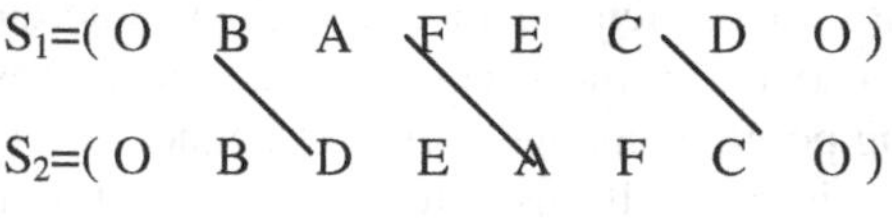

S_1=(O B A F E C D O)

S_2=(O B D E A F C O)

After the cross, get two new chromosomes:

$$S_1' = (O, D, A, A, E, C, D, O)$$

$$S_2' = (O, B, B, E, F, F, C, O)$$

Two chromosomes are not effective solution, the first S_1' have two A, two D, and lack of B and F, after revising, then get :

$$S_1''=(O, D, A, E, C, B, F, O)$$

Also, revise the S_2', and get:

$$S_2''=(O, B, E, F, C, A, D, O)$$

After revising, get two effective solution, then crossover success.

Therefore, cross algorithm as follows:

(1) Work out the total number K of chromosome according to crossover probability;

(2) Set a variable as cross statistics k, and give the initial value ;

(3) Number chromosomes population in 1~M, randomly generated two numbers among 1~M, selected two chromosomes, according to the above method to cross;

(4) Cross number k add 1, if 2 k >K, then cross over, otherwise, return(3) ;

⑥ Mutation operator

The exchange of using simple method as a mutation operator, carried out in a given mutation rate variation. The method is as follows:

(1) According to the mutation probability of total variation to work out a chromosome number L;

(2) Set a variable l as number of variation statistics, and give the initial value ;

(3) Randomly generated a number of 1~M, Selected a chromosome, then produce two random numbers, and set the variation of two genes bits;

(4) Exchange the position of the two genes which determined in (3);

(5) Variation number l add 1, if l >L, then ending variations, otherwise, return(3);

For example,

$$S_1=(O, A, E, F, G, B, C, D, O),$$

Randomly select two receiving points, assuming you have selected F and C, exchange the position of both and get the mutated chromosomes

$$S_1'=(O, A, E, C, G, B, F, D, O)$$

⑦ The optimal preservation strategy

After the selection, crossover and mutation, compare the best individual of new generation of fitness with the best individual of previous generation, if reduce, then the best individual of the above generation replace the new generation of worst individual. the strategy can ensure that so far the best individual won't be damaged by genetic operations of cross, variation and so on, it is an important guarantee of genetic algorithm convergence.

5 Application Cases

Set 51 receiving points of transportation problem for an example, suppose the 51 receiving points are 1,2,3,…,51, among them 1 is transportation center, each point coordinate position as is shown in table 1.

Table 1. 51 receiving point coordinates

1(37,52)	2(49,49)	3(52,64)	4(20,26)	5(40,30)	6(21,47)	7(17,63)
8(31,62)	9(52,33)	10(51,21)	11(42,41)	12(31,32)	13(5,25)	14(12,42)
15(36,16)	16(52,41)	17(27,23)	18(17,33)	19(13,13)	20(57,58)	21(62,42)
22(42,57)	23(16,57)	24(8,52)	25(7,38)	26(27,68)	27(30,48)	28(43,67)
29(58,48)	30(58,27)	31(37,69)	32(38,46)	33(46,10)	34(61,33)	35(62,63)
36(63,69)	37(32,22)	38(45,35)	39(59,15)	40 (5, 6)	41(10,17)	42(21,10)
43(5,64)	44(30,15)	45(39,10)	46(32,39)	47(25,32)	48(25,25)	49(48,28)
50(56,37)	51(30,40)					

Through the improved genetic algorithm for the transportation of 51 receiving some solving path, then will get the optimal loop as shown in figure 1 below:

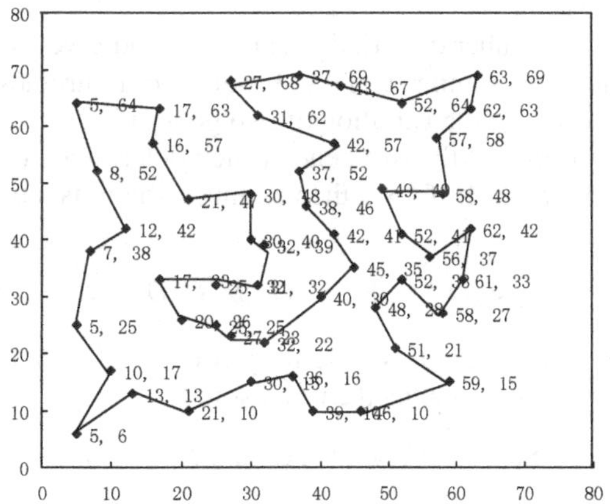

Fig. 1. The best circuit of this paper obtained.

Improvement plan of the values of the genetic algorithm after the optimal path is 422.2, but using the traditional genetic algorithm and carry out the optimal path is 555[6], then get the conclusion that the improved genetic algorithm in performance has improved significantly, thus we can obtain better results.

6 Conclusion

The paper puts forward a new method which using the genetic algorithm to solve the TSP problem.This method neatly improvements to the intersection method, and through the uniform variation, not only makes the population evolution, but also expand the search space. Using the data in TSPLIB of eil51 for experiments, the proposed genetic algorithm is better than the traditional method for the loop length get cut.

References

1. Wang, X., Cao, L.: Genetic Algorithms: Theory, Applications and Software, pp. 10–14. Xi'an Jiaotong University Press, Xi'an (2002)
2. Zhang, W.: The mathematical basis of the genetic algorithm. Xi'an Jiaotong University Press, Xi'an (2003)
3. Wang, Y.P., Han, L.X., Liy, H.: A new encoding based genetic algorithm for the traveling salesman an problem. Engineering Optimization 38(1), 1–13 (2006)
4. Ma, X., Zhu, S., Yang, P.: An Improved Genetic Algorithm of Traveling Salesman Problem. Computer Simulation 20(4), 36–37 (2003)
5. An Improved Genetic Algorithm of Traveling Salesman Problem 43(6), 65–68 (2007)
6. Luo, L., Huang, K.: A new method of Genetic algorithm solving traveling salesman problem. Jiaying College Journal 28(5), 18–21 (2010)

Study on Establishing Public Traffic Network of Qianjiang City

Haijiang Ye

College of Arts and Science, Jilin Agricultural Science and Technology University, Jilin 132109, Jilin Province, China
yehaijiang@sohu.com

Abstract. This paper presents the nodes which are appropriate for bus stations in Qianjiang city with the use of satellite map and field investigation, as well as the fig.1-1 of the distribution of nodes and main traffic lines with layer-by-layer analysis and mathematical model. The methods and results provided in this paper are applicable in service systems for public access computers in medium and small cities, design of public traffic lines and the evaluation and reformation of public traffic network maps, etc.

Keywords: urban public traffic communication, shortest route, network map.

1 Background

The current researches on public traffic network optimization, which mostly concern the qualitative and quantitative traffic network design, as well as the discrete and continuous variables of designing, have accomplished outstanding results [1]; some researches put forward the design of mixed traffic network and the model of double-layer network. The public traffic network shall be designed according to the continuous development and demand changes of passenger volume with regular adjustment and improvement. The current issues concerning the public traffic network are the long traffic lines, high degree of overlap among different lines and unscientific network designWe consider the following programming problem:

2 Hypothesis on the Issue

1) The nodes in the map can be taken as the bus station;
2) The node can be taken as a spot on the public traffic line which travels along the city road;
3) The road between two nodes is a straight line with a linear rate of 100 percent;
4) The passengers will not consider the time length while taking bus; they may take taxi or other vehicles if they are in a hurry, the above assumption accords with the reality.
5) The fare is irrelevant with the distance of traveling and the number of bus;

D. Jin and S. Lin (Eds.): Advances in CSIE, Vol. 1, AISC 168, pp. 311–315.
springerlink.com © Springer-Verlag Berlin Heidelberg 2012

3 The Establishment of Urban Public Traffic Network

3.1 The ultimate goal of establishing the public traffic bus is to provide convenience for passengers, who mainly concern the minimum bus transfers and if there is any straight bus, which is the inertial thinking, so the establishment of urban public traffic system shall put the convenience of passengers as the main concern and the shortest route is presumed to be a travel line with minimum number of bus transfers. Citizens always have a single destination, while the tourists have multiple destinations, and this paper mainly concerns the single destination. The public traffic lines and bus stations construct the complex network map, and the bus stations can be taken as the nodes in the network map and are connected with public traffic lines with the nodes connected (on a traffic line) or unconnected (not on any traffic line). Therefore, the public traffic network map is very complicated with many nodes and crossed public traffic lines, which is different from other regular network map. We take the small developing Qianjiang city as the subject and obtain the main traffic roads and network nodes in the city from the electronic map.

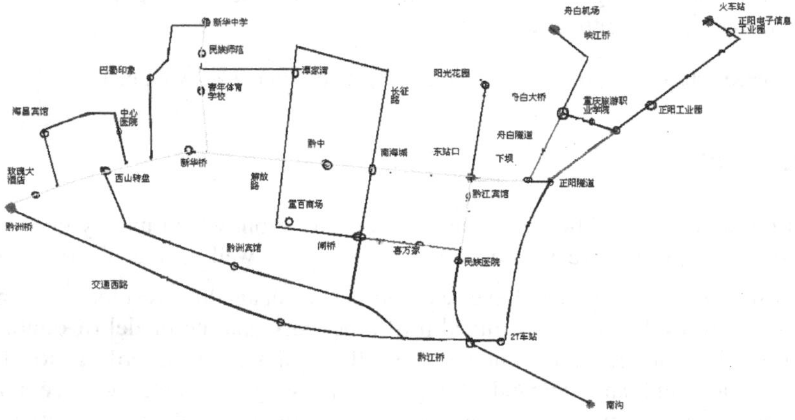

Fig. 1. Distribution map of main traffic roads and nodes in Qianjiang city

We use s_i and v_i to represent the arcs between these nodes and the adjacent ones

Table 1. The actual distances of the main traffic roads and those between adjacent stations in traffic network

Bus Station	Name of Bus Station	Arc between Adjacent Stations	Two Ends of Arc (Actual Stations)	Length between Adjacent Stations (km)
s_1	Qianzhou Bridge	v_1	Qianzhou Bridge - Rose Hotel	0.228
…	…	…	…	…

3.2 Mathematical Model

The establishment of public traffic network with the shortest route mainly concerns the rate of direct destination without the need of bus transfer, and the rate of overlap between different public traffic lines (the total length of the public traffic lines / the total length of urban roads available in the public traffic network), the optimized public traffic network can be obtained from the minimum function of overlap rate between public traffic lines.

The key to establish the public traffic network is to choose the proper start point and the end, which can be change into single direction with one adjacent arc. Based on the field investigation, the geographical condition of the place is not good with many mountains and rivers, and the actual area of urban land is small with many nodes positioned at the narrow roads which will obstruct the traffic if they were taken as the bus stations. In conclusion, the stations which are appropriate to be taken as the start points and the end are Qianzhou Bridge, Dongzhankou, Yangguang Neighbourhood, Nangou, Xiaba, Railway Station, and Zhoubai Airport [2], which can construct the following traffic lines, including Qianzhou Bridge – Dongzhankou, Qianzhou Bridge – Yangguan Neighbourhood, Qianzhou Bridge – Nangou, Qianzhou Bridge – Xiaba, Qianzhou Bridge – Railway Station, Qianzhou Bridge – Zhoubai Airport, Dongzhankou - Yangguan Neighbourhood, Dongzhankou - Xiaba, Dongzhankou –Nangou, Dongzhankou – Railway Station, Dongzhankou – Zhoubai Airport, Yangguan Neighbourhood – Nangou, Yangguan Neighbourhood – Xiaba, Yangguan Neighbourhood –Railway Station, Yangguan Neighbourhood – Zhoubai Airport, Nangou – Xiaba, Nangou –Railway Station, Nangou – Zhoubai Airport, Xiaba – Railway Station, Xiaba – Zhoubai Airport, Railway Station – Zhoubai Airport, L_{ij} is used to represent the route between s_i and s_j, and the length of each public traffic line, according to the standard, is controlled between 5km and 10km with the min L_{ij} after calculation as the following:

$$L_{1,15} = 2.817km, \quad L_{1,19} = 3.664km,$$
$$L_{1,16} = 3.323km, \quad L_{1,29} = 4.753km,$$
$$L_{1,27} = 12.898km, \quad L_{1,24} = 9.081km,$$
$$L_{29,19} = 2.783km, \quad L_{29,16} = 3.289km,$$
$$L_{29,27} = 12.066km, \quad L_{29,24} = 8.249km,$$
$$L_{19,15} = 1.353km, \quad L_{15,16} = 0.847km,$$
$$L_{15,16} = 0.506km, \quad L_{19,24} = 6.732km,$$
$$L_{19,27} = 9.234km, \quad L_{27,24} = 9.833km,$$
$$L_{15,27} = 10.081km, \quad L_{15,24} = 7.579km$$

The lengths which meet the standard are $L_{1,29} = 4.753km$, $L_{29,24} = 8.249km$, $L_{19,24} = 6.732km$, $L_{19,27} = 9.234km$, $L_{27,24} = 9.833km$, $L_{15,24} = 7.579km$, $L_{1,24} = 9.081km$. According to the actual demand of the local passengers, $L_{1,19} = 3.664km$ and $L_{1,16} = 3.323km$ are also appropriate to be taken as the main urban public traffic lines, that is, Qianzhou Bridge – Xiaba, Qianzhou Bridge – Yangguang Neighbourhood, Qianzhou Bridge – Nangou, Nangou – Zhoubai Airport, Xiaba – Railway Station, Xiaba – Airport, Dongzhankou – Airport, Railway Station – Zhoubai Airport, Qianzhou Bridge – Zhoubai Airport. The public traffic network is established on the basis of these main lines. First, we locate all the public traffic lines with the use of computer, and then find out a group, which shall cover all the nodes in the public traffic network, with minimum overlap rate. The target function is obtained as follows:

$$\min z = \sum (L_{ij} \Big/ \sum_{i=1}^{39} v_i)$$

Control condition:

The l_{ij} is the combination of all the nodes on the public traffic line from s_i to s_j, that is, $l_{ij} = \begin{bmatrix} s_i & \cdots & s_j \end{bmatrix}$, and V_i is the arc v_i which contains the two end points $V_i = [v_{i_1}, v_{i_2}]$

$$\begin{cases} 5km \le L_{ij} \le 10km \\ \bigcup l_{ij} = \bigcup V_i \end{cases}$$

The public traffic lines from Xiaba and 27 Bus Terminal to Xinmin Hospital can only pass through the Zhenyang tunnel, and so is the line from Zhoubai Bridge to Xiaba through Zhoubai tunnel, and both tunnel are single direction.

All the public traffic lines along the main roads are obtained from the calculation with MATLAB software with the minimum overlap rate as follows:

$$\min z = \sum (L_{ij} \Big/ \sum_{i=1}^{39} v_i) = 1.797665$$

$$s_1 \to s_3 \to s_4 \to s_9 \to s_{11} \to s_{12} \to s_{15} \to s_{19}$$

$$s_1 \to s_5 \to s_{30} \to s_{18} \to s_{29}$$

$$s_1 \to s_5 \to s_6 \to s_7 \to s_{10} \to s_{11} \to s_{12} \to s_{15} \to s_{17} \to s_{18} \to s_{29}$$

$$s_1 \to s_5 \to s_6 \to s_7 \to s_{10} \to s_{12} \to s_{15} \to s_{16}$$

$$s_1 \to s_{18} \to s_{29}$$

$s_1 \to s_{15} \to s_{30} \to s_{14} \to s_{12} \to s_{15} \to s_{19}$

$s_{29} \to s_{28} \to s_{20} \to s_{21} \to s_{24}$

$s_{19} \to s_{20} \to s_{21} \to s_{22} \to s_{23} \to s_{24}$

$s_{19} \to s_{20} \to s_{21} \to s_{25} \to s_{26} \to s_{27}$

$s_{27} \to s_{26} \to s_{25} \to s_{21} \to s_{22} \to s_{23} \to s_{24}$

4 Conclusion

This paper provides a design of public traffic lines by using many methods of operational research to establish a mathematical model for the optimization design of public traffic network, and obtain the optimized solution through mathematical software and the corresponding public traffic network with the overlap rate of public traffic lines as the target function and covering rate and length standard of lines as the control condition.

References

[1] Liu, C.: Modern Transportation Planning. China Communications Press, Beijing (2001)
[2] GB50220-95. Urban Traffic Planning and Design Standards
[3] Liu, W.: Analysis of evaluation indexes quantum used in the yard layout of highway hub. China Journal of Highway and Transport 16(2), 86–89 (2003)
[4] Xiao, W.: Graph theory and its Algorithm. Aviation Industry Press (1992)
[5] Du, D.: Graph-based Operational Research. Beijing University of Aeronautics and Astronautics Press (1990)
[6] Ye, X.: Practical Operational Research - using Excel to establish the model and find the solution. China Renmin University Press (2007)
[7] Wang, W., Xu, J.: Theory and Practice of Urban Transportation Planning. Southeast University Press (1998)
[8] Zhou, X., Yang, J.: Study of dispatching at minimum waiting time of public transportation transfer under the condition of ITS 17(2), 82–84 (2004)
[9] Liu, W.: MATLAB Program Design Course. China Waterpower Press (2005)

Vehicle Routing Problem with Soft Time Windows

Jixian Xiao and Bing Lu

College of Science, Hebei United University, Tangshan Hebei 063009, China
xiaojix@yahoo.com.cn, lubing311@126.com

Abstract. Vehicles start from the logistics center, they do not repeat to finish all the customer points, and return to the logistics center. on the way, traffic jams may be encountered at any time, you can choose to wait or a new route, if the service vehicles can not reach the customer points within the specified time, we need to implement some of the punishment, if the service vehicles select a new path, it will increase the journey distance, resulting in new additional costs, so we establish soft time window of the vehicle routing problem, and design the algorithm, get the minimum transportation costs as objective function.

Keywords: soft time windows, genetic algorithm, vehicle routing problem.

1 Introduction

Vehicles start from the logistics center, they delivery goods to all the customer points under demanding every customer point's goods requirements and certain time window requirements, every vehicle has limited capacity, the vehicles return the logistics center after they complete the delivery mission. if traffic congestion occurs when the vehicles are in the road. they can choose to wait or a new path, based on the type of vehicle routing problem, at the premise of minimum transportation costs as the optimization objectives, we make the decision whether to adjust the running line. based on the above, the paper creates a dynamic vehicle routing optimization model.

2 Modeling

Assuming collection points of customers are $C=\{C_1, C_2, \cdots, C_N\}$, the distance between two points is $d(C_i, C_j)$, $C_i, C_j \in C$ $(1 \le i, j \le N)$ among $C_0 = C_{N+1}$ is logistics center, $I(1), I(2), \cdots, I(N)$ is a full array of $1, 2, \cdots, N$. vehicles start from the transportation center C_0 and delivery goods to N customer points, at last return the logistics center after finish the task.

It solves the problem based on the following assumptions:

1) The demands of customer points will not exceed the vehicle capacity;

2) The travel distance of vehicles does not exceed the maximum travelling distance;

3) Vehicles can get instantaneous information on traffic condition, that is the vehicles get the travel time of all sections in a short time passed;

D. Jin and S. Lin (Eds.): Advances in CSIE, Vol. 1, AISC 168, pp. 317–322.
springerlink.com © Springer-Verlag Berlin Heidelberg 2012

4) After the initial line, it adjust the line, it does not change the vehicles to access the customer's collection, only to change the vehicle to access the customer's order;

Explanation of symbols:

N—the number of customer points;

C_0—logistics center;

W—the total cost;

c_{ij}—the time travel from customer point i to customer point j;

δ—unit penalty costs of vehicles arrive earlier than the earliest specified time;

θ—unit penalty costs of vehicles arrive later than the latest specified time point;

Q_k $(k=1,2,\cdots,K)$—the maximum loading capacity of the number k vehicle;

K—number of vehicles in distribution center;

$F(C_i)$ $(i=1,2,\cdots,N)$—the earliest service time of the customer point i;

$L(C_i)$ $(i=1,2,\cdots,N)$—the latest service time of the customer point i;

$T(C_i)$ $(i=1,2,\cdots,N)$—the time of the vehicle arrives at the customer point i;

$q_k(C_j)$ $(k=1,2,\cdots,K)$—the number j customer's demand of the number k vehicle;

$d_k(C_{I(i)},C_{I(i+1)})$ $(k=1,2,\cdots,K)$—the distance between the number $I(i)$ customer and the number $I(i+1)$ customer serviced by the number k vehicle in the order of service;

$$x_{ijk}=\begin{cases}1, & vehicle\ k\ travels\ from\ customer\ point\ i\ to\ customer\ point\ j\\ 0, & or\end{cases}$$

$$y_{ik}=\begin{cases}1, & customer\ point\ i\ serviced\ by\ vehicle\ k\\ 0, & or\end{cases}.$$

The soft time window problems objective function is as follows:

$$\min W=\sum_{i=1}^{N}\sum_{j=1}^{N}\sum_{k=1}^{K}c_{ij}x_{ijk}+\sum_{i=1}^{N}\max\{0,(F_i(C_i)-T_i(C_i))\}\delta$$
$$+\sum_{i=1}^{N}\max\{0,(T_i(C_i)-L_i(C_i))\}\theta. \tag{1}$$

Constraints :

$$\sum_{j=1}^{l_k}q_k(C_j)\le Q_k\,. \tag{2}$$

$$\sum_{k=1}^{K}l_k=N\,. \tag{3}$$

$$\sum_{k=1}^{l_k}d_k(C_{I(i)},C_{I(i+1)})\le D_k\,. \tag{4}$$

$$\sum_{k=1}^{N} y_{ik} = 1,\ (i = 1,2,\cdots,N)\ . \tag{5}$$

$$\sum_{i=1}^{N} x_{ijk} = y_{jk},\ (j = 1,2,\cdots,N),\ \forall k\ . \tag{6}$$

$$\sum_{j=1}^{N} x_{ijk} = y_{ik},\ (i = 1,2,\cdots,N),\ \forall k\ . \tag{7}$$

$$x_{ijk} = 1\text{或}0\ . \tag{8}$$

$$y_{ik} = 1\text{或}0\ . \tag{9}$$

(1) The least-cost objective function;

(2) The demand of the customers in loop will not exceed the vehicle capacity;

(3) The number of customer points serviced by all participating vehicles is exactly equal to the total number of customer points, that is every customer get serviced;

(4) The length of the vehicle in every loop will not exceed the maximum mileage;

(5) The customer point i only serviced by the vehicle k;

(6) and (7) ensure that every customer is visited exactly once;

(8) The vehicle k travels from the customer point i to the customer point j when $x_{ijk} = 1$; if not, it is $x_{ijk} = 0$;

(9) The customer point i serviced by the vehicle k when $y_{ik} = 1$; if not, $y_{ik} = 0$;

3 Algorithm Design

Step1 It determines encoding method of the variables.

It uses natural number coding method, it traverses the customer point of order as coded character set, that is $\{C_0, C_{I(1)}, C_{I(2)}, \cdots, C_{I(N)}, C_0\}$, $I(1), I(2), \cdots, I(N)$ is a full array of $1,2,\cdots,N$, C_0 is the Logistics Center.

Step2 It defines fitness function.

$$f_i = \frac{1}{W_i}$$

f_i is the fitness value of chromosomes i; W_i is the objective function value and penalty function of chromosome i.

Step3 Initialized groups.

The genetic algorithm population size is generally 20 to 200.

Step4 Selection operator.

1) It calculates the fitness of every individual as the fitness formula, and then calculates the sum $\sum f_i$ of the individual's fitness of the population;

2) It calculates the probability of every individual selected, its value is the fitness of every individual divided by the sum of fitness;

$$p_i = \frac{f_i}{\sum f_i}$$

3) It selects with roulette mode.

Step5 Crossover operator

It figures out the number K of chromosome according to the crossover probability, it sets a variable k as the cross number and gives initial value, they are numbered 1~N to the chromosomes population, it randomly generated two numbers among 1~N.

Step6 Variation operator

At first, it calculates variation of total chromosome number L according to the mutation probability, and then gives the variation of this statistic number l, and gives the initial value, it generates a number among 1~M randomly, select a chromosome, and then generates two random numbers, as a variant of the two genes, the selected two genes on chromosome position exchange, variation number l plus 1, if $l > L$, then variation over, Otherwise, choose to variation of chromosome.

Step7 Optimal preservation strategy

At first, it calculates the best individual's fitness value of the previous generation, and then calculates the best individual's fitness value of the new generation, if the individual's fitness value of the new generation is smaller than the previous generation, the worst individual's fitness value of the new generation is replaced by the best individual's fitness value of the previous generation.

4 Case Study

The logistics center provides service to 10 customer points using a car, the average speed is 50km/h of the vehicle, assuming the unit transportation cost is 1 yuan/min of the vehicle, if the vehicle reaches earlier than the earliest time of the customer point requires, the vehicle must wait until the earliest service time starts, the unit penalty cost is 0.5 yuan/min, if the vehicle reaches later than the latest time of the customer point requires, the unit penalty cost is 2 yuan/min, coordinates location of 10 customers and the earliest service time and the latest service time of every customer point as the following table (1) below.

1) How to design appropriate transportation lines making costs least under meeting the conditions of the soft time window;

2) The vehicle travels according to the question 1), when it travels from the customer point C_{10} to the next customer point, traffic jam occurs, the traffic jam time is about 40 minutes, how to arrange the line making total transportation costs least.

Table 1. Coordinates and time windows requirements of customer points.

client-point number	x-coordinate (km)	y-coordinate (km)	the earliest service time (min)	The latest service time (min)	service time (min)
C_0	35	35	0	280	0
C_1	41	49	120	140	20
C_2	35	17	20	40	20
C_3	55	45	155	175	20
C_4	55	20	185	205	20
C_5	15	30	50	70	20
C_6	25	30	40	60	20
C_7	20	50	71	91	20
C_8	10	43	60	80	20
C_9	55	60	130	150	20
C_{10}	30	60	95	115	20

Solution process is as follows:

1) It is to cost at least in the condition of the soft time window, the vehicle travels in the line below (1) as follows:

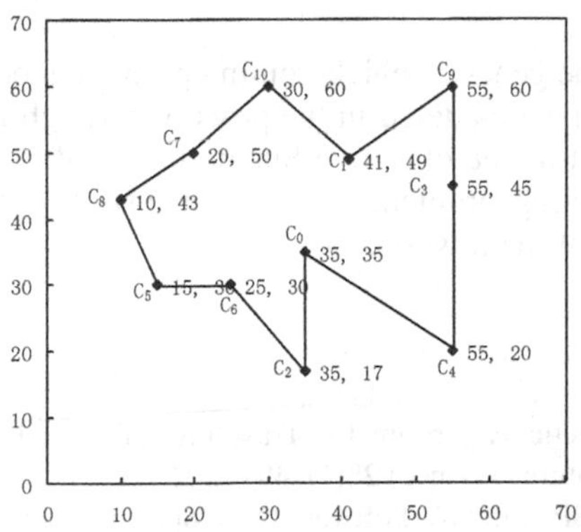

Fig. 1. The traffic routes under soft time window required.

It completes tasks following the above line, transportation costs to 219.647yuan.

2) When the vehicle travels from the customer point C_{10} to the next customer, a traffic jam about 40 minutes occurred intersection, if the vehicle travels following the original route, the total cost of at least

$$219.647+101.2226*2=422.0922\text{yuan}$$

If changing the traffic routes at this time, the adjusted line as shown (2) as follows:

After the traffic jam, it adjusts the line, the total cost of at least

$$236.5348+5.386*0.5+83.6429*2=406.5136\text{yuan}$$

It can be seen, the adjusted line cost less than waiting through, so I chose to continue to adjust the line after delivery tasks.

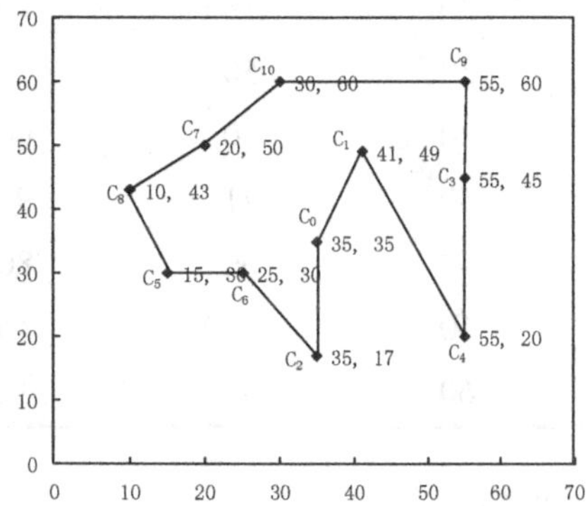

Fig. 2. The adjustment line after traffic jam.

5 Conclusion

The article is based on the general vehicle routing problem model, when the vehicle in road traffic jam situation encountered in the process, through increasing the soft time window, to the least cost as the objective function, the soft time window required to establish the vehicle routing problem, through analysis of the model and algorithm design, making the cost of the lowest cost.

References

1. Li, F., Bai, Y.: A Genetic Algorithm for Traveling Saleman Problem. Journal of North University of China (Natural Science) 28(1), 49–52 (2007)
2. Lu, T.: A Solution to TSP with GA. Software Development and Design, 29–30 (2010)
3. Luo, L., Huang, K.: A New Genetic Algorithm for TSP Problem. Journal of Jiaying University 28(5), 18–21 (2010)
4. Wang, Y.P., Han, L.X., Liy, H.: A new encoding based genetic algorithm for the traveling salesman an problem. Engineering Optimization 38(1), 1–13 (2006)
5. Dantzig, G., Ramser, J.: The truck dispatching problem. Management Science (6), 81–89 (1959)
6. Clarke, G., Wright, J.W.: Scheduling of vehicles from a central depot to a number of delivery points. Operations Research 12, 568–581 (1964)

Research on Monitoring Data Processing and Fractional Predictive Method of the Phreatic Line*

Xiaoxia Wang[1], Fengbao Yang[2], Linna Ji[1], and Qingying Fan[1]

[1] Information and Communications Engineering College
[2] Science College
North University of China, Taiyuan Shanxi 030051, China
wangxiaoxia@nuc.edu.cn

Abstract. It is a key issue of security for the tailing dam that how to deal with monitoring data and predict the phreatic line. In this paper, the empirical mode decomposition method is employed to purify the data. According to the periodicity and self-similarity of monitoring data, the data is divided into different stages. The fractal interpolation is utilized to obtain the iterated function systems of the various stages, and weighted iterative function systems of phreatic line changes are attained at prediction moment through its weighted sum, then the attractor of weighted iterative function system is obtained by the random iterative algorithm. Finally, the fractional prediction model of the phreatic line is established, and some examples are verified its validity. The predicting outcomes show the average prediction error of the No. 2 monitoring point is 1.50%, which expresses the results are ideal.

Keywords: tailing dam, phreatic line, fractal theory, iterated function system.

1 Introduction

With the development of automatic monitoring technology, online monitoring system of phreatic line has developed successfully and has been put into service [1,2]. Phreatic line is taken as "lifeline" of tailing dam, and its predicting outcomes are closely linked to the safety of the dam. In the actual monitoring, predicting outcomes of phreatic line are mainly affected by two aspects: (i) trend items of the errors, which are resulted from sensor errors and condition mutation, reduce the accuary of monitoring data, so the data must be excluded. (ii) There are many existing predictive methods, such as grey system, neural network, and support vector machine etc, and some achievements are obtained. These methods are dependent on the length of the data, also less data have greater effect on predicting outcomes. Therefore, it is of great significance to purify monitoring data and research on new predicting method for phreatic line.

---

* Biography: Xiaoxia Wang(1980.08-), instructor, doctoral candidate, majoring in signal and information processing and pattern recognition. Fengbao Yang(1968.11-), professor, doctoral supervisor, majoring in signal and information processing and image fusion.
Correspondence Address: Information and Communications Engineering College, the North University of China, Taiyuan Shanxi 030051.

D. Jin and S. Lin (Eds.): Advances in CSIE, Vol. 1, AISC 168, pp. 323–328.
springerlink.com © Springer-Verlag Berlin Heidelberg 2012

Empirical mode decomposition method can decompose any type of signal, and it has a distinct advantage in dealing with nonlinear data. Therefore, the revision of pauta criterion and the empirical mode decomposition method are employed to purify the data, in order to improve the accuracy of the monitoring data.

The tailing dam consists of the porous media, and the skeleton structures and pores of the dam are formed into fractal space. When the fluid flow in the fractal space, it surely has the fractal characteristic and some other special properties [4,5]. However, phreatic line is the free line which forms in the seepage process, so in different positions, the change law of phreatic line can be taken as complex nonlinear system which has self-similarity and cyclical. The attractor of iterated function systems can be described the change characteristics of the line. Therefore, the fractional prediction model of the phreatic line is established by the fractal interpolation, then the proposed method is verified its effectiveness by some examples, and it provides theoretical basis and the scientific basis for predicting precision of phreatic line.

2 Monitoring Data Processing

The trend items are frequency components of the monitoring data in which vibration cycles are greater than the sampling length, because some reasons of measurement system cause the slow change trend error in the time series [6]. In the actual monitoring data, the trend items are produced for many reasons, such as the zero drift which is resulted from the temperature change of sensors; when we monitor phreatic lines, the sensors are inertial measurement bases, so in the dynamic measurement, the displacement change of dam will cause data migration; the trend is caused by time delay which is resulted from the seepage velocity and so on. These trend items will produce large error on the correlative analysis of the data, even make the data completely lost authenticity. Therefore, the data that do not exclude trend items is used to predict phreatic line directly, which may cause large errors, so extraction and exclusion of the trend items is an essential link to forecast the phreatic line.While empirical mode decomposition (EMD) method does not need any prior knowledge, and it is convenient to utilize, also it has good removal efficiency [7]. Therefore empirical mode decomposition method is used to remove trend items of monitoring data.

Empirical mode decomposition method is used for data decomposition according to the time scale features of data itself, and it decomposes processed data $y(t)$ into:

$$y(t) = \sum_{i=1}^{n} c_i(t) + r_n(t) \tag{1}$$

Where c_i is a natural function, and r_n is residual item (trend item).

The effect of original data and the data extracted trend items is compared to the data of No. 2 monitoring point for phreatic line, which is collected in June 2011, as shown in figure 1:

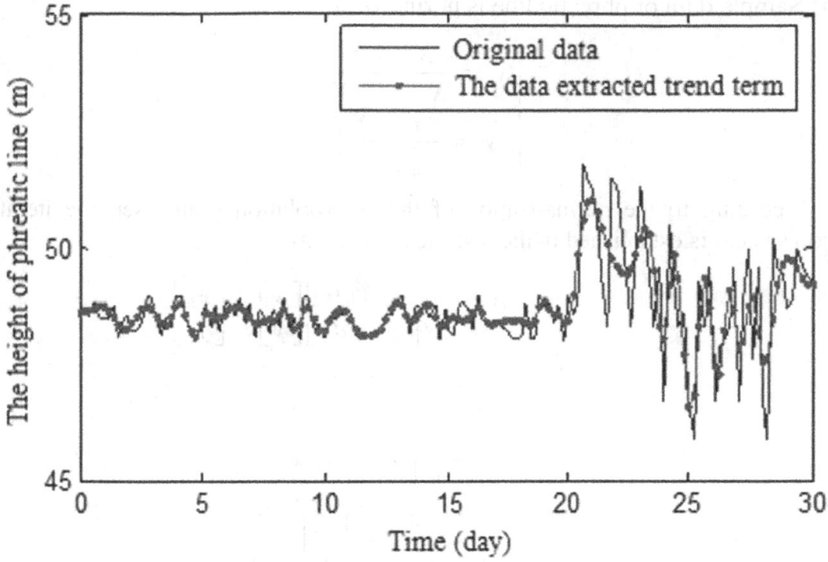

Fig. 1. The comparison of original data and the data extracted trend items

3 Fractal Prediction of Phreatic Line

Because phreatic line has the characteristics of randomness and nonlinearity, it is difficult to accurately give quantitative function relation which describes the position of phreatic line by the monitoring data. Therefore, fractal interpolation method is used to forecast the phreatic line, according to self-similarity and periodicity of phreatic line, and interoperability of time and scale rule in fractal time series. Take No. 2 monitoring data after purification for example, the specific prediction procedure is as follows (other monitoring the same):

(i) Purification data of every six days is combined with each other as a time period, then five time periods can be obtained, and take the period that is nearest to July 1th to 5th (i.e. the combination of June 25th to 30th) as a benchmark stage.

(ii) For each time period, the data of phreatic line is regarded as sample data, then interpolating point set is establish.

$$\Omega \stackrel{\Delta}{=} \{(X_i, Y_i) | i = 1, 2, \cdots 6\} \tag{2}$$

Where (X_i, Y_i) is interpolation point, X_i is monitoring time, and Y_i is monitoring data after purification.

(iii) Sample data of phreatic line is normalized:

$$\begin{cases} x_i = \dfrac{X_i - X_{min}}{X_{max} - X_{min}} \\ y_i = \dfrac{Y_i - Y_{min}}{Y_{max} - Y_{min}} \end{cases} \quad (i = 1,2,\cdots,6) \tag{3}$$

(iv)According to the normalization of the interpolation points set, the iteration function system is established in the historical moments:

$$w_i \begin{bmatrix} x \\ y \end{bmatrix} = \begin{bmatrix} a_i & 0 \\ c_i & d_i \end{bmatrix} \begin{bmatrix} x \\ y \end{bmatrix} + \begin{bmatrix} e_i \\ f_i \end{bmatrix} \tag{4}$$

and

$$\begin{cases} w_i \begin{bmatrix} x_0 \\ y_0 \end{bmatrix} = \begin{bmatrix} x_{i-1} \\ y_{i-1} \end{bmatrix} \\ w_i \begin{bmatrix} x_M \\ y_M \end{bmatrix} = \begin{bmatrix} x_i \\ y_i \end{bmatrix} \end{cases} \tag{5}$$

The parameters a_i、c_i、e_i、f_i、d_i are calculated by formula (4) and (5), and the affine transformation is determined.

(v) The distances among the parameters of iteration function system are obtained through using the Euclidean distance, and the weights of parameters are derived through its weighted sum by the method in [8], then weighted parameters of iteration function system are determined. Weighted parameters of the curve for No.2 phreatic line are calculated in July 1th to 5th, as shown in table 1:

Table 1. Weighted parameters of the curve for No.2 phreatic line in July 1th to 5th

Serial number	a_i	c_i	e_i	f_i	d_i
1	0.1820	-0.2100	0.0000	0.4720	0.2000
2	0.1820	-0.0186	0.2000	0.2451	0.3400
3	0.1820	0.1036	0.4000	0.1278	0.4860
4	0.1820	0.1788	0.6000	0.2450	0.4700
5	0.1820	-0.0848	0.8000	0.4026	0.4620
Weighted parameters	0.1820	-0.0039	0.4000	0.3010	0.3793

(vi) The attractor of weighted iterative function system is obtained by the random iterative algorithm. According to the attractor, y_i corresponding to x_i is calculated, and the change of the phreatic line is predicted based on the monitoring data, finally the prediction value is obtained by formula (3), the results as shown in table 2:

Table 2. Prediction value of phreatic line in July 1th to 5th

Serial number	comparison of the results		
	real value （m)	prediction value (m)	relative error (%)
July 1th	49.60	49.20	0.81
July 2th	48.32	48.96	1.32
July 3th	49.98	51.20	2.44
July 4th	47.90	48.50	1.25
July 5th	48.80	47.98	1.68

From the table 2, the maximal relative error is 2.44%, the minimum relative error is 0.81% and the average relative error is 1.50% through the fractal theory to predict phreatic line. The outcomes are ideal, so the model is feasible in this paper, and the results reflect the change trend items of the phreatic line perfectly.

4 Conclusion

(i) Because an amount of noise data exist in the actual monitoring of phreatic line, the empirical mode decomposition method is employed to purify the data based on analyzing the noise characteristics of the data, which ensures the authenticity of the predicting data.
(ii) According to randomness, nonlinearity, periodicity and self-similarity of phreatic line, the data is divided into different stages. The fractal interpolation of the fractal theory is utilized to obtain the iterated function systems of the various stages, and weighted iterative function systems of phreatic line are determined at predicting moment through its weighted sum, then the attractor of weighted iterative function system is obtained by the random iterative algorithm, finally the fractional prediction model of the phreatic line is established. The predicted results show the average prediction error of the No. 2 monitoring point is 1.50%, and the predictive value reflects the trends of the phreatic line.
(iii) Space fractal interpolation method is employed to determine the overall change of phreatic surface of tailings dam, according to different predictions of monitoring

points. Therefore, this study not only extends the application field of fractal theory, but also provides theoretical basis and predict the scientific basis for data processing and prediction of other indicators of the dam.

Acknowledgements. This paper is supported by the National Natural Sicence Foundation of China (No.61171057), the Natural Sicence Foundation of North University of China, and International Office of Shanxi Province Education Department of China.

References

1. Zhu, C.: Study of data warehouse and data mining application on geohazard in three gorges reservior area. Degree thesis of doctor (2010)
2. Corominas, J., Moya, J.: A review of assessing landslide frequency for hazard zoning purposes. Engineering Geology 102, 193–203 (2008)
3. Wang, Y., Qin, P., Qin, B.: Forecasting model of monitored data of arch dam's temperature based on improved variable dimension fractal theory. Journal of Yangtze River Scientific Research Institute 12(26), 33–35 (2009)
4. Lin, P., Tang, L., Song, H., et al.: Fractal geometry in the past, present and future research of rock and soil mechanics. Northwestern Seismological Journal 33, 24–29 (2011)
5. Liu, Y., Sheng, J., Ge, X., et al.: Evaluation of rock mass quality based on fractal dimension of rock mass discontinuity distribution. Rock and Soil Mechanics 28(5), 971–975 (2007)
6. Duan, H., Xie, F., et al.: Signal trend extraction of road surface profile measurement. In: The 2nd International Conference on Signal Processing System, vol. 07, pp. 694–698 (2010)
7. Huang, N.E., Zheng, S.H., Steven, R.L., et al.: The empirical mode decomposition and the Hilbert spectrum for nonlinear and non-stationary time series analysis. Pro. Royal Society London A, 903–995 (1998)
8. Wang, X., Yang, F.: A kind of evidence combination rule in conflict. Journal of Projectiles; Rockets; Missiles and Guidance 27(5), 255–257 (2007)

High-Speed Train Adaptive Control Based on Multiple Models

Yonghua Zhou, Chao Mi, and Yaqin Yin

School of Electronic and Information Engineering, Beijing Jiaotong University, Beijing 100044, China
{yhzhou,10120304,10125072}@bjtu.edu.cn

Abstract. High-speed train puts up stricter requirements on its control strategy. One requirement is to make the controller quickly adapt to the large-scope external disturbances and system parameter variations. This paper develops the control model of train movement with maximum torque control for adaptive control strategy. The maximum torque control is converted to the state following control around the optimal slip speed. The method of multiple model adaptive control with second level adaptation is introduced into the train control. This method avoids the observer design and the estimation of unobservable states. The simulation results demonstrate that the control strategy has superior performance when the system parameters are time-variant with uncertain disturbances. The control scheme is as an alternative of the multiple controllers based on multiple models to improve the reliability of train control with large-scope prompt adaptation.

Keywords: adaptive control algorithm, multiple models, second level adaptation, train, high speed.

1 Introduction

High-speed train construction has attained great progress in recent years. Under high-speed transport architecture, fast and adaptive response to the uncertain disturbances plays a critical role in ensuring the safe and comfortable operation of trains. Compared with traditional PID control, adaptive control technique can provide train with flexible and versatile control efforts to deal with the system parameter variations and the external disturbances. Thus, more favorable transient performance can be achieved such as high safety and comfortableness.

Anti-slip or -skid control is a basic problem for train control. The anti-slip train control has been discussed with various observers [1-4]. The adhesion control system based on the observer is proposed to determine the maximum traction torque [1]. The disturbance observer is put up for the dynamic adhesion force based on the mathematical model of the mechanical structure of transmission unit [2]. A new adhesion control is brought forth based on the torque feedback control using the first order disturbance observer [3]. The observer is proposed to estimate the friction force and the derivative, and thus the optimal slipping velocity is estimated to realize the maximum adhesive force [4]. The single model is utilized in those methods, which

D. Jin and S. Lin (Eds.): Advances in CSIE, Vol. 1, AISC 168, pp. 329–334.
springerlink.com © Springer-Verlag Berlin Heidelberg 2012

covers the limited scope of system parameter variations and external disturbances. This paper attempts to propose a train control strategy based on multiple model adaptive control [5-7]. Because multiple models are employed, the high adaptability to the large scope of uncertainty and the quick response can be achieved. In addition, multiple model adaptive control has theoretic guarantee of control stability.

2 Model

Each train has several cars, and partial cars are equipped with several driving motors. This paper only addresses the model of one wheel driven by one motor. The similar analysis can be extended to the total train. Without loss of generality, the inner force q_i between cars is omitted. Based on the force analysis between wheel and rail, we have

$$F_i = \mu(v_{si})m_i g = m\dot{v}_i + f_i = m\dot{v}_i + c_{vi}v_i^2 , \tag{1}$$

$$T_{Li} = F_i r , \tag{2}$$

$$J_i \frac{d\omega_i}{dt} = T_i - T_{Li} - b_\omega \omega_i \tag{3}$$

where v_{si} is the slip speed of car i, which is the difference between the train wheel speed v_{wi} and the train body speed v_i , i.e. $v_{si} = v_{wi} - v_i$. F_i is the traction force. m_i is the mass. f_i is the resistance. c_{vi} is the resistance coefficient. T_{Li} is the load torque. ω_i is the angular velocity of wheel. J_i is the wheel moment. r is the wheel radius. T_i is the torque of electrical motor. b_ω is the friction coefficient. And $\mu(v_{si})$ is the viscosity friction coefficient. There exists the optimal slip velocity v_{si}^m related to the maximum viscosity friction coefficient around which the torque is imposed on the wheel so that the maximum traction force can be achieved.

Define $x_1 = v_i$ and $x_2 = v_{\omega i} = \omega_i r$. Equations (1)-(3) are linearized as

$$\begin{bmatrix} \dot{x}_1 \\ \dot{x}_2 \end{bmatrix} = \begin{bmatrix} -(f+a) & f \\ bf & -(bf+c) \end{bmatrix} \begin{bmatrix} x_1 \\ x_2 \end{bmatrix} + \begin{bmatrix} d \\ e \end{bmatrix} + \begin{bmatrix} 0 \\ 1 \end{bmatrix} u \tag{4}$$

where $a = 2c_{vi}v_0 / m_i$, v_0 is the initial speed, $f = F'(v_{si}^m)$, $F(v_{si}) = \mu(v_{si})g$, $v_{si} = x_2 - x_1$, $b = r^2 / J_i$, $c = b_w / J_i$, $d = d' + F(v_{si}^m)$, $d' = c_{vi}v_0^2$, $e = -r^2 F(v_{si}^m) / J_i$, and $u = T_i r / J_i$.

Define $A = \begin{bmatrix} -(f+a) & f \\ bf & -(bf+c) \end{bmatrix}$ and $\begin{bmatrix} \bar{x}_1 \\ \bar{x}_2 \end{bmatrix} = \begin{bmatrix} x_1 \\ x_2 \end{bmatrix} + A^{-1} \begin{bmatrix} d \\ e \end{bmatrix}$. It follows that

$$\begin{bmatrix} \dot{\bar{x}}_1 \\ \dot{\bar{x}}_2 \end{bmatrix} = \begin{bmatrix} -(f+a) & f \\ bf & -(bf+c) \end{bmatrix} \begin{bmatrix} \bar{x}_1 \\ \bar{x}_2 \end{bmatrix} + \begin{bmatrix} 0 \\ 1 \end{bmatrix} u . \tag{5}$$

The output is

$$y = [\bar{x}_1 \; \bar{x}_2]^T . \tag{6}$$

The companion form of equations (5) and (6) is as follows:

$$\begin{bmatrix} \dot{\tilde{x}}_1 \\ \dot{\tilde{x}}_2 \end{bmatrix} = \begin{bmatrix} 0 & 1 \\ -(ac+cf+abf) & -(a+c+f+bf) \end{bmatrix} \begin{bmatrix} \tilde{x}_1 \\ \tilde{x}_2 \end{bmatrix} + \begin{bmatrix} 0 \\ 1 \end{bmatrix} u , \tag{7}$$

$$y = \begin{bmatrix} f & 0 \\ a+f & 1 \end{bmatrix} \begin{bmatrix} \tilde{x}_1 \\ \tilde{x}_2 \end{bmatrix} = C \begin{bmatrix} \tilde{x}_1 \\ \tilde{x}_2 \end{bmatrix} . \tag{8}$$

Assuming control objectives $\begin{bmatrix} x_1 \\ x_2 \end{bmatrix} = \begin{bmatrix} D \\ D+v_{si}^m \end{bmatrix}$, $\begin{bmatrix} \tilde{x}_1 \\ \tilde{x}_2 \end{bmatrix} = C^{-1} \left(\begin{bmatrix} D \\ D+v_{si}^m \end{bmatrix} + A^{-1} \begin{bmatrix} d \\ e \end{bmatrix} \right)$.

3 Multiple Model Adaptive Control with Second Adaptation

The controlled plant is described by the state equations

$$\textstyle\sum_p: \quad \dot{x}_p(t) = A_p x_p(t) + bu(t) . \tag{9}$$

The elements of the last row of matrix A_p are $[\theta_{p1} \, \theta_{p2} \cdots \theta_{pn}] = \theta_p^T$ and are assumed to be time-variant.

The reference model for adaptive control is described by

$$\textstyle\sum_m: \quad \dot{x}_m(t) = A_m x_m(t) + br(t) . \tag{10}$$

The last row of matrix A_m is denoted as $[\theta_{m1} \, \theta_{m2} \cdots \theta_{mn}] = \theta_m^T$. The objective of adaptive control is to determine the input $u(t)$ such that $\lim_{t\to\infty}[x_p(t) - x_m(t)] = 0$.

The indirect adaptive control based on multiple models is utilized to control the plant. The N identification models are described by

$$\textstyle\sum_i^{'}: \quad \dot{x}_i(t) = A_m x_i(t) + [A_i(t) - A_m] x_p(t) + bu(t) - \lambda e_i(t) \tag{11}$$

with the initial condition $x_i(t_0) = x_p(t_0)$. The last row of matrix $A_i(t)$ is represented as $[\theta_{i1} \, \theta_{i2} \cdots \theta_{in}] = \theta_i^T(t)$. The identification errors are $e_i(t) = x_i(t) - x_p(t)$ with $e_i(t_0) = 0$.

Define the Lyapunov function as $V_i(\xi_i,\theta_i)=\xi_i^T P\xi_i+\theta_i^T\theta_i$ where P is the positive definite matrix solution of the Lyapunov equation $A_m^T P+PA_m=-Q$ with $Q=Q^T>0$, and $\xi_i=e_i/\sqrt{1+x_p^T x_p}$. Using the adaptive law $\dot{\theta}_i=-e_i^T Pbx_p/(1+x_p^T x_p)$, it follows that $\dot{V}_i(\xi_i,\theta_i)<0$ if $\lambda>\lambda_0$. Correspondingly, to ensure the stability of the plant, the feedback input is $u(t)=-k^T(t)x_p(t)+r(t)$ where $k(t)=\theta_i(t)-\theta_m$.

In order to make the system have faster response, the following virtual model [6, 7] is defined using $N=n+1$ identification models:

$$\hat{\theta}_p=\sum_i \alpha_i\theta_i(t). \tag{12}$$

The weighting coefficients α_i are iteratively estimated as $\dot{\overline{\alpha}}(t)=-M^T(t)M(t)\overline{\alpha}(t)+M^T(t)\ell(t)$ where the ith column of M is $M_i(t)=e_i(t)-e_{n+1}(t)$ and $l(t)=-e_n(t)$.

4 Simulation Results

Two kinds of disturbances are considered in the simulation. One is the variation of load torque, and the other is the change of adhesion coefficient. Fig. 1 demonstrates the plant parameter variations caused by these two disturbances. Fig. 2 displays the results of tracking control using the traditional adaptive control. From Fig. 2, we can learn that there exist big tracking and identification errors, and the performance is unsatisfactory. Fig. 3 shows the results of trajectory following and model identification utilizing the multiple model adaptive control with second level adaptation. Fig. 3 reveals that the control strategy can achieve favorable tracking performance and convergent identification errors although the initial values of the virtual model greatly deviate from the true ones.

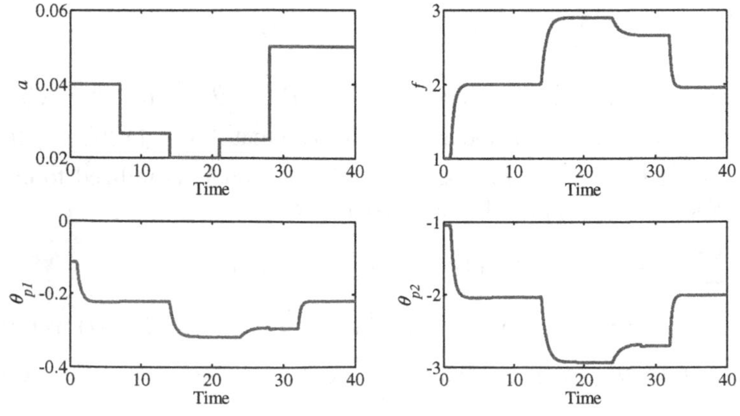

Fig. 1. The plant parameter variations caused by the disturbances.

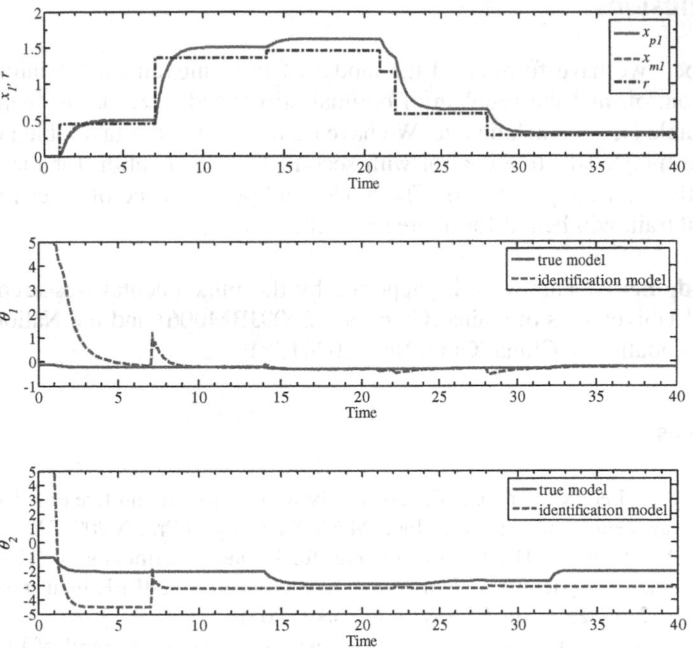

Fig. 2. The traditional adaptive control of train

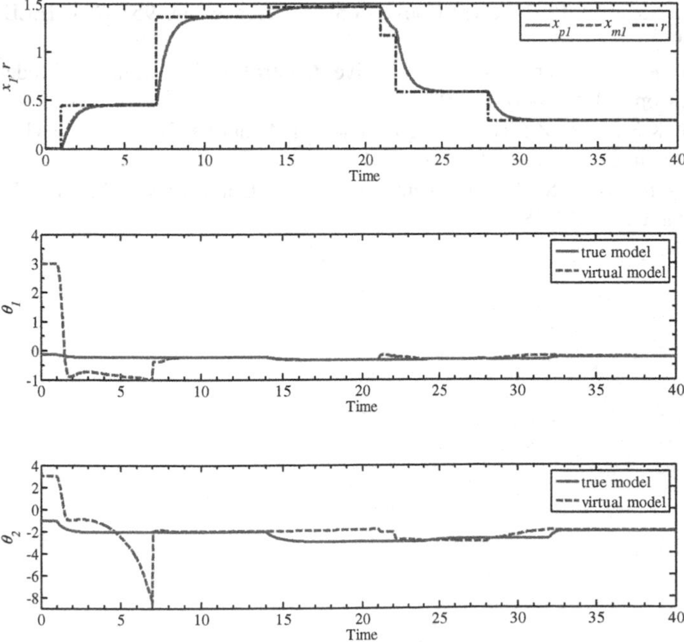

Fig. 3. The multiple model adaptive control with second level adaptation of train.

5 Conclusions

In this paper, we have formulated the model of train movement for multiple model adaptive control, and the problem of optimal slip speed control has been converted into detectable state's tracking one. We have demonstrated the favorable performance of multiple model adaptive control with second level adaptation for the input track control with uncertain parameters. The model and performance of inner force control for the total train will be left for future research.

Acknowledgments. This work is supported by the Fundamental Research Funds for the Central Universities of China (Grant No. 2009JBM006) and the National Natural Science Foundation of China (Grant No. 61074138).

References

1. Spiryagin, M., Lee, K.S., Yoo, H.H.: Control System for Maximum Use of Adhesive Forces of a Railway Vehicle in a Tractive Mode. Mech. Syst. Signal Pr. 22, 709–720 (2008)
2. Iannuzzi, D., Rizzo, R.: Disturbance Observer for Dynamic Estimation of Friction Force in Railway Traction Systems. In: 29th Annual Conference of the IEEE Industrial Electronics Society, pp. 2282–2979. IEEE Press, New York (2003)
3. Ohishi, K., Ogawa, Y., Miyashita, I., Yasukawa, S.: Adhesion Control of Electric Motor Coach Based on Force Control Using Disturbance Observer. In: 6th International Workshop on Advanced Motion Control, pp. 323–328. IEEE Press, New York (2000)
4. Ishikawa, Y., Kawamura, A.: Maximum Adhesive Force Control in Super High Speed Train. In: Nagaoka 1997 Power Conversion Conference, pp. 951–954. IEEE Press, New York (1997)
5. Narendra, K.S., Balakrishnan, J.: Adaptive Control Using Multiple Models. IEEE T. Automat. Contr. 42, 171–187 (1997)
6. Narendra, K.S., Han, Z.: The Changing Face of Adaptive Control: The Use of Multiple Models. Annu. Rev. Control 35, 1–12 (2011)
7. Han, Z., Narendra, K.S.: New Concepts in Adaptive Control Using Multiple Models. IEEE T. Automat. Contr. 57, 78–89 (2012)

Generalized Simulation Models of Vehicle Movements for Road and Railway Traffic Using Cellular Automata with Braking Reference Distance

Yonghua Zhou, Chao Mi, and Yaqin Yin

School of Electronic and Information Engineering, Beijing Jiaotong University, Beijing 100044, China
{yhzhou,10120304,10125072}@bjtu.edu.cn

Abstract. The common characteristic of automobile and train movements is autonomous, corporative and restrictive. This paper attempts to establish the generalized simulation models to describe the movements of road and railway vehicles under the modeling framework of cellular automata. The models distinguish the acceleration, deceleration and speed holding according to the speed limit and the difference between the estimation of the distance from the vehicle to its front target point and the braking reference distance. They describe various control strategies with adjustable acceleration and deceleration parameters. The tempo-spatial constraints such as space and speed ones are converted into the limitations caused by real or dummy cell states on the state transition of cells within a time step. The proposed models can reproduce the realistic phenomena such as stop-and-go wave and human-machine collaborative feedback adjustment.

Keywords: computer simulation, traffic flow, train, cellular automata, modeling.

1 Introduction

Computer simulation of traffic flow is a powerful tool of supporting traffic operation planning and real-time feedback adjustment. The various simulation models have been separately proposed for road and railway vehicles. Actually, the movements of vehicles in road and railway network have the common places, i.e. autonomous, corporative and restrictive. No matter for cars on roads or trains on railway, humans (drivers) and machines (vehicles) form the unified automatic control equipment, although the automation degree of trains is greater than that of automobiles in terms of speed control. The movements among vehicles are mutually coordinated to avoid collision. The front vehicle restricts the speed variation of the hinder adjacent one.

The cellular automaton (CA) model was introduced in the 1940s by von Neumann and Ulam, which is one kind of generalized description approach of the movements of complex physical and biological processes. The cellular automata (CAs) have been utilized to model the road [1] and railway [2, 3] traffic. Correspondingly, various improvement models have been put up to make up the disadvantages of already existing models, for example in [4]. This paper aims at establishing the unified modeling framework for the movements of road and railway vehicles.

D. Jin and S. Lin (Eds.): Advances in CSIE, Vol. 1, AISC 168, pp. 335–340.
springerlink.com © Springer-Verlag Berlin Heidelberg 2012

2 Models

Three kinds of models are established in this paper. One is for the vehicle movements of road traffic, and the other two are for the train movements under the fixed- and moving-block systems. These models have the similar mechanism of acceleration, deceleration and speed holding, but the model for road traffic has the random characteristic. All of them include two main parts, i.e. speed and position update.

2.1 Vehicle Movement of Road Traffic

The microscopic simulation model for road traffic describes the speed v_{n+1} at next instant as the function of the current speed v_n of vehicle, the distance d_t to the front adjacent vehicle excluding the safety margin distance *SM*, the acceleration a, the deceleration b and the braking reference distance d_r. The acceleration and deceleration locate between their respective maximum and minimal ones, i.e. $a \in [a_{\min}, a_{\max}]$ and $b \in [b_{\min}, b_{\max}]$. Define two kinds of braking reference distances, i.e. $d_{r1} = v_n^2 /(2b_{\max}) + v_n$ and $d_{r2} = v_n^2 /(2b_{\min}) + v_n$. The main framework of the model is as follows.

(1) speed update
if $d_t \geq d_{r2}$, $v_{n+1} = \min(v_n + d_t - d_{r2}, v_n + rand(a_{\min}, a_{\max}), v_{\max})$
elseif $d_t < d_{r2}$ and $d_t \geq d_{r1}$,
begin
$p = (d_t - d_{r1})/(d_{r2} - d_{r1})$
if $rand(0,1) \leq p$, $v_{n+1} = \min(v_n + d_t - d_{r1}, v_n + rand(a_{\min}, a_{\max}), v_{\max})$
else $v_{n+1} = \min(d_t, \max(v_n - rand(b_{\min}, b_{\max}), 0)$
end
else $v_{n+1} = \min(d_t, \max(v_n - b_{\max}, 0))$
(2) position update
$x_{n+1} = x_n + v_{n+1}$

In the above model, *rand*(a_1, a_2) means producing a number between a_1 and a_2 according to the driving characteristic.

2.2 Train Movement for Fixed-Block System

In order to establish the generalized model of train movement for the fixed-block system, we define three kinds of target points:

(1) p_a: the start point of the block section that the front adjacent train is occupying, which can be detected utilizing signaling equipment. The speed limit v_a at that point is generally set to be 0.

(2) p_b: the end point of the block section the train is occupying where there is speed limit v_b determined by the signal color denoted by the signal lamp or the codes sent by track circuits.

(3) p_c: the end point specified by the scheduling command where the speed limit v_c can be specified as certain value.

Suppose that the distances from the current train to the above three kinds of target points are d_a, d_b, and d_c, respectively. p_a and p_b always exist. Let $LC(sc)$=1 denote the scheduling command is still active and $LC(sc)$=0 otherwise. The allowable distance that train can move, i.e. *movement authority (MA)*, is determined by

$$d_m = \min(d_a, d_c) \cdot LC(sc) + d_a \cdot (1 - LC(sc)). \tag{1}$$

With regard to the current instant, the distance of a train from its current position to the nearest target point p_t, called *instantaneous MA*, is defined as

$$d_t = \min(d_a, d_b, d_c) \cdot LC(sc) + \min(d_a, d_b) \cdot (1 - LC(sc)). \tag{2}$$

Correspondingly, p_t is called *instantaneous target point*.

Given the speed v_t at the target point p_t, the distance d_t to p_t, and the deceleration b, in order to avoid exceeding the speed limit at p_t, the current speed v_n of train should be controlled below or equal to

$$v_r = \sqrt{v_t^2 + 2bd_t}. \tag{3}$$

We call the curve engendered by Equation (3) the basic speed limit one. If train's current speed v_n locates on the basic speed limit, the train can arrive at p_t with the target speed v_t at the deceleration of b when there is no other speed restriction.

Define the braking reference distance $d_r = (v_n^2 - v_t^2)/(2b) + v_n$. Given the distance d_t to the nearest target point p_t, if $d_t > d_r$, train can accelerate; if $d_t = d_r$, train can keep the current speed v_n; if $d_t < d_r$, train will decelerate to v_t.

There exist various speed limits, denoted as $v_{\lim}(x)$ in a unified way where x is the position of train. v_t at p_t adopts the minimum value of all kinds of speed limits.

Define that x_n and x_{n+1} are the positions of train at current and next instants, respectively; v_{n+1} is the velocity at next instant; a is the acceleration rate; w is the overspeed degree (≥1) and b_w is the deceleration corresponding to w. The proposed model using CAs is as follows.

(1) Speed update

if $v_n > wv_{\lim}(x_n)$, $v_{n+1} = \max(v_n - b_w, 0)$

elseif $v_n = v_{\lim}(x_n)$ AND $d_t \geq (v_n^2 - v_t^2)/(2b) + v_n$, $v_{n+1} = v_n$

else

begin

if $d_t > (v_n^2 - v_t^2)/(2b) + v_n$, $v_{n+1} = \min(v_n + a, v_{\max})$

elseif $d_t = (v_n^2 - v_t^2)/(2b) + v_n$, $v_{n+1} = v_n$

else

begin

if $v_n = v_t \neq 0$, $v_{n+1} = v_n$

else $v_{n+1} = \min(\max(v_n - b, v_t), d_m)$
end
end
(2) Position update
$x_{n+1} = x_n + v_{n+1}$

In the above model, the acceleration a, related to d_t, d_r, v_n and so on, is determined by the control strategies such as time-optimal and energy-efficient control. The deceleration b_w is determined by the braking control mode such as conventional or emergent braking according to the degree of the difference between v_n and $v_{\lim}(x_n)$. a and b are the adjustable parameters for various types of trains.

2.3 Train Movement for Moving-Block System

In the moving-block system, there exist no fixed block sections. We dynamically decompose the railway lines into a series of sections according to the *most restrictive speed profile* (MRSP). Three kinds of target points are also defined:

(1) p_a: the desired stop point of a train if its preceding train comes to a sudden halt considering the safety margin distance. Especially, if the front is station, the desired stop point is just the station. The speed limit v_a at that point is generally set to be 0.
(2) p_b: the end point of a speed-restriction segment in MRSP where the speed limit v_b is the minimum value among all kinds of speed restrictions at that point.
(3) p_c: the end point specified by the scheduling commands (scheduled *MAs*) where the speed limit v_c can be specified as certain value.

p_a and p_b always exist, and moreover dynamically update. Suppose that the distances from a train to p_a, p_b and p_c are d_a, d_b, and d_c, respectively. Similar to those in fixed-block system, d_m and d_t can be determined using equations (1) and (2), respectively. Different from that in fixed-block system, the target speed v_t at *instantaneous target point* p_t in moving-block system is determined by:

$$\begin{aligned} v_t = {} & \min(v_{\lim}(x_t), \sqrt{2b(d_a - d_t)}, \sqrt{2b(d_c - d_t) + v_n^2}, \\ & \sqrt{2b(d_b - d_t) + v_b^2}, v_{\max}) \cdot LC(sc) + \min(v_{\lim}(x_t), \\ & \sqrt{2b(d_a - d_t)}, \sqrt{2b(d_b - d_t) + v_b^2}, v_{\max}) \cdot (1 - LC(sc)) \end{aligned} \tag{4}$$

where x_t is the position of p_t, v_n is the velocity at current instant, and $v_{\max}$ is the maximum speed of train.

The proposed CA model for moving-block system has the same description as that for fixed-block one, however, due to the different meanings of d_a and d_b, d_m, d_t and v_t are calculated according to the mechanism of moving-block system.

3 Simulation Results

Fig. 1 demonstrates the macroscopic characteristic of traffic flow using the proposed model and the NaSch model, respectively. Fig. 2 shows the plots of speed and position or time of four vehicles. From Fig. 1 and 2, we can learn that the proposed model has more speed and density diversities. Because the speed update depends on the difference between d_t and d_r, there is no abrupt speed fall when deceleration.

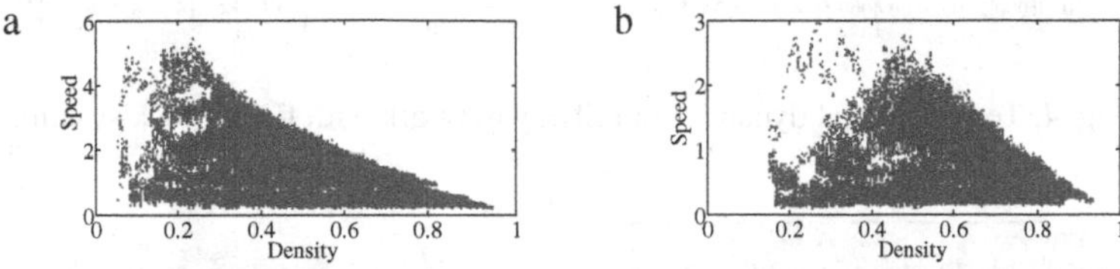

Fig. 1. Speed-density diagram. (a) the proposed model. (b) the NaSch model.

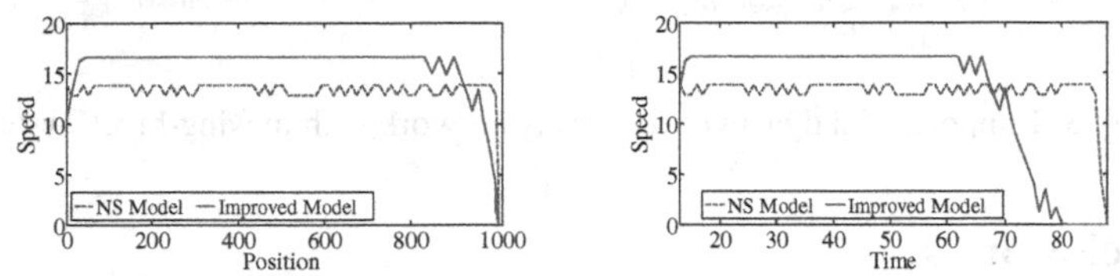

Fig. 2. Diagram of speed versus position or time.

The simulated railway network is shown in Fig. 3. The network has five stations, i.e. *A*, *B*, *C*, *D* and *E*. Line 1 and 2 are from station *A* to *E* and *B* to *D*, respectively, passing junction section *F* and *G* and stopping at station *C*. Track segment *AF*, *BF*, *FC*, *CG*, *GD* and *GE* are composed of 50, 30, 30, 30, 30, 50 block sections. Each block section has 1000 *cells* with each cell 1*m*. The speed limits at *F* and *G* are 22*cells*/*s*. The simulation step is 1*s*. Fig. 4 displays the tempo-spatial dynamics of line 1 in the fixed-railway network. The 4-time main accelerating and decelerating processes can be observed, respectively.

The railway network shown in Fig. 3 is still utilized for the simulation of moving-block system. However, the length of track segment *AF*, *BF*, *FC*, *CG*, *GD* and *GE* is 2500, 1500, 3000, 3000, 1500 and 2500 *cells*, respectively, with each *cell* 1*m*. The speed limits at *F* and *G* are 11*cells*/*s*. Fig. 5 demonstrates the tempo-spatial dynamics of line 1 in the moving-block system. The 4-time main accelerating and decelerating processes can also be observed, respectively.

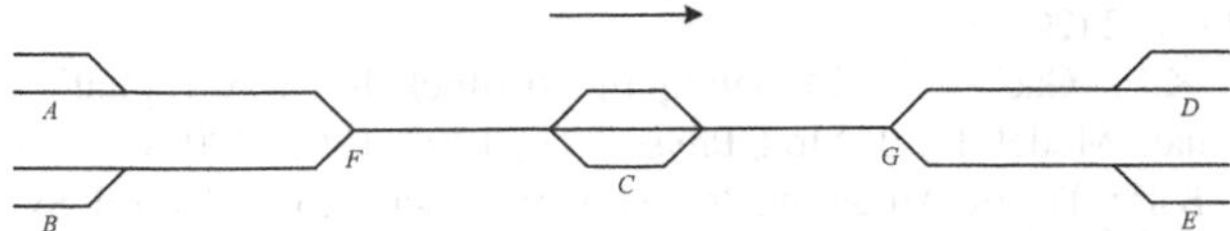

Fig. 3. The railway network.

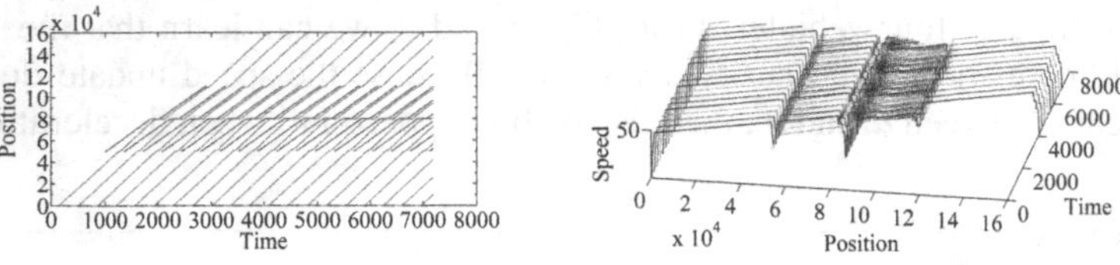

Fig. 4. Tempo-spatial dynamics of railway network with fixed-block system.

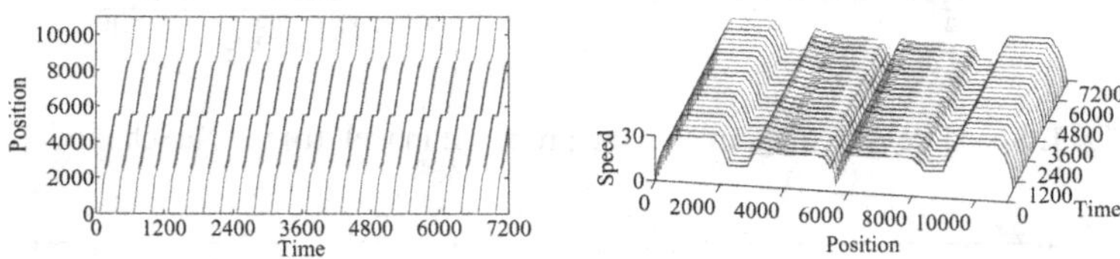

Fig. 5. Tempo-spatial dynamics of railway network with moving-block system.

4 Conclusion

This paper purposes to reveal the common places of road and railway traffic simulation models. The basic idea of all models is firstly to judge whether the vehicle requires braking or not, and then to update the speed. We regard the stop points specified by the scheduling commands as the space constraints caused by the dummy cell states. The speed constraints are converted into the space ones also by the dummy cell states in one time step. In this sense, the proposed models describe the vehicle movements using CAs under tempo-spatial constraints.

Acknowledgments. This work is supported by the Fundamental Research Funds for the Central Universities of China (Grant No. 2009JBM006) and the National Natural Science Foundation of China (Grant No. 61074138).

References

1. Nagel, K., Schreckenberg, M.: Cellular Automaton Models for Freeway Traffic. Phys. I 2, 2221–2229 (1992)
2. Li, K.P., Gao, Z.Y., Ning, B.: Cellular Automaton Model for Railway Traffic. J. Comp. Phys. 209, 179–192 (2005)
3. Ning, B., Li, K.P., Gao, Z.Y.: Modeling Fixed-block Railway Signaling System Using Cellular Automata Model. Int. J. Mod. Phys. C 16, 1793–1801 (2005)
4. Tang, T., Li, K.P.: Traffic Modeling for Moving-block Train Control System. Commun. Theor. Phys. 47, 601–606 (2007)

Multi-objective Test Paper Evaluation in the Process of Composing with Computer

Yan Li[1,2], Jiqiang Tang[3], and Min Fu[1,2]

[1] College of Mechanical Engineering, Chongqing University, 400030, China
[2] Engineering Research Center of Mechanical Testing Technology and Equipments, Chongqing University of Technology, 400054, China
[3] Information & Education Technology Center, Chongqing University of Technology, 400054, China
{ly,tjq,fumin}@cqut.edu.cn

Abstract. Since traditional statistics evaluation can't completely be used to evaluate test paper composing because it is hardly to get score sample for no testing held, the multi-objective test paper evaluation is proposed to evaluate test paper composing with computer. The key method of proposed evaluation is to use the constraints defined in the outline of examination to establish the multi-objective of test and to use the absolute distance between temporary paper and final test paper to compute the degree of approximation. The experiments show that the proposed evaluation can evaluate test paper composing with computer, and the tradeoff can be got between the objective evaluation with computer and subjective evaluation with person.

Keywords: multi-objective, test paper evaluation, test paper composing with computer.

1 Introduction

Generally, test paper evaluation is considered as to compute the value of test's statistics with score sample and compare them with normal distribution. To evaluate a test, many statistics are computed, such as test difficulty, test discrimination, test reliability, test validity, the degree of knowledge coverage, etc. The computed statistics should be fall in a certain range to meet the requirements of normal distribution. However, the statistics evaluation can't completely be used to the process of test paper composing with computer because it is hardly to get score sample for no testing held. To evaluate the test paper composing with computer, the multi-objective which be defined in the outline of examination should be considered, such as item type distribution object, test paper difficulty distribution object, knowledge coverage object, etc[1]. Furthermore, the approximation between the multi-objective measure computed from test paper composing with computer and the multi-objective requirements defined in the outline of examination should also be considered.[2]

In this thesis, the theory of multi-objective test paper evaluation is proposed to evaluate the quality of test paper composing with computer. Firstly, the model of paper and the definition of multi-objective measures are introduced. Secondly, the

D. Jin and S. Lin (Eds.): Advances in CSIE, Vol. 1, AISC 168, pp. 341–348.
springerlink.com © Springer-Verlag Berlin Heidelberg 2012

theorems based on the definition are proved. Finally, the experiments show that the proposed multi-objective test paper evaluation can compute the degree of approximation between the multi-objective measure and the multi-objective requirements.

2 Test Paper Model and Multi-objective Constrains

To figure the characters of test item's attributes, the test paper is denoted by test paper matrix, so test paper matrix P is a m×n matrix whose m rows denote the m items of test paper and n columns denote n attributes of one item. Similarly, the item bank is denoted by item bank matrix too, so item bank matrix B is a sxn matrix whose s rows denote the items of item bank and n columns denote n attributes of one item.[3-6]

$$P = \begin{bmatrix} p_{11} & p_{12} & \cdots & p_{1n} \\ p_{21} & p_{22} & \cdots & p_{2n} \\ \vdots & \vdots & \ddots & \vdots \\ p_{m1} & p_{m2} & \cdots & p_{mn} \end{bmatrix} \quad B = \begin{bmatrix} b_{11} & b_{12} & \cdots & b_{1n} \\ b_{21} & b_{22} & \cdots & b_{2n} \\ \vdots & \vdots & \ddots & \vdots \\ b_{s1} & b_{s2} & \cdots & b_{sn} \end{bmatrix} \quad S = \begin{bmatrix} s_{11} & s_{12} & \cdots & s_{1n} \\ s_{21} & s_{22} & \cdots & s_{2n} \\ \vdots & \vdots & \ddots & \vdots \\ s_{m1} & s_{m2} & \cdots & s_{mn} \end{bmatrix}$$

Moreover, to generate a test paper with computer, the temporary paper(matrix) should satisfy many constraints from the outline of examination. If the test paper P denotes the final test paper user demanded, the temporary paper S composing with computer should have the same statistics characteristics as the test paper P.

3 Multi-objective Test Paper Quality Evaluation

Generally speaking, the quality evaluation of temporary paper S can only take into account three constrains including item type distribution constrains, knowledge coverage constrains, difficulty distribution constrains. [3]The three constrains can also be regard as three objectives, each objective corresponding to one attribute of test item, so the test paper matrix P, the item bank matrix B and the temporary matrix S should be redefined.

$$P = \begin{bmatrix} p_{11} & p_{12} & p_{13} \\ \vdots & \ddots & \vdots \\ p_{m1} & p_{m2} & p_{m3} \end{bmatrix} \quad B = \begin{bmatrix} b_{11} & b_{12} & b_{13} \\ \vdots & \ddots & \vdots \\ b_{s1} & b_{s2} & b_{s3} \end{bmatrix} \quad S = \begin{bmatrix} s_{11} & s_{12} & s_{13} \\ \vdots & \ddots & \vdots \\ s_{m1} & s_{m2} & s_{m3} \end{bmatrix}$$

The redefined matrix P, B and S has only three columns, the first column denotes item types, the second column denotes knowledge, and the third column denotes difficulty. From the matrix angle of view, each column is not only a objective vector but also a kind of discrete distribution, all approximation computing and compare computing will be based on the objective vector.

Definition 1: The degree of item type distribution approximation between temporary paper S and final test paper P is defined as the absolute distance[7] of item type statistics vector between matrix S and matrix B.

In the definition, the item type statistics vectors are denoted as T_s, T_p separately. The vector T_s is classified statistic from the first column of temporary matrix S, each element of T_s denotes corresponding item type count of the first column of temporary matrix S, such as t_1 of T_s means that the count of item type 1 is t_1. Correspondingly, the vector T_p is classified statistic from the first column of final matrix P, each element of T_p has the same meaning as vector T_s.

$$T_p = [t_1 \quad t_2 \quad \cdots \quad t_b]$$

$$T_r = [t_1 \quad t_2 \quad \cdots \quad t_b]$$

Based on above definition of item type statistics vector, the absolute distance F_t can be denoted as:

$$F_t = |T_s - T_p| = \sum_1^b |t_i^s - t_i^p|, (t_i^p \le t_i^b) \tag{1}$$

From the function, the following theorems can be deduced.

Theorem 1: The range of F_t is [0,m].
Prove:

$$\because \sum_1^b t_i^s = m,\ 0 \le |t_i^s - t_i^p| \le t_i^s,\ \sum_1^b t_i^p = m \text{, and } 0 \le |t_i^s - t_i^p| \le t_i^p$$

$$\therefore 0 \le \sum_1^b |t_i^s - t_i^p| \le m \text{, and } F_t \in [0,m]$$

Theorem 2: Smaller is the value of F_t, more approximate is the item type distribution between temporary paper S and final test paper P.

Prove:

$$\because F_t(T_x, T_x) = 0,\ F_t(T_x, T_y) = F_t(T_y, T_x),\ F_t(T_x, T_y) \ge 0,$$

and $F_t(T_x, T_z) \le F_t(T_x, T_y) + F_t(T_y, T_z)$

$\therefore$Theorem 2 can be draw.

From above two theorems, the function F_t is a bounded function which means it can find the best temporary paper S in finite time, and when temporary paper S is the best one, the value of F_t will be zero.

Definition 2: The degree of knowledge coverage approximation between temporary paper S and final test paper P is defined as the absolute distance[7] of knowledge union collection between matrix S and matrix B.

In the definition, the knowledge union collections are denoted as K_s, K_p separately. The collection K_s is a union from the second column of temporary matrix S, each element of K_s denotes the key of knowledge. Correspondingly, the collection K_p is a union from the second column of final matrix P, each element of K_p has the same meaning as collection K_s.

$$K_p = \{k_1, \ k_2, \ \cdots \ k_p\}$$

$$K_s = \{k_1, \ k_2, \ \cdots \ k_s\}$$

Based on above definition of knowledge union collection, the absolute distance F_k can be denoted as:

$$F_k = |K_p \cup K_s - K_p \cap K_s| \tag{2}$$

It should be noted that the operator I K I means compute the count of collection K.From the function, the following theorems can be deduced.

Theorem 3: The range of F_k is [0,p+s].

Prove:

$\because 0 \le |K_p \cup K_s| \le p+s$, and $0 \le |K_p \cap K_s| \le p+s$

$\therefore 0 \le |K_p \cup K_s - K_p \cap K_s| \le p+s$

Theorem 4: Smaller is the value of F_k, more approximate is the knowledge coverage between temporary paper S and final test paper P.

Prove: The steps of prove are the same as theorem 2.

From above two theorems, the function F_k is also a bounded function which means it can find the best temporary paper S in finite time, and when temporary paper S is the best one, the value of F_k will be zero.

Definition 3: The degree of difficulty distribution approximation between temporary paper S and final test paper P is defined as the absolute distance of difficulty rank statistics vector between matrix S and matrix B.

In the definition, the difficulty rank statistics are denoted as D_s, D_p separately. The vector D_s is classified statistic from the third column of temporary matrix S, each element of D_s denotes corresponding difficulty rank's count of the third column of temporary matrix S, such as d_1 of D_s means that the count of difficulty rank 1 is d_1. Correspondingly, the vector D_p is classified statistic from the third column of final matrix P, each element of D_p has the same meaning as vector D_s.

$$D_p = [d_1 \quad d_2 \quad \cdots \quad d_p]$$

$$D_s = [d_1 \quad d_2 \quad \cdots \quad d_p]$$

Based on above definition of difficulty rank statistics vector, the absolute distance F_d can be denoted as:

$$F_d = |D_s - D_p| = \sum_1^p |d_i^s - d_i^p| \tag{3}$$

From the function, the following theorems can be deduced.

Theorem 5: The range of F_t is [0,m].

Prove:

$$\because \sum_{1}^{p} d_i^s = m,\ 0 \le \left|d_i^s - d_i^p\right| \le d_i^s, \text{ and } \sum_{1}^{p} d_i^p = m,\ 0 \le \left|d_i^s - d_i^p\right| \le d_i^p$$

$$\therefore 0 \le \sum_{1}^{p} \left|d_i^s - d_i^p\right| \le m, \text{ and } F_d \in [0,m]$$

Theorem 6: Smaller is the value of F_d, more approximate is the difficulty distribution between temporary paper S and final test paper P.

Prove: The steps of prove are the same as theorem 2.

From above two theorems, the function F_d is also a bounded function which means it can find the best temporary paper S in finite time, and when temporary paper S is the best one, the value of F_d will be zero.

Above definitions and theorems explain the measurement of objective item type distribution, knowledge coverage, and difficulty distribution separately. To evaluate the degree of approximation between temporary paper S and final test paper P thoroughly, the relationship among them should be considered.

Definition 5: The degree of approximation between temporary paper S and final test paper P is defined as the dot product of normalized function vector and weight vector.

In the definition, the normalized function vector and the weight vector are denoted as F, W separately. The normalized function vector is composed by above normalized function F_t, F_k, and F_d. The weight vector is the weight of function F_t, F_k, and F_d separately.

$$F = \left[{F_t}/{m} \quad {F_k}/{(p+s)} \quad {F_d}/{m} \right]$$

$$W = \left[w_t \quad w_k \quad w_d \right]$$

Based on above definitions, the dot product F_O can be denoted as:

$$F_o = F \cdot W = \frac{F_t}{m} \cdot w_t + \frac{F_k}{p+s} . w_k + \frac{F_d}{m} . w_d \tag{4}$$

According to theorem 1 to theorem 8, the function F_O is a multi-objective function, and it is also a bounded function which means it can find the best temporary paper S in finite time, and when temporary paper S is the best one, the value of F_O will be zero. If F_O is zero, the value of F_t, F_k, and F_d must be zero too, or the weight vector W will coordinate approximation among F_t, F_k, and F_d.

4 Compose and Evaluate Temporary Paper with Multi-objective Test Paper Evaluation

To use genetic algorithm, the coding scheme, selection operation, crossover operation and mutation operation should be established firstly. [8]In our experiment, binary coding scheme is used, the binary vector X_s is defined as:

$$X_s = [x_1 \quad x_2 \quad \cdots \quad x_s], (x_i \in \{0,1\} and \sum_1^s x_i = m)$$

To establish the selection operation, the multi-objective function F_0 is employed as the fitness function and the proportion is adopted to form the individual selection probability. The selection probability P_{is} is defined as:

$$P_{is} = \frac{\sum_{k=1}^{s} F_o^k - F_o^i}{\sum_{k=1}^{s} F_o^k}$$

From above equation, the part $\sum_{k=1}^{s} F_o^k$ is the total fitness in the population of one generation, and the part F_o^i is the fitness of the i'th individual in the population of one generation. To compute the multi-objective function value F_o^i, the other objective function values will be computed.

5 Experiments and Evaluation

In the experiments, 1000 test items are created with computer randomly, and the parameters of the final test paper P are defined in table 1, the parameters of genetic algorithm are defined in table 2.

Table 1. The parameters of final test paper P

Multi-objective	Parameter value
Test item type distribution vector Tp	[15,5,10]
Knowledge collection Kp	{1,2,3,4,5,6,7,8,9,10,11,12,13,14,15}
Difficulty distribution vector Dp	[10,10,10]
Weight vector W	[0.5,0.3,0.2]

Table 2. The parameters of genetic algorithm

Parameter name	Parameter value
Population size M	200
Crossover probability PC	0.6
Mutation probability PM	0.01
Generation T	50

The experiments' result can be draw in a figure. From the figure, the vertical coordinates represent absolute distance value of F_0, F_t, F_k, and F_d, the horizontal coordinates represent the generations.

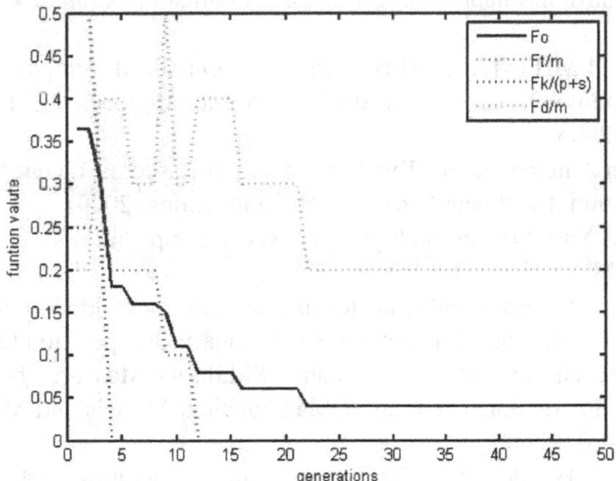

Fig. 1. The absolute distance value and generation

From the experiments, with the increase of generation, the absolute distance becomes more and more smaller, and the best situation is the absolute distance is zero. From this point of view, the multi-objective test paper evaluation can apply to test paper evaluation in the process of test paper composing with computer. The user can choose a tradeoff test paper through the evaluated test paper.

6 Conclusion and Future Work

In this thesis, multi-objective test paper evaluation is proposed to solve the problem of temporary paper evaluation created by the process of test paper composing with computer. Statistics evaluation can't be completely used to evaluate temporary paper because it is hardly to get score sample for no testing held, so it has to use the multi-objective constraints defined in the outline of examination to evaluate it. The multi-objective constraints are the framework of multi-objective test paper evaluation which includes item type distribution constrains, knowledge coverage constrains, and difficulty distribution constrains. It supplies the solution of multi-object test paper evaluation through the definitions and theorems of the degree of approximation between temporary paper and final test paper. The experiments show that the multi-objective can be applied to the process of test paper composing with computer, and the benefits of tradeoff between test paper quality and test paper composing efficient can be got too. In the future, it should pay more attention how effects the process of test paper composing with computer by varies of F_0, F_t, F_k, and F_d.

References

1. Hwang, G.-J.: A test sheet generating algorithm for multiple assessment requirements. IEEE Trans. Educ. 46(3), 329–337 (2003)
2. Liu, B.: Quality control mechanism of self-taught examination. China Examinations 04, 14–17 (2006)
3. Hwang, G.J., Lin, B.M.T., Tseng, H.H., Lin, T.L.: On the development of a computer-assisted testing system with genetic test sheet-generating approach. IEEE Trans. SMC-Part C 35(4), 590–594 (2005)
4. Gao, H., Li, Y.: An Autogenerating Test Paper Method Based on Latent Semantic Analysis and Genetic Algorithm. Intelligent Systems and Applications (2009)
5. Guo, P., Liu, L.-L., Yao, Q.: An intelligent test sheet composition system. IT in Medicine and Education (2008)
6. Hu, X.-M., Zhang, J.: An intelligent testing system embedded with an ant colony optimization based test composition method. Evolutionary Computation (2009)
7. Cha, S.-H.: Comprehensive Survey on Distance Similarity Measures between Probability Density Functions. International Journal of Mathematical Models and Methods in Applied Sciences 1(4), 300–307 (2007)
8. Huang, W., Wang, Z.-H.: Design of Examination Paper Generating System from Item Bank by Using Genetic Algorithm. Computer Science and Software Engineering (2008)
9. Tsai, K.H., Wang, T.I., Hsieh, T.C., et al.: Dynamic computerized testlet-based test generation system by discrete PSO with partial course ontology. Expert Systems with Applications 37(1), 774–786 (2010)
10. Guan, E.J., Han, X.B., Zhou, Q., Shen, W.H.: The item bank management system for university common courses: design, software development and application. China Education Inf. 12, 30–33 (2008)
11. Hwang, G.J., Lin, B.M.T., Lin, T.L.: An effective approach for test-sheet composition from large-scale item banks. Computers and Education 2(46), 122–139 (2006)

Breakout Prediction System Based on Combined Neural Network in Continuous Casting

Xin Jin, Tingzhi Ren, Xueliang Shi, Rong Jin, and Dawei Liu

School of Mechanical Engineering, Yanshan University, 066004 Qinhuangdao, China
jinxin@ysu.edu.cn

Abstract. A breakout prediction system based on combined neural network in continuous casting is developed. It adopts the radial basis function (RBF) neural network for single-thermocouple temperature pattern pre-diagnosis, and logic judgment unit for multi-thermocouple temperature pattern recognition at first, then uses fuzzy neural network based on Takagi-Sugeno (T-S) model to make final decision. In the RBF network, the maximum entropy function is used to normalize input data. According to the law of crack growth in the bonding location, the horizontal network prediction model is adopted to logically judge multi-thermocouple temperature pattern, the prediction time is shortened. In the T-S fuzzy neural network model, the overall influencing factors of breakout are considered. The results show that the breakout prediction system based on combined neural network can effectively decrease the false alarm rate and improve the prediction accuracy.

Keywords: breakout prediction, continuous casting, pattern recognition, radial basis function network, fuzzy neural network.

1 Introduction

Breakout is the most hazardous production accident in continuous casting process. The practice has proved that the breakout prediction system is effective to prevent breakout of molten steel.

At present, the most widely application of the breakout prediction system is the real-time detection of copper plate temperature using the thermocouples which are installed in the copper plate [1]. According to the thermocouple temperature pattern to judge whether bonding occurred, it can decrease the breakout accident and reduce economic loss and casualties, however, various systems also have several defects. The traditional prediction system based on logical judgment depends on specific equipments and technological conditions, so its development is restricted because of lack of robustness and poor tolerance [2]. Thermocouple temperature pattern is often affected by some factors, thus, the breakout prediction system has high false alarm rate if it judges breakout or not only according to the thermocouple temperature pattern [3]. The breakout prediction expert system is an effective prediction system, but its expert knowledge is bad portability, and it needs the re-accumulation experience for a new set of production line [4].

D. Jin and S. Lin (Eds.): Advances in CSIE, Vol. 1, AISC 168, pp. 349–355.
springerlink.com © Springer-Verlag Berlin Heidelberg 2012

In this paper, a new type of breakout prediction system based on combined neural network for continuous casting is proposed, which consists of the RBF neural network, a logic judgment unit and the fuzzy neural network. This system takes full advantages of complementarity between logic judgment and fuzzy neural network, and the learned system has the advantages of self-study and the adjustability of parameters. A curve fitting method based on the maximum entropy function is used to process input data [5]. The result shows that it can not only reduce the interference of bad data but improve the system accuracy.

2 Breakout Prediction Model

This breakout prediction system consists of three parts. The first part is that the RBF neural network recognizes single-thermocouple temperature pattern; the second part is that the logical judgment program judges multi-thermocouple temperature pattern; the third part is that the fuzzy neural network makes final decisions according to the scene information. The process of breakout prediction model is shown in Fig. 1.

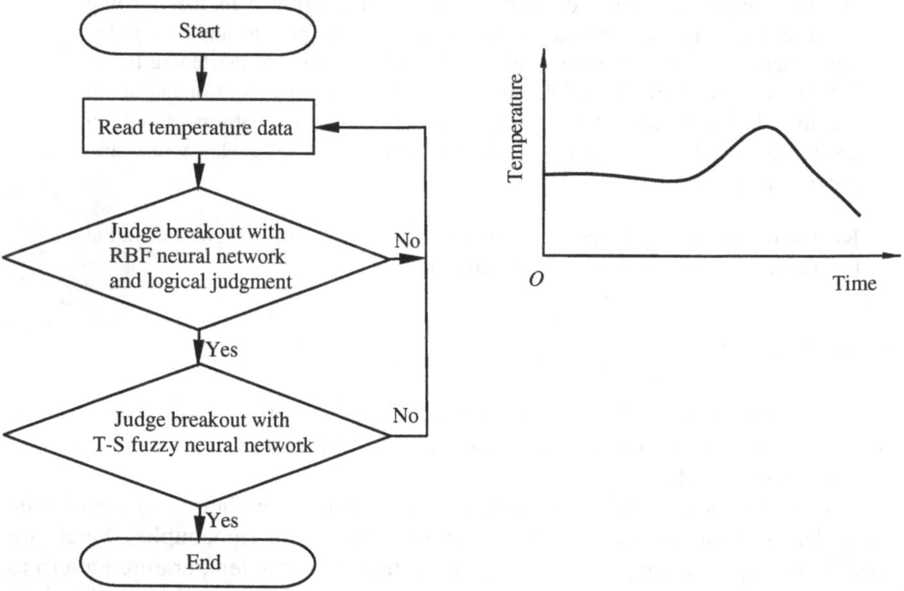

Fig. 1. Flow chart of breakout prediction system

Fig. 2. Thermocouple temperature curve

2.1 Pattern Recognition of the Single-Thermocouple Temperature

When bonding happened in mould, the molten steel will flow into the fractured position where it directly contacts with the mould wall, causing the nearby thermocouples temperature to rise. As the fractured position gradually goes away, the temperature of thermocouples will decrease, so the temperature pattern corresponds to a curve increased firstly and then decreased. The pattern recognition function of RBF neural network is adopted to recognize thermocouples temperature pattern.

When bonding happened, the thermocouple temperature change curve is shown in Fig. 2. It needs about 30 seconds of temperature change from increase to decrease. The data acquisition system should select an appropriate sampling period, if the sampling period is short, some data are easily interfered, causing the desired waveform not to be recognized. However if the sampling period is long, the prediction time will be extended, causing not to get the desired prediction effect. In this paper, the sampling period is 2s, so the network has 15 input vectors. The thermocouple temperature data are a dynamic and continuous sequence, but the RBF neural network is a kind of static networks, thus the TDNN (time delay neural network) is adopted [6], then sampling data are put into a shift register to change the dynamic sequence to static input.

The thermocouple temperature pattern reflects its temperature change tendency. Because of being influenced by bad factors, the collected temperature data are not completely accurate. For just need to recognize the temperature change tendency of thermocouples in a period, the data fitting method is adopted in the data processing. To obtain the regression parameters, the least square method is often used, which possesses good statistical properties, such as unbias, consistency, the least variance. However, when some abnormal data exist for gross error, or when the data generality distribution deviates from normal distribution, for example, thermocouple temperature data inevitably have some abnormal samples. In this case, the regression analysis result will lose its good statistical properties if the least square method is adopted, so the robust criterion function—I_1 norm criterion is adopted. $y = f(x, \Theta)$ is defined as a fitting function, which is a nonlinear function of Θ, and the problem is transformed into an optimization problem of minimization residual.

$$\min_{\Theta} err(\Theta) = \sum_{i=1}^{n} |err_i(\Theta)| \tag{1}$$

Where, Θ is the unknown parameter vector, err is the residual vector.

In numerous algorithms, the entropy function method is adopted.

$$err(\Theta) = \sum |f(x_i) - y_i| = \sum \max\{f(x_i) - y_i, y_i - f(x_i)\} \tag{2}$$

Suppose

$$F_p(\Theta) = \frac{1}{p} \sum \ln\left\{\exp\left[p\left(f(x_i) - y_i\right)\right] + \exp\left[-p\left(y_i - f(x_i)\right)\right]\right\} \tag{3}$$

Where, $p>0$ is a control parameter, Fp (Θ) is the maximum entropy function, and when $p>0$, Fp (Θ) and $f(x_i)$—y_i have the same order smoothness.

As shown in Fig. 3, the maximum entropy method is better than the least square method when they are used to fit data because of the existing of anomalous points.

The 15 temperature values of thermocouples are fitted to a curve by the maximum entropy method, and then one point is selected at regular intervals from 1, and the selected 8 points are used as input data.

The network performance is improved by this method which greatly simplifies the network structure and filter out some bad factors at the same time, the number of the final RBF network input layer nodes is reduced to 8. The improved nearest neighbor clustering classification algorithm is used to select cluster center points. The number of hidden nodes is 5. The output layer randomly output a number between 0 and 1 which represents the probability of breakout, the closer it is to 1, the bigger the probability is, on the contrary, the closer it is to 0, the smaller the probability is.

The training speed of network processed with curve fitting method is obviously improved. The network is tested by 20 test samples. The result in Fig. 4 shows that the network performance is good.

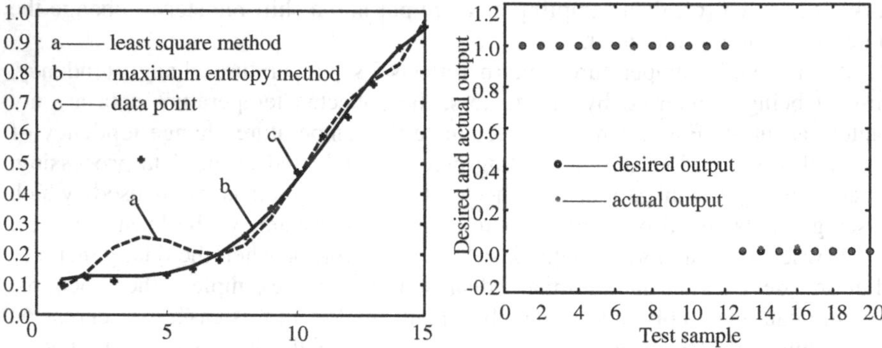

Fig. 3. Curves fitting

Fig. 4. Test result of RBF network performance

2.2 Pattern Recognition of the Multi-thermocouple Temperature

The multi-thermocouple temperature pattern is judged after the recognition of single-thermocouple temperature pattern. As a matter of experience, cracks in the bonding location extend not only along casting direction, but also along transverse direction at an angle between 40° and 60° in the slab surface. The bigger the casting speed is, the greater the angle is. Because the horizontal distance of thermocouple is 0.15 m, and the vertical spacing is 0.20 m, combining with the law of crack growth in the bonding location, the horizontal network is used to predict breakout in multi-thermocouple temperature pattern.

Fig. 5 shows that the bonding point is close to the point B when the bonding happened. So the thermocouple in point B can detect bonding firstly, and then other two thermocouples A and C will detect the occurrence of bonding in a space of time. The temperature patterns of the three thermocouples are shown in Fig. 6.

In horizontal network prediction model, it bases on a thermocouple to judge whether the temperature of the adjacent thermocouple is in dangerous range in specified time or not. If it is in dangerous range, the system will alarm.

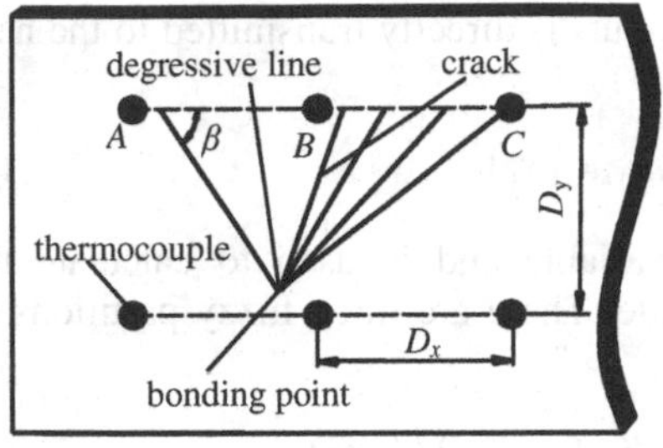

Fig. 5. Sketch of crack growth

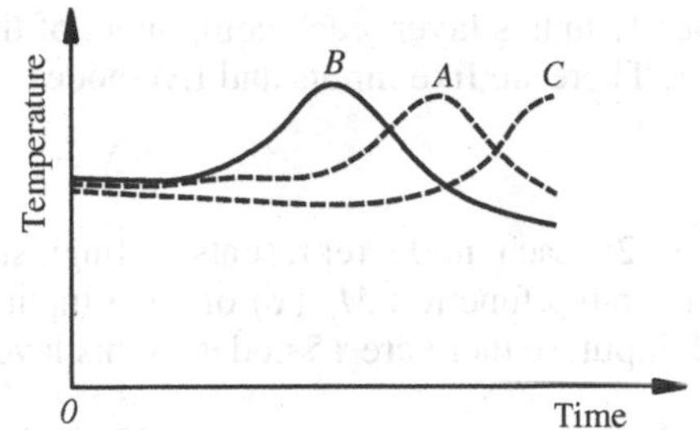

Fig. 6. Multi-thermocouple temperature pattern

2.3 The Fuzzy Decision

Because of the complexity of continuous casting process, thermocouple temperature pattern is only one basis of breakout judgement, but not the all bases. In actual operation, workers judge if breakout would happen according to practical situations, such as casting speed, liquid level position in mould. However, these bases are difficult to be expressed exactly because they have certain fuzziness. Based on T-S model[7], a fuzzy breakout prediction model is developed, as shown in Figure 7.

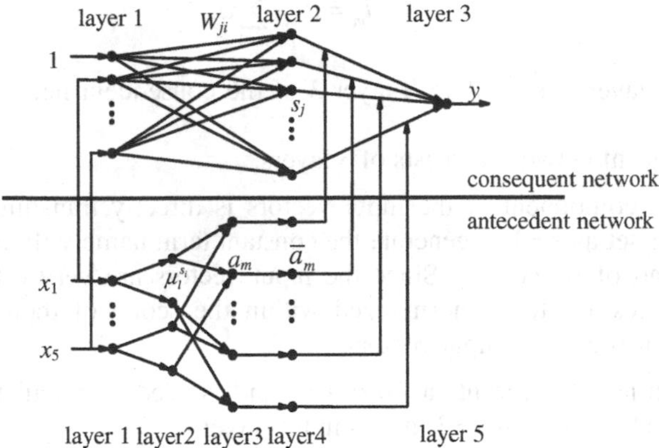

Fig. 7. Sketch of crack growth

The model is established based on the five factors: casting speed—S, casting speed change rate —a, the average temperature of thermocouples—T, the height of molten steel level in tundish—H, pouring temperature—W. The membership functions of the five inputs correspond to the definition of "high", "medium", "low" in fuzzy language. All of them adopt regularized Gaussian function μ_{ij}=exp $(-((x_i-c_{ij})/\sigma_{ij})\ 2)$.The number between 0 and 1 outputted in the system reflects the probability of breakout, the closer it is to 1, the bigger the probability is. On the contrary, the closer it is to 0, the smaller the probability is.

The antecedent network consists of 5 layers.

Layer 1: In this layer, each component of the inputs is directly transmitted to the next layer. There are five inputs and five nodes.

$$X = (x_1, x_2, x_3, x_4, x_5) \tag{4}$$

Layer 2: Each node represents a linguistic variable and is used to calculate the membership function $M_{ji}(x_i)$ of each input value. There are three fuzzy partitions of each input, so there are 15 nodes in this layer.

$$\mu_{ji} = M_{ji}(x_i) = \exp\left(-(x_i - c_{ij})^2 / \sigma_{ij}^2\right) \tag{5}$$

Where, c_{ij} and σ_{ij} respectively correspond to the central value and the width of the regularized Gauss membership function, i=1, 2, 3, 4, 5; j=1, 2, 3.

Layer 3: Each node represents a fuzzy rule and is used to calculate the fitness of each rule. There are $k=3^5$ nodes in this layer.

$$a_m = \mu_{1i} \times \mu_{2j} \times \mu_{3p} \times \mu_{4k} \times \mu_{5l} \tag{6}$$

Where, m=1, 2, k; i, j, p, k and l are 1, 2, 3.

Layer 4: This layer is used for a normalized computing.

$$\overline{a}_m = a_m \Big/ \sum_{m=1}^{k} a_m \tag{7}$$

Layer 5: This layer is shared with layer 3 of the consequent network, as the system output.

The consequent network consists of 3 layers.

Layer 1: Each component of the input vectors is directly transmitted to layer 2. A node's input is set as x_0=1 to generate the constant term namely threshold value of the consequent part of fuzzy rule. Since the input vectors are very different from each other, it is necessary to be normalized within the scope of their own in order to achieve the consistency of input vectors.

Layer 2: Each node represents a fuzzy rule and is used to calculate the consequent part of fuzzy rule. There are $k=3^5$ nodes in this layer.

$$s_j(m) = w_{j0} + w_{j1}x_1 + \cdots + w_{j5}x_5 \tag{8}$$

Where, j=1, 2,…, k.

Layer 3: This layer is shared with layer 5 of the antecedent network used to calculate y which is the final output.

$$y = \overline{a}_1 s_1 + \overline{a}_2 s_2 + \cdots + \overline{a}_k s_k \tag{9}$$

3 Assessment of the Model Performance

286000 records of a factory are used to test this model offline. As shown in table 1, the two prediction models are different in the accuracy of prediction. The prediction model based on combined neural network predicts more accurately than the prediction model based on single-stage RBF neural network. It proves that the fuzzy neural network is important and necessary in breakout prediction model.

Table 1. Prediction result of two models

Model type	Quote rate/%	Prediction rate/%
RBF neural network model	100	89.7
Combined neural network model	100	97.6

4 Conclusion

In this paper, maximum entropy function is used to deal with the inputs of RBF neural network. As a result, the interference of bad data is reduced and the network structure is simplified.

The prediction speed of the network is increased by using the horizontal network prediction model and combining with the law of crack growth in the bonding area.

The T-S fuzzy neural network is introduced in the prediction model. The thermocouple temperature pattern is used as anterior information of breakout judgement, and the false alarm rate of this model is reduced and the prediction accuracy is improved by synthetically considering the factors that affect breakout in continuous casting. The analysis results show that the performance of the breakout prediction system based on combined neural network is superior to the performance of the breakout prediction system based on RBF neural network.

Acknowledgments. This work is supported by Natural Science Foundation of Hebei Province of China (E2011203078).

References

1. Li, Y., Zhai, Y., Wang, Z., et al.: The Application of Breakout Prediction System with Thermal Imaging. Advanced Materials Research 402, 386–389 (2012)
2. Langer, M., Arzberger, M.: New Approaches to Breakout Prediction. Steel Times International 26, 23–26 (2002)
3. Li, Y., Wang, Z., Ao, Z.-G., et al.: Optimization for breakout prediction system of BP neural network. Kongzhi yu Juece 25, 453–456 (2010) (in Chinese)
4. Mills, K.C., Billany, T.J.H., Normanton, A.S., et al.: Causes of Sticker Breakout during Continuous Casting. Ironmaking and Steelmaking 18, 253–265 (1991)
5. Liu, Y., Liu, X., Song, Y.: Curve Fitting Based on Maximum Entropy Method and Its Application. Statistics and Decision 3, 133–144 (2007) (in Chinese)
6. Guh, R.-S., Shiue, Y.-R.: Fast and Accurate Recognition of Control Chart Patterns Using a Time Delay Neural Network. Journal of the Chinese Institute of Industrial Engineers 27, 61–79 (2010)
7. Han, M., Sun, Y., Fan, Y.: An Improved Fuzzy Neural Network Based on T-S Model. Expert Systems with Applications 34, 2905–2920 (2008)

Prediction of House Trade Based on Enthusiasm and Purchasing Power of Buyers

Miao Xu and Qi Meng

School of Architecture, Harbin Institute of Technology, Harbin, P.R. China
sylviaxm1009@gmail.com, mengq@hit.edu.cn

Abstract. Present, more and more people focused on the house trade in China. It has been demonstrated in previous studies that the successful house trade is depend on both subjective and objective factors of buyers. In this study, a large scale questionnaire survey has been undertaken from spring 2011 to winter 2011 in Beijing and Harbin, China, to determine what factors influence buyers' enthusiasm and purchasing power in house trade. Taking three types of flats as an example, the prediction models have been developed based on these factors. The above results can enhance the influencing factors of house buyers' enthusiasm, and they also form a good basis for further prediction models for house trade.

Keywords: Prediction model, enthusiasm of buyer, house trade, questionnaire survey.

1 Introduction

Nowadays, the house trade became a big problem to people in China. Taking Harbin as an example, 50 km^2 residences were built in 2010 [1]. It has been demonstrated in previous studies that the successful house trade is depend on both subjective and objective factors of buyers [2, 3, 4]. From them, the objective factors are mainly about the income of buyers and types of flats, and the subjective factors are mainly about the enthusiasm of buyers. Sometimes, the objective factors are easy to evaluation, while the subjective factors are hard to evaluation. In this study, a large scale questionnaire survey has been taken from spring 2011 to winter 2011 in capital cities of China, such as Beijing and Harbin to determine what factors influence the enthusiasm of buyers in house trade. The methods in psychology of consumption and marketing have been also used in this paper to analyze psychology of buyers in house trade [5]. A Prediction model has been final developed based on some main factors, such as three types of flats, number of floors, waiting time for the houses, income and enthusiasm of buyers.

In this paper, the methodology has been given first, say questionnaire survey[6,7,8] and Delphi method[9,10,11,12]; Then, some objective and subjective factors in house trade have been analyzed, from them, the details about enthusiasm of buyers has been introduced and discussed. At last, a model has been developed to predict house trade based on enthusiasm of buyers and other objective factors.

D. Jin and S. Lin (Eds.): Advances in CSIE, Vol. 1, AISC 168, pp. 357–363.
springerlink.com © Springer-Verlag Berlin Heidelberg 2012

2 Methodology

In this section, a large scale questionnaire survey was undertaken firstly for buyers to determine how many factors influence buyers in house trade. Then, Delphi method was used for experts to determine the weights of these factors to enthusiasm of buyers.

2.1 Questionnaire Survey

A field study was conducted through a questionnaire survey at selected case study sites in Beijing and Harbin, China. In these cities, six residential areas were chose as survey locations. In terms of subjective investigation, nearly 600 valid questionnaires were recovered from spring of 2011 to winter of 2011 on these residential areas. Some details were shown in Table 1. Around 80 to 100 interviews were conducted at each survey location using the same questionnaire. The statistics results are that 8 factors which have influence their enthusiasm in house trade were pointed out by interviewees, say type of flat, number of the flats, price of flats, waiting time for the houses, numbers of floors, buyers' income, bank savings, and loan rate. In purchasing power investigation, buyers were divided into four age groups, say 20-30 years, 31-40years, 41-50 years and more than 50 years. The objective factors influence them in house trade are different, say income is the main factor for persons from 20 to 30 years old, while bank savings is the main factor for persons from 41 to 50 years old.

Table 1. Some details of questionnaire survey

Survey sites	Cities	Number of interviews
Bin Cai residential area	Harbin	79
Yuan Da residential area	Harbin	102
Min Jiang residential area	Harbin	88
Sheng He residential area	Beijing	75
Wan Ke residential area	Beijing	62
Lv Di residential area	Beijing	34
Jian Wai SOHO residential area	Beijing	77

2.2 Delphi Method

Delphi method was also used for 10 experts to get the weights of the factors for enthusiasm of buyers. They are 2 project managers in property companies, 2 managers in insurance companies, 2 managers in banks, 2 professors whose research field is mainly about building or urban design, and 2 government officials. The weights of the factors were given by them though three times grade. Some details were shown in Table 2.

Table 2. Some details of process in Delphi method

Grade	The most important factor (Number of experts)
First time	waiting time for the houses (3)
	numbers of floors (3)
	price of flats (4)
Second time	waiting time for the houses (2)
	numbers of floors (3)
	price of flats (5)
Third time	waiting time for the houses (2)
	numbers of floors (2)
	price of flats (6)

3 Methodology Results and Analysis

3.1 Relationships between Waiting Time and Enthusiasm of Buyers

Taking the survey results in Harbin as an example, the relationships between the waiting time for houses and the enthusiasm of buyers to buy the houses (1, very low; 2, low; 3, neither low nor high; 4, high; and 5, very high), referred below as waiting time and enthusiasm of buyers, are shown in Fig. 1, where linear regressions and coefficients of determination R^2 are also given. To minimise personal bias, the enthusiasm of buyers which have less than 5 responses are not included in this figure. In Fig. 1, it can be seen that there is generally a strong positive correlation between the waiting time and the enthusiasm of buyers ($p<0.001$), with coefficients of determination R^2 as 0.625. With the increase of waiting time, the mean score of enthusiasm of buyers becomes lower.

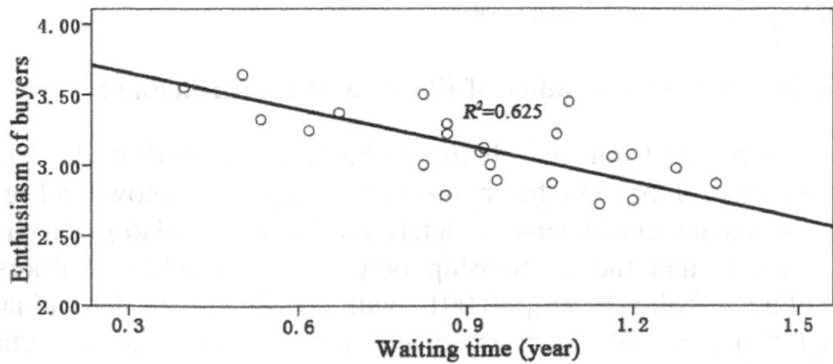

Fig. 1. Relationships between the waiting time and the mean evaluation score of enthusiasm of buyers, with linear regressions and coefficients of determination R^2.

3.2 Relationships between Price of Houses and Enthusiasm of Buyers

The relationships between the price of houses and the enthusiasm of buyers are shown in Fig. 2, where linear regressions and coefficients of determination R^2 are also given. In Fig. 2, it can be seen that there is also a strong positive correlation between the price of houses and the enthusiasm of buyers ($p<0.001$), with coefficients of determination R^2 as 0.442. With the increase of price of houses, the mean score of enthusiasm of buyers becomes lower.

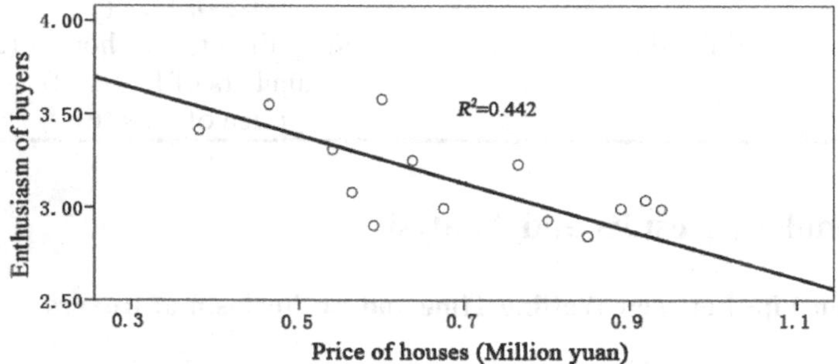

Fig. 2. Relationships between the price of houses and the mean evaluation score of enthusiasm of buyers, with linear regressions and coefficients of determination R^2.

3.3 Relationships between Amount of Houses and Enthusiasm of Buyers

The relationships between the amount of houses and the enthusiasm of buyers are shown in Fig. 3, where linear regressions and coefficients of determination R^2 are also given. In Fig. 3, it can be seen that there is also a strong positive correlation between the amount of houses and the enthusiasm of buyers ($p<0.001$), with coefficients of determination R^2 as 0.415. With the increase of price of houses, the mean score of enthusiasm of buyers becomes higher too.

3.4 Relationships between Number of Floors and Enthusiasm of Buyers

Taking the survey results from some high-rise buildings in Harbin, the relationships between the number of floors and the enthusiasm of buyers are shown in Fig. 4, where quadratic regressions and coefficients of determination R^2 are also given. In Fig. 4, it is interesting to note that the relationship between the number of floors and the enthusiasm is of a parabolic shape ($p<0.001$), with coefficients of determination R^2 as 0.792. When the number of floors is lower than or more than a certain value, approximately 20, the enthusiasm of buyers becomes lower.

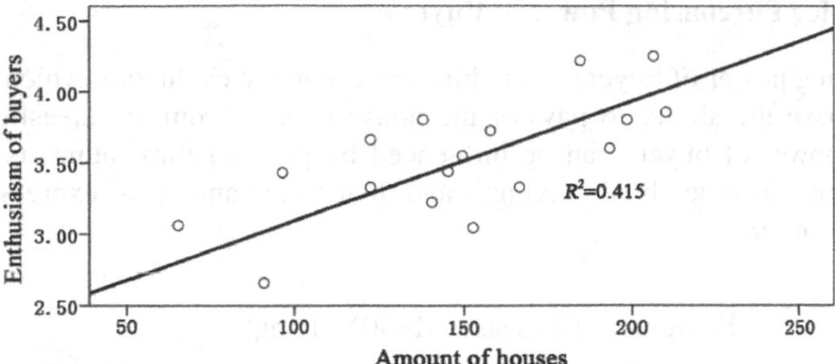

Fig. 3. Relationships between the amount of houses and the mean evaluation score of enthusiasm of buyers, with linear regressions and coefficients of determination R^2.

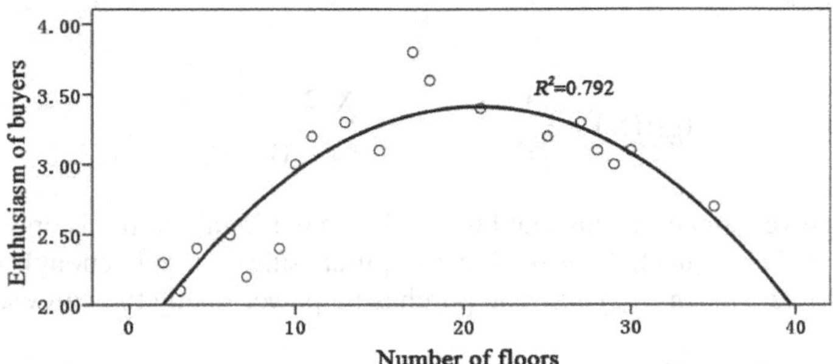

Fig. 4. Relationships between the number of floors and the mean evaluation score of enthusiasm of buyers, with quadratic regressions and coefficients of determination R^2.

4 Prediction Models

4.1 Model for Enthusiasm of Buyers

The concept of enthusiasm of buyers is the buyers' comprehensive expected for price waiting time and characteristics of houses. A model for enthusiasm was developed firstly. In this model which combines objective and subjective factors together, the subjective factors were affected by the changes of objective factors. From the investigation, the enthusiasm of buyers can be influenced by the price of houses, the number of floors, amount of houses and the waiting time for houses, and it is expressed in the following equation:

$$e_s = t_s (a_s + b_s\eta_s) + l_s (c_s + d_s\eta_s) + m_s (f_s + g_s\eta_s) \quad (1)$$

Where e_s is the enthusiasm of buyers, t_s is the price of houses, l_s is the number of floors, m_s is the amount of houses, and η_s is the waiting time for houses. The variables such as a_s, b_s, c_s, d_s, f_s, g_s are weights that given by the experts. It should be pointed out that these weights are different in vary cities.

4.2 Model for Purchasing Power of Buyers

The purchasing power of buyers is an objective economic evaluation which can judge the people have the ability to pay for the house or not, From the investigation, the purchasing power of buyers can be influenced by price of flats, numbers of floors, buyers' annual income, bank savings, and loan rate, and it is expressed in the following equation:

$$F=\mathrm{sign}(M-0.2\ t_i)\ \mathrm{sign}(M+30X-\ t_i\ -ht_i) \tag{2}$$

Where M is bank savings, h is loan rate, X is the buyers' annual income, t_i is the price of houses, 0.2 is the percentage of the first payment in total price of the house, 30 is the deadline of the loan.
And,

$$\mathrm{sign}(x)=\begin{cases}1 & x\geq 0\\ 0 & x<0\end{cases} \tag{3}$$

For x, all positive number means that buyers' income or bank saving is enough to pay for the houses. For sign(x), 1 means buyers' purchasing power is enough to pay for the houses, while 2 means they have no purchasing power to buy the houses.

4.3 Model for House Trade

First of all, the flats were divided into three types, say luxury flats (80-120m^2), general flats (50-80m^2), and small flats (30-50m^2). The price of these three kinds of flats, is t_1, t_2 and t_3, respectively, and the amount of houses' supply is m_1, m_2 and m_3 respectively. With all these objective factors, we can develop a prediction model for buyers, combined the two factors which is mentioned in 3.1 and 3.2, and it is expressed in the following equation:

$$D=\mathrm{sign}(\max(e_i-t_i))\mathrm{sign}(M-0.2t_i)\mathrm{sign}(M+30X-t_i-ht_i) \tag{4}$$

Where the longest waiting time that the buyers can patient is η, the price of flats is t_i, buyers' annual income is X, buyers' bank savings is M.
And,

$$\mathrm{sign}(x)=\begin{cases}1 & x\geq 0\\ 0 & x<0\end{cases} \tag{5}$$

In this function, 1 means the buyers can pay for the houses, and 2 means the buyers can't pay for the houses.

5 Conclusions

In this study, a large scale questionnaire survey has been undertaken from spring 2011 to winter 2011 in Beijing and Harbin, China. The factors which influence to enthusiasm and purchasing power of buyers have been analyzed. From them price of houses, the number of floors, amounts of houses and the waiting time for houses are main factors for enthusiasm of buyers, while buyers' annual income, bank savings, and loan rate are main factors for purchasing power of buyers. Taking three types of flats as an example, a prediction model has been developed based on these factors. The above results can enhance the influencing factors of house buyers' enthusiasm, and they also form a good basis for further prediction models for house trade.

Acknowledgements. This research is funded by school of architecture, Harbin Institute of Technology.

References

1. Information on, `http://www.hlj.gov.cn/zxxx/tjxx/`
2. Qingxia, R.: Study on the Current Power of Chinese Residents' Purchasing the Merchantable House——Taking Beijing, Shanghai, Guangzhou and Shenzhen as the Examples. Sci-Tech Information Development & Economy 16(22), 130–131 (2006)
3. Li, C., Bin, L., Guang, L.: Study on Prediction of Commercial Housing Sale Volume Based on BP Neural Network. Optimization of Capital Construction 26(04), 56–58 (2005)
4. Shouliang, Z.: Sharp Decreases of Commercial House Dealing Quantity: The Change of Commercial House Demand. Journal of Harbin University of Commerce (Social Science Edition) (03), 62–65 (2009)
5. Shuangshuang, C.: A Study on the Psychology and Management Strategies of Customers Waiting in Line. The Guide of Science & Education (04), 134, 172 (2011)
6. Lkezawa, T., Sunaga, N.: Questionnaire survey on the actual conditions of eco-schools funded by mext - Environmental adjustment technique and energy consumption. Journal of Environmental Engineering 74, 783–788 (2009)
7. Kuwano, S., Morimoto, M., Matui, T.: A questionnaire survey on noise problems with elderly people. Acoustical Science and Technology 26, 305–308 (2005)
8. Carayon, P., Schoepke, J., Hoonakker, P.L.T., Haims, M.C., Brunette, M.: Evaluating causes and consequences of turnover intention among IT workers: The development of a questionnaire survey. Behaviour and Information Technology 25, 381–397 (2006)
9. Kanama, D., Kondo, A., Yokoo, Y.: Development of technology foresight: Integration of technology roadmapping and the Delphi method. International Journal of Technology Intelligence and Planning 04, 184–200 (2008)
10. Landeta, J.: Current validity of the Delphi method in social sciences. Technological Forecasting and Social Change 73, 467–482 (2006)
11. Okoli, C., Pawlowski, S.D.: The Delphi method as a research tool: An example, design considerations and applications. Information and Management 42, 15–29 (2004)
12. Elmer, F., Seifert, I., Kreibich, H., Thieken, A.H.: A Delphi method expert survey to derive standards for flood damage data collection. Risk Analysis 30, 107–124 (2010)

A 3D Acoustic Temperature Field Reconstruction Algorithm Based on Tikhonov Regularization

Hua Yan, Kun Li, and ShanHui Wang

School of Information Science and Engineering,
Shenyang University of Technology, Shenyang, China
yanhua_01@163.com, likun0606@126.com

Abstract. A new 3D temperature field reconstruction algorithm based on Tikhonov regularization is proposed. A space surrounded by 32 acoustic sensors is divided into pixels. Four typical temperature field models are reconstructed using the new algorithm for verifying its reconstruction ability. The influence of regularization parameter on detail reconstruction and anti-noise ability are investigated. 3D displays of reconstructed temperature fields and reconstruction errors are given, which shows that the new algorithm is good at reconstructing complex fields.

Keywords: reconstruction algorithm, acoustic tomography, 3D temperature field, Tikhonov regularization.

1 Introduction

Temperature field reconstruction based on acoustic travel-time tomography [1,2,3] uses the dependence of sound speed in materials on temperature along the sound propagation path. In this paper, a new reconstruction algorithm of 3D temperature field based on acoustic travel-time tomography is proposed. Unlike the least square method, a widely used reconstruction algorithm, the new algorithm doesn't require the number of pixels less than that of the sound paths. As a result, it is more suitable for the reconstruction of complex temperature fields.

2 The Principle of Acoustic Temperature Field Reconstruction

Temperature measurement by acoustic method is based on the principle that the sound velocity in a medium is a function of the medium temperature. The sound velocity c in a gaseous medium at an absolute temperature T is given by

$$c = Z\sqrt{T} \tag{1}$$

where Z is a constant decided by gas composition. The value of Z for air is 20.05.

In this paper, the space to be measured is 10m×10m×10m and 32 acoustic sensors are set on its periphery. Fig.1 shows the position of 32 sensors and the 172 effective sound wave paths. On the basis of sensor positions and the travel-time measurements

D. Jin and S. Lin (Eds.): Advances in CSIE, Vol. 1, AISC 168, pp. 365–370.
springerlink.com © Springer-Verlag Berlin Heidelberg 2012

along each effective sound wave path, the temperature field can be reconstructed by using suitable reconstruction algorithms.

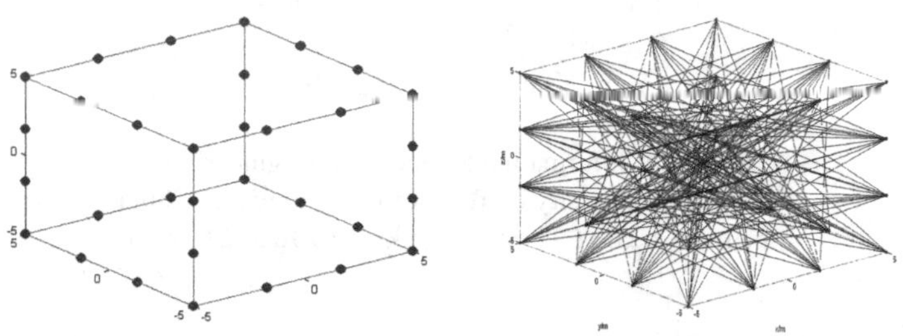

Fig. 1. A space surrounded by 32 sensors and the 172 effective sound wave paths

3 The Theory of the New Reconstruction Algorithm

Supposing that the distribution of temperature is T(x,y,z), and the distribution of reciprocal of sound speed is f(x, y, z), then T(x, y, z) can be presented as:

$$T(x,y,z)=\frac{1}{Z^2 f(x,y,z)^2} \quad . (2)$$

The acoustic travel-time gk along path p_k can be expressed as:

$$g_k = \int_{p_k} f(x,y,z)dp_k \,, \quad k=1,2,...,K \quad . (3)$$

where K is the number of the effective sound travel paths, that is, the number of the effective sound travel-times.

Divide the space to be measured into M pixels (cells). Let the center coordinates of pixel m is (x_m, y_m, z_m). Expand f(x,y,z) over a finite set of basis functions as follows:

$$f(x,y,z)=\sum_{m=1}^{M}\varepsilon_m \phi_m(x,y,z) \quad . (4)$$

$$\phi_m(x,y,z)=e^{-\beta\sqrt{(x-x_m)^2+(y-y_m)^2+(z-z_m)^2}} \quad . \quad (5)$$

where β is the shape parameter of basis function $\phi_m(x,y,z)$, β is related to the size of the space and the layout of the sound sources/receivers.

Combining Eq. 3, Eq. 4 and Eq. 5, we have:

$$g_k=\sum_{m=1}^{M}\varepsilon_m \int_{p_k}\phi_m(x,y,z)dp_k=\sum_{m=1}^{M}\varepsilon_m a_{km}\,,\, a_{km}=\int_{p_k}\phi_m(x,y,z)dp_k \quad (6)$$

Define $A=(a_{km})_{k=1,...,K;m=a,...,M}$, $g=(g_1,...,g_K)^T$, $\varepsilon=(\varepsilon_1,...,\varepsilon_M)$, Eq. 6 can be expressed as:

$$g = A\varepsilon \tag{7}$$

Since matrix A is ill-conditioned, we solve Eq. 7 by means of Tikhonov regularization. The basic theory of the regularization is adding apriori information of solution to the original problem and making the norm of solution minimum on the condition of data-fitting. Since the sound velocity is bounded and fluctuant, $\|\varepsilon\|_2^2$ is used as apriori information, and solving Eq.(7) is changed to $\min\left\{\|A\varepsilon-g\|_2^2+\mu\|\varepsilon\|_2^2\right\}$.

Using the singular value decomposition of matrix[4], matrix A can be written uniquely as:

$$A = U\begin{bmatrix}\Sigma & 0\\ 0 & 0\end{bmatrix}V^T, \Sigma = \mathrm{diag}(\sigma_1,\sigma_2,...,\sigma_\gamma) \tag{8}$$

where $\sigma_1 \ge \sigma_2 \ge ... \ge \sigma_\gamma \ge 0$ are the non-zero singular values of matrix A, U and V are orthogonal matrixes and their columns are eigenvectors of AA^T and A^TA respectively. Then the solution of Eq. (7) based on Tikhonov regularization can be expressed as:

$$\varepsilon = A^+g,\quad A^+ = V[\begin{matrix}\Sigma^{-1} & 0\\ 0 & 0\end{matrix}]U^T,\quad \Sigma^{-1} = \mathrm{diag}\left(\frac{\sigma_1}{\sigma_1+\mu},\frac{\sigma_2}{\sigma_2+\mu},...\frac{\sigma_\gamma}{\sigma_\gamma+\mu}\right) \tag{9}$$

Where μ is a non-negative regularization parameter. Combining Eq. 4 and Eq. 9, the sound speed reciprocal of each pixel can be gained. And then the temperature of each pixel can be obtained by using Eq. 2. Finally, the detailed temperature distribution of the space can be calculated by 3D interpolation method.

4 Simulation Research on the New Reconstruction Algorithm

For the space surrounded by 32 sensors shown in Fig.1, the shape parameter β of 0.0001 is suitable, and the space is divided into 10×10×10=1000 pixels. Four typical temperature field models are used. They are:

Model 1 : One-peak symmetry model[5] defined as:

$$T(x,y,z) = \frac{2000}{0.05(x^2+y^2+z^2)+1}$$

Model 2 : One-peak bias model defined as:

$$T(x,y,z) = \frac{350}{0.05[(x+2)^2+(y-2)^2+(z-2)^2]+1}$$

Model 3 : Two-peak symmetry model defined as:

$$T(x,y,z) = \frac{350}{0.2[(x+3)^2+y^2+(z-1.6)^2]+1} + \frac{350}{0.2[(x-3)^2+y^2+(z-1.6)^2]+1}$$

Model 4 : Four-peak symmetry model defined as:

$$T(x,y,z)=\frac{350}{0.5[(x+2.5)^2+y^2+(z-2.5)^2]+1}+\frac{350}{0.5[(x-2.5)^2+y^2+(z-2.5)^2]+1}$$
$$+\frac{350}{0.5[(x+2.5)^2+y^2+(z+2.5)^2]+1}+\frac{350}{0.5[(x-2.5)^2+y^2+(z+2.5)^2]+1}$$

To evaluate the quality of the reconstructions, the root-mean-squared percent error of the reconstructed field is used, which is defined as follows.

$$Er_{ms}=\frac{\sqrt{\frac{1}{M}\sum_{j=1}^{M}[T(j)-T(\hat{j})]^2}}{T_{ave}}\times 100\%$$

where $T(j)$ is the model temperature of pixel j, $\hat{T}(j)$ is the reconstruction temperature of pixel j.

The reconstruction errors and the reconstruction temperature fields under different μ when the sound travel-times are noise-free and noisy are given in Table 1, Table 2, Table 3, Fig.2 and Fig.3. The noise added is white Gauss noise with a mean of 0 and a standard deviation of σ.

Table 1. The reconstruction errors when the sound travel-times are noise-free

μ	E_{rms} [%]			
	Model 1	Model 2	Model 3	Model 4
0	1.32	1.38	2.32	5.12
10^{-7}	1.52	1.56	2.81	6.92
10^{-6}	1.99	2.18	4.60	12.52

Table 2. The reconstruction errors when the sound travel-times are noisy (σ=0.001)

μ	E_{rms} [%]			
	Model 1	Model 2	Model 3	Model 4
0	1.65	1.78	2.72	5.49
10^{-7}	1.62	1.67	2.98	7.21
10^{-6}	2.04	2.23	4.69	12.68

Table 3. The reconstruction errors when the sound travel-times are noisy (σ=0.01)

μ	E_{rms} [%]			
	Model 1	Model 2	Model 3	Model 4
0	9.52	10.39	11.57	12.68
10^{-7}	5.42	5.52	7.03	10.95
10^{-6}	3.81	4.02	6.43	14.43

5 Discussion and Conclusion

Following can be concluded form the simulation researches given in section 4.

1) The regularization parameter μ controls the weight of measured data and apriori information in solution. If μ is too small, the error of measured data can't be well restrained, while if μ is too large, the solution will lose much detail information.

2) When the travel-times are noise-free, the reconstruction errors are smallest when μ is equal to 0. If a small positive μ is used, the reconstruction errors will increase obviously. The larger the value of μ , the larger the reconstruction errors.

3) When the travel-times are with noise, the reconstruction results are not ideal when μ is equal to 0, and the reconstruction accuracy can be improved significantly by using a small positive μ. The larger the value of μ is, the stronger the anti-noise ability of reconstruction fields is; and meantime, the more detail information in solution is eliminated, which decreases the reconstruction accuracy too. Therefore, for a complex temperature field, such as four-peak symmetry temperature field, the value of μ should be small to preserve more detail information in the reconstruction field; while for a simple temperature filed, such as one-peak symmetry or one-peak-bias temperature filed, a bigger value of μ can be used to eliminate more noise.

Unlike the well-known least square method, the number of pixels of a reconstruction field obtained by the new algorithm is much greater than the number of travel-times. As a result, the new algorithm is more suitable for reconstruction of complex temperature fields. Simulation results show that the new algorithm is good at reconstructing complex fields too.

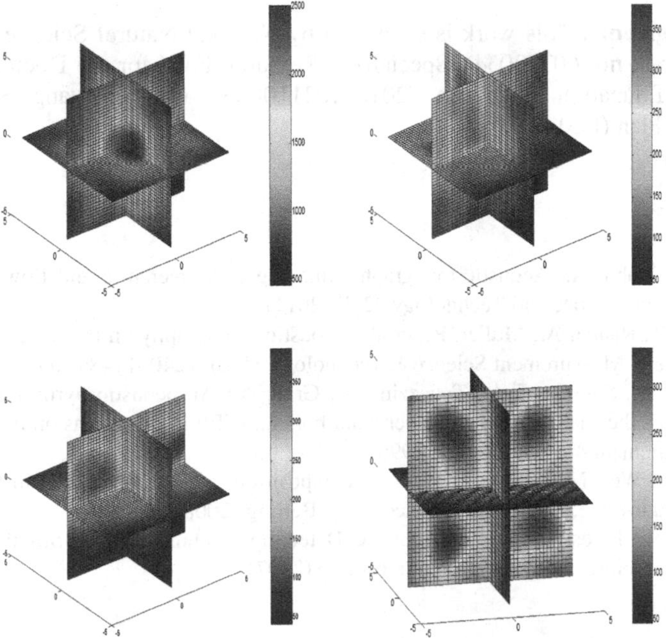

Fig. 2. Reconstruction temperature fields with noise-free sound travel-times and μ of 0

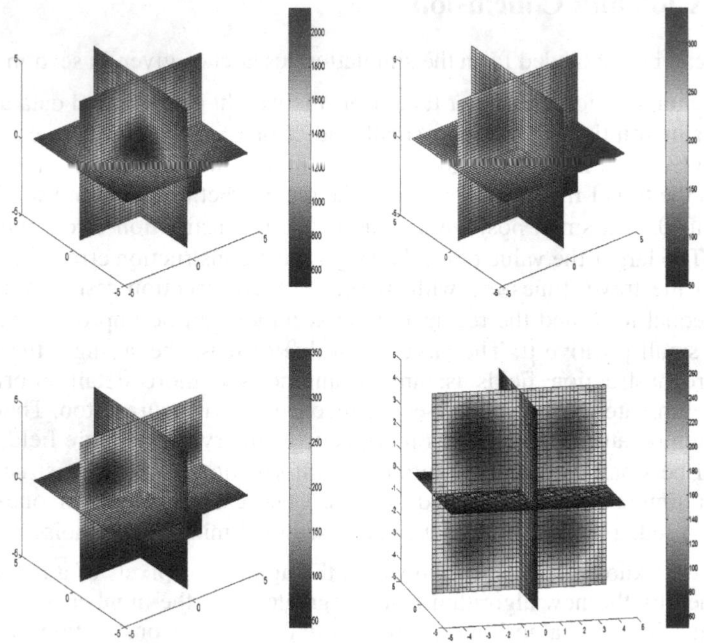

Fig. 3. Reconstruction temperature fields with noisy sound travel-times (σ= 0.01) and μ of 10^{-6}

Acknowledgments. This work is supported by National Natural Science Foundation of China (grant no. 60772054), Specialized Research Fund for the Doctoral Program of Higher Education of China (20102102110003) and Shenyang Science and Technology Plan (F10213100).

References

1. Barth, M., Raba, A.: Acoustic topographic imaging of temperature and flow fields in air. Measurement Science and Technology 22(3) (2011)
2. Holstein, P., Raabe, A., Muller, R., et al.: Acoustic tomography on the basis of travel-time measurement. Measurement Science & Technology 15(6), 1240–1248 (2004)
3. Bramanti, M., Salerno, E.A., Tonazzini, A., Gray, A.: An acoustic pyrometer system for tomographic thermal imaging in power plant boilers. IEEE Transactions on Instrumentation and Measurement 45(1), 159–161 (1996)
4. Guo, W.B., Wei, M.S.: Singular value decomposition and its application in the theory of generalized inverse, pp. 14–15. Science Press, Beijing (2008)
5. Wang, R.X.: Investigations on Acoustic Detecting Techniques and Simulation of 3-D Temperature Field. Daqing Petroleum institute (2007)

Intelligent Identification of High Rock-Filled Embankment Elastic-Plastic Model Parameters and Application*

Chengzhong Yang[1], Xianda Xu[1,2], and Shufang Wang[1]

[1] Institute of Civil and Architecture, East China Jiaotong University, Nanchang China 330013 China
[2] Huahai Engineering Co., Ltd, Shanghai Engineering Bureau of China Railway Engineering Corporation., Shanghai 200436 China

Abstract. Based on the elastic-plastic model and the orthogonal design table, the ANSYS finite element method was combined with the BP neural network to establishe high rock-filled embankment constitutive model parameters inverse analysis method. According to the field monitoring settlement data of Hurongxi highway, the elastic-plastic model parameters of high rock-filled embankment were identified and the inversion parameters were used to calculate and analyze. Compared with the measured settlement, it was found that the calculated settlement were in good agreement with it. Therefore, the method is higher reliability.

Keywords: High rock-filled embankment, elastic-plastic model, ANSYS finite element, BP network, back analysis, settlement.

1 Introduction

In the western mountainous in China, the terrain geological conditions are complex. The high rock-filled embankment and digging tunnel has become inevitable. People often use high-quality tunnel gravel as filler in the actual construction, which not only can solve the problem of lack of gravel in the mountains, but also can reduce the damage to the ecological environment and geological disasters along highways, so the high rock-filled embankment has become more common, economic and environmental subgrade type in the mountainous highway.

But the settlement of highway embankment is very strict, it is very important issue which can determine the project success and failure, the settlement calculation and the settlement prediction are also very complex. In general, choosing or establishing a appropriate constitutive model can solve these problems. However, the constitutive model parameters is not accurate, it has become the "bottleneck" of restricting the

* Supported by : National Nature Science Foundation of China (51168016), Scientific Research Foundation of Jiangxi Provincial Science and Technology Department (2010 BSA17000) and Scientific Research Foundation of Jiangxi Provincial Education Department (GJJ11442).

D. Jin and S. Lin (Eds.): Advances in CSIE, Vol. 1, AISC 168, pp. 371–377.
springerlink.com © Springer-Verlag Berlin Heidelberg 2012

problem analysis [1]. Traditionally, the constitutive model parameters often derive from indoor or field test [2], because of environmental and artificial factors, these parameters errors are too large and can't meet the engineering actual requirements. In order to solve the mechanical parameters, people turn to study the model theory and method, which is based on back analysis of field monitoring settlement, to keep the existing theoretical calculations can be used in engineering practice.

Artificial neural network theory has quickly developed to become a kind of intelligent scientific theory nearly 20 years, it has powerful and incomparable advantages in the simulation and learning, highly nonlinear recognition ability and self-learning ability are the most important aspects. It can ascertain the constitutive model of soil and the non-linear relationship of other parameters through studying the experimental and field monitoring data, and new examples and accumulated data can improve accuracy of the model.

Therefore, based on BP neural network theory and the ANSYS simulation technology, we establish BP neural network method of the displacement back analysis, and the method is used to identify the elastic-plastic model parameters of the high rock-filled embankment in Hurongxi highway.

2 BP Neural Networks Realizing of the Displacement Back Analysis

2.1 Basic Principle of the Displacement Back Analysis

The displacement back analysis method based on the site monitoring displacement to ascertain the required parameters with the optimization method, and will these parameters look up as the input to calculate these measured displacement ,then the calculated displacement is compared with the site monitoring displacement, the smallest amount of error is solution of back analysis. It is the key of back analysis that establishing the objective function and finding the optimal solution, the objective function is that:

$$F(P)=\sum_{i=1}^{n}\sum_{j=1}^{k}\left(FEM_{ij}(P)-u_{ij}\right)^{2} \tag{1}$$

Where, (P) represents inversion parameters, $FEM_{ij}(P)$ represents the numerical value of the i measured displacement components j ; u_{ij} represents the corresponding displacement measured weight; n represents the measured total number; k represents displacement components number.

Once establish the objective function, back analysis is transformed into a tiny value problems for the objective function. Through a series of optimization algorithm to get minimum of objective function $F(P)$, the corresponding parameters (P) are the optimal parameters of back analysis at this time.

2.2 BP Neural Network Theory and the Realization of the Analysis

The BP network is a multilayer feed-forward neural network based on error back propagation, typical the BP network is three feed-forward classic networks. It contains the input layer, hidden and output layer, between adjacent layers are all connections. The main idea is, in the model, the samples are input firstly, spread forward to hidden nodes, activate sigmoid function, and then transfer hidden nodes information to the output node. Finally output the results. If the output information fails to meet the target, returning the error along the original connection path to reduce errors by modifying the connection weights of each node between different layers. This is repeated until the errors meet the set requirements. The learning process of BP network can be divided into "forward propagation mode, error back-propagation, memory training, learning convergence".

3 Engineering Example

The high rock-filled embankment locates in between Zhujiayan tunnel and fire live rock tunnel of Hurongxi highway, in structure dissolution to erosion mountain regions, its karst landforms is developable. The surface slope angle of the slope is about 30 ° ~ 75 °, belongs to the typical mountains area. Long is about 130 m, width is about 64.4 m, The maximum filling height of the slope is about 72 m, the maximum center height about is 65 m, filling number is about 370000 m3, filling height is the first in Asia, and the second in the world [3-5].In order to solve the gravel packing problem of mountain and reduce the damage to the ecological environment, we use slag of the two tunnels to fill the high rock-filled embankment.

3.1 Numerical Calculation Principle and Observe Scheme

For the high rock-filled embankment in Hurongxi highway, choose typical slope embankment section ZK51+725 as the research object, use settlement data of the section. Buried settlement pipes at vertical direction in different parts of embankment ZK51+725, set settlement ring outside settlement pipe, as shown in Fig. 1. Utilize Plane42 plane strain element and DP elastic-plastic model to simulate the embankment. When the embankment is completed filling, the model diagram has 1985 nodes, 1857 quadrilateral elements , the settlement point Ⅰ1、Ⅰ2、Ⅰ3、Ⅰ4 , Ⅱ1、Ⅱ2、Ⅱ3、Ⅱ4、Ⅱ5、Ⅱ6、Ⅱ7、Ⅱ8、Ⅱ9 and measured point1 correspond to the model node, 30, 42 ,147, 288, 262, 279, 426, 509, 782, 1039, 1287, 1561, 1810 and 1830.Consider the influence of construction and the accuracy of the measured data, the vertical displacements of 42, 147, 279, 288, 509, 536, 782, 1039, 1287, 1561, 1810, 1830 are used in this article. The lower part of the foundation is rock, it is thought to have reached a steady state, and the rock strength is very high, so transverse displacement of the finite element node on the contacting surface of the subgrade and foundation u is zero, vertical displacement v is zero too. Don't consider the impact of vehicle loads in calculation. Finite element mesh and the constraints are shown in Fig.2.

In order to simulate dynamic construction of the embankment, the embankment is divided into nine layers. After one-time modeling and loading, kill the filling layer with element command "kill" (all kill, the initial stress is zero.), then according to principles of layered hierarchical loading, as shown in Fig.3. (When the first layer is completed, the structure is the first layer; the load is the dead weight of the layer. When the second layer is completed, the structure is the first layer and the second layer, the load incremental is the dead weight of the second layer, interred by the same rules until the completion of the whole embankment, Adopt element "live" command from the first layer, activate the killed element layer by layer, thus simulate the construction process of each layer. Each filling layer, counters once, and records the accumulation settlement of embankment.

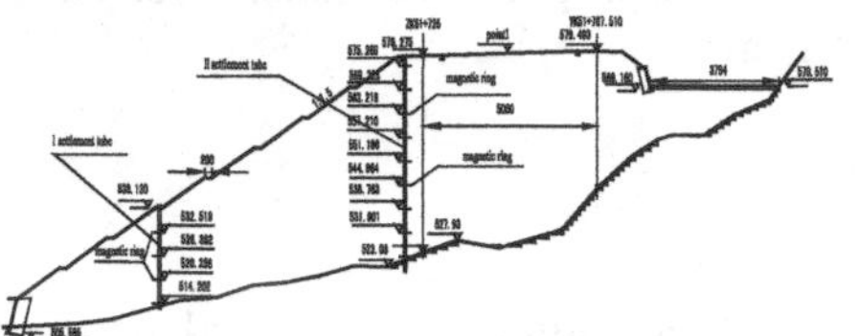

Fig. 1. The schematic drawing to lay settlement tube

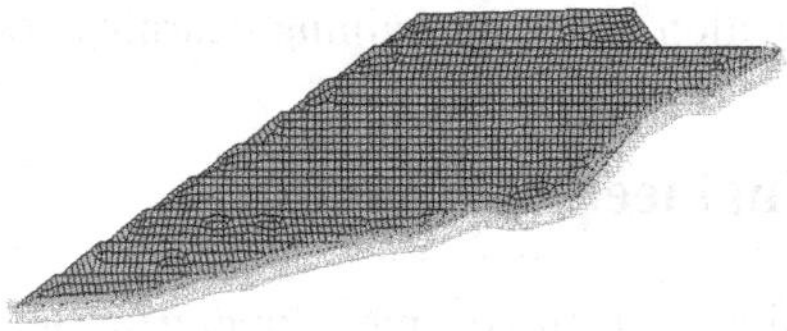

Fig. 2. Finite element mesh and the constraints

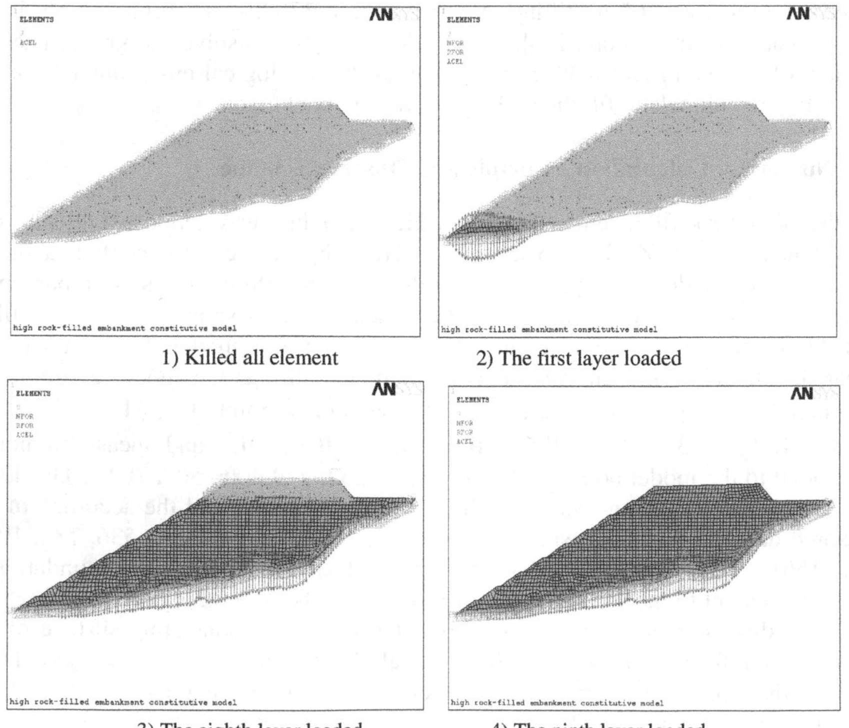

1) Killed all element

2) The first layer loaded

3) The eighth layer loaded

4) The ninth layer loaded

Fig. 3. The load diagram layer by layer (note: space limitations, not list.)

3.2 Elastic-Plastic Model Parameters Orthogonal Simulation

According to the part test contents of site material and the experiment results of similar materials, take into account the actual engineering characteristics, the calculated parameters in table 1. The selection scheme of parameter is ascertained by orthogonal table which is 4 level 5 factors (L16 (45), such as table 2. The calculated displacement of table 3 corresponds to the parameters of table 2.

Table 1. Parameter value level of elastic-plastic constitutive model parameters

value level	inversion parameters				
	γ/KN/m³	E/MPa	c/kPa	μ	φ/°
1	18	60	8	0.20	25
2	21	80	10	0.25	30
3	23	100	12	0.30	35
4	25	120	15	0.45	40

Table 2. Orthogonal design table of inversion parameters

inversion parameters	Parameters seletion scheme															
	1	2	3	4	5	6	7	8	9	10	11	12	13	14	15	16
γ/KN/m³	23	21	25	18	25	18	23	21	21	23	18	25	18	25	21	23
E/MPa	60	100	80	120	60	100	80	120	60	100	80	120	60	100	80	120
c/kPa	12	8	12	8	8	12	8	12	15	10	15	10	10	15	10	15
μ	0.25	0.45	0.45	0.25	0.30	0.20	0.20	0.30	0.20	0.30	0.30	0.20	0.45	0.25	0.25	0.45
φ/°	30	30	35	35	40	40	25	25	35	35	30	30	25	25	40	40

3.3 Network Design Parameters and Training

Establish an ANN structure with 12 node of the input layers, 28 node of the hidden layers and 5 nodes of the output layer. Activate the structure between input layers and hidden layers by Sigmoid function while pure linear function between input layers and hidden layers and nodes of layers is connected to each other. The input of the structure is the displacement values of table 3, the goal output is the inversion parameter values of table 2. Because of S function as BP network transfer function, the target vector which is made up of the parameters of table 2 need be normalized, in order to make it in the [0-1] interval. With the following: severe γ takes the original one percent, the elastic modulus E takes the original one over two hundred, cohesion C takes the original one over twenty, Poisson ratio μ is unchanged, the internal friction Angle φ takes the original one percent.

Table 3. Simulation displacement value (note: "-" calculation conver gence)

simulation scheme	node vertical displacement value (mm)											
	u_{42}	u_{147}	u_{279}	u_{288}	u_{509}	u_{536}	u_{782}	u_{1039}	u_{1287}	u_{1561}	u_{1810}	u_{1830}
1	-22.4	-84.4	-84.8	-105.2	-181.8	-80.8	-245.2	-269.5	-250.3	-189.1	-95.39	-124.6
2	-3.8	-21.3	-19.1	-23.6	-42.9	-14.7	-60.9	-70.9	-69.6	-34.4	-27.3	-37.6
3	-8.6	-31.4	-28.3	-38.0	-63.7	-21.6	-90.4	-105	-103	-80.4	-40.3	-55.7
4	-7.7	-29.7	-31.9	-37.8	-68.1	-28.5	-90.8	-98.7	-91.1	-68.2	-34.2	-5.9
5	-	-	-	-	-	-	-	-	-	-	-	-
6	-9.6	-37.2	-40.9	-47.7	-87.4	-36.2	-116.2	-126.1	-116.1	-86.9	-43.8	-58.7
7	-	-	-	-	-	-	-	-	-	-	-	-
8	-10.1	-37.7	-34.7	-46.2	-75.2	-35.4	-102.7	-114.7	-108.3	-82.8	-42.2	-53.4
9	-20.6	-78.4	-83.9	-98.7	-178.8	-76.1	-239.2	-259.9	-239.8	-180	-90.8	-120
10	-	-	-	-	-	-	-	-	-	-	-	-
11	-10.6	-40.9	-43.4	-52.1	-92.8	-38.6	-124.2	-135.7	-125.8	-94.5	-47.3	-63.6
12	-14.6	-55.7	-54.8	-69.4	-117.3	-53	-158.3	-174.1	-161	-119.8	-59.7	-78.8
13	-8.3	-30.6	-27.2	-37.	-61.3	-22.8	-87.2	-101.5	-99.6	-77.9	-39.2	-53.8
14	-17.6	-66.0	-60.4	-82.1	-130.6	-63.9	-178.6	-199.8	-188.4	-141.5	-70.5	-91.6
15	-12.7	-50.9	-55.7	-65.3	-118.9	-49.1	-158.5	-172.3	-159.1	-119.	-59.78	-80.1
16	-5.3	-19.2	-17.4	-23.3	-39.1	-13.4	-55.5	-64.4	-63.1	-49.3	-24.7	-34.2

Create the BP neural network by MATLAB toolbox "newff" function, train the BP neural network by "traingdx" function. When finished the training, test by "sim" function, finally the measured settlement of ZK151+725 section looks upon as the input of BP network structure above. The learning rate 0.1, training momentum factor 0.9, set goal error 1e-4, the training stepinterval maximum 70000.

After 555534 steps, BP neural network finishes the training, error is 9.99989e-5, as is shown in Fig.1.Table 4 gives the inversion results.

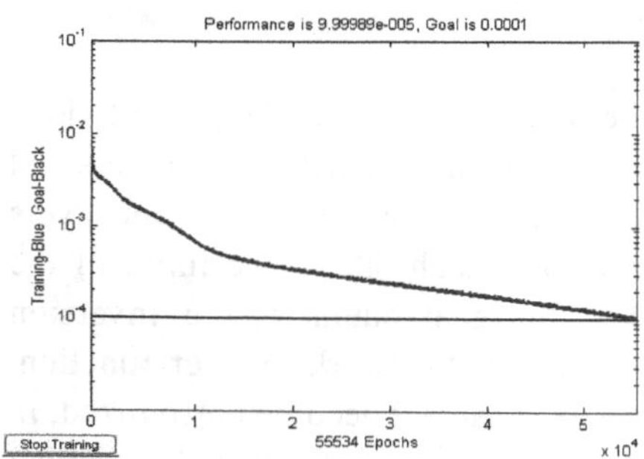

Fig. 4. Error performance curve

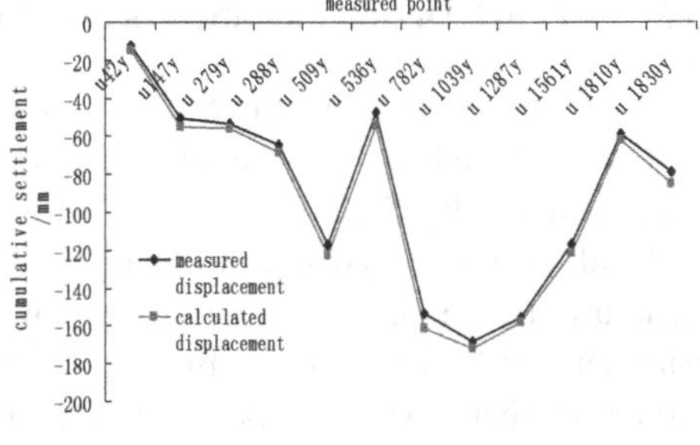

Fig. 5. The comparison curves between inversion and measured value

Table 4. Inversion value of identification parameters

γ/KN/m^3	E/MPa	c/kPa	μ	φ /°
21.44	77.2	11.914	0.2934	35.44

3.4 Inversion Results Analysis

Through BP neural network identify value of table 4, calculate the displacement of measuring point and draw comparison curves between calculated and measured. Fig.5 gives the contrast of inversion and measured value .As we can be seen from the graph, the calculated values and the measured values are in good agreement. So inversion parameters calculated settlement of the high rock-filled embankment which can meet the project accuracy requirement. It shows that inversion method has higher reliability.

4 Conclusions

By use of DP elastic-plastic constitutive model, introducing the BP neural network algorithm and ANSYS finite element method, identifying the high rock-filled embankment model parameters, some conclusions are followed as:

(1) BP neural network has a strong nonlinear mapping ability, it is fit for solving the embankment indefinite displacement back analysis problems without mathematical expressions between the input and output;

(2) The parameters back analysis based on orthogonal design, finite element method and BP neural network to determine rock-filed embankment is feasible and relatively accurate. This method can also fit for Settlement prediction;

(3) By the orthogonal design method to arrange simulation calculation scheme of ANSYS is representative. It can reduce the experiments number and save time;

(4) By the method of the parameters identification is more reliable. It can meet the engineering accuracy and provide reference with the design and construction.

References

1. Yang, L., et al.: Geotechnical problems of inversion theory and engineering practice. Science Press, Beijing (1999) (in Chinese)
2. Gao, W., Zheng, Y.: Back analysis in geotechnical engineering based on fast-convergent genetic algorithm. Chinese Journal of Geotechnical Engineering 23(1), 120–122 (2001) (in Chinese)
3. Yang, C., Liu, X., Wang, S., et al.: Dynamic stabllity analysis and site monitoring of reinforced high rockfill embankment during construction. Chinese Journal of Rock Mechanics and Engineering S1, 3321–3328 (2010) (in Chinese)
4. Yang, C.: Analysis of dynamic response of high rockfill embankment under seismic loads. Advanced Materials Research 163-167, 4363–4366 (2011)
5. Yang, C., Xu, X.: Intelligent identification of high rock-filled embankment constitutive model parameters. Advanced Materials Research 183-185, 2139–2142 (2011)

Distributed Detection Approach for Wormhole Attack in Wireless Sensor Networks

Haofei Wang and Tongqiang Li

School of Information & Electronic Engineering, Zhejiang GongShang University, Xuezheng Street 18, Jianggan District, 310018 HangZhou, China

Abstract. Wormhole attacks put severe threats to wireless sensor networks since its immunity to cryptology and independence of MAC protocols, which makes it difficult to detect. Most existing methods have some drawbacks, no matter the tight time synchronization or the dedicated hardware units, are difficult to achieve in reality, which limits their use in practice. In this paper, we present a robust distributed detection approach without special hardware or time synchronization. The simulation shows that with proper parameters this method has very high detection ratio and low false alarm probability.

Keywords: wormhole attack, distributed algorithms, sensor network, local topology information.

1 Introduction

The Wireless Sensor Network (WSN) is made up of lots of miniature and independent nodes with thick density and self-organizing. Since the node is source limited, the WSN faces serious security challenges and wormhole attack is a severe one. A wormhole attack is a particularly severe attack on MANET (Mobile Ad-hoc Network) routing where two attackers connected by a high-speed off-channel link called the wormhole link [1]. For its independence of MAC protocols and immunity to cryptology techniques, the existing detection methods almost take other means, for example, the method proposed by Y.Hu requests the constraints of geography and time [2]. The geography constraints method needs additional location information while the time constrains method requires the tight time synchronization [3]. L.Hu proposed using directional antennas to find the impossible communication links between nodes to prevent wormhole attacks [4]. All those methods have impractical requirements, which limit their applicability in practice.

Recently, the topological detection has been developing quickly and some new approaches have been proposed. W. Wang and B. Bhargava presented method of combining graphics and the topological detection [5]. Ritesh proposed using the connectivity information [6], which gives us a lot of inspiration.

In this paper, we analyze the effect of wormhole attack on the local connectivity of the network [7] and find distinct changes. We get our algorithm by analyzing the changes and simulate it. The results show with proper parameters this method has stable performance and high detection accuracy.

D. Jin and S. Lin (Eds.): Advances in CSIE, Vol. 1, AISC 168, pp. 379–384.
springerlink.com © Springer-Verlag Berlin Heidelberg 2012

2 Assumptions and Principles

2.1 The Basic Assumptions

Firstly, we assume the sensor network is static or quasi-static, the nodes are distributed uniformly and randomly, the density of the nodes and the communication radius can ensure the connectivity of network. There are no obvious divided points and edge borders. Secondly, the communication radius of the attacker is larger than that of the node and the distance between the collusion attackers is large enough so that the attacked areas on both sides will not overlap.

2.2 The Detection Principles

Firstly, we consider the normal condition, the node is uniform and random distribution and its neighbor nodes is $\pi r^2 S / A$, r is the communication radius, S is the total number of nodes, A is the area of the network. The neighbor nodes shared by two internal nodes should satisfy $0 \le |N_X \cap N_Y| \le D$, N_X and N_Y are the neighbor nodes collection of the node X and Y. Especially, when the distance between the two nodes is more than 2r, deriving $N_X \cap N_Y \approx 0$.

Now considering the collusion attack, we assume E_1 and E_2 are the collusion attackers and X and Y are internal nodes. If X locates in the attack area of E_1, it will mistakes all the nodes locates within the area of E_2 as its neighbor nodes, that is to say $N'_X = N_X \cup N_{E2}$, and vice versa. N'_X is the false neighbor collection of X. Here we will get an important conclusion:

$$|N'_X| = |N_X \cup N_{E2}| \ge |N_X \cup N_X| = 2N_X$$

That is to say, if a node suffers wormhole attack, its neighbor nodes will double at least, which can be used as the preliminary basis to judge whether a node suffers wormhole attack or not. Further discussion, if X and Y locate the opposite side of the attack tunnel, their mutual neighbor nodes are:

$$|N'_X \cap N'_Y| = |(N_X \cup N_{E2}) \cap (N_Y \cap N_{E1})| = |N_X \cup N_Y| \approx 2D$$

Else if X and Y at the same side:

$$|N'_X \cap N'_Y| = |(N_X \cup N_{E2}) \cap (N_Y \cap N_{E2})| = |(N_X \cup N_Y) \cup N_{E2}|$$

It is obvious that $D \le |N_{E2}| \le |N'_X \cap N'_Y| \le D + |N_{E2}|$, so it is easy to derive:

$$|N'_X \cap N'_Y| = |(N_X \cup N_{E2}) \cap (N_Y \cap N_{E2})| > D$$

When the distance between X and Y is larger than 2r, we can get $N_X \cap N_Y = \Phi$, on this condition, we get:

$$N'_X \cap N'_Y = (N_X \cup N_Y) \cup N_{E2} = N_{E2}$$

So we can get the attacked nodes' collections of the other side by the not adjacent attacked nodes of this side. Then we can remove them respectively. Z is the attacked node, and if Z lies in the same side of X and Y:

$$N_Z^{'} - \left(N_X^{'} \cap N_Y^{'}\right) = \left(N_Z \cup N_{E2}\right) - N_{E2} = N_Z$$

If Z lies in the opposite side of *X* and *Y*:

$$N_Z^{'} \cap \left(N_X^{'} \cap N_Y^{'}\right) = \left(N_Z \cup N_{E1}\right) \cap N_{E2} = N_Z$$

We can verify $N_{E2} \subset N_Z^{'}$ is true or false to judge whether Z lies the same side with X and Y or not.

3 Detection Algorithm

The new approach doesn't need the stations' participation and is completed by the nodes cooperation. Each node will firstly judge itself, if its neighbor nodes is more than $K_{ab}D$, it should be regard as anomaly node. K_{ab} is the threshold ratio of the anomaly nodes. This approach can be divided into three steps; competing for the sponsor node; competing for the response node; removing the wormhole attack;

3.1 Competing for the Sponsor Node

The anomaly node will compete for the sponsor node as the following steps.

- Anomaly nodes broadcast request, every neighbor node receives the request should return their anomaly marks, which will be used to calculate the ratio of the normal nodes to anomaly nodes, and record it as R_{na}.
- If one anomaly node with not zero R_{na} doesn't receive any competitive messages, or receives but R_{na} is less than itself, or equals but the ID is bigger than itself, it will broadcast its own message $\{ID, R_{na}\}$. Otherwise it will broadcast the competitive messages of others and record the last message. If the R_{na} of itself is zero, the message will be recorded as $\{ID, 0\}$.

After this process, the node with the max R_{na} and less ID will become the sponsor node marked as Z_I and its message will be broadcasted over the area attacked.

3.2 Competing for the Response Node

The response node should respond the request and complete the wormhole detection as following steps.

- The request message $\{ID_{ZI}, N_{ZI}^{'}, ID_R, |N_{ZI}^{'} \cap N_R^{'}\}$will be broadcasted by Z_I and recorded. R: non-neighbor node of Z_I.

- If the neighbor node of Z_I receives the request the first time, forwarding it and recording the message; else if the $|N'_{ZI} \cap N'_R|$ in the message is less than records, or equals but the ID_X is less, forwarding the message and updating the record; none of the above, discarding it.
- If the non-neighbor node X receives the request the first time and R_{na} is not zero, it will calculate the common neighbor nodes with Z_I, if $|N'_{ZI} \cap N'_X|$ is less than $|N'_{ZI} \cap N'_R|$, or equals but ID_X is less, it will broadcast the new request message $\{ID_{ZI}, N'_{ZI}, ID_X, |N'_{ZI} \cap N'_X|\}$; else forwarding the received message and updating the sending records.
- If X receives the request on the second time or the R_{na} is zero, besides, if $|N'_{ZI} \cap N'_R|$ in the message is less than records , or equals but ID_R is less, it will forwards the original request message and update the sending records; else discarding the received requests

After this process the node has least shared neighbor nodes with Z_I and less ID will become the response node, marked as Z_R.

3.3 Removing the Wormhole Attack

If Z_R doesn't receive more competitive request message during the time intervalΔ, it can be affirmed as the response node and will execute the detection as follows:

- Z_R calculates its shared neighbor nodes with Z_I, recording the collection as $N'_{ZI} \cap N'_{ZR}$. If $N'_{ZI} \cap N'_{ZR} < K_{ab}D$, no wormhole attack exists, removes all the anomaly marks and the detection process is over. Else the neighbor nodes collection should be update as $N'_{ZR} - (N'_{ZI} \cap N'_{ZR}) = N_{ZR}$ and the detection result is $\{ID_{ZI}, ID_{ZR}.N'_{ZI} \cap N'_{ZR}\}$, all detection results after this will be ignored.
- If other anomaly nodes X receive the detection result on the first time, it will judge $X \in N'_{ZI} \cap N'_{ZR}$ is true or false. If true shows X, Z_I, Z_R not in the same side and updating the neighbor collection as $N'_X \cap (N'_{ZI} \cap N'_{ZR}) = N_X$; else shows they in the same side and calculate $N'_X - (N'_{ZI} \cap N'_{ZR}) = N_X$ then forwards the detection results, ignoring all the results after this.

If $(N'_{ZI} \cap N'_{ZR}) = N_E$, E is the wormhole attacker lies in the opposite side of Z_I and Z_R; if $|N'_{ZI} \cap N'_{ZR}| \approx 0$, the node density of some area is too thick but no wormhole attack exists; if $|N'_{ZI} \cap N'_{ZR}| < K_{ab}D$, the distance between Z_I and Z_R is less than 2r, no wormhole attack exists ; otherwise, wormhole attack exists.

4 Simulations

The parameters of the simulation model set as follows: the communication radius is 10m; the attack radius is 25m; the density expectation is 8 and the anomaly threshold ratio is 3, the others parameters will be generated automatically.

Firstly, the relationship between performance and attack radius, showing as figure 1:

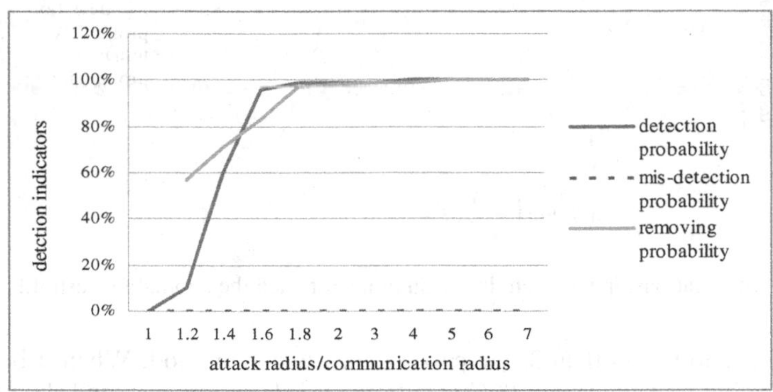

Fig. 1. The relationship between performance and attack radius

We can see the probability of the detection and the removing are increasing with the ratio and when the ratio is bigger than 1.8, they are both keep 100%. The misdetection probability is unrelated with the ratio and keeps 0.

Secondly, we consider the relationship between the performance and the density expectation of the nodes, showing in figure 2:

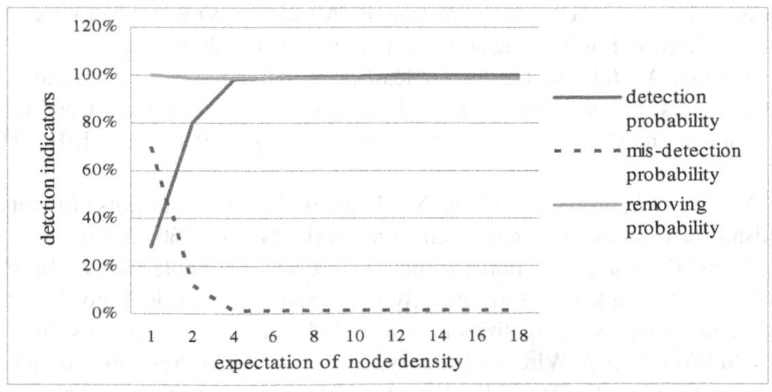

Fig. 2. The relationship between detection indicators and the node expectation

With the expectation increasing the detection probability increases while the misdetection decreases. When it is bigger than 4, the detection probability is nearly to 100% and misdetection keeps 0. The removing probability is unrelated with the expectation and keeps 100%.

Lastly, let us discuss the relationship of the performance and the anomaly threshold.

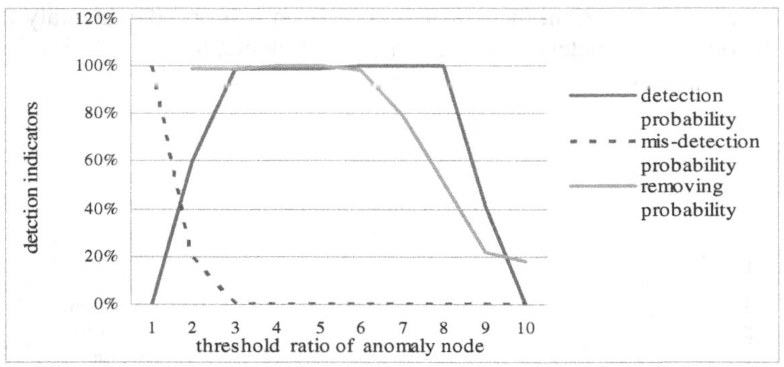

Fig. 3. The relationship between detection indicators and the anomaly threshold ratio.

When the ratio is less than 3, the performance is not so good; When it belongs [3 6], the detection probability is 100%, while the misdetection is 0, and the removing detection keeps 100%. But when the threshold exceeds 6, the indicators become worse. So the threshold set [3 5] is appropriate.

The simulation results show the algorithm has a lower detection threshold and a stable performance in a wide range of parameters with high detection accuracy. So it is more feasible and practical.

References

[1] Win, K.S.: Analysis of Detecting Wormhole Attack in Wireless Networks. In: Word Academy of Science, Engineering and Technology, vol. 48 (2008)

[2] Hu, Y., Perring, A., Johson, D.: Packet leashes: A defense against wormhole attacks in wireliss networks. In: Proceeding of the Twenty-second Annual Joint Conference of the IEEE Computer and Communications Societies, vol. 3, pp. 1976–1986. IEEE, Pisttsburgh (2003)

[3] Zhang, Y., Liu, W., Lou, W., Fang, Y.: Location-based compromise-tolerant security mechanisms for wireless for wireless sensor networks 24, 247–260 (2006)

[4] Hu, L., Evans, D.: Using directional antnnas to prevent wormhole attacks. In: Proceeding of the Eleventh Network and Distributed System Security Symposium, pp. 131–141 (2004)

[5] Wang, W., Bhargava, B.: Visualizition of wormhole s in sensor networks. In: Proceeding of the ACM Workshop on Wireless Security, pp. 51–60. ACM Press, Philadelphia (2004)

[6] Maheshwari, R., Gao, J., Das, S.R.: Detecting Wormhole Attacks in Wireless Networks Using Connectivity Information. In: Proc. of the 26th Annual IEEE Conference on Computer Communications (INFOCOM 2007), Anchorage, AK, pp. 107–115 (2007)

[7] Yu, B., Li, T.: Study of wormhole attack detecting based on lcoal connectivity information in wireless sensor networks. In: International Conference on Computer Science and Service System (CSSS), pp. 3585–3588 (2011)

Routing Optimization Design of Wireless Ad Hoc Network Based on Electric Energy Data Acquire System

GangJun Gong, JuanXi Zhu, XingChuan Wang, and Chen Xiong

School of Electrical and Elcctronic Engineering,
North China Electric Power University, Beijing 102206, China
{948553964,546582715,421488273,4753199}@qq.com

Abstract. This paper puts forward an improved AODV routing algorithm after analyzing the characteristics of wireless ad hoc network desired by electric energy data acquire system in smart grid and the topological structure of intelligent power consumption community. The proposed algorithm decides delay time of node access and means of link maintenance by residual energy and buffer queue length. Thus nodes which have more energy and fewer loads can access route more quickly. The theoretical analysis and simulation results show the improved algorithm effectively improves network throughput, reduces packet loss rate and end-to-end delay.

Keywords: Electric Energy Data Acquire System, Routing Algorithm, AODV.

1 Introduction

The special environment of intelligent power consumption community ask its communication network must be with the convenience, speediness, flexibility, economy, reliability and other characteristics. Wire communication have weaknesses such as wiring limit, low networking flexibility and difficulty for repair and maintenance. Yet wireless ad hoc network is a kind of self-organizing multi-hop network without infrastructure which is in favour of building a community network with simple operation, flexible networking and low power consumption [1].

Wireless ad hoc network is object-oriented, so different application object and environment determine different network topology structure and routing algorithm in design practice. The traditional routing algorithm have DSDV, DSR, AODV and so on[2], but most of these algorithms often lead to overuse small center nodes and make its consumption of energy resource too fast. Consequently the network will be partitioned and reduced life and throughput.

There are most of communications nodes in residential quarter in electric energy data acquire system. So plentiful data information may cause congestion during transmission. In consequence, this paper will work on improving traditional AODV putting forward a routing algorithm which decides delay time of node access and means of link maintenance by residual energy and buffer queue length. The proposed algorithm can dramatically resolve congestion control problem and extend the life cycle of whole network.

D. Jin and S. Lin (Eds.): Advances in CSIE, Vol. 1, AISC 168, pp. 385–390.
springerlink.com © Springer-Verlag Berlin Heidelberg 2012

2 Topology Analysis and Routing Requirements

2.1 Network Topology Analysis

By the reason of different patterns of small residential buildings and building layouts, we should place concentrations and gateway nodes according to different geographic environment so that the collected data can be saved and transmitted remotely through gateway node. Because of the environment complexity of residential areas, its transmission distance will be compact to some degree. In our scheme design, the indoor distance standard is shortened to 50 m and the outdoor distance standard is shortened to 500 m. We obtain community topology structure according to the current mainstream community distribution characteristics, as shown in Fig. 1.

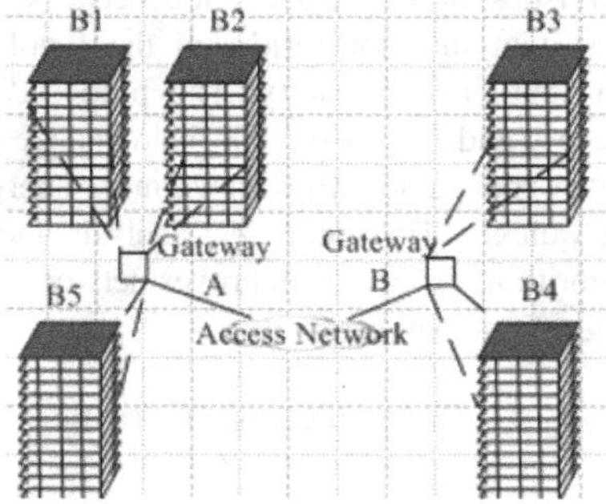

Fig. 1. Community topology structure

In the community shown by Fig.1, building 1 and building 2 are located side by side, building 1 and building 5 are located face to face and building 3 and building 4 are located face to face but from building 2 on a certain distance. To use this kind of structure, we can place gateway A during building 1, building 2 and building 5, and place gateway B between building 3 and building 4. In this case, we both can satisfy indoor and outdoor transmission distance of sensor network to ensure all collected data be transmitted to gateway by a hop or more hop, also can make access terminals of access network less as far as possible to reduce the access network burden.

2.2 Routing Requirements of Acquire System

Existing remote meter reading collects a data every 15 minutes, so the real time requirement is not high. The position of electric meters and collectors is fixed, so we don't consider the mobility of nodes. But the flow is bigger during transferring meter reading data. This may produce a short congestion and cause packet loss rate and time delay to rose. So global channel quality will decrease. For this reason, congestion control is needed in routing design. Electric energy data acquire system also have certain requirements for fault-tolerant and robust. When some of node in network failed, newly adjusting and calculating routing are needed. Local repair technology is mainly used at present. In addition, because the energy of node is limited, we should consider to make energy consumption of whole sensor network balanced when choosing route to extend the life cycle of network.

3 Routing Algorithm

3.1 Brief Introduction of AODV

AODV completes routing function through the route setup and route maintenance[3]. In the process of setting up route, it is the source node that starts the process of route discovery. The basic process of the algorithm is as follows: When source node wants to send information to destination node, if the routing table of itself has not route to destination node, it will send route request packet RREQ to all its neighbors nodes to request them help to find route to destination node. Each node receiving to RREQ packet records node information at last hop. If the node itself is not the destination node, it will broadcast the RREQ packet. Through this flooding way, RREQ packet will be broadcasted to the destination node. After the destination node receives the RREQ packet, it oppositely sends the route reply packet RREP to the source node through the coming path of RREQ in the way of unicast. Over the course, intermediate nodes record the next-hop-node to destination node, so when the source node receives the RREP packet, t electric energy data acquire system he route from source node to destination node has been set up.

AODV reduces frequency of broadcasting. So the algorithm is simple and practicable. But AODV doesn't consider the conditions of residual energy and congestion of nodes, so packet loss rate is high and the energy consumption is in imbalance[4]. Under the circumstances, we design a improved AODV algorithm according to the characteristic of routing requirements of electric energy data acquire system. This algorithm will be suitable for the environment of wireless ad hoc network desired by the acquire system..

3.2 Improved AODV Algorithm

Routing by Residual Energy

In order to realize load balance, intermediate nodes calculate the route delay weight w_e according to residual energy E after receiving RREQ node and get the delay time. Each node in route routes in the delay time. Limiting top limitation E_H=3/4E_0 and bottom limitation E_L=1/4E_0 for residual energy, we determine the delay time according to the limitations:

1. If $E>E_H$, setting W_e=0, node forwards RREQ directly
2. If $E_L \le E \le E_H$, $W_e=E_0/E, T_e=W_e T_B$

In types: E_0 is the primary energy; T_B is delay constant; T_e is the delay time. Node forwards RREQ in T_e

3. If $0 \le E \le E_L$, node stops to response the route request

This algorithm regarding residual energy make nodes which have less residual energy access route at a lower pace, so it balances the energy consumption of nodes.

Routing with Congestion Control

We use buffer queue protocol to measure congestion state of route. In the process of route setup, intermediate nodes choose route according to its buffer queue after receiving RREQ packet. In AODV, the maxlength of group buffer queue is 64. Limiting top limitation $L_H=3/4L_0$ and bottom limitation $L_L=1/4L_0$ for group buffer queue similarly, we determine the delay time according to the limitations:

a. If $L>L_H$, setting $W_L=0$, node forwards RREQ directly
b. If $L_L\le L\le L_H$, $W_L=L/L_0, T_L=W_L T_B$

In types: L is the current length of buffer queue; L_0 is the maxlength of buffer queue; T_L is the delay time; W_L is the delay weight. Node forwards RREQ in T_L.

c. If $0\le L\le L_L$, node is in a serious state of congestion, so it will stop to response the route request.

Through congestion control, nodes which have less load can preferentially access route.

Process of Route Setup

Regarding synthetically residual energy and buffer queue, our algorithm improves on the basis of AODV. Intermediate node routes according to delay time after receiving RREQ. A detailed description of routing delay algorithm is as follows:

a. If $E>E_H$, delay time is $T=T_L$
b. If $E_L\le E\le E_H$: if $L>L_H$, delay time is $T=T_e$; if $L_L\le L\le L_H$, delay time is $T=(aW_e+bW_L)$; if $0\le L\le L_L$, node stops to response the route request
c. If $0\le E\le E_L$, node stops to response the route request

Through the improvement, the algorithm makes nodes to calculate local delay time and forward RREQ in the delay time. Consequently, in that delay time, RREQ is transmitted through path whose nodes have more energy and less load. So preferable route is preferentially set up. If there is no other better route, the original route will still be set up.

Process of Route Maintenance

In electric energy data acquire system, because of outage and equipment failure, routes which have been set up will be disabled. This means link fracture communication interrupt. For this reason, we must maintain and repair the routes in acquire system after each round of information collection. Only in this way can we ensure acquire system normal operation.

As with AODV, our algorithm broadcasts HELLO packet to detect whether the route is failure. When node detect link disconnection, it does not directly conduct local routing repair, but make a choice between local routing repair and reporting to the upstream node according to its buffer queue. If the length of buffer queue is greater than 48 or residual energy is less than 25% of primary energy, node directly report mistake. In this way, source node will newly set up a new route which has less load and more energy. Conversely, node conducts local routing repair. One example of process of route maintenance is shown in figure 2.

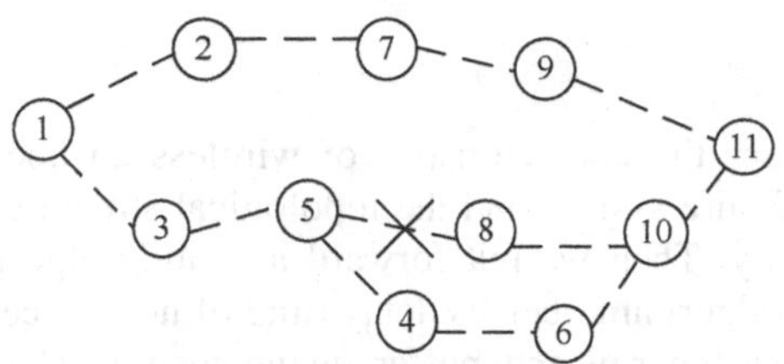

Fig. 2. Process of route maintenance

As shown in figure 2, source node 1 transmits data to destination node 11 through route 1-3-5-8-10-11. Node 5 realizes that path 5-8 is failed by means of broadcasting HELLO packet. So faulted node 5 checks its routing table and makes the following judgment: If finding the length of buffer queue is greater than 48 or residual energy is less than 25% of primary energy, node 5 sends RERR packet to its upstream node 3. Next node 3 sends the RERR packet to source node 1. After receiving the RERR packet, node 1 newly sets up route 1-2-7-9-11. Conversely, node 5 conducts local routing repair. In this, node 5 broadcasts itself a RREQ packet to destination node 11 and set up a new path 5-4-6-10-11 according to judge residual energy and buffer queue of nodes.

4 Simulations and Analysis

We use OPNET 14.5 as simulation platform in this paper. 50 nodes are arrayed randomly in a 1000 m by 1000 m square area. All nodes are fixed. Simulation time is one hour. Transmitting interval of data is 2 s. FHSS. T_B=0.005s , a=b=0.5.

We use AODV as our contrast algorithm. Fig. 3 is comparison of packet loss rates of two algorithms. In the opening phase of route-setting, there is no failed node because of high energy. So packet loss rates of both AODV and improved AODV are 0. In the 1500 s, node failure appears. So packet loss appears and gradually increases. However, by the reason of high stability, packet loss rate of improved AODV is less than AODV.

Fig. 4 is comparison of end-to-end delay of two algorithms. Though the delay time of improved AODV is less than AODV invariably in one hour, as time goes on, disparity of two algorithms is smaller and smaller. Reason of this phenomenon is that improved algorithm selects nodes which have more residual and less load to set up new route.

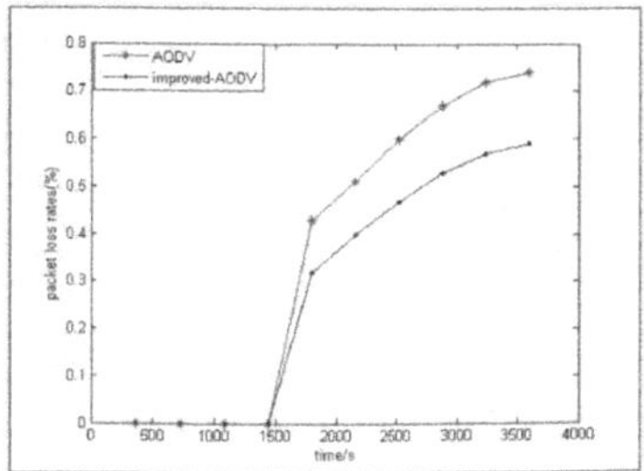

Fig. 3. Comparison of packet loss rates

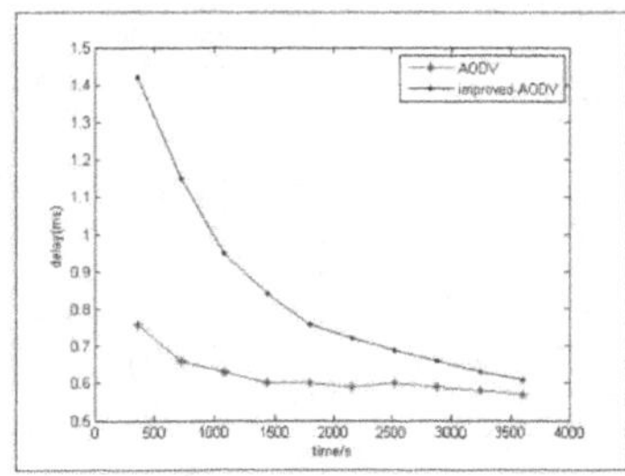

Fig. 4. Comparison of end-to-end delay

5 Summary

This paper first analyses the characteristics of wireless ad hoc network desired by electric energy data acquire system and the topological structure of intelligent power consumption community. Then we put forward a routing algorithm on the basis of AODV. The proposed algorithm decides delay time of node access and means of link maintenance by residual energy and buffer queue length. Thus nodes which have more energy and less load can more quickly access route. The theoretical analysis and simulation results show the improved AODV algorithm effectively improves network throughput, balances the energy consumption of nodes and reduces packet loss rate and end-to-end delay. The proposed algorithm can not only dramatically resolve congestion control problem and extend the life cycle of whole network, but also bring about better tolerance and robustness, so it meets with the requirements of electric energy data acquire system. Besides, as a new means of communication, wireless ad hoc network expands the functions of communication network, improves business level of electric energy data acquire system and promotes intelligent construction of remote meter reading. So we should attach importance to the research and applications of wireless ad hoc network.

References

1. Yu, F., Hu, J., Zhang, Q., et al.: Research of data transfer routing for low power customers wireless meter reading system. Power System Protection and Control 38(1), 66–69 (2010)
2. Wang, B.: Implementation of AODV Routing Protocol for Wireless Sensor Networks. Computer and Modernization 1, 86–89 (2009)
3. Liu, Y., Guo, L., Ma, H.: An energy efficient and load balanced Ad hoc routing protocol. Journal of Harbin Engineering University 29(3), 294–297 (2008)
4. Liu, G., Wang, H., Wei, G.: Congestion adaptive protocol based on aided-routing for multi-hop wireless Ad hoc networks. Journal of Systems Engineering and Electronics 32(05), 1070–1076 (2010)

Prediction Research Based on Improved Intelligent Algorithm Model

Mingjun Wang[1,3] and Shuxian Deng[2]

[1] Zhengzhou University, 450007 Zhengzhou, China
[2] Henan Engineering Institute, 450007 Zhengzhou China
[3] Zhejiang Vocational College of Commerce, 310053 HangZhou, China
mingjun_w@126.com, dshuxian@163.com

Abstract. This paper is to focus on the prediction and application of tourism human resources by the intelligent algorithm. In this paper, an improved gray prediction algorithm is applied to analyze the tour human resources data of every city in Zhejiang province from 2005 to 2010 and obtained the predict data on 2011 and 2012. The gray prediction algorithm is fit to study these problems of small sample and poor information and these objects of extension clear and meaning not clear.

Keywords: intelligent, algorithm, prediction.

1 Introduction

This paper is mainly about the application of intelligence algorithm in the prediction and development of tourism human resources demand in the cities of Zhejiang Province. This research analyzes and predicts the statistics provided by Zhejiang Provincial Tourist Bureau. This paper employs the improved grey prediction algorithm, which is suitable for the research of "small sample, inadequate information" and "clear extension and unclear intension" objects.

(1) Base on the statistics. The establishment of prediction model is based on a certain number of sample statistics and conforms to the rules reflected by the data. Generally speaking, the first move of establishing prediction model is not about the choosing of prediction method but the changing of the sample data which is served as the basis of the appropriate prediction method.

(2) Set the prediction accuracy standard. Even for the same prediction object, prediction methods vary.

The employment of different parameter estimation will lead to a prediction result with different accuracy.

(3) Set meeting the actual needs as a prerequisite. For one, the prediction method should be in accord with the facts and easy to be understood by the application staff. Secondly, cost consumption during the application of the method can be afforded. Finally, the error range from prediction result to actual result should fall into a certain limit.

D. Jin and S. Lin (Eds.): Advances in CSIE, Vol. 1, AISC 168, pp. 391–397.
springerlink.com © Springer-Verlag Berlin Heidelberg 2012

2 Grey Prediction Algorithm

Prediction can be done from two perspectives, namely, quantitative and qualitative. Since the demonstration of this paper is based on the statistics of Zhejiang Provincial Tourist Bureau, the grey prediction algorithm is applied to the qualitative research.

The prediction of future development is done through setting up a mathematical model. In view of the characteristics of grey prediction, this paper uses GM (1, 1) to predict the tourism human resources demand. Grey system theory is for the research of "small sample, inadequate information" and "clear extension and unclear intension" objects [3].

2.1 Basic Principles of Grey System Theory

Grey system theory includes the following basic principles: information gap principle, i.e. gap is information, where there is information, there is gap; non-uniqueness of solutions, i.e. information is incomplete, the solution is not unique; less information principle, the amount of available information is the turning point of identifying grey and non-grey, making full use of the available "less information" is the basic idea of grey system theory; cognitive basis principle, i.e. information is the basis of cognition, with complete, certain information one can get complete and certain recognition, with incomplete, uncertain information serving as the basis, one can only get incomplete, uncertain grey recognition; forever grey principle, i.e. incomplete information, grey is absolute[4-9]·

2.2 Model Construction and Solving Process

Theorem 1 [4]. Set that $Y^{(0)}$ is original sequence: $Y^{(0)} = (y^{(0)}(1), y^{(0)}(2), ..., y^{(0)}(n))$, let that the stepwise ratio of $Y^{(0)}$ is $\sigma^{(0)}(k)$, $\sigma^{(0)}(k) = \dfrac{Y^{(0)}(k-1)}{Y^{(0)}(k)}$, $k \geq 3$, then, when $\sigma^{(0)}(k) \in (0.1353, 7.389)$, $Y^{(0)}$ can become a model of non-malformation GM(1,1).

Theorem 2[4]. Set that $Y^{(0)}$ is non-negative sequence: $Y^{(0)} = (y^{(0)}(1), y^{(0)}(2), ..., y^{(0)}(n))$, where $y^{(0)}(k) \geq 0$, $k = 1,2,...,n$; $Y^{(1)}$ is the accumulating generation sequence of $Y^{(0)}$ (1-AGO); $Y^{(1)} = (y^{(1)}(1), y^{(1)}(2), ..., y^{(1)}(n))$, where $y^{(1)}(k) = \sum_{i=1}^{k} y^{(0)}(i)$, $k=1,2,...,n$; $Z^{(1)}$ is the compact adjacent mean generation sequence of $Y^{(1)}$: $Z^{(1)} = (z^{(1)}(1), z^{(1)}(2), ..., z^{(1)}(n))$, where $z^{(1)} = \dfrac{(e-2)y^{(1)}(k+1) + y^{(1)}(k)}{e-1}$, $k = 1,2,...,n$.

If $\hat{a} = [a, \mu]^T$ is the parameter column, and

$$Y = \begin{pmatrix} y^{(0)}(2) \\ y^{(0)}(3) \\ \cdots \\ y^{(0)}(n) \end{pmatrix} \quad B = \begin{pmatrix} -z^{(1)}(2) & 1 \\ -z^{(1)}(3) & 1 \\ \cdots & \cdots \\ -z^{(1)}(n) & 1 \end{pmatrix} \tag{1}$$

Then the least square estimation sequence of $y^{(0)}(k) + az^{(1)}(k) = u$ of GM(1,1) model satisfies $\hat{a} = [B^T B]^{-1} B^T Y$. The application scope of GM(1,1) model is related to the developing coefficient a , Only when $|a| < 2$, GM(1,1) model is meaningful; Otherwise GM(1,1) model loses its significance.

Construct the differential equation in accord with GM (1, 1) model:

$$\frac{dy^{(1)}(t)}{dt} + ay^{(1)}(t) = \mu . \tag{2}$$

Where $y^{(1)}(t)$ is the accumulated value of one time of the system status variable $y^{(0)}(t)$, y(t) is the t-th year demand quantity of tourism human resources, a and μ are undetermined parameters.

Construct B matrix and Y matrix:

$$B = \begin{pmatrix} -\frac{1}{2}[y^{(1)}(1) + y^{(1)}(2)] & 1 \\ -\frac{1}{2}[y^{(1)}(2) + y^{(1)}(3)] & 1 \\ \cdots & \cdots \\ -\frac{1}{2}[y^{(1)}(n-1) + y^{(1)}(n)] & 1 \end{pmatrix} \quad Y = \begin{pmatrix} y^{(0)}(2) \\ y^{(0)}(3) \\ \cdots \\ y^{(0)}(n) \end{pmatrix}$$

Then a and μ can be obtained by the following method:

$$[a, \mu]^T = [B^T B]^{-1} B^T Y \tag{3}$$

Time response equation is that:

$$\hat{y}(k+1) = \left[y^{(1)}(0) - \frac{\hat{\mu}}{\hat{a}} \right] e^{-\hat{a}k} + \frac{\hat{\mu}}{\hat{a}} \tag{4}$$

When $k = 1, 2, 3, \ldots, n-1$, get the fitted value by the above equation; when $k \geq n$, $y(k+1)$ is the prediction value. The solving process is the following:

(1) Construct the accumulated generating sequence. Set original sequence is

$$y^{(0)}(k) = \{y^{(0)}(1), y^{(0)}(2), \ldots, y^{(0)}(k)\}, k = 1, 2, \ldots n . \tag{5}$$

The generating sequence is

$$y^{(1)}(k) = \{y^{(1)}(1), y^{(1)}(2), ..., y^{(1)}(k)\}, k = 1, 2, ...n \,. \tag{6}$$

And, between (4) and (5) the following relationship is satisfied:

$$y^{(1)}(k) = \sum_{k=1}^{n} y^{(0)}(k) \,. \tag{7}$$

According to the above method to accumulate these data, then, obtain the generation sequence (n expresses the sample space) [7-9].

(2) Apply the least square method to solve the parameters a, μ. By (2) and (6), we can get the vectors $\hat{a}$, $\hat{\mu}$ then, obtain the parameter response vector $\hat{y}(k+1)$ by (3).

2.3 Error Checking

There are three methods about the error checking of GM(l, l) such as residual size checking, relational coefficient checking, posterior variance checking[5]. The checking of residual size is to use the method of point to test. The checking of relational coefficient is to test the approximation between the model constructed and the function appointed. The checking of posterior variance is to test the statistical characteristics of residual distribution.

3 Grey Predictions about Brain Drain in Tourism Industry

According to the static from Zhejiang Provincial Tourism Bureau, till the end of 2010, there were 109 tourism colleges in Zhejiang Province with 53902 undergraduates, among which there were 43 vocational tourism colleges (including those with tourism department or tourism major) with 23071 undergraduates and 66 vocational schools (including those with tourism department or tourism major) with 30831 undergraduates.

The Table 1 show that the distribution of tourism human resources of main city of Zhejiang province.

Table 1. The data of tourism human resources from 2005 to 2010

City	2005	2006	2007	2008	2009	2010
Hangzhou	7812	9246	9814	11071	12685	13508
Ningbo	2451	3210	3492	4105	4832	5435
Wenzhou	1294	1630	1840	2080	2431	2706
Jiaxing	440	561	696	852	1087	1257
Shaoxing	728	1076	1127	1223	1332	1397
Jinhua	751	942	1106	1340	1604	1736
Quzhou	457	603	653	728	792	845
Zhoushan	1079	1310	1424	1629	1865	1987
Taizhou	549	762	898	1078	1261	1342
Lishui	330	419	454	520	653	690
Total	16098	20271	23186	25332	29430	31948

3.1 Program Code Prediction and Analysis in 2011

In order to introduce the modeling process of grey model carefully, next to apply the data of the Table 1 to construct the grey model such as the following:

Step 1 : Original data is non-negative, so we directly construct the model.

Step 2 : From Table 1 we know that it includes the statistical data of six years, which is in accord with $y^{(0)}(1) \dots y^{(0)}(6)$. By Table 1 we may get $\sigma^{(0)}(k)$.

According to Theorem 1, we know that the value of $\sigma^{(0)}(k)$ is valid, so the data of Table 1 is fit to the demand of constructing the grey model.

Step 3 : From Step 2 we know that the data of Table 1satisfies the condition of constructing the grey model. We accumulate $Y^{(0)}$ of Table 1 and then get $Y^{(1)}$ of $Y^{(1)} = (y^{(1)}(1), y^{(1)}(2), \dots, y^{(1)}(n))$

Step 4 : From Step 3 and Theorem 2 we may get $Z^{(1)}$ such as the following. $Z^{(1)} = (z^{(1)}(1), z^{(1)}(2), \dots, z^{(1)}(n))$

Step 5 : According to Theorem 2, (3) and (4) we may obtain the prediction data of main cities of Zhejiang province in 2011 and developing coefficient a in 2011such as the following.

$$y^{(1)}(6) = (y^{(1)}(0) - \frac{\mu}{a})e^{-5a} + \frac{\mu}{a} \tag{8}$$

Computing and get the result of prediction such as the Prediction data in 2011.

Program listings or program commands in the text are normally set in typewriter font, e.g., CMTT10 or Courier.

3.2 Prediction and Analysis in 2012

Similar to the method to predict the data in 2011, we may obtain its result such as the following figures.

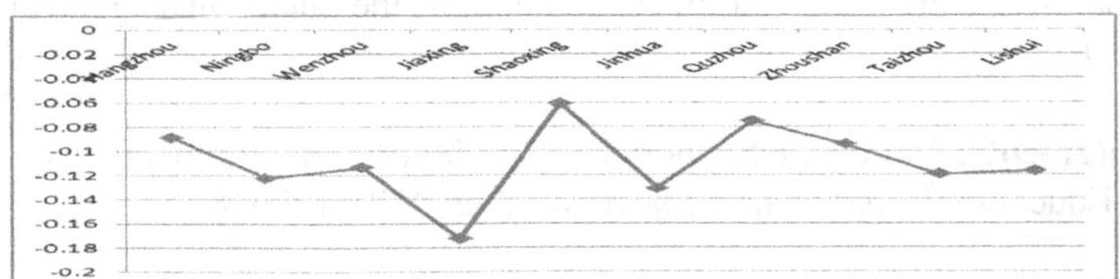

Fig. 1. Developing coefficient a in 2012

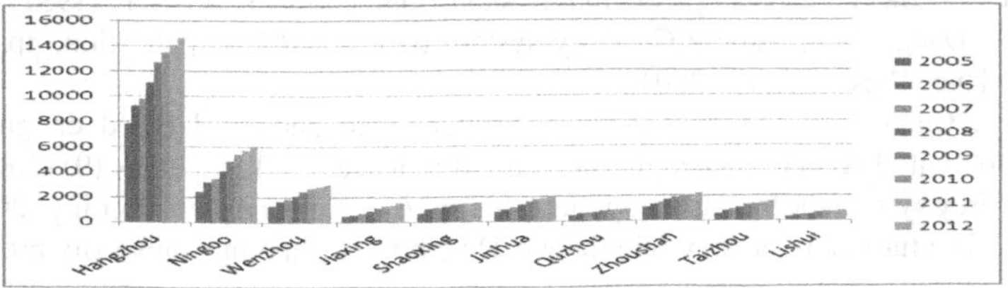

Fig. 2. Prediction data in 2012

4 Conclusions

Talents are the most active factor of the productive force and the most precious resource. The work about talent is the strategic and foundational one to the development. From the above data we know that in order to the rapid comprehensive coordinated development a powerful organization guarantee has to be provided by positive innovation talents working mechanism, continuously improving quality talent team and giving full play to the role of all kinds talent.

4.1 The Current Situation of Tour Human Resource

The news from Zhejiang Tourism Bureau, in 2010, there are 258276 person-times in Zhejiang tourism industry employees, including the job training of 238128 person-times, where these trained personnel are from the tourist hotel, travel agencies, travel and tourism administrative department. The hotel and tourism account for 77% and 17% respectively. Including adult education 20148 person-times, secondary education 12424 person-times and higher education 7724 person-times, they are from the tourist hotel, travel agencies, travel and tourism administrative department . Zhejiang province has organized all kinds of examination the 16 times such as tour guide examination, hotel/travel agency manager post training examination and outbound manager position. There are more 15000 persons to attend these tests.

4.2 Countermeasure and Measure

The work on the talent is a system engineer touched upon all aspects of society. Pay special attention to do the construction of the talent team, improve the talents working mechanism, innovate continuously the way which the Party supervises the talents and improve the work of the talents to a new level. Firstly, improve positively introduction of talents and perfecting the policy measures to attract talent. Secondly, strengthen the training of personnel. Talent growth rules tell us, the link of cultivating is very important. Thirdly, perfect the mechanism of the talents management to play a role of talent.

Acknowledgments. A Project Supported by Scientific Research Fund of Zhejiang Provincial Education Department. Project Number: Y201122618

References

1. Yi, D.: Grey model and talent prediction. System Engineering 5(1), 36–43 (1987)
2. Liu, S.F., Dang, Y.G., Fang, Z.G.: Grey system theory and Its Application, pp. 126–128. China Science Press, Beijing (2004)
3. Rong, X., Chao, A.: Prediction demand for technical personnel based on grey system theory. Journal of Anhui Engineering Institute of Science and Technology (9) (2006)
4. Liu, S.: Grey system theory and its application, pp. 6–9. Science press, Beijing (2004)
5. Deng, J.: The tutorial of grey system theory. Huazhong technology university press, Wuhan (1990)

6. Deng, S., Wang, M.: Researches on a Class of Reaction-diffusion Thermo-plaMaterial Equations. Advanced Materials Research 219-220, 1022–1025 (2011)
7. Wang, M., Deng, S.: Research to E-commerce customers losing predict based on rough set. Applied Mechanics and Materials 58-60, 164–170 (2011)
8. Deng, S., Wang, Z.: Researches on a class of thermo-viscous-elastic anisotropic dissipative material system equation. Advanced Materials Research 219-220, 126–129 (2011)
9. Deng, S., Yang, M.: Analysis and Simulation to A Class of Parabolic-Elliptic System. In: ICIC 2010, June 4-6, vol. I, pp. 274–278 (2010)

The Research of Information Search System Based on Formal Concept Analysis Theory

GuoFang Kuang* and HongSheng Xu

College of Information Technology, Luoyang Normal University, Luoyang, 471022, China
xhs_ls@sina.com

Abstract. As a branch of applied mathematics, FCA (formal concept analysis) comes of the understanding of concept in philosophical domain. It is to describe the concept in formalization of symbol from extent and intent, and then realize the semantic information which can be understood by computer. This paper describes current situations of search, analyzes the features and shortcomings of existing information search systems. The paper presents the development of information search system based on formal concept analysis theory. The method provides query expansion function for the system.

Keywords: Formal concept analysis, concept lattice, information search.

1 Introduction

With the Internet the information platform in people's daily life and work is playing an increasingly important role, as the Internet search engine, information retrieval tools become more and more usage of the system. However, with the explosive development of Internet information, ordinary Internet users want to find the essence of accurate information just like needle in a haystack. In order to satisfy the needs of users, the need for traditional search engine technology to compare a big improvement [1]. Therefore, the study of a new generation of search engines with a lot of new search technology, a number of other related areas of research have also been introduced into retrieval techniques which improve. Search engine to a certain strategy on the Internet to gather, find information, understand the information, extract, organize, and process and provide users with access services, to serve as an information navigation purposes.

Currently the majority of search engines on the Internet are using keyword-based query techniques; the typical representative of Google and Baidu, the content can cover the vast majority of content on the Internet. Keyword-based search engines generally search engine, indexer, crawler and user interfaces of four parts.

Semantic correlation reflects the correlation between text or word level. Semantic relevance calculation can be applied to information retrieval, syntactic disambiguation, and text classification, text clustering and other fields. In information retrieval, semantic relatedness is more text or to reflect the user's query in the sense of line with the degree of inter-related.

---

* Author Introduce: Kuang GuoFang (1974-), Male, lecturer, Master, College of Information Technology, Luoyang Normal University, Research area: concept lattice, information search.

D. Jin and S. Lin (Eds.): Advances in CSIE, Vol. 1, AISC 168, pp. 399–404.
springerlink.com © Springer-Verlag Berlin Heidelberg 2012

Corpus-based approach is more objective, comprehensive reflection of the words in the syntax, semantics, pragmatics, etc. of the similarities and differences. However, this method is used depends on the training corpus, compute-intensive, complex calculations, in addition, data sparse and data noise by the interference of large, sometimes obvious errors. In this paper, first, describe and demonstrate dentally which used the FCA technology. Calculated by the method of semantic relevance of research, this paper proposes a concept lattice-based semantic relevance calculation. This method is out of the traditional methods of mindset, to a certain extent, break the limitations of traditional methods and can provide a system for search engines semantic processing support, to support the user's query expansion.

2 Information Search Terms of Semantic Relatedness

For Search Engine web page information stored in the database or text message, we can think of it by one or more background composition. So, we can put the knowledge database organized into one or more formal context. Through the application of concept lattice construction algorithm, we can get the knowledge of the corresponding concept lattice. As a result, we can use to generate the concept lattice of information retrieval, clustering and semantically related mentioned in this calculation.

From computer science, especially from the standpoint of artificial intelligence, natural language understanding task is to create a computer model, this computer model can give people so as to understand, analyze and answer natural language. Natural language understanding is a technology-based next-generation search engine, called intelligent search engine. Because it information retrieval based on keywords from the current level to knowledge-based (or concept) level, there is a certain understanding of knowledge and processing capability, enabling segmentation techniques, synonyms technology, concept searches, phrase recognition and machine translation technology [2]. Therefore, this search engine with intelligent information services, user-friendly feature, allowing users using natural language information retrieval, really make the search easier and more precise.

For two Chinese words $W_1$and W_2, if W1 has n meaning items (concepts): S_{11}, S_{12}, …, S_{1n}, W_2 of m-defined items (concept): S_{21}, S_{22}, …, S_{2m}, this article provides, $W_1$and $W_2$is equal to the similarity of the similarity of the maximum value of each concept, that is equation 1:

$$Sim(W_1, W_2) = \max_{i=1...n, j=1...m} Sim(S_{1i}, S_{2j}) \tag{1}$$

In this way, put the issue of similarity between two words attributed to the similarity between two concepts the problem. Of course, this consideration is the similarity of two words in isolation. If you are in a certain context, among the two words, it is best to first make meaning disambiguation, the concept of the word mark and then calculate the similarity of concepts. Therefore, the calculation of similarity to the word, we must first compute the similarity of each meaning of the original. Suppose the original meaning of this two-level system in the path distance d, you can get the original meaning of these two semantic distances between.

$$Sim\ (p_1, p_2) = \frac{\alpha}{d + \alpha} \tag{2}$$

$p_1$and $p_2$represent two of which meaning the original (Primitive), d is the original meaning of $p_1$and $p_2$in the hierarchy of the path length, is a positive integer, α is an adjustable parameter. Thus, the two concepts of semantic expressions denoted by the overall similarity: equation 3.

$$R_{u,i}^{strength} = \frac{1}{|F(u,i)|} R_{u,i} = \frac{1}{|F(u,i)|} \sum_{j \in F(u,i)} I(V_i; V_j) \tag{3}$$

Algorithm is implemented: the first item on the item and the mutual information between the evaluation value calculated for each target item i, using equation (3) Ti in the calculation of the user set all the user's reasonable strength, according to from the high to low order, and then select the user a certain proportion of r (the ratio will affect the prediction accuracy and efficiency of operation), as a predictive measure of similarity when selecting the size of the user's nearest neighbors based on the final prediction formula based on the use of 2 calculations to predict.

Based on the face of the lexical semantics of the statistical model description of the semantic similarity between two words $Sim(x, y)$can be calculated in three different words feature space (L_x, R_x, C_x) in the distance to be. The similarity is calculated as follows equation 4.

$$p_{k,m}(x_{a,b}) = x_{a,b} - (\overline{x_a} - \overline{x_k}) - (\overline{x_b} - \overline{x_m}) \tag{4}$$

3 Formal Concept Analysis Theory

The concept of formal concept analysis is a mathematical concept of philosophy that is on people's cognitive knowledge of a mathematical description. In formal concept analysis, the concept of extension of this concept is understood as belonging to the set of all objects, while the content was considered to be all the characteristics common to these objects or set of attributes, which implements the concept of a formal understanding of the philosophy. Between them all together with the generalization of the concept / example of the relationship you form a concept lattice. Mathematics is based on concept lattice theory and lattice theory of order. Concept lattice structure model is the theory of formal concept analysis of the core data structure, which essentially describes the link between objects and features that the concepts of generalization and example of the relationship. Concept lattice corresponding Hasse diagram is to achieve visualization of data.

Mathematics is based on concept lattice theory and lattice theory of order. The first concept lattice is a lattice; it must meet the definition of and relationship between grid computing.

Given formal context $K = (G, M, I)$, if formal context $K_1 = (G_1, M_1, I_1)$and $K_2 = (G_2, M_2, I_2)$meet the$G_1 \subseteq G$, $G_2 \subseteq G$, $M_1 \subseteq M$, $M_2 \subseteq M$, then says $K_1$and $K_2$is the same domain formal context, they are all the son formal contexts of K, also says the concept lattice L (K_1) of formal context $K_1$and the concept lattice L (K_2) of formal K_2 are the same domain concept lattice [3].

Definition 1. Set (A, ≤) is a partial order set, forB ⊆ A,if there is a,b∈A, and for any x ∈ B, all meet x ≤ a, then called a is the upper bound of the subset B. Similarly, if for any x ∈ B, b ≤ x is satisfied, then says b is the lower bound of subset B

Definition 2. Set (A, ≤) is a partial order set, B ⊆ A, a is any one upper bound of B, if all of the upper bound are using a ≤ y, then say a is the minimum upper bound on B (that supremum), recorded sup (B). Also, if b is any one of the lower bound of B, if all of the lower bound z are using z ≤ b, then say b is the biggest lower bound of B (infimum), recorded inf (b).

Below respectively, simply introduce two extension independent domains concept lattice of the vertical combating operation and the two connotation independent domains concept lattice horizontal combating operation. First, the vertical combain operaions of the domain concept lattice have the following definitions.

Definition 3. If (M, ≤) is a partial order set, a, b, c and d are the elements of M and b < c. Then set[b, c] : = {x ∈ M | $b \le x \le c$ } called interval (interval), collection (a] : = {x ∈ M | $x \le a$ } called principal ideal (principal ideal), set [d) : = {x ∈ M | $x \ge d$ } called principal sub filter (principal filter). Besides, $a \prec b \Leftrightarrow a<b$ and $[a, b]=\{a, b\}$.

Definition 4. If L (K_1) and L (K_2) are two extension independent domain concept lattices, define L (K_1) ∪L (K_2) is concept lattice L, if L meets:

(1) one lattice node C_1 of $L (K_1)$ and one lattice node C_2 of $L (K_2)$, make $C_3=C_1+C_2$, if any lattice node C'_1 bigger than C_1, has no C'_1 equals to or less than C_3; Likewise, any lattice node C'_2 more than C_2, are not has C'_2 equals to or less than C_3, then $C_3 \in L$.

(2) for any lattice node C_1 of L (K_1), if there is no lattice node equal to or less than C_1 in $L(K_2)$, then $C_1 \in L$.

(3) for any lattice node C_2 of L (K_2), if there is no lattice node equal to or less than C_2 in $L(K_1)$, then $C_2 \in L$.

Theory 1. Let K = (G, M, I) is a formal context,)≤B (K) = ((G, M, I), is the formal context of concept lattice of K, then B (K) is a complete lattice, for (G, M, I) of any non-empty set, the least upper bound sup (B (K)) and the greatest lower bound inf (B (K)) were equation5 and equation6.

$$\bigvee_{i \in I}(X_i, g(X_i)) = (g(f(\bigcup_{i \in I} X_i)), \bigcap_{i \in I} g(X_i)) \tag{5}$$

$$\bigwedge_{i \in I}(X_i, g(X_i)) = (\bigcap_{i \in I} X_i, f(g(\bigcup_{i \in I} g(X_i)))) \tag{6}$$

Concept lattice can be graphically represented as a labeled graph (labelled line diagram). We call this Hasse diagram. Generated graph as follows: If C1 <C2, C3, and no element grid makes C1 <C3 <C2, then there is an edge from C1 to C2. Line diagram, we use the concept of a black circle indicates the form, use the line-up that

sub-concept - the concept of parent relationship. For an object, if C is a concept that contains the smallest object, the object's name is attached to the corresponding circle on C. For a feature, if C is a concept that contains the characteristics of the largest, the feature's name is attached to the corresponding circle on C. Concept lattice of a label is often used as a communication line graph mode, which makes the concept of a given data structure of the background became clear and easy to understand, enabling visualization of concept lattices show.

Applied in the concept lattice or research, it must first be completed using construction algorithm forms the background to the concept lattice from the construction process. Currently, the concept lattice construction algorithm can be divided into three categories: batch algorithms, incremental construction algorithm and parallel algorithm. Currently, the concept lattice as an effective tool for knowledge analysis, got more and more researchers' attention, and has been in information retrieval, software engineering and other fields of knowledge discovery has been successfully applied.

4 The Research of Information Search System Based on FCA

For Search Engine web page information stored in the database or text message, we can think of it by one or more background composition. So, we can put the knowledge database organized into one or more formal context. Through the application of concept lattice construction algorithm, we can get the knowledge of the corresponding concept lattice. As a result, we can use to generate the concept lattice of information retrieval, clustering and semantically related mentioned in this calculation.

We select the user's query words as a collection of objects. The selection of objects, where the search engine borrowed from the FCA system attributes extraction module part of the results. The module will query words from the text information out of a property, we have adopted certain rules extracted from its properties in terms of words only as words - words form the background characteristics of the object set.

Selected words from the text, it needs Chinese word segmentation. Chinese and English are different between English words and words separated by spaces, while the Chinese is the word as a unit, the sentence in order to link all the words to describe a meaning. The sequence of Chinese characters cut into meaningful words, is the Chinese word segmentation, sometimes also known as Chinese segmentation. For the search engine system, the most important is not to find all the results, because the billions of web pages to find all the results do not have much meaning, no one can view all of the results.

Located in the original concept lattice inf(L) elements C1, C1 and add special handling properties of the object x * set f (x *) the intersection of computing f(x*)∩Intent(C1); then the lattice L the content of the remaining elements of the set content from small to large number of elements to sort, followed by removing elements and C_j, for computing Intent(C_j)∩f(x*), to determine the appropriate type C_j, for appropriate action. Sorting and use of database error handling, reducing the number of search and determine the process to improve the performance of the algorithm, thus forming the information search result as shown in figure 1.

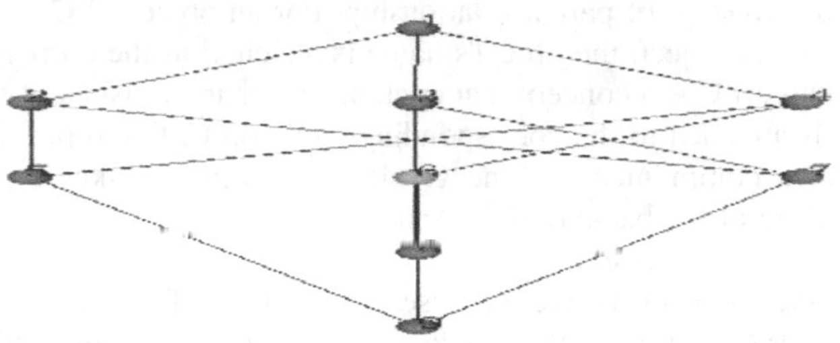

Fig. 1. The result of information search system based on FCA

FCA information search system to support keyword search, based on increased understanding of the semantics of the query keywords and the search results clustering. This system is our search engine development and application of a FCA to explore and study. This paper to address the core problem is the establishment of concept lattice-based semantic relevance calculation model and the use of the model support system for keyword search engine FCA semantic understanding, which ultimately is the system with the expansion of the query keyword query capabilities.

5 Summary

The purpose of this paper is to build an information search system based on FCA model, the theory through the introduction of formal concept analysis, clustering of search results increases as well as the semantics of the query processing power of words to make up some of the existing search engine system insufficient.

References

1. Ganter, B., Wille, R.: Formal Concept Analysis: Mathematical Foundations. Springer, Berlin (1999)
2. Godin, R., Missaoui, R., Alaoui, H.: Incremental concept formation algorithms based on Galois (concept) lattices. Computational Intelligence 11(2), 246–267 (1995)
3. Godin, R., Mili, H., Mineau, G.W., et al.: Design of class hierarchies based on concept (Galois) lattices. Theory and Application of Object Systems 4(2), 117–134 (1998)

Correlation Analysis among International Soybean Futures Markets Based on D-vine Pair-Copula Method

Jianbao Chen, Ting Yang, and Huobiao Zhou

Department of Statistics, School of Economics,
Xiamen University (XMU), Xiamen 361005, P.R. of China
yangting1010@gmail.com

Abstract. This paper uses D-vine pair-Copula, T-Copula and Gaussian-Copula to study the correlations among three soybean futures markets in Tokyo Grain Exchange (TGE), Chicago Board of Trade (CBOT) and Dalian Commodity Exchange (DCE). The obtained results show: (1) the correlation between soybean futures markets in TGE and DCE is the highest no matter which method is used; (2) the correlations obtained with pair-Copula method is lower than the other corresponding correlations obtained by T-Copulas and Gaussian Copula, which indicates D-vine pair-Copula method gives net correlation between two variables by getting rid of the impacts of other variables.

Keywords: Correlation, GARCH model, D-vine pair-Copula function.

1 Introduction

With the development of economic globalization, different economies are connected more and more closely. Especially in financial markets, a shock in one market often induces some fluctuation in other markets. Therefore, it's necessary to understand the relevance among financial markets to reduce investment risks. Usually, Pearson correlation is applied to study the linear relationship between two variables. However, distributions of financial variables are commonly nonlinear and asymmetric, so Pearson correlation is inappropriate to describe the relations among different financial markets. As a new tool to measure the correlations, Copula functions can capture the characteristics of nonlinearity and asymmetry between variables. Moreover, marginal distributions in Copula functions are not restricted. For these reasons, people often use Copula theory other than Pearson correlation to study the correlations among financial markets.

There are many published papers in Copula theory and its empirical study, such as Sklar (1959) [5], Bedford and Cooke (2001) [1], Schirmacher and Schirmacher (2008) [4], Czado et al (2008) [2], Zhang (2002) [7], Ouyang and Wang (2008) [3] and etc. However, most empirical researchers focused on stock markets and paid little attention to futures markets, and they only employ bivariate Copula functions, multivariate Gaussian Copula or t Copula to do empirical analysis. In this paper, we will try to apply D-vine pair-Copula construction method to study the correlation among three soybean futures markets in TGE, CBOT and DCE.

D. Jin and S. Lin (Eds.): Advances in CSIE, Vol. 1, AISC 168, pp. 405–408.
springerlink.com © Springer-Verlag Berlin Heidelberg 2012

2 Methodology

In order to study the correlation among three futures markets, we will use the D-vine pair-Copula construction method, which consists of the following three steps:

Step 1. Estimate the marginal distribution functions with appropriate model.

Step 2. Construct D-vine pair-Copula decomposition and estimate Copula functions and correlations. Specifically, the joint density function of an n-dimensional random variable $X = (X_1, \cdots, X_n)$ can be decomposed as

$$f(x_1, x_2, ..., x_n) = \prod_{k=1}^{n} f(x_k) \prod_{j=1}^{n-1} \prod_{i=1}^{n-j} c_{i,i+j|i+1,...,i+j+1}\left(F\left(x_i \mid x_{i+1}, ..., x_{i+j+1}\right), F\left(x_{i+j} \mid x_{i+1}, ..., x_{i+j-1}\right)\right)$$

where $f(x_k)$ is the marginal density function; $c_{i,j|k}(\cdot,\cdot)$ is the pair-Copula density function; $F(\cdot \mid \cdot)$ is the conditional distribution function. MLE method can be used to estimate the unknown parameters in above equation.

Step 3. Test the goodness of fitting for the pair-Copula decomposition through probability integration transformation (PIT). The PIT of X is

$$T(X) = (T_1(X_1), \cdots, T_n(X_n)) \triangleq (Z_1, Z_2, \cdots, Z_n)$$

where $Z_i = F_{i|1,...,i-1}(x_i \mid x_1, ..., x_{i-1}), i = 1, 2, \cdots, n,$ are iid variables with uniform distribution on $[0,1]$.

3 Empirical Analysis

In this paper, the selected samples consist of the daily return rates of soybean futures in TGE, CBOT and DCE over the period of April 01, 2003 to September 18, 2009. The return rates $r_{i,t} = 100 \ln(P_{i,t}/P_{i,t-1}), i = 1, 2, 3, t = 1, \cdots, 1451,$ where $P_{i,t}$ is the closed price of soybean futures on t-th day in i-th futures market.

According to the steps given in section 2, D-vine pair-Copula construction is employed to explore the correlation among the three futures markets.

1. Estimate the marginal distribution functions. The autocorrelation function (ACF) and partial autocorrelation function (PACF) tests show that there exists autocorrelation in series $\{r_{1,t}\}$ and $\{r_{3,t}\}$ but no autocorrelation in series $\{r_{2,t}\}$. The ACF tests of squared return rates in three markets indicate GARCH effect. According to the suggestion of Wei and Zhang (2008) [6], we apply AR(R)-GARCH(1,1) model to fit the marginal distribution function of each return rates series, where R=1 for TGE and DCE and R=0 for CBOT. The AR(R)-GARCH(1,1) model can be mathematically expressed as

$$r_{i,t} = \mu_{i,t} + \sum_{k=1}^{R} \lambda_{i,t} r_{i,t-k} + \varepsilon_{i,t}, \varepsilon_{i,t} = h_{i,t}^{1/2} \xi_{i,t}$$

$$h_{i,t} = \omega_i + \alpha_i \varepsilon_{i,t-1}^2 + \beta_i h_{i,t-1}$$

Moreover, according to the descriptive statistical analysis of the three return rates series, their distributions are not normally distributed with high peaks and fat tails. Therefore, we can suppose that $\xi_t \sim t(v)$, where v is an unknown parameter. All the estimated results are list in Table 1. We find that all the estimates are significant at 5% significance level.

Table 1. The estimation of marginal distribution functions (with t-values in the parentheses)

	λ	μ	ω	α	β	v
TGE	0.050911	-	0.071861	0.86557	0.11874	15.912
	(1.8209)		(2.6559)	(33.4490)	(4.9620)	(6.3603)
CBOT	-	0.1376	0.085817	0.92428	0.059185	4.7187
		(3.1984)	(2.7634)	(59.3707)	(4.7325)	(8.0558)
DCE	-0.10097	0.048368	0.49098	0.59157	0.40843	2.5113
	(-4.0087)	(2.2232)	(2.6753)	(9.1778)	(2.6215)	(10.9944)

2. Construct pair-Copula and estimate the correlations. The Pearson correlations between soybean futures in TGE and the other two markets are the highest, so fix soybean futures in TGE as the common node in D-vine pair-Copula construction. Make the scatter diagram between the every two residuals obtained in GARCH model in step 1, and compare scatters with the simulation diagrams of known Copula functions, so we take Clayton Copula between soybean futures in TGE and CBOT, Frank Copula between soybean futures in TGE and DCE, and Gaussian Copula between soybean futures in CBOT and DCE under the condition of soybean futures in TGE. Estimates of correlations are obtained through MLE method. To compare with pair-Copula method, the correlations derived with Pearson, Gaussian-Copula and T-Copula method are given in Table 2 as well.

Table 2. The correlations among the soybean futures in three markets with different methods

Method	TGE-CBOT	TGE-DCE	CBOT-DCE
Pearson	0.2706	0.4232	0.2240
Gaussian-Copula	0.2914	0.4178	0.2370
T-Copula	0.4238	0.5615	0.5132
Pair-Copula	0.1734	0.3770	0.1336

3. Test the pair-Copula construction. Three new variables Z_1, Z_2, Z_3 are obtained through the PIT of the residuals in GARCH model in step 1. With PIT, the p-value of S statistics is 0.0124, which indicates that Z_1, Z_2, Z_3 are iid. So pair-Copula construction is appropriate in measuring the correlations studied in this paper.

4 Discussion and Conclusion

Comparing each column in Table 2, we can see that the correlation between soybean futures in TGE and DCE is highest, no matter which method is used. So soybean futures market in Japan affects that in China more strongly than soybean futures market in America does.

Compared with the correlations obtained with other methods, the corresponding correlations obtained with pair-Copula method is the lowest. The possible reason is the correlation obtained by pair-Copula construction method is not affected by other variables, this correlation is the local and net relationship. As we know that the correlations derived through other three methods do not get rid of the effect of the other variables.

In summary, the nonlinearity and asymmetry of financial variables make that Copula theory is more appropriate to study their correlations than Pearson theory. Compared with common multivariate Copula functions (Gaussian Copula and T-Copula), pair-Copula construction uses Copula functions depending on the specific situations, so pair-Copula is more realistic. Moreover, the correlations derived through pair-Copula are the local relationship between two variables and get rid of the effect of other variables.

Acknowledgments. This work was supported by social sciences research grants 08&zd034 09AZD045, 2009b051 and 11JZD019.

References

1. Bedford, T., Cooker, M.: Probability Density Decomposition for Conditionally Dependent Random Variables Modeled by Vine. Annals of Mathematics and Artificial Intelligence 32, 245–268 (2001)
2. Czado, C., Min, A., Baumann, T., Dakovic, R.: Pair-Copula Constructions for Modeling Exchange Rate Dependence, http://www-m4.ma.tum.de/Papers/Czado/czado-min-baumann-dakovic.pdf
3. Ouyang, Z.S., Wang, F.: Modeling Dependence Risk of Treasury with Copulas in China. Statistical Research 25, 82–85 (2008)
4. Schirmacher, D., Schirmacher, E.: Multivariate Dependence Modeling using Pair-Copulas. The Society of Actuaries (January 31, 2008)
5. Sklar, A.: Fonctions de répartition à n dimensions et leurs marges. Publ. Inst. Statist. Univ. Paris (1959)
6. Wei, Y.H., Zhang, S.Y.: Copula Theory and Its Application in Financial Analysis. Tsinghua University Press (2008)
7. Zhang, R.T.: Copula Technique and Financial Risk Analysis. Statistical Research 4, 48–51 (2002)

The Research for the Algorithm of Digital Image Watermark Based on Decomposition of Wavelet Packet

ZhenLei Lv* and HaiZhou Ma**

Information Technology Engineering Yellow River Conservancy Technical Institute, Kaifeng, China
lvzhenlei_82@126.com, kfsea@126.com

Abstract. This paper adopt the Shannon entropy algorithm to search the optimal wavelet packet from bottom to up, and decompose the original image and watermark respectively with wavelet packet, and finally embed the watermark data into the selected basis of wavelet packet of the original image. The experiments against attack show that the watermark created by the algorithm is invisible, and image quality have been well preserved, and strong robustness have been shown at same time.

Keywords: digital watermarking, wavelet packet decomposition, image processing, robustness, algorithm.

1 Introduction

The image digital watermark has been classified into two categories, visible and invisible. Visible watermark is easy to be attacked, so its application is limited. Invisible watermark as a new technology of protection of digital products needs to meet two basic requirements: the invisibility and the robustness. Based on these two requirements, the domestic and foreign researchers have done lots of deep studies and put forward many algorithms. These watermark algorithms can be divided into two main categories: spatial-domain method[1] and frequency-domain (or transform domain) method[2-4].

Spatial-domain watermark is to directly modify the image pixel with various methods. But the frequency-domain watermark is to embed watermark into the images after the transformation of frequency-domain, such as discrete cosine transform (Discrete Cosine Transform, DCT), discrete Fourier transform (DFT), wavelet transform (DWT), etc.[5-7].

The JPEG standard is generally to use DCT transformation to implement the image lossy compression. The JPEG2000 abandoned the encoding format adopted by JPEG which mainly is based on block discrete cosine transform, and adopt multi-analytic coding method based Wavelet Transform. One of the core technologies of the

---

* ZhenLei Lv, male, lecturer, Master Degree, his research interest include Image Processing and artificial information processing.

** HaiZhou Ma, male, lecturer, Master Degree, his research interest include Image Processing and artificial information processing.

D. Jin and S. Lin (Eds.): Advances in CSIE, Vol. 1, AISC 168, pp. 409–414.
springerlink.com © Springer-Verlag Berlin Heidelberg 2012

JPEG2000 is based on wavelet transform algorithm[8]. The Embedded Zerotree Wavelet (EZW)[9-11] was put forward by Shapiro in 1993, is a milestone in the study of wavelet image coding. Its essence is that convert the original image into the wavelet system through the wavelet transform, and then do quantization coding for wavelet coefficients.

This paper presents a algorithm based on wavelet watermark, which respectively does wavelet packet decomposition for the original image and watermark, sufficiently makes use of the characteristics of wavelet packets with the visual perception model of wavelet-domain, and embeds the watermark data into low-frequency sub-band of wavelet domain of the original image, ensures that the watermark are robustness. Experimental results show that the algorithm can embed a large amount of watermark into images and watermark are more invisible and more robust to the compression of JPEG2000.

2 Algorithm of Optimal Wavelet Packet Basis

Because the wavelet packet decomposition have a variety of forms and different wavelet packet decomposition have different effects, therefore, for a given signal, the selection of suitable wavelet packet is very important.

Wavelet packet decomposition is a more sophisticated analysis method. Wavelet packet decomposition can not only do binary partition for scale-space V_i but also do similar things for wavelet space W_i, therefore, wavelet packet decomposition is more flexible. It can select the optimal wavelet packet according to certain criteria (or cost function). The packet made the cost function smallest in all the wavelet packets is known as the optimal wavelet packet. We find the optimal wavelet packet based on the Shannon entropy principle in this paper[12,13].

The Shannon entropy of the sequence x = (xi) is defined as:

$$M(x) = -\sum_i P_i \log_2 P_i \tag{1}$$

Where $P_i = \frac{x_i^2}{\|x\|^2}$, and $P_i \log P_i = 0$ when $P_i = 0$.

The algorithm of selection the optimal packet generally is a bottom-up searching method, which seeks the optimal packet based on Shannon entropy through selecting a panel of wavelet packet, and make M (x) smallest.

3 Watermark Algorithm Based on Decomposition of Wavelet Packet

In this paper the algorithm intent to embed the size of 32 × 32 image, that is to say, two-dimension watermark into images, which has the advantages that it can carry more amount of information and be more visible than the one-dimension watermark.

A. Decomposition of Wavelet Packet and Embedding Watermark

There are similar between the process of human's recognition to image and the one of wavelet transform. Orthogonal wavelet decomposition is to decompose the image into a simple multi-level framework, i.e. the multi-resolution representation of the image. Each component of the framework has the features of different frequency characteristics and different spatial orientation, which become a basic foundation to further analysis and process images.

Decompose the original image and the watermark based on the optimal wavelet packet basis with two-ordered wavelet. In the mode of wavelet decomposition, the low-frequency components has been further decomposed, while the wavelet packet transform is to divide frequency band into multi level and the high-frequency part which hasn't been divided by the multi-resolution would be further decomposed, that lead to a increase of time-frequency resolution, and improvement of insensible to watermarked image. According to our experiments, as compared with the orthogonal wavelet transform, wavelet packet transform has a better ability to image reconstruction.

Set f (i, j), f '(i, j) represent the pixels of the original image and ones of the watermarked image; w is value of the amplitude of watermark, then embedding watermark can be expressed as:

$$f'(i,j) = bf(i,j) + aw \tag{2}$$

Where, a is amplitude factor, the choice of its value need to consider the nature of images and visual systems (Human Visual System, HVS) characteristics. b is the modulation coefficient, which is aimed to improve the feature of insensible of watermark. After embedding watermark, the sub-band coefficients of wavelet are inversely transformed to get watermarked image.

B. Extraction and Detection of Watermark

Extraction of watermark is the reverse process of embedding the watermark. First, the image with watermark to be decomposed based on the basis of optimal wavelet packet, and the information of watermark is extracted by the data of orthogonal wavelet decomposition at same level of the original image. The corresponding watermark signal decoding formula is:

$$w(i,j) = [f'(i,j) - bf(i,j)]/a \tag{3}$$

The watermarked image can be constructed again from the extracted information of watermark based on the basis of optimal wavelet packet.

Because the watermark information of the image is two dimensions, which can be viewed and be determined whether it exist intuitively. But in other situation a computer is needed to help us automatically detect the watermark. Here, the method that calculating the correlation coefficient between the extracted watermark image and the original watermark image has been adopted.

Set the size of original watermark image as K × L, then the correlation coefficient between the extracted watermark and the original watermark can be expressed as [14]:

$$Cov(F,F') = \sum_{i=0}^{K-1}\sum_{j=0}^{L-1} w(i,j)\cdot w'(i,j) / (\sum_{i=0}^{K-1}\sum_{j=0}^{L-1}(w'(i,j))^2)^{1/2} \tag{4}$$

Where, w(i, j), w'(i, j), are pixels of the original watermark image and extracted watermark image respectively.

Set the threshold T: if Cov (F, F ')> T, then the decision that a watermark has been embedded into the tested image can be made, otherwise, the tested image has no watermark. The choice of the threshold T should also take the both probability of false alarm and missed alarm into account. Because if T decreases, the missed alarm probability decreases too and the false alarm probability increases conversely; while when T increases, the false alarm probability decreases and the missed alarm probability also increases.

4 Experiment and Result Discussion

The experiment uses grayscale images with the size of 512 × 512 pixels, Lena; and the watermark is 32 × 32 grayscale images, Camera. Do two order wavelet decomposition to the original image Lena and the watermark image Camera, and according to formula (2) embed watermark information into the sub-belt of the wavelet decomposition of Lena. Amplitude factor a and modulation coefficient b are usually determined based on experience. In this paper, the experiment set amplitude factor a as 0.1, and the modulation coefficient b as 0.9. Then obtain watermarked image by wavelet reconstruction to the watermark embedded data. The original image of Lena and the watermarked image of Lena is shown in Figure 1 and Figure 2. Figure 3 is relevant results of the experiment.

Fig. 1. Original Lena image

Fig. 2. Watermarked Lena image

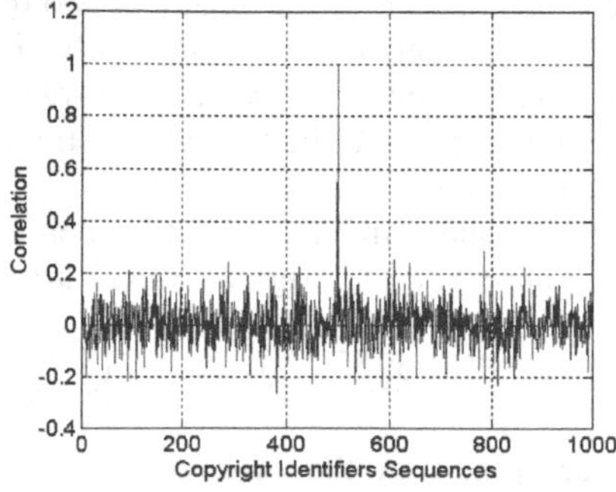

Fig. 3. Experimental result of relevant detection

In order to evaluate the robustness of the watermark algorithm, we simulate some attacks against the Lena image with watermark, then extract information of watermark image which is used to reconstruct the watermark from the attacked Lena image. The experimental results are shown in figure 4. The figure 4 (b) and (c) is the extracted watermark from the image with JPEG lossy compression. Where, figure 4 (b) come from the JPEG compression image with quality of 75% and compression ratio is 5.909; Figure 4 (c) come from the JPEG compression image with quality of 50% and compression ratio is 8.125. Figure 4 (d) and (e) are the extracted watermark from the image which went through the process of Gaussian blur and sharpening. Figure 4 (f) is the extracted watermark from the image with the noise intensity of 0.05.

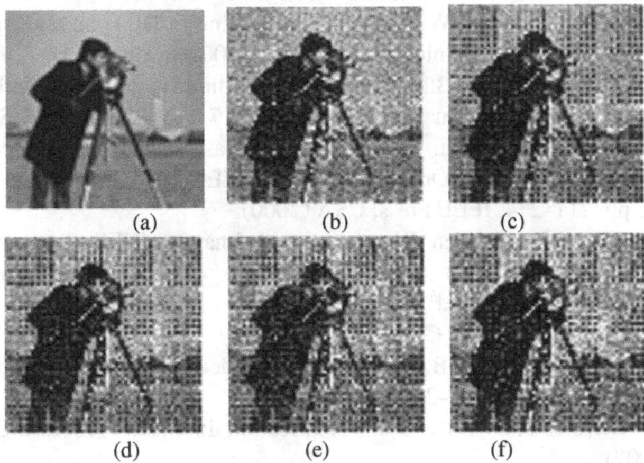

Fig. 4. Extracted watermark from attacked Lena image.

Experiments show that among the range of wavelet this algorithm not only can embed abundant information of watermark into image, but also effectively guarantee the watermark is insensible.

As usually, the visual importance of coefficients of low-frequency sub-band images is more significant than the one of high-frequency sub-band images. So the algorithm in the paper chose to embed information of watermark into the coefficients of low-frequency sub-band image. And thus, if the watermark was destructed, the image with it would inevitably lead to serious distortion. So the robustness of the watermark has been enhanced. Experiments show that the algorithm is robust for some attacks against watermark, especially for JPEG compression.

5 Conclusion

This paper presents a watermark algorithm based on decomposition of wavelet packet with two-dimension watermark information. The algorithm search the optimal wavelet packet basis based on the principle of Shannon entropy through wavelet packet transformation to both the original image and the watermark image respectively. Experimental results show that the algorithm can embed into images large amount of information of watermark with better insensible and more robust against JPEG compression.

References

1. Nikolaidis, N., Pitas, I.: Robust image watermarking in the spatial domain. Signal Processing 66(3), 385–403 (1998)
2. Podilchum, C.I., Zeng, W.: Image-adaptive watermarking using visual models. IEEE Journal on Selected Areas in Communications 16(4), 525–539 (1998)
3. Cox, J., Kilian, J., Leighton, F.: Secure spread spectrum watermarking for multimedia. IEEE Transactions on Image Processing 6(12), 1673–1687 (1997)
4. Huang, J., Shi, Y.Q., Cheng, W.: Image watermarking in DCT: an embedded strategy and algorithm. Acta Electronica Sinica 28(4), 57–60 (2000) (in Chinese)
5. Eggers, I.J., Su, J.K., Girod, B.: Robustness of a Blind Image Watermarking Scheme. In: International Conference on Image Processing, pp. 17–20. IEEE Press, USA (2000)
6. Piva, A., Barni, M., Barrolini, F., et al.: DCT-Based Watermark Recovering Without Resorting to the Uncorrupted Original Image. In: IEEE International Conference on Image Processing, pp. 211–214. IEEE Press, USA (2000)
7. Hsu, C.T., Wu, J.L.: Hidden Watermarks in Images. IEEE Transactions on Image Processing 8(1), 58–68 (1999)
8. Li, J.: Wavelet Analysis and Signal Processing - Theory, Application and Software Realization. Chongqing Press, Chongqing (1997)
9. Vidakovic, B., Johnstone, C.B.: On Time Dependent Wavelet Denoising. IEEE Trans. Signal Processing 46(9), 2549–2551 (1998)
10. Chen, W.: Wavelet Analysis and Its Application in image processing. Science Press, Beijing (2002)
11. Zhang, Y.: Image Engineering (first volume) image processing. Qinghua University Press, Beijing (2006)
12. Peng, Y.: Wavelet Transform and Engineering Application. Science Press, Beijing (1999)
13. Mallat, S.: A theory for multiresolution signal decomposition: the wavelet representation. IEEE Transactions on PAMI 11(7), 674–693 (1989)
14. Cox, I.J., Killian, J., Leighton, F.T., et al.: Secure Spread Spectrum Watermarking for Multimedia. IEEE Trans. Image Processing 6(12), 1673–1687 (1997)

A Technology and Implementation of Embedding Watermark to Digital Pictures Dynamically

HaiZhou Ma* and ZhenLei Lv**

Information Technology Engineering Yellow River Conservancy Technical Institute, Kaifeng, China

{kfsea,lvzhenlei_82}@126.com

Abstract. This paper introduces the technology that dynamically and automatically embedding watermark into digital pictures in ASP.NET web site without changing the original website code, and meanwhile gives the primary code which implements the function. The technology proposed in this paper is an efficient, cheap and feasible solution.

Keywords: Image digital watermark, Adding dynamically, ASP.NET, HttpHandler.

1 Introduction

The image digital watermark has been classified into two categories, visible and invisible. As we all know, the visible image watermark has a lot of advantages, such as simple implementation, lower cost, fast deployment, but it is vulnerable against attack and its ability to protect the work is limited. However, as a compromise, it is widely applied in such fields that less demand for copyright. And now most images in website use the technology. Figure 1 shows a website the picture is added watermark by the technology.

Fig. 1. Watermark on the picture

* HaiZhou Ma, male, lecturer, Master Degree, his research interest include Image Processing and artificial information processing.

** ZhenLei Lv, male, lecturer, Master Degree, his research interest include Image Processing and artificial information processing.

D. Jin and S. Lin (Eds.): Advances in CSIE, Vol. 1, AISC 168, pp. 415–420.
springerlink.com
© Springer-Verlag Berlin Heidelberg 2012

There are three methods to implement image digital watermark:

1) Edit the pictures manually and add watermark. We edit each original image and embed watermark into it by using photo-editing tools. This method is laborious and time-consuming, and only fit for handling little images.

2) Edit the pictures and embed watermark automatically by a program. Programs embed a digital watermark into each picture automatically. However, this method has a drawback that the watermark will overwrite the corresponding part of the original images, and those would be corrupted.

3) Embed digital watermark into the pictures dynamically when it need to display. The pictures were embed digital watermark temporarily before they were send to the client, then the pictures which were embedded watermark were send to the client. This approach is a good way to make up deficiencies in the previous two methods.

The paper is intend to illustrate the third method implemented by ASP.NET technology.

2 HTTP Request Procedure

We firstly look at the process of HTTP request before introducing the implement of embedding the watermark to image digital dynamically.

The procedure to handle ASP.NET request is based on PipleLines model which consists of many HttpModules and one HttpHandler. HTTP requests has been passed by ASP.NET to each HttpModule in the pipeline one by one, and finally been tackled by HttpHandler. After finished it, ASP.NET returns the results to the client through the pipeline again. We can intervene with the procedure in every Http module. The workflow of HTTP request in HttpHandler and HttpModule is shown in Figure 2.

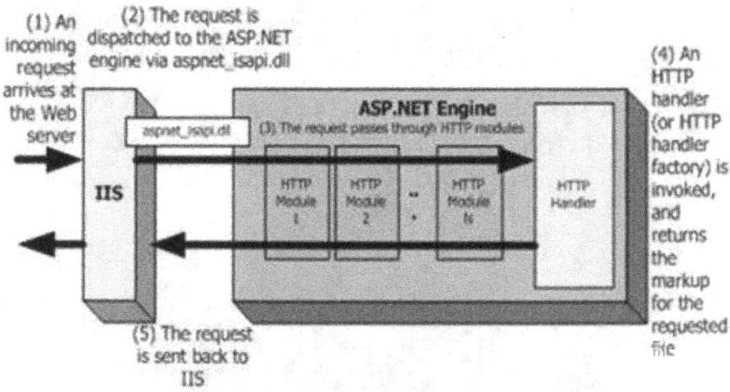

Fig. 2. The workflow of HTTP request in HttpHandler and HttpModule

As can be seen from Figure 2, in the HTTP request processing, the request can pass through (call) several HttpModule, but in the end just only calls a HttpHandler.

When a request reaches HttpModule, ASP.NET engine doesn't really deal with this request, means that, in terms of the HTTP request, HttpModule is the only way between the client sending a request and the client receiving the response from the server. So that we can add some other additional information to the request before it reach to the processing center (HttpHandler), or intercept the request and do some extra work, or terminate the request.

The HTTP request processing center is HttpHandler, in which the server page client requested is compiled and executed. After the HttpHandler finishes the request, the processed information is attached to the HTTP request information flow and returns back to the HttpModule. Thus we can reprocess the results of request processing in the corresponding HttpModule. HttpModule will continue to deliver the processed HTTP request information to every level until the information return to the client.

How does IHttpHandler deal with HTTP requests? When a HTTP request is delivered to HttpHandler container through HttpModule container, ASP.NET Framework can call the HttpHandler's ProcessRequest member method to deal with this HTTP request. For example with .aspx page, it is here one .aspx page is resolved by systems, and the results processed by the HttpModule is passed along continuously by HttpModule, until it reaches the client.

3 Create HttpHandler Program

To put it simply, HttpHandler handler is a server processing program and the its primary role is to make right response for the client request.

After learning about an HttpHandler, let's look at how to achieve our own HttpHandler. To achieve an HttpHandler, must implement the IHttpHandler interface, and any class which implements this interface can be used to handle the input HTTP request.

IHttpHandler interface has a method ProcessRequest, and one attribute IsReusable.

IsReusable attribute: used to set whether the current IHttpHandler instance can be reused, default returns False.

ProcessRequest Method: complete response to user requests, this method's only parameter is an HttpContext instance, and through this parameter the request, response and other objects in the the current request can be accessed.

In the ASP.NET 2005/2008, you can conveniently create a HttpHandler application. As follow, we implement dynamically add image digital watermark through the design of HttpHandler applications

A. The Idea of Adding Digital Watermark

The process of embedding a digital watermark process is somewhat similar to Photoshop's layer processing:

1) New one canvas which has the same size with the picture you want to add a watermark to.

2) Copy the picture properties, such as the information of resolution, dates and so on.

3) First, paint the layer white.

4) Second paint the original picture on it.

5) Then paint the watermark on it.

6) Save.

B. Automatically Add Watermark to JPG Images on the Site

We hope that the code does not make any changes to the codes in original site, and the displaying JPG images was added the digital watermark automatically. To achieve this way, only by capturing the request for the picture to realize, needs the treatment as follows:

1) Modify the configuration information in the file web.config, turn all request for access to .jpg file to HttpHandler processing program.

2) Catch the request, and get the path of images to which the user wants access.

3) Find the corresponding picture according to the request path.

4) Put the digital watermark image on the lower right corner of the picture you want access.

5) Modify output type of the program, and put out the new assembled pictures.

- Modify the configuration file

In web.config file add the following code:

```
<httpHandlers>
        <add verb="*" path="Photo/*.jpg" type="PicHandler"/>
</httpHandlers>
```

the meaning of above various attributes are as follows:

verb: means the predicate, can be written as "GET, POST ","*", etc.," * "as wildcard, here identifies all requests.

path: access path, for example, "Photo / *. jpg" the path means that all the request for accessing to the. jpg format images under this path will be handed over to PicHandler.cs class.

type: specifies class / program set which are separeated by comma. PicHandler specified here is to be programed for HttpHandler.

- Create HttpHandler program

As follow, we implement the HttpHandler to achieve adding watermark in this way. Create a PicHandler.cs class derived from class IHttpHandler by default, the class is added to the App_Code folder. PicHandler.cs code is shown as follow:

```
using System.Drawing;
using System.Drawing.Imaging;
using System.IO;
public class PicHandler : IHttpHandler
{
    private const string WATERMARKADDR = "~/Images/watermark.jpg";
    private const string DEFAULTIMAGE = "~/Images/noperson.jpg";
    public void ProcessRequest(HttpContext context)
{
        System.Drawing.Image image;
        if (File.Exists(context.Request.PhysicalPath))
        {
          image = Image.FromFile(context.Request.PhysicalPath);
            Image watermark =
Image.FromFile(context.Request.MapPath(WATERMARKADDR));
            Graphics g = Graphics.FromImage(image);
```

```
                    g.DrawImage(watermark, new Rectangle(image.Width -
watermark.Width, image.Height - watermark.Height, watermark.Width,
watermark.Height), 0, 0, watermark.Width, watermark.Height, GraphicsUnit.Pixel);
                    g.Dispose();
                    watermark.Dispose();
                }
                else
                {
                    image =
Image.FromFile(context.Request.MapPath(DEFAULTIMAGE));
                }
                context.Response.ContentType = "image/jpeg";
                image.Save(context.Response.OutputStream,
System.Drawing.Imaging.ImageFormat.Jpeg);
                image.Dispose();
                context.Response.End();
            }
            public bool IsReusable
            {
                get
                {
                    return false;
                }
            }
        }
```

4 Create Test Page

To evaluated the technology, we created a web page ShowPhoto.aspx, whose "design" view in Visual Studio is shown in Figure 3. We can see all pictures without any watermark.

Fig. 3. The reality original picture in "design" view

The code in "source" view is shown as follows:

```
<img src="photo/2.JPG"   alt="bb" width="260" height="195" />
```

The figure 1 shows the situation when the web site is running. From the figure we can see all the pictures to been added a watermark.

5 Conclusion

This paper described the technology of dynamically adding watermark in the ASP.NET web site without damaging the original image and modifying the existing website's code, and the whole work we need to implement it is that just embedding a HttpHandler program and a line of code in the web. It is a simple, low-cost, efficient and effective solution of embedding watermark into the pictures, when we reform an old sites or create a new Web site.

References

1. Wang, T., Yang, J., Qian, P.: An Investigation On Low-Level Request-Process Mechanism Of Asp.Net Webforms. Computer Applications and Software 24(10), 120–121, 125 (2007)
2. Gu, Q.: The method of extending ASP.NET Applications with custom Http Handlers. Computer Knowledge and Technology (05), 3–5 (2005)
3. Zhang, Y.: ASP.NET Almighty Quick Fact Collection. Posts&Telecom Press, Beijing (2009)
4. Microsoft Corporation. MSDN Library for Visual Studio

Practice of Bringing Up Mechatronics All-Round Innovation Talents with Measure, Quality and Standardization Characteristics

Xiaolu Li and Bi Sun*

School of Electromechanical Engineering, China Jiliang University, 310018 Hangzhou, China
sunbi1113@163.com

Abstract. Bringing up all-round innovation talents with speciality characteristic is one side of Chinese higher education reform in nowadays. In order to train spirit of innovation and characteristics of measure, quality and standardization for mechatronics undergraduates, our school educates theories and techniques of measure, quality, and standardization for students. Meanwhile, the main aim is bringing up high-caliber personnel with innovation and practice abilities, and the main methods are training innovation ability, practice education, patent application, attending extracurricular practice and setup student associations. The school forms the talents training characteristic of fostering mechatronics high-caliber all-round personnel with soild quality sense, explicit standardization consciousness and high measure ability.

Keywords: Innovation Ability, Mechatronics, Undergraduate, Characteristics of Measure, Quality and Standardization.

1 Introduction

Situated in Hangzhou, the capital city of Zhejiang Province and founded in 1978, China Jiliang University (CJLU), formerly China Institute of Metrology, is the only university qualified to offer bachelor and master degrees in the sectors of Quality Supervision, Inspection and Quarantine in China. School of Electromechanical Engineering was founded in 2000, and has four undergraduate programs as follows: Automation, Electrical Engineering & Automation, Machine Design and Mechatronics Engineering (Automation is state-level key discipline and Zhejiang Province key discipline). It offers two master programs as follows: Detection Technology & Automatic Equipment and Control Theory & Control Engineering. With 98 full-time faculty members, the school enrolls around 2,000 full-time students. It owns a strong teaching and research team. Faculty members with master and doctoral degrees account for 87%. According to the characteristics of CJLU and the social needs of all-round innovation talents, School of Electromechanical Engineering has practiced bringing up mechatronics all-round innovation talents with measure, quality and standardization characteristics.

*Corresponding author.

D. Jin and S. Lin (Eds.): Advances in CSIE, Vol. 1, AISC 168, pp. 421–425.
springerlink.com © Springer-Verlag Berlin Heidelberg 2012

This paper introduces School of Electromechanical Engineering bringing up mechatronics all-round innovation talents with its characteristics.

2 Background

2.1 Social Needs Analysis of Graduates

In order to educate more high-level talents for economic development, China has been promoting the higher education since 1999. The gross enrollment rate rose from 9.8 percent in 1998 to 23 percent in 2008, and it reached 38 percent in 2008 in Zhejiang and Jiangsu provinces, which marks that the transformation from elite education to mass education. Meanwhile, the number of colleges and universities is increasing quickly, there are 742 institutions of higher learning, over 70 percent of them are teaching universities, almost similar type schools have same specialities, like Electrical Engineering & Automation and Machine Design. This raises two questions: one is same speciality in different schools, which have different histories, different technical backgrounds, different academic styles, and different occupational direction, if they should have same teaching program, same teaching material, and same evaluation criteria. The other one is how to fulfill socity and companies for special talents. The first question is a universal existing problem in teaching universities. They have same curricula and same teaching program, so that we shall be educated all alike. This training pattern faces contradiction with varity of talents need. Therefore, same speciality with characteristic education improves and perfects the wide scope education, and this talents training pattern will achieve important results.

According to the educational departments, high education of China paid more attention to mass higher education, quality education, and wide scope education in previous years. In recent years, major characteristics, training innovation talents, and innovation education have been the research objects. It actually expresses the needs of education and talents in the socity. Nowadays, most high schools have taken the initiative in trying to research major characteristic and traing innovation talents.

2.2 Law of Training Talents

Automation, Electrical Engineering & Automation, Machine Design and Mechatronics Engineering are basis majors of industry, which set up by every engineering college in China. Therefore, the students of these majors, who study at CJLU, must have their merits and characteristics in comparison with other colleges. According to academic accumulation and traditional advantage of CJLU, we present the principle of "characteristic and innovation".

2.2.1 Characteristic

The majors of electromechanical engineering fit in with the needs of the society, according to characteristics of CJLU, the speculative knowledge and specialized skill of measure, quality, and standardization are intrduced into traditional education pattern, and forms the talents training characteristic of fostering mechatronics high-caliber all-round personnel with soild quality sense, explicit standardization consciousness and high measure ability.

2.2.2 Innovation

Classroom education combined with scientific and technological activities after class, which enlighten and arouse the potential innovation ability of the students. At last, we accomplish bringing up mechatronics all-round innovation talents with measure, quality and standardization characteristics.

3 Practice of Training Innovation Talents

School of Electromechanical Engineering has trained the student innovation talents, and formed the training pattern of "major characteristic and innovation".

3.1 Pattern of Bringing Up Mechatronics All-Round Innovation Talents with Measure, Quality and Standardization Characteristics

According to the company needs of mechatronics talents with theory and skill of measure, quality, and standardization, its teaching experience has been introduced into the major education through establishing teaching plan, scientific and technological activities, graduation project, and academic forum since 2002. In this way, the mechatronics graduates of CJLU not only have the professional knowledge and skills, but also accept the theory and skills of measure, quality, and standardization. These build up mechatronics all-round innovation talents with its characteristics, and have obvious superiority in entrepreneurship and employment.

The majors are added with characteristic courses as follows: fundamental of metrology, modern quality engineering, standardization generality, interchangebility and base of measuring technology, measurement technique. Meanwhile, every major presents and develops the distinctive personnel-training scheme.

3.2 Training Innovation Ability through Developing Scientific and Technological Activities

In recent years, School of Electromechanical Engineering has developed scientific and technological activities to train innovation ability and reserve innovation skills. The students have achieved more than 200 provincial-level and state-level awards in different academic contests. An "automatic fish killer" received the first prize of China Graduate Machine Design Competition 2008, this work caused great interest and many companies wanted to buy this technology.

3.3 Setting Up Patent Pending Course to Train Creative Thinking

Prof. Jialing Liang in Institute of Electromechanical Engineering, who has the most patents in the Chinese colleges, offers a course of "patent writing and legislation". Under his instruction, the students of School of Electromechanical Engineering have been over 80 patents in 2007 and 2008. In recent years, these students have been the selective preference graduates to companies, and some of them manage firms through their patents.

3.4 Developing the Research of Education and Practice

In recent years, teachers have paid more attentions to the research of innovation education, developed the reformation of education. The college has offered innovation courses to train the students since 2001 as follows: innovation and practice on synthetical training, innovation design of electronics, product innovation design, and so on. Meanwhile, the college invites the professors, postgraduates, and undergraduates to pass on their own experiences.

4 Cultivation Effects

4.1 Theory and Specialist Knowledge

In order to introduce the practice ability and innovative spirit into theory education, School of Electromechanical Engineering has launched a variety of educational reforms and practices. They encourage students to take part in teachers' research activities, develop experimental class education, work out diverse training plans to different students, integrate of undergraduate and postgraduate education, and so on. All these contribute students to consolidate the specialist knowledge and strengthen the ability of analysing and solving problems.

4.2 Practice Ability and Innovative Spirit

The school of eletromechanical engineering exerts specialized subject advantages to promote the special environment and system that help the students develop personality and make the creative talents grow, foster the talents who have high quality and creative ability [1]. In recent years, the students have achieved more than 10 state-level awards in different academic contests and 50 patents.

4.3 Employment and Entrepreneurship

The graduates are well adopted by the society because of solid basis, distinct characteristics, and high practical ability. Moreover, the employment rate of School of Electromechanical Engineering is always the first one in CJLU.

4.4 Social Evalution

On February 26, 2008, Guangming Daily carried an article titled CJLU encourages students to invent and create [2]. The news reported the attempts and achievement in training innovation talents through encouraging student to apply for patents. Meanwhile, an opinion piece on the front page titled training curiosity and encouraging creativity commented on the significance of training the innovation ability.

5 Conclusion

In order to train spirit of innovation and characteristics of measure, quality and standardization for mechatronics undergraduates, School of Electromechanical Engineering educates theories and techniques of measure, quality, and standardization for undergraduates. Meanwhile, the main aim is bringing up high-caliber personnel with innovation and practice abilities, and the main methods are training innovation ability, practice education, patent application, attending extracurricular practice and setup of student associations. The school forms the talents training characteristic of fostering mechatronics high-caliber all-round personnel with solid quality sense, explicit standardization consciousness and high measure ability.

Acknowledgments. This research is supported by 2011 Education Sciences Research of Zhejiang Province (Research of Training the Innovation Ability of Mechatronics Undergraduates Based on TRIZ Theory, No.SCG130) and 2009 Innovative Experimental Zone for Training Mode of Talents in China (Training experimental zone for Mechatronics creative talents with characteristics of measurement, quality and standardization, No.38).

References

1. Li, X., Li, Y., Gu, L.: Training Innovation Ability for Mechatronics Undergraduates Based on TRIZ Theory. In: Lin, S., Huang, X. (eds.) CSEE 2011, Part III. CCIS, vol. 216, pp. 402–406. Springer, Heidelberg (2011)
2. China Jiliang University Encourages Students to Invent and Create. Guangming Daily (February 26, 2008)

Research on Frequency Assignment in Battlefield Based on Adaptive Genetic Algorithm

Libiao Tong, Hongjun Zhang, Wenjun Lu, and Jie Zhou

New Star Research Institute of Applied Technology, Hefei 230031, China
libtong@sohu.com, {nuptzhj,lwj2963}@sina.com,
962854399@qq.com

Abstract. Based on analyze the actuality of frequency assignment when the communication radio works in training and dry run , a method of frequency assignment based on Genetic Algorithm is proposed , and the Genetic Algorithm is also improved. The simulation validates that the method of frequency assignment is real-time and high-efficient.

Keywords: Genetic Algorithm, Communication radio, Frequency assignment.

1 Introduction

Nowadays the method of frequency assignment of our military communications radios is still the traditional method such as artificial sub-division method、group-classification method and cross-classification method etc. Although these method can solve the problem of electromagnetic compatibility to some extent when many radios work together on the battlefield, these traditional methods can't meet the needs on real-time, high-efficiency and stability, and can't adapt to the high-tech war and the increasing complexity battlefield electromagnetic environment in the face of increasing levels of information. We must plan the frequency of battlefield communications equipments through scientific, orderly and high-efficient method of frequency assignment. So this paper proposes a widely-used optimal genetic algorithm to solve the problem of frequency assignment of our military communications radio.

2 To Establish the Mathematical Model of Frequency Assignment

In this paper we study the problem of frequency assignment of N communication radios working in the same area. We assume there are N communication radios working in the same area, and the available band of the i-th radio is (f_L^i , f_H^i), of which f_L^i is starting and f_H^i is termination frequency. If the frequency interval of the radio is Δf in this band, the selectable frequency of the i-th radio is (f_L^i , $f_L^i+\Delta f$, $f_L^i+2\Delta f$,..., f_H^i). So we can get a combination of all radios

D. Jin and S. Lin (Eds.): Advances in CSIE, Vol. 1, AISC 168, pp. 427–432.
springerlink.com © Springer-Verlag Berlin Heidelberg 2012

frequency (f^1 , f^2 ,..., f^i ,..., f^N), of which $f^i \in \{f_L^i, f_L^i + \Delta f, f_L^i + 2\Delta f, ..., f_H^i\}$ and f^i is the frequency assigned to the i-th radio. The purpose of the frequency assignment is to find such a frequency combination, and these radios can work compatibly with the assigned frequency.

Therefore, after we analyze the electromagnetic compatibility of the radios working simultaneously, and set constraints to avoid the occurrence of interference, and count the number of violating the constraint, we establish the following mathematical model:

(1) The judgment of co-channel interference between the i-th and j-th radio

$$A(f_i, f_j) = \begin{cases} 1 & f_i = f_j \text{ and } d_{i,j} < D \\ 0 & \text{others} \end{cases} \qquad (i, j \in [1, N])$$

In which D is the minimum separation distance without co-channel interference between two radios, and $d_{i,j}$ is the actual distance between two radios, and N is the total number of channels.

(2) The judgment of adjacent channel interference between the i-th and j-th radio

$$H(f_i, f_j) = \begin{cases} 1 & |f_i - f_j| \le l \\ 0 & |f_i - f_j| > l \end{cases} \qquad (i, j \in [1, N])$$

In which l is the minimum distance between two adjacent channels.

(3) The judgment of intermediation interference between the i-th and j-th radio

$$M(f_i, f_j) = \begin{cases} 1 & 2f_i - f_j = f_k \text{ or } f_i + f_j - f_k = f_t \\ 0 & 2f_i - f_j \ne f_k \text{ and } f_i + f_j - f_k \ne f_t \end{cases} \qquad (i, j, k, t \in [1, N])$$

We can get the objective function from the above three constraints and that is the expression of the total number of interference:

$$E(f) = \sum_{i=1}^{N} \sum_{j=1}^{N} [\alpha A(f_i, f_j) + \beta H(f_i, f_j) + \gamma M(f_i, f_j)]$$

In which α , β and γ are the weight coefficient.

3 The Implementation and Improvement of the Algorithm

3.1 Design of the Fitness Function

The objective function $E(f)$ has been designed in the previous content. As the design requirements of the fitness function is that the individual with large interference will be inherited to the next generation with smaller probability, and the individual with little interference will be inherited to the next generation with greater probability, we should define the fitness function $F(E)$ and the objective function $E(f)$ as the inverse relationship in order to the individual with less interference will make greater fitness. As the number of the individual' interference may be zero, we must set a fitness constant C. The relationship between the fitness function and the objective function can be expressed by the following formula:

$$F(E) = \frac{1}{E(f) + C}$$

In which C is the fitness constant. The best value of C is 3.2 which are verified by the author in the 2nd literature.

3.2 Chromosome Representation and Coding Schemes

There are usually two coding way of the symbol coding and the binary coding in the Genetic Algorithms. We use the symbol coding in this paper. We assign numbers to the available frequency of the radio, The frequency assigned to each radio can be expressed by the number, so we can get one individual' coding string after every group radios are assigned one frequency number. For example when the available band of the i-th radio is (f_L^i , f_H^i), and the frequency interval of the radio is Δf , so there are $n+1$ available frequency points (f_L^i , $f_L^i + \Delta f$, $f_L^i + 2\Delta f$, ..., $f_L^i + n * \Delta f$). We can encode them into $(0,1,2,...,n)$.

3.3 Design Population Initialization

There are usually two ways to generate initial population. One is completely random method, and the others is the method added some priori knowledge. In this paper we use the 2nd way. Firstly we assign numbers to available frequency points of each radio according to the radio' available band and the frequency interval Δf . Then we randomly assign numbers to every radio one by one within their frequency numbers until all radios are assigned.

3.4 Design and Improvement of Genetic Operation

3.4.1 Design and Improvement of Selection

In this paper we use the roulette wheel method in selection and improve the method. Supposed the selection probability of the k-th individual is P_k and the fitness value is F_k , the relationship between P_k and F_k can be expressed by the following formula:

$$p_k = \frac{F_k}{\sum_{k=1}^{N} F_k}$$

In the traditional selection method p_k is divided in the range of (0,1), then the random number r is selected in the range of (0,1). The range of the r is judged and the individual are saved according to its site. In this paper the average fitness value $\overline{F}$ is quoted. We want the minimum value of the objective function and the maximum value of the fitness function in this paper. So the individual is saved when $F_k \geq \overline{F}$, then we calculate its selection probability p_k. The computational efficiency will be improved greatly.

3.4.2 Design and Improvement of Crossover

In crossover the two parent individuals of populations cross each other according to the crossover probability p_c and the new generation is produced. So the crossover probability must be set before the crossover is performed. The best value is 0.91 which is verified by the author in the 2nd literature. But the solution set will gradually approach the optimal solution with the increase in the number of the evolution in the later period of evolution. Now if we still use the bigger crossover probability, there will be many new individuals scattering in the entire search space. The bigger crossover probability will damage to the proportion of the best individual and slow down the convergence process. So we improve the crossover probability p_c in this paper, and use the adaptive crossover probability which change with the evolution. The formula of p_c is:

$$p_c = \frac{p_{c0} - (p_{c0} - p_{c\min}) * d}{D}$$

In which p_{c0} is the initial crossover probability, and $p_{c\min}$ is the minimum of the crossover probability, and d is the current evolution algebra, and D is the overall evolution algebra. Usually, p_{c0} equals 0.95, and $p_{c\min}$ equals 0.4.

3.4.3 Design of Mutation

Mutation can provide the genes which are not contained in the initial population, or recover the genes lost during selection, and provide new content for the population. So the selection of mutation probability is very important. If the mutation probability is too big, a lot better model may be destroyed. But if the mutation probability is too small, the ability of generating new individuals and inhibiting the premature will be less in mutation. Usually the mutation probability is between 0.01 and 0.1. In this paper we use the best mutation probability 0.095 which is verified by the author in the 2nd literature.

4 Simulation and Analysis

In this paper the objects of study are 30 short-wave radios which work in between 1.5MHz and 30MHz, and the frequency interval is 25 KHz. Then we assign their work frequency. After mathematical modeling, we simulate by the MATLAB7.0. The result is shown in figure 1. From the figure we can see that the results of the original algorithm and the improved algorithm all converge to the best fitness value, but the convergence speed of the improved algorithm is faster than the original algorithm'. So this will greatly improve the speed of operation.

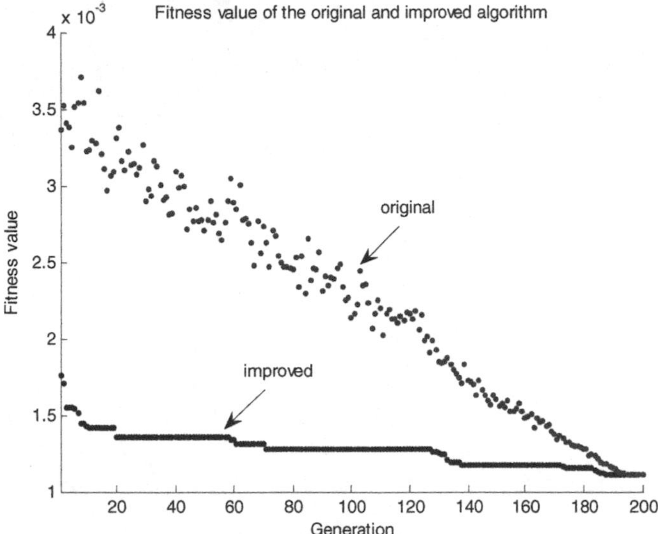

Fig. 1. Fitness value of the original and improved algorithm

5 Conclusions

Based on analyze the actuality of our military radio' frequency assignment, a method of frequency assignment based on Genetic Algorithm is proposed, and the Genetic Algorithm is also improved. The adaptive Genetic Algorithm is used in this paper. The simulation result shows that the adaptive Genetic Algorithm can effectively deal with the problem of the frequency assignment, and improve the convergence rating and the convergence speed.

References

1. Li, M., Kou, J., Lin, D., Li, S.: Basic Theory of Genetic Algorithms and Applications. Science Press (2002)
2. Hao, C.: Application of Genetic Algorithms in Frequency Assignment Issue. Beijing Jiaotong University (2009)

3. Lei, Y., Zhang, S., Li, X., Zhou, C.: MATLAB Genetic Algorithm Toolbox and its Application. Xi'an University of Electronic Science and Technology Press (2004)
4. Cui, S.: Some Improvements of the Genetic Algorithm and their Applications. China University of Science and Technology (2010)
5. Huo, J., Yi, X., Gu, C.: Shipboard Electronic Countermeasure 33(6), 49–53 (2010)

Research on Model of Circuit Fault Classification Based on Rough Sets and SVM

Fu Yu, Zheng Zhi-song, and Wu Xiao-ping

College of Electronic Engineering, Naval Univ. of Engineering, Wuhan 430033 China

Abstract. Aiming at the characteristic of lacking swatches and paroxysmal faults, A fault classification model based on rough sets and SVM is put forward. The pretreatment of diagnosis data is constructed by attribute reduction in rough sets. Redundancy attribute is deleted from the diagnosis decision-making table without losing useful information, and the reduced diagnosis decision-making table is used as original training sets of classification sub-system. The dimension of fault symptom and the capability of classification is balanced. Finally an example shows the model is effective and reasonable.

Keywords: rough sets, support vector machine, fault classification.

1 Introduction

In recent years, support vector machine (SVM) has been successfully applied to data mining, machine learning and pattern recognition by virtue of solid theoretical base, good generalization performance and broad application prospects of widespread concern. SVM could solve the problem of small samples of machine learning, which based multi-classifier has high classification accuracy and good generalization ability, So SVM has been as a new method of intelligent diagnosis classification used in circuit fault with high research value [1-3].

Circuit fault diagnosis in most cases is a small sample of machine learning problem. There are two reasons: on the one hand, the circuit fault occurs with a certain surprise which is not repeated and simulation. On the other hand, compared with the speech recognition and image recognition and other issues, which have thousands of samples and hundreds of feature dimension, the circuit failure mode is not only a relatively small number of samples, but also the nodes which could obtain diagnostic information are limited. Therefore the dimensions of fault samples are relatively small, so SVM could fast fault classification[4-5]. In addition, the size of fault samples and normal samples is imbalance, which causes "a bias" using the common effect of classifier. While SVM can take effective measures to reconcile such “a bias” to ensure the accuracy of diagnosis with the minority class samples based on the minimization structure risk[6]. Therefore, the article chose SVM to create a circuit fault classification model, which is a good solution to the problem of circuit fault classification. In the model, the different points of failure could produce the same fault characterization, which could bring the uncertainty. To solve this problem, the method of rough set theory and machine learning technology is introduced.

D. Jin and S. Lin (Eds.): Advances in CSIE, Vol. 1, AISC 168, pp. 433–439.
springerlink.com © Springer-Verlag Berlin Heidelberg 2012

2 A Fault Classification Model Based on SVM

Support vector machines applied to the circuit fault classification, which is established in the diagnostic process model of support vector machine classification process. The modeling process is: support vector machine classifier corresponds to the fault classifier, training sample can be corresponded to fault circuit, and the voltage or current value of the node consists of test vectors (usually obtained through fault simulation); test samples and the corresponding circuit to be diagnosed node consist of voltage and current values of test vectors. Classifier training process fault diagnosis device corresponds to the establishment and initialization process, the sample testing process to be diagnostic fault classifier corresponds to the fault diagnosis process [7]. The model is shown in Fig. 1.

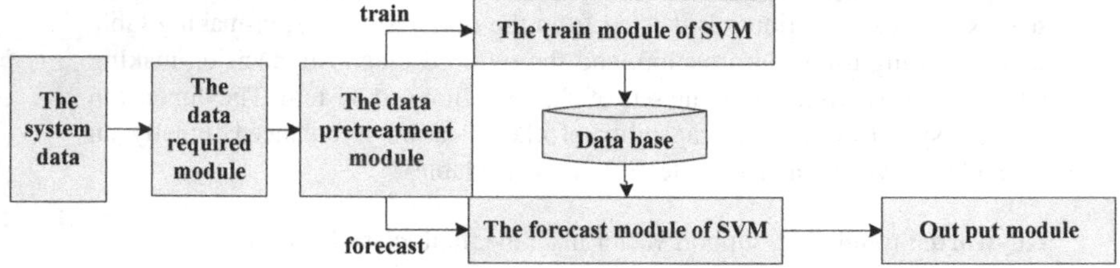

Fig. 1. The frame of circuit fault diagnosis based on SVM

Model of treatment process: firstly to capture the signal flow of circuit system by the data acquisition module, so the characteristic parameters could be extracted by the parameters of the signal; secondly the characteristic parameters will be pretreated as the input vector of the SVM. If it is in state of training, characteristic parameters will be trained through the support vector machine modules, and the training results, ie a set of support vector, will be storaged into the database; If it is in the state of predicting, the input vector will be predicted through module of SVM to get the output value. When the circuit is abnormal the failure event has occurred.

3 Major Steps in Fault Classification

The main steps of system fault classification are showed in figure 2 based on SVM and rough set.

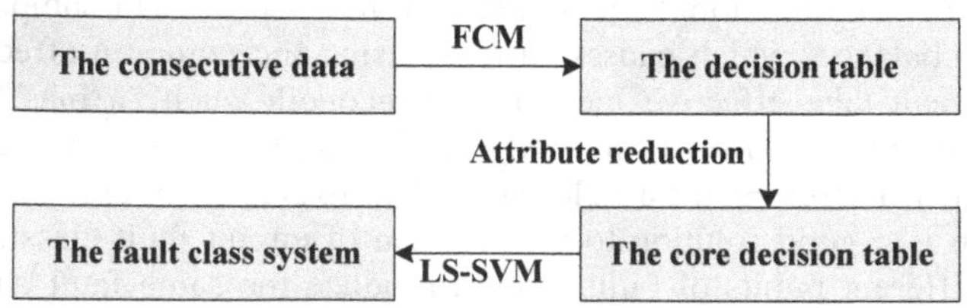

Fig. 2. The realization of fault classification based on SVM and rough set

Step 1 Data discrimination

The system parameters of circuit fault diagnosis of state are usually numerical bariables collected by the data acquisition module, which is not easy to be dealed with by the rough sets. This paper uses fuzzy C means (FCM) clustering algorithm as a pretreatment algorithm before reduction of rough set. FCM algorithm was first proposed by the Bezkek, which is obtained by the iterative algorithm to approximate the optimal value of the objective function. Consider a sample set $X = \{x_1, x_2, ..., x_n\}$, where $xi = (x_{i1}, x_{i2}, ..., x_{ik})$ is a k-dimensional vector, which will be divided into c fuzzy subsets based on certain criteria where c is the number of clusters given by the user, and the result of clustering center with a cluster membership matrix vector and expressed as:

$$\begin{aligned} V &= (v_1, v_2, \cdots, v_c) \\ u &= [u_{ij}]_{i=1,\cdots,c;\, j=1,\cdots,n;} \end{aligned} \tag{1}$$

Where, u_{ij} is the membership which means the degree that x_{ij} belong to the category i, u_{ij} meet:

$$\begin{cases} \sum_{i=1}^{c} u_{ij} = 1 \quad j = 1,\cdots,n \\ 0 < \sum_{j=1}^{n} u_{ij} < n \quad i = 1,\cdots,c \end{cases} \tag{2}$$

Objective function of FCM algorithm is:

$$J(u_{ij}, v_k) = \sum_{i=1}^{c}\sum_{j=1}^{n} u_{ij}^{m} \left\| x_j - v_i \right\|^2 \tag{3}$$

Equation (3), m is index weight for the matrix of fuzzy degree

$$\begin{aligned} v_i &= \left(1 / \sum_{j=1}^{n} (u_{ij})^m \right) x_j, i = 1,\cdots,c \\ u_{ij} &= \left(1 / \left\| x_j - v_i \right\|^2 \right)^{1/m-1} / \sum_{k=1}^{c} \left(1 / \left\| x_j - v_k \right\|^2 \right)^{1/m-1} \end{aligned} \tag{4}$$

Some membership values data can be gotten after the FCM clustering and the greatest value of the corresponding membership category is selected as the sample corresponding to discrete categories, which is easy to handle for rough set decision table.

Step 2 Rough Set Theory

To make each input as the condition attributes and each output as the decision attributes for the decision table. The discernibility matrix based on logical operation on rough set attribute reduction algorithm is as follows.

① calculated the discernibility matrix M of decision table system S = <U,C∪D>:

$$M(i,j)=\begin{cases}\{a_k \mid a_k \in C \wedge a_k(x_i)\neq a_k(x_j)\}, d(x_i)\neq d(x_j)\\ 0 \quad d(x_i)=d(x_j)\end{cases} \tag{5}$$
$$i,j=1,2,\cdots,n$$

② establish the appropriate expression of disjunctive logic

$$L_{ij}, L_{ij}=\vee_{a_i\in M_{ij}} a_i, M_{ij}\neq 0, M_{ij}\neq \Phi \tag{6}$$

③ calculation of CNF L

$$L=\overset{M_{ij}\neq 0, M_{ij}\neq\Phi}{\wedge} L_{ij} \tag{7}$$

④ make L convert to CNF then $L=\vee_i L_i$;

Every item on the results of CNF is corresponding to the attribute reduction, which contains the least number of attributes are asking for the minimum set of attributes C [8].

Step 3 Least squares Support Vector Machine (LS-SVM)

The quadratic programming has been required for training solutions SVM. Although the solution obtained is the only optimal solution, the complexity of the algorithm depends on the number of sample data. An effective solution is to use least squares support vector machine (LS-SVM) which can improve the convergence speed by solving a set of linear equations [9].

Let the training data set $\{x_t, y_t\}_{t=1}^{N}, x_t \in R_m$ is the t sample of the input mode, $y_t \in R$ is the expected output corresponds to the t sample, while N is the number of training samples. LS-SVM to take the following form:

$$y(x)=w^T\phi(x)+b \tag{8}$$

To map input data into a high dimensional feature space in the equation (8). The dimension of w is not pre-specified (which can be infinite dimensional.)

In the LS-SVM, the objective function is described as:

$$\min J(w,e)=(1/2)w^T w=\gamma(1/2)\sum_{t=1}^{N} e_t^2 \tag{9}$$

$$\text{constraints are: } y(x)=w^T\phi(x_t)+b+e_t, t=1,\cdots,N \tag{10}$$

4 Example Calculation

The specific process of the method can be illustrated by the differential circuit.

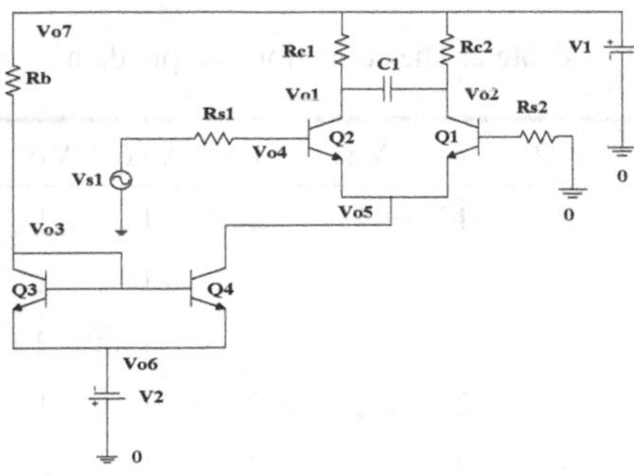

Fig. 3. Differential circuitry

In the fig. 3 the resistance and the tolerance of capacitance components is 5%, when there is a single soft fault, we can set the nominal value of resistance fluctuations in R_b fluctuate to 25%, Rc1 fluctuate to -30%, Rc2 fluctuate to 30% . When a component fluctuates beyond its tolerance, not only it will cause changes in the voltage across the device, but also it will lead the other node voltage change, what is more the voltage values are different in different failure mode. The seven test points are selected as samples which is Vo1 Vo2 Vo3 Vo4 Vo5 Vo6 and Vo7 shown in Table1.

Table 1. Sample data

Vo1	Vo2	Vo3	Vo4	Vo5	Vo6	Vo7	S
0.082	0.014	0.088	0.112	0.021	6.215	0.015	normal
0.195	0.035	0.062	0.137	0.026	5.242	0.168	normal
0.104	0.101	0.006	0.072	0.051	6.101	0.385	failure
0.015	0.026	0.033	0.092	0.018	5.001	0.069	failure
0.206	0.014	0.128	0.032	0.188	6.284	0.110	failure
0.184	0.087	0.071	0.074	0.029	5.301	0.106	failure
0.274	0.012	0.011	0.165	0.006	3.329	0.076	normal
0.279	0.038	0.010	0.187	0.007	6.469	0.094	normal
0.206	0.022	0.128	0.197	0.171	2.968	0.170	failure
0.082	0.014	0.088	0.112	0.021	6.215	0.015	normal
0.206	0.022	0.128	0.197	0.171	2.968	0.170	failure

The data obtained by FCM is shown in Table 2, then rough set attribute reduction algorithm can be used for attribute reduction the result after which is that the inputs are leaving only the {Vo1, Vo2, Vo3, Vo4}. In other words, through the reduction of rough set input have not been greatly reduced. Simulation results show that the algorithms of the model can be reduced greatly.

Table 2. The collection sample data

Vo1	Vo2	Vo3	Vo4	Vo5	Vo6	Vo7	S
1	2	1	2	2	1	1	1
2	1	1	2	2	1	2	1
1	1	2	2	2	1	2	2
1	2	2	2	2	1	1	2
2	1	2	1	2	1	2	2
2	2	1	2	2	1	2	2
2	1	2	2	2	2	1	1
2	1	2	2	2	1	1	1
2	1	1	2	1	2	1	2

5 Conclusion

The fault classification model is established based on rough set theory and SVM in this paper. In addition, this paper uses the rough set theory get the pre-processor without loss of valid information, which can remove the diagnostic decision-making table to solve the problem of dimension and the classification.

References

1. Han, H., Gu, B., Ren, N., et al.: Fault Diagnosis for Refrigeration Systems Based on Principal Component Analysis and Support Vector Machine. Journal of Shanghai Jiaotong University 45(9), 1355–1361 (2011)
2. Ma, C., Chen, X., Xu, Y., et al.: Analog circuit fault diagnosis based on attribute reduct ensemble of support vector machine. Chinese Journal of Scientific Instrument 32(3), 660–665 (2011)
3. Li, F., Tang, B., Zhang, G.: Fault diagnosis model based on least square support vector machine optimized by leave-one-out cross-validation. Journal of Vibration and Shock 29(9), 170–174 (2010)
4. Shen, Y., Meng, C.: Research on Method of Analog Circuit Fault Diagnosis Based on Information Fusion and SVM. Computer Measurement & Control 17(11), 2177–2180 (2009)
5. Zhu, B., Hu, S.J.: A New Method of Support Vector Machine (SVM) for Power Electronic Fault Diagnosis. Control Engineering of China 16(S), 206–208 (2009)

6. Lin, H., Wang, D.: Optimizing Support Vector Machine Parameters and Application to Fault Diagnosis. Computer Science 37(4), 255–257 (2010)
7. Gu, X., Zhao, J., Wang, S., et al.: Fault Diagnosis Technology Based on Multi-resolution Analysis and Support Vector Machine in Power Electronic Circuit. Journal of Henan University (Natural Science) 41(3), 300–303 (2011)
8. Cao, C.F., Yang, S.X., Zhou, X.F.: Fault Diagnosis of Rotating Machinery Based on an Improved Support Vector Machines Model. Journal of Vibration, Measurement & Diagnosis 29(3), 270–274 (2009)
9. Jiang, H., Wu, C., Ma, J.: Design of Cognitive Engine for CR Based on Least Squares SVM. Journal of University of Electronic Science and Technology of China 40(1), 41–45 (2011)

Research on Model of Circuit Fault Diagnosis Based on Bayesian Inference

Fu Yu, Zheng Zhi-song, and Wu Xiao-ping

College of Electronic Engineering, Naval Univ. of Engineering, Wuhan 430033 China

Abstract. According to the problem of information uncertainty during the process of fault diagnosis, a model of circuit fault diagnosis is proposed based on Bayesian inference. The definition of probability is extended in this model, which is explained as the subjective faith degree of experts. Besides, the knowledge process is added to the process of fault diagnosis, and the determinant rules of fault are given. Finally, the model is used in some circuit fault diagnosis, and qualitative analysis and quantitative calculation show that the model is effective and reasonable.

Keywords: Bayesian inference, circuit, fault diagnosis.

1 Introduction

There is much uncertainty in the circuit fault diagnosis which appears in the form of probability. And the uncertainty also exists in the connections between the fault and symptoms [1-3]. Bayesian theory can deal with such uncertainty. Bayesian uses the method of probability theory to diagnose the circuit fault in which the probability can be viewed as the relationship of trust [4,5].The basic problem of statistical inference is judging the unknown aspects of the overall distribution by using information provided by the overall model and samples. The basic feature of Bayesian theory is using the information provided by samples as well as prior information to deal with statistical problems.

2 Bayesian Inference Method

Bayesian is an inference method firstly applied to uncertain information fusion, and its basic idea is to calculate the posterior probability by making use of the prior probability, according to which to make the decision. Thus the uncertainty problem can be dealt with [6,7]·

Let Ω be a set of basic events and P be the probability measure on Ω, and ω be an assumed random variable on Ω. Suppose there is an N class ω_j ($j = 1,2, ..., N$), if the prior probability of each class is P(ω_j), and for the D-dimensional random vector $X=(x_1,x_2,\ldots,x_D)$, the conditional probability of which is P($X|\omega_j$), according to Bayesian theory:

D. Jin and S. Lin (Eds.): Advances in CSIE, Vol. 1, AISC 168, pp. 441–447.
springerlink.com © Springer-Verlag Berlin Heidelberg 2012

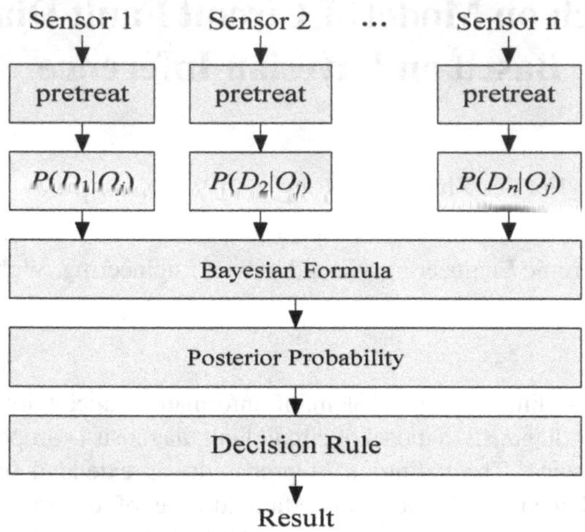

Fig. 1. The application of Bayesian theory in the information fusion

$$P(\omega_j|X) = \frac{P(X|\omega_j)P(\omega_j)}{\sum_{i=1}^{N} P(X|\omega_j)P(\omega_j)} = \frac{P(X|\omega_j)P(\omega_j)}{P(X)} \tag{1}$$

Then the posterior probability P(ω_j|X) can be gotten. Assuming the elements of *X* are independent, that is $x_i \cap x_j = \phi$, (ϕ is an empty set), then, substitute $P(X|\omega_j) = \prod_{i=1}^{D} P(X|\omega_j)$ into equation (1), we have:

$$P(\omega_j|X) = \frac{\prod_{i=1}^{D} P(x_i|\omega_j)P(\omega_j)}{\sum_{i=1}^{N}\prod_{k=1}^{D} P(x_k|\omega_i)P(\omega_i)} \tag{2}$$

Equation (2) transforms a multi-dimensional decision-making problem into the combination of each elements' conditional probability and prior probability. Besides,

$$P(x_i|\omega_j) = \frac{P(\omega_j|x_i)P(x_i)}{P(\omega_j)}$$

so

$$P(\omega_j|X) = \frac{\prod_{i=1}^{D} P(\omega_j|x_i)}{P(\omega_j)^{D-1}} \tag{3}$$

Equation (3) transforms the posterior probability of each elements into the total posterior probability, which integrates the decision-making results of each elements into the total results. This is a typical decision-fusion method. Bayesian decision rule is: if $X_0 \subset X$, and $P(\omega_{j0}|X_0) = \max\{P(\omega_j|X)\}\ (j=1,2,\cdots,N)$, then ω_{j0} is validated under the evidence X_0 .

In the above equations, prior probability, as the known conditions, is an uncertainty measure of Ω, and conditional probability is an uncertainty measure of the rules, both of which are given based on human experience. The results are posterior conditional probability. Therefore, probability measure is actually regarded as the uncertainty measure, which uses the probabilistic methods to solve the problem of uncertain reasoning. The multi-sensor fusion process for Bayesian theory is shown in Fig. 1.

3 Fault Diagnosis Model

3.1 Modeling

A fault diagnosis system is a triple group, including: S_{SD}、S_{COMPS}、S_{OBS}. S_{SD} is the system description, which is a set of predicate formulas; S_{COMPS} is the component of the system; S_{OBS} is a finite set of constants which is an observation set, consisting of a finite set of predicate formulas. The circuit in Figure 2 is composed of three multipliers, e.g., m_1, m_2 and m_3, and two adders e.g., a_1 and a_2, and S_{COMPS} is composed of m_1, m_2, m_3, a_1 and a_2; S_{SD} describes the normal input and output of the circuit. And a measurement S_{OBS} in the circuit is

$$\text{in}_1(m_1)=3,\ \text{in}_2(m_1)=2,\ \text{in}_1(m_2)=3,\ \text{in}_2(m_2)=2,\ \text{in}_1(m_3)=3,\ \text{in}_2(m_3)=2,$$
$$\text{out}(a_1)=10, \text{out}(a_2)=12.$$

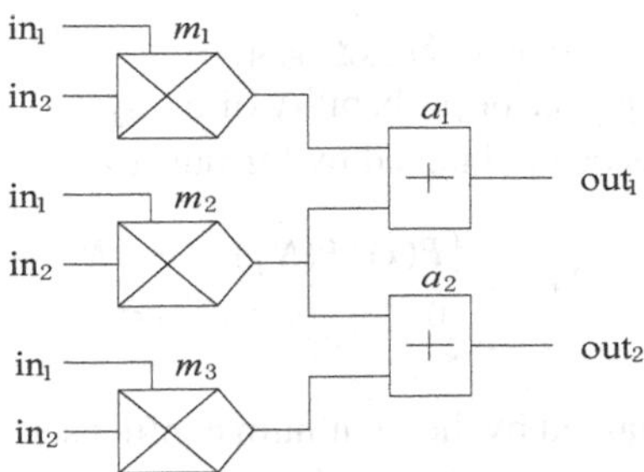

Fig. 2. The diagnosed circuit

The behavior of the system means a set of all measureable points under various normal working conditions which is expressed as S_{BEH}. Thus, while there are no faults, $S_{OBS} \in S_{BEH}$. In Figure 2, when the inputs in_1=3 and in_2=2, the normal circuit outputs should be out_1=12 and out_2=12; but the actual observation is that the out(a_1)=10 and out(a_2)=12. Therefore the conclusion that some components are at fault can be drawn.

The circumstance $S_{OBS} \not\subseteq S_{BEH}$, is called a symptom. The process of diagnose is guided by symptoms, each of which indicates that the assumption is invalid. Intuitively, a conflict is a hypothesis set that supports certain symptoms. That is to say, if one symptom is present, there is at least one element is at fault. Considering the symptom of " out_1 turned out to be 10, not 12", forecasting out_1 = 12 depends on a_1、m_1 and m_2. That is, if a_1、m_1 and m_2 are working normally, then out_1=12. However the out_1 is not 12, at least one of the a_1,m_1 and m_2 are at fault. Therefore, set $\{m_1, m_2, a_1\}$ is a conflict for the symptoms . In addition, you can come to another conflict $\{m_1, m_3, a_1, a_2\}$. The so-called minimal conflict set is one that has conflicts even if it has no subsets, and any other conflicts are a superset of it. For example, $\{m_1, m_2, m_3, a_1, a_2\}$ is a conflict, because the subset $\{m_1, m_2, a_1\}$ is a conflict, so $\{m_1, m_2, m_3, a_1, a_2\}$ is not the smallest subset of the conflict.

3.2 Determining the Faults

The probability theory is introduced based on the above work, and a Bayesian model is established based on diagnosis model. Assuming the system includes a set of components C = $\{c_1, c_2, ..., c_n\}$, for each component, there is a prior probability value without any prior observations. Assuming that P_i is probability value which means components c_i is normal. Then, for each component c_i, we can give a corresponding Boolean variable x_i, when c_i is working normally, x_i=1; when c_i is working abnormally, x_i=0. Before any observation on the system, a particular system state $x = (x_1, x_2, ..., x_n)$ can be expressed as the probability of

$$P(x) = \sum_{i=1}^{n} P_i^{x_i} (1 - P_i)^{1-x_i} \tag{4}$$

When we get a measurement and $S_{OBS} \not\subseteq S_{BEH}$, some of the system state is not a reasonable diagnosis. The posterior probability of the system state can be gotten by the prior probability which has been obtained by formula (4)

$$P'(x) = \begin{cases} P(x) / P(N_d) & x \in N_d \\ 0 & x \in N_c \end{cases} \tag{5}$$

N_c and N_d can be easily obtained by the minimum conflict set. Take N_c for example, it is actually a collection of all conflicts[8], in other words, N_c can be expressed as minimal conflict sets

$$N_c = (m_1 \wedge m_2 \wedge a_1) \vee (m_1 \wedge m_3 \wedge a_1 \wedge a_2) \tag{6}$$

In equation (6), m_1 means the multiplier m_1 is in the normal state, and m_1' means the multiplier m_1 is not in normal state. Since N_c and N_d are complementary, N_d can be expressed as:

$$N_d = m_1^{'} \vee a_1^{'} \vee (m_2^{'} \wedge m_3^{'}) \vee (m_2^{'} \wedge a_2^{'}) \quad (7)$$

The posterior probability P_i^* of component c_i can be expressed as:

$$P_i^*(c_i) = P(c_i \mid N_d) \quad (8)$$

The prior probability can convert into the posterior probability [9,10]. According to maximum posterior probability logic if $P_i^*(c_i) = \max_i \{P(c_i \mid N_d)\}$ then the component c_i is fault.

4 Example Calculation

In Fig. 2, assuming that the prior probability of a certain device which is in normal state is $P(m_1)= P(m_2)= P(m_3)=0.95$; $P(a_1)= P(a_2)=0.98$. When the inputs $in_1=2$ and $in_2=3$, the output observation is $out_1=10$ and $out_2=12$. The minimum conflict set can be obtained according to the preceding discussion $\{m_1, m_2, a_1\}$ and $\{m_1, m_3, a_1, a_2\}$, so the different states of the system can be gotten (1,1,1,1,1), (1,1,1,1,0), (1,1,0,1,1), (1,1,0,1,0), (1,0,1,1,1), which is indicated by Table1. Probability of occurrence of the system state is:

$$P = (m_1, m_2, m_3, a_1^{'}, a_2) = 0.95 \times 0.95 \times 0.95 \times 0.02 \times 0.98 = 0.0168$$

Similarly, other possible states can be obtained. Table1 shows that in the all 27 possible states, there are 16 states at fault and 11 states at normal. At last we can get:

$$P^*(m_1^{'}) = 0.69, P^*(m_2^{'}) = 0.09, P^*(m_3^{'}) = 0.08, P^*(a_1^{'}) = 0.27, P^*(a_2^{'}) = 0.03 \cdot$$

From the result we can see in all impossible 5 states the state of multiplier m_1 is normal which can be judged intuitively that the possibility of m_1 at normal is low. The prior probability of multipliers in failure is 0.05, and the posterior probability is 0.69. According to the fault determination rules, you can determine the component is failure, which is consistent with the intuitive judgments.

Table 1. The system state and the possibility

System state	Possibility P*	System state	Possibility P*
m_1, m_2, m_3, a_1, a_2	impossibly	m_1', m_2, m_3, a_1, a_2	0.04333805
m_1, m_2, m_3, a_1, a_2'	impossibly	$m_1', m_2, m_3, a_1, a_2'$	0.00088445
m_1, m_2, m_3, a_1', a_2	0.01680455	$m_1', m_2, m_3, a_1', a_2$	0.00088445
$m_1, m_2, m_3, a_1', a_2'$	0.00034295	$m_1', m_2, m_3, a_1', a_2'$	0.00001805
m_1, m_2, m_3', a_1, a_2	impossibly	$m_1', m_2, m_3', a_1, a_2$	0.00228095
$m_1, m_2, m_3', a_1, a_2'$	impossibly	$m_1', m_2, m_3', a_1, a_2'$	0.00004655
$m_1, m_2, m_3', a_1', a_2$	0.00088445	$m_1', m_2, m_3', a_1', a_2$	0.00004655
$m_1, m_2, m_3', a_1', a_2'$	0.00001805	$m_1', m_2, m_3', a_1', a_2'$	0.00000095
m_1, m_2', m_3, a_1, a_2	impossibly	$m_1', m_2', m_3, a_1, a_2$	0.00228095
$m_1, m_2', m_3, a_1, a_2'$	0.00088445	$m_1', m_2', m_3, a_1, a_2'$	0.00004655
$m_1, m_2', m_3, a_1', a_2$	0.00088445	$m_1', m_2', m_3, a_1', a_2$	0.00004655
$m_1, m_2', m_3, a_1', a_2'$	0.00001805	$m_1', m_2', m_3, a_1', a_2'$	0.00000095
$m_1, m_2', m_3', a_1, a_2$	0.00228095	$m_1', m_2', m_3', a_1, a_2$	0.00012005
$m_1, m_2', m_3', a_1, a_2'$	0.00004655	$m_1', m_2', m_3', a_1, a_2'$	0.00000245
$m_1, m_2', m_3', a_1', a_2$	0.00004655	$m_1', m_2', m_3', a_1', a_2$	0.00000245
$m_1, m_2', m_3', a_1', a_2'$	0.00000095	$m_1', m_2', m_3', a_1', a_2'$	0.00000005

5 Conclusion

The diagnosis based on model establishes the system model according to the connections between the components of the system, the main idea of which is to reduce the uncertainty of the system. Before observation, if all of the system states are possible, then the uncertainty is maximum. After observation, some states can be excluded by the model-based diagnosis, so the uncertainty can be reduced.

References

1. Liu, Z., Gao, Y.: Research of Fault Diagnosis Based on Grey Theory of Multi-wavelet Entropy. Computer Measurement & Control 19(6), 1318–1324 (2011)
2. Xu, X., Wang, Y., Wen, C.: Information-fusion method for fault diagnosis based on reliability evaluation of evidence. Control Theory & Applications 28(4), 504–510 (2011)

3. Zhao, S., Liu, F.: Cross-correlation fault diagnosis in control loop based on Bayesian network. Journal of Southeast University (Natural Science Edition) 40, 277–281 (2010)
4. Zhou, Z., Ma, C., Dong, D., et al.: Auto-study diagnosis method based on Bayesian fusion. Application Research of Compute 27(5), 1764–1766 (2010)
5. Li, Y., Lu, Q., Su, W., et al.: Learning Bayesian Network from Small Scale Dataset and Application. Computer Science 38(7), 181–184 (2011)
6. Zhao, W.: Transformer fault diagnosis based on selective Bayes classifier. Electric Power Automation Equipment 31(2), 44–47 (2011)
7. Cai, Z., Sun, S., Yannou, B., et al.: Conditional Bayesian network classifier and its application in product failure rate grade indentifying. Computer Integrated Manufacturing System 16(2), 417–423 (2010)
8. IIe, X., Tong, X., Sun, M.: Distributed power system fault diagnosis based on Bayesian network and Dempster-Shafer Evidence Theory. Automation of Electric Power Systems 35(10), 42–47 (2011)
9. Zhang, D., Wu, S., Luo, X., et al.: Research on rapid diagnosis algorithm for complex system based on Bayesian theory. Engineering Journal of Wuhan University 44(1), 128–132 (2011)
10. Wu, Q., Wu, S., Liu, J.: Mechanical fault diagnoses approach based on Fv-SVM. Systems Engineering – Theory & Practice 30(7), 1266–1271 (2010)

Research on the Alkalinity of Sintering Process Based on LS-SVM Algorithms

Rui Wang[1], Ai-min Wang[2], and Qiang Song[3]

[1] Mechanical and Electrical Engineering Xinxiang University, China
[2] Computer Science Department, Anyang Normal University, China
[3] Mechanical Engineering Department of Anyang Institute of Technology, China
songqiang01@126.com

Abstract. The measurement of R in sintering process is difficult to control , on the other hand, it is easily to be disturbed by almost process steps. A prediction model of R in sintering process based on LS-SVM is proposed to judge the trend of R. The application result shows that the prediction with this method can achieve higher robust, better utility and expensive value. It was concluded that the LS-SVM model is effective with the advantages of high precision, less requirement of samples and comparatively simple calculation.

Keywords: Alkalinity of sinter, LS-SVM, Prediction, The sintering process.

1 Introduction

In the modern steel enterprises, the sintering process of blast furnace material is one of the best important production process. The chemical composition of sintering alkalinity has a direct effect on production and economic benefits of the whole steel enterprise. Therefore almost every steel factory is equipped with many instruments and automatic control systems in its sintering plant for its production process control. But the complexity of sintering production process makes it difficult to be described by a set of mathematic models. Since this process often has large time delay and dynamic time-variability. It is hard to perform control tasks of total sintering process by using conventional control models. Sintering process is a complex physical chemistry process, which relates to a lot of characteristics, such as complicated mechanism, high non-linear, strong coupling, high delay-time and etc. So we can not establish its mathematical model, but performance index of sintering process decides the policy of blending process[1]. Because of the restriction of the detecting means, chemical examination of sinter alkalinity generally needs forty minutes. In the whole craft process, its time sometimes can even exceed one hour. Obviously, such a long delay-time can not meet the needs of actual productivity, therefore we must detect sinter alkalinity and establish the prediction model. In this research, we provide a nonlinear model method, which are least squares-support vector machine (LS-SVM). SVM has the capability of dealing with linear and nonlinear multivariate calibration and resolving these problems in a kernel function scheme. The prediction ability of nonlinear model methods will be compared with other methods..

D. Jin and S. Lin (Eds.): Advances in CSIE, Vol. 1, AISC 168, pp. 449–454.
springerlink.com © Springer-Verlag Berlin Heidelberg 2012

2 Least-Squares Support Vector Machine Algorithm Modeling

2.1 Least-Squares Algorithm Support Vector Machine

Recently, least squares support vector machine (LS-SVM) has been applied to machine learning domain successfully. It is a promising technique due to its successful application in classification and regression tasks. It is established based on the structural risk minimization principal rather than the minimized empirical error commonly implemented in the neural networks. LS-SVM achieves higher generalization performance than the neural networks in solving these machine learning problems. Another key property is that unlike the training of neural networks which requires nonlinear optimization with the danger of getting stuck into local minima, training LS-SVM is equivalent to solving a set of linear equation problem. Consequently, the solution of LS-SVM is always unique and globally optimal. In this paper, we discuss the application of LS-SVM in the prediction of the alkalinity in sintering process[2].

Given a training set $\{x_t, y_t\}_{t=1}^{N}$,with $x_t \in R^n$, $y_t \in \mathrm{R}$, $x_t \in R^n$ is input vector of the first t samples , $y_t \in \mathrm{R}$ is the desired output value of the first t corresponds to samples, N is the number of samples, the problem of linear regression is to find a linear function y(x) that models the data. In feature space SVM models take the form:

$$y(x) = w^{\mathrm{T}} \varphi(x) + b \tag{1}$$

Where the nonlinear function mapping $\varphi(\bullet): R^{\mathrm{n}} \to R^{n_{\mathrm{h}}}$ maps the high-dimensional space into the feature space. w dimension is not a pre-specified (which can be infinite dimensional).

The least squares approach prescribes choosing the parameters (w,b) to minimize the sum of the squared deviations of the data, the square loss function described as:

$$\min J(w,e) = \frac{1}{2} w^{\mathrm{T}} w + \gamma \frac{1}{2} \sum_{t=1}^{\mathrm{N}} e_t^2 \tag{2}$$

subject to the restrictive conditions, $y(x) = w^{\mathrm{T}} \varphi(x_t) + b + e_t$, for t = 1,…, N ;

$$\begin{bmatrix} 0 & \boldsymbol{1}^T \\ \boldsymbol{1} & \varphi(x_t)^T \varphi(x_t) + \boldsymbol{D} \end{bmatrix} \begin{bmatrix} b \\ \boldsymbol{\alpha} \end{bmatrix} = \begin{bmatrix} 0 \\ \boldsymbol{y} \end{bmatrix} \tag{3}$$

where $y=[y_1,\cdots,y_N]$, $1=[1,\cdots,1]$, $\alpha=[\alpha_1, \ldots, \alpha_N]$,

$D=diag[\gamma_1,\cdots,\gamma_N]$,Select $\gamma>0$, and guarantee matrix $\varphi=\begin{bmatrix}0 & 1^T \\ 1 & \varphi(x_t)^T\varphi(x_t)+D\end{bmatrix}$

$$\begin{bmatrix}b\\ \alpha\end{bmatrix}=\varphi^{-1}\begin{bmatrix}0\\ y\end{bmatrix}$$

Important differences with standard SVM are thc equality constrains and the squared error term, which greatly simplifies the problem.

Only equality constraints, and the optimization objective function is the error loss, which will simplify the problem solving.

Finally, The solution is obtained after constructing the lagrangian , Lagrange function definition can be written as,

$$L(w,b,e,\alpha)=J(w,e)-\sum_{t=1}^{N}\alpha_t\{w^T\varphi(x_t)\tag{4}$$

Where α_t is Lagrange multipliers. By Karush-Kuhn-Tucker (KKT) optimal conditions, the conditions for optimality are

$$\begin{cases}\dfrac{\partial L}{\partial \omega}=0\rightarrow\omega=\sum_{i=1}^{n}\alpha_i\varphi(x_i)\\ \dfrac{\partial L}{\partial b}=0\rightarrow\sum_{i=1}^{n}\alpha_i=0\\ \dfrac{\partial L}{\partial e_i}=0\rightarrow\alpha_i=\gamma e_i, i=1,\ldots,n\\ \dfrac{\partial L}{\partial \alpha_i}=0\rightarrow\omega^T\varphi(x_i)+b+e_i-y_i=0, i=1,\ldots,n\end{cases}$$

After elimination of e_t and w , the solution is given by the following set of linear equations

Reversible, the LS-SVM algorithm optimization problems would be transformed by least-square method .

2.2 Choice of the Kernel Function

By the KKT-optimal conditions, obtains w, and thus gains the training sets of nonlinear approximation

$$y(x)=\sum_{t=1}^{N}\alpha_t k(x,x_t)+b\tag{5}$$

where

$$k(x_t, x_k) = \varphi(x_t)^T \varphi(x_k) \; t, k = 1, \cdots, N \tag{6}$$

The choice of the kernel function $\varphi(\cdot): R^n \to R^{n_h}$ has several possibilities. It can meet arbitrary symmetry theorem Mercer function. In this work, the radial basis function (RBF) is used as the kernel function of the LS-SVM, because RBF kernels tend to give good performance under general smoothness assumptions ,for we commonly used Gaussian RBF (Radial Basis Function, RBF)function as a kernel function,

$$K(x, x_t) = \exp\{-\| x - x_t \|_2^2 / 2\sigma^2\} \tag{7}$$

Where σ is a positive real constant. By (8) - (10), gains object nonlinear model as follows:

$$y(x) = \sum_{t=1}^{N} \alpha_t \exp\{-\| x - x_t \|_2^2 / 2\sigma^2\} + b \tag{8}$$

The LS-SVM prediction involves two parameters to be optimized, which are σ (the width of the Gaussian kernel function which cover the input space) and γ (the regularization factor, allowing to avoid obtaining a too local model).

This LS-SVM regression leads to solving a set of linear equations, which is for many practitioners in different areas. Especially, the solution by solving a linear system is instead of quadratic programming. It can decrease the model algorithm complexity and shorten the computing time greatly. All the calculations were performed using MATLAB 7.0 (The Math Works, Natick, USA). The free LS-SVM toolbox (LS-SVM V 1.5, Suykens, Leuven, Belgium) was applied with MATLAB 7.0 to develop the LS-SVM models.

2.3 LS-SVM Design Steps

Based on the above analysis, we will put forward a new idea or algorithm. we adopt the least squares support vector machine to process and prediction.

The algorithm design steps as follows: 1) Firstly, input the original sequence;2)Secondly, Select Kernel function ; 3) Support Vector Machine method of solving optimization problems formula (8) ; 4) build up regression function; 5)Finally, model test.

3 Simulation Based on LS-SVM

3.1 System Input Parameters Choice

This paper adopts LS-SVM to predict the alkalinity, aiming at this important output index. In the whole craft process, synthesizes the variables related to the alkalinity and make sure that ten important input variables as the input of grey neural network, such as

①the layer thickness②the trolley speed③the first mixing water rate④the mixing temperature⑤the content of SiO2 in the mineral⑥the content of CaO in the mineral ⑦the content of FeO in the mineral⑧the second mixxing wate rate⑨the proportion of CaO⑩the proportion of Coal.

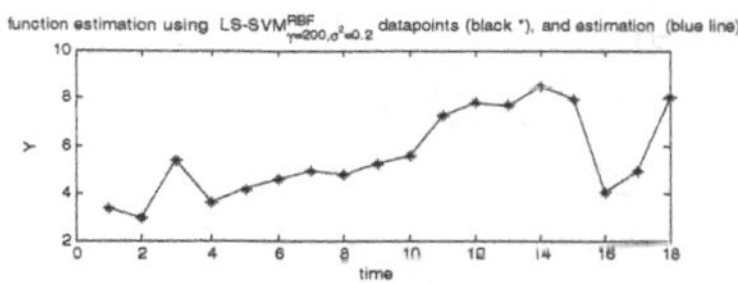

Fig. 1. The prediction of the alkalinity based on LS-SVM

3.2 Assessment of the Prediction Performance

The assessment of the prediction performance of the different soft computing models was done by quantifying the prediction obtained on an independent data set. The mean absolute percentage error (MAPE) was used to study the performance of the trained forecasting models for the testing data. MAPE is defined as follows:

$$MAPE = \frac{1}{N}\sum_{i=1}^{N}\left(\frac{P_{actuali} - P_{predictedi}}{P_{actuali}}\right) \tag{11}$$

Where $P_{actuali}$ is the actual value on day i and $P_{predictedi}$ is the forecast value of the load on that day.

As can be seen from Fig.1, Support vector machines show good timing prediction and generalization performance, and significantly accelerate the convergence rate. which is also capable of more accurate forecasts. In this paper, based on LS-SVM to identify the parameters of the forecasting model using the structure to minimize the risk criteria can be good to avoid this problem, so that the model not also can get the smaller fitting error, but also has good generalization ability and robustness.

4 Conclusion

The Support Vector Machine in resolving a lot of problems such as the small sample, high-dimensional nonlinear problems and pattern recognition etc, it show many unique advantages, so it becomes a current hotspot. This article proposes a new forecasting model of grey support vector machine. These results fully demonstrate the prediction accuracy of new model is superior to a single model, the theoretical analysis and simulation are fully presented the validity of the forecast model. We obtained significantly better results when using a simulation of the prediction model of grey support vector machine. We observe that the use of prediction model allows to improve the quality of the alkalinity.

References

1. Fan, X.-H., Wang, H.-D.: Mathematical model and Artificial Intelligence of sintering process. Central South University Press (2002)
2. Song, Q., Wang, A.-M.: Simulation and Prediction of Alkalinity in Sintering Process Based on Grey Least Squares Support Vector Machine. Journal of Iron and Steel Research, International 16(5), 1–6 (2009)

Leveling Accuracy Research Based on Trimble 5700 GPS Antenna

Tianzi Li[1], Huabin Chai[1], and Yuanyuan Xu[2]

[1] School of Surveying and Land Information Engineering, Henan Polytechnic University, Jiaozuo ,China
[2] Jiefang District Committee, Jiaozuo City Jiaozuo, China

Abstract. At present, the research on GPS elevation is mostly based on the analysis of the sources errors and the studying of the phase center deviation in different antennas and the methods of collecting the abnormalelevation values. After conducting experiences, the key factor which influent the accuracy is found,is how the phase center deviation in different antennas influent the elevation errors.And then the method of GPSelevation fitting can meet the four standards ofmeasurement accuracy of GPS.

Keywords: elevation errors, the deviation of the phase center, GPS elevation fitting, accuracy.

1 Introduction

The refraction of Ionosphere and troposphere affects the accuracy of GPS elevation measurement. During the recent 10 years, people in the surveying and mapping have reached a very excellent level at home and abroad[1]. As for the influence of the GPS receiver antenna to elevation, people have got the conclusion that the value of the deviation of the phase center $\triangle$h is very small, about 1mm [2], as for the same types of the same model of GPS receiver antenna combination. Then I reach a different conclusion and find that the deviation of the phase center of the same types of the same type of GPS receiver antenna is still the key factor to affect the elevation measurement by conducting a reach with Trimble 5700 receiver.

The WGS-84 geodetic height system is measured by GPS elevation, but we use the normal height system, which is based on quasigeoid. So if we want to calculate the normal height of GPS points accurately, we must do some transformations. At present, the methods of calculating the current high ground point are the normal standards of height GPS, GPS gravity height, GPS triangular elevation, the conversion parameters, the overall adjustment and neural network method and other methods.

2 The Effects of the Antenna Phase Center to Elevation

During the recent 10 years, people in the surveying and mapping have reached a very excellent level at home and abroad, which the article will no longer repeat again.

D. Jin and S. Lin (Eds.): Advances in CSIE, Vol. 1, AISC 168, pp. 455–459.
springerlink.com © Springer-Verlag Berlin Heidelberg 2012

The influence of the deviation of the phase center to the accuracy of elevation mainly focuses on the GPS receiver antennas, which have the different models for the same type or the different types. Because of the differences of the antenna designing, construction, manufacturing processes, the deviation of the phase center differs a lot [2]. Here are some kinds of calibrating methods: such as the indoor microwaves calibration, the bias of vertical component of the outdoor GPS receiver antenna phase center, switching antenna method, phase center variation correction model, etc. [2] The same types of the same model of GPS receiver antenna combination, which are designed same and have the same production processes and materials, are believed that$\triangle$h, the deviation value of vertical component of antenna phase center is small. And the error can be regarded as the accidental error [3]. So in this regard, I carried on a static observation with trimble 5700GPS receiver in one city in Henan. And the point accuracy, density design, reference design and graphic design, and the observation time and the number of repeat station measurements have reached the E-Class GPS network requirements. The graph of adjustment of network is as follows figure 1:

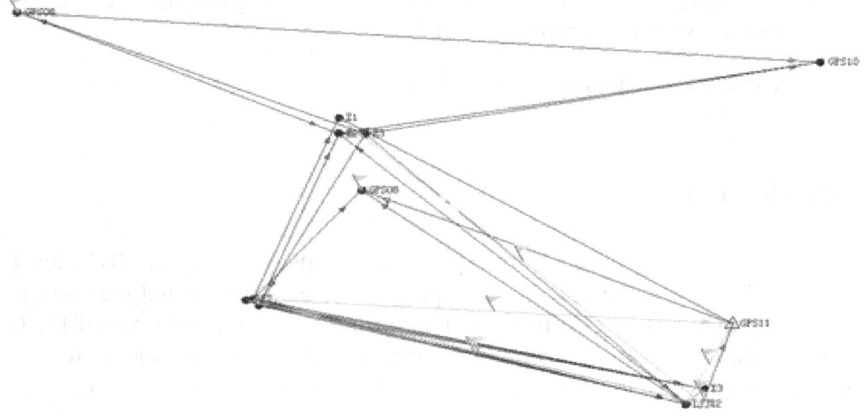

Fig. 1. The graph of GPS adjustment of network

Data are resolved. In restraint adjustment, in order to reduce the influence which the constraints have on the GPS antenna phase center, we use one point of GPS11 to conduct the adjustment with the software TGO comes with a trimble 5700GPS receiver and universal EGM96 geoid and GPS11 controlling point elevation is a fourth-level measurement result. In accordance with Fourth-level measurement, we conduct the level survey again to all other points except GPS06, and minus the result which is calculated with the TGO software in EGM96 geoid surface and compare with it. The results are as the following table 1:

Table 1. The comparison of different antennas between height differences

Point	x(N)	y(E)	H(EGM96)	H normal	△H	Antenna(S/N)
Z3	****098.208	***388.815	272.889	272.884	-0.005	
X3	****070.411	***764.931	325.847	325.844	0.003	9036
△GPS11	****581.734	***970.224	327.679	327.679	0	
GPS06	****062.559	***949.683	257.748			
F1	****730.461	***637.116	253.243	253.389	-0.146	5228
F2	****779.496	***592.888	251.824	251.973	-0.149	
GPS10	****661.886	***590.927	320.67	320.664	0.006	
X1	****946.003	***641.952	325.473	325.475	-0.002	
X2	****945.775	***755.761	326.08	326.058	0.022	5232
LJJ	****944.372	***637.774	325.575	325.571	0.004	
Z2	****101.219	***204.791	260.079	260.250	-0.171	
F3	****772.600	***546.649	250.8	250.948	-0.148	
Z1	****224.677	***206.307	252.89	253.062	-0.172	4274
GPS08	****645.283	***364.809	257.734	257.872	-0.138	

From the table we can see that the height difference is largely due to receiver bias and antenna. The influences which the receiver host has on the measurement accuracy of elevation are mainly the receiver clock error, the signal path delay, the delay locked loop error and machine error of noise and so on [4]. And the error in the receiver has been largely eliminated when the receiver has been produced. At this point the main factors affecting the accuracy of elevation are the deviation of receiver's antenna phase center.

3 The Project of Reducing the Error

We use antenna 9036 to observe Z3, X3, GPS11. Because the controlling point is GPS11, the antenna deviation is close to 0. The table 2 shows the error statistics of several antennas been very mature; we will not repeat them.

Table 2. Error statistics of several antennas

Antenna	9036	5228	5232	4274
Average (mm)	-0.001	-0.148	+0.008	-0.157
Maximum error (mm)	-0.004	+0.002	+0.014	+0.019
Middle error (mm)	±0.004	±0.002	±0.010	±0.017
Corrected value (mm)	△undetermined	△-0.147	△+0.009	△-0.156

Middle error: $\sigma = \pm\sqrt{VV/n-1}$

V: error values , n: The numbers of observations.

According to the table, the average value of every antenna does not represent the real value of the deviation of the phase center. However, it just represents the deviation value of the antenna 9036 in the controlling point, for the elevation of GPS11 has been fixed. Then, the table 1 shows the corrected values of the antennas and △ is an undetermined value.

So in the data processing, when completing the adjustment of the free net, we can get a highly accurate normal height of GPS to meet the four standards of measurement accuracy by getting an adjustment of the free net of a fixed point in the local coordinate system to calculate the elevations of different points under testing, and correcting it using the corrected value of antenna height difference.

EGM96 geoid is a global geoid. To get a further accuracy improvement of geoid measurement, we can use these methods such as GPS level elevation, GPS gravity height, GPS triangular elevation, the conversion parameters, the overall adjustment and neural network method and so on.

4 The Conclusion

The deviation of the phase center of GPS antenna plays a leading role in the error of the elevation measurement. The relationship of mutual deviation plays a very important role in developing the accuracy of GPS elevation measurement, when you use GPS antenna. This article also shows the data settlement of GPS when using the mutual error of GPS antenna, which plays a leading role in engineering applications.

Acknowledgement

1.Henan Polytechnic University Youth Fund Project : Q2010-17

2.Henan provincial key research project : 112102210193.

References

1. Hu, W., Li, M.: Key techniques of precise GPS height surveying. Journal of Hohai University (Natural Sciences) 36(5), 659–662 (2008)
2. Gao, W., Yan, L., Xu, S., Jiang, Y.: Research of influence and correction of GPS antenna phase center deviation on GPS height. Chinese Journal of Scientific Instrument 28(9), 2053–2057 (2007)
3. Beutler, G., Brockmann, E., Fankhauser, S., et al.: The bemese GPS software version 4.0. Astronomical Institute, University of Berne, Switzerland (1996)
4. Xu, S., Zhang, H., Yang, Z., Wang, Z.: Principle and Application of GPS Surveying, pp. 110–111. Wuhan university press

Feature Learning Based Multi-scale Wavelet Analysis for Textural Image Segmentation

Jing Fan

Department of Mechanical & Electrical Engineering, Xi'an University of Arts and Science, Xi'an, 710065, China

Abstract. In order to increase the edge accuracy and the areas consistency, and to reduce the partition error rate in textural image segmentation, we propose a new multi-scale wavelet analysis based on feature learning in this paper. It improves the textural image segmentation by reducing the effect of redundant features on segmentation results. This method includes three stages as follows: feature extraction, optimizing the feature vectors and feature space clustering. In the stage of filtrating valid features, we optimize the feature vectors by feature learning. The experimental results demonstrate that the improved algorithm is effective for textural image segmentation.

Keywords: Texture Image Segmentation, Wavelet Transform, Feature Learning.

1 Introduction

Texture is an important characteristic used for identifying objects or regions of interest in an image, and it exists extensively in each kind of images. Most of the images obtained by all kinds of observation system are textural images, for instance, aviation and satellite remote sensing images, hydrology and geology images, medical micrograph, material microgram etc., even a lot of natural sceneries images can be regarded as they are constituted by much tiny texture [1]. We can acquire a lot of useful macroscopic and microcosmic information by analyzing these textural images. So the research of the textural image has important academic meaning and wide application foreground.

Most textural image segmentation techniques can span several classes, including contour based techniques, feature threshold based techniques, region based techniques, clustering, template matching, etc [10,12]. Derraz *et al.* [11] presented an unsupervised segmentation of textural images based on integration of texture descriptor in formulation of active contour, which described the geometry of textural regions using the shape operator defined in Beltrami framework. Wang *et al.* [12] proposed an improved watershed segmentation algorithm combined with texture features to construct a well segmentation of textural images. Gaetano *et al.* [13] introduced a recently developed hierarchical model into textural image segmentation for reducing the computational burden and preserving contours at the highest spatial space. Paragios *et al.* [14] proposed a supervised texture segmentation method,

D. Jin and S. Lin (Eds.): Advances in CSIE, Vol. 1, AISC 168, pp. 461–466.
springerlink.com © Springer-Verlag Berlin Heidelberg 2012

applying a Gabor filter-bank to the pattern image. The filter-bank responses are represented as multi-component conditional probability density functions and a textural feature vector encoding boundary information is generated. Compared to gray-level based approaches, these texture-based segmentation methods are faced with more difficulties and many methods available can not consistently and accurately segment textural images.

In this paper, we use multi-scale wavelet analysis as a tool to study the algorithm in textural image segmentation and actualize it, based on feature learning. When we extract the features using wavelet analysis, we can obtain texture features in abundance from an image, but the feature vectors include a great deal of redundant information. The existence of the redundant features which can affect the final segmentation is not effective for textural images. So we improve it through adding a stage called optimizing feature vectors between feature extraction and clustering to reduce the effect of redundant features on segmentation results. With this method, we can obtain a better segmentation result with good area consistency and give attention to good edge accuracy at the same time.

The rest of this paper is organized as follows: In section 2, the proposed method is presented, where the three stages of the proposed method are discussed respectively. Finally, we give the experimental results in section 3 and conclusions in section 4.

2 Improved Algorithm

In this paper an algorithm contains three stages for textural image segmentation is proposed. The first stage provides us with texture features to be analyzed which is derived from multi-scale wavelet transform. The stage, named optimizing the feature vectors, is practiced by feature learning to reduce the redundant features. Then, the K-means clustering algorithm is applied to the chosen features to achieve a segmentation map.

2.1 Feature Extraction

The wavelet transform is first proposed by a France geophysicist Morlet at the beginning of 80's of 20th century as a kind of mathematics tool on signal analysis when analyzing the physical geography signal. The formation of theory of wavelet is the production of the scientists' joint efforts in the numerous realms such as mathematician, physicist and the engineers etc. Developed for several decades, it not only made the breakthrough progress on the theories and methods, but also acquired an extensive application in the realms just as the signal and image analysis, the physical geography signal processing, the computer vision and code and speech recognition etc. [2-4]. The texture image analysis methods based on wavelet have received more and more attention in that the energy measures of the channels of the wavelet decomposition were found to be very effective as features for texture analysis.

The aim of feature extraction is to obtain a set of texture features that can distinguish efficiently between the different texture regions. The texture feature set is usually made up of l_1 norm of subband [5], and its expression is as follow:

$$e = \frac{1}{MN}\sum_{m=1}^{M}\sum_{n=1}^{N}|x(m,n)| \tag{1}$$

Where $x(m,n)$ denotes the wavelet coefficients of a subband in the M×N local neighborhood centered at pixel (m, n).

The dimension of the feature vectors is $D=(3\times L+1)$. L is the decompose level. The steps of extracting wavelet texture feature vectors from textural image are shown as follows [6-7]:

- From top left corner of the image, select a proper local square area as a gliding window.
- Use the wavelet of Daubechies3 to decompose the window, the decompose level is L ,then calculate the l_1 norm of each export frequency band image using formula(1), and give this result to the center pels of the area image as its feature vector.
- Glide the window way and the step length is 1 every time , then calculate the feature vector of center pels of the next image's local area, one by one until the feature vectors of the whole image are calculated.

2.2 Optimizing the Feature Vectors

When we extract the features using wavelet analysis, we can obtain texture features in abundance from an image, but the feature vectors include a great deal of redundant information. The existence of the redundant features which can affect the final segmentation is not effective for textural images.

Due to this problem, we adopt a method optimizing feature vector by feature learning, and make the feature vectors express the feature of different textures more availably and accurately.

We carry out this idea by the following steps. First choose each texture swatch random from the image, clustering them on each dimension and calculate the partition right rate of each dimension. Then wash out the dimensions (make the weight as 0) that redundant features focus on and whose partition right rate is lower than 80%~90% (this parameter could be adjusted appropriately according to actual circumstance). At last give the proper weight to the dimensions left pro rata according to their partition right rate, namely give the dimensions whose right rate is higher a bigger weight and the lower ones a smaller weight [8]. Follow the steps above, the effect on segmentation results of the dimensions that redundant features focus on can be removed, and the effect on results of the redundant features can be reduced. It can make the dimensions whose right rate is higher play a more important role, thus obtain a more accurate segmentation result.

Clustering is a process for dividing a data set into several sets or clusters. It should make the data in a same set have a higher comparability and that in different sets has a lower.K-means clustering has a widely applied realm in numerously clustering algorithm[9]. In this paper, we cluster the feature vectors optimized in feature space with K-means.

3 Experimental Results and Discussion

Experiment 1: An experiment is done in this paper to validate the necessity of reducing the effect of redundant features on segmentation. In this experiment, for the image feature vectors extracted by wavelet transform, we cluster each dimension of feature vectors and the results are shown by figure map. Take the decompose level as 3 and 10 results is obtained after respectively clustering. The experimental results are displayed as follows (see figure 1):

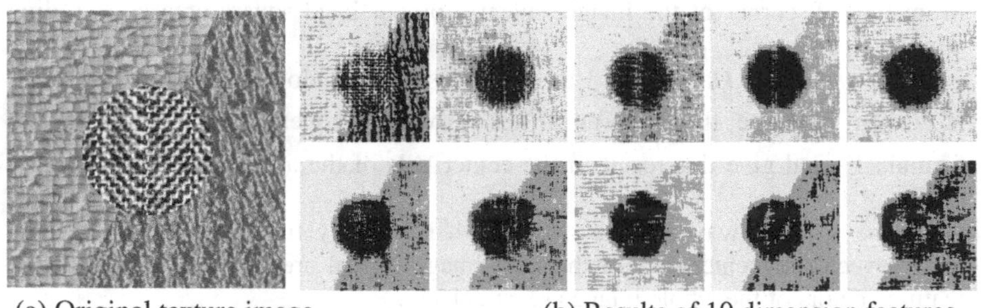

(a) Original texture image (b) Results of 10 dimension features

Fig. 1. Textural image clustering results of each dimension

Easily observed from the experimental results above, the partition error rate in some dimensions is very low, which can express the difference and the boundary among actual textures, but the partition error rate in others is high, which can't express the difference among actual textures and affect the segmentation results veracity (the edge accuracy and the areas consistency) greatly on the contrary. The dimensions whose partition error rate is high include less beneficial information for the segmentation but more redundant features focused on. For the sake of the improvement of segmentation results, reducing the effect of redundant features on segmentation results is really necessary.

Experiment 2: Some experiments are done on several different images to contrast the results after improvement to that of before in this paper. In these experiments, we partition the images by the traditional algorithm and the improved algorithm proposed above. The experimental results are shown by figure map as follows (see figure 2 and figure 3):

In figure 2, (a) is a textural image constituted with two different textures, and in figure 3, (a) is a textural image constituted with four different textures. In figure 2 and 3, (b) is the segmentation result without optimizing feature vectors; (c) is that with optimizing feature vectors in which use the method mentioned in this paper.

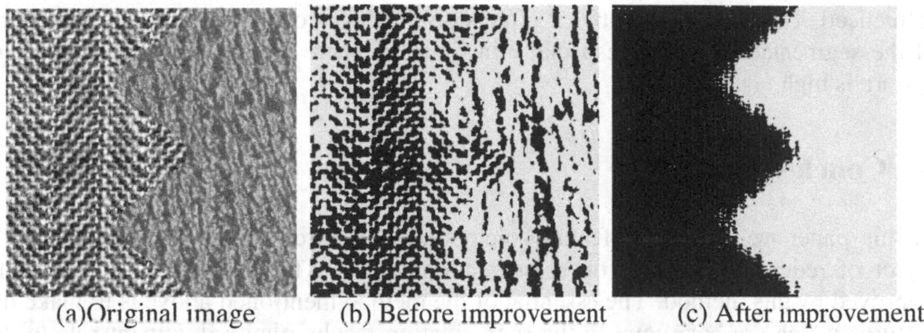

(a)Original image (b) Before improvement (c) After improvement

Fig. 2. Textural image segmentation results (two texture regions)

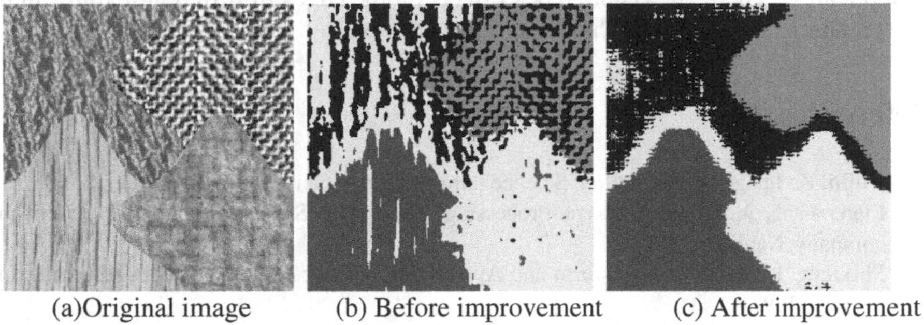

(a)Original image (b) Before improvement (c) After improvement

Fig. 3. Textural image segmentation results (four texture regions)

Seen from figure 2 and 3 clearly, the edge accuracy and the area consistency are improved, and the performance of segmentation is better after improvement. We compare our algorithm with that before improvement by two parameters--the number of the dimensions filtrated and the partition error rate, which are shown in table 1 below.

Table 1. The parameters of experiment 2

Image	**Feature number**		**Error rate/%**		**The size of Square region**
	Before improvement	After improvement	Before improvement	After improvement	
Fig.2	10	7	36.56	2.12	27
Fig.3	10	4	30.19	10.67	27

Seen from the table above, we use lesser feature dimensions but obtain lower partition error rate. To make a comprehensive view to the four parameters analyzed above, our method is valid to the textural image segmentation. It can improve the area consistency and obtain low partition error rate. But because of the dismission of some features, some detail information lost at the same time, so the edge accuracy is influenced. There are visible block effects on the edges of different textures in some of the segmentation results, especially in figure 3 where the number of the different texture is high.

4 Conclusions

In this paper we optimized feature vectors with feature learning, and reduced the effect of redundant features on segmentation results. The segmentation results are improved by this method. The essential of the method mentioned above is to make the features which are beneficial to the segmentation results play a dominant role to the results, and to avoid or lessen the effect of the features which are adverse to the segmentation results. However, this method proposed in this paper has its localization in application. There are visible block effects on the edges of different textures in the segmentation results. Due to this disadvantage, we should do a lot of work in future.

References

1. Yujin, Z.: Image Segmentation. Science publishing company, Beijing (2001)
2. Liangzheng, X.: Digital Image Processing (Revision). Southeast university publishing company, Nanjing (1999)
3. Shixiong, L.: Wavelet Transform and Application. Higher education publishing company, Beijing (1997)
4. Zhengxing, C.: Wavelet Analysis Algorithm and Application. Xi'an jiaotong university publishing company, Xi'an (1998)
5. Yujin, Z.: Image Project-—Image Processing and Analysis. Qinghua university publishing company, Beijing (1999)
6. Zhaoling, H., Dazhi, G., Yehua, S.: Extracting Texture Information of Satellite SAR Image Based on Wavelet Decomposition. Journal of Remote Sensing 5(6), 423–427 (2001)
7. Yu, X., Zaiming, L.: Study on Extension and Statistics in Image WT Coding. Journal of UEST of China 28(3), 223–227 (1999)
8. Gaohong, W., Yujin, Z., Xinggang, L.: Texture Segmentation with Wavelet Transform and Feature Weighting. Journal of Image and Graphics 6A(4), 333–337 (2001)
9. Xiaoyan, D., Zhongyang, G., Qinfen, L., Jianping, W.: An Overview of Spatial Clustering Analysis and Its Application. Shanghai Geology (4), 41–46 (2003)
10. Mukhopadhyay, S., Chanda, B.: Multiscal Morphological Segmentation of Gray-Scale Images. IEEE Transactions on Image Processing 12(5), 533–549 (2003)
11. Derraz, F., Taleb-Ahmed, A., Peyrodie, L., Pinti, A., Chikh, A., Bereksi-Reguig, F.: Active Contours Based Battachryya Gradient Flow for Texture Segmentation. In: 2nd International Congress on Image and Signal Processing (CISP), Tianjin, pp. 1–6 (2009)
12. Shuang, W., Xiuli, M., Xiangrong, Z., Licheng, J.: Watershed-based Textural Image Segmentation. In: International Symposium on Intelligent Signal Processing and Communication Systems (ISPACS), Xiamen, pp. 312–315 (2007)
13. Gaetano, R., Scarpa, G., Poggi, G.: Hierarchical Texture-Based Segmentation of Multiresolution Remote-Sensing Images. IEEE Transactions on Geoscience and Remote Sensing 47(7), 2129–2141 (2009)
14. Paragios, N., Deriche, R.: Geodesic Active Regions and Level Set Methods for Supervised Texture Segmentation. Int'l J. Computer Vision 46(3), 223–247 (2002)

Magneto-electric Generator Based on Automotive Vibration Energy

Jian Hu, Jian Sun, and Yanfei Zhou

School of Mechanical and Electronic Engineering, Wuhan University of Technology, 122, Luoshi Road, 430070, Wuhan, Hubei Province, P.R. China
hujian@whut.edu.cn, {sj517794780,feifei90315}@163.com

Abstract. Due to the uneven road, the unstable quality of transmission system and so on, the automotive rear axle will vibrate randomly, which decreases the energy transformation efficiency of automobile in a certain extent. In order to use automotive vibration, a kind of magneto-electric vibration generator is put forward in this paper. The generator can collect the automotive vibration energy, which is transformed into usable electrical energy. The device is designed based on the principle of magnetic induction, and permanent magnet installed on the mandrel can feel the automotive vibration incentive. Therefore, it makes reciprocating motion in axial direction and relative movement with the coil, eventually cuts the magneto-tactic lines to generate electricity. The structure and three-dimensional model, the design of parameters and experiment platform are researched. The generator's advantages include small volume, high efficiency, saving energy and environmental friendly.

Keywords: Automotive Vibration, Permanent Magnet, Energy Saving, Generator.

1 Introduction

The vibration is inevitable during the automotive working. Quantities of energy are missed and noise and vibration are also brought at the same time, which decrease the comfortable and stable quality of cars. As the cars are becoming universal, the energy loss caused by vibration has been great. Since the global energy becomes less and less and Chinese government is putting emphasize on the construction of conservation-minded society, the full use of energy is a problem that we have to solve. The vibration of cars comes from quantities of reasons including the uneven roads, the change of speed's value and direction, the bumpy operation of wheel, the engine and transmission system, gear impact and so on. All of these will cause the vibration of cars and it's rather difficult to solve it by technique means. The vibration causes many unfavorable impacts for the operation of car, which include partially using of energy, bad economy, bad passing, steering stability and quiet running, damaging components and parts, shortening the using span, and so on. Therefore, we research on a kind of generator device for collecting the vibration, transforming the energy of vibration into electronic power, then offering the electrical system extra electronic power, thus achieving the goal of raise the using efficiency of car energy's fuel.

D. Jin and S. Lin (Eds.): Advances in CSIE, Vol. 1, AISC 168, pp. 467–473.
springerlink.com © Springer-Verlag Berlin Heidelberg 2012

Since most of road surface in China is B (Cement and asphalt) or C degree. We assume that the car is running on the B degree's road. When the car is running, due to roads of different degrees and different speed, road's roughness incentive is varying. Road model is a math description, which reflects how the road roughness varies with the change of speed and road degree. Using the Simulink, the output curve of spring load angle and spring load quality mass displacement can be got under the system parameters and road incentive. Adding the average car speed of 36-108 Km/h, relating wheel input time frequency range of 0.1-75Hz, the suspension system inherent frequency of 1Hz and the opposite of suspension system's inherent frequency of 10-16Hz, we can get that in the 10-16Hz range of frequency, absorber's spring load quality mass displacement amplitude lies mostly on -40dB. After counting, we can get that the vibration amplitude of rear frame is about 2cm.

Therefore, the design working environment for magneto-electric vibration generators are: the car speed vary from 36 to 108 Km/h, on the road of B degree, with road surface and car vibration's incentive on the rear frame, the vibration frequency's range is 10-16 Hz, vibration amplitude is about 2cm.

2 Selection of the Magneto-electric Vibration Generator Structure

Based on the design environment of system, we can design the magneto-electric vibration generator into straight line vibration generator. During the design of straight line vibration generator, there are two main problems: (1) the selection of permanent magnet's internal or external structure; (2) the selection of stator.

In the generator, permanent magnet's place can be internal or external as shown in Fig.1.

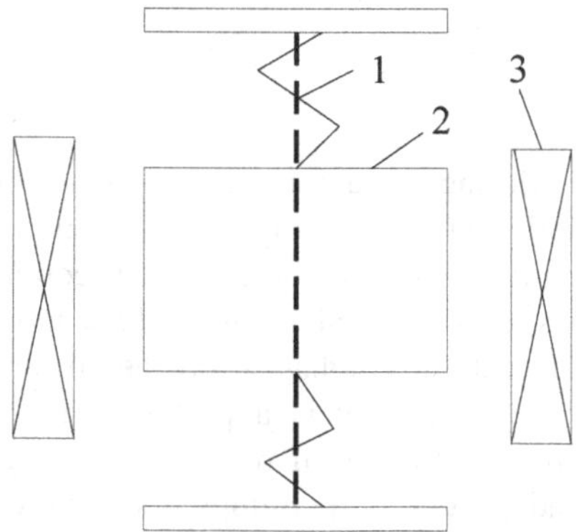

(a) Internal type of the permanent magnet

Fig.1. The place scheme of coil and permanent magnet

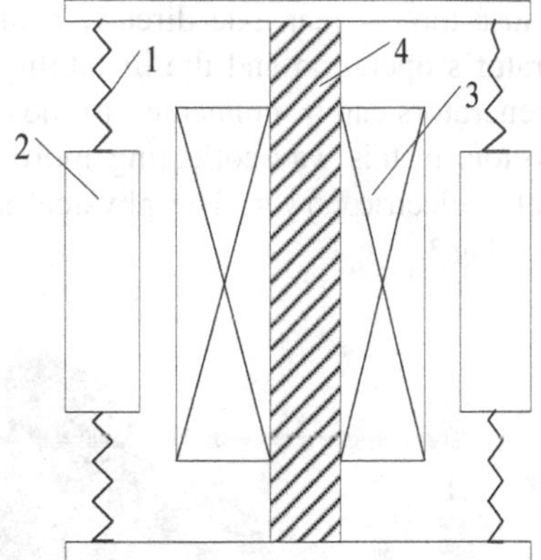

(b) External type of the permanent magnet
1.spring 2.permanent magnet 3.coil 4.core figure

Fig. 1. *(continued)*

According to literatures, compared to permanent magnet built-in generator, the permanent magnet built-out generator has many advantages including current-limiting, small degauss effect, strong stability, and high efficiency. The experiment result demonstrates that the permanent magnet built-out generator cost less than the other one, but its no-load induction EMF amplitude is higher about 0.6V, and its efficiency is raised by 1.64% or so. The simulation analysis and the experiment result both show that the permanent magnet built-out generator is more supreme and has bigger development potential. Considering comprehensively, we choose the generating type of permanent magnet built-out place.

In the aspect of stator selection, considering the character of automotive vibration, we use the way of coil fixed and permanent magnet's respond to the automotive vibration to make the permanent magnet and coil move relatively. The selection can make the manufacture of vibration generator more convenient.

3 Design of Three-Dimensional Model of Magneto-electric Vibration Generator

The three-dimensional model of magneto-electric vibration generator is shown in Fig.2. Coil is winding on the permanent magnet, fixed in the generator shell though conduction materials. Permanent magnet of sphere-shaped is put on the central spindle, bonding with the shell though spring and shell. The joint of central spindle and shell is line bearing. With this way, the efficiency can be raised by decreasing the friction. The work theory is: when the vibration generator receives the outside vibration, central spindle drives the permanent magnet and bilateral spring to do line reciprocating motion in the axial direction, thus makes relative motion with coil, and produces the movement of cutting the magneto tactic lines, forming electric currency. Then the electric currency is put into rectification circuit though head, outputting stable direct current. Because the car rear axle directly bears the vibration incentive from uneven roads, generator's operation and the instability of transmission system, more than one vibration generators can be connected to the car rear axle by screws or magnet in the real application, in this way collecting more car vibration energy and transforming it into available electric power. The physical sample of magneto-electric vibration generator is shown in Fig.3.

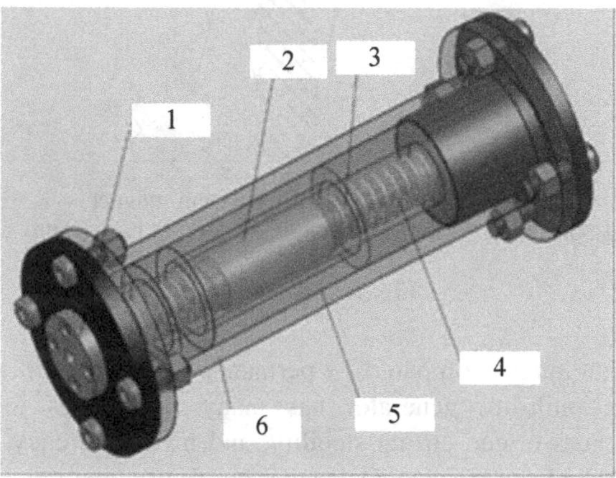

1. linear bearing 2. permanent magnet 3. mandrel 4. spring 5. coil 6. shell

Fig. 2. The three-dimensional model of magneto-electric vibration generator

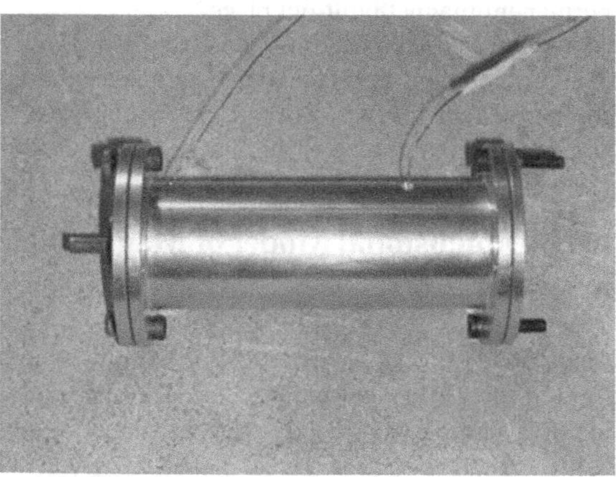

Fig. 3. The physical prototype of magneto-electric vibration generator

4 Design of the Parameters of Magneto-electric Vibration Generator

4.1 Design of the Size of Permanent Magnet

According to the permanent magnet formula, the size of permanent magnet motor is as follow:

$$V_{\mathrm{m}} = 51 \frac{P_N \sigma_0 K_{ad} K_{\mathrm{dF}}}{fK_g K_{\mathrm{u}} C(BM)_{\max}} * 10^9 \mathrm{mm}^3 \tag{1}$$

Where:

P_N is the rated capacity of the generator. We assume that $P_N = 5W$;

Based on calculation, the magnetic leakage factor $\sigma_0 = 1.05$;

Axial magnetic potential armature reaction straight into the coefficient $K_{ad} = 0.797$;

The permanent magnet magnetic potential and straight shaft magnetic potential armature reaction when the generator is in short circuit $K_{df} = 0.95$;

The vibration frequency of the system is f=15Hz;

The biggest energy utilization coefficient Kg=0.89;

Voltage waveform coefficients $K_u = 1.414$;

Boron for the biggest energy of the product C=0.9;

$(BM)_{max} = 240KJ/M^2$.

Therefore, $V_m = 15100mm^3$.

After calculating, the Outer diameter is φ=20, the inner diameter is φ=6, the magnets of L=40 can meet the needs. The hole of φ=6 can be used as the guide mounting holes.

4.2 Calculations of the Generator Armature Winding Circle Number

The vibration generator armature winding circle number can be determined by formula(2).

$$N = E_0/(4.44fBS) \tag{2}$$

Where:

E_0 is No-load emf;

F is the frequency of the generator;

B is the residual magnetism induction intensity of the permanent magnets;

S is the cross-sectional area of the permanent magnets;

Then, we can calculate the vibration generator armature winding circle number as following: N=820.

Based on the actual winding occasion, 900 turn on aggregate. As to the small permanent magnet motor, current density is $3.5\text{-}5A/mm^2$. Assuming line diameter is 0.2mm, as a result, the requirements can be satisfied. Winding 200 turns in every layer, then the length of the winding line is 0.2mm*300=60mm, coherent with the dimension of the magnets. Consequently, we know there are 3 layers of winding lines in total.

The total length of the wire is:

L=900*π*d=900*π*26=73m.

5 Experiments and Analysis of Magneto-electric Vibration Generator

We have adopted the permanent magnet built-in generators in our analysis process. However, there are still a lot of advantages in the permanent magnet built-in generators: The permanent magnet built in generators don't contain a Iron core, the permanent placed in the internal winding can vibrated by fluctuating. Compared with the other vibration generator, this kind of generator have simpler configuration, lighter active cell, therefore the vibration frequency is much more higher, the trip is longer.

The simulation platform of magneto-electric vibration generator is shown in Fig.4. Placing the manufactured vibration generator in vibration experiment table ,which is used for simulating the actual road situation, to test the theoretical results and collect data. In the first round, LED can flashed normally, which demonstrate that the vibration system is advisable. However, after a period of duration, we noticed that the electric energy production is little, far from our initial expectation, accounting for the low energy transforming frequency. By analyzing, we concluded that it owes to the unreasonable design of some parts, as well as the disqualification of the manufacturing craft, leading to many harmful frictions, which can furthermore result in low efficiency of the system.

Re-produced the brand new vibration system by improving the manufacturing craft and tested again, LED flashed normally again, what's more, after period of duration, the electric energy production is enhanced. Approaching the design aims, the energy transform efficiency is high.

All from above, we summarized that the design of the system is reasonable. It can transform vibration into useful electricity efficiently. Nevertheless, there are still some defects, such as we should adopt in vitro for permanent magnet type power generation, thus the power efficiency can be improved.

Fig. 4. Simulation platform of magneto-electric vibration generator

6 Conclusions

The automotive vibration uses the permanent magnet built-in generators, line cutting the magnetic induction line and transforming the vibration mechanic energy into electric power. Through theoretical design, we manufactured the actual automotive vibration generator, did actual experiments, which demonstrates that magneto-electric vibration generator can transform the mechanic energy of automotive vibration into electric power. Automotive vibration generator also has the advantages of small volume, low cost, convenient manufacturing. Based on these advantages, magneto-electric vibration generator will be promoted well. Promoting the automotive vibration system will raise the fuel efficiency of cars, donating to the sustainable development of the world.

Acknowledgment. We would like to give our acknowledgments to Wuhan University of Technology and Wuhan City, Hubei Province, China because the research work reported here has been supported by "2010 Teaching Research and Innovation Project of Wuhan University of Technology" and "2012 International Science and Technology Cooperation Plan Project of Key Science and Technology Problem Project of Wuhan City"-"Research on Passenger Car Pneumatic Braking Control Technology for Active Safety".

References

1. Guo, L., Lu, F., Ye, Y.: Performance of Novel Linear Reciprocating Generator with External Permanent Magnet. Journal of Zhejiang University (Engineering Science) 41(9), 1604–1608 (2007) (in Chinese)
2. Zhen, Y.: Electrodynamics, 3rd edn. Science Press, Beijing (2010) (in Chinese)
3. Sun, G., Qiang, W.: Magnetic Functional Materials. Chemical Industry Press, Beijing (2007) (in Chinese)
4. Wen, B.: Mechanical Vibration. Metallurgical Industry Press (2000) (in Chinese)
5. Xian, H., et al.: Study on Internal Permanent Magnet Linear Reciprocating Generator. Micro Generator 43(1), 16–17 (2010) (in Chinese)
6. Zhou, S.: Electromagnetic Field and Mechanical and Electrical Energy Conversion. Shanghai Jiaotong University Press, Shanghai (2008) (in Chinese)
7. Cheng, D.: Mechanic Design Manual, 5th edn. Chemical Industry Press, Beijing (2007) (in Chinese)

Handwriting Digit Recognition Based on Fractal Edge Feature and BP Neural Net

YingChao Zhang, TaiLei Liu, and XiaoLing Ye

College of Information and Control,
Nanjing University of Information Science & Technology, Nanjing 210044, China
{yc.nim,peggy_ltl,xyz.nim}@163.com

Abstract. Handwritten digit recognition (HDR) is one of the difficult research areas on pattern recognition, Hence, evaluation a performance of algorithms on HDR problem is of great importance. In this paper, a method of the recognition of the handwriting digits based on fractal edge feature and BP neural net is proposed. First, extracting the edge feature of the character by using the fractal edge detect method; then, combining the fractal edge feature with three other features (ring zones, projection histograms, moments); finally, taking BP neutral net as the classifier and getting the result. The experimental results show that the proposed method has good anti-noise and high accuracy performance.

Keywords: pattern recognition, handwriting digit recognition, fractal edge feature, BP neural net.

1 Introduction

Handwritten digit recognition (HDR) is an important branch of optical character recognition technology (OCR), its object of study is how to make use of computer automating recognition of Arabic handwriting on the paper. HDR systems typically include two steps: feature extraction and classification. The features can be classified into two major categories: statistical and structural features [1]. Some previous works on recognition of isolated characters have used structural features [2], moment features [3] wavelet features [4] and fractal features [5].

At present, lots of researches have been done on the aspect that fractal dimension is used as image feature in image analysis. [6], [7] have presented the edge detect method base on fractal dimension, and proved the anti-noise performance of this method. In this paper the fractal edge feature, together with three other features (ring zones, projection histograms, and moments), is introduced into the HDR system.

The paper is organized as follows. Firstly, describes the image preprocessing algorithms. Secondly, presents the extraction method of the fractal edge feature and three other features. Finally, the experiment results are provided, and a conclusion is given.

D. Jin and S. Lin (Eds.): Advances in CSIE, Vol. 1, AISC 168, pp. 475–481.
springerlink.com © Springer-Verlag Berlin Heidelberg 2012

2 Image Preprocessing

Image preprocessing refers to the original image is converted into a recognizer that can accept binary form. The preprocessing stage here includes two major steps: the processing of image binarization and the processing of digit slant correction.

2.1 Binarization

The image binarization means convert the image into gray image that only contain two kinds of values (0, 1). Its expression is as follows:

$$g(i,j)=\begin{cases}1, & f(i,j)<T\\ 0, & f(i,j)\geq T\end{cases} \tag{1}$$

Where $f(i,j)$ is the gray level of the pixel (i,j), T is binarization threshold.

2.2 Slant Correction

The slant correction method is, first, estimated the slant angle as the inclination of the line connecting the gravity centers of the top 25% part and the bottom 25% part of the image (Fig. 1), then, a sub-pixel precision shear transformation is performed in order to remove the estimated inclination.

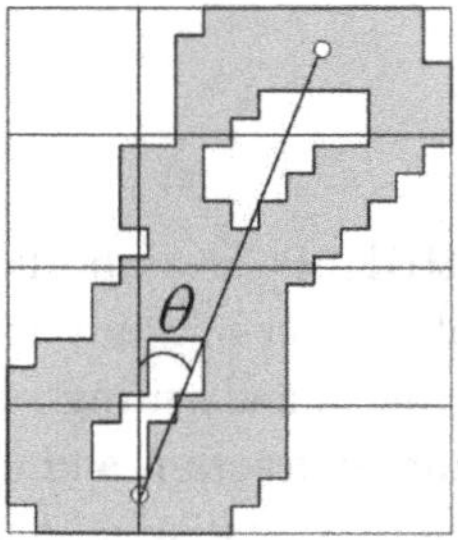

Fig. 1. Slant correction

2.3 Feature Extraction

Fractal Edge Feature. Mandelbrot defined the behavior of a fractal surface as follows[8]:

$$Area(\varepsilon)=K\varepsilon^{2-D}. \tag{2}$$

Where D is the fractal dimension of the surface, and K is a constant. Here K reflects the variety ratio of gray surface area of the image, the rougher the image gray surface is, the much bigger K is, so it can be taken as the edge feature [6]. Taking the logarithm of the both sides of Eq.2 yields:

$$\log Area(\varepsilon) = (2 - D)\log \varepsilon + \log K . \tag{3}$$

In this paper, the blanket method [9] has been introduced to calculate the coefficient K of local window, and each local window has size $w \times w$ .The algorithm is described as follows:

Covering the image surface $g(i, j)$ by using a blanket with top μ_ε and bottom b_ε surfaces:

$$u_\varepsilon(i, j) = \max\left\{u_{\varepsilon-1}(i, j) + 1, \max_{|(m,n)-(i,j)|\le 1} u_{\varepsilon-1}(m, n)\right\} \tag{4}$$

$$b_\varepsilon(i, j) = \min\left\{b_{\varepsilon-1}(i, j) - 1, \min_{|(m,n)-(i,j)|\le 1} b_{\varepsilon-1}(m, n)\right\} \tag{5}$$

Where $u_0(i, j) = b_0(i, j) = g(i, j)$, if ε is the number of blankets, the area of the blanket $Area(\varepsilon)$ is computed by:

$$Area(\varepsilon) = \frac{\sum_{i,j}\left(u_\varepsilon(i, j) - b_\varepsilon(i, j)\right)}{2\varepsilon} \tag{6}$$

Then the K value may be estimated from the linear fit of Eq.3 and Eq.6 with the blanket's scale range from 1 to ε .

The fractal edge feature extract method can be described as follows:

1) Estimating the fractal signature K pixel by pixel based on the above algorithm.
2) Computing the maximal value $K_{\max}$ of image and choose an appropriate coefficient a , here $0.5 < a < 1$. Let $Th = a \times K_{\max}$. Scanning the image, the pixel are considered as real edge if their K are larger than Th . Otherwise is not real edge.
3) Getting the pixel coordinate (i, j) of the edge points (pixels which $K > Th$), and calculating the centroid C of the digit.
4) Computing the distance d between the centroid C and all edge points, here we take the top left pixel as the initial point (Fig.2). Then the fractal edge feature will be $F_{edge} = [d_1, d_2, d_3, \cdots d_n]$, where n is the number of edge points.

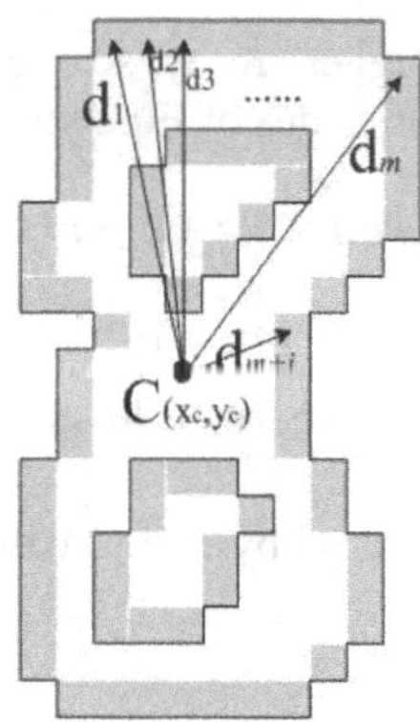

Fig. 2. Fractal edge feature

Ring Zones. According to this feature, features are extracted as pixel counts in rings zones around the gravity center of the image (Fig. 3a). We have used three rings, each divided in different number of equal zones. The outmost ring has a radius r equal to the distance from the gravity center to the furthest black pixel of the image. The radius of the innermost ring and the second ring is 0.2r and 0.5r, respectively.

Projection Histograms. According to the projection histograms, image is scan along a line from one side to another side and number of fore ground pixel on the line is counted. In this experiment, we calculate horizontal, vertical and diagonal projections (Fig.3b).

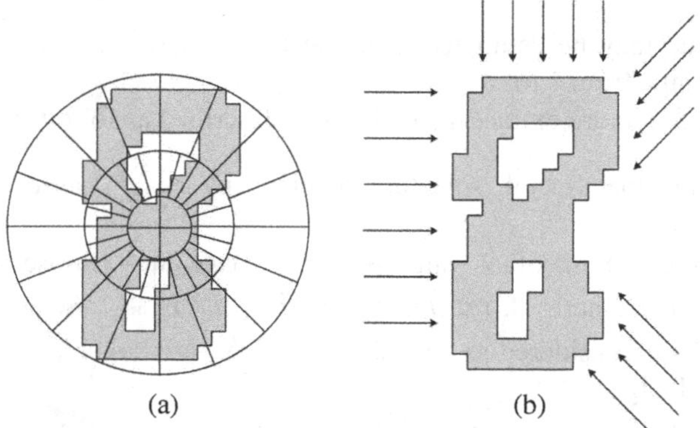

Fig. 3. (a)Ring zone (b) Projection histograms

Moments. Moment invariants have been widely applied to image pattern recognition in a variety of applications due to its invariant features on image translation, scaling and rotation [10]. In our case we use the seven Hu-moments for our experiments.

3 Experiments

3.1 Dataset

Handwritten digits database that has been used in the experiments is derived from the original MNIST database [11]. The MNIST database has a training set of 60,000 examples, and a test set of 10,000 examples, the digits have been size-normalized and centered in a fixed-size image. Here, we used MATLAB to extract the isolated handwritten images from each file. Fig. 4 shows a sample of the dataset.

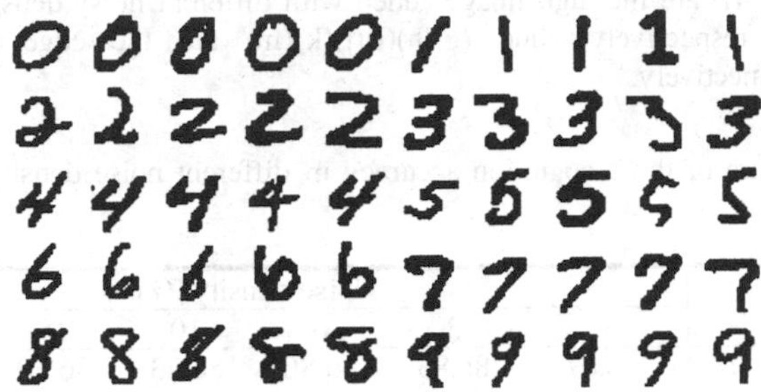

Fig. 4. The MNIST database

3.2 Experiments and Results

In this experiment the digits database has a training set of 20000 examples, and a test set of 5000 examples derived from the MNIST, then the experiment steps are described as follows:

1) Image preprocessing, since the MNIST has already been size-normalized, we can just do the other two steps: binarization and slant correction.
2) Feature extraction, in our experiment 4 kinds of features will be extracted: 72 fractal edge features, 44 ring zone features, 18 projection features (both 5 features from the horizontal projections and from the vertical projections, and 4 features from each of the two diagonal projections), and 7 moments features.
3) Classification, BP neural net is introduced in this experiment, here we set three layers: input layer, one hidden layer, and the output layer. The input is the feature vector mentioned above in (2), the output is the binary code of the digits (e.g. the binary code of 9 is 1001).

Experiment A. Testing the anti-noise performance of the fractal edge feature.
First, we add different density of noise in the handwriting digit image, and detect their edge by using the fractal edge method (Fig. 5). Then we only take the fractal edge feature as the input of the BP neural net, and test the recognition accuracy under different noise density (Table 1).

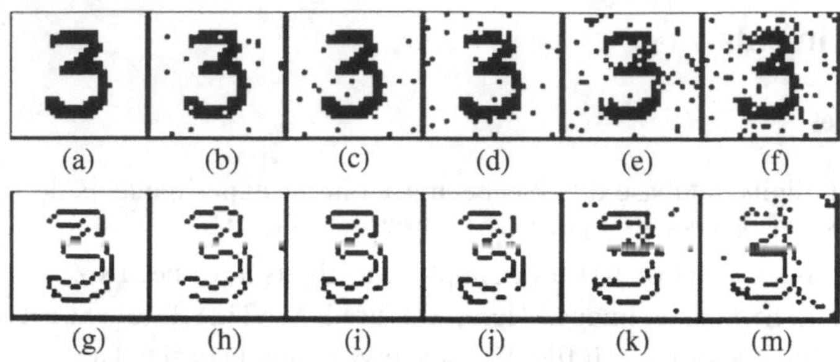

Fig. 5. (a)(b)(c)(d)(e)(f) are the digit image added with different noisy density (0%, 3%, 6%, 10%, 15%, 20%, respectively), and (g)(h)(i)(j)(k)(m) are the edge detect result of (a)(b)(c)(d)(e)(f), respectively.

Table 1. The compare of the recognition accuracy in different noise density with the fractal edge feature only

Digits		Noise density [%]					
		0	3	6	10	15	20
Accuracy [%]	0	86.87	86.83	86.80	86.73	86.60	86.46
	1	82.45	82.41	82.38	82.31	82.18	82.01
	2	78.47	78.43	78.40	78.36	78.22	77.98
	3	80.13	80.08	80.03	79.96	79.83	79.69
	4	80.71	80.67	80.62	80.56	80.42	80.27
	5	79.69	79.64	79.59	79.55	79.50	79.39
	6	77.17	77.14	77.12	77.08	77.02	76.91
	7	81.37	81.34	81.29	81.21	81.12	80.98
	8	76.01	75.97	75.94	75.91	75.85	75.76
	9	77.09	77.06	77.04	77.01	76.94	76.86
Average		79.50	79.46	79.42	79.37	79.27	79.13

Experiment B. Testing the recognition accuracy of the multi-feature input.
In this experiment, all of the four features which has introduced above are taken as the input of the BP natural net, the accuracy result is showed in Table 2.

Table 2. The recognition accuracy with multi-feature input

Digits	Accuracy [%]	
	Training	Testing
0	97.89	92.56
1	98.95	93.27
2	95.69	91.13
3	97.52	92.08
4	97.96	92.46
5	97.13	91.98
6	95.24	90.02
7	96.93	92.18
8	94.89	90.45
9	96.26	91.29

4 Conclusions

In this paper, a handwriting digit recognition method based on fractal edge feature and BP neutral net was proposed. First, we test the anti-noise performance of the fractal edge feature, and then we test the accuracy performance with the multi-feature.

The present results show that the fractal edge feature has a high anti-noise performance, and we also proved that the recognition accuracy can be improved by using multi-features.

Acknowledgments. This work was supported by the National Project of Public Welfare Industry (Meteorology) Research (GYHY201106040), the Project Funded by Prospective Joint Study of Jiangsu Province (BY2011111), the Project Funded by Priority Academic Program Development of Jiangsu Higher Education Institutions (Project name: Sensor Networks and Modern Meteorological Equipment).

References

1. Mozaari, S., Faez, K., Ziaratban, M.: A Hybrid structural/statistical classifier for handwritten Farsi/Arabic numeral recognition. In: Proceedings of IAPR Conference on Machine Vision Application, Japan (2005)
2. Gorgevik, D., Cakmakov, D.: Handwritten digit recognition by combining SVM classifiers. In: IEEE Proceeding of International Conference on Computer as a Tool, pp. 1383–1396 (2005)
3. Hossian, M.Z., Amin, M.A., Hong, Y.: Rapid feature extraction for Bangla handwritten digit recognition. In: IEEE Proceeding of International Conference on Machine Learning and Cybernetics, pp. 1832–1837 (2011)
4. Mowlaei, Faez, K., Haghighat, A.T.: Feature extraction with wavelet transform for recognition of isolated handwritten Farsi/Arabic characters and numerals. In: Proceedings of 14th International Conference and Digital Signal Processing, pp. 923–926 (2002)
5. Mozaffari, S., Faez, K., Rashidy Kanan, H.: Recognition of isolated handwritten Farsi/Arabic alphanumeric using fractal codes. In: IEEE Proceeding of Southwest Symposium on Image Analysis and Interpretation, pp. 104–108 (2004)
6. Qiong, L., Jun, G., Long, G., et al.: A novel edge detection method based on fractal theory. In: IEEE Proceeding of International Conference on Neural Networks and Signal Processing, pp. 1105–1108 (2003)
7. Jipeng, D., Yanming, G., Peng, Q., et al.: Multi-feature edge extraction for gray-scale images with local fuzzy fractal dimension. In: IEEE Proceeding of International Conference on Fuzzy Systems and Knowledge Discovery, pp. 583–587 (2010)
8. Mandelbrot, B.B., Passoja, D.E., Paullay, A.J.: The fractal character of fracture surface of metals. Nature, 721–722 (1984)
9. Peleg, S., Naor, J., Hartley, R., et al.: Multiple resolution texture analysis and classification. IEEE Trans. Pattern Anal. Mach. Intell., 518–523 (1984)
10. Zhihu, H., Jinsong, L.: Analysis of Hu's Moment Invariants on Image Scaling and Rotation. In: IEEE Proceeding of 2nd International Conference on Computer Engineering and Technology (ICCET), pp. 476–480 (2010)
11. Information on, `http://yann.lecun.com/exdb/mnist/`

Software Development of Crude Oil Storage System Energy Consumption Analysis and Comprehensive Optimization

Jian Zhao[1], Yang Liu[1], Hang Dong[1], and XueQing Sun[2]

[1] Northeast Petroleum University
[2] The Designing Institute of Daqing Oil Field, Daqing, P.R. China
soulkissing@163.com

Abstract. The mathematical model of energy consumption analysis and execution plan optimization of crude storage system was established. On the basis of .Net platform and Access 2003 DB tool, With the help of OLEDB database connection and GDI+ graphics applications interface, The software of energy consumption analysis and comprehensive optimization of crude storage system was developed, by which the function of static and dynamic data management, graphics modeling, energy consumption analysis and execution plan optimization can be realized.

Keywords: crude oil storage, software, energy consumption analysis, optimization.

1 Introduction

According to the twelfth five-year plans, the developed aim of low-carbon economy makes higher demands on energy consumption level of crude oil storage and transmission system. On the basis of mathematical model of crude oil storage energy consumption analysis and execution plan optimization, according to the software engineering, the common software of energy consumption analysis and comprehensive optimization of crude oil storage system was developed which can be used as a tool to find out the energy consumption level of crude oil storage and provide the decision support for managers. Finally, the efficient and lower energy consumption aim of crude oil storage system will be achieved.

2 Mathematical Model

2.1 Energy Consumption Analysis

By three-link modal, the energy consumption pattern of crude oil storage system can be divided into three kinds, the energy conversion process, transmission process and technological utilization process. The energy use efficiency and lost energy ration were taken as evaluation index, the mathematical model of crude oil storage energy consumption analysis was established.

D. Jin and S. Lin (Eds.): Advances in CSIE, Vol. 1, AISC 168, pp. 483–488.
springerlink.com © Springer-Verlag Berlin Heidelberg 2012

(1)Energy conversion process

The major energy conversion units are the pump which is used to convert electric energy into pressure of oil and the furnace which is used to convert chemical energy of fuel into thermal energy.

①Furnace

Energy Balance Modal: $E_{suph} + E_{supe} + E_{jin} - E_{jout} + (E_{l1} \mid E_{l2} \mid E_{l3}) + E_{lin}$ (1)

Positive Balance Efficiency: $$\eta_{jz} = \frac{Q_j}{3600BQ_r} \times 100\% \tag{2}$$

Counter Balance Efficiency: $$\eta_{jf} = 100\% - (q_1 + q_2 + q_3 + q_4 + q_5) \tag{3}$$

②Pump

Energy Balance Modal: $$E_{pin} + E_{supe} = E_{pout} + E_{lp} \tag{4}$$

Positive Balance Efficiency: $$\eta_{pz} = \frac{N_e}{N_c} \times 100\% \tag{5}$$

Counter Balance Efficiency: $$\eta_{pf} = \frac{H}{H + 102C_p(\Delta t - \Delta PS_p)} \times 100\% \tag{6}$$

(2)Energy transmission process

Pipeline is the major energy transmission unit, the energy input and output of pipeline are the pressure and heat energy of starting point and terminus. The lost energy during transmission process mainly includes the pressure loss caused by friction and topography and thermal dissipation.

Transmission efficiency : $$\eta_g = \frac{Q_{se}''}{Q_{se}'} \times 100\% \tag{7}$$

Lost energy: $$E_{Ll} = \Delta E_{Lh} + \Delta E_{Lp} = (E_{Lhin} + E_{Lpin}) - (E_{Lhout} + E_{Lpout}) \tag{8}$$

Lost energy per unit: $$M_L = \frac{\Delta E_{Lh} + \Delta E_{Lp}}{L \cdot G_L} \tag{9}$$

Energy loss ratio: $$\eta_L = \frac{\Delta E_{Lh} + \Delta E_{Lp}}{E_{Lhin} + E_{Lpin}} \times 100\% \tag{10}$$

Heat loss ratio: $$\eta_{Lh} = \frac{\Delta E_{Lh}}{E_{Lhin} + E_{Lpin}} \times 100\% \tag{11}$$

Pressure loss rati $$\eta_{Lp} = \frac{\Delta E_{Lp}}{E_{Lhin} + E_{Lpin}} \times 100\% \tag{12}$$

(3)Energy technological utilization process
Generally, the tank needs to be heated to insure the oil is stored safely in winter. So, the tank can be taken as an energy utilization unit.

Energy balance model: $$E_{\text{ghin}} + E_{\text{suph}} + E_1 = E_{\text{ghout}} + E_{\text{lgh}} + E_2 \quad (13)$$

Energy loss ratio: $$\xi_{\text{g}} = \frac{E_{\text{lgh}}}{E_{\text{suph}} + E_1 + E_{\text{ghin}}} \times 100\% \quad (14)$$

Heat energy efficiency: $$\xi_{\text{g}} = \frac{\max(E_2 \text{-} E_1, 0) + E_{\text{ghout}}}{E_{\text{suph}} + E_1 + E_{\text{ghin}}} \times 100\% \quad (15)$$

2.2 Oil Scheme Optimization

(1)Optimization model
The energy consumption of crude storage system includes the power consumption of pumps and fuel consumption of furnaces. On condition that the system perform oil storage and transportation safely, the purpose for optimizing scheme is to formulate schemes and parameters of oil pump and heating furnace to make the comprehensive energy consumption fees of crude storage system lowest, The mathematical model[2] of the optimization problem can be established.

(2)Solving algorithm
In the problem of scheme optimization, the variable which characteristic the open positions for oil pump and heating furnace is discrete variable, the equipment running parameters are continuous variable. For both the target function and constraint conditions include nonlinear functions, so the optimization problem belong to nonlinear mixed variables optimization problem[2,3]. In this paper, the problem can be solved by using hierarchical optimization method[3].

3 Software Running Mechanism

Based on the mathematical model, crude oil storage system energy analysis and comprehensive optimization software is developed on the platform of .Net. Based on object-oriented program design idea, the data management, graphics management, system energy consumption analysis and comprehensive optimization modules of software were established by combining the DAManager, GRManager, ENEvaluate and PIOptimize custom class with OLE DB database connection technology and GDI+ graphics application interface. The running mechanism is shown bellow.

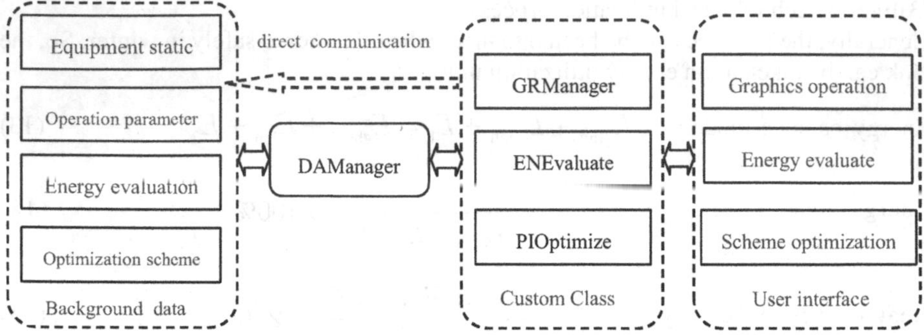

Fig. 1. Software running mechanism

4 Main Function Module

4.1 Data Management

Based on the function of software, The data management module can be divided into three parts, the background database, application program interface and data management custom class.

(1)Database Design. Based on the Access2003 system, the software database was established which was divided into equipment static database, system operation parameters database, system energy using database and optimized plan database..

①Equipment static database. This database includes tables of pipeline, furnace and other equipment's static data which used to store equipment's inherent information. Taking pipelines for example, the properties include length, diameter, buried depth, etc.

②System operation parameters database. This database includes data tables of pipelines, pumps or other equipment's operation parameters data which used to store system operation data of different date.

③System energy consumption analysis database. This database includes pipeline, pump or other equipment's energy consumption results. Taking pipelines for example, the properties include pipe transmission efficiency, energy loss rate, heat loss rate data, etc.

④Optimized plan database. This database includes tables of pumps, furnaces and other equipment's optimized parameters used to store optimized operation parameters.

(2)DAManager Class. In order to realize the communication between application and backend database, Based on OLEDB technology, The DA Manager class was created, which can be used to realize the communication with backend database, data reading and data edit. In DA Manager class, The backend database is taken as OLE DB provider, the application is taken as OLEDB data user, and the database is operated by OLEDB Connection, OLEDB Command and OLEDB Data Reader methods etc.

(3)Data management interface. The data management interface of application is the interactive channel between users and database. With it, Users can directly manipulate the database without understanding its implementation and principles. There are two ways to accomplish this function:

①Data management interface. On the basis of visual programming technology of C#.Net, the data management interface of software was designed which is used to implement the communication between users and backstage database.

②Data manipulation by graphic management module. Not Only dose the graphic management module have the function of system modeling, but also by the graphic interface, the communication between users and database is achieved by the custom class of this module.

4.2 Graphics Management Module

Graphical interface is a visualized friendly and efficient man-machine interaction method, which enables users to have an intuitive understanding to the internal process flows of storage. Graphics management class GR Manager is set based on the drawing application programming interface GDI+, In the meantime, the database connection technology OLE DB is embedded into GR Manager class, making users handle equipment data information in a graphical interface except operating the graphical interface. The realization mechanism of the module is as follows:

(1)The Draw Line•Draw Rectangle methods in the Graphic class etc of the GDI+ technology was applied to realize the drawing of the pipeline and the nodes such as joints or branch points;

(2)The professional drawing software AutoCAD was applied to create figure unit files of the equipments with complex structure such as pumps, furnace etc, the FromImage method of graphic class is used to realize the creating and operating of graphics file;

(3)The TImage components was applied to act as the drawing pad of the graphic operating interface, with the attributes of event function and drawing buttons' properties of the TImage component, the users' communication through the application interface and GR Manager class was realized;

(4)By embedding the OLEDB technology into GRManager class and editing programming code, the operating processes such as reading and writing and inquiring of the background date through the graphics management module were achieved.

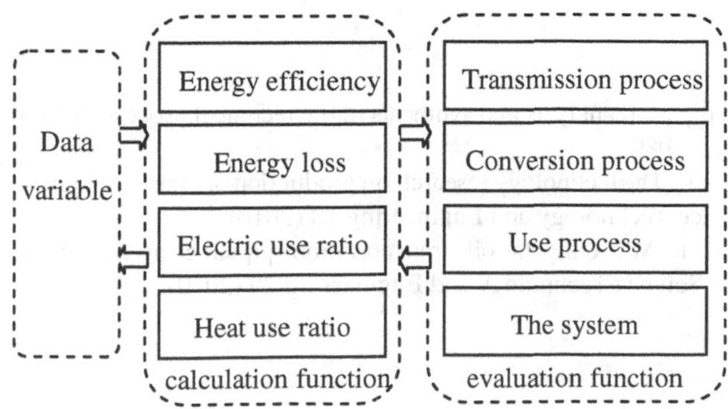

Fig. 2. The realization mechanism of ENEvaluate class

4.3 Energy Consumption Analysis Module

Oil storage system energy consumption analysis is the process of evaluating integral energy used level for energy used equipments and the system, on the basis of energy efficiency, energy loss in different energy used model and comparing with the industry standards and national norms. In order to achieve this function, based on the mathematic model, the ENEvaluate class was created which mainly includes data variables used for calculation and functions representing energy used analysis model. The realization mechanism of the ENEvaluate class is as shown in figure 2.

4.4 Comprehensive Optimization Module

This module mainly includes two components: the interface to realize man-machine interaction and the PIOptimize class to realize the optimization mathematical model and solving algorithm with computers. By this module, when solving a specific problem, Users do not need in-depth knowledge about the realization mechanism of the mathematical model and solving algorithm, as long as the fundamental parameters and control parameters of optimization algorithm were applied and sent to the PIOptimize class, then the optimizing operation scheme will be obtained.

5 Conclusion

(1)On the basis of three link module, the energy consumption process of crude oil storage system was divided into energy conversion process, transmission process and technological utilization process. By taking the energy efficiency and energy loss ratio as evaluation indexes, the energy consumption analysis model was established.
(2)Based on object-oriented program design concept, with the C# language in the .Net platform, four custom classes were created: DAManager, GRManager, ENEvaluate and PIOptimize class. The crude oil storage system energy consumption analysis and comprehensive optimization software was developed to realize energy analysis and scheme optimization in the system on the foundation of database developed in Access2003, With OLEDB database connection technology and the drawing application programming interface GDI+.

References

1. Hua, B.: Energy used analysis and synthesis in the technical process. hydrocarbon processing Press, Beijing (1989)
2. Liu, Y., Zhao, J.: The technology research on production operation scheme optimization in oil storage. Science Technology and Engineering 10 (2010)
3. Wei, L., Chen, M.: QingHa oil transportation pipeline production operation scheme optimization. Science Technology and Engineering 29 (2010)

A Strategy of Genetic Operations Based on Schema

Fachao Li[1], Jie Yang[2], and Chenxia Jin[1]

[1] School of Economics and Management, Hebei University of Science and Technology, 050018, Shijiazhuang Hebei, China
[2] School of Science, Hebei University of Science and Technology, 050018, Shijiazhuang Hebei, China
lifachao@tsinghua.org.cn, hebeikedayj@163.com, jinchenxia2005@126.com

Abstract. Genetic algorithm (GA) is an important kind of intelligent computing technique. In order to solve the low computing problem, this paper proposes a protecting strategy based on schema. According to this strategy, we establish a GA based on schema (BS-GA). Finally, we proof the convergence of BS-GA by the theory of Markov chains. All the results indicate that, BS-GA is essentially the extension of SGA, and it is better than SGA obviously in convergence performance.

Keywords: Genetic algorithm, schema theorem, excellent schema, Markov chain.

1 Introduction

As an intelligence optimization method, genetic algorithm(GA), with the features of easy structure and strong adaptability, has been successfully applied in many fields. In 1975, Holland first proposed genetic algorithm and schema theory [1]. After that, many scholars studied schema theory in different ways. For instances, Goldberg discussed the influence of excellent schema by different selection methods [2]; Tang et al. discussed the schema theory of real code [3]; Yang et al. discussed the effect of building blocks to GA [4]; You et al. discussed several crossover operators based on schema theory [5]. But it is worth noting that the computation efficiency of GA is low, and GA usually doesn't converge to global optimal solution. For this problem, many scholars improved genetic operation strategies through designing different GA. For example, Cervantes et al. researched the GA under the influence of mutation rate, then designed some improved methods based on mutation rate [6]; Chang et al. studied the function of selection operator, and designed a new compact algorithm [7]; paper [8] proposed an algorithm which searches for the refine estimates in the subspace; paper [9] studied the property and main factor of parallel genetic algorithm; Xu et al. studied the combination of the parallel genetic algorithm, neural network and fuzzy system [10].

The literatures above mainly focused on the design of genetic parameters, but the extraction and protection of excellent schema is less involved. In this paper, we mainly

D. Jin and S. Lin (Eds.): Advances in CSIE, Vol. 1, AISC 168, pp. 489–494.
springerlink.com © Springer-Verlag Berlin Heidelberg 2012

discuss schema protecting problem for genetic algorithm, and we establish a new GA based on schema (BS-GA). our main contributions are: 1) we first give a selection method for excellent schema. 2) Then we design a schema protecting method by different strategies of genetic operations. 3) Finally, we analyze the global convergence of BS-GA by Markov chain.

2 Preliminaries

Schema is a set of individuals with some similar characteristics. For a binary code string $a_1a_2\cdots a_{l-1}a_l$ ($a_i \in \{0, 1\}$), schema L can be represented by a string with 0, 1 or *, and * represents 0 or 1. And the number of the first gene loci to last one is defined as the length of schema L (denoted by $\delta(L)$). The number of all the fixed gene loci is the order of schema L (denoted by $O(L)$). For example, for a binary code with the length 7, schema L=0*10*11 can be represented as {0010011, 0010111, 0110011, 0110111}, obviously, $\delta(L)$=7-1=6, $O(L)$=5. Using the above concepts, Holland proposed the conclusion reflecting the characteristics of GA (schema theorem) as follows:

Theorem 1 (Schema Theorem). Suppose that $\vec{X}(t)$ is the population at generation t, $\bar{f}(H,t)$ is the average individual fitness value of schema H in $\vec{X}(t)$, $\bar{f}(\vec{X}(t))$ is the average individual fitness value of $\vec{X}(t)$, $m(H,t)$ is the expected number of individuals which belong to schema H in $\vec{X}(t)$. Then

$$m(H,t+1) \geq m(H,t)\frac{\bar{f}(H,t)}{\bar{f}(\vec{X}(t))}[1-p_c\frac{\delta(H)}{(l-1)}-O(H)p_m]. \qquad (1)$$

Where, p_c is the crossover probability, p_m is the mutation probability, $\delta(H)$ is the length of schema H, $O(H)$ is the order of schema H, l is the length of individual string.

3 A Selection Method for Excellent Schema

According to schema theorem, the schema with short length, low-order, high average fitness (called the excellent schema) reflects the iteration trends of genetic algorithm. Thus, searching for and protecting excellent schema will improve the efficiency of GA. We will give a selection method for excellent schema based on the following example.

Suppose that the length of individual string is 10, the size of the population is 40. The genetic iteration is t. we get 5 different individuals which the fitness are higher than the others, we can regard the five individuals as the basic start for excellent schema searching.

Table 1. Genetic statistical chart of five different individuals

			No. 1	No. 2	No. 3	No. 4	No. 5	No. 6	No. 7	No. 8	No. 9	No. 10
Part I		1	1	1	0	1	1	0	0	1	1	0
		2	1	0	0	1	0	1	0	0	1	1
		3	1	1	0	1	1	0	1	1	1	1
		4	1	0	0	1	0	1	0	1	1	0
		5	1	1	0	1	0	1	1	1	1	1
		Σ_i	5	3	0	5	2	3	2	4	5	3
Part II	①	E_{1i}	1	0.6	1	1	0.6	0.6	0.6	0.8	1	0.6
	②	E_{2i}	1	0	1	1	0	0	0	0.8	1	0
	③	E_{3i}	2	2.5	2.5	1	0	0.64	2.04	2.04	1	0
	④	E_{3i}	3.5	2.5	2.5	1	0.64	2.04	2.04	2.04	1	0

From the table above, the sum Σ_i represents the similarity of gene loci in the same column; the genes of No.1, No.3, No.4 and No.9 are same, which reflects the same common features; so they could be regarded as the fixed gene loci of excellent schema and called quasi-excellent gene loci.

Because the length of schema is too long and the order is too high, the principle of short length and low order can't be reflected, then, the gene loci with small number and strong intensive from excellent schema can be extracted. For example, 1) if the length and order are both not more than 3, then 1*01****** is regarded as an excellent schema; 2) if the length is not more than 2 and order is not more than 3, then **01****** is regarded as an excellent schema. But it is worth noting that the quasi-excellent genetic loci with the same value does not always exist. In real operation, we can determine quasi-excellent gene loci through the similarity rate. Based on the above analysis, the excellent schema can be obtained by the following principles:

Principle 1. Rank the individuals in t^{th} population according to the fitness, and select the first m individuals by some proportion of the size of population scale.
Principle 2. Compare the similarity of each gene value of the first m individuals, then determine the quasi-excellent gene loci combining with a certain threshold $\beta \in (0.5, 1]$.

Principle 3. According to the length of individuals, determine the length L and order K of schema, then select the quasi-excellent gene loci with strong intensive degree, proper length and order to construct excellent schema.

Obviously, $K \leq L+1$. In order to reflect characteristics of short length and low order of schema, we assume that $K = L+1$, and K is regarded as the upper limit of the order of excellent schema. β and K are parameters. When β is larger (or smaller), the common features of excellent individuals decided by excellent schema are stronger (or weaker). Usually, we use $\beta \in [0.8, 1]$. When K is too large, it doesn't satisfy the requirement; when K is too small, it can't reflect the common features of the excellent individuals. Generally, $K \in \{2, 3, 4, 5\}$.

The excellent schema by principle 3 may be more than one. In the following, we will give the concrete steps for selecting excellent schema from Table 1.

Step 1. Calculate the similarity of each gene value (as ① in Table 1): $E_{1i} = \max\{\frac{1}{5}\Sigma_i, 1-\frac{1}{5}\Sigma_i\}, i = 1, 2, \cdots, 10$;

Step 2. According to the given threshold β, the quasi-excellent gene loci should be revised by the similarity of each gene as follows (if β=0.8, the result is as ② in Table 1):

$$E_{2i} = \begin{cases} E_{1i}, & E_{1i} \geq \beta \\ 0, & E_{1i} < \beta \end{cases}, \quad i = 1, 2, \cdots, 10. \tag{2}$$

Step 3. According to the order K of excellent schema and (3), calculate the intensive degree E_{3i} for each column of quasi-excellent gene locus. 1) if $j > 10$, then $E_{2j} = 0$; 2) $\alpha \in (0, \infty)$ is a importance parameter reflecting the similarity of each gene loci; 3) $\lambda \in [0, 1]$ is a importance parameter reflecting the connection of each gene loci; 4) if $K = 2, \alpha = 2, \lambda = 0.5$, the calculation is as ③ in Table 1; 5) if $K = 3, \alpha = 2, \lambda = 0.5$, the calculation is as ④ in Table 1):

$$E_{3i} = E_{2i}^{\alpha} + (1 + \lambda E_{2i} E_{2(i+1)}) E_{2(i+1)}^{\alpha} + \cdots + (1 + \lambda E_{2(i+K-1)} E_{2(i+K)}) E_{2(i+K)}^{\alpha}, \quad i = 1, 2, \cdots, 10. \tag{3}$$

Step 4. if $E_{3i_0} = \max\{E_{3i} \mid i = 1, 2, \cdots, 10\}$, select the fixed genetic loci No.i_0, No.$(i_0+1), \cdots$, No.$(i_0 + \mathrm{K})$ to construct excellent schema (according to ③ and ④ in Table 1, we know that: 1) if $K = 2, \alpha = 2, \lambda = 0.5$, the excellent schema is **01******; 2) if $K = 3, \alpha = 2, \lambda = 0.5$, the excellent schema is 1*01******).

Obviously, if the quasi-excellent gene loci exist, the excellent schema with the order is lower than K can be always obtained according to step 1~4, but the excellent schema with the order is equal to K may not. For a binary coding with length 10,when the quasi-excellent gene loci is No.1, No.5, No.9, then the 2 or 3 orders excellent schema doesn't exist.

4 The Protection of Excellent Schema

Based on the above discussions, if the excellent individuals have some similar characteristics, we always find the excellent schema satisfying the requirements. Thus, in the process of genetic operation, we can try to protect the excellent schema by different crossover and mutation strategies. The specific steps are as follows:

1) The Protection Strategy in Crossover Operation. The essential significance of crossover operation is to produce new individuals by genetic recombination of elder generation individuals. So the protection of excellent schema can be realized through different crossover probability or selection probability for crossover points. In the sequel, Crossover point k denotes the right of the kth gene loci is segment point.

For the single point crossover operation, the specific operation can be set according to the following process: i) randomly produce crossover point in same probability; ii) if the crossover point is in the position between the fixed genetic loci of excellent schema, crossover operation can be operated by $p_c' \in [0, p_c'']$, otherwise, operated by p_c''. For instance, for an excellent schema 1**1*** with the length 7, if the crossover point is between 1 and 3, the crossover probability is p_c=0.6, if it is between 4 and 7, the crossover probability is p_c=1.

For the multipoint crossover operation, the specific operation can be set according to the following process: i) randomly produce crossover point in same probability; ii) operate crossover operation in same probability. For example, for an excellent schema ***1**1*** with the length 10, the proportion of the length of excellent schema in the length of individual is 0.3, so we can select a threshold r_0 less than 0.3. If the random number $r \le r_0$, and $r \in [0,1]$, produce crossover points between 4 and 6, else produce crossover points between 1 and 3 or between 7 and 10.

2) The Protection Strategy in Mutation Operation. Mutation operation can expand searching range, so we can protect the excellent schema in different mutation probability. For the schema mutated bit by bit, the specific operation can be set according to the following: i) change the value of non-fixed gene loci of excellent schema in mutation probability p_m'; ii) change the value of fixed gene loci of excellent schema in mutation probability $p_m'' \in [0, p_m']$. For a schema 1**1*** with the length 7, we can change the gene value between 1 and 4 in mutation probability 0.01 bit by bit, and change the other in mutation probability 0.05.

5 Convergence of BS-GA

In this section, we mainly analyze the global convergence of BS-GA combining with Markov chain.

Definition 1. let $\vec{X}(t) = \{X_1(t), X_2(t), \cdots, X_N(t)\}$ be the tth population of genetic algorithm in GA, $Z_t = \max\{f(X_i(t) \mid i = 1, 2, \cdots, N\}$, $f^* = \max\{f(X) \mid X \in S\}$ denotes the global optimal value of the individuals. If $P\{|Z_t - f^*| < \varepsilon\} \to 1\ (t \to \infty)$ for any $\varepsilon > 0$, then we say the genetic sequence $\{\vec{X}(t)\}_{t=1}^{\infty}$ converges.

Theorem 2. BS-GA using the elitist preserving strategy in replication process is global convergent.

Proof. For the sake of simplicity, in the following, we only discuss the Markov chain $\{X(t)\}_{t=0}^{\infty}$ of BS-GA using double individuals reserved strategy. Denote $p_{ij}^{(n)}$ be transition probability of state i to j after n steps. Suppose the state of contemporary population (for example generation t) is j, the double individuals reserved strategy is used, then some a individual in state j of generation t (for instance the individual at position k) is the most superior individual of previous generation (generation j-1), or, some two individuals in state j of generation t (for instance the individual at position k_1 and k_2) are the most superior and sub-optimal individuals of previous generation (generation j-1), which indicate the most superior individual is more excellent than or equal to that of generation t-1. Adding the strategy of schema protection (BS-GA), the fitness value of excellent schema is higher than the average at generation t. We suppose that generation t' before generation t, and i be the population state of generation t', and a more superior new individual is produced in the evolution process from generation t' to generation t (namely the most superior individual of generation t is more outstanding than the most superior individual of generation t'). that $p_{ij}^{(n)} > 0$ from the ergodic of BS-GA, it is obviously that we may obtain that BS-GA is a irreversible evolution process, so the genetic sequence $\{X(t)\}_{t=1}^{\infty}$ of BS-GA will finally converge to the global optimal solution in probability 1.

6 Conclusion

According to the schema theory, this paper proposes a strategy of schema protection (BS-GA) to improve the disadvantages of the schema protection of GA. Through the theory analysis, it is easy to see that this protection strategy can increase the performance of algorithm. The result shows that the strategy of schema protection is valuable and it improves the performance of genetic algorithm.

Acknowledgment. This work is supported by the National Natural Science Foundation of China (71071049) and the Natural Science Foundation of Hebei Province (F2011208056).

References

1. Holland, J.H.: Adaptation in Nature and Artificial Systems. Univ. of Michigan, USA (1975)
2. Goldberg, D.E.: A practical Schema Theorem for Genetic Algorithm Design and Tuning, pp. 328–335. Morgan Kaufmann, San Francisco (2001)
3. Tang, F., Teng, H.F., Sun, Z.G., Wang, W.Z.: Schema Theorem of the Decimal-Coded Genetic Algorithm. Mini-Micro Systems 21(4), 364–367 (2000) (in Chinese)
4. Yang, H.J., Li, M.Q.: Schema Theorem and Building Blocks in Evolution Algorithms. Chinese Journal of Computers 26(11), 1550–1554 (2003) (in Chinese)
5. You, X.X., Hu, S.L.: Schema theory of Genetic algorithm. Journal of Hubei Normal University 1(1), 24–28 (2007) (in Chinese)
6. Cervantes, J., Stephens, C.R.: Limitations of Existing Mutation Rate Heuristics and How a Rank GA Overcomes Them. IEEE Transactions on Evolutionary Computation 13(2), 369–397 (2009)
7. Chang, W.A., Ramakrishna, R.S.: Elitism-Based Compact Genetic Algorithms. IEEE Transactions on Evolutionary Computation 7(4), 367–385 (2003)
8. Ali, H., Doucet, A., Amshah, D.I.: GSR: A New Genetic Algorithm for Improving Source and Channel Estimates. IEEE Transactions on Circuits and Systems I 54(5), 1088–1098 (2007)
9. Guo, X., Shi, X.H.: Performance Analysis of Parallel Genetic Algorithm. Aeronautical Computer Technique 28(3), 86–89 (1998) (in Chinese)
10. Xu, J., Wang, M.H.: Integrations of Parallel Genetic Algorithms and Neural Networks and Fuzzy Systems. Mini-Micro Systems 18(7), 1–7 (1997) (in Chinese)

Facial Expression Recognition via Fuzzy Support Vector Machines

Xiaoming Zhao[1] and Shiqing Zhang[2]

[1] Department of Computer Science, Taizhou University, 318000 Taizhou, China
[2] School of Physics and Electronic Engineering, Taizhou University, 318000 Taizhou, China
{tzxyzxm,tzczsq}@163.com

Abstract. In this paper a new facial expression recognition method via fuzzy support vector machines (FSVM) is presented. We extract the local binary patterns (LBP) features for facial representation, and then employ FSVM to conduct facial expression recognition. The experimental results on the popular JAFFE facial expression database demonstrate that the FSVM with the Gaussian kernel obtains the best accuracy of 83.25%.

Keywords: Facial expression recognition, Local binary patterns, Fuzzy support vector machines.

1 Introduction

An automatic facial expression recognition system involves two crucial parts: facial feature representation and classifier design. Facial feature representation is to extract a set of appropriate features from original face images for describing faces. Mainly two types of approaches to extract facial features are found: geometry-based methods and appearance-based methods [1]. So far, Principal Component Analysis (PCA) [2], Linear Discriminant Analysis (LDA) [3], and Gabor wavelet analysis [4] have been applied to either the whole-face or specific face regions to extract the facial appearance changes. Recently, Local Binary Patterns (LBP) [5], originally proposed for texture analysis [6] has been successfully applied as a local feature extraction method in facial expression recognition [7, 8]. Classifier design is to use the extracted facial features to develop a promising classifier to recognize different expressions. The support vector machines (SVM) [9] classifier is a powerful tool for solving classification problems, and has been widely used for facial expression recognition [10] [11]. However, SVM still has one limitation. That is, in SVM each training point is enforced to belong to either one class or the other. In many real-world applications, input samples may not be exactly assigned to one class and the effects of the training samples might be different. Some are more important to be fully assigned to one class so that SVM can separate these samples more correctly. Some samples might be noisy and less meaningful and should discard them. Equally treating every data samples may cause the unsuitable over-fitting problem.

To overcome the mentioned-above limitation of SVM, in recent years an improved SVM algorithm, called fuzzy support vector machines (FSVM) [12], has been

D. Jin and S. Lin (Eds.): Advances in CSIE, Vol. 1, AISC 168, pp. 495–500.
springerlink.com © Springer-Verlag Berlin Heidelberg 2012

proposed. FSVM combines fuzzy logic and SVM to make different training samples have different contributions to their own class. Motivated by the deficiency of studies on FSVM for facial expression recognition, in this work we aim to illustrate the potential of FSVM on facial expression recognition tasks. To verify the effectiveness of FSVM for facial expression recognition, we conduct facial expression recognition experiments on the popular JAFFE [4] facial expression database.

2 Fuzzy Support Vector Machines

Fuzzy support vector machines (FSVM) [12] applies a fuzzy membership to each input point and reformulate the SVM algorithm such that different input points can make different contributions to the learning of decision surface.

For a binary classification problem, a set S of l training samples, each represented are given as (x_i, y_i, μ_i) where x_i is the feature vector, y_i is the class label, and μ_i is the fuzzy membership. Each training sample belongs to either of two classes. These samples are given a label $y_i \in \{-1, +1\}$, a fuzzy membership $\sigma < \mu_i \leq 1$ with $i = 1, \ldots, l$, and sufficient small $\sigma > 0$. Decision functions are simple weighted sums of the training samples x_i plus a bias are called linear discriminant functions, denoted as

$$D(x) = w \cdot x + b \tag{1}$$

where w is the weight vector and b is a bias value. Let $z_i = \varphi(x_i)$ denote the corresponding feature space vector with a mapping function φ from R^N to a feature space Z. The hyperplane can be defined as

$$w \cdot z + b = 0 \tag{2}$$

The set S is said to be linearly separable if there exists (w, b) such that the inequalities

$$\begin{cases} w \cdot z_i + b \geq +1 \Rightarrow y_i = +1 \\ w \cdot z_i + b \leq -1 \Rightarrow y_i = -1 \end{cases} \tag{3}$$

To deal with data that are not linearly separable, the previous analysis can be generalized by introducing some non-negative variables $\xi_i \geq 0$ such that Eq. (3) is modified to

$$y_i(w \cdot z_i + b) \geq 1 - \xi_i, \ i = 1, \ldots, l \tag{4}$$

the non-zero ξ_i in Eq. (4) are those for which the data samples x_i does not satisfy Eq. (3). Since the fuzzy membership μ_i is the attitude of the corresponding sample x_i toward one class and the parameter ξ_i is the measure of error in the SVM, the term $\mu_i\xi_i$ is a measure of error with different weighting. The optimal hyperplane problem is then regarded as the solution to

$$\begin{aligned} &\text{minimize } \frac{1}{2}\|w\|^2 + C\sum_{i=1}^{l}\mu_i\xi_i \\ &\text{s.t. } y_i(w\cdot z_i + b) \ge 1-\xi_i,\ i=1,\ldots,l \end{aligned} \tag{5}$$

where C is a constant. The parameter C can be regarded as a regulation parameter. The optimization problem (5) can be solved by introducing Lagrange multiplier α and transformed into:

$$\begin{aligned} &\text{minimize } W(\alpha) = \frac{1}{2}\sum_{i=1}^{l}\sum_{j=1}^{l}\alpha_i\alpha_j y_i y_j (z_i\cdot z_j) - \sum_{i=1}^{l}\alpha_i \\ &\text{s.t. } \sum_{i=1}^{l} y_i\alpha_i = 0,\ 0\le\alpha_i\le\mu_i C,\ i=1,\ldots,l \end{aligned} \tag{6}$$

The mapping ϕ is usually nonlinear and unknown. Instead of calculating φ, the kernel function K is used to compute the inner product of two vectors in the feature space Z and thus implicitly defines the mapping function, which is

$$K(x_i, x_j) = \phi(x_i)\cdot\phi(x_j) = z_i\cdot z_j \tag{7}$$

Kernel is one of the core concepts in SVMs and plays a very important role. The following are three types of commonly used kernel functions:

$$\text{linear kernel: } K(x_i, x_j) = x_i\cdot x_j \tag{8}$$

$$\text{polynomial kernel: } K(x_i, x_j) = (1 + x_i\cdot x_j)^p \tag{9}$$

$$\text{Gaussian kernel: } K(x_i, x_j) = \exp(-\|x_i - x_j\|^2/2\sigma^2) \tag{10}$$

The decision function can be expressed by using the Lagrange multiplier:

$$D(x) = sign(w\cdot z + b) = sign(\sum_{i=1}^{l}\alpha_i y_i K(x_i, x) + b) \tag{11}$$

3 Experiments

The popular JAFFE facial expression database [4] used in this study contains 213 images of female facial expressions. Each image has a resolution of 256×256 pixels. The head is almost in frontal pose. The number of images corresponding to each of the 7 categories of expression (neutral, happiness, sadness, surprise, anger, disgust and fear) is almost the same. As done in [7] [8], we normalized the faces to a fixed distance of 55 pixels between the two eyes. Automatic face registration can be achieved by a robust real-time face detector based on a set of rectangle haar-like features [13]. From the results of automatic face detection, such as face location, face width and face height, two square bounding boxes for left eye and right eye are created respectively. Then, two eyes location can be quickly worked out in terms of the centers of two square bounding boxes for left eye and right eye. Based on the two eyes location, facial images of 110×150 pixels were cropped from original frames. No further alignment of facial features such as alignment of mouth was performed in our work.

As LBP tolerates against illumination changes and operates with its computational simplicity [5], we adopt LBP for facial image representations for facial expression recognition. The LBP operator is applied to the whole region of the cropped facial images of 110 × 150 pixels. For better uniform-LBP feature extraction, two parameters, i.e., the LBP operator and the number of regions divided, need to be optimized. We selected the 59-bin operator, and divided the 110 × 150 pixels face images into 18 × 21 pixels regions, giving a good trade-off between recognition performance and feature vector length. Thus face images were divided into 42 (6 × 7) regions, and represented by the LBP histograms with the length of 2478 (59 × 42).

To reduce the length of the extracted LBP features, the most popular linear PCA method [2] is used to perform dimensionality reduction. The reduced dimension of LBP features is confined to the range of [10, 100] with an interval of 10. We randomly choose 70% samples from the JAFFE database for training, the remaining 30% samples for testing. We investigate the performance of FSVM with three typical kernels, including the linear, polynomial ($p = 2$), and Gaussian kernels ($\sigma = 1$).

Table 1. The best accuracy for different kernels of FSVM with corresponding reduced dimension

Methods	Linear kernel	Polynomial kernel	Gaussian kernel
Dimension	100	20	100
Accuracy (%)	83.01	77.83	83.25

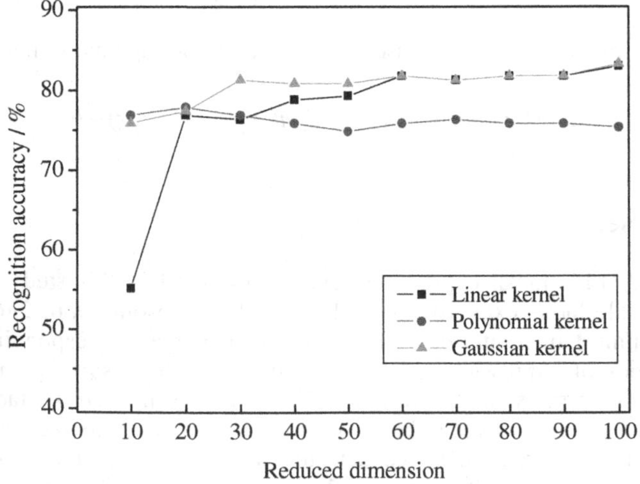

Fig. 1. Performance comparisons of different used kernels of FSVM

The different recognition results of three kernels of FSVM are given in Fig.1. The best accuracy for different kernels with corresponding reduced dimension is presented in Table 1. The results in Fig.1 and Table 1 indicate that the Gaussian kernel of FSVM obtains the highest accuracy of 83.25% with 100 reduced dimension, outperforming the linear and polynomial kernels. Nevertheless, the linear kernel of FSVM still yields the best accuracy of 83.01%, highly close to the performance of the Gaussian kernel of FSVM. The polynomial kernel of FSVM achieves the lowest accuracy of 77.83%.

4 Conclusions

In this paper, we presented a new method of facial expression recognition based on FSVM. The experiment results on the popular JAFFE facial expression database indicate that FSVM with the Gaussian kernel obtains the highest accuracy of 83.25%. This confirms that FSVM is suitable for facial expression recognition.

Acknowledgments. This work is supported by Zhejiang Provincial Natural Science Foundation of China under Grant No.Z1101048 and Grant No. Y1111058.

References

1. Tian, Y., Kanade, T., Cohn, J.: Facial expression analysis. In: Handbook of Face Recognition, pp. 247–275 (2005)
2. Turk, M.A., Pentland, A.P.: Face recognition using eigenfaces. In: IEEE Conference on Computer Vision and Pattern Recognition (CVPR), Maui, USA, pp. 586–591 (1991)
3. Belhumeur, P.N., Hespanha, J.P., Kriegman, D.J.: Eigenfaces vs. fisherfaces: Recognition using class specific linear projection. IEEE Transactions on Pattern Analysis and Machine Intelligence 19(7), 711–720 (1997)
4. Lyons, M.J., Budynek, J., Akamatsu, S.: Automatic classification of single facial images. IEEE Transactions on Pattern Analysis and Machine Intelligence 21(12), 1357–1362 (1999)
5. Ojala, T., Pietikïnen, M., Mënpä, T.: Multiresolution gray scale and rotation invariant texture analysis with local binary patterns. IEEE Transactions on Pattern Analysis and Machine Intelligence 24(7), 971–987 (2002)
6. Ojala, T., Pietikïnen, M., Harwood, D.: A comparative study of texture measures with classification based on featured distributions. Pattern Recognition 29(1), 51–59 (1996)
7. Shan, C., Gong, S., McOwan, P.: Robust facial expression recognition using local binary patterns. In: IEEE International Conference on Image Processing (ICIP), Genoa, pp. 370–373 (2005)
8. Shan, C., Gong, S., McOwan, P.: Facial expression recognition based on Local Binary Patterns: A comprehensive study. Image and Vision Computing 27(6), 803–816 (2009)
9. Vapnik, V.: The nature of statistical learning theory. Springer, New York (2000)
10. Kotsia, I., Pitas, I.: Facial expression recognition in image sequences using geometric deformation features and support vector machines. IEEE Transactions on Image Processing 16(1), 172–187 (2007)

11. Xu, Q., Zhang, P., Yang, L., Pei, W., He, Z.: A Facial Expression Recognition Approach Based on Novel Support Vector Machine Tree. In: Liu, D., Fei, S., Hou, Z., Zhang, H., Sun, C. (eds.) ISNN 2007, Part III. LNCS, vol. 4493, pp. 374–381. Springer, Heidelberg (2007)
12. Lin, C.-F., Wang, S.-D.: Fuzzy support vector machines. IEEE Transactions on Neural Networks 13(2), 464–471 (2002)
13. Viola, P., Jones, M.: Robust real-time face detection. International Journal of Computer Vision 57(2), 137–154 (2004)

The Performance Analysis of Genetic Algorithm Based on Schema

Chenxia Jin[1], Jie Yang[2], and Fachao Li[1]

[1] School of Economics and Management, Hebei University of Science and Technology, 050018, Shijiazhuang Hebei, China
[2] School of Science, Hebei University of Science and Technology, 050018, Shijiazhuang Hebei, China
jinchenxia2005@126.com, hebeikedayj@163.com, lifachao@tsinghua.org.cn

Abstract. In this paper, for a new genetic algorithm (GA) based on schema (BS-GA), we mainly analyze the performance of BS-GA through simulation. Through two examples, we verify the effectiveness of our algorithm. All the results indicate that, BS-GA is better than standard genetic algorithm (SGA) obviously in computation efficiency and convergence performance.

Keywords: genetic algorithm, schema theorem, excellent schema, evolution operation, Markov chain.

1 Introduction

As an intelligence optimization method, genetic algorithm, with the features of easy structure and strong adaptability, has been successfully applied in many fields. In 1975, Holland first proposed genetic algorithm (GA) and schema theory [1]. But it is worth noting that the computation efficiency of GA is low, and GA usually doesn't converge to the global optimal solution. For this problem, many scholars improved genetic operation strategies through designing different GA. For example, Hajabdollahi et al. studied multi-objective optimization of pin fin to determine the optimal fin geometry by GA [2]. Goncalves et al. applied a parallel multi-population biased random-key GA for a container loading problem [3]. Kumar et al. analyzed the reliability of waste clean- up manipulator by GAs and fuzzy methodology [4]. Chakraborty proposed a branching process model for GAs [5]. Annicchiarico et al. proposed a approach to optimize finite element bidimensional models based on GAs [6]. Srinivasa et al. proposed a self-adaptive migration model GA for data mining applications [7]. Cervantes et al. researched the GA under the influence of mutation rate, then designed some improved methods based on mutation rate [8]; Chang et al. studied the function of selection operator, and designed a new compact algorithm [9]; Ali et al. proposed an algorithm which searches for the refine estimates in the subspace [10].

D. Jin and S. Lin (Eds.): Advances in CSIE, Vol. 1, AISC 168, pp. 501–506.
springerlink.com © Springer-Verlag Berlin Heidelberg 2012

In previous work, we mainly discuss schema protecting problem for genetic algorithm, and propose a new GA based on schema (BS-GA). In this paper, we mainly analyze the performance of BS-GA through simulation. Through two examples, we also verify the effectiveness of our algorithm.

2 Performance of BS-GA

2.1 A General Introduction to BS-GA

According to schema theorem, the schema with short length, low-order, high average fitness (called the excellent schema) reflects the iteration trends of genetic algorithm. Thus, searching for and protecting excellent schema will improve the efficiency of GA. Paper [11] establishes a new genetic algorithm based on schema (BS-GA). It can find the excellent schema and protect it through designing different crossover and mutation operators. The excellent schema can be obtained by the principles of reference [11], and the strategy of schema protecting in crossover and mutation operators are also involved. Please see the details in [11]. Next we mainly analyze the performance of BS-GA through the following two examples.

2.2 Application Examples

Example 1. Consider the following global extreme value problem.

$$\begin{cases} \min f(x) = x_1^2 + 2x_2^2 - 10\sin(2x_1)\sin(x_3) + 0.5\cos(x_1 + 2x_2) + x_1^2 x_3^2 - 5\sin(2x_1 - x_2 + 3x_3), \\ \text{s.t.} \quad -10 \le x_i \le 10, \ i = 1, 2, 3. \end{cases}$$

It is a usual multiple humped function optimization problem for analyzing the performance of GA. The global minimum point is x_1=-0.676256, x_2=-0.381512, x_3=-1.282868. The corresponding optimum value is -12.765474. In the following, we use standard genetic algorithm (SGA) and BS-GA to test the function respectively, and all the calculations are based on Matlab 7.0 and 2.00 GHz Intel core 2 processor and worked out under WindowsXP Professional Edition platform. The parameters are as follows: the coding length is 30; the size of population is 80; the maximum generation is 100; for SGA, the crossover probability is 1 and the mutation probability is 0.001; for the excellent schema of BS-GA, the crossover probability is 0.8, the mutation probability of the fixed gene loci is 0.0001; for the non-excellent schema of BS-GA, the crossover probability is 1, the mutation probability is 0.001. The results are shown as Fig.1, Fig. 2, and, the full curve indicates the maximum value, chain dotted curve indicates the average value, dotted curve indicates the minimum value. The calculated results and performance data of 10 experiments are as Table 1 and Table 2.

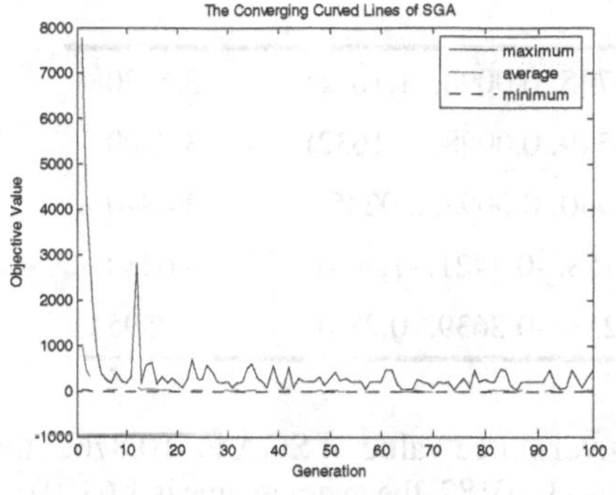

Fig. 1. 100 iterations results of SGA

Fig. 2. 100 iterations results of BS-GA

Table 1. The results of 10 experiments of Example 1for SGA

	Convergence value	Optimum solution (x_1, x_2, x_3)	Time (s)
1	-10.4378	(0.6354, 0.9482, 0.5963)	1.4220
2	-11.2189	(0.6940, -0.4594, 1.8866)	1.7650
3	-10.6679	(0.6158, -0.2639, 2.0430)	1.3910
4	-10.0182	(0.6158, 0.0684, 2.1994)	1.5310
5	-10.6035	(0.7136, 0.0098, 1.9844)	1.6720
6	-10.5896	(0.6940, 0.0098, 2.0430)	2.0470
7	-10.9311	(0.7136, -0.3812, 2.2190)	1.4060
8	-10.8767	(0.6158, 0.0098, 2.0626)	1.5310
9	-10.9402	(0.7136, -0.4790, 1.8671)	1.7960
10	-12.4425	(-0.6745, 0.0098, -1.1632)	1.7970
average	-10.8726	(0.6744, -0.0537, 1.5738)	1.6350

Table 2. The results of 10 experiments of Example 1for BS-GA

	Convergence value	Optimum solution (x_1, x_2, x_3)	Time (s)
1	-12.6535	(-0.6158, -0.4790, -1.3392)	4.0780
2	-12.7170	(-0.7136, -0.3030, -1.2414)	3.7190
3	-12.6158	(-0.6158, -0.2248, -1.2414)	4.0000
4	-12.6015	(-0.6940, -0.6354, -1.3392	3.6250
5	-12.7651	(-0.6745, -0.3812, -1.2805)	3.6410
6	-12.7401	(-0.6940, -0.3030, -1.2414)	3.9380

Table 2. (*continued*)

7	-12.4425	(-0.6745, 0.0098, -1.1632)	3.6720
8	-12.4347	(-0.6549, 0.0098, -1.1632)	3.7500
9	-10.6060	(0.6940, 0.0098, 2.0235)	3.4840
10	-12.6840	(-0.6158, -0.3421, -1.3001)	3.6880
average	-12.4260	(-0.5258, -0.2639, -0.2861)	3.7595

According to Table 1 and 2, the average convergence value of SGA is -10.8762, the maximum value is -12.4425, the minimum value is -10.0182, the average time is 1.6350 (s). The average convergence value of BS-GA is -12.4260, the maximum value is -12.7651, the minimum value is -10.6060, the average time is 3.7595 (s). Compared with Fig. 1 and Fig. 2, it is easy to see that the convergence speed and stability of BS-GA is better than that of SGA.

Example 2. Consider the following global extreme value problem

$$\min f(x) = \frac{\sin^2 \sqrt{x_1^2 + x_2^2} - 0.5}{[1+0.001(x_1^2 + x_2^2)]^2} - 0.5, \quad |x_i| \leq 10, \ i = 1, 2 .$$

The global optimum solution is -1, and the global minimum point is x_1=0, x_2=0. There are many different local minimum points rounding the global minimum point around the range of 3.14, and the function shocks strongly. It is hard to get the optimum solution by analytical methods. We use standard genetic algorithm (SGA) and BS-GA to test the function respectively. The parameters are as follows: the coding length is 20, the size of population is 80, the maximum generation is 100, for SGA, the crossover probability is 1, and the mutation probability is 0.001; for the excellent schema of BS-GA, the crossover probability is 0.8, the mutation probability is 0.0001. The results are shown as Fig. 3, Fig. 4, Table 3 and Table 4 in the following.

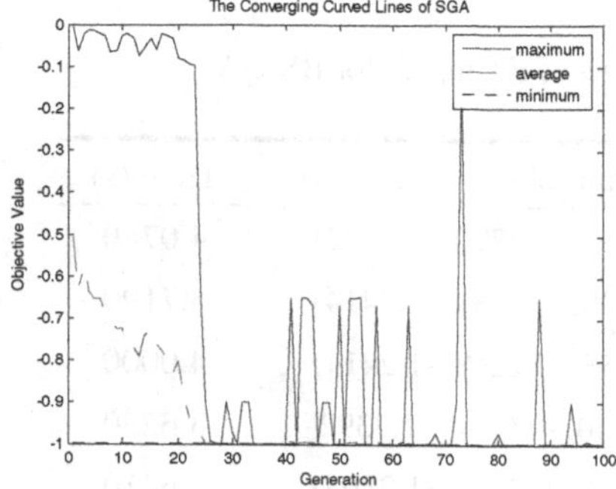

Fig. 3. 100 iterations results of SGA

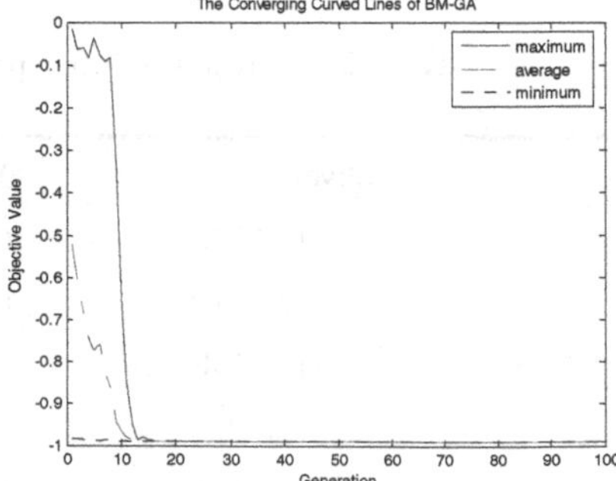

Fig. 4. 100 iterations results of BS-GA

Table 3. The 10 experiments results of Example 2 for the SGA

	Convergence value	Optimum solution (x_1, x_2)	Time (s)
1	-0.9903	(0.4985, 3.0987)	1.6560
2	-0.9903	(-3.0205, -0.8504)	1.6400
3	-0.9903	(-3.0401, -0.7331)	1.9380
4	-0.9903	(-2.2776, 2.1603)	1.3750
5	-0.9903	(0.4985, 3.0987)	1.5630
6	-0.9903	(3.0987, -0.4985)	1.5630
7	-0.9903	(0.0880, -3.1378)	1.8750
8	-0.9903	(0.4985, -3.0987)	1.4530
9	-0.9903	(-2.8837, 1.2414)	1.5940
10	-0.9903	(1.8475, -2.5318)	1.3440
average	-0.9903	(-0.4692, -0.1251)	1.6001

Table 4. The 10 experiments results of Example 2 for the BS-GA

	Convergence value	Optimum solution (x_1, x_2)	Time (s)
1	-0.9903	(-2.9228, -1.1437)	3.0320
2	-0.9998	(0.0098, 0.0098)	3.2970
3	-0.9903	(2.9423, -1.0850)	3.9380
4	-0.9903	(3.0205, 0.8504)	3.0940
5	-0.9998	(-0.0098, 0.0098)	3.4680
6	-0.9903	(-2.5318, -1.8475)	3.2030
7	-0.9998	(-0.0098, -0.0098)	3.3910
8	-0.9903	(1.9062, 2.4927)	3.3900
9	-0.9903	(0.9091, -3.0010)	3.0470
10	-0.9998	(0.0098, -0.0098)	3.3750
average	-0.9941	(0.3323, -0.3734)	3.3235

According to Table 3 and Table 4, the average convergence value of SGA is -0.9903, the maximum value is -0.9903, the minimum value is -0.9903, the average time is 1.6001 (s). The average convergence value of BS-GA is -0.9941, the maximum value is -0.9903, the minimum value is -0.9998, the average time is 3.32355 (s). From Fig. 3 and Fig. 4, the convergence speed and stability of BS-GA is better than that of SGA.

Combined with the experiments and analysis, we know that, 1) the convergence precision of BS-GA is much better than that of SGA; 2) the optimal solution of BS-GA is nearer to the global solution than SGA; 3) the convergence time of BS-GA increases slightly compared with SGA; 4) from the convergence figures, the convergence speed and stability of BS-GA are better than SGA.

3 Conclusion

In this paper, we mainly analyze the performance of BS-GA through simulation. Through two examples, we verify the effectiveness of our algorithm. Experiments show that, though the convergence time of BS-GA increases slightly compared with SGA, the convergence precision of BS-GA is better than that of SGA obviously. The result shows that the strategy of schema protection is valuable and feasible. It improves the performance of genetic algorithm to a certain degree.

Acknowledgment. This work is supported by the National Natural Science Foundation of China (71071049) and the Natural Science Foundation of Hebei Province (F2011208056).

References

1. Holland, J.H.: Adaptation in Nature and Artificial Systems. Univ. of Michigan, USA (1975)
2. Hajabdollahi, F., Rafsanjani, H.H., Hajabdollahi, Z., Hamidi, Y.: Multi-Objective Optimization of Pin Fin to Determine the Optimal Fin Geometry Using Genetic Algorithm. Applied Mathematical Modelling 36, 244–254 (2011)
3. Goncalves, J.F., Resende, M.G.C.: A Parallel Muti-Population Biased Random-Key Genetic Algorithm for a Container Loading Problem. Computers & Operations Research 39, 179–190 (2011)
4. Kumar, N., Borm, J.H., Kumar, A.: Relibility of Waste Clean-Up Manipulator by Genetic Algorithms and Fuzzy Methodology. Computers & Operations Research 39, 310–319 (2011)
5. Chakraborty, U.K.: A Branching Process Model for Genetic Algorithms. Information Processing Letters 56, 281–292 (1995)
6. Annicchiarico, W., Cerrolaza, M.: Optimization of Finite Element Bidmensional Models: an Approach based on Genetic Algorithms. Finite Elements in Analysis and Design 29, 231–257 (1998)
7. Srinivasa, K.G., Venugopal, K.R., Patnaik, L.M.: A self-adaptive migration model genetic algorithm for data mining applications. Information Sciences 177, 4295–4313 (2007)
8. Cervantes, J., Stephens, C.R.: Limitations of Existing Mutation Rate Heuristics and How a Rank GA Overcomes Them. IEEE Transactions on Evolutionary Computation 13(2), 369–397 (2009)
9. Chang, W.A., Ramakrishna, R.S.: Elitism-Based Compact Genetic Algorithms. IEEE Transactions on Evolutionary Computation 7(4), 367–385 (2003)
10. Ali, H., Doucet, A., Amshah, D.I.: GSR: A New Genetic Algorithm for Improving Source and Channel Estimates. IEEE Transactions on Circuits and Systems I 54(5), 1088–1098 (2007)
11. Li, F.C., Yang, J., Jin, C.X.: A Strategy of Genetic Operations based on Schema. In: 2012 Second International Conference on Computer Science and Information Engineering (CSIE 2012) (2012)

The Research of E-commerce Recommendation System Based on Collaborative Filtering Technology

ZhiQiang Zhang[1,*] and SuJuan Qian[2]

[1] Henan Occupation Technical College, Zhengzhou, 450046, China
[2] Zhengzhou Vocational College of Economics and Trade, Zhengzhou, 450006, China
zhangzhiqiang70@sina.com

Abstract. Recommendation System is the use of statistical and knowledge discovery techniques to solve the interaction with the target customers to provide products recommended problem. This paper presents the method of e-commerce recommendation system based on collaborative. The recommendation algorithm based on Collaborative filtering is proposed in order to better solve the collaborative filtering recommender system implementations exist in the data sparse and synonyms. The compared experimental results indicate that this method has great promise.

Keywords: recommendation system, collaborative filtering, e-commerce.

1 Introduction

Recommended system (Recommender Systems) is the use of statistical and knowledge discovery techniques to solve the interaction with the target customers to provide products recommended problem. Now widely used recommendation systems (Recommender System) is defined Resnick & Varian given in 1997: "It's in e-commerce system to provide customers with product information and recommendations to help customers decide what products to buy, analog sales recommend products to customers to complete the purchase process."Recommended system recommended in the e-commerce site which is commodity purchases of merchandise, customer demographics or purchase history of customers on the analysis generated [1]. Recommended activities are prevalent in our daily life, according to the different objects and recommended method recommended, recommended activities in various forms, such as supermarkets, shopping Purchasing Guide users to recommend the user's favorite products and improve sales ability; real life friends love each other recommended film; salesman selling products to the users, are recommended activities, it can be said, as long as there are multiple candidates, there is choice, there are recommendations for selecting a problem.

In the virtual environment of e-commerce, businesses and types of goods offered by a very large number, the user can not by a small computer screen at a glance to know all the goods, as users can not check in the physical environment as the

* Author Introduce: ZhiQiang Zhang(1970), male, associate professor, Master, Hehan Occupation Technical College, Research area: web mining, data mining.

D. Jin and S. Lin (Eds.): Advances in CSIE, Vol. 1, AISC 168, pp. 507–512.
springerlink.com © Springer-Verlag Berlin Heidelberg 2012

selection of goods. Therefore, businesses need to provide some intelligent purchasing guidance, recommendations based on user interests and hobbies, or user satisfaction merchandise may be of interest, so that users can easily get what they need to get the goods. Then, if the business be able to meet the needs of the user vague recommendation of goods to the user, the user can put the potential demand into real demand, thus improving product sales.

In this context, the recommendation system (Recommender Systems) came into being, it is based on the user's characteristics, such as hobbies, it is recommended to meet the user requirements of the object, also known as personalized recommendation system (Personalized Recommender Systems) [2]. The recommended characteristics of the object, there are mainly two types of recommendation systems, a web-based search system recommended objects, mainly Web data mining methods and techniques for users to recommend web pages in line with their interests and hobbies, such as Google etc.; the other is online shopping (especially B2C type) environment, commodity-recommended target of personalized recommendation system for users to recommend products meet the interests and hobbies, such as books, audio-video, saying that this recommendation system for electronic Business personalized recommendation system, referred to as e-commerce recommendation system (recommender system in E-commerce).

2 Recommend Method of E-commerce Recommendation System

E-commerce recommendation system, currently used techniques are: Bayesian network (Bayesian Network), association rules (Association Rules), clustering (Clustering), Horting map (Horting Graph), collaborative filtering (Collaborative Filtering) and so on. One collaborative filtering recommender system technology is the application of the earliest and one of the most successful technologies.

Obviously, the recommended method is the recommended system is the most crucial and most critical part, largely determines the merits of the recommended system performance. Currently, the main recommendations include: content-based recommendation, collaborative filtering, recommendation based on association rules, based on the effectiveness of recommendations, recommendation and portfolio recommendations based on knowledge.

(1) Content-based Recommendation: Content-based recommendation (Content-based Recommendation) is a continuation of information filtering technology and development, it is based on the contents of the information to make recommendations, without the need for items based on user evaluations, more need to use machine learning approach from the characterization of examples of the content of the user's interest to get information [3]. The user's data model depends on the learning method, commonly used decision trees, neural networks and other vector-based representation.

(2) Collaborative filtering recommendation: Collaborative filtering (collaborative filtering recommendation) is the most studied personalized recommendation technology.

(3) Association Rule-based Recommendation: Recommendation based on association rules (Association Rule-based Recommendation) is based on association rules, the rules of the purchased goods as head rules the body for the recommended

object. Association rule mining can be found in the sales of different commodities in the process of relevance in the retail sector has been successfully applied.

(4) Utility-based Recommendation: Utility-based recommendation (Utility-based Recommendation) is based on the user entry on the utility of computing, the core problem is how to create for each user a utility function, so the user data model is largely system used by the utility function of the decision. Recommended based on utility is that it can non-product attributes.

(5) Knowledge-based Recommendation: Knowledge-based recommendation (Knowledge-based Recommendation) is to some extent can be seen as a kind of reasoning (Inference) technology; it is not based on user needs and preferences based on recommendations. Knowledge-based methods because they use different functional knowledge are apparently different.

3 The Research of Collaborative Filtering Recommendation

Collaborative filtering recommendation systems in e-commerce website and widely used, it is by far the most successful information filtering technologies. Collaborative filtering, also known as social filtering (Social Filtering), the basic idea by comparing the user's interests and past behavior of the degree of similarity, to identify and target users with the same or similar interests of user groups, according to their resources assessment to predict the target user's interest, to recommend to the target user of resources.

In general, this requires the establishment of user information, with its large data base, which is difficult to do precisely the same time it also makes the field of collaborative filtering technology is relatively narrow (almost concentrated in the entertainment: music, movies), in the broader area (such as content-based filtering has been quite successful in the relevant areas of text) application is still not enough [4].

To the first idea, for example, under normal circumstances, collaborative filtering systems typically require three steps: first, to obtain user information (user entries for the evaluation of certain information, etc.); Second, analysis of the similarity between users, the formation of the recent neighbors; finally, the resulting recommendations is shown in Figure 1.

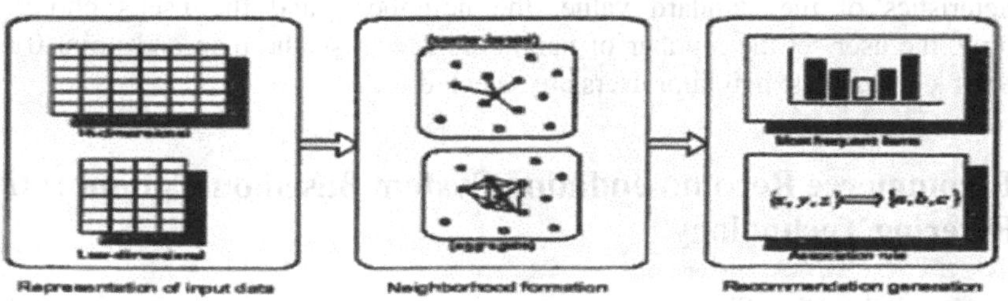

Fig. 1. Collaborative filtering of the implementation steps

(1) Cosine similarity: the user-item ratings as n-dimensional vector space, the similarity between users by the cosine of the angle between the vector measure, the greater the cosine value, indicating that the higher the degree of similarity of two users. Let user *i* and user *j* in the n-dimensional space on the item scores were expressed as a vector, the user *i* and user *j* is the similarity between.

$$sim(i, j) = \cos(\vec{i}, \vec{j}) = \frac{\vec{i} \bullet \vec{j}}{\|\vec{i}\| \|\vec{j}\|} \tag{1}$$

(2) the amendment of the cosine similarity (Adjusted Cosine): Since the cosine similarity measurement method does not take into account different user ratings scale problems, the modified cosine similarity measure method by subtracting the average user score of items to improve the defect, set by the user *i* and user *j* with a common set of scoring items that I_{ij}, I_i and I_j, respectively, by the score of user *i* and user *j* collection of items, the user *i* and user *j* is the similarity between.

$$sim(i, j) = \frac{\sum_{c \in I_{i,j}} (R_{i,c} - \overline{R_i})(R_{j,c} - \overline{R_j})}{\sqrt{\sum_{c \in I_i} (R_{i,c} - \overline{R_i})^2} \sqrt{\sum_{c \in I_j} (R_{j,c} - \overline{R_j})^2}} \tag{2}$$

Which, R_i, c means that the user i's score on item c, i, and represent user i and user j, the average score of the items. Figure 2 shows a neighbor collaborative filtering in a user-set formation process.

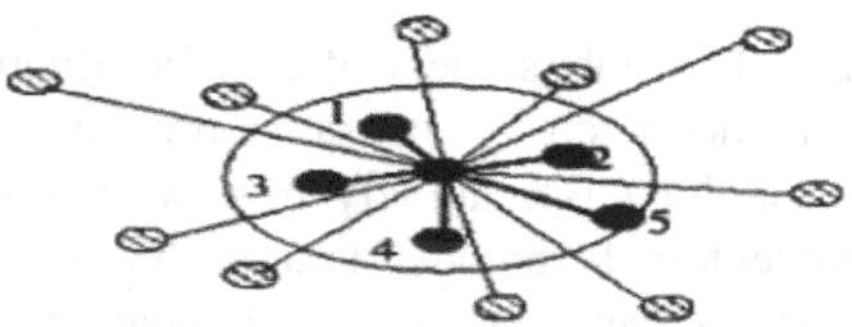

Fig. 2. The formation of collaborative filtering

First, According to the user between one or a few related properties to calculate the current user 0 (with the figure black spots) and the similarity between other users. Then, according to the set of neighbors or the user sets the size range of relevant characteristics of the standard value, the neighbors, and the user's choice. For example, the user set the number of neighbors k = 5, so the map and point 0 as the center of k = 5 nearest neighbor users are selected as.

4 E-commerce Recommendation System Based on Collaborative Filtering Technology

Although collaborative filtering recommendation systems in e-commerce application to obtain a greater success, but with the site structure, content, complexity and the increasing number of users, based on collaborative filtering recommendation system development faces two major challenges.

Collaborative filtering algorithm can easily provide for the thousands of good recommendations, but for e-commerce sites often need to provide recommendations to millions of users, which on the one hand the need to improve response time requirements, to provide users real-time to be recommended; the other hand should also take into account the storage space requirements, recommended to minimize the burden of system operation.

In order to better solve the collaborative filtering recommender system implementations exist in the data sparse, synonyms (similar products with different names to describe, but can not find this correlation) and other issues, the current proposed use is widely used in information retrieval, and used to solve the problem of synonyms and polysemy dimension reduction techniques - hidden semantic indexing (Latent Semantic Indexing, LSI). Dimensionality reduction can improve the data by the density; find more hidden information about the user evaluation. LSI uses singular value decomposition (Singular Value Decomposition, SVD) as matrix factorization algorithms. SVD can be well combined with the collaborative filtering technology to effectively reduce the data noise, found a potential association, and SVD calculation can be carried out offline.

S VD can be a m n matrix R is decomposed into three matrices. Algorithm: based on singular value decomposition (SVD) of the recommendation algorithm.

Input: matrix R ', the user U, the corresponding set of options has been Iu.

Output: correlation matrix T, S, D.

(1) Decomposition of the matrix with the SVD method R 'be the matrix T0, S0, D0 the S0 reduced to dimension K of the matrix, to get S (k <r, r is the rank of matrix R)

(2) Simplified the matrix T0, D0 to be T, D

(3) Calculate the square root of S to be Sl / 2

(4) Calculate two correlation matrices TS1 / 2, S1/2D '

(5) Compare whether the similarity between the feedback connection weight vector and the input pattern vector is greater than ρ (pre-defined vigilance parameter) or not, if$| T_g U_k | / | U_k | > \rho$①

then transferr to the step (7), otherwise turn to the step (6)

(6) cancel the recognition results and exclude the neuron g from the scope of identification, then return the step (4) .

(7) return to the step (3), and recognize the next input pattern.

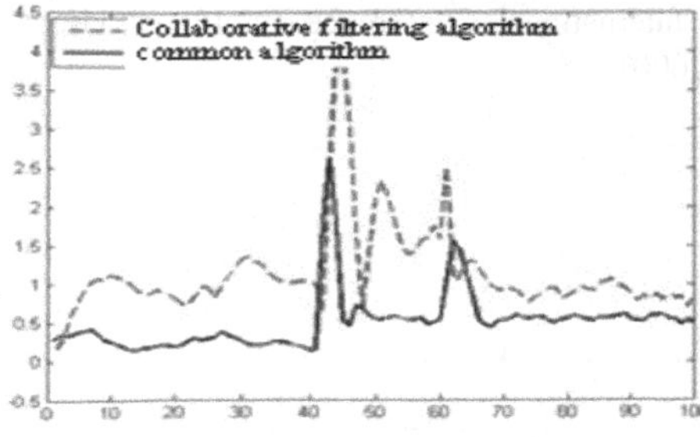

Fig. 3. The compare result of collaborative filtering algorithm and common algorithm

Dimension based on a simplified algorithm to solve the data better sparseness problem, and because k <<n, to predict or TopN recommended for calculating a corresponding reduction in consumption, but also help solve the scalability problem, but the amount of data larger effect is not very satisfactory. Meanwhile, although the SVD can be calculated off-line, but very expensive training required. Taking into account the dynamic change of data elements, the recommended system should also consider re-calculated the frequency of SVD or use Incremental SVD algorithm(but to consider whether to use the algorithm to obtain better accuracy). Generally speaking, the method in solving the user-based collaborative filtering algorithm issues, than on the accuracy of the results on the effect of significantly improving operational efficiency.

5 Summary

This paper presents E-commerce recommendation system based on Collaborative filtering technology. Collaborative filtering recommendation systems in e-commerce application to obtain a greater success, but with the site structure, content, complexity and the increasing number of users, based on collaborative filtering recommendation system.

Acknowledgement. This paper is supported by Education Department of Henan Province, 2011 Natural Science Research Program (2011C520019).

References

1. Fu, X., Budzik, J., Hammond, K.I.: Mining navigation history for recommebdatuib. In: Proceedings of 2000 International Conference Intelligent User Interfaces, pp. 106–112. ACM, New Orleans (2000)
2. Karypis, G.: Evaluation of Item-Based Top-N Recommendation Algorithms. In: Proceedings of the Tenth International Conference on Information and Knowledge Management, Atlanta, Georgia, USA, pp. 247–254 (2001)
3. Sarwar, B., Karypis, G., Konstan, J., Riedl, J.: Analysis of recommendation algorithms for E-commerce. In: ACM Conference on Electronic Commerce, pp. 158–167 (2000)
4. Fu, X., Budzik, J., Hammond, K.I.: Mining navigation history for recommebdatuib. In: Proceedings of 2000 International Conference Intelligent User Interfaces, pp. 106–112. ACM, New Orleans (2000)

Autoregressive Model for Automobile Tire Pressure and Temperature Based on MATLAB

RuiRui Zheng, Min Li, and BaoChun Wu*

Information & Communication Engineering College,
Dalian Nationalities University, Dalian, China
{zhengruirui,wubaochun2007}@yahoo.cn

Abstract. This paper proposed the autoregressive model to forecast pressure and temperature inner aumobile tire for promoting the reliability of tire burst forewarning. The appropriate order for autoregressive model was optimized based on MATLAB. This paper also discussed the relationship between pressure and temperature. MATLAB stimulations show that AR(1) model can forecast the pressure and temperature quickly with a high accuracy, and the pressure and temperature inner tire has a linear relationship which seems fits the ideal gas law.

Keywords: Autoregressive model, Tire pressure monitoring system, Tire temperature monitoring system, MATLAB.

1 Introduction

The statistics from the Ministry of Public Security and Ministry of Transport of the People's Republic of China show that in 2008 China, 70 per cent of highway traffic accidents were caused by tire burst, and 100 per cent people died in roll-over accident caused by front tire burst when the its speed was faster than 160Km/h [1]. According to the American Society of Automotive Engineers statistics, more than 260 thousand accidents were caused by faulty tires, 75 per cent of the faulty tires had problems like lower pressure compared to the standard one, tire gas leakage and over high tire temperature [2]. In a ward, automobile tire faults result in series accidents, and make huge losses to both society and economic.

Thus, research on automobile tire burst early-warning system get more and more attention. Thus tire burst early-warning system gradually becomes one of the essential devices of car electronics. Both U.S. government and European Union adopted mandatory standards of installing the tire pressure monitoring system (TPMS) in cars [3]. In September 2008, the National Development and Reform Commission of the People's Republic of China proposed a draft named "automobile tire pressure monitoring system standard", which stipulated performance requirements of TPMS [4]. There are some licensed automobile tire burst early-warning system products on the market currently, these devices alarm mainly according to monitored pressure inner tire. Because statistics show that if tire pressure is lower than the rated values,

* Corresponding author.

D. Jin and S. Lin (Eds.): Advances in CSIE, Vol. 1, AISC 168, pp. 513–520.
springerlink.com © Springer-Verlag Berlin Heidelberg 2012

0.03MPa for example, the service life of tier will be reduced by 25 per cent, meanwhile, if the tire pressure is 25 per cent higher than the default value, tire service life will be declined by 15 to 20 per cent [5]. Although TPMS devices on the market can alter when the pressure inner tire is more or less than the default value, the best time may be missed and cause an accident, because normally the speed of pressure growing is very fast, and the driver needs time to react to the alarm.

In this paper, we propose an idea of short-time forecast for the pressure and temperature inner tire to give forewarning to the driver, so that the driver may have much more time to react to the future state of car, the reliability of forewarning will be promoted. Because the pressure and temperature values are time series, so we decide to use time series method, autoregressive model, to analyze experiments data based on MATLAB.

2 Autoregressive Model

For a single-output signal $\{x_t\}$, if the relationship of a certain moment t and its former p output values is given by the following Eq. (1).

$$x_t = \varphi_1 x_{t-1} + \varphi_2 x_{t-2} + \ldots + \varphi_p x_{t-p} + \varepsilon_t \,. \tag{1}$$

If the random sequence $\{\varepsilon_t\}$ is white noise and not related to x_t , $k<t$, then this model is called p order autoregressive model, denoted as AR (p). If x_t mainly relative to x_{t-1}, it is 1 order autoregressive model denoted as AR (1) and represented by Eq. (2).

$$x_t = \rho x_{t-1} + + \varepsilon_t,\ t=1, 2, 3\ldots \,. \tag{2}$$

AR (1) satisfied the weak dependence condition if $|\rho|<1$, and then there is Eq. (3).

$$\mathrm{Corr}(x_t, x_{t+h}) = \frac{\mathrm{Cov}(x_t, x_{t+h})}{\sigma_y^2} = \rho^h \,. \tag{3}$$

Correlation coefficient declines quickly as h getting bigger. AR (p) model reflects the influence of historical observations of series itself on predicted objects, without the assumption that model variables are independent. Thus, AR (p) model is capable of eliminating the argument selection and multicollinearity problems of common autoregressive methods [6] and [7].

3 Data Acquisition

We search for information about pressure and temperature data inner running tires on the internet and CNKI database, and notice that there is almost no public pressure and temperature data for some reasons. So to analysis these data we must acquire them first. We carefully design a device and a series experimental program to acquire these data in circumstances closer to the real situation. Each experiment differs from each other on its ambient temperature, ambient pressure, car load, car speed and duration.

The acquisition device we designed to monitor the pressure and temperature inner tire mainly constructed with MPXY8300A, which had already been widely used in measurement of pressure and temperature inner tire. The MPXY8300A chip contains the pressure, temperature and 2-axis accelerometer sensors, a microcontroller, and an RF output all within a single package [8]. The pressure range of our data acquisition device is -40°C to +125°C, and the temperature range is 100Kp to 800Kp. The device can be installed on all vehicles except large trucks.

The experiments were done in the National Tire and Rubber Products Quality Supervision and Inspection Center in November 2011, Qingdao, China. The center is one of the national laboratories for tire testing. We use 3 in use tires during the experiments because statistics indicate that the brand new tire is rarely burst and the old or faulty one may burst in certain conditions. Like we mentioned in former paragraph, these experiment data are obtained in different experiment circumstances. Although subject to the constraints such as test time, site, condition and funds limitation, we only gained 9 datasets, but each dataset has thousands of value points because the experiments last more than hours. Table 1 and Table 2 show part data of two different experiments. Table 1 is part of dataset2, which ambient temperature, ambient pressure, car load and speed of dataset2 are 40°C, 240Kp, 500kg and 60Km/h. Table 2 is part of dataset7, which ambient temperature, ambient pressure, car load and speed of dataset2 are 40°C, 240Kp, 672kg and 90Km/h.

Table 1. Data in dataset 2

Data	Date	Time
[FFFE0FIA3B3C3D3M06P40T68VB9Z1CX23K00]	2011/11/14	11:27:57
[FFFE0FIA3B3C3D3M06P40T69VB9Z1DX26K00]	2011/11/14	11:27:59
[FFFE0FIA3B3C3D3M06P40T69VB9Z1CX25K00]	2011/11/14	11:28:02
[FFFE0FIA3B3C3D3M06P40T68VB9Z1DX27K00]	2011/11/14	11:28:04
[FFFE0FIA3B3C3D3M06P40T69VB9Z1CX26K00]	2011/11/14	11:28:09
[FFFE87ID1D9E1E9M83P1FTB4V5CZ8EX12KBC]	2011/11/14	11:28:13
[FFFE0FIA3B3C3D3M06P40T68VB9Z1DX25K00]	2011/11/14	11:28:16
[FFFE87ID1D9E1E9M83P20T34VDCZ8EX13KBC]	2011/11/14	11:28:25
[FFFE0FIA3B3C3D3M06P40T69VB9Z1DX25K00]	2011/11/14	11:28:27
[FFFE0FIA3B3C3D3M06P40T69VB9Z1CX25K00]	2011/11/14	11:28:32

In the data stream, codes "PXX" and "TXX" represent pressure code and temperature code respectively. After the data stream with a space is the sampling time. The transfer function for pressure versus an 8-bit result from the REIMS_PMEAS firmware is:

$$\text{Pressure}=2.767\times\text{PCODE}+97.233 \tag{4}$$

The transfer function for temperature versus an 8-bit result from the REIMS_PMEAS firmware is:

$$\text{Temperature}=\text{TCODE}-55 \tag{5}$$

Fig. 1 is the raw pressure values calculated by Eq.(4) of dataset 2. Data in Fig.1 are sampled every 30 seconds to reduce the display points to make figure clearly.

Table 2. Data in dataset 7

Data	Date	Time
[FFFE0FIA3B3C3D3M06P37T5CVAFZ0DX0FK00]	2011/11/15	11:47:56
[FFFE07ID1D9E1E9M83P1BTADVD7Z87X88KBC]	2011/11/15	11:48:03
[FFFE0FIA3B3C3D3M06P38T5CVB0Z0FX12K00]	2011/11/15	11:48:15
[FFFE0FIA3B3C3D3M06P38T5CVAFZ0FX11K00]	2011/11/15	11:48:17
[FFFE0FIA3B3C3D3M06P39T5DVAFZ0EX11K00]	2011/11/15	11:48:19
[FFFE0FIA3B3C3D3M06P39T5CVAFZ0EX11K00]	2011/11/15	11:48:22
[FFFE0FIA3B3C3D3M06P39T5CVAFZ0EX10K00]	2011/11/15	11:48:26
[FFFE0FIA3B3C3D3M06P39T5CVAFZ0FX10K00]	2011/11/15	11:48:29
[FFFE0FIA3B3C3D3M06P39T5CVAFZ0FX11K00]	2011/11/15	11:48:31
[FFFE0FIA3B3C3D3M06P39T5CVB0Z0FX11K00]	2011/11/15	11:48:33

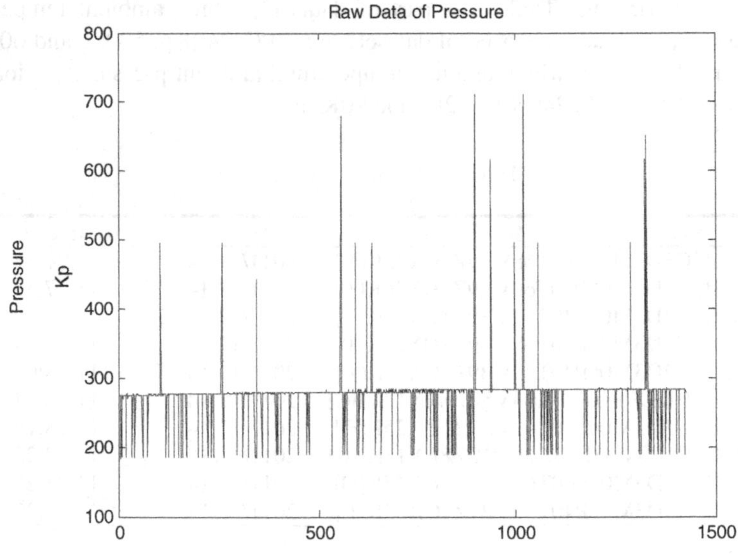

Fig. 1. Raw pressure data

4 AR Model for Forecast Pressure and Temperature

4.1 Data Preprocess

From Fig. 1, we can see the raw data have gross errors, and these gross errors must be deleted to prevent the measurement disturb. Here, we choose the 3 σ criteria [9] to determine a data whether a gross error or not. Autoregressive model require equally spaced series for forecasting, so the raw data must be interpolated first, and then sampled with an equal space. In order to facilitate the calculation, we choose the linear interpolation method, "interp1q" in MATLAB. The sampling space is 30 seconds, because the statistics show that the drivers' average response time to emergency is 1 second to 3 seconds, and if the sampling space is too short, like 1 second, the forecast is meaningless, meanwhile, if the sampling space is too long, like 1 minute, the forecast can't reflect the rapid changes inner tire, and may lead to missing alarm, besides, many researcher choose 30 seconds as their sampling space too [10]. The interpolated and sampled pressure data of dataset 2 is shown in Fig. 2.

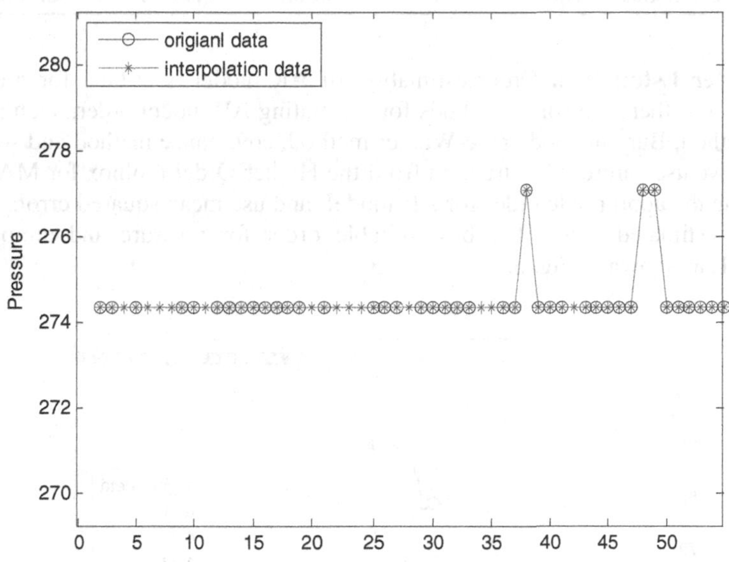

Fig. 2. Pressure data interpolation

4.2 Autoregressive Model

Models Selection. Obviously, the pressure and temperature data are time series, so we can use the time series analysis theory to analyze and forecast data. There are lots of time series analysis theories, like autoregressive model (AR), moving average model (MA), autoregressive and moving average model (ARMA), grey model (GM) and artificial neural network (ANN). ANN is too complex to implement on hardware. GM can't forecast every point in the series. They are not suitable for forecasting pressure and temperature series. Then we analyze whether AR, MA and ARMA are suitable for this problem. Sample autocorrelation computes the sample autocorrelation function of a univariate, stochastic time series with confidence bounds, and sample partial autocorrelation computes the partial autocorrelation function of the series. Analyzing sample autocorrelation and partial autocorrelation can help us to determine the model type. Fig. 3 show the autocorrelation function (AF) and partial autocorrelation function (PAF) of pressure sequences from dataset 2 respectively. The "tail off" and "cut off" properties of autocorrelation function and partial autocorrelation function are the foundation to estimate time series model type. Table 3 demonstrates the autocorrelation and partial autocorrelation functions characteristics of a time series for AR, MA and ARMA models. According to Table 3, the pressure and temperature sequences could be modeled by AR model.

Table 3. ARMA Model Series Characteristics

Model types	AR	MA	ARMA
Autocorrelation function	tail off	cut off	tail off
Partial autocorrelation function	cut off	tail off	tail off

Model Order Estimation. Order estimation of AR model is critical for modeling accuracy. Thus, there are some methods for estimating AR model order, such as least squares method, Burg method, Yule-Walker method, covariance method and so on. In this paper, we use "arorder" instruction from the Higher-Order toolbox for MATLAB to determine the appropriate order for AR model, and use mean squared errors (MSE) to test the estimated order. The best suitable order for pressure and temperature forecast is 1, as shown in Fig. 3.

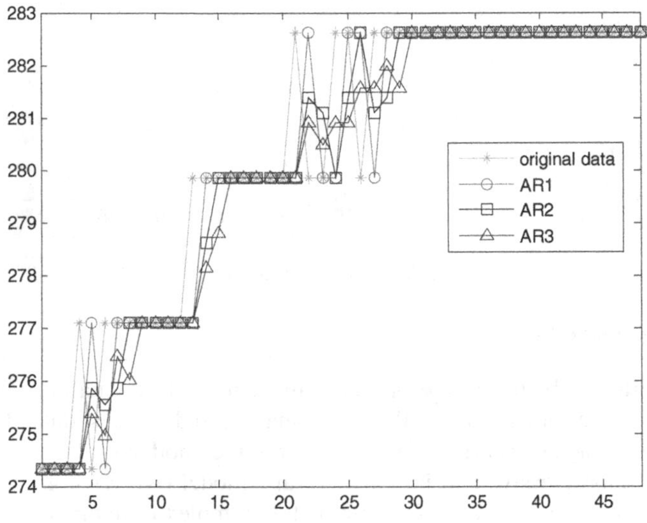

Fig. 3. Pressure forecast using AR models with different orders

5 Pressure and Temperature Relationship Analysis

People believe that the gas pressure and temperature inner tire fit the ideal gas law (6). Where, p means pressure, V represents volume, n denotes the amount of gas, R is a constant, and T means temperature. For gas inner tire, n and R are the same, and V change a little. Suppose V do not change, because its change is too small to measure, then p and T has a proportional relationship. Under this assumption, the pressure values calculated by (6) are close to the measurement values of pressure. The polynomial curve fitting with MATLAB indicates that p and T seems have a linear relationship, which shows the polynomial curve fitting result of dataset 2. But, the polynomial coefficients of the different datasets differ from each other, shown in Table 4.

$$pV=nRT \tag{6}$$

Table 4. Polynomial coefficients

Dataset number	Gradient	Intercept
1	1.7934	-306.8400
2	2.0589	-389.7258
3	2.3087	-520.7472
4	1.9471	-407.4407
5	1.7226	-278.1361
6	1.3693	-163.1378
7	2.5297	-533.6105
8	0.9855	115.0040
9	1.9079	-211.4884

6 Conclusions

To promote the reliability of tire burst forewarning, we utilize autoregressive model to forecast the values of pressure and temperature inner tire, analyze the appropriate order for AR mode, and analyze the relationship between pressure and temperature. MATLAB stimulations show that AR(1) model can forecast the pressure and temperature quickly with a high accuracy, and the pressure and temperature inner tire has a linear relationship which seems fits the ideal gas law. There are still some problems need to be solved too. For example, why the different dataset has different gradients and intercepts. And we will study those issues in our future work.

Acknowledgment. This work is supported by "the Fundamental Research Funds for the Central Universities" under Grant No. DC10020117, "the Foundation of Department of Education of Liaoning Province of China" under Grant No. L2010094, and "the Foundation of Scientific Research for Doctors of Liaoning Province of China" under Grant No. 20111050.

References

1. Li, M., Zhao, J.-Y., Chen, X.-W., Yang, Y.-N.: An Overview of Automobile Tire Burst Early-warning System. Journal of Dalian Nationalites University 13(5), 454–457, 464 (2011)
2. Zang, H.-Q., Tian, C., Zhao, B.-J.: Design of Tire Pressure Monitoring System Based on Embedded Operating System. Transactions of Beijing Institute of Technology 28(10), 870–874 (2008)

3. Liao, W.: Viewing Automobile Electronic Safety Industry from FORYOU TPMS. Automobile Parts (9), 50–52 (2010)
4. Ren, J.-Q.: Research on a novel online testing system for TPMS. Modern Manufacturing Engineering (12), 122–124, 136 (2010)
5. Wang, G.: Simple analysis of tire tyre burst early warning system technical application of the car. Journal of Changchun Institute of Technology (Nature Science Edition) 10(3), 109–112 (2009)
6. Zhang, S., Lei, Y., Feng, Y.: MATLAB Applications in Time Series Analysis. XiDian University Press, Xi'an (2007)
7. Yang, S., Wu, Y., Xuan, J.: Time Series Analysis in Engineering Application. Huazhong University of Science and Technology Press, Wuhan (2007)
8. Freescale Semiconductor Product Specification MPXY8300. Rev 4, 05 (2009)
9. Fei, Y.-T.: Error Theory and Data Processing. China Machine Press, Beijing (2011)
10. Zang, H., Tian, C., Zhao, B.: Design of Tire Pressure Monitoring System Based on Embedded Operating System. Transactions of Beijing Institute of Technology 28(10), 870–874 (2008)

The Feature Selection Method of Radar Modulation Signal Based on Particle Swarm Optimization Algorithm

Libiao Tong, Fei Ye, Hongjun Zhang, and Wenjun Lu

New Star Research Institute of Applied Technology, Hefei 230031, China
libtong@sohu.com, nuptzhj@sina.com, lwj2963@sina.com

Abstract. Feature selection is the important component of pattern recognition. Particle Swarm Optimization (PSO) is sample and has excellent performance. Through adding improvement strategy to standard PSO algorithm, the feature selection of radar modulation signal based on PSO algorithm is proposed, and the simulation experiment validates the algorithm is effective.

Keywords: PSO algorithm, feature selection, modulation type recognition.

1 Introduction

PSO is a bionic optimization algorithm. It is sample and easy to realization and has not many adjusted parameters and needn't grads information. So PSO has great potential in the application of nonlinear programming, TSP, system distinguish, system design, dynamic multi-target optimization, pattern recognition, schedule and so on[1][2]. The paper proposes an improved PSO algorithm and applies it to feature selection of radar emitter signal modulation type recognition.

2 The Basic Thought of Algorithm

Different feature subset is regarded as a particle of feature space, and then some particles are put in feature space. Every particle has one place and flies in the space in order to flying the best place. After period of time, they change places continually and communicate each over, then surround possible local or full best places, and at last they find the optimal place. For the search ability POS algorithm can be used to perform feature selection and find optimal feature subset. For N features, there have 2^N feature subset with different length and feature, and the optimal position searched by algorithm is the feature subset with the minimal length and highest classification quality.

In the late of the search period, Standard PSO has low search strategy, gets into local extremum easily, optimizes performance inadequately and so on shortcoming[3]. In order to improve the quality of solution, the follow improvement strategy is adopted:

(1) In order to prevent falling local extremum, the life force is given to particle all the time. If the speed increment is close to zero in the iteration process, particles are put random speed in certain scope, and then particles are promoted to move.

D. Jin and S. Lin (Eds.): Advances in CSIE, Vol. 1, AISC 168, pp. 521–526.
springerlink.com © Springer-Verlag Berlin Heidelberg 2012

(2) Introduce mutation mechanism. The algorithm can get on with search in the late of iteration, so its ability of full search is improved, and the possibility got into local optimization can be decreased, and finally the quality of search result is improved.

(3) The phenomenon of "birds of a feature flock together" is common and it is called niche. According to the theory of niche[5], if the place of two particles approach excessively, then the position and speed of particle with less fitness is evaluated over again.

An improved PSO is formed by adding improvement strategy. Some problems in the algorithm are illuminated as follow.

(1) The expression of particle position

The particle position is expressed as binary code with length N, and N is the number of feature. Each code symbol expresses one feature, 1 means it's not selected and 0 means it's selected. Every position (binary code) is a feature subset.

(2) Particle speed and position updating

In binary system space, the movement of particle can be achieved by reversing bit value. The updating formula of speed and position:

$$v_{ij} = v_{ij} + c_1 \cdot rand(\cdot) \cdot \left| p_{ij} - x_{ij} \right| + c_2 \cdot Rand(\cdot) \cdot \left| p_{gj} - x_{ij} \right| \tag{1}$$

$$\begin{cases} v_{ij} = V_{\max} & v_{ij} \geq V_{\max} \\ v_{ij} = -V_{\max} & v_{ij} < -V_{\max} \end{cases} \tag{2}$$

$$x_{ij} = \begin{cases} !x_{ij} & rand_2(\cdot) < S(v_{ij}) \\ x_{ij} & rand_2(\cdot) > S(v_{ij}) \end{cases} \tag{3}$$

$S(v_{ij}) = \dfrac{1}{1+\exp(-v_{ij})}$ is sigmoid function. $rand(\cdot)$、 $Rand(\cdot)$ and $rand_2(\cdot)$ is the stochastic number of [0,1]. $!$ is the NOT operation, that is to say $!1=0$, $!0=1$. Speed v_{ij} decides the probability whether the x_{ij} will be reversed, and v_{ij} is bigger then the reversed possibility of x_{ij}. At the same time, $V_{\max}$ is kept to restrict the final probability whether the x_{ij} will be reversed.

(3) Fitness function

Suppose $index$ is the index to evaluate feature subset, if big $index$ expresses its classification ability is good, then fitness function is defined that:

$$Fitness = \alpha \cdot index + \beta \cdot \frac{N-L}{N} \tag{4}$$

If small $index$ expresses its classification ability is bad, then fitness function is defined that:

$$Fitness = \frac{\alpha}{index} + \beta \cdot \frac{N-L}{N} \tag{5}$$

N is the dimension of feature set, L is the code number of 1 in the particle position and is the length of feature subset selected. α and β are two variable parameter, and express the importance of the classification quality and the subset length, and $\alpha \in [0,1]$, $\beta = 1-\alpha$.

3 The Feature Selection Based on Improved PSO Algorithm

Assume N is the dimension of candidate feature set, the procession of feature selection algorithm based on PSO is as follow:

(1) N candidate features are sorted randomly;

(2) Population initialization. $m(20 \le m \le 40)$ particles are produced randomly, particle position is binary code with the length of N. 0 means the feature of corresponding position is not selected, and 1 means the feature of corresponding position is selected. Particle speed is a N-dimension vector, and the speed in every dimension is a random count in $(0, V_{\max})$.

(3) The selected feature subset is used to classify the sample set (there need clustering for sample set without class sign), and the classification result can be evaluated with $index$. Supposed classification quality is more important than the length of feature subset, so set $\alpha = 0.9$, $\beta = 0.1$, the every particle's fitness is computed according formula (4) or (5).

(4) The distance of two particle can be defined the number of difference binary code symbol. According to niche phenomenon, whether there have the distance between particles in population less than ε (threshold), if there exist, then new particle is produced to replace the particle with small fitness while all the distance of particles in population are bigger than ε, and the fitness of new particle is computed.

(5) Update the optimal position of every particle $\mathbf{P}_i$ and population optimal position $\mathbf{P}_g$.

(6) According formula (1), (2) and (3), the speed and position of particle is updated.

(7) If the increment of the particle iteration speed is less than Δv (threshold), then this particle loses life force, and the new particle speed is produced randomly over again in $(0, V_{\max})$.

(8) Mutation operation. For the $j(j = 1, 2, \cdots, N)$ part of $i(i = 1, 2, \cdots, m)$ particle, a binary count $rand\,\mathrm{int}(0,1)$ and stochastic count $rand(\cdot)$ of [0,1] are produced randomly. If $rand\,\mathrm{int}(0,1) = 1$ and $rand(\cdot) < (1 - \frac{iter}{iter_{\max}})^2$, then

position x_{ij} upturn. $iter$ is the iteration times and $iter_{max}$ is the maximal iteration times.

(9) Repeat step (3) to (8), until the iteration times are equal to the maximal iteration times.

(10) The feature which corresponds to the "1"code symbol of population optimal position is the feature selected.

4 Simulation Experiment

Experiment one. The performance of GA (Genetic Algorithm) and improved PSO is compared by six-apex humpback function. In order to let two algorithms in the same condition, immigrant strategy is added in GA for the sake of keeping population diversity. Six-apex humpback function can be expressed as:

$$\begin{cases} f(x,y) = (4 - 2.1x^2 + \frac{x^4}{3})x^2 + xy + (-4 + 4y^2)y^2 \\ -3 \le x \le 3 \\ -2 \le y \le 2 \end{cases} \tag{6}$$

The function has six local minimal values, $(-0.0898, 0.7126)$ and $(0.0898, -0.7126)$ is the full minimal values, the minimal value is $f(-0.0898, 0.7126) = f(0.0898, -0.7126) = -1.031628$. Let the number of original population is 20, iteration times is 20, the algorithm is repeated 10 times, then the search process for function minimal value shows Fig 1. GA algorithm gets minimal value -1.0316 at 12 times and PSO gets minimal value -1.0316 at 9 times averagely.

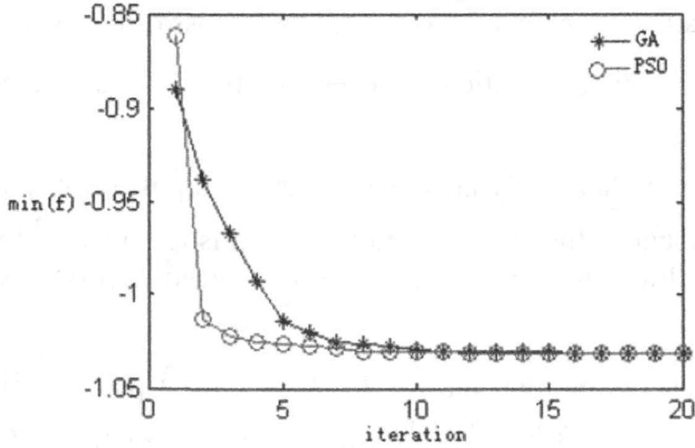

Fig. 1. The process of searching function minimal value by GA and PSO

Experiment two. The 20 kinds feature samples of 12 kind radar emitter signals are adopted. The radar signals of 12 kinds are fixed frequency, linearity frequency modulation, V type frequency modulation, tangent frequency modulation, arctangent frequency modulation, hyperbola frequency modulation, equal progression frequency modulation, linearity step frequency modulation, frequency coding, two-phase coding and four-phase coding signal. The signals are decomposed by four-class wavelet and four box dimensions, four information dimensions, four correlation dimensions, four multi-fractal entropy and four lacunarity measure variation rate can be get, so 20 kinds features is D_{B1}, D_{B2}, D_{B3}, D_{B4}, D_{I1}, D_{I2}, D_{I3}, D_{I4}, D_{C1}, D_{C2}, D_{C3}, D_{C4}, $H_{D_q 1}$, $H_{D_q 2}$, $H_{D_q 3}$, $H_{D_q 4}$, $E_{\Lambda 1}$, $E_{\Lambda 2}$, $E_{\Lambda 3}$, $E_{\Lambda 4}$ [6]。 In feature selection, the number of various signal samples is 100, and the number of original population is 20 and the maximal iteration times in PSO feature selection algorithm. The feature selection algorithm in the paper is used to select and using nearest classifier achieves the recognition of radar emitter signal modulation type, so the runtime, feature subset selected and the last recognition rate of feature selection is show as Tab1.

From the result of Tab1, it can be found that the feature subset selected by PSO feature selection algorithm is 5 dimensions and less the dimension of original feature set greatly, but the algorithm runtime has quite long. Because the algorithm is a feature selection without teacher, its search space is great bigger than the feature selection algorithm with teacher.

Table 1. Font sizes of headings. Table captions should always be positioned *above* the tables.

runtime(s)	dimension of feature subset selected	feature subset	recognition rate
9.8493×103	5	D_{B4}, D_{I3}, D_{C4}, $H_{D_q 3}$, $E_{\Lambda 2}$	88.17%

5 Conclusions

Particle Swarm Optimization is sample and has excellent search performance, but in the late of search process its search efficiency becomes low and is easy to get into local optimization. So through adding improved strategy to standard PSO algorithm improved PSO algorithm is formed. Comparing the improved PSO algorithm and GA, the result of simulation experiment one indicates that PSO has more excellent search performance. Using improved PSO algorithm to select feature, simulation experiment two shows that the feature selection algorithm can greatly reduce the dimension of recognition and get high recognition rate.

References

1. Duan, H., Wang, D., Yu, X.: Research on some novel bionic optimization algorithm. Computer Simulation 24(3), 169–172 (2007)
2. Lu, B.: The research and development of particle swarm algorithm. Business Technology (501), 25–26 (2007)
3. Cui, Z., Zeng, J.: Dynamic adjusting modified particle swarm algorithm. Journal of Systems Engineering 20(6), 657–660 (2005)
4. Zheng, X., Qian, F.: A modified particle swarm optimization algorithm. Computer Engineering 32(15), 25–27 (2006)
5. Wang, J.-N., Shen, Q., Shen, H., Zhou, X.: A clustering-based niching particle swarm optimization. Information and Control 34(6), 680–684 (2005)
6. Ye, F., Yu, Z., Luo, J.: Analysis of Radar Emitter Signal Feature Based on Multifractal Theory. In: Proceedings of 2007 8th International Conference on Electronic Measurement & Instruments, Xi an, China, vol. I, pp. 14–17 (2007)

Dynamic Trust-Based Access Control Model for Web Services

YongSheng Zhang[1,2], YuanYuan Li[1,2], ZhiHua Zheng[1,2], and MingFeng Wu[1,2]

[1] School of Information Science and Engineering, Shandong Normal University, Jinan 250014, China
[2] Shandong Provincial Key Laboratory for Novel Distributed Computer Software Technology, Jinan 250014, China
zhanys@sdnu.edu.cn, lizzy.01@163.com, Zhengzhi121@sina.com, wumingfeng310@126.com

Abstract. The paper presents a new trust-based access control model. In this model, the service provider registers trust threshold and trust weight on the trust server. Only the service requester's trust value exceeds the threshold, he can get the service provider's trust and their services. Within the trust, direct trust is regularly updated, and both implement trust evaluation through direct interaction of experience and recommendation trust information. Combine the Security Assertion Markup Language (SAML) and Extensible Access Control Markup Language (XACML) together and embed SAML authentication module and XACML access control module into the model. The model is flexible and reliable because it not only takes into account the trust value of service requester, but also implements the authentication mechanism.

Keywords: direct trust, recommendation trust, trust threshold, affair impact factor (AIF).

1 Introduction

Access control technology plays a significant role in addressing the subject access to the object, and it can restrict user's access to system resources, prevent unauthorized users' access to the system, and avoid the legitimate users using system resources by illegal means [1]. Reference[2] proposed the concept of trust management, based on which researchers expanded some traditional access control models, such as access control matrix, access control list (ACL), etc[3]. Reference[4] put forward a certification credibility-based access control model, which introduced the concept of trust threshold based on the CAS model. However, the weight of each rule needed to be changed dynamically with the network in authentication system rule base of CAS server. When the resource providers have greater autonomy, how to set this threshold needs to be improved. Reference[5] came up with a model WS-TBAC, in which calculation of trust was through the algorithm of regret system, the direct trust and recommended trust were known as individual latitude and social latitude. The calculation of individual latitude needed to set different weights according to different

D. Jin and S. Lin (Eds.): Advances in CSIE, Vol. 1, AISC 168, pp. 527–534.
springerlink.com © Springer-Verlag Berlin Heidelberg 2012

events, but it did not consider the requester's recent performance, by which updated the trust. So this article presents a model that is a set of credible, certification, trust threshold, at the same time, it takes trust update into account and makes it more perfect in order to better adapt to the dynamic, open Web Service environment.

2 Improved Trust Model

A new trust-based access control model is proposed on the basis of the existing research in this paper. The work including three aspects is as follows:

(1)Introduce confidence threshold and certified weight.

(2)Combine SAML authentication module and XACML access control module.

(3)Assign the corresponding appropriate weights to history trust and recent trust to update the trust according to the individual's recent performance.

There are four entities in the trust domain: service requester, service provider, trust server and witnesses that provide trust information.

Whether service requester is able to obtain access authorization or not needs to go through two tests: one is the recommendation trust returned by witnesses, the other is synthesis trust synthesized by trust synthesis module. If the two are standard, requester is granted access to resources, otherwise, the requester can not get the resource authorization.

Definition 1: Certification credibility is a confidence value owned by users after authenticated by authentication system rule base in trust server, represented by $A(u)$, and $A(u) \in [0,1]$.

Definition 2: Trust threshold e is the boundary used to distinguish whether to trust. Users must achieve minimum confidence value to get access authorization, represented by e, and $e \in [0,1]$.

Definition 3: Certification weight factor λ is determined by the resource providers. After authenticated by trust server, users obtain certification credibility $A(u)$ needed to be multiplied by λ.

Definition 4: Consistency verification is a kind of mechanism used to test the legality of the certificate submitted by the user, whether the certificate has been revoked by certification authority or not.

Definition 5: Authentication system rule base includes some attribute information of service requesters, such as identity, passwords, voice, fingerprint, etc. Authentication system rule base is regularly updated by the server.

Overall institutions and process of model are shown in Fig.1.

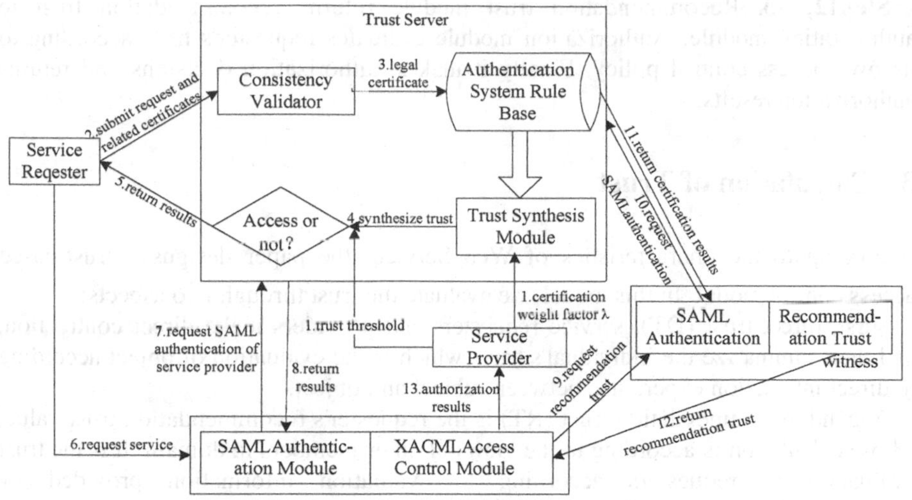

Fig. 1. A new trust-based access control model

Step1. Service providers register resources to trust server, meanwhile submit certification weight factor λ of all resources and trust threshold e of corresponding authorities to trust server.

Step2. Service requester inquiries service from the system in order to find corresponding service providers. Requesters select service strategy and service providers according to their needs, and submit service requests and associated server certificate to the trust server.

Step3. Consistency verifier is used to verify the legality of the certificates.

Step4, 5. In the authentication system rule base, the calculation of certification credibility $A(u)$ is according to the weight of the certificate, its value gets from $T = \lambda A(u) + (1-\lambda)T_0$ in the trust synthesis module, T_0 is the original trust of requester, whose value gets from history table of trust. Compare T with e, if $T \geq e$, allow users to make requests for access to resources, otherwise, and do not allow users to make requests for access to resources.

Step6. Service requesters propose service requests to XACML authorization module, and request service providers to give the corresponding service.

Step7. After XACML authorization module receives the service requests from service requesters, SAML authentication module sends SAML requests to trust server to complete SAML authentication.

Step8. Authentication results are returned by the trust server.

Step9. After SAML authentication, XACML access control module needs to obtain the recommendation from other members in the trust domain to decide the final authorization result.

Step10, 11. Recommendation trust module requests SAML authentication of service requesters and providers to trust server, and gets authentication results from trust server.

Step12, 13. Recommendation trust module returns recommendation trust to authorization module. Authorization module evaluates requester's trust according to its own access control policy. Finally it makes authorization decisions and returns authorization results.

3 Calculation of Trust

According to the characteristics of Web Service, the paper designs a trust-based access control model. In this model, we evaluate the trust through two aspects:

First, direct trust (DT), service requesters and providers make direct contraction, and then summarize the individual's trust, which is the evaluation of object according to direct interaction experience between subject and object.

Second, recommendation trust (RT) is the requester's recommendation trust value, whose calculation is according to the views of other members in domain. It is the trust evaluation of requesters according to evaluation information provided by recommendation nodes and credibility of witnesses [6].

3.1 Credit Assessment of Witness

We introduce the concept of affair impact factor (AIF) to identify the impact extent of affairs so that malicious members in domain show good trust condition in the ordinary affairs to improve their trust in more important events to make serious damage. The greater the value of AIF is, the greater the impact of individual trust, for example, AIF of transmission of confidential data is bigger than transmission of common data; AIF of writing access is bigger than reading access.

In order to cope with different trust requirements of different access services, individual trust values should be distributed in a range, for example, trust's range is 0 to 1.0 is absolutely incredible; 1 is credible entirely. The higher the trust is, the more credible the individual is [7].

3.2 The Calculation and Update of Direct Trust

(1) The calculation of direct trust

Direct trust is the subjective expectations of future behavior of the target entity according to experience in a specific environment and time.

If there are m times interaction between service requester q and provider p, the satisfaction of the i-th is $Sa(q,p,i)$, $Sa(q,p,i) \in [0,1]$.The satisfaction increases with $Sa(q,p,i)$. $AIF(q,p,i)$ is the affair impact factor of the i-th interaction, According to direct experience, the calculation formula of service requester's trust is as follows:

$$D(q)=\frac{\sum_{i=1}^{m} Sa(q,p,i)*TCF(q,p,i)}{\sum_{i=1}^{m} TCF(q,p,i)} \quad (1)$$

The value of $Sa(q,p,i)$ is determined by service provider according to the performance of requester, and each direct trust is recorded in the history table to be easy to query and update.

The value of $AIF(q,p,i)$ is determined by service provider according to the importance of the event, $AIF(q,p,i)\in[0,1]$.If individuals' performance is absolutely satisfied in all events, $Sa(q,p,i)$ for all i is equal to 1,so $D(q)=1$,we will consider q is absolutely credible.

(2) The update of direct trust

The behavior of the network is changing constantly, therefore, the recent behavior of the requester is more valuable, better reflect the current status of the trust. The update formula is as follows [8]:

$$\begin{cases} D_i(q)=\sigma\times D_{i-1}(q)\oplus(1-\sigma)\times D_i(q) \\ Sa(q,p,i)=\sigma\times Sa(q,p,i-1)+(1-\sigma)\times Sa(q,p,i) \\ AIF(q,p,i)=\sigma\times AIF(q,p,i-1)+(1-\sigma)\times AIF(q,p,i) \end{cases} \quad (2)$$

In above-mentioned formula, $\oplus$ is used to combine direct trust, $D_i(q)$ is updated direct trust, $D_{i-1}(q)$ is history trust. Similarly, $Sa(q,p,i-1)$ is the trust of the i-th interaction between requester and provider, $AIF(q,p,i-1)$ is the affair impact factor of i-th interaction. σ is the corresponding weight of historical events, and its value is determined according to a recent performance of the service requester.

$$\sigma=\begin{cases} \sigma_h, s\ge f \\ \sigma_l, s\le f \end{cases} \quad (3)$$

In the formula above, $0<\sigma_h<0.5<\sigma_l<1$. s is the number of successful interaction, similarly f is the number of unsuccessful interaction, the greater the value of σ_h is, the greater proportion owned by history state, that can prevent malicious individuals from increasing trust by taking specific actions in the short term after they found out they have too low trust recently; the greater the value of σ_l is, the greater proportion owned by recent state, that can supervise requester's behavior opportunely to punish the consequences created by malicious acts severely.

3.3 Recommendation Trust

Service requester q obtains recommendation trust from other members, n is the total number of events, $r(q,p,j)$ is the recommender of the j-th event, $Sa(q,p,j)$ is the satisfaction of $r(q,p,j)$ to q in the j-th event, $T(r(q,p,j))$ is the trust of the recommender, the affair impact factor is $AIF(q,p,j)$ in the j-th event, then the formula of recommendation trust is as follows:

$$R(q)=\frac{\sum_{j=1}^{n} Sa(q,p,j)*T(r(q,p,j))*AIF(q,p,j)}{\sum_{j=1}^{n} T(r(q,p,j))*AIF(q,p,j)} \tag{4}$$

If all recommenders are satisfied with individuals absolutely, $Sa(q,p,i)$ for all j is equal to 1, so $R(q)=1$, we will consider q is absolutely credible.

3.4 Comprehensive Trust

In the paper, for comprehensive trust, we consider three aspects:

- Direct trust of service provider;
- Recommendation trust of other members;
- Service requester's own original trust T_0.

Select weights $\alpha \in (0,1)$, $\beta \in (0,1)$:

$$T(q)=\alpha*\frac{\sum_{i=1}^{m} Sa(q,p,i)*AIF(q,p,i)}{\sum_{i=1}^{m} AIF(q,p,i)}+\beta*\frac{\sum_{j=1}^{n} Sa(q,p,j)*T(r(q,p,j))*AIF(q,p,j)}{\sum_{j=1}^{n} T(r(q,p,j))*AIF(q,p,j)}+(1-\alpha-\beta)T_0 \tag{5}$$

4 Authorization

The workflow of XACML access authorization model showed in Fig.2 is as follows:

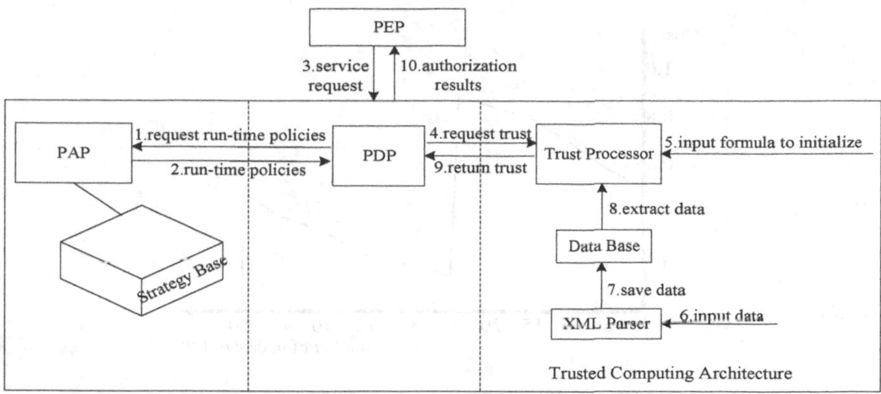

Fig. 2. XACML access authorization module

PAP generates security strategy by strategy base, the security strategy is described by XACML language. PAP was requested by PDP for operation strategy to make the decision. User sends the requests depending on the application, these access control requests are intercepted then unified into XACML format by PEP, and then sent to the PDP. PDP returns the eventual decision results to PEP according to the trust of requester and strategy in domain.

Trusted Computing Architecture: Service providers input the formula to initialize the trust processor and get the way of calculating trust. The necessary data for the calculation of trust is sent to the XML parser, which is stored in a corresponding database table. Data transmission is in the form of XACML file. The trust processor gets the corresponding data from the database when trust degree is needed [5].

5 Authorization

The layout of the scene is as follows: there is 500 interaction between service requester and provider, initialize the following values: the weight of requester's certificate is 0.6;Initial trust T_0 gets from the history table, the value is 0.5;The weight of DT is 0.3,the weight of RT is 0.4,that is $\alpha = 0.3$, $\beta = 0.4$;Weight factor of initial resource is 0.3,that is $\lambda = 0.3$;Initial trust threshold is 0.7,that is $e = 0.7$;Initial affair impact factor is 0.2,that is $AIF = 0.2$;When there is 10 interaction, DT will be updated once. The update weight of DT is as follows: $\sigma_h = 0.8$, $\sigma_l = 0.2$.

In the experiment, we assume that malicious individuals make malicious attacks at the time of the 24th update of DT, stop attacks at the time of the 25th update of DT. The result is shown in Fig.3.

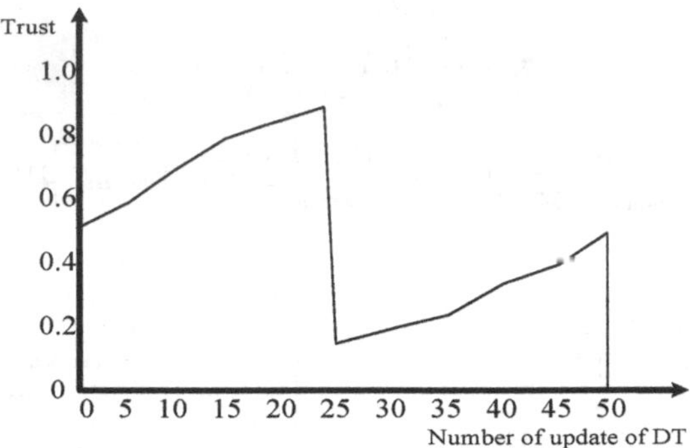

Fig. 3. Changes of trust as individuals make malicious attacks

6 Conclusion

A trust model and improved trust evaluation algorithm are proposed in this paper, this model is like certification agency between service requester and provider, it helps service provider to complete the requester's trust assessment, and submits the final authorization decision to service provider, that is credible for service provider and against deliberate attacks of malicious individuals.

Acknowledgments. This work was supported by Natural Science Foundation of Shandong Province of China under Grant No. ZR2011FM019.

References

1. Luo, H.: Research on Several Access Control Models of Automated Trust Negotiation. Network Security Technology & Application, 77–80 (2010)
2. Blaze, M., Feigenbaum, J., Lacy, J.: Decentralized trust management (EB/OL) (October 10, 2008), `http://www.cs.utsa.edu/winsboro/teaching/CS6463-S06/Papers/BFL96.pdf`
3. Zhao, R., Song, G.: An improved method for white noise reduction based on wavelet transform. Journal of Xidian University (Natural Science) 27, 619–622 (2000)
4. Wang, Y., Chang, C., Liu, C., Fu, X.: Design and analysis of an access control model based on credibility. Network and Computer Security, 3–5 (2009)
5. Ma, X., Feng, Z., Xu, C.: Trust-Based Access Control in Web Service. Computer Engineering 36, 10–12 (2010)
6. Hu, J., Zhou, B., Wu, Q.: Research on Incentive Mechanism Integrated Trust Management for P2P Networks. Journal on Communications 32, 22–31 (2011)
7. Liu, Y.: Trust-based P2P access control model. Computer Engineering and Applications 44, 145–147 (2008)
8. Chen, Z., Zi, B., Jiang, G., Liu, Y.: Trust evaluation for wireless sensor networks based on trust-cloud. Journal of Computer Applications 30, 3346–3348 (2010)

Predicting Reservoir Production Based on Wavelet Analysis-Neural Network

Zhidi Liu[1], Zhengguo Wang[2], and Chunyan Wang[3]

[1] School of Petroleum Resources, Xi'an Shiyou University, China
[2] Trans-Asia Gas Pipeline Company Limited, China
[3] Exploration Department, Yanchang Oilfield Company, China
lzdtxy@sina.com

Abstract. During oil field development, production prediction is related to effectively develop oil reservoirs. In the process of prediction production commonly using modular dynamics testing (MDT), it will introduce larger error that MDT data is directly used to predict production. Considering this issue, the wavelet coefficients that are extracted from the MDT data using wavelet analysis method, then the neural network method is used for establishing production predicting model that use drill stem testing (DST) production and wavelet coefficients. The set of MDT production predicting method is applied to predict production in Karamay oil field. The results show that it can obtain good accuracy.

Keywords: Wavelet Analysis, Neural Network, Productivity Prediction.

1 Introduction

Oil and gas field development often show the non-repetitive and non-experimental characteristics [1-3], so it has an important role in guiding that predict reservoir productivity. Formation testing is the main data to study reservoir productivity. Because MDT test time is short, it will lead to increase errors that directly use the data to predict productivity. The detailed features of pressure curve reflect the relationship of productivity and pressure changes in the pressure test process, and it is the foundation of productivity estimation. Based on this understanding, the relationship between DST data and MDT wavelet coefficient are established using wavelet analysis and neural network method, and the predicting method of reservoir productivity is formed.

2 MDT Wavelet Analysis

The MDT pressure data is a set of time-varying signal which is a sequence of discrete signals. In order to conduct wavelet transform for MDT pressure signal, the numerical approximation integration replace the integral in wavelet transform.

D. Jin and S. Lin (Eds.): Advances in CSIE, Vol. 1, AISC 168, pp. 535–539.
springerlink.com © Springer-Verlag Berlin Heidelberg 2012

$$WT_x(a,\tau) = \frac{\Delta T}{\sqrt{a}} \sum_n x(n) \psi\left(\frac{n-\tau}{a}\right) \quad (1)$$

Where, $\psi\left(\frac{n-\tau}{a}\right)$ is continuous wavelet function which is generated by wavelet generating function; a is scale factor, a >0; τ is displacement factor; ΔT is sampling interval; $x(n)$ is signal sequence.

Equation (1) shows that wavelet coefficient of each resolution is a set of data which reflect the change information of formation fluid pressure and production. Therefore, the productivity information which is scaled using DST test data can be extracted, and can be used to study the relationship between productivity and wavelet coefficients of MDT pressure. Figure 1 is a MDT pressure curve of wavelet analysis (scale factor a = 2). Since then, the wavelet coefficients can see such characteristics for high-frequency wavelet coefficients of high-performance MDT pressure curve, the distribution is wide, the time is long duration, and it is distributed more evenly.

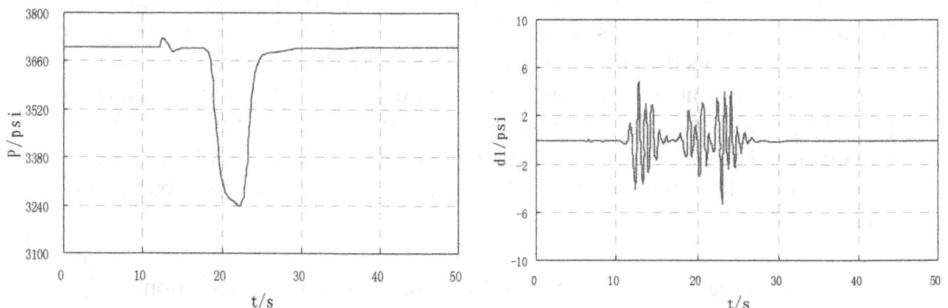

Fig. 1. The relationship chart of wavelet coefficients and productivity in S303 well

From the wavelet analysis of the pressure curve, the wavelet analysis can reveal the tiny pressure disturbance of the whole formation test pressure curve. Therefore, the wavelet coefficients of fluid flow section can reflect the fluid productivity information. In this figure, the wavelet coefficient which is equal to 2 obviously fluctuates in the nearby 18s and 30s. But contrasted on pressure test curve, the fluid section is confined to between 25s and 80s. Thus, the wavelet coefficients within this time can be selected as neural network's input nodes. Wavelet coefficient's fluctuation only reflects the mud pressure disturbance before 25s; the disturbance is irrelevant to formation productivity, so it can't be used to study the productivity evaluation.

3 Productivity Predicting Model

3.1 Basic Principle of Productivity Predicting

In the process of MDT formation test, due to the test condition and operation personnel set of conditions are different, MDT data is actually the production data of different pressure. In order to eliminate the influence of production pressure, all of the wavelet coefficients were divided by production pressure. So there is equation as follows.

$$a_{ij} = \frac{a'_{ij}}{\Delta P} \tag{2}$$

Where, a_{ij} is wavelet coefficient after standardizing production pressure; a'_{ij} is the first j wavelet coefficients of the first pressure test data; ΔP is the production pressure during MDT testing.

MDT is usually single point test data, DST is generally single-layer test and co-layer test. In most cases, the relationship between MDT and DST data is not the one-to-one. Generally, a DST test section includes several MDT test points, so each test point is the productivity of specific depth segment. Hypothesis the scope of depth segment is 1 meter, DST is multi-layer combined test, so DST productivity of multi-layer combined test decompensate to homogeneous single-layer. The function is expressed as follows according to Darcy's law.

$$Q = K\frac{\Delta P \times S}{\mu \times L} \tag{3}$$

Where, Q is test production; K is permeability; S is flow cross-sectional area; μ is flow viscosity; L is flow length in the ΔP differential pressure.

Hypothesis DST multi-layer combined test contains n layers. Each layer's thickness and permeability are respectively Δh_i, k_i. Then the following equation is expressed.

$$\begin{cases} Q = \sum_{i=1}^{n} q_i \\ q_1 : q_2 : \ldots q_n = \frac{k_1 h_1}{\mu_1} : \frac{k_2 h_2}{\mu_2} \ldots\ldots \frac{k_n h_n}{\mu_n} \end{cases} \tag{4}$$

The above-mentioned equation suggests that we know the permeability, viscosity and thickness of single-layer, the multilayer DST test data can be decomposed into a single productivity. For physical obvious non-uniform and only a testing point, it needs to decomposite DST productivity data according to layer. Then, make a correspondence analysis for the nearest of DST testing data and the MDT testing data.

3.2 The Neural Network Model of Production Predicting

After DST production is decomposed to the corresponding measuring point, the productivity prediction model of hydrocarbon reservoir can construct using neural network to train the wavelet coefficients (Fig.2). The MDT wavelet coefficients are normalized for production pressure when a is equal to 2, the previous 22 wavelet coefficients act as the neural network input nodes, the number of hidden nodes is 5. The production predicts using DST testing data by the neural network training method.

4 Application Case Studies

According to the above method, the productivity prediction is conducted for SN31 and other wells in Karamay oil filed using the program of wavelet analysis and neural network. Table 1 is the compared between DST test productivity and reservoir productivity which prediction using the wavelet analysis and neural network in work area. We can see that the difference between prediction productivity and DST test productivity are large from this table. For MDT prediction data of single-point, the difference between the prediction production and DST test data are lager, the error between MDT's single point productivity and distribution DST productivity is lager because in the case of low productivity. However, compared with the same layer of MDT predict production and DST tests production, the errors is small, and the precision can completely meet the actual requirements to prediction accuracy.

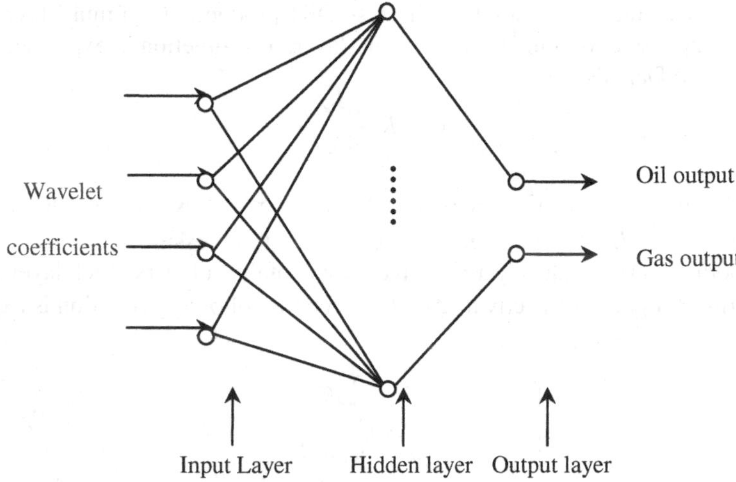

Fig. 2. The wavelet -neural network structure

Table 1. Compared with the DST tests and prediction production

Well	Test point	Actual oil production (m^3/d)	Prediction oil production (m^3/d)	Actual gas production (m^3/d)	Prediction gas production (m^3/d)	Oil production prediction error(%)	Gas production prediction error (%)
SN31	34	2.42	2.22	310.69	337.37	8.38	8.59
SN31	35	7.13	6.59	913.79	915.34	7.51	0.17
SN31	39	15.25	15.65	1955.52	1966.55	2.62	0.56
SN31	Total	24.80	24.46	3180.00	3219.26	1.37	1.23
S304	65	0.21	0.22	642.52	671.73	5.77	4.55
S304	71	0.10	0.11	321.26	290.99	5.77	9.42
S304	67	18.89	18.27	58276.23	58287.87	3.26	0.02
S304	Total	19.20	18.60	59240.00	59250.58	3.12	0.02
MN1	67	2.60	2.81	24997.33	26134.74	8.08	4.55
MN1	69	2.83	3.01	27177.37	26294.09	6.47	3.25
MN1	70	44.61	44.60	428877.31	426397.44	0.03	0.58
MN1	Total	50.04	50.42	481052.01	478826.27	0.76	0.46

5 Conclusions

The wavelet analysis - neural network method which predicted the reservoir production were detailed studied by theoretical study and practical production test data analysis. The oil and gas production were predicted using the conducted model of predicting production. Actual application results show that the method can more accurately predict reservoir production. This method can be used productivity prediction in other Karamay areas.

References

1. Daungkaew, J., Harfoushian, S., Cheong, B.: Mini-DST Applications for Shell Deepwater Malaysia. SPE 109279 (2007)
2. Orban, N., Ayan, C., Ardila, M.: Development and Use of Improved Wireline Formation Tester Technologies in Challenging Deltaic Reservoirs. SPE 120691 (2009)
3. Kuang, L.C.: Introduction to MDT Data Application. Petroleum Industry Press, Beijing (2005)
4. Xu, L., Chen, D., Lai, F., et al.: Study on Method of Nervous Network about Productivity Prediction in New Gas Well. Well Testing 17(3), 12–14 (2008)
5. Peter, H., Dong, C., Nikos, V., et al.: Application of Artificial Neural Networks to Downhole Fluid Analysis. SPE 123423 (2009)

Predicting Formation Fracture Pressure Based on BP Neural Network

Zhidi Liu[1], Xiaoyan Tang[2], and Chunyan Wang[3]

[1] School of Petroleum Resources, Xi'an Shiyou University, China
[2] College of Geology and Environment, Xi'an University of Science & Technology, China
[3] Exploration Department, Yanchang Oilfield Company, China
lzdtxy@sina.com

Abstract. Presently, the theoretical models of calculating formation fracture pressure are generally used, but there are few statistical models that can obviously reflect the formation fracture pressure and the rock mechanics parameters. In this paper, taking into account the relation of formation fracture pressure and rock mechanics parameters, the four parameters which have a close connection to formation fracture pressure, such as Young's modulus, bulk modulus, poison's ratio and depth are selected. Thus based on the BP neural network, the statistical models of formation fracture pressure are established to calculate fracture pressure of carbonate formation. The research shows that BP neural network model is complex, and its model is difficult to set up, but the error of estimate's formation fracture pressure is little, the precision is high.

Keywords: Formation fracture pressure, BP neural network, Rock mechanics parameter.

1 Introduction

Currently, the theoretical model of calculating fracture pressure (*Fp*) are more, such as Eaton formula, Anderson formula [1], Huang formula [2,3] and Feng formula [4~6], but the direct statistical prediction models between measured fracture pressure and rock mechanics parameters are rarely reported. These theoretical models have limitations when calculate formation fracture pressure. Calculating pore pressure, formation tensile strength, structural stress factor and Biot's coefficient are complicated process, and can not be accurately calculated. Furthermore, the model can not adapt to the actual formation situation, it need to correct in order to use. With the rapid development of modern computing technology, the modeling technology of neural networks brings the gospel for detailed pressure interpretation. The study find that the statistical model, which is established using the measured formation fracture pressure data and rock mechanical parameters, can be effectively used to predict the formation fracture pressure. This paper aims to establish the statistical model of predicting formation fracture pressure using BP neural network.

D. Jin and S. Lin (Eds.): Advances in CSIE, Vol. 1, AISC 168, pp. 541–545.
springerlink.com © Springer-Verlag Berlin Heidelberg 2012

2 Relationship between the Rock Mechanical Parameters and Formation Fracture Pressure

By analyzing the existed theoretical models of formation fracture pressure, we find that the formation fracture pressure is a function of overburden pressure (P_0), pore pressure (P_p), Poisson's ratio (μ), tensile strength (σ_t) and the tectonic stress coefficient (ξ). Depth, rock density and the acoustic logging reflects the overburden pressure and pore pressure value. Rock Poisson's ratio is the inherent flexibility of underground rock, and mainly reflects the plasticity of rocks. Formation fracture pressure increases with Poisson's ratio. Young's modulus (E) reflects the rock of the tensile strength and rigidity, and decreases with the formation fracture pressure. Bulk modulus (K) reflects the formation of tensile and compressive deformation characteristics, and increases with the formation fracture pressure. Therefore, the predicting model of formation fracture pressure can be established using the parameters which are μ, E and K, etc... Logging data may well reflect the rock mechanical properties, these rock mechanics parameters can easy to extract from compression wave and shear wave slowness. Its calculating equations are follows.

$$\mu = \frac{\Delta t_s^2 - 2\Delta t_c^2}{2\left(\Delta t_s^2 - \Delta t_c^2\right)} \tag{1}$$

$$E = 2\rho_b \beta (1+\mu) / \Delta t_s^2 \tag{2}$$

$$K = \frac{\rho_b \left(3\Delta t_s^2 - 4\Delta t_c^2\right)}{3\Delta t_s^2 \cdot \Delta t_c^2} \tag{3}$$

Where, μ is dimensionless; E, K units are MPa; Δt_c, Δt_s are the formation compression wave, shear wave time slowness, $\mu s / ft$; ρ_b is the rock bulk density, g / cm^3; β is a unit conversion factor, its value is 9.290304×10^7.

3 BP Neural Network Prediction Model of Formation Fracture Pressure

Prediction of formation fracture pressure involves the two processes of learning model and parameters predict using BP neural network. In this paper, three layers BP network selected by the input layer, hidden layer and output layer (Fig.1). The algorithm parameters include the learning rate, impulse factor, absolute error, the global cost function value and the number of iterations. Sample set of independent variables data pretreat, and the dependent variable standardize. The weights W (1, i, j) and W (2, j, k) can be given random number which range between zero and one as initialization value.

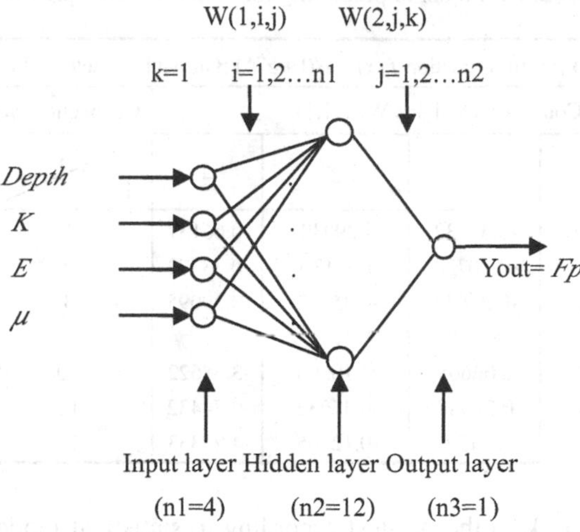

Fig. 1. BP neural network structure in this study

Numerous studies show that any linear and nonlinear functions are infinitely close to three layers network, only increase the number nodes of neurons (n2) in hidden layer. *Depth*, *E* *K* and μ act as s input information (n1=4), the measured formation fracture pressure act as desired output value (n3=1). The 12 hidden layer nodes (n2=12) are chosen through experience, the formation fracture pressure (*Fp*) of the BP neural network topology model is formed, it shows in Fig.1.

Fp acts as the dependent variable(y). *Depth*, *E* *K* and μ act as the independent variable (equivalent to x1, x2, x3, x4) in Table.1. These samples will be loaded into the neural network structure, repeated training and learning model. The BP neural network mathematical statistical models of predicting fracture pressure is ultimately determined, it shows in Table.2.

Table 1. Rock mechanics parameters and fracture pressure data of carbonate rocks in area

Depth(m)	*E* (10^4MPa)	*K* (10^4MPa)	μ	*Fp*(MPa)
4065.0	6.957	3.690	0.186	45.406
4104.0	4.807	2.624	0.195	43.051
4133.0	9.089	5.319	0.215	47.488
4141.0	8.314	3.289	0.079	24.432
...	...	...	...	...
4446.5	8.329	6.329	0.281	74.746
4458.0	8.165	5.587	0.257	66.380
4476.5	8.488	5.952	0.263	68.132
4479.0	9.002	4.098	0.134	33.458
4541.0	9.131	6.494	0.265	68.978

Table 2. BP model of predicting formation fracture pressure

Three layer network; Action function $f(x) = 1/(1+e^{-x})$;Output parameter values or forecast *Fp*						
Connection weights W(1,i,j)					Connection weights W(2,i,j)	
j \ i	1	2	3	4	k \ j	1
1	-0.92199	-0.37732	4.29039	-0.66137	1	8.25108
2	-0.29277	-0.20728	1.82130	-0.33372	2	-3.66998
3	-0.28402	-0.33673	-0.35972	-0.29095	3	0.88297.
.	.	.	.	.	..	.
10	-0.19693	2.06605	-2.10606	-3.68622	10	-3.55724
11	-0.42094	0.26587	-0.45632	-0.74432	11	1.33110
12	0.16494	-0.31723	-0.12405	-4.22353	12	-3.88647

The model data back to the sub-test according to statistical model in Table.2. We find the predicting fracture pressure error is small using BP neural network, and has high accuracy. Maximum absolute error is less than ±1MPa. So, it shows that the formation fracture pressure prediction is feasible applying BP neural network model.

4 Applications Analysis

The carbonate formation fracture pressure is predicted using the above model for DU3 well, its results show in Table.3. This table shows that the actual error between the predicting formation fracture pressure and measured fracture pressure is small, the maximum relative error is 1.6 percent, and the smallest relative error is only 0.17%. Thus, it indicates that the method can more accurately predict the formation fracture pressure, has high precision, and meet the precision requirements of fracture pressure prediction in petroleum engineering.

Table 3. Comparison between measured and predicted formation fracture pressure in DU3 well

Depth (m)	*Fp* (MPa)	*Fpan* (MPa)	*Relative error*(%)	*Depth* (m)	*Fp* (MPa)	*Fpan* (MPa)	*Relative error*(%)
4140.000	78.440	78.307	0.17	4272.125	79.618	78.982	0.80
4146.750	77.522	76.743	1.00	4294.750	71.523	70.438	1.52
4154.250	81.046	80.225	1.01	4306.750	61.140	61.192	0.09
4167.875	85.521	84.935	0.69	4319.250	62.641	62.177	0.74
…	…	…	…	…	…	…	…
4180.375	69.439	68.722	1.03	4334.375	58.889	58.468	0.71
4190.750	73.843	73.353	0.66	4348.125	89.593	88.157	1.60
4200.500	85.853	85.543	0.36	4349.875	90.893	89.446	1.59
4211.375	84.469	83.824	0.76	4353.875	90.780	89.342	1.58
4253.625	89.096	88.337	0.85	4354.250	92.886	91.473	1.52

5 Conclusions

The BP neural network model of predicting formation fracture pressure can be established. Its prediction of formation fracture pressure has small error, and high precision.

The μ、E、K parameters can be fast, continuously, and accurately extracted, the BP neural network model to calculate the formation fracture pressure is more practical. Relative to thc theoretical model is concerned, the predicted formation fracture pressure has also important significance for drilling mud density and fracturing design.

References

1. Stephen, R.: Prediction of Fracture Pressure for Wildcat Wells. JPT 34(4), 863–872 (1982)
2. Draou, A., Osisanya, S.: New Methods for Estimating of Formation Pressures and Fracture Gradients from Well Logs. SPE 63263 (2000)
3. Huang, R.: A Model for Predicting Formation Fracture Pressure. East China Petroleum Institute Journal 4(4), 335–346 (1984)
4. Zhou, N., Yang, Z.: Overview on Pressure Prediction of Formation Fracture. Chongqing Institute of Technology (Natural Science) 13(1), 36–39 (2011)
5. Qi, B., Xie, G., Zhang, S., et al.: Application of Fracture Pressure Log Interpretation Methods in the LG area. Nature Gas Industry 29(10), 38–41 (2009)
6. Li, M., Lian, Z., Chen, S., et al.: Rock Mechanical Parametric Experiments and the Research of Formation Fracture Pressure Prediction. Oil Drilling & Production 31(5), 15–18 (2009)

Multi-layer Background Subtraction Based on Multiple Motion Segmentation in Sport Image

WenHui Ma

Physical Department, North China Institute of Science and Technology,
Langfang 065201, China

Abstract. Segmentation of motion in an image sequence is one of the most challenging problems in image processing, while at the same time one that finds numerous applications. In this paper, we propose a robust multi-layer background subtraction technique which takes advantages of local texture features represented by local binary patterns (LBP) and photometric invariant color measurements in RGB color space. Due to the use of a simple layer-based strategy, the approach can model moving background pixels with quasiperiodic flickering as well as background scenes which may vary over time due to the addition and removal of long-time stationary objects. The use of a cross-bilateral filter allows to implicitly smooth detection results over regions of similar intensity and preserve object boundaries.

Keywords: multi-layers, motion segmentation, level sets.

1 Introduction

Foreground objects detection and segmentation from a video stream captured from a stationary camera is one of the essential tasks in video processing, understanding and visual surveillance. A commonly used approach to extract foreground objects consists of performing background subtraction. Despite the large number of background subtraction methods that have been proposed in the past decade and that are used in real-time video processing, The task remains challenging when the background contains moving objects (e.g. waving tree branches, moving escalators) as well as shadows cast by the moving objects we want to detect, and undergoes various changes due to illumination variations, or the addition or removal of stationary objects.

In this paper, we propose a layer-based method to detect moving foreground objects from a video sequence taken under a complex environment by integrating advantages of both texture and color features. Compared with the previous method proposed by Heikila, several modifications and new extensions are introduced. First, we integrate a newly developed photometric invariant color measurement in the same framework to overcome the limitations of LBP features in regions of poor or no texture and in shadow boundary regions. Second, a flexible weight updating strategy for background modes is proposed to more efficiently handle moving background objects such as wavering tree branches and moving escalators. Third, a simple layer-based background modeling/detection strategy was developed to handle the

D. Jin and S. Lin (Eds.): Advances in CSIE, Vol. 1, AISC 168, pp. 547–550.
springerlink.com © Springer-Verlag Berlin Heidelberg 2012

background scene changes due to addition or removal of stationary objects (e.g. a car enters a scene and stay there for a long time). It is very useful for removing the ghost produced by the changed background scene, detecting abandoned luggage, etc. Finally, the fast cross bilateral filter was used to remove noise and enhance foreground objects as a post-processing step.

2 Background Subtraction Algorithm

In this section, we introduce our approach to perform background modeling subtraction. We describe in turn the background model, the overall algorithm, the distance used to compare image features with modes, and the foreground detection step.

2.1 Background Modeling

Background modeling is the most important part of any background subtraction algorithms. The goal is to construct and maintain a statistical representation of the scene to be modeled. Here, we chose to utilize both texture information and color information when modeling the background. The approach exploits the LBP feature as a measure of texture because of its good properties, along with an illumination invariant photometric distance measure in the RGB space. The algorithm is described for color images, but it can also be used for gray-scale images with minor modifications.

Let $\mathrm{I}=\{\mathrm{I}^t\}_{t=1,.....N}$ be an image sequence of a scene acquired with a static camera, where the superscript t denotes the time. Let $M^t=\{M^t(x)\}_x$ represent the learned statistical background model at time t for all pixels x belonging to the image grid. The background model at pixel x and time t is denoted by

$$M^t(x)=\{K^t(x),m_k^t(x)_{k=1,.....k^t(x)},B^t(x)\} \tag{1}$$

and consists of a list of $K^t(x)$ modes $m_k^t(x)$ learned from the observed data up to the current time instant, of which the first $B^{t(x)}\le K^t(x)$ have been identified as representing background observations. Each pixel has a different list size based on the observed data variation up to the current instant. To keep the complexity bounded, we set a maximal ode list size $K_{\max}$. In the following unless explicitly stated or needed, the time superscript t will be omitted to simplify the presentation. Similarly, when the same operations applies to each pixel position, we will drop the x notation.

2.2 Foreground Detection

Foreground detection is applied after the update of the background model. First, a background distance map $D^t=\{D^t(x)\}_x$ is built, which can be seen as the equivalent of the foreground probabilities in the mixture of Gaussian approach. For a given pixel x, the distance is defined as $D^t(x) = Dist(m_k^{t-1}(x))$ which is the distance to the closest, unless we have $k > B^t(x)$ and $L_k(x) = 0$. It is defined as:

$$D^t(x) = \frac{1}{W(x)} \sum_v G_{\sigma s}(\| v - x \|) G_{\sigma r}(| I^{g,t}(v) - I^{g,t}(x) |) D^t(v) \tag{2}$$

Where $W(x)$ is a normalizing constant, $I^{g,t}$ denotes the gray-level image at time t, σ_s defines the size of the spatial neighborhood to take into account for smoothing, σ_r controls how much an adjacent pixel is down weighted because of its intensity difference and G_σ denotes a Gaussian kernel. As can be seen, the filter smoothes values that belong to the same gray-level region, and thus prevents smoothing across edges.

3 Experimental Results

In this section, we examined the performance of our proposed method on both simulated and real data.

To evaluate the different components of our method, we performed experiments on simulated data, for which the ground truth is known:

Background Frames (BF): For each camera, 25 randomly selected background frames containing no foreground objects were extracted from the recorded video stream.

Background and Shadow Frames (BSF): In addition to the BF frames, we generated 25 background frames containing highlight and (mainly) shadow effects. The frames were composited as illustrated in Figure2, by removing foreground objects from a real image and replacing them with background content. The experiment result is shown in fig.1 and fig.2.

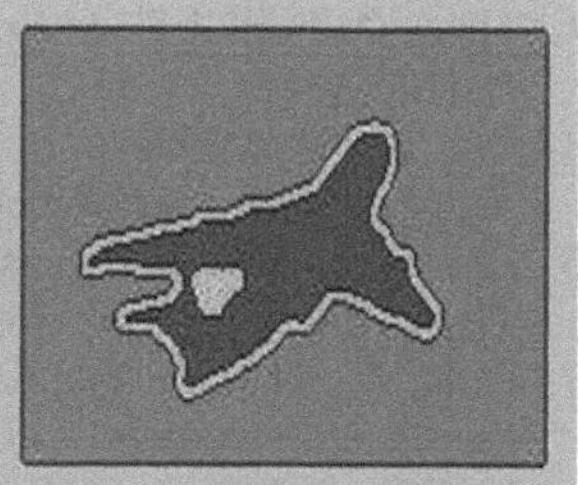

Fig. 1. Image processing results

  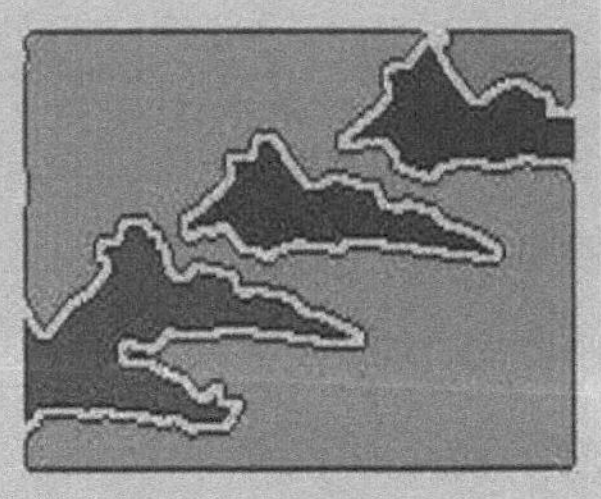

Fig. 2. Image processing results

4 Conclusions

A robust layer-based background subtraction method is proposed in this paper. It takes advantages of the complementarity of LBP and color features to improve the performance. While LBP features work robustly on rich texture regions, color features with an illumination invariant model produce more stable results in uniform regions. Combined with an 'hysteresis' update step and the bilateral filter (which implicitly smooth results over regions of the same intensity), our method can handle moving background pixels (e.g., waving trees and moving escalators) as well as multi-layer background scenes produced by the addition and removal of long-time stationary objects.

On both simulated and real data with the same parameters show that our method can produce satisfactory results in variety of cases.

References

1. Heikkila, M., Pietikainen, M.: A texture-based method for modeling the background and detecting moving objects. IEEE Trans. Pattern Anal. Machine Intell. 28(4), 657–662 (2006)
2. Jacques, J.C.S., Jung, C.R., Musse, S.R.: Background subtraction and shadow detection in grayscale video sequences. In: SIBGRAPI (2005)
3. Hu, J.-S., Su, T.-M.: Robust background subtraction with shadow and highlight removal for indoor surveillance. EURASIP Journal on Advances in Signal Processing, 14 pages (2007)
4. Tuzel, O., Porikli, F., Meer, P.: A bayesian approach to background modeling. In: CVPR, p. 58 (2005)
5. Kim, K., Chalidabhongse, T.H., Harwood, D., Davis, L.: Real-time foreground-background segmentation using codebook model. Real-Time Imaging 11(3), 172–185 (2005)
6. Lee, D.-S.: Effective gaussian mixture learning for video background subtraction. IEEE Trans. Pattern Anal. Machine Intell. 27(5), 827–832 (2005)
7. Paris, S., Durand, F.: A Fast Approximation of the Bilateral Filter Using a Signal Processing Approach. In: Leonardis, A., Bischof, H., Pinz, A. (eds.) ECCV 2006, Part IV. LNCS, vol. 3954, pp. 568–580. Springer, Heidelberg (2006)

Design and Implementation of Intelligent Distribution Management System for Supply Chain DRP

Aimin Wang[1], Qiang Song[2], and Jipeng Wang[1]

[1] Computer and Information Engineering School, Anyang Normal University, Anyang, China
[2] Mechanical Engineering Department of Anyang Institute of Technology, Anyang City of Henan Province, China

Abstract. This system is designed basing on WEB and the criterion of J2EE. The Java programming language, XML, Internet/Intranet and the Unified Modeling Language (UML) are used in the implementation of this system. The object oriented software engineering model is also involved. So the system has no limitation on the usage in different software or hardware environment. And the performance will even be more splendid when having the integration with other task management systems or Enterprise Resource Planning (ERP) systems.

Keywords: Internet technique, multilateral architecture, supply chain management (SCM), distribution system design.

1 Introduction

China's enterprise management operation is connecting with international practice after joining the WTO. The popularity and thorough application of the Internet brings profound influence. The competition among enterprises has evolved into the competition among the corresponding enterprise supply chain. Enterprise management's scope, including not only their own resources (such as improving productivity and reducing operational costs), this concept is extended to the dealers and end-user session. The embodiment is to help middle links sell and reduce the logistics cost of marketing aspects.

At present, the distribution channel's inventory level of the world-class supply chain management (SCM) is lower 50% than the same industry opponents which do not carry on SCM. In China, the SCM software's application environment is not yet mature. The SCM software's market is focused on large enterprises and medium-sized enterprises. Its industry users are mainly in the manufacturing, distribution, and energy industries. The main pulling power of the SCM software market is the logistics management module adopted by corresponding enterprises and the management software adopted by a number of the third-party logistics enterprises. From the customer's application, the existing products are not relatively perfect and their functions not complete.

With the supply chain development from the supply-driven mode (for stocks) to the demand pull mode (for demand), the future supply chain products should have four major characteristics: real-time visibility (across the whole supply chain), flexibility (supply and sources choice), responsiveness (for the changing customer demand and the shorten delivery cycle), and rapid new product going on the market (based on

D. Jin and S. Lin (Eds.): Advances in CSIE, Vol. 1, AISC 168, pp. 551–555.
springerlink.com © Springer-Verlag Berlin Heidelberg 2012

market trends and new design).The next generation SCM system should include the supply chain process management and event management capabilities. This allows real-time information on event to be gradually submitted to the appropriate person in the enterprise, to carry out effective decision-making. The products have higher visibility and more accessible real-time information to enhance the predictability of the implementation of decision-making.

2 Design Principle and Technical Analysis of the DRP Management System

2.1 System Development Environment and Functional Analysis

This DRP management system based on WEB and J2EE used the object oriented software engineering model. The Java programming language, XML, Internet/Intranet and UML are used in the implementation of the system, using the Unified Modeling Language model and design the system. Taking the three-tired structure and Browser/Server model, integrating the internal workflow effectively, the system enhances its flexibility and greatly improves company's working efficiency.

The DRP management system based on the component technology can meet the personalized needs of enterprises. The System can achieve effectively the integration with other systems that are constructed with standard components, such as the enterprise internal management systems, e-commerce system, logistics and distribution system, cross-enterprise supply chain management system, customer relation management system. And it also linked closely ERP system and financial system of major vendors, supporting the business process reengineering.

2.2 The Three-Tired Computing Model

The DRP management system uses a new structure-the three-tier computing model as shown in Figure 1.

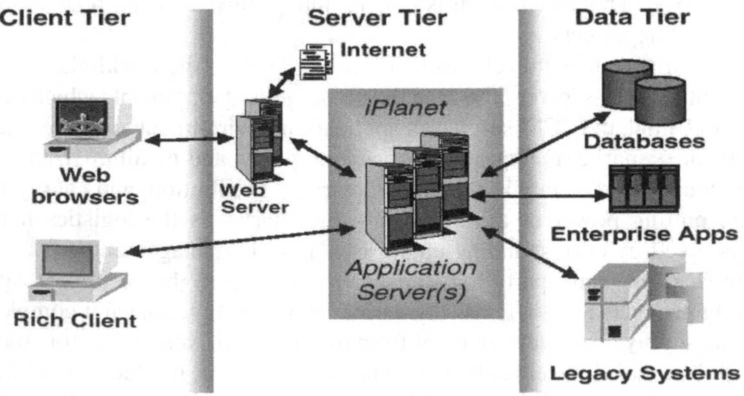

Fig. 1. The three-tier computing model

The structure of model is: ① the client tier, as the user interface and the place to send out a user's request, the typical application is a web browser and fat clients (such as Java programs); ② the server tier, the typical application is a web server and application server running business code; ③ the data tier, a typical application is relational databases and other back-end data resources, such as Oracle and SAP R/3, etc.

In the three-tier structure, customers (request for information), procedures (process the request) and data (operated) are physically separated. The flexible three-tier architecture separates the display logic from the business logic. This means that the business code is independent and users need not care how to display codes. The middle business logic tier does not need to concern the type of customers requesting data, and can maintain the relative independence with back-end systems. The three-tier structure has better portability, may across different types of platforms, and allows users to request load balancing conducted in a number of servers. The three-tier structure has better security, because the application procedures have been separated with customers.

In the enterprise application, the application server is the middleware between enterprise data (as well as other legacy systems) and customers accessing business data. The application server provides the environment storing and running the business code, and separates the business logic with the client side and data resources physically. The application server enables a commercial system to be developed and deployed simply and rapidly, and can adapt to the increase of commercial system users without restructuring the system, because it is in a relatively independent layer.

2.3 The J2EE (Java 2 Platform, Enterprise Edition) Technical Standard

The leading J2EE standard is used in the DRP management system. J2EE provides a good mechanism to build important business systems with scalability, flexible, easy to maintain.

J2EE is one kind of technical architecture that uses the Java 2 platform to simplify complex problems related to the development, deployment and management of enterprise solution. The basis of J2EE technology is the core Java platform or J2SE (Java 2 Platform, Standard Edition). J2EE not only consolidate many advantages in the standard edition, such as the feature of "Write once, Run anywhere", JDBC API (Application Programming Interface) and CORBA technologies facilitate to access database, the security mode to protect data in Internet application and so on, but also provides the full support to EJB (Enterprise JavaBeans), Java Servlets API, JSP (Java Server Pages), as well as XML technology.

The J2EE architecture provides a middle layer integration framework to meet the application needs of high usability, high reliability and extensibility. By providing a unified development platform, J2EE reduces the cost and complexity of multi-tier application development, while it provides strong support to the existing applications integration and Enterprise JavaBeans. So it speeds up the enterprise application development.

J2EE also has the following characteristics. ① Cross-platform deployment: J2EE provides a simple unified platform, programmers do not need to consider the kind of application servers, and the application systems based on J2EE can be deployed to any service platform. ② Integration with enterprise existing heterogeneous systems easily: J2EE applies the unified naming technology, and it can access the database and interoperate with CORBA. ③ High Scalability: through a series of complete enterprise middleware service, scalable application systems based on J2EE can support millions of transaction processing, as well as tens of thousands of different users' concurrent request. ④ High security: J2EE is designed as a high-security model, and provides a series of access control.

J2EE is a development framework based on standardized, modular, reusable components. Its API processes a number of applications automatically, so that programmers concentrate on the business logic and speed up the project development.

3 Technology Innovation of the DRP Management System

3.1 Technical Tools

The project technical innovation is that we have adopted an advanced and creative software development component platform, and achieve the ability of "Develop once, Deploy more". In other words, this allows developing only once in technology, but supports wide software application ways through different deployment modes of the component platform. Moreover, it increases the software development efficiency substantially.

In this project, we have used the self-developed component development platform Junco on the Java application server to achieve the following purposes. The platform mainly solves the secondary problems of software activities (express these abstract entities with programming languages, map them into the machine language in defined space and time), so that software developers can concentrate their efforts on solving fundamental tasks (create the complex concept structure that can be the basis of these abstract software entities).

3.2 Support Multiple Deployment Modes

In the project, the self-developed component development platform Junco has been used to realize the approach of "Develop once, Deploy more".

① The single host model, shown in Figure 2, is suitable for the small-scale, single-point application.

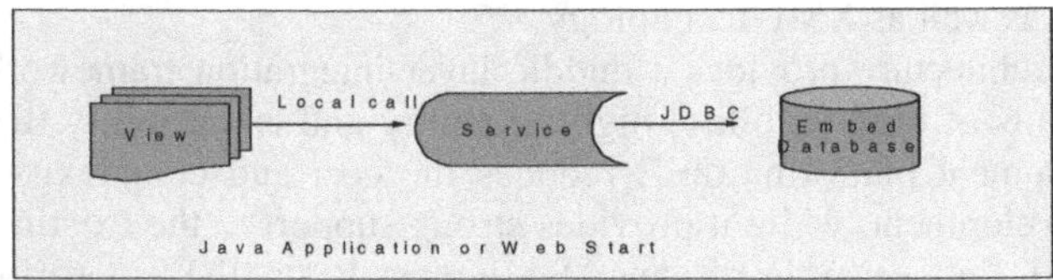

Fig. 2. The single host model

② The C/ S mode in LAN (Local Area Network), is shown in Figure 3. Deploy the interface interpreting engine separately in this mode.

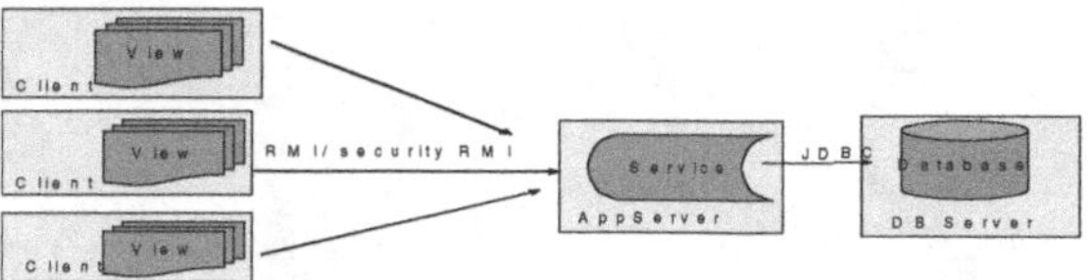

Fig. 3. The C/ S mode in LAN

③ The inter-network operation mode. Large, distributed retailing and manufacture enterprises with internal private lines can deploy the three-tier structure application with fat clients, as Figure 4 displays.

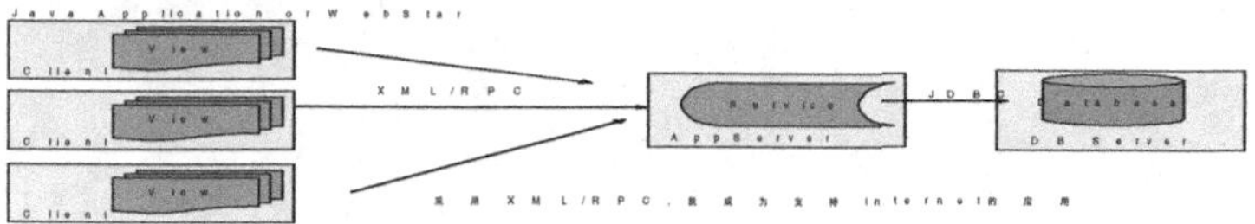

Fig. 4. The inter-network operation mode

4 Conclusion

The enterprise using the system has achieved better economic and social benefits. The effects that can be directly quantified are as follows:①reduce the average inventory loss 10-20%;②reduce the average inventory amount of distribution centers 10-30%; ③rise the stock turnover ratio from the originally approximate 20 times/year to 25-30 times/year;④shorten the time of accounts receivable from the originally approximate 30 days to the average of 10-15 days;⑤shorten the processing period of single order from the original 7-8 days to 1-3 days.

References

1. Bit Culture. JAVA Applet Design Practice. China Railway Press (November 2007)
2. Hong, W.-E.: JVJA 2 Object-Oriented Programming. China Railway Press (August 2008)
3. Wang, A.-M.: The research of large system decision theory and its Intelligent Decision Support System. Computer Engineering and Applications 39(33), 75–78 (2008)

Fig. 3 [illegible]

Key Technologies Research of Network Remote Control

Aimin Wang[1], Qiang Song[2], and Jipeng Wang[1]

[1] School of Computer and Information Engineering, Anyang Normal University, Anyang, China
[2] Mechanical Engineering Department of Anyang Institute of Technology, Anyang City of Henan Province, China

Abstract. This Paper introduces the design and implementation of an intelligent, real-time controlling management system which has already been deployed in real use and proves high effect in practice. Based on Socket network communication and API programming, the key control functions of network communication, remote screenshots, mouse and keyboard locking, screen-darkening, open and shut down computers are all realized in this remote-control system. Trojan horse technology is involved and the system is developed by Delphi 6.0.

Keywords: Remote control, Intelligence, Trojan horse, LAN, key technologies.

1 Introduction

With the growing popularity and popularization of the network technology, how to supervise LAN scientifically and effectively has become more and more important. Basically, the current management method is that managers specially inspect users back and forth; users executing illegal operations are given some tips immediately, so that the effective and scientific management are greatly reduced. This paper studies the network monitoring technology, applies the remote control technology in LAN management, and has achieved unattended function securely and reliably. Schedule the ventilation time in computer rooms visually and conveniently. Control the users' computers effective (key control of mouse and keyboard locking, blank screen, open and shutdown computers). Monitor users' computer screen in the real-time (Remote Screenshots). The system has advantages of full, flexible statistical analysis of data and rapid, accurate accounting management. Here the mechanism of remote control is that computers controlled running a program to intercept port and receive data packets, and the controlling side sends data packets through its port to the port of controlled side. According to this principle, prepare two procedures, one is the controlling side, and the other side is the controlled side. Controlled computers wait for instructions send by the controlling side, and then implement the corresponding operation. This paper studies and has achieved the key technologies including the file system management (Including files and directories upload, download and delete operations, etc.),registration form management, screen monitor, all present tasks management and control(Including the keyboard, mouse control, restart the computer, shutdown, etc.) to controlled computers.

D. Jin and S. Lin (Eds.): Advances in CSIE, Vol. 1, AISC 168, pp. 557–561.
springerlink.com © Springer-Verlag Berlin Heidelberg 2012

2 Implementation of Network Communication

Delphi has two types of controls that can achieve the purpose of communicating. One type is client socket used by the control side and server socket (all in Internet pages) used by the controlled side, and the other type is NMUDP (in the Fast Net pages) used by both sides [1]. The difference between the two types of controls is that the former uses TCP transmission control protocol, the latter uses the UDP (User Datagram Protocol) protocol. TCP is a connection-oriented protocol, two sides establish a connection (or disconnection) through three interactions, it is more time-consuming, but achieves reliable data transmission. UDP isn't a connection-oriented protocol, no confirmation to the data is needed, so its transmission speed is faster than TCP's, but data could be lost, and it is therefore not reliable. The amount of transmission data required by remote control is little and higher reliability is needed, so the former type of controls is generally used.

Delphi uses the client Socket component and Tserver Socket component to achieve connectivity and communication between the client Socket and servers Socket respectively. The two components are used to manage the connection of client and server side. Therefore, a TServersocket component should be placed in the server procedures and a Client TClientsocket component placed in the client procedures.

The control to different controlled computers is achieved through setting dynamically the Active and Address attributes of Clientsocket in procedures. Port attribute specifies the server port, the number should be larger than 1024. Here the number is 10738. The data send by the server can be read in the On Read incident of Clientsocket. The port number of Serversocket is set for 10738 (it should be according with the port number of Clientsocket) and the Active value is set to "True", so the client's instructions can be accepted when starting the server and read in the On Client Read incident (to be explained in "the definition of communication protocols" in details).The system communication settings are completed.

Each instruction that Client sends should meet the following format: send instructions through Clientsocket.socket.sendtext () method. For example: send Short Message "Good morning", Clientsocket1.socket.sendtext ('103 Good morning').

When Server received the instruction that Client sends out, Server analyzes the instruction, fetches function codes and implements the corresponding module. This function is completed by the OnClientRead incident of Serversocket.

```
ReceiveMessage:=socket.ReceiveText;        // Fetch instruction
MessageHead:=copy(ReceiveMessage,1,3); // Extract function codes
case.strtoint(MessageHead) of
101: ...
102: ...
103: ...
... ...                                     // Implement the corresponding
module
end;
```

3 Trojan Technology Applications

This part discussed the Trojan applications in Server procedure and the technical problems such as automatic operation during starting computers, prevention to run multiple copies, installation and hide, covert operation and so on [2].

3.1 Automatic Operation of Server Procedure during Starting Computers

Achieve automatic operation of Server procedure during starting computers through the way of modifying the registry. Join the Registry unit; add the following code in the OnCreate incident of the window.

```
Const
  K='\Software\Microsoft\Windows\CurrentVersion\Run';
… …
with TRegistry.Create do
  try
RootKey:=HKEY_LOCAL_MACHINE;
OpenKey(K,TRUE );
WriteString('ODBC32', Path);   // 'ODBC32'is key name,  Path is procedure path
  Finally
    free;
end;
```

3.2 Prevent Running Multiple Copies of Server Procedure

To avoid running more than one copy of Server procedure (saving system resources), the procedure is usually set to run one each time. General, there are two ways: one way is that Server procedure searches the same operation when running, if any, to immediately withdraw the process; the other is that Server procedure searches the same operation when running, if any, to close existing procedure and then start. The second way facilitates the upgrade. The old version will be automatically covered by the new version. The system in this paper used the first way.

The modified Server project documents are as follows:

```
var
  ReLoadHandle:Thandle;
begin
ReLoadhandle:=windows.CreateMutex(nil,true,'ODBC32');
    if ReLoadhandle<>0 then
      begin
        if       getlasterror=windows.ERROR_ALREADY_EXISTS
then
          begin
            closehandle(ReLoadhandle);
            halt;
          end;
      end;
```

3.3 Installation and Hide of Server Procedure

Such procedures usually run themselves hidden through the following methods: self-copy, resources file, web-installation, binding like virus. The self-copy method is applicable to an entire document itself; install many documents at the same time in the resources file method, pay to MS firstly for security certificates in the web-installed method, the method of binding like virus uses the principle of virus. Here the self-copy method is used.

The principle of this approach is that the procedure checks if it is in a specific directory when running or not, continues to run itself if true, copy itself to a specific directory if not, then run the new procedure and withdraw from the old procedure.

Here Server procedure is installed in the Windows System folder (the path in Win9X is \Windows\system, the path in WinNT is \Windows\system32). To hide it better, it is generally named a file name similar to system file names. For example, some Trojan horse is named kernel32.exe, shield by the operating system kernel 'kernel32.dll'.

In this system, the file name after Server is installed is entitled ODBC32.exe, which is shielded by ODBC32.dll.

First, the Windows API function GetSystemDirectory ( ) acquires the current Windows system folder, and compares with the current folder. If it is not in the Windows system folder, it will copy itself to the system folder.

3.4 Covert Operation of Server Procedure

In order to achieve the purpose of covert surveillance, a monitoring software should meet some conditions. No interface is in window and nothing is show in taskbar during normal operation. Its process doesn't appear in the task list when pressing Ctrl + Alt + Del. It is essential to change the current process to a system process to achieve the purpose.

(1) Hide interface

Add the following codes in the project files of Server procedures:

Application.ShowMainForm :=false;

(2) Hide in the taskbar

Call the Windows API function SetWindowLong () to achieve hiding in taskbar. Add the following function in the OnCreate incident of the Server procedure main form:

SetWindowLong (Application.Handle, GWL_EXSTYLE, WS_EX_TOOLWINDOW);

(3) Hide in the task list

Call the Windows API function RegisterServiceProcess () to achieve hiding in the task list. Add declaration after the implementation:

Function RegisterServiceProcess (dwProcessID, dwType: Integer): Integer; stdcall; external 'KERNEL32.DLL';

Then, add the following content in the OnCreate incident of the Server window:

RegisterServiceProcess (GetCurrentProcessID, 1);

4 Conclusion

The main innovation of this paper is to solve the LAN remote control problem with the software-only mode. The practice shows that this technology is safe and reliable. The most important feature is that the system can work closely with the teaching and learning activities. In addition to meet the management requirement of the computer room, it can also assess the attendance of courses on computer system. Through the statistical analysis on the students' computer-using and experimental teachers' work, the system not only reflects students' study situations which help strengthen the management and control of experimental course, also monitors at any time teachers' work which provides effective data to evaluate experimental teachers' work. The system has good interfaces, high-level intelligence and strong fault-tolerant capability. The system that applied in 12 colleges or universities in China helps obtain a very good teaching effect during the past two years. The research achievement was appraised in Science and Technology Department of Henan Province In June 30, 2006. All experts agreed that the overall technologies of the achievement come up to internal advanced level of similar systems.

References

1. Cui, H., Wang, A.: Advanced Programming with J2EE Web Services. Tsinghua University Press (April 2005)
2. Li, G., Li, X., He, X.: Application in aero-engine fault diagnosis based on reformative ART1 neural network. Control & Automation 9(1), 156–158 (2005)
3. Jin, C., Wu, H.: A neural networks training algorithm based on adaptive genetic algorithm. Control & Automation 10(1), 49–51 (2008)

Extraction of Ideas from Microsystems Technology

Dirk Thorleuchter[1] and Dirk Van den Poel[2]

[1] Fraunhofer INT, Appelsgarten 2, D-53879 Euskirchen, Germany
[2] Ghent University, Faculty of Economics and Business Administration, B-9000 Gent, Tweekerkenstraat 2, Belgium
Dirk.Thorleuchter@int.fraunhofer.de,
dirk.vandenpoel@ugent.be
http://www.crm.UGent.be

Abstract. In literature, idea mining is introduced as an approach that extracts interesting ideas from textual information. Idea mining research shows that the quality of the results strongly depends on the domain. This is because ideas from different domains consist of different properties. Related research has identified the idea properties for the medical domain and the social behavior domain. Based on these results, idea mining has been applied successfully in these two domains. In contrast to previous research, this work identifies the idea properties from a general technological domain to show that this domain differs from the two above mentioned domains and to show that idea mining also can applied successfully in a technological domain. Further, idea properties are identified by use of backward selection as main approach in stepwise regression, which is in contrast to previous research. Predictive variables are selected considering their statistical significance and a grid search is used to adapt the parameters of the idea mining algorithm. Microsystems technology is selected for a case study. It covers a wide range of different technologies because it is widely used in many technological areas. The case study shows that idea mining is successful in extracting new ideas from that domain.

Keywords: Idea Mining, Microsystems, Technology, Textmining, Knowledge Discovery.

1 Introduction

Today, a large amount of textual information is accessible in the internet covered different topics. This information could be a valuable source for decision makers because it consists of many interesting ideas, which possibly are relevant for a current decision problem. In scientific literature, a well known question is how to extract these interesting ideas from the large amount of textual information. The most commonly used way (an intuitive extraction of these ideas by a human expert) is very time consuming. This is because the expert has to scan all the relevant documents for the occurrence of interesting ideas.

Literature suggests idea mining as granular view [1] on data to extract ideas using different sources (internet [2], scientific publications [3], patents [4], research projects [5], databases [6], and internet blogs [7, 8]). Idea mining divides an idea in a means

D. Jin and S. Lin (Eds.): Advances in CSIE, Vol. 1, AISC 168, pp. 563–568.
springerlink.com © Springer-Verlag Berlin Heidelberg 2012

and an end, creates representative textual pattern for means and end, and identifies new ideas by searching for unknown means appearing together with known ends or vice versa. Idea mining is successfully applied in case studies where ideas from the medical [9] and from the social behavior domain [10] are extracted. In contrast to previous research, this work applies idea mining on the domain of Microsystems technology that covers a wide range of different technologies and it is widely used in many technological areas. Thus, Microsystems technology can be seen as representative for a general technological domain.

Although ideas from Microsystems technology also consists of means and ends, literature has shown [11] that the way an idea is formulated differs from domain to domain. Considering these results, the standard idea mining approach that is proven in the medical and in the social behavior domain has to be adapted to the Microsystems technology domain. Different parameters or at least different parameter values are needed for a successful extraction of new ideas.

Existing parameters from literature are taken over one by one considering their statistical significance in the field of Microsystems technology. Their values are adapted by use of a 'backward selection procedure' [12], a grid search [13], and a 5-fold cross-validation [14].

A case study shows that extracting ideas from Microsystems technology is successful by use of the newly selected and adapted parameter values. The use of the selected parameters together with the adapted parameter values lead to a successful extraction of ideas from Microsystems technology comparing to the baseline as shown by the results of a case study. These parameter values outperform those values that are commonly used to extract ideas from the medical and from the social behavior domain.

2 Background – The Idea Mining Classifier

The algorithms standing behind idea mining is taken over from technique philosophy where an idea can be divided into two components: a means and an end [15]. If a combination of means and end is new then the means and end represent a new idea. To distinguish new ideas from known ideas, a given context is needed that describe existing ideas represented as already known combinations of means and ends. For each idea in a given text, idea mining [9] extracts textual information that represents a means and an appertaining end. Means and ends themselves consist of domain-specific terms. Idea mining also extracts terms representing means and ends from the context. These terms are compared to the terms extracted from a given text by use of similarity measures. As a result, a new idea can be found in a given text where terms representing the means and the appertaining end do not occur together in the context.

The parameters that are used for calculating the sub measures are α as a set of terms from a given text and β as a set of terms from the context. Both set of terms represent textual patterns and stop words are excluded. They are compared only if one term is in both sets. The first sub measure m_1 is a function based on the number of terms in α ($p = |\alpha|$) and the number of terms in both sets ($q = |\alpha \cap \beta|$). m_1 has its maximum where the number of terms describing a known (or unknown) means is equal to the number of terms describing an unknown (or known) end.

The second sub measure m_2 is of a high value if terms describing the known means or the known end occur frequently in the given text and in the context. The parameter that is used for calculating the second sub measures is z as percentage to create δ as a set of most z % frequent terms in context. Then, $r = |\alpha \cap \beta \cap \delta|$ as the number of frequent terms existing in both sets is divided by the number of terms in both sets to calculate m_2.

The third sub measure m_3 is of a high value if terms describing the unknown means or the unknown end occur frequently in the given text. Based on φ as a set of z % frequent terms in new text, $s = |\alpha \cap \bar{\beta} \cap \varphi|$ is the number of frequent terms in α and it is divided by the number of terms in α that are not in β to calculate m_3.

Sub measure m_4 identifies the occurrence of characteristic terms (e.g. advanced, improved etc.) in α that may lead to a new and useful idea. λ is the set of these terms and $t = |\alpha \cap \lambda|$ is the number of these terms in α. If $t>0$ then m_4 equals 1. The idea mining measure m is the sum of all sub measures multiplied by weighting factors $g_1, \ldots, g_4 \geq 0$ with $g_1+g_2+g_3+g_4 = 1$ for each sub measure. A further parameter is $\hat{a}$ as a threshold for the idea mining measure m to ensure a binary classification decision: if $m \geq \hat{a}$ then α represent a new and useful idea otherwise it does not. Further parameters are the term weight u for stop words and the term weight v for non-stop words as well as l as a parameter to determine the length of a textual pattern (number of terms). Terms are added to α or β one by one until the sum of all added term weights is greater than or equal to l.

3 Methodology

The methodology of this approach is based on a text that consists of new ideas and on a text that consists of existing ideas from the used context. The idea mining classifier is used to extract the new ideas from the first text however; all nine parameters (l, u, v, g_1, g_2, g_3, g_4, $\hat{a}$, and z) have to be optimized first to increase the performance of the classifier using unseen data. Reducing the number of parameters is done by use of a parameter selection procedure. Using too many parameters is computationally expensive and it results in a less comprehensible resulting model. Thus, backward-selection is used where parameters with low predictive performance are discarded. Based on their χ^2-statistic (lowest first), they are excluded one by one from the resulting model while their performance are below a specific threshold [16]. Calculating an optimized parameter set is done by use of grid search [13] on the selected parameters that uses the training data and is based on an n-fold cross-validation approach [14]. Discrete sequences of the parameters are defined to reduce the number of parameter combinations. The grid search results in the highest performance using unseen data. It is evaluated by the F-measure where precision and recall are equally important as suggested by [10].

4 Case Study

A case study focuses on Microsystems technology as a representative for a technological domain. Microsystems technology is a very interesting research area, because it is involved in nearly all technological areas e.g. mobility, medical, information and communication. The fast development of new technologies leads to an increasingly importance of Microsystems technologies in future [17].

To identify new and useful ideas from that domain, a new text that contains ideas and a context is necessary. A review paper is used as context that contains a detailed description of existing research founding in this technology [18]. It is written by the research founding department of the German Ministry of Education and Research. Many existing ideas can be found in this document. As new text, a forecasting paper is selected also written by the same organization. It consists of thoughts about future research directions in Microsystems technology [19].

The parameter selection procedure identifies u, v, and g_4 as parameters with low predictive performance. Thus, we use standard values for them (u = v = 100 % and g_4 = 0). The parameter l is identified with high predictive performance that is in contrast to related research where l is of low predictive performance for identifying ideas from the medicine or social behavior domain [10]. Further, domain independent recommendations from [9] are considered to select the sequences of the parameters g_1, g_2, g_3, â, l and z in the proposed grid search approach ($g_1 \in$ [20%, ..., 80%], $g_2 = g_3 \in$ [10%, ..., 40%], â ∈ [0.20, ... 0.50], l = [5, …, 15], z = [0.05, ..., 0.30]). As a result, different value combinations are evaluated to identify the parameter value results based on the highest performance (g_1 = 0.40, $g_2 = g_3$ = 0.20, â = 0.50, l = 7, and z = 0.06). Using these results instead of the optimized parameter values as calculated for the social behavior domain (g_1 = 0.20, $g_2 = g_3$ = 0.40, â = 0.20, l = 10, and z = 0.10) and for the medical domain (g_1 = 0.40, $g_2 = g_3$ = 0.20, â = 0.50, l = 10, and z = 0.30) leads to an increase in performance e.g. concerning the medical domain precision increases from 40% to 45%, recall increases from 25% to 30%, and F-measure increases from 32,5% to 37,5%. Further, these results outperform the precision (40%) and recall (25%) values as calculated for the extraction of ideas from the medical domain. To compare the results to a baseline, we use a baseline as introduced in the evaluation of [8] that consists of heuristic similarity measures replacing the idea mining measure. The baseline results (30% for precision at 20% recall) is lower than the 45% (precision) at 30% (recall) as calculated by use of the optimized parameter values.

5 Results from the Case Study (Examples)

As true positive result, it was correctly identified that Microsystems technology in the future will done decisions with foresight and communicate with their surroundings. Further, it is correctly identified that Microsystems technology will have an auto diagnosis function and will operate largely autonomously - characteristics which come very close to cognitive abilities. A third example for a true positive result is that Microsystems technology enables sensors that measure the movement and recognize that the laptop is being tilted. The hard drive is switched off instantly and the data are secured. A relevant idea that is not identified correctly (false negative) is the intelligent integration of the individual components to form more and more complex systems. This is relevant because it enables completely novel functions that lead to new intelligent products.

Furthermore, this approach identified an idea that innovative Microsystems work as tiny invisible helpers in countless areas of our everyday lives. This idea is similar to a true positive idea identified some sentences above: Virtually unnoticed, Microsystems are taking over more and more sensitive tasks in our everyday lives. This idea is not a new one, thus, it is a false positive idea. A second false positive idea is that the German Ministry of Research plans to implement a funding of EUR 10 million for Microsystems technology in 2010. This might be a correct statement in politics however it is not relevant as technological idea.

6 Conclusion

A new methodology based on the use of backward parameter selection as main approach in stepwise regression as well as the use of grid search and n-fold cross-validation is presented. It optimizes the parameter values for the use of idea mining classifier in a specific domain. A case study shows that the performance of the idea mining classifier in the field of Microsystems technology is improved. The F-measure of the newly created parameter values increases compared to the optimized parameter values from the social behavior domain, the medical domain, and the baseline.

In general, ideas can be found in all domains; however ideas in the Microsystems technology domain are formulated in a different way than ideas in the social behavior and the medical domain. This is shown in this paper because the calculated parameter values for Microsystems technology are different to that of the social behavior and the medical domain. In detail, the parameter l is identified with high predictive performance that is in contrast to the social behavior and the medical domain. Further, the parameter value of z in the Microsystems technology domain is much smaller than the value of z in the social behavior and the medical domain.

This result leads to three interesting aspects. First, ideas in the Microsystems technology domain are described by a smaller number of terms ($l = 7$) than ideas in the social behavior and the medical domain ($l = 10$). Thus, a concisely wording is used in texts from Microsystems technology. Second, the total number of domain-specific terms in Microsystems technology is smaller than the number of domain-specific terms in the other domains. This means that in Microsystems technology, each domain-specific term occur more frequently than they occur in other domains. Thus, in texts from Microsystems technology it is sufficient to consider the 6% most frequently terms as domain-specific terms in contrast to the social behavior (10%) and the medical (30%) domain. This also shows the difference between the Microsystems technology and the medical domain where a large number of different domain-specific terms normally is used. Third, the parameter values of g_1, g_2, g_3, and $\hat{a}$ are similar to the values in the medical domain; however they are in contrast to that of the social behavior domain. This confirms research results as published in [10].

References

1. Thorleuchter, D., Van den Poel, D.: High Granular Multi-Level-Security Model for Improved Usability. In: System Science, Engineering Design and Manufacturing Informatization (ICSEM 2011), pp. 191–194. IEEE Press, New York (2011)
2. Wang, C., Lu, J., Zhang, G.: Mining key information of web pages: A method and its application. Expert Syst. Appl. 33(2), 425–433 (2007)

3. Thorleuchter, D., Van den Poel, D.: Semantic Technology Classification. In: Uncertainty Reasoning and Knowledge Engineering (URKE 2011), pp. 36–39. IEEE Press, New York (2011)
4. Thorleuchter, D., Van den Poel, D., Prinzie, A.: A compared R&D-based and patent-based cross impact analysis for identifying relationships between technologies. Technol. Forecast. Soc. Change 77(7), 1037–1050 (2010)
5. Thorleuchter, D., Van den Poel, D., Prinzie, A.: Mining Innovative Ideas to Support new Product Research and Development. In: Locarek-Junge, H., Weihs, C. (eds.) Classification as a Tool for Research, pp. 587–594. Springer, Berlin (2010)
6. Park, Y., Lee, S.: How to design and utilize online customer center to support new product concept generation. Expert Syst. Appl. 38(8), 10638–10647 (2011)
7. Thorleuchter, D., Van den Poel, D.: Companies Website Optimising concerning Consumer's searching for new Products. In: Uncertainty Reasoning and Knowledge Engineering (URKE 2011), pp. 40–43. IEEE Press, New York (2011)
8. Thorleuchter, D., Van den Poel, D., Prinzie, A.: Extracting Consumers Needs for New Products. In: Knowledge Discovery and Data Mining (WKDD 2010), pp. 440–443. IEEE Computer Society, Los Alamitos (2010)
9. Thorleuchter, D., Van den Poel, D., Prinzie, A.: Mining Ideas from Textual Information. Expert Syst. Appl. 37(10), 7182–7188 (2010)
10. Thorleuchter, D., Herberz, S., Van den Poel, D.: Mining Social Behavior Ideas of Przewalski Horses. In: Wu, Y. (ed.) Advances in Computer, Communication, Control and Automation. LNEE, vol. 121, pp. 649–656. Springer, Heidelberg (2011)
11. Stumme, G., Hotho, A., Berendt, B.: Semantic Web Mining: State of the art and future directions. J. Web Semant. 4(2), 124–143 (2006)
12. Van den Poel, D., Buckinx, W.: Predicting Online-Purchasing Behavior. Eur. J. Oper. Res. 166(2), 557–575 (2005)
13. Jiménez, Á.B., Lázaro, J.L., Dorronsoro, J.R.: Finding optimal model parameters by deterministic and annealed focused grid search. Neurocomputing 72(13-15), 2824–2832 (2009)
14. Thorleuchter, D., Van den Poel, D., Prinzie, A.: Analyzing existing customers' websites to improve the customer acquisition process as well as the profitability prediction in B-to-B marketing. Expert Syst. Appl. 39(3), 2597–2605 (2012)
15. Thorleuchter, D.: Finding New Technological Ideas and Inventions with Text Mining and Technique Philosophy. In: Preisach, C., Burkhardt, H., Schmidt-Thieme, L., Decker, R. (eds.) Data Analysis, Machine Learning and Applications, pp. 413–420. Springer, Berlin (2008)
16. Coussement, C., Van den Poel, D.: Churn prediction in subscription services: An application of support vector machines while comparing two parameter-selection techniques. Expert Syst. Appl. 34(1), 313–327 (2008)
17. Fluitman, J.: Microsystems technology: objectives. Sensors and Actuators A: Physical 56(1-2), 151–166 (1996)
18. VDI/VDE Innovation + Technik GmbH: Mikrosystemtechnik, Innovation, Technik und Trends. Technologie & Management 1(2), 22–23 (2007)
19. VDI/VDE Innovation + Technik GmbH: Fortschritt mit System (June 09, 2010), `http://www.mstonline.de/mikrosystemtechnik/mikrosystemtechnik`

K-means Optimization Clustering Algorithm Based on Particle Swarm Optimization and Multiclass Merging

Youcheng Lin, Nan Tong, Majie Shi, Kedi Fan, Di Yuan, Lincong Qu, and Qiang Fu

College of Science and Technology
Ningbo University
Ningbo, China
fuqiang@nbu.edu.cn

Abstract. Be aimed at the existing defects of traditional K-means, which is the clustering results is related to the input data sequence, heavily dependent on the initial mass center, and easy to trap into the local minimum, this paper presents a new K-means optimization clustering algorithm based on particle swarm optimization and multiclass merging, referred to as PM-Kmeans algorithm. The algorithm first selects the initial cluster center by improving particle swarms clustering algorithm under default number of clustering, then optimizes the clustering, and last carries out cluster merging based on multiclass merging condition, in order to obtain the best clustering results. The experimental results show that, the algorithm can effectively solve the defects of K-means algorithm, and has a faster convergence rate and better global search ability, as well as better cluster category effect.

Keywords: Particle Swarm Optimization (PSO), Multiclass Merging, K-means algorithm, Fitness variance.

1 Introduction

Since K-means algorithm has advantages of simple structure and fast convergence rate, it has been widely applied in many fields such as data mining, image segmentation and pattern recognition. K-means algorithm is a typical distance-based clustering algorithm, using Euclidean distance as the evaluation of similarity, that is, the closer the two objects, the greater the similarity. This algorithm regards cluster as data points in proximity.

However, the traditional K-means algorithm has several drawbacks: sensitive to the initial value, and easy to fall into local optimum. In order to solve this problem, some scholars have improved it with combination of PSO algorithm. For example, literature [1] and [2] proposed a K-NM-PSO algorithm, a hybrid algorithm based on K-means algorithm, Nelder-Mead and PSO, which has favorable global search capability. Literature [3] proposed a hybrid clustering algorithm based on particle swarm, which can effectively avoid falling into local optimal solution. But the algorithms all literature [1], [2] and [3] proposed are not ideal for clustering effect of complex data. This paper presents a multiclass merged hybrid clustering algorithm based on particle swarm and

D. Jin and S. Lin (Eds.): Advances in CSIE, Vol. 1, AISC 168, pp. 569–578.
springerlink.com © Springer-Verlag Berlin Heidelberg 2012

K-means. This algorithm combines PSO and K-means algorithm, then improves and optimizes PSO algorithm to make full use of their own advantages. The experimental tests prove that the algorithm can achieve better clustering results.

2 K-means Algorithm

K-means clustering algorithm can minimize the sum of squares of distance from each data point to the cluster center. The principle is, first to select K initial cluster centers, calculate the distance from very data point to each of the K centers, find the minimum distance and classify it into the nearest cluster center, modify the center value to the mean value of all data in this class, then calculate again the distance from very data point to each of the K centers, reclassify and modify the new centers. End until the new distance center equals to the previous center.

K-means Algorithm realization steps are as follow:

Step1: Arbitrarily select K data points from the data set as initial cluster centers;

Step2: calculate the distance from very data point to each of the K centers; classify them into the nearest cluster;

Step3: Again calculate the mean value of each cluster as the cluster center;

Step4: If the division results no longer change or reach the maximum number of iterations, end it; if not, return to Step2.

K-means algorithm is relatively simple to implement, it has low computational complexity and scalability. However, K-means algorithm also has the following disadvantages: it is only applicable to the situation that all the mean value of each cluster object is meaningful; the clusters number K must be given in advance, its heavily dependence on the initial mass center and sensitivity to the order of input data; it is also very sensitive to noise and abnormal data because these data may affect the mean of the cluster object. In addition, there is a big limitation of the division rules based on Euclidean distance.

3 Particle Swarm Optimization (PSO)

PSO is an effective global optimization algorithm, first proposed by Kennedy and Eberhar from United States in 1995. It is an optimization algorithm based on the theory of swarm intelligence, which is through swarm intelligence generated by cooperation and competition between particles in the group to guide the search optimization. In PSO algorithm, each solution of an optimized problem can be regarded as a bird in the search space, namely "particle". First it generate the initialized population, that is, to randomly initialize a group of particles in feasible solution space, each particle is a feasible solution of optimized problems, and the objective function will determine a fitness value to each of them. All the particles are moving in the solution space, determined by the moving velocity their flight direction and distance. Usually particles search in the solution space following the current optimal particle. In each iteration, every particle can remember the optimal solution it searched, marked as $gBest$, and the best position the whole particle swarm experienced, that is, the current optimal solution the entire group searched, marked as $pBest$. Those particles constantly adjust their position in accordance with $pBest$ and $gBest$ to search for new solutions. It specific update their speed and position according to the following two formulas.

$$v_i(n+1) = \omega v_i(n) + c_1 \cdot rand_1() \cdot (pBest - x_i(n)) + c_2 \cdot rand_2() \cdot (gBest - x_i(n)) . \tag{1}$$

$$x_i(n+1) = x_i(n) + v_i(n+1) . \tag{2}$$

In which, $v_i(n)$ is current speed of the particlc, $x_i(n)$ is the current position of the particle. i = 1, 2.... N, N is the dimension of the current space; $\text{rand}_1()$ and $\text{rand}_2()$ is random number between [0, 1]; c_1 and c_2 is the learning factor, usually taken $c_1 = c_2 = 2$; ω is the weighting coefficient, usually between 0.1 and 0.9.

PSO realization steps are as follow:

Step1: Initialize the particle swarm, each particle is randomly set with the initial position X and the initial velocity V;

Step2: Calculate the fitness value of each particle;

Step3: For each particle, compare its fitness value and the fitness value of the best location it experienced, say $pBest$; if better, update $pBest$

Step4: For each particle, compare its fitness value and the fitness value of the best location entire group experienced, say $gBest$; if better, update $gBest$

Step5: Adjust the speed and position according to formula (1) and (2);

Step6: If reach the end conditions (good enough position or the maximum number of iterations), end it; if not, return to Step2.

4 K-means Optimization Clustering Algorithm Based on Particle Swarm Optimization and Multiclass Merging

PSO was originally used to search globally to find the global optimal solution, need try to avoid the emergence of local optimum. In this paper, the requirement has been modified for the usage of its local search function to find the optimal initial mass center require by K-means algorithm, and improve PSO.

The whole idea of this algorithm is to increase several times the given optimal number of clusters K and assign it to G, then bring G as the number of classes into improved PSO to optimize the center point. That is, first divide the data into a more categories, then use K-means algorithm to optimize cluster, getting better pre-clustering results; and finally according to a particular merging principle to merge the pre-divided categories, until the merged category reach the optimal number of cluster K and output the best cluster.

4.1 Measures of Algorithm Optimization

This paper aims at improving the defects of PSO such as easy to fall into local optimization and optimal solution oscillation, introduce fitness variance to speed up the convergence rate. Besides, it also brings in the thought of cluster merging, effectively solve the defects of K-means algorithm and PSO based on Euclidean distance divided clustering.

4.1.1 Weighting Factor

In order to eliminate vicinity near the optimal solution in PSO optimization process, as well as larger ω leads to particle swarms' strong global search capability and smaller ω tends to local search; we improve the weighting factors as follow. The weighting factor ω in speed update formula is from the maximum ω_{min} to the minimum ω_{max} .

$$\omega = \omega_{max} - run \frac{(\omega_{max} - \omega_{min})}{runMax} . \quad (3)$$

In which, run is the current number of iterations, $runMax$ is the total number of iterations algorithm.

4.1.2 Fitness Variance

No matter premature convergence global convergence in PSO, the particles will have "gathering" phenomenon, whether gather in a specific location or in a few specific locations, depending on the characteristics of the problem itself and the choice of fitness function. Consistence of the location is equivalent to the same fitness of each particle. Therefore, study the overall fitness change of all the particles can track the state of particle swarm, determining whether the algorithm converges.

Formula of fitness variance as follow:

$$\sigma^2 = \sum_{i=1}^{N} \left(\frac{F_i - F_{avg}}{f} \right)^2 . \quad (4)$$

In which, N is the number of particles in particle swarm; F_i is the i particle's fitness; F_{avg} is the current average fitness of the particle swarm; σ^2 is the group fitness variance; f is a normalized scaling factor, its role is to limit the size of σ^2 . f can take any value, just pay attention to two conditions: i) after be normalized, the maximum value of the $|F_i - F_{avg}|$ is no larger than 1; ii) f changes with the evolution of the algorithm. In this algorithm, the value of f adopts the following formula:

$$f = \begin{cases} \max\{|F_i - F_{avg}|\}, \max\{|F_i - F_{avg}|\} > 1 \\ 1, others \end{cases} . \quad (5)$$

Group fitness variance reflects the convergence level of all particles in the particle swarm. The smaller of σ^2, the particle swarm tends to converge; otherwise, the particle swarm is in the random search phase.

Therefore, in this paper, the algorithm adopts fitness variance in the program, if σ^2 less than the set threshold, it jumps out of the particle swarm algorithm part into the next program.

4.1.3 Cluster Merging

As clustering with the direct use of the improved particle clustering algorithm, the choice of fitness function is very harsh, so it is difficult to choose the appropriate function to get the optimal clustering division. In addition, the criteria of PSO and

K-means algorithm based on Euclidean distance for clustering division has serious limitations, the division effect of two very close clusters is far from satisfactory. Therefore, we propose a multiclass merging idea for an effective solution to these shortcomings, but the selection of merging conditions must be rational in order to achieve the optimal clustering division.

This algorithm select a merging condition based on density, consolidation steps and specific criteria are as follow:

Step1: Calculate the distance between every two centers in the pre-clustering results, and store them into a matrix in ascending order to generate a new matrix, namely $range$;

Step2: Take out the two centers which have the minimum center distance under the current clustering results (the first group of values in matrix range);

Step3: Take the mid-value of the two centers as pre-merging center;

Step4: Set 1/2 of the minimum center distance as the density range, calculate the density of the three centers selected Step2 and Step3 by the density range, the density of original center are recorded as Da and Db, the density of pre-merging center as Dc;

Step5: Determine whether the density of pre-merging center (Dc) is greater than or equal to 1/4 of the original center density, the formula is as follow:

$$Dc \geq \frac{1}{4}(Da + Db) . \tag{6}$$

If not meet the above condition, to choose another two original centers corresponding to the next set of values in the center distance matrix (range), and turn to Step3; Otherwise, if meet the conditions, to merge the two original clusters, re-divide the merged cluster;

Step6: If the merged cluster number reach the optimal number of clusters K, jump out of the merging step, save the best clustering results, or else turn to Step1 to continue merging.

4.2 Algorithm Realization

4.2.1 Parameter Encoding

In this algorithm, the part of PSO is adopting the encoding method based on cluster centers, the location of each particle is composed of the m cluster centers, in addition to the particle position, there are speed V and the fitness value F. As the sample vector dimension is d, so the location of the particle is m×d-dimensional variable, the speed of the particles should be m×d-dimensional variables, each particle has a fitness value F_i as well.

Therefore, the particle using the following coding structure:

$C_{11}C_{12}\cdots C_{1d}C_{21}C_{22}\cdots C_{2d}\cdots C_{m1}C_{m2}\cdots C_{md}$	$V_1V_2\cdots V_{m*d}$	F_i

In which, $C_{i1}C_{i2}\cdots C_{id}$ represents the i th d-dimensional cluster center, 1≤i≤m, V_{m*d} represents the m*d-dimensional velocity.

Then, the whole particle swarm using the following encoding format:

$$POP = \begin{Bmatrix} X_{11}, X_{12}, X_{13}, \cdots, X_{1m}, V_{11}, V_{12}, V_{13}, \cdots, V_{1m}, F_1 \\ X_{21}, X_{22}, X_{23}, \cdots, X_{2m}, V_{21}, V_{22}, V_{23}, \cdots, V_{2m}, F_2 \\ \cdots\cdots \\ X_{N1}, X_{N2}, X_{N3}, \cdots, X_{Nm}, V_{N1}, V_{N2}, V_{N3}, \cdots, V_{Nm}, F_N \end{Bmatrix}.$$

In which, $X_{ij} = C_{j1}C_{j2} \cdots C_{jd}$ represents its belonging to which particle.

When the cluster center is determined, the cluster decision is decided by the following nearest-neighbour rule, if x_i, c_j satisfied:

$$\|x_i - c_j\| = \min\|x_i - c_k\| \ , k = 1,2, \cdots, m. \quad (7)$$

x_i belongs to the j th class. For a certain particle, calculate its fitness according to the following formula:

$$F_i = \sum_{i=1}^{L}\sum_{j=1}^{m}\|x_i - c_{ij}\|^2 . \quad (8)$$

In which, L is the number of samples, x_i is the input samples, F_i represents the fitness value of the i th particle.

4.2.2 Algorithm Specific Process

Step1: Determine the best number of clusters K, set $G = 3*K$, and bring it as pre-clustering number into PSO for center optimization;

Step2: Population initialization: when initializing particles on, first randomly assign each data point to a certain class as the initial clustering division; then calculate the cluster centers of each class as the initial position encoding of particles; at the same time, calculate the particle fitness as the optimal location of the individual particles, and randomly initialize the particle velocity. Repeat N times, generating a total population of N initial particles;

Step3: For each particle, compare its fitness and the fitness of the best location it experienced; if better, update the particle's best position of the;

Step4: For each particle, compare its fitness and the best position the group experienced; if better, update the global best position;

Step5: According to formula (1) and (2) to adjust the speed and position of the particle;

Step6: For the new generation of particles, optimize in accordance with the following methods;

1. On the basis of cluster center encoding of the particle and the nearest neighbor rule to determine clustering division of the according particles;
2. Calculate the new cluster center according to clustering division, update the fitness value of particle, and replace the original code values. Classify clustering according to this way can make the convergence of the algorithm much faster.

Step7: Calculate the fitness of PSO variance;

Step8: Determine whether the fitness variance is less than the set threshold, or the algorithm reaches the maximum number of iterations, if so, end particle swarm optimization and output the best center, or else turn to Step3;

Step9: Regard the center obtained by PSO as the initial center of K-means, enter K-means algorithm for clustering division, output the pre-clustering classification results;

Step10: Combine the pre-merging division by steps described in 4.1.3;

Step11: Output the best clustering results.

In step6, when reclassify clustering categories for a new generation individuals, empty cluster appear. If there is an empty cluster, settle the center of the class as follows.

$$C_k = \text{minsam} + (\text{maxsam} - \text{minsa m}) \times \text{rand}(1,d) . \quad (9)$$

In which, C_k is the empty cluster center; minsam is the minimum value in coordinates of the sample data; $\text{rand}(1, d)$ is the maximum data in coordinates of the sample data; maxsam is a randomly generated point with the same dimension of the center coordinate.

5 Experimental Testing and Results Analysis

Three standard data sets were chosen for testing which can be divided into six, three and seven cluster number respectively. The structures are shown in Figure 1:

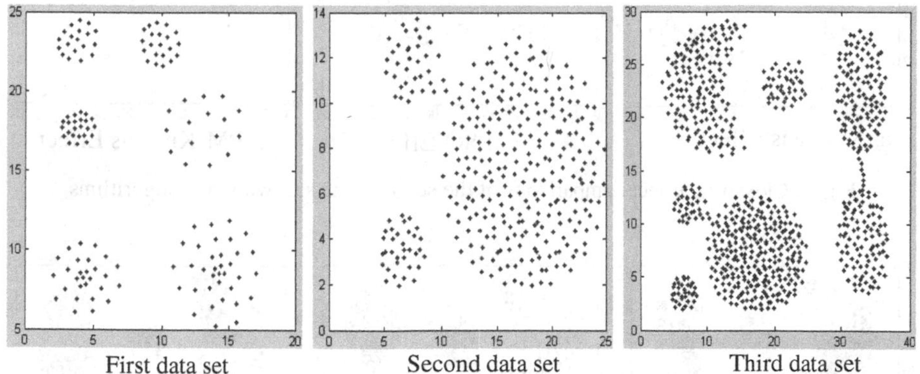

Fig. 1. Structures of three data sets

The algorithm parameter setting is as follows, particle number N is set to 50, the best cluster number K take value according to the data set, $c_1 = c_2 = 2$, ω changes linear as formula (3); $\omega_{max} = 0.9$, $\omega_{min} = 0.4$, fitness variance threshold threo is set to 0.07, cycles times M of hybrid clustering in particle swarm set to 100.

In this experiment, in order to compare the better clustering effect and wider range of applications, under the MATLAB environment, we run the program 50 times for these three data sets respectively with K-means algorithm, PSO algorithm and the algorithm stated in this paper. The experimental results are summarized as follows shown in Figure 2, 3, 4.

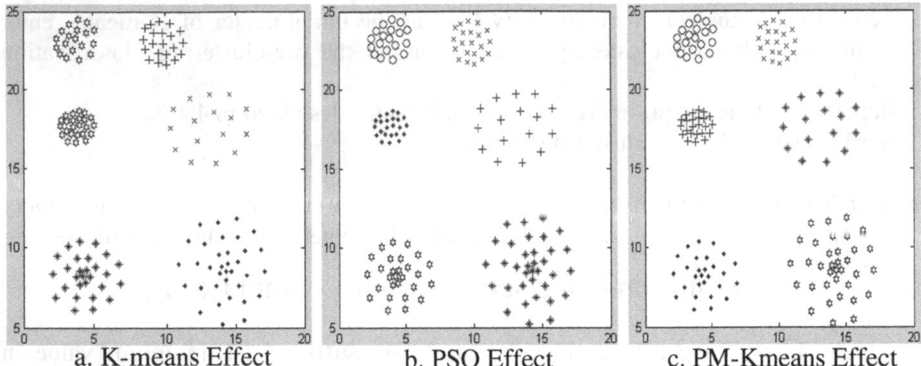

a. K-means Effect b. PSO Effect c. PM-Kmeans Effect

Fig. 2. Clustering effect comparisons of the first data sets with three algorithms

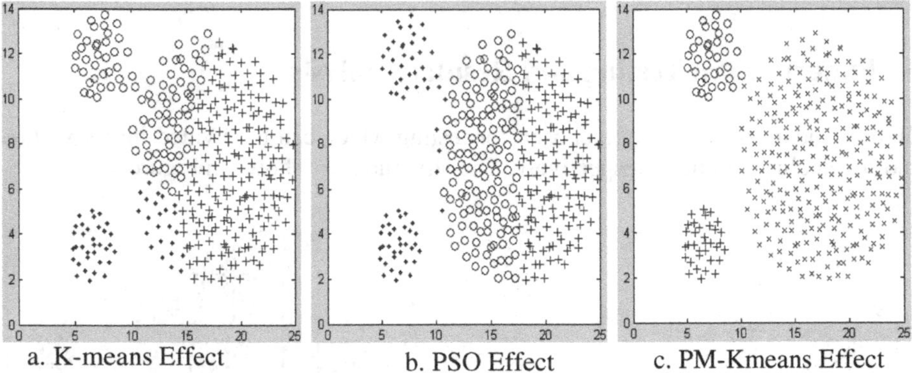

a. K-means Effect b. PSO Effect c. PM-Kmeans Effect

Fig. 3. Clustering effect comparisons of the second data sets with three algorithms

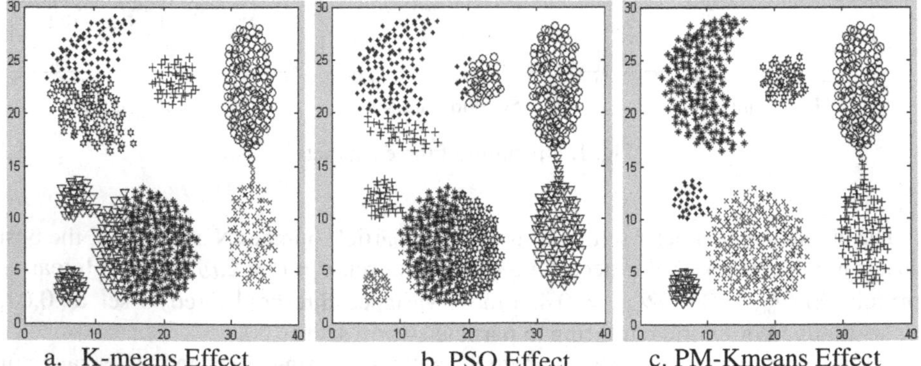

a. K-means Effect b. PSO Effect c. PM-Kmeans Effect

Fig. 4. Clustering effect comparisons of the third data sets with three algorithms

Figure 2 shows that, for clustering data sets with data points distributed evenly and within the categories has larger distance between categories, clustering results are satisfactory whether it adopts K-means algorithm, PSO algorithm, or algorithm in this paper. K-means algorithm results a relatively fast convergence speed but there is instability, occasionally appears poor clustering results, and the clustering results are easily affected by the initial value. While the algorithm stated here in this paper and the PSO algorithm are more stable and accurate.

From Figure 3 and 4, we can see K-means algorithm and PSO algorithm have clustering result not satisfactory enough for the second and third group of data sets, while the algorithm of this paper is much more advantageous on the clustering effect of more complex data sets. When dealing with data sets where there are smaller categories beside bigger categories, K-means algorithm and PSO algorithm are easy to divide the smaller into bigger categories, as shown in Figure 3.a, cannot effectively carve out of the small categories, while the algorithm of this paper does well in this aspect. This is because K-means algorithm and PSO algorithm are based on Euclidean distance criteria, there are big defect clustering results when clustering data sets with smaller distance such as the second group and third group of data sets. However, using the multiclass merging method can solve this problem well. The algorithm of this paper, compared with K-means algorithm and PSO algorithm, has better clustering effect and more stable when dealing with more complex data sets, it can well integrate local optimization and global optimization with faster convergence rate.

6 Conclusion

K-means Optimization Clustering Algorithm Based on Particle Swarm and Multiclass Merging (PM-Kmeans) stated in this paper, is mainly to integrate K-means clustering algorithm and improved PSO, and introduce the thought of multiclass merging to optimize clustering. It uses PSO algorithm to find the initial cluster centers, and select best K-means algorithm operation time according to their fitness variance, realizing an effective combination of PSO algorithm and K-means algorithm and speeding up the pace, and finally achieve the optimal clustering division by multiclass merging.

The results of theoretical analysis and experimental data show that the algorithm of this paper overcomes the problems existing in traditional K-means clustering algorithm, its global optimization ability is superior to the current K-means algorithm and PSO with a faster convergence rate and more superior clustering effect for complex data sets.

Acknowledgment. The work was supported by "New talent" Project of Zhejiang Province, "A Project Supported by Scientific Research Fund of Zhejiang Provincial Education Department (Y201119877)", China.

References

1. Kao, Y.-T., Zahara, E., Kao, I.-W.: A hybridized approach to data clustering. Expert Systems with Applications 34(3), 1754–1762 (2008)
2. Tsai, C.-Y., Kao, I.-W.: Particle swarm optimization with selective particle regeneration for data clustering. Expert Systems with Applications 38(6), 6565–6576 (2011)

3. Cui, X., Potok, T.E.: Document Clustering Analysis Based on Hybrid PSO+K-means Algorithm. Journal of Computer Science (2005)
4. Xia, S., Li, W., Zhou, Y., Zhang, L., Niu, Q.: Improved k-means clustering algorithm. Journal of Southeast University 23(3) (2007)
5. Bagirov, A.M.: Modified global k-means algorithm for minimum sum-of-squares clustering problems. Pattern Recognition 41(10) (2008)
6. Steinley, D.: K-means clustering: A half-century synthesis. The British Journal of Mathematical and Statistical Psychology 59(1), 1–34 (2006)
7. Faber: Clustering and the continuous k-means algorithm. Los Alamos Science 22, 138–144 (1994)
8. Kennedy, J., Eberhart, R.: Particle Swarm Optimization. In: Proc IEEE International Conference on Neural Networks, vol. IV, IEEE Service Center, Piseataway (1995)
9. Van der Merwe, D.W., Engelbrecht, A.P.: Data clustering using particle swarm optimization. Evolutionary Computation 1, 215–220 (2003)
10. Zhang, X., Wang, J., Zhang, H., Guo, J., Li, X.: Spatial clustering with obstacles constraints using particle swarm optimization. In: Proc. of the 2nd International Conference on Scalable Information Systems, ICST, Brussels, Belgium (2007)
11. Shi, Y., Eberhart, R.C.: Parameter Selection in Particle Swarm Optimization. In: Porto, V.W., Waagen, D. (eds.) EP 1998. LNCS, vol. 1447, pp. 591–600. Springer, Heidelberg (1998)

Discrimination Method of the Natural Scenery and Camouflage Net Based on Fractal Theory

Tao Zou, Wen-bing Chen, Han Ding, and Fei Ye

New Star Research Institute of Applied Technology, Hefei 230031

Abstract. Fractal theory can be applied to distinguish the natural scenery and artificial scenery, because it can describe the irregularity of nature scenery. According to the characteristic of the camouflage net in grass, an improved computation method of fractal dimension is used to segment image in order and extract fractal dimension. According to contrastive analysis of experiments, the result shows that this method can distinguish camouflage net and grass and provide a new way of surface features of the model identification.

Keywords: camouflage net, fractal dimension, remote-sensing image, blanket methods.

1 Introduction

The camouflage net is a very important equipment of camouflage, called "umbrella" of the military target which include the armament and military establishment etc. Camouflages are various methods of cheating and illusive concealment which hide the true and show the fake. It's an important content for battle safeguards. The basic theory of the camouflage is decreasing the difference of scattering or radiating properties between the target and background in the visible, infrared or microwave. By concealing the real objective, reducing its explorative character, simulating or extending the difference of target and background, camouflage makes up a fake target to cheat enemies. So how to identify the camouflage objective exactly within the shortest time is the research hotspot for military experts all over the word.

Fractal geometry can describe and analyze the natural phenomena which are ubiquitous, anomalistic and random more profoundly based on the self-similarity of the natural scenery.

This paper provides an extraction method of fractal dimension about the camouflage net of natural vegetation and woodland. The result indicates that this method could distinguish the camouflage of the natural vegetation and woodland effectually.

2 Fractal and Fractal Dimension

Mandelbrot finds the fractal geometry theory based on the description of the self-similarity of the complex natural scenery. There are two definitions of fractal[1]:the

D. Jin and S. Lin (Eds.): Advances in CSIE, Vol. 1, AISC 168, pp. 579–585.
springerlink.com © Springer-Verlag Berlin Heidelberg 2012

aggregation A of D(A) > d(A),called fractal aggregation, D(A) is called Hausdorff dimension (or fractal dimension) of A, d(A) is called topology dimension. Generally, D(A) is not integer but fraction.

The part and integer have self-similarity according to the mode called fractal. By the inspection of theoretic and application, people find that it is difficult to express the complex content of the fractal using these two definitions. In fact, there aren't exact definitions about what's the fractal at present. People usually explain these by a series of characters of fractal. The characters of the fractal are as follows [2]:

(1) Fractal aggregation has elaborate structure, namely the random small proportional detail;

(2) Fractal aggregation is so anomalous that its part and integer can't be described by traditional geometry language;

(3) Fractal aggregation usually has self-similarity, may be approximate or statistical;

(4) Generically, the fractal dimension (defined according to some mode) of fractal aggregation is bigger than its topology dimension;

(5) In the interested condition, fractal aggregation is defined according to a simple method and may be engendered by iteration.

Mandelbrot puts forward that when we calculate the fractal dimension of a pictures, we actually measure the irregularity [3] of natural phenomenon using fractal dimension D. First, we should choose a calculation methods, it's based on a fractal model. The fractal model of the picture has to be adapted for every kind of the method of calculating fractal dimension. There is a very important model in the random fractal theory, called FBR(Fractional Brownian Random field). Pentland had proved that under the illumination, the external gray images of 3d fractal also satisfy the FBR[4].So Peleg puts forward the blanket method[5] used for calculating the fractal dimension of gray images. Here, we apply an improved blanket method[6]:

1) Let's regard the picture as a massif, the altitude as the gray value, and spread the blanket with 2ε in thickness on both sides that the distance from the exterior is ε. The area of the blanket is its cubage divided 2ε. For different ε, the fractal area $A(\varepsilon)$ of gray surface is : $A(\varepsilon) = v_{\varepsilon} / 2\varepsilon = \sum_{i,j} (u_{(\varepsilon)}(i,j) - b_{\varepsilon}(i,j)) / 2\varepsilon$

In this way, the blanket cubage divided 2ε equals its area. In that formula, we just use one variable v_{ε} which has error to decrease in resultant computing error.

2) Because the fractal model just simulates the complex structure characters of natural object in special scale range. Here, "special scale range" reflects different veins characters. The excellence of computing fractal dimension is different in different scales. Just in a special scale range, we could reflect the fractal characters better. So we must choose different scales among different veins structures. This improved method which applies multi-scale fractal dimension is used to simulate dimension. $c(\varepsilon)$ is the linearity fit linear slope of five pairs of points which are $(\log(\varepsilon), \log A(\varepsilon))$, $(\log(\varepsilon+1), \log A(\varepsilon+1))$, $(\log(\varepsilon+2), \log A(\varepsilon+2))$, $(\log(\varepsilon+3), \log A(\varepsilon+3))$, $(\log(\varepsilon+4), \log A(\varepsilon+4))$ .According to the size order of ε, the linear fits the result of the

first, second, third, fourth, fifth pair of point as fractal dimension of scale 1. The result of the second, third, fourth, fifth, sixth pairs of point be regarded as fractal dimension of scale 2, and so on. Comparing with every fractal dimension among different scales, we use the steady fractal dimension as the picture's dimension.

3) $\log\varepsilon$ and $\log(A(\varepsilon))$ aren't rigorous linearity. If we choose different $\log\varepsilon_i, \log(A(\varepsilon_i))$, they would affect the final result. So we use fit character of every scale corresponding point based on the fit error [7-8]as standard for eliminating high error. In this way, we eliminate the points outside the limited error range, and compute the fractal dimension of every scale.

If the coordinates of n points are $(x_1, y_1), (x_2, y_2), \cdots, (x_n, y_n)$ in the double logarithm reference frame $\{\log A(\varepsilon), \log\varepsilon\}$, then the equation fits the regressed line:

$$y = \alpha + \beta x$$

In this formula :

$$\beta = l_{xy} / l_{xx}, \quad \alpha = \bar{y} - \beta\bar{x}, \quad \bar{x} = \frac{1}{n}\sum_{i=1}^{n} X_i, \quad \bar{y} = \frac{1}{n}\sum_{i=1}^{n} y_i, \quad l_{xy} = \sum_{i=1}^{n}(x_i - \bar{x})^2,$$

$l_{xy} = \sum_{i=1}^{n}(y_i - \bar{y})^2$, $l_{xy} = \sum_{i=1}^{n}(x_i - \bar{x})(y_i - \bar{y})$, β is an estimative fractal dimension.

If R is the correlation coefficient between y and x: $R^2 = \beta^2 l_{xx} / l_{xy}$

We can get the result from the fit analysis theory: $R_\alpha(n-2) = t_{\alpha/2}(n-2) / ((n-2) + (\tau_{\alpha/2}(n-2))^2)^{1/2}$

Here, $R_\alpha(n-2)$ is called the critical value of correlation coefficient R whose degree of freedom is (n-2) in marked level α. In the above formula, $\tau_{\alpha/2}(n-2)$ is the probability consistency value of t distribution whose degree of freedom is (n-2), and marked level is α.If $R > R_\alpha(n-2)$, y and x have better linearity connection. We could calculate the fractal dimension using β, otherwise we need to analyze and deal with the observation data once more. So, the value of R is bigger, the fit degree of regression line and observation data point is better. Therefore, R could used to be a measurement of linearity connection degree between y and x. In the range of a certain scale, the linearity correlation degree between y and x is the statistic self-comparability of the fractal aggregation. When R=1, the linear connection between y and x is certain, thereby the fractal aggregation S is self-comparable. When R tends towards 0, y and x don't have linear connection. We think S isn't statistic self-comparable. So R describes the statistic self-comparability degree of S quantification ally. In this algorithm, we confirm the reliability of this character according to R which is a weight used for calculating the fractal dimension.

When $R > R_\alpha(n-2)$ isn't right, we need to analyze the fit points and eliminate abnormity points. In order to get exact estimate value of the fractal dimension, we need to find the data points which fall short of request and eliminate them.

The residual of the i-th point is: $e_i = y_i - y'_i, \quad i = 1, 2, \cdots, n$

Here, y'_i is the calculation result of the fit linear formula by x_i.

The standard residual definition of e_i is: $d_i = e_i / \delta_i$,

Here, $\delta_i = ((1+1/n+(x_i-\bar{x})^2/l_{xy})(1-R^2)l_{xy}/(n-2))^{1/2}$

If one point's residual absolute value is bigger than other points, it is regarded as abnormity point and need to be eliminated.

3 The Identification of Camouflage Net Based on Fractal Character

Generally, it is difficult to find the human body and military target covered with some camouflage in background of the nature. The differences between the camouflage net and vegetation have two aspects: one is the character of radicalization, it represents the gray difference in the picture; the other is veins difference. The character of the imitated wildwood camouflage net designed for vegetation is similar with the character spectrum of vegetation on the visible near infrared wave band. The determination of this kind of imitated vegetation camouflage net is more difficult because of the low dimension of picture.

By analysis of the theory research and experiment data for many years, we find that the fractal model can be inosculated with exterior and space structure of nature background like sky, offing, ground and others in certain scale range. The reason is that the physics process can engender fractal exterior directly. On the other hand, we couldn't identify the fractal exterior and nature exterior, because they are quite similar. So the fractal model could be regarded as a math model of the nature background. The comprehensive applicability of the fractal model for the nature background is realized by changing the character parameter of model. We could describe different nature backgrounds using the aggregate of fractal model which has the different parameters, namely fractal model genus. There are connatural differences between the exterior and space structure of artificiality and the expressive discipline of fractal model. So the objects described by the fractal model don't include artificial object. We could inspect the artificial object in the background of nature according to the connatural differences [9]

4 The Analysis and Results of Experiment

Here, we analyze the camouflage net of the natural vegetation and woodland as following: choose two pictures (from literature10), put the camouflage net into the natural vegetation, and choose grass and dead grass as the natural vegetation separately. In this experiment, the wave band of CCD camera is 443nm~865nm, the spectrum bandwidth is 30nm~50nm, the precision is 12bit, the size is1024×1024. We choose 64 samples from the sequential partition image of grass and the camouflage net in the picture. The size of these samples are 64×64, extract the fractal dimension using an improved blanket algorithm.

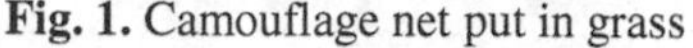

Fig. 1. Camouflage net put in grass

Fig. 2. Camouflage net put indeed grass

Fig. 1 and Fig. 2 are camouflage net pictures which background are grass and dead grass, Fig. 3 and Fig. 4 are partition samples extracted from Fig.1 and Fig. 2 from left to right and from top down. We just choose a few samples because of the limited space. The Fig.3 (a) and Fig.4 (a) are samples of the camouflage net. Fig.3 (b) and Fig.4 (b) are samples of the background picture of grass and dead grass extracted from Fig. 1 and Fig. 2.We could see from the pictures that the difference between the background picture of Fig. 1 and the camouflage net is apparent, but the difference between the background picture of Fig. 2 and the camouflage net isn't apparent very much.

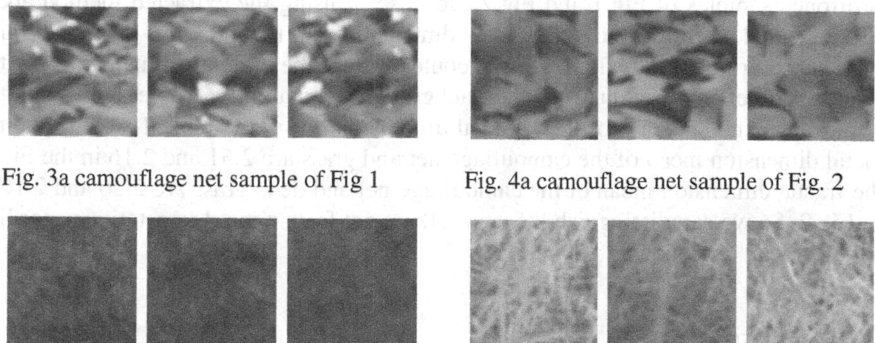

Fig. 3a camouflage net sample of Fig 1

Fig. 4a camouflage net sample of Fig. 2

Fig 3b grass sample of Fig 1

Fig. 4b dead grass sample of Fig2

Fig. 3. Sample of Fig 1

Fig. 4. Sample of Fig 2

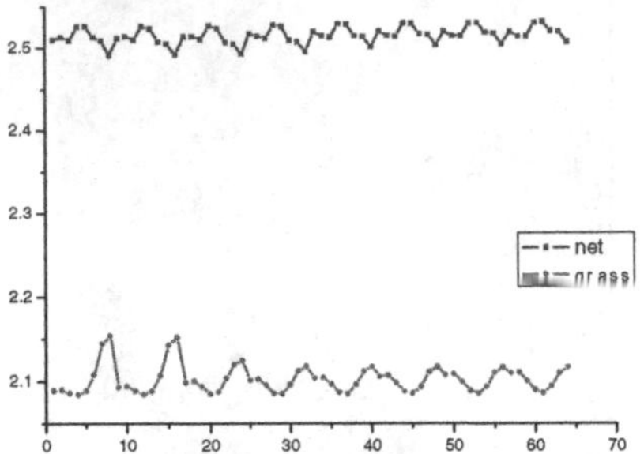

Fig. 5. Fractal dimension static diagram of Fig 1

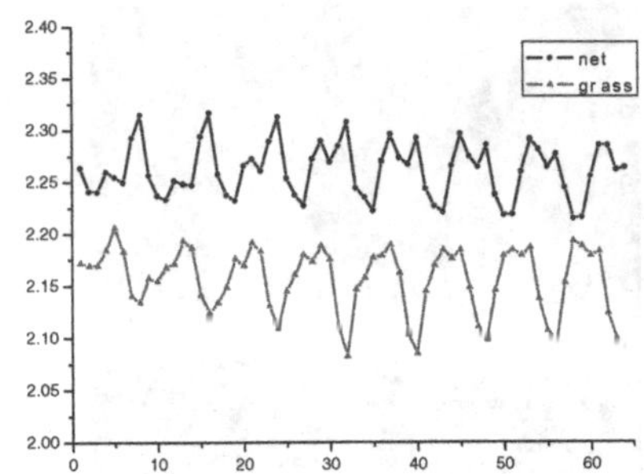

Fig. 6. Fractal dimension static diagram of Fig 2

Table 1. Fractal dimension static chart of the background and the camouflage net

target	mean	max	min	std
grass in Fig 1	2.1556	2.1962	2.0945	0.0359
Camouflage net in Fig 1	2.5142	2.5292	2.4916	0.0094
Dead grass in Fig 2	2.1555	2.2060	2.0825	0.0318
Camouflage net in Fig 2	2.2609	2.3162	2.2145	0.0261

Fig 5 and Fig 6 show the fractal dimension statistical results of the sequentially partitioned samples of Fig.1 and Fig.2. Because of using the extracted method about the sequential partition, so the fractal dimensions of the pictures have the same characters. From these figures, we could know the fractal dimension of the camouflage net is different from the backgrounds' apparently. Table 1 presents the mean, range and standard of their fractal dimension. From the table 1,we can see the fractal dimension mean of the camouflage net and grass are 2.51 and 2.16 in the Fig.1. The fractal dimension mean of the camouflage net and dead grass are 2.26 and 2.16 in the Fig.2. So we can distinguish the camouflage net from natural vegetation according to their fractal dimension.

5 Conclusion

Fractal geometry is an important math embranchment in non-linear science. Fractal theory provides an effective method for solving some realistic problems. In this paper, an extraction experimentation of fractal feature was carried out by applying an improved blanket method. We find easily that there are many differences in the fractal dimension of the vegetation and woodland camouflage nets. The vegetation satisfies the fractal model, and has obvious self-similarity.

The result from integrated analysis and experimentation validating shows that the method has better effect in distinguishing two kinds of scenery. Fractal dimension is a convenient and exact method in distinguishing vegetation camouflage net from the woodland camouflage net, and this method can also be applied to other different physiognomy sorting of remote sensing images.

References

1. Chen, Y.Q., Lu, A.S., Hu, H.P.: Summary of image analysis method based on fractal. Computer Engineering and Design 26(7) (2005)
2. Falcone, K.: Fractal geometry: The foundation of Mathematics and it application. Northeast Engineering Institute Press, Shengyang (1991)
3. Mandelbrot, B.B.: Fractals: from chance and dimension, pp. 189–192. Freeman, San Francisco (1977)
4. Mandelbrot, B.B.: Fractional brownianmotions, fractional noises and application. SIAM Review (10), 422–437 (1968)
5. Peleg, S., Narorand, J., Hartley, R.: Multiple resolution texture analysis and classification. IEEE Trans. PAMI 6(4), 518–523 (1984)
6. Zhang, T., Yang, Z.B., Huang, A.M.: Improved Extracting Algorithm of Fractal dimension of Remote Sensing Image. Journal of ordnance Engineering College
7. Wu, Z.: Image segmentation based on fractal theory.Nanjing University of Aeronautics and Astronautics (2002)
8. Zhang, X.G.: Data analysis and experimental optimum design, 156–244 (1986)
9. Zhao, Y.G.: Fractal Models of Natural Background and Automatic Identification of Man-Made Objective
10. Zhang, C.Y., Cheng, H.F., Chen, C.H., Zheng, W.W., Cao, Y.: A Study of Polarization Degree and Imaging of Camouflage Net in Natural Background. Journal of National University of Defense Technology 30(5), 34–37 (2008)

The Construction of the Embedded System Development Platform Based on μC/OS-Ⅱ

Hong He[1,*], BaoSheng Yu[1], ZhiHong Zhang[2], and YaNan Qi[1]

[1] Tianjin Key Laboratory for Control Theory and Application in Complicated Systems, Tianjin University of Technology, China
[2] Tianjin Broadcast and Television Development Company Ltd, Tianjin, China
heho604300@126.com

Abstract. Focusing on solving the problem that the traditional super-loops can not meet the requirement of Modern Control Systems no matter on resource management and system control, this paper introduces a way to construct an Embedded System Development Platform, combining the strength of the S3C2440A, a high-performance 32 bits ARM9 MCU and μC/OS-Ⅱ,a real time operation system(RTOS)[1].By providing the API and the scheme of system management, the Embedded System Development Platform can provide a good solution to solve the problem and make the portability of the program based on it much stronger.

Keywords: ARM9, MMU, RTOS, API.

Introduction

The Embedded System Development Platform construction mainly refers to three parts, the RTOS transplant, the source management scheme and the API. RTOS transplant is to make the RTOS run on the different MCU by modifying its source code [1].And the API is the function provided by the system. The μC/OS-Ⅱ is a very famouse, portable, ROMable and scalable Kernel. The S3C2440A is a high performance MCU produced by the SAMSUNG [2].

1 System Initialization

The system initialization is to design the startup code to fulfill the task of processor mode initialization, stack initialization and system clock initialization [3]. Different MCUs have different start-up codes. Fig. 1 shows the process of the initialization of the S3C2440A offering an outstand feature with its CPU core, a 16/32-bit ARM920T RISC processor.

* Hong He, School of Electrical Engineering, Tianjin University of Technology (China), 300384.

D. Jin and S. Lin (Eds.): Advances in CSIE, Vol. 1, AISC 168, pp. 587–590.
springerlink.com © Springer-Verlag Berlin Heidelberg 2012

1.1 Interrupt Vectors Initialization

The interrupt is a mechanism that the program is halted by the CPU for some fixed reasons. There are seven kinds of interrupt controlled by the interrupt vector table stored at the first 32 bytes of the memory, containing 8 vectors, a reserved vector included. Every vector contains an instruction to copy the address of the corresponding ISR to the PC register in the S3C2440A. Fig. 2 shows the process of the interrupt service.

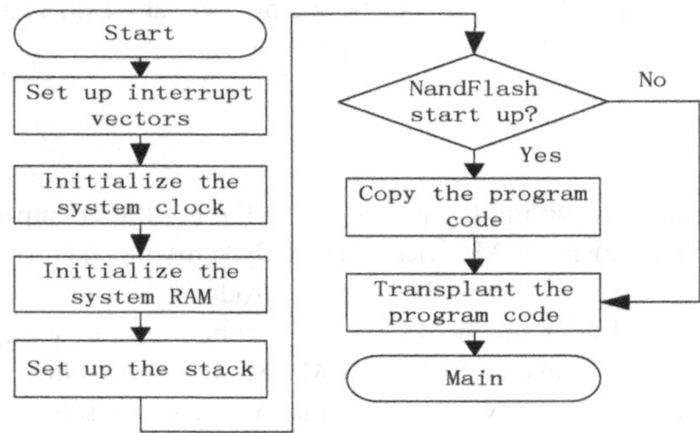

Fig. 1. There are seven parts of the start process. This shows the relationship of these seven parts. After this process the program jump to the main function.

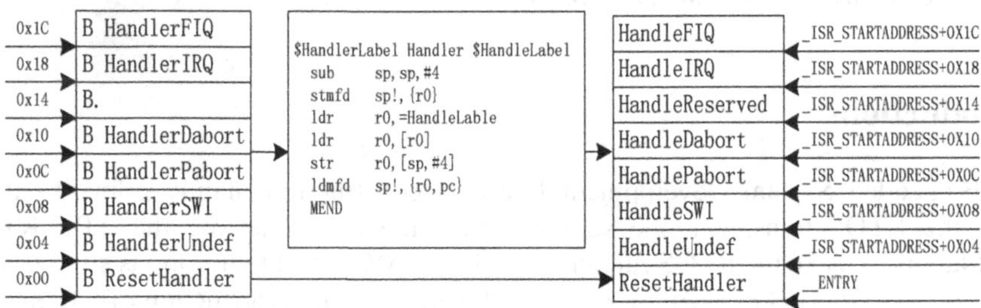

Fig. 2. When a interrupt occurs, the macro will be called. This macro will lead the program to jump to the corresponding codes to execute the ISR.

1.2 System Clock

The default basic frequency of the S3C2440A is 12MHZ or 16.9344 MHZ and the 12MHZ is preferred. With the PLL circuit, system can generate higher basic frequency for the CPU and its peripheral circuit. S3C2440A has two phase locked loops (PLL), the MPLL and the UPLL. MPLL generate three kinds of frequency, FCLK, HCLK and PCLK.

In order to make the PLL generate the needed clock signal, the parameter P M and S, related to the PLL must be set properly and so does the LOCKTIME register to determine the lock time. Here is how it dose. The Equation 1 shows the relationship between the frequency and the PLL parameters.

$$FCLK = MPLL = ((M + 8) * Fin) / ((P + 2) + 2 ^ S) \quad (1)$$

When the FCLK determined, system will generate the needed HCLK and PCLK through frequency division. Different FDIVN: HDIVN: PDIVN determine different FCLK: HCLK: PCLK. Table.1 shows the common proportion relationships.

Table 1. This table shows the different relationship between different divisions.

FDIVN	*HDVIN*	*PDIVN*	*Divide Ratio*
0	0	0	1:1:1
0	0	1	1:1:2
0	1	0	1:2:2
0	1	1	1:2:4
1	0	0	1:4:4

2 Import μC/OS-Ⅱ into the S3C2440A

Considering the portability, most codes of the μC/OS-Ⅱ is written in C. However there are still some codes must be written in assembly to manipulate the hardware.

2.1 Memory Remap

The physical address of the interrupt vectors have been changed after the code being copied and transplanted, with virtual address still unchanged .So before the importing, the virtual address of the system must be mapped again, making the virtual address of the interrupt vectors be mapped to the new physical address[4].

In the S3C2440A, he address mapping is done by the MMU. There is a table in the RAM, whose start address is stored in the register C2 of the coprocessor CP15.There are 4096 entries stored in the table, and each entries represent 1M address mapping [5]. So the remap can be done by changing the entries of the virtual address.

2.2 Modify the OS_CPU_A.ASM

The OS_CPU_A.ASM contains four assembly functions. OSStartHighRdy, OSIntC-txSw, OSCtxSw , and OSTickISR. The OSIntCtxSw is called in the ISR to do a task switch. The OSStartHighRdy is called in the OSStart to make the task with the hig-hest priority into the ready mode.

The μC/OS-Ⅱ needs users to provide a system clock to realize delay and timeout. The Clock Tick of the μC/OS-Ⅱ occurred from 10 to 100 times every second. When the tick occurred the system call the OSTickISR. In this function the system will deal with the task delay, task timeout and task switch. After the delay, the system will return back to the original place or switch to a higher priority task.

3 API and System Management

The system management of μC/OS-Ⅱ mainly contains task management and clock management. There are totally five task modes in the μC/OS-Ⅱ, the task dormant, the task running, the task ready, the ISR running and the task waiting. The μC/OS-Ⅱ provides the task control block, the event control block, the semaphore,the mutex, the event flag and the Message Mail box, totally five kinds of data structure to manage the system. Each data structure has their own API and through task switch, users can deal with complex applications.

4 Summary

With the development of the embedded system, the system control is becoming more and more complex. The appearance of the high-performance MCU could meet the hardware demand of the control system, and bring a series of problem at the same time. The RTOS can divide the complex application into different tasks and by the task switch users can accomplish the complex application. With the API, the users no longer need to know the inside details of the function, and can simply design a program by calling them. These make it easy to do a transplant. The Embedded System Development Platform can combine the strong point of the RTOS and the high performance MCU, efficient for hardware resource using and easy for program development.

Acknowledgment. This paper is supported by Tianjin Social Development key Foundation of China (NO. 09ZCGYSF00300) and Tianjin Key Laboratory for Control Theory and Application in Complicated Systems, Tianjin University of Technology, Tianjin 300384, China.

References

1. Labrosse, J.J.: The real-Time kernel, 2nd edn. Beijing aerospace university press, Beijing (2002)
2. Samsung's official website, `http://www.samsungsemi.com`
3. Sun, G.: The Embedded System Development Based on S3C2440A. Xi'an university of electronic science and technology press, Xi 'an (2010) (in Chinese)
4. Chen, G.: The μC/OS-Ⅱ operation system and stack. Electronic Component & Device Application 03, 38–46 (2011) (in Chinese)
5. Pan, X.: Digital Oscilloscope Based on ARM9 and μC/OS-Ⅱ. Microcomputer Information 25, 108–110 (2009)

Based on the of FFT Algorithm the Minimum Mean Square Error Linear Equilibrium Algorithm

Hong He[1,*], Xin Yin[2], Tong Yang[3], and Lin He[3]

[1] Tianjin Key Laboratory for Control Theory and Application in Complicated Systems, Tianjin University of Technology, China
[2] School of Computer and Communication Engineering, Tianjin University of Technology, China
[3] Tianjin Mobile Communications Co., Ltd
heho604300@126.com

Abstract. The main purpose of the Joint Detection (JD) techniques is an accurately estimating the user's signal, the difficulty lies in the system matrix inversion, so to find the fast inversion algorithms is the main key to make better the JD algorithms. This paper, it studies the new improved MMSE-BLE algorithm based block FFT, and through simulation that test and verify the algorithm's performance. The improved fast algorithm higher operational efficiency, the improved fast algorithm reduces the computing magnitude and the computational complexity and better improving the system performances.

Keywords: Inter symbol interference, Joint Detection, Fourier transformation.

1 Introduction

Using the code orthogonal properties, the system CDMA make the capacity of the whole system get great increase when it transmit different user data in the same frequency band . In the actual TDD mode, because wireless mobile channel has time-varying properties and multipath effect that it exist Inter Symbol Interference (ISI) between the same user data and exist Multiple Access Interference (MAI) between different user data. ISI interference limits the rate of the symbols transfer. MAI interference cause near-far effect. Both of they are indirectly suppress the increase of system capacity.

Joint Detection technology is a kind of multiple code channel data detection, and it has better performance than a single code channel receiver (such as RAKE receiver). Theoretical analysis, Joint Detection can improve three times the system performance [1]. But the calculated amount of linear Joint Detection algorithm is too big [2], and they need fast algorithm to simplify .This paper mainly studies an improved algorithm based on block FFT algorithm.

* Hong He, School of Electrical Engineering, Tianjin University of Technology (China), 300384.

D. Jin and S. Lin (Eds.): Advances in CSIE, Vol. 1, AISC 168, pp. 591–594.
springerlink.com © Springer-Verlag Berlin Heidelberg 2012

2 The Improved MMSE-BLE Algorithm (Based on Block FFT Algorithm)

By the analysis of the Joint Detection, it is known that if do not consider the noise, and the solution of MMSE-BLE is showed in the formula (1), which also is the solution of the linear equation $e = Ad + n$.

$$\hat{d}_{MMSE-BLE} = A^{+} e \tag{1}$$

However, system matrix $A \in C^{M(NQ+W-1)\times NK}$ is not Square matrix, but the first block row of A is row cycle when it in the case of array antenna multi-user. So it can expand A to a block cyclic matrix that it can use block FFT algorithm to get the solution.

After expanding A to a block cyclic matrix, the block cyclic matrix will have $D \times D$ blocks. The D is known in $D = N + [(Q+W-1)/Q] - 1$, and in every block $P \times K (P = MQ)$.

The block cyclic matrix ($\overline{A}$) of which the line and row are widen, so the corresponding vector x should be expanded by using zero to fill in. Because only the first block row of A are used to calculate Λ, so only fill V with zero to make it reach $DP \times K$ that is ok.

Through the block FFT transformation, the $\overline{A}$ can be realized the block diagonalization in the formula (2).

$$\overline{A} = F_{D\otimes P}^{-1} \cdot \Lambda \cdot F_{D\otimes K} \tag{2}$$

Similarly, the correlation matrix $\overline{A}^H \overline{A}$ of $\overline{A}$ also can be diagonalized in the formula (3).

$$\overline{A}^H \overline{A} = F_{D\otimes P}^{-1} \cdot T \cdot F_{D\otimes K} \tag{3}$$

And

$$T = diag_{(K,K)} \left\{ F_{D\otimes K} \cdot \left[\overline{A}^H \overline{A} \right]_{,1:K} \right\} \tag{4}$$

Fill the data e with zero and expand it to DP dimension. So it can be showed in the formula (5).

$$\overline{\hat{d}}_{MMSE-BLE} = (I + (R_d A^H R_n^{-1} A)^{-1})^{-1} (F_{D\otimes P}^{-1} \cdot T \cdot F_{D\otimes K})^{-1} (F_{D\otimes P}^{-1} \cdot \Lambda \cdot F_{D\otimes K})^H \overline{e} \tag{5}$$

After cholesky decomposition [3], the block matrix T would be $R^H R$ (R is upper triangular matrix). The formula (5) can be simplified to:

$$\overline{\hat{d}}_{MMSE-BLE} = (I + (R_d A^H R_n^{-1} A)^{-1})^{-1} [F_{D\otimes K}^{-1} R^{-1} (R^H)^{-1} \Lambda^H F_{D\otimes P} \overline{e}] \tag{6}$$

Based on the block FFT Algorithm the Minimum Mean Square Error Linear Equilibrium algorithm avoid the time domain inverse operation of correlation matrix, and it only involves FFT、IFFT and cholesky decomposition which greatly reduces the complex rate of computation [4].

3 Process Analysis

Suppose that the receive power of each user is the same and the time delay and incident angle of each user are evenly distributed, when it simulates the performance of multi-user.

Figure.1 is the simulation result of performance comparison of the MMSE-BLE algorithm based on the block FFT algorithm and the MMSE-BLE algorithm when the user number is 4 and the MS speed is 30km/h.

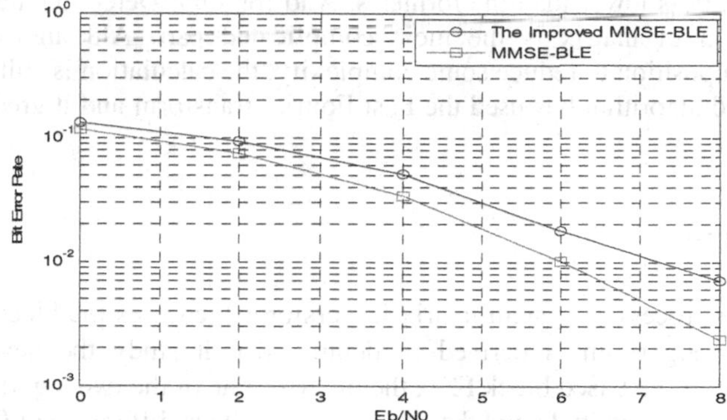

Fig. 1. Curve: Comparison of the two algorithm's performance. From the simulation results we can see that the performance of the improved MMSE-BLE algorithm has a little decline compared with the primary MMSE-BLE algorithm. Although this has 0.2dB decline it is in the permissible range of the system. The bit error rate of the two algorithms is still in the same order of magnitude, so it can be concluded that the performance of the improved algorithm and the primary algorithm are basically consistent.

4 Complexity Analysis

Compare the complexity of the MMSE-BLE algorithm based on the block FFT algorithm with the primary MMSE-BLE algorithm, and it as listed below:

Table 1. Complexity comparison

The Primary Algorithm		The Improved Algorithm	
Operation	Complex Rate	Operation	Complex Rate
$H^H * e$	$O(NM)$	$FFT(H(:,1))$	$O(N\log_2 N)$
$H^H * H$	$O(NM^2)$	$FFT(e)$	$O(N\log_2 N)$
$CholH^H * H$	$O(M^3)$	$E = FFT(e)./FFT(H(:,1))$	$O(N)$

According to the order of magnitude increasing, the time complexity of calculation complexity is arranged as follows: constant tatarian $O(1)$, logarithm tatarian $O(\log_2 N)$, linear tatarian $O(N)$, linear logarithm tatarian $O(N\log_2 N)$, square tatarian $O(N^2)$, cubic tatarian $O(N^3)$, ... , k power tatarian $O(N^k)$ and exponent tatarian $O(2^N)$. With the N is increasing, the time complexity above is continuously increasing, while the efficiency of the algorithm is more and more low.

So from the table.1 we can know that the complexity of the primary MMSE-BLE algorithm is square tatarian and cubic tatarian, while the highest order of the improved MMSE-BLE algorithm is linear logarithm tatarian. Obviously that the latter's complexity is lower than the former's. And for Joint Detection algorithm, the inverse operation of matrix is the most complicated part. Although it uses the Cholesky decomposition to reduce some complexity, the calculation is still very big.

The improved algorithm has used the Fast Fourier transform and it greatly reduces the complexity.

5 Conclusions

This paper, the process of system matrix in version which uses the block FFT (Fast Fourier change) algorithm is derived in detail. Also it study the new improved MMSE-BLE algorithm based block FFT, the bit error rate of the two algorithms is still in the same order of magnitude and the simulation results is difference of 0.2 dB , that can be received by the system. Therefore it can be concluded that the performance of the improved algorithm and the primary algorithm are basically consistent, but the MMSE-BLE algorithm based on FFT greatly reduces the computational complexity and has more advantages.

Acknowledgment. The title selection is mainly originated from Tianjin science and technology innovation special funds project (10FDZDGX00400) and Tianjin Key Laboratory for Control Theory and Application in Complicated Systems, Tianjin University of Technology.

References

1. Nigam, I., Mallik, R.K.: A Joint Performance Measure for the Deeorrelating Multi-user Detector. IEEE Trans. Wireless Commune. 3(4), 1024–1030 (2004)
2. Kang, S., Qiu, Z., Li, S.: Analyses and comparison of the performance of linear joint detection algorithms in TD-SCDMA system. Journal of China Institute of Communications 23 (2002)
3. Kim, S., Lee, J., Kim, Y.: Adaptive Cholesky based MMSE equalizer in GSM. Circuits and Systems 109, 886–889 (2008)
4. Wu, G., Dai, W., Tang, Y., Li, S.: Performance comparison of channel estimation algorithms for TD-SCDMA PIC receiver. In: IEEE International Conf. of Communications, Circuits and Systems, Chengdu, China, vol. 2 (July 2002)

Research on Embedded Intrusion Detection System Based on ARM9

Shiwei Lin[1] and Yuwen Zhai[2]

[1] College of Information and Control Engineering, Jilin Institute of Chemical Technology, Jilin, China
13704406003@126.com
[2] Mechanical & Electrical Engineering College, Jiaxing University, Jiaxing, China
wanglei_new814@126.com

Abstract. An embedded intrusion detection system based on ARM9 is proposed to overcome problem existing in the current network intrusion detection systems. The system is designed as ARM9 microprocessor core, using the Linux-2.4 kernel as the underlying operating system; with misuse detection and protocol analysis techniques to complete the data packets of real-time detection and extraction using unsupervised clustering algorithm for intrusion characteristics, expansion of the existing rule base invasion. Experimental results show that the design of the system has good detection capability and high stability under certain conditions.

Keywords: intrusion detection system, embedded techniques, clustering algorithm, protocol analysis techniques.

1 Introduction

Existing intrusion detection system of the following deficiencies: intrusion detection system can not detect all of the good packets; attack signature database updates are not timely; detection method single; different intrusion detection systems are not interoperable; and other networks can not interoperable security products; structural problems, in other aspects of the architecture can not meet the distribution, for openness [1]-[3].

As embedded systems with high performance, low power consumption and price advantage, the embedded systems used in intrusion detection systems can significantly increase the detection performance of the system. Therefore, the proposed and initial implementation of Intrusion detection system based on ARM9 embedded.

2 The Design of Hardware System

System hardware uses a modular design approach, including eight modules: control module, FLASH module, SDRAM modules, LCD and keyboard module, Ethernet interface module, SMS alarm module, the external circuit module of, power modules[4], such as shown in Figure 1.

D. Jin and S. Lin (Eds.): Advances in CSIE, Vol. 1, AISC 168, pp. 595–599.
springerlink.com © Springer-Verlag Berlin Heidelberg 2012

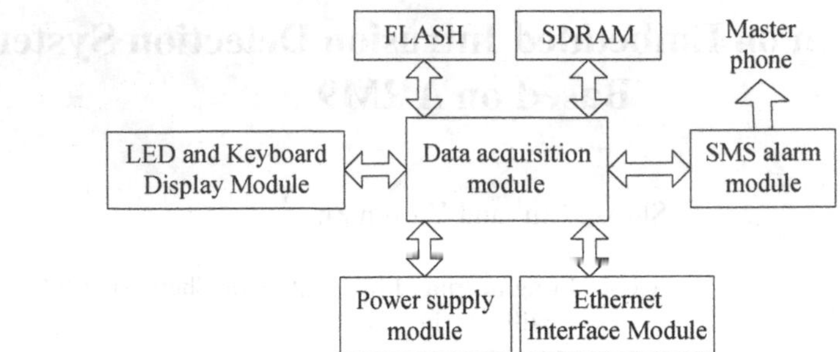

Fig. 1. The design of hardware system

3 The Design of Software System

IDS system design software design is the difficulty; we use embedded Linux system as the underlying operating system. In addition, to ensure system stability, chose Linux-2.4kernel for software development [5].

To overcome the shortcomings of traditional IDS systems, the software design will misuse detection technology combined with protocol analysis. The overall structure of the collector, protocol analysis module, misuse and intrusion alarm detection module composed of modules, shown in Figure 2.

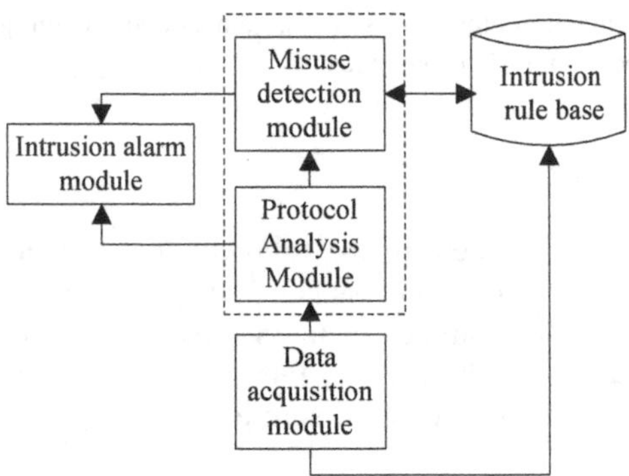

Fig. 2. The overall structure

4 Key Technology

4.1 Data Acquisition

Data acquisition module is system-wide basis, in accordance with predefined policies for network data packets flowing through the acquisition. Ethernet data transfer process, the data packet destination address matches will follow the way of transmission, the network interface to accept only their own data packets, while the packets are not discarded their own. In order to enable them to receive all the data flowing through the network interface packet interface mode can be set to promiscuous mode limpkin provides a separate user-level network

Network packet capture interface, and full consideration to the application's portability, can be easily ported to multiple operating systems; Web application development is very convenient. Use limpkin for network packet capture from the host can limit the agreement; it is used here to capture data packets. The main code is as follows:

```
/ * Create multi-threaded data reported capture * /
  int m_threadHandle;
  pthread_t p_thread;
  m_ThreadHandle    =    pthread_create    (&p_thread,    NULL,
(void*)
  myCaptureThread, NULL);
  /* Packet capture thread */
  char* inter; /* Network interface name */
  u_char *pr; /* Handle packets to be captured */
  pcap_t* p1;
  struct bpf_program bpf1; /* Filter code structure*/
  struct pcap_pkthdr hr;
  device = pcap_lookupdev(errbuf);
  void myCaptureThread (void){
  inter = pcap_lookupdev (errbuf);
  p1 = pcap_open_live (inter, 9000, 1, 400, errbuf);
  pcap_compile(p1, &bpf1, filter_string, 0, netmask);
  pcap_setfilter(p1, &bpf1);
  while ((pr = (char *) (pcap_next(p1, &hr))) = =NULL);
  e_net = (struct libnet_ethernet_hdr *)pr;
  pcap_close(p1);
  }
```

4.2 Intrusion Rules Extraction

Intrusion rules misuse detection signatures is an important part of the module, it directly affects the overall system performance. Typically, the source IP, destination IP, protocol type, source port, destination port and attack signatures should be included in the rules. System has been included as part of the rules can be used for misuse intrusion detection. In addition, to ensure that the latest attack signatures can be detected in time, the system can capture data from Ethernet packets for analysis by cluster analysis algorithm to extract the invasion characteristics.

Cluster analysis algorithm uses the network's normal behavior is much larger than thinking abnormal behavior, unsupervised clustering (UC) algorithm to data sets into different groups, within groups have a higher similarity between objects, whereas the similarity between groups low [6]

Euclidean distance can be used to detect the distance between objects within the group

$$d(x_i, x_j) = \left(\sum_{k=1}^{n} \left| x_{ik} - x_{jk} \right|^2 \right)^{\frac{1}{2}} = \sqrt{\left| x_{il} - x_{jl} \right|^2 + \left| x_{il} - x_{jl} \right|^2 \ldots \left| x_{ik} - x_{jl} \right|^2} \quad (1)$$

① $d_{ij} \le d_{ik} + d_{kj}, \quad \forall i, j, k;$

② $d_{ij} \succ 0, \quad \forall i, j, X_i \ne X_j;$

③ $d_{ij} = 0, \quad \forall i, j, X_i = X_j;$

④ $d_{ij} = d_{ji}, \quad \forall i, j.$

Clustering using the following steps:

Step 1 Choose an arbitrary number of objects, and build the cluster center.
Step 2 to read a new object, and use the Euclidean distance formula dollars
Considered the minimum distance. If it does not exceed a given threshold value, it places the formation of cluster centers. On the contrary, in order to build it out of a cluster of nine new and repeat step 2.
Step 3 class of objects to sort, filter out the noise and outliers. After setting the proportion of the number N, determine the class of data objects is greater than N, if N is greater than the class can be considered normal, whereas that is the exception class. On the exception class data.

Extraction, and data with the existing rules to compare, delete duplicate rules, expansion of the existing rule base invasion.

Table 1. Experiment result

Threshold	Intrusion types							
	DOS		Probing		U2R		U2L	
	Detection rate (%)	False alarm rate (%)	Detection rate (%)	False alarm rate (%)	Detection rate (%)	False alarm rate (%)	Detection rate (%)	False alarm rate (%)
10	72.35	4.62	73.19	2.61	73.02	3.58	63.08	6.05
20	71.32	4.50	72.85	2.48	70.89	2.89	68.70	5.02
30	68.22	6.51	90.02	3.18	86.59	2.62	37.89	5.36
40	75.01	3.25	88.06	1.03	88.48	2.23	87.56	4.52
50	65.48	4.02	82.42	2.45	80.05	3.25	61.22	5.12
60	74.56	4.21	82.03	2.48	79.18	3.21	65.45	4.35
70	74.98	3.98	84.41	2.50	79.53	2.10	68.22	4.40

5 Experiments

Use KDDCUP 99 intrusion data set [6] to evaluate the system. In the experiment, using a different threshold, when the threshold value of 40 when, Pro-bing and U2R detection rate was higher false alarm rate low; When the threshold is 10, the emergence of a higher false positive rate; when threshold is 70, the low false alarm rate, the experimental results shown in Table 1.

6 Conclusions

This article will advanced ARM embedded technology and intrusion detection technology combined with high performance, low-power S3C2440 microprocessor as the core of their networks in real-time packet capture and detection. System design, the introduction of unsupervised clustering algorithm for intrusion rules library expansion, making the intrusion rules can update, good to make up for misuse detection of defects.

References

1. Wei, X.: A semi-supervised clustering algorithm for network intrusion detection. Tiedao Xuebao/Journal of the China Railway Society 32, 49–53 (2010)
2. Hubballi, N.: Network specific false alarm reduction in intrusion detection system. Security and Communication Networks 4, 1339–1349 (2011)
3. Kachurka, P., Golovko, V.: Neural network approach to real-time network intrusion detection and recognition. In: The 6th IEEE International Conference on Intelligent Data Acquisition and Advanced Computing Systems: Technology and Applications, pp. 393–397. IEEE Press, New York (2011)
4. Sun, S., Wang, Y.: A weighted support vector clustering algorithm and its application in network intrusion detection. In: 1st International Workshop on Education Technology and Computer Science, pp. 352–355. IEEE Press, New York (2011)
5. Chen, R.-C., Hsieh, C.-F., Huang, Y.-F.: An isolation intrusion detection system for hierarchical wireless sensor networks. Journal of Networks 5, 335–342 (2010)
6. Wang, J., Zhang, X.: KDDCUP99 Network Intrusion Detection Data Analysis and Pretreatment, pp. 407–408. Science & Technology Information (2008)

Application of Rough Set in Highway Construction Program

Xiao-jun Lu[1] and Ping Zhang[2]

[1] School of Transportation, Wuhan University of Technology, Wuhan, China
[2] Wiuhan Zhenghua Architectural Design CO., LTD, Wuhan, China
lxjzp@126.com, 47079389@qq.com

Abstract. It is always of concern in highway and bridge construction to effectively succeed past highway construction programs and guide new highway construction. A kind of knowledge model resolution method based on rough data analysis was presented in the paper, the key point of which is reduction algorithm on decision table. With decision table, it can automatically extract typical example from original programs and obtain higher classification accuracy. It can also better process interference from noise and perform attributes reduction, so as to reduce computation complexity and generate relatively simple rule form. Using the method to establish knowledge model of highway construction program based on a group of classic highway construction data, simulation result illustrates the effectiveness of the method.

Keywords: rough set, highway and bridge construction, attribute reduction.

1 Introduction

In the case-based reasoning (CBR) systems, it is an important step to automatically extract useful and typical cases from database in case reasoning, the target of which is to eliminate interference from noise data while obtaining high quality cases. It can also reduce time complexity of case reasoning while decreasing space complexity of case storage. Promoted by this target, peoples proposed various algorithm of case reasoning in recent years. Ref. [1] researched a method to build case database from meteorological data with data mining technique. Ref. [2] brought out a CBR system instance indexing model based on AHP. The shortcoming of these methods is that it can not process noise data and need many pre-determined parameters. The rough set (RS) case extraction algorithm presented can better solve these deficiencies.

With the continuous expansion of fixed asset investment on transportation, the highway mileage has increased rapidly. At the same time, the highway construction also has accumulated various construction programs, rule understandings and design experiences. It has been a concern in highway and bridge construction to effectively succeed past highway construction programs and guide new highway construction. The paper utilized rough set reduction to study on how to optimize highway construction program parameters and perform discussion on intelligent important attributes indexing method according to diagnosis rules of rough set reduction. The paper is organized as follows: section 2 introduces basic theories and methods of

D. Jin and S. Lin (Eds.): Advances in CSIE, Vol. 1, AISC 168, pp. 601–607.
springerlink.com © Springer-Verlag Berlin Heidelberg 2012

rough set reduction; section 3 gives case database extraction algorithm; section 4 presents highway construction program reasoning example based on rough set and section 5 concludes our work.

2 Basic Theories and Method of Rough Set

In the rough set theory, knowledge is considered as ability to classify objective things, program development technology based on which is seen as capability to classify actual or abstract objects. For standardization, the objective things to be investigated are always called objects and been described by attributes and their values. All objects are called as set domain. To convenient data processing, it needs to perform symbol representation on knowledge. The information system can be seen as a system consisted of a set of 4 points that denoted as $S=(U,R,V_r,f), r\in R$. Where, U is non-empty closed domain constituted by n research objects $\{x_1,x_2,\cdots,x_n\}$; $R=C\bigcup D$ is attributes set; $C=\{c_i,i=1,2,\cdots,m\}$ is conditional attributes set; $D=C=\{d_i,i=1,2,\cdots,n\}$ is decision attributes set; Vr is attribute set and f is total decision function; $f_r=U\rightarrow V_r$ so that for any $x\in U, r\in R$, there is $f(x,r)\in V_r$.

The data table of knowledge representation system can be divided into decision table and non-decision table. The former on is a special and important knowledge representation system, which plays vital role in the decision applications.

The resolution matrix can be used for attribute reduction of decision table. It condenses information about attribute classification in the decision table into a matrix. In the information system S, if $a_k(x_j)$ is value of sample x_j on attributes, we can define the resolution matrix of system as $M(S)=\lfloor m_{ij}\rfloor_{n\times n}$, the element of i-th row j-th column is

$$m_{ij}=\begin{cases} a_k\in C, a_k(x_i)\neq a_k(x_j)\wedge D(x_i)\neq D(x_j) \\ \varnothing, D(x_i)=D(x_j) \quad i,j=1,2,\cdots .n \end{cases}. \tag{1}$$

Therefore, the element m_{ij} in matrix is set of all attributes that can distinguish object x_i and x_j. If x_i and x_j belong to same decision class, the element m_{ij} take empty set $\varnothing$. Clearly, the matrix is an n-order symmetric matrix according to main diagonal, so we can only consider its upper triangular or lower triangular in the computation.

For each resolution matrix $M(S)$, it corresponds to a merely resolution function $f_M(s)$. It defines the resolution function of information system S is a boolean function with m variables $a_1,a_2,\cdots,a_m$. It is the conjunction of $(\vee m_{ij})$, while $(\vee m_{ij})$ is disjunction of each element m_{ij} in the matrix, namely

$$f_M(s)(a_1,a_2,\cdots,a_m)=\wedge\{\vee m_{ij}, 1\leq j<i\leq n, m_{ij}\neq\varnothing\}. \tag{2}$$

Each disjunction normal form in the resolution function corresponds to a reduction. Although the reduction based on resolution is simple, at the circumstance of complex decision table and too many conditional attributes, as the requirements on storage space is too big, it is difficult to perform attribute reduction with pure resolution matrix. Here we will present a heuristic minimum reduction algorithm based on resolution matrix to solve the problem better.

The importance of attributes and reduction on decision table are all in the table. Different conditional attribute has difference importance and some attributes provide a wealth of information to play vital role on production decision, while other attributes are seem to be nothing. Therefore, we can reduce conditional attributes while keep correct classification ability of decision table, so as to decrease unnecessary attributes. The importance of reduced attributes can be measured by its dependence $R_E(C) - R_{E-F}(C)$, where E and F are subset of conditional attributes and F is subset of E.

Thus, it is necessary to sort conditional attributes according to m_{ij}. In the generation of resolution matrix, the emergency frequency of each attribute will be record at same time. These frequencies can be used to assess importance of attribute and optimal selection of attribute. It is because if an attribute emerge more frequent, its potential classification ability will be greater. In case of attribute frequency computation, it is not simple count, but weighted. The value of weight is determined according to length of attribute in the resolution matrix. In this way, for a resolution matrix $M(S) = \lfloor m_{ij} \rfloor_{n\times n}$, the importance computation formula of corresponding attribute a is

$$f(a) = \sum_{i=1}^{n} \sum_{j=1}^{n} \frac{\lambda_{ij}}{|m_{ij}|}, \quad \lambda_{ij} = \begin{cases} 0, a \notin m_{ij} \\ 1, a \in m_{ij} \end{cases}. \tag{3}$$

Sort all attributes according to importance weight and then delete redundant samples introduced by conditional attribute reduction and non-consistent sample, the reduction condition is

$$p_1 / p_0 < \alpha . \tag{4}$$

Where, p_0 is sample number in information table before reduction operation; p_1 is number of non-consistent sample introduced after reduction; α is threshold, which is determined according to actual needs.

3 Case Database Extraction Algorithm

Definition 1: Attribute similarity. There are two different program C_1 and C_2 in the decision table $(U, C \cup \{d\})$. Set V_{1i} and V_{2i} are values of C_1 and C_2 on attribute a and $V_{\max}$ and $V_{\min}$ are maximum and minimum of attribute a. The similarity of programs C_1 and C_2 on attribute $a \in A$ is

$$sim(v_{1i}, v_{2i}) = 1 - \frac{|v_{1i} - v_{2i}|}{|v_{\max} - v_{\min}|}. \tag{5}$$

Definition 2: Program similarity and similarity classification capability. The similarity S between C_1 and C_2 in decision table $(U, C \cup \{d\})$ is product of similarity of conditional attribute $a \in C$ and its importance:

$$sim(C_1, C_2) = \sum_{i=1}^{n} f(a) \times sim(v_{1i}, v_{2i}) . \quad (6)$$

Therefore, according to basic concept of similar rough set, the basic process of CBR case extraction is as follows: compute similarity among different attributes with (5), the specific description of which can refers to [3, 4]. The similarity of each program can be computed with object similarity (6) so as to perform extraction on typical cases. The case generation algorithm is as follows:

Input: decision table $(U, A \cup \{d\})$, wehre $A = \cup a_i, i = 1,2,\cdots,n$;
Output: extracted program
Begin
Open decision table

a) Initializing p_0, p_1, α, i;
b) Set reduced attribute set equal to conditional attribute set, namely *Reduct*=*R*;
c) Compute resolution matrix *M* and find attribute set *S* that not contain core attributes;
d) Describe attribute combination not containing core attributes as form of disjunction normal form, namely $P = \wedge\{\vee a_{ik}, i = 1,2,\cdots,s; k = 1,2,\cdots,m\}$.
e) Compute importance $f(a)$ of attribute according to (3);
f) While ($p_0/p_1 < \alpha$){Select the attribute *a* with minimum importance so that *Reduct*=*Reduct*-{*a*};Delete redundant case and non-consistent cases introduced with conditional attributes; *i*=*i*+1;}
g) Compute attribute similarity after indexing based on attribute importance;
h) Compute similarity of index cases;

End

4 Highway Construction Reasoning Case Based on Rough Set

Highway construction program design is vital part in the whole highway construction cycle. It's scientific and level directly relates to whole construction effect of highway, final practice and economic benefits. Currently, the work amount of highway construction design is huge. Taking task of some Highway and Bridge Design Institute as example, there needs to conduct about 10 construction programs each year, evolving more than 10,000 km mileage. Due to different geological, geomorphological and feature, the highway construction programs are vary from each other, involving plains, hills and deserts, so it is difficult to design. In the highway construction program design, there are always problems of many using materials, long design period and subject to human effects, which affect the design quality and working efficiency to some extent.

The highway construction program design system is a program developing system based on many years of program design and approval process, which can be divided into 4 items and each one contains index for its evaluation, as shown in Table 1.

Table 1. Evaluation indexes of highway construction program design system.

Design program	Indexes
Program foundation	1. Prospect; 2. Research situation; 3. Program foundation
Research basis	4. Related work accumulation; 5. Qualifications; 6. Implementation team
Program content	7. Title; 8. Keywords; 9. Content; 10. Implementation method and technique route; 11. budget
Program evaluation	12. Comprehensive evaluation; 13. Leadership attitude; 14. Program weight

Based on practical status of highway construction program design, we built program database according to large amount of practices and the decision table (U, R) for knowledge representation. Since 1999, the Institute recorded 1,520 programs for highway construction. We selected 7 randomly as program close domain $U = \{x_1, x_2, \cdots, x_7\}$, where the conditional attribute $C = \{c_1, c_2, \cdots, c_{13}\}$ represents 13 indexes respectively in program development.

Decision attribute is the weight of program. Form orderly hierarchical structure according to domination relationship using AHP. The upper one is target layer; middle is criteria layer and lower program layer. Determine total order of relative importance of all factors in the layer by the way of multiple comparisons [5-7]. Its solution process is as follows: In order to apply rough set theory, we firstly discretize continuous sample data, namely divide value range of each attribute into several sections and each one represents different discrete value, as shown in Table 2.

Table 2. Importance division of decision attribute.

Decision factor	Attribute value
Absolutely important	9
Very important	5
Slightly important	3
As important as	1

The paper uses Chi-merge discretization method to regard sample value as break point. Verify each attribute with same significance level a to determine whether merge adjacent sections. Gradually reduce significance level and loop till the termination condition is met. The discrete program parameter decision table is shown in Table 3.

Table 3. Parameter decision table of highway construction program.

U	c_1	c_2	c_3	c_4	c_5	c_6	c_7	c_8	c_9	c_{10}	c_{11}	c_{12}	c_{13}	d
x_1	2	1	2	2	2	2	1	3	2	2	2	2	2	1
x_2	3	1	3	2	2	1	1	1	2	1	1	1	1	5
x_3	2	1	1	1	3	2	2	3	2	3	2	2	1	3
x_4	2	1	2	2	2	2	1	3	2	2	2	2	2	1
x_5	3	2	2	2	2	2	1	1	2	2	1	1	3	3
x_6	2	1	1	3	2	1	3	2	2	2	2	3	2	9
x_7	2	1	2	2	1	2	3	2	1	1	1	2	2	1

Refresh the decision table and merge some rows in it. The reduced decision table may have many, but they all have same cores. The core of decision table refers to intersection of each reduced decision tables. It is the set of characteristic attributes that can not be reduced and the computation basis of all reduction. We can arrive at resolution matrix *M*(*S*) and resolution function based on (1) and (2). Therefore, the attribute combination not containing core attributes is as follows:

$m_{12} = c_1 \vee c_3 \vee c_6 \vee c_8 \vee c_{10} \vee c_{11} \vee c_{12} \vee c_{13} \vee d$

$m_{13} = c_3 \vee c_4 \vee c_5 \vee c_7 \vee c_{10} \vee c_{13} \vee d$

$m_{15} = c_1 \vee c_2 \vee c_8 \vee c_{11} \vee c_{12} \vee c_{13} \vee d$

$m_{16} = c_3 \vee c_4 \vee c_6 \vee c_7 \vee c_8 \vee c_{12} \vee d$

$m_{17} = c_5 \vee c_7 \vee c_8 \vee c_9 \vee c_{10} \vee c_{11}$

$m_{23} = c_1 \vee c_3 \vee c_4 \vee c_5 \vee c_7 \vee c_{10} \vee c_{11} \vee c_{12} \vee d$

$m_{24} = c_1 \vee c_3 \vee c_6 \vee c_8 \vee c_{10} \vee c_{11} \vee c_{12} \vee c_{13} \vee d$

$m_{25} = c_2 \vee c_3 \vee c_6 \vee c_{10} \vee c_{11} \vee c_{13} \vee d$

$m_{26} = c_1 \vee c_3 \vee c_4 \vee c_7 \vee c_8 \vee c_{10} \vee c_{11} \vee c_{12} \vee c_{13} \vee d$

$m_{27} = c_1 \vee c_3 \vee c_5 \vee c_6 \vee c_7 \vee c_8 \vee c_9 \vee c_{12} \vee c_{13} \vee d$

$m_{34} = c_3 \vee c_4 \vee c_5 \vee c_7 \vee c_{10} \vee c_{13} \vee d$

$m_{35} = c_1 \vee c_2 \vee c_3 \vee c_4 \vee c_5 \vee c_7 \vee c_8 \vee c_{10} \vee c_{11} \vee c_{12} \vee c_{13}$

$m_{36} = c_4 \vee c_5 \vee c_6 \vee c_7 \vee c_8 \vee c_{10} \vee c_{12} \vee c_{13} \vee d$

$m_{37} = c_3 \vee c_4 \vee c_5 \vee c_7 \vee c_8 \vee c_9 \vee c_{10} \vee c_{11} \vee c_{13} \vee d$

$m_{45} = c_1 \vee c_2 \vee c_8 \vee c_{11} \vee c_{12} \vee c_{13} \vee d$

$m_{46} = c_3 \vee c_4 \vee c_6 \vee c_7 \vee c_8 \vee c_{12} \vee d$

$m_{47} = c_5 \vee c_7 \vee c_8 \vee c_9 \vee c_{10} \vee c_{11}$

$m_{56} = c_1 \vee c_2 \vee c_3 \vee c_4 \vee c_6 \vee c_7 \vee c_8 \vee c_{11} \vee c_{12} \vee c_{13} \vee d$

$m_{57} = c_1 \vee c_2 \vee c_5 \vee c_7 \vee c_8 \vee c_9 \vee c_{10} \vee c_{12} \vee c_{13} \vee d$

$m_{67} = c_3 \vee c_4 \vee c_5 \vee c_6 \vee c_9 \vee c_{10} \vee c_{11} \vee c_{12} \vee d$

Based on heuristic reduction algorithm of resolution matrix, the importance of 6 attributes based on (3) is as:

$f(c_1) = 1.1009$ $f(c_2) = 0.7104$ $f(c_3) = 1.6405$ $f(c_4) = 1.1957$

$f(c_5) = 1.3433$ $f(c_6) = 1.0639$ $f(c_7) = 1.6088$ $f(c_8) = 1.8199$

$f(c_9) = 0.8242$ $f(c_{10}) = 1.4417$ $f(c_{11}) = 1.5882$ $f(c_{12}) = 1.6088$

$f(c_{12}) = 1.3866$ $f(d) = 2.0465$

So the importance order of attributes is:

$$d - c_8 - c_7 - c_{12} - c_3 - c_{11} - c_{10} - c_{13} - c_5 - c_4 - c_1 - c_6 - c_9 - c_2$$

However, in the classification sense, not all conditional attributes are necessary. The reduction of excess attributes will not affect result of program design. The reduction is defined as minimum condition attribute set that can ensure correct classification but not containing redundant attributes.

Based on pre-determined accuracy requirement of highway construction design program, $\alpha = 5\%$. Perform attribute reduction with (4) and the result is $\alpha = 3.34\% < 5\%$ after reduction two indexes, which can meet the accuracy needs. After index reduction, the original sample index has changes and learning efficiency has been improved.

5 Conclusion

The paper gave algorithm to extract typical case with rough set and introduced algorithm with rough set decision table. The advantages of rough set decision table reduction algorithm in case extraction was also provided. The experiment result based on some Highway and Bridge Design Institute proved that the algorithm has better capability to deal with noise compared with others classification algorithm. It also has higher classification accuracy to decreases dimension of information space and attributes number, which is an analytical method with fundamental sense in the knowledge representation system. Of course, the generated case number in the extraction is determined by threshold α and related to specific field knowledge, which is our research focus in the next future for threshold optimization and automatic settings.

References

1. Zhao, P., Ni, Z.-W., Jia, Z.-H.: Using Data Mining Technique to Establish Case Base from Weather Database. Computer Technology and Development 12, 67–70 (2002)
2. Qu, X.-L., Du, J., Sun, L.-F.: A Model for Cases Indexing in CBR System Based on AHP. Application Research of Computers 22, 33–34 (2005)
3. Pal, S.K., Mitra, P.: Case generation: a rough-fuzzy approach. In: Proceedings of Workshop Program at the 4th International Conference on Case-base Reasoning, pp. 236–242 (2001)
4. Chmielewaki, M.R., Grzymala-Busse, J.W.: Global discretization of continuous attributes as preprocessing for machine learning. International Journal of Approximate Reasoning 15, 319–331 (1996)
5. Shen, Y.-H., Wang, F.-X.: Attribute discretization method based on rough set theory and information entropy. Application Research of Computers 44, 221–224 (2008)
6. Ji, S., Yuan, S.-F.: Algorithm in Case Generation Based on Similarity Rough Set. Journal of Chinese Computer Systems 28, 1072–1075 (2007)
7. Beynon, M.: Reducts within the variable precision rough sets model: a further investigation. European Journal of Operational Research 134, 592–605 (2001)

Web Design and Analysis of Usability Issues and Recommendations

Ying Liu[1] and Zhen-ya Liu[2,*]

[1] Netease Proper Information Technology Company
[2] Jiangxi Institute of Education, Nanchang, Jiangxi, China 330029
liuzhenya@126.com

Abstract. This paper studies focused on web design, visual, interactive process of usability design frequently asked questions, for web design, visual, interactive design process, the availability of common problems were analyzed and compared, and gives a solution to the problem and recommendations on the web visual design, interaction usability design has a certain reference value.

Keywords: web design, usability, problem analysis, the proposed.

1 Text Question

1.1 Issues and Analysis

In the original system, there is no uniform text of the page, naming irregularities, giving users a semantic confusion, in violation of the consistency of usability problems. For example, the query in the original system in both records, there are records check that the menu structure. Meanwhile, in the navigation bar of this menu, sub menu names at all levels is also a lack of norms, for example, both query a certain item, but there are certain items query.Form the main navigation bar, there are two naming, noun + verb or verb-noun, for example, pre-existing card management and query name. Noun + verb simple way for the functionality of the page, such as the naming of the various functional modules, and specific to the specific operation of the process, then the appropriate form of the verb + noun, in order to enhance the image of the user a sense of action, given the psychological implications.

1.2 Proposal

The specific terms set up the page specification, unified name, individual professional term needs an explanation.

Navigation
To improve the system, according to a menu name menu, sub menu with the noun + verb, three sub-menus and the end of the operation, the use of verb + noun form.

* Corresponding author.
LIU Zhen-ya, Jiangxi Institute of Education, Nanchang, Jiangxi, China 330029.

D. Jin and S. Lin (Eds.): Advances in CSIE, Vol. 1, AISC 168, pp. 609–615.
springerlink.com © Springer-Verlag Berlin Heidelberg 2012

2 Log on Problems

2.1 Issues and Analysis

In the user survey indicated that most of the users that login page for the horizontal arrangement of four in the need to enter the verification code in English composition, this approach is not convenient and useful. Login button and the Cancel button in the center of the page, and the Cancel button usage is very low, the majority of user input error is rarely used when the Cancel button, but directly in the text box one by one to delete.

There are two main methods of Pages:

Page horizontally, track user's attention span, but a small proportion of page space occupied.

Vertical arrangement of the page, the user's line of sight path is a straight line, but the page occupy a large space.

The main form verification code:

The main role is to verify the code to prevent a malicious site attacks, improve system security. However, usability point of view, the verification code and the user's actions not directly related, are additional operations, thus needs to consider the design verification code.

Chinese validation code

Difficult to identify, and inconvenience the user to input

Alphanumeric verification code:

Safe, but also to users is difficult to identify, difficult issues such as input

Digital verification code:

Enter the simple, easily recognizable, and has a certain security

Contrast

Yahoo China's login page is very simple, only two input ID and password, using the horizontal layout of the login button at the bottom right corner, user moves the mouse right-hand operation habit. When the user input errors will be prompted to register and display a red error message in the input box top.

2.2 Proposal

Login page using vertical approach

Login page elements generally small, as the system only has two user name and password input. Therefore, the use of longitudinally way, in order to facilitate user input.

Remove validation code

The system is intended for business, an internal application site, so a lower security requirements, in order to simplify the process of user input, consider removing the verification code.

3 Issue of the Query Form

3.1 Issues and Analysis

Query page the original system (see Figure 1) uses a vertical arrangement of the way, the query options accounted for half of the page, this approach highlights the query

conditions, and reduce the query results, apparently more concerned about the user query results should be . After the user submits a query to get results, but because half of the Query Options page, each page after the operation the user must drag the scroll bar below the query results will be moved to the middle of the screen, increasing the burden on the user's operation, but also cause a bad user experience.
Check the layout of the page there are three main ways:
Query on the right side (see Figure 2)
The first layout of the query terms are more appropriate when placed in the left side of the page query terms, the right to place the results of tabular data for the query leaving sufficient space.
Query at the top (see Figure 2)
The second layout for the query item less when placed in the top of the page query, place the bottom of the query results. This approach will take up some space on the page, to a certain extent, weaken the concern of the user query results. So the query term is best between 4-6, if more than this range, consider adding more links to expand the query page.
Query using the label layout (see Figure 2)
The third layout for the query of the more simple time, such as classification inquiries. This approach eliminates the need for user input in the process, simplify the operation process, but also reduced the query functionality.

3.2 Proposal

Business support systems for the main field of inquiry, such as card number, name, time, query items less. Therefore, using the second layout, the query conditions placed on the top of the page, to facilitate the user's actions, while the impact of the results less attention. (See Figure 3)

Application name : Please input
Product name : Wanshang card
The start card number : - -
Number : 1
nominal value : US $ To US $
Expiration date : Month To Month
Application instructions : Please input

.Query Cancel

Application name	Product name	start card number	Number	nominal value
xxxxxxxxx	xxxxxxxxx			

Fig. 1. Query form

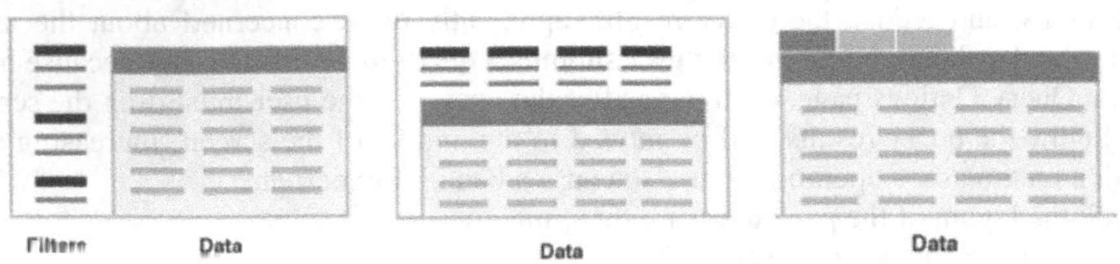

Fig. 2. The three structures the query form

Application name : Please input
nominal value : US$ To US$
Product name : Wanshang card
Expiration date : month To month
start card number : - -
Application instructions : Please input
Number : 1
Query
End of card number : xxxx-xxxx-xxxxxxxx

Application name	Product name	start card number	Number.	nominal value	application time	Applicant
xxxxxxxxx	xxxxxxxxx				xxxxxxxxx	xxxxxxxx:

Fig. 3. The modified query form

4 Page Issue

4.1 Issues and Analysis

The page number for the original system drop-down box, shown in Figure 4, users select the appropriate drop-down box after the page number to jump to the corresponding page. This page allows users to design in the form of the number of pages at a glance, easy to jump quickly. However, this method is not suitable for pages more page, when more than 10 pages when the drop-down list becomes too long, likely to cause the user's misuse, and users need to scroll to see all page numbers, is not conducive to choice. Use the drop-down list can not remember the way the user has visited the page, the user survey results found that page when the user selects the misuse rate is very high.

Fig. 4. Page

Contrast
Google (see Figure 5)
Google's page design is very creative, top of the large font of Google, in the middle with o instead of page numbers, giving users a new visual experience, while the larger picture also easy for users to click. Google's page provides links to previous and next, while the red shows the location of the current page. But for visited pages, Google does not give the corresponding color change, when users visit a lot of information, is not conducive to the user memory of filtered information.
Sina (see Figure 5)
Sina page design in the upper left corner shows the message number, upper right corner of the page gives the traditional and the current popular use of Ajax technology, users can drag the scroll bar to select the page number, when the mouse cursor will Displays the current page.

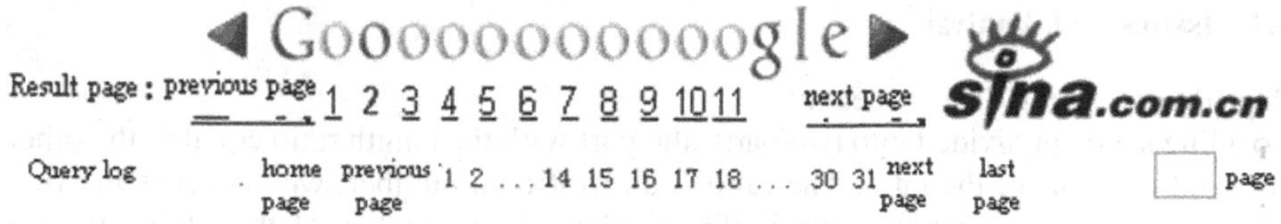

Fig. 5. Google and Sina page

4.2 Proposal (See Figure 6)

Need to provide the "Previous" and "Next" link, and designed according to the state distinction. When the user is located when the first page, do not provide links to the previous page, the same token, if the user is in the last page, no links to display the next page;

Highlight the current page number, give the user a clear display, and to distinguish between visited and did not show visited pages.

Highlight the current page number, and try to make the page easier to click (do not tightly close together), all the pages at once and do not show up;

Show the "first page" and "last page" link, under the circumstances to consider given page number to jump box;

Query log home page previous page 1 2 ... 14 15 16 17 18 ... 30 31 next page last page page

Fig. 6. Revised page

5 Control Problems

5.1 Issues and Analysis

No initial value of radio button group
The initial state of radio button group has not been selected items. User may not select this radio button group, there is no default value, then increases the chance of user error.

Error-prone input text box
There are strict requirements for the format of the table item, use the text box to accept user input, such as time. Authority over an open text box is likely to cause the user input errors. What is the user input data format should be designed by the developer, rather than requiring the user in form of input.

5.2 Proposal

Initial value for the radio button set up, the time input box to select the calendar form, this method reduces the user's error rate and improve the operational efficiency and improve the user experience.

6 Layout Problem

6.1 Issues and Analysis

Page disproportion, lack of beauty
To a line segment divided into two parts, the part with the length ratio equal to the other part and this part of the ratio. The ratio is an irrational number, whichever is the first three digits of the approximation is 0.618. Meet the aesthetics of this style, gives a feeling of harmony. Golden Section in all walks of life at home and abroad are widely used in Web pages as well.
Golden Section
The original page will be important things to put in the center, said the emphasis on trying to draw users' attention. However, research shows that through the user, the center part of the user's blind spot area. That is according to human visual habits is in the center is often overlooked. And placed in the center will not lead to visual stimulation. According to the golden ratio study, the visual center should be located as shown on page four points, (see Figure 7) in the layout design process, the need to take this into account.

6.2 Proposal

Large layout
Basically, the layout is divided into pages, on the lower, left, right, left and right on the mixed types. Currently, most computers are more than 1024 * 768 resolution, between the page and leave some blank display, the ratio is closer to the golden ratio. Taking into account the system's page content, use of the structure on the left and right, and use the golden ratio to be divided. Left 40 percent, accounting for about 60 percent of the middle, and above 31%.

Fig. 7. Jiugong Figure

Small layout

This system, the upper navigation bar to the left for the shortcuts, the right is the system the main content area. Based on the above study can know the contents of the core should not be placed on top, end, or the middle, top and middle is the middle position, while from a usability point of view, add a small navigation bar prompts the user to the current in which the page position.

References

1. Cooper, A.: About Face 2.0, America, p. 11 (2003)
2. Price, J.: Interaction Design beyond Human-Computer Interaction, pp. 10–13. John Wiley & Sons, America (2003)
3. Nielsen, J.: Usability Engineering. Academic press, America (1993)
4. Chewer, C.M.: User-Centered Critical Parameters for Design Specification, Evaluation, and Reuse (2005)
5. Herb, C.L.: Website usability evaluation using sequential analysis (2005)
6. Livingston, A.: A comparative analysis of style of user interface look and feel in a synchronous computer supported cooperative work environment (2005)
7. Beatnik, G.: Automatic web usability evaluation: what needs to be done? (DB/OL) (2003), `http://www.tri.sbc.com/hfweb/brajnik/hfweb-brajnik/hfweb-brajnik.html`

Design of Digital Signal Transmission Performance Analyzer

Yanyi Xu, Ming Yin, and Jianlin Xu

Department of Electronic Engineering
Naval University of Engineering
Wuhan, China
kevinyinming@yahoo.com.cn

Abstract. The system consists of digital signal producing module, low pass filtering module, pseudo random signal producing module and digital signal analysis module. Pseudo random signals and digital signal analysis module are produced by the FPGA. Digital signal analysis module consists of active filter, level conversion module .This system perfectly completed the digital signal extraction synchronous signal. In the signal-to-noise ratio (SNR)-9.54 dB, the system also can regenerate the signal information, and show the eye chart. Through the oscilloscope we can get the signals eye chart and analysis of the digital signal transmission performance.

Keywords: Synchronized signal extraction, FPGA, VHDL, phase locked loop.

1 Introduction

Currently there is an increasing interest in developing and applying monitoring system for electrical equipment like transformers, Generator and Fire Control system[1].The monitoring system has the potential to diagnose failure and verify the design.

During the development stages of Monitoring system , an ancient control and guide equipment was used. Although it is out of commission, some function can be custom made. In this experimental scenario the behavior models were designed and validete. They were fed with the need of all data.

This paper focuses in hardware and software description of the Data Acquisition based on Virtual Instrument, developed. Although, a brief description of the complete architecture of monitoring system is also done. It is responsible for acquisition ,processing and storing of measurement. The System is being monitored for a certain control and guide equipment with satisfactory results[2].

2 System Architecture

This system based on FPGA technology, producing an M series signal and an Manchester yards signal which bit rates are from 10 to 100 Kpbs step for 10 Kpbs. Another M series pseudo-random signal(using for simulation of the noise) is bit rate

D. Jin and S. Lin (Eds.): Advances in CSIE, Vol. 1, AISC 168, pp. 617–622.
springerlink.com © Springer-Verlag Berlin Heidelberg 2012

for 10 Mpbs . Coding signal which is through low-pass filter (analog transmission channel characteristics) added with the simulated noise. The eye pattern of M sequence is displayed in an oscilloscope , and the range of it is shown in LCD [3].The system can get the synchronized signal of Manchester yards in the signal-to-noise ratio-13 db using Phase Locked Loop technique . The test shows that the system works steadily, and can get good performance. Figure 1 illustrates this system chart [4].

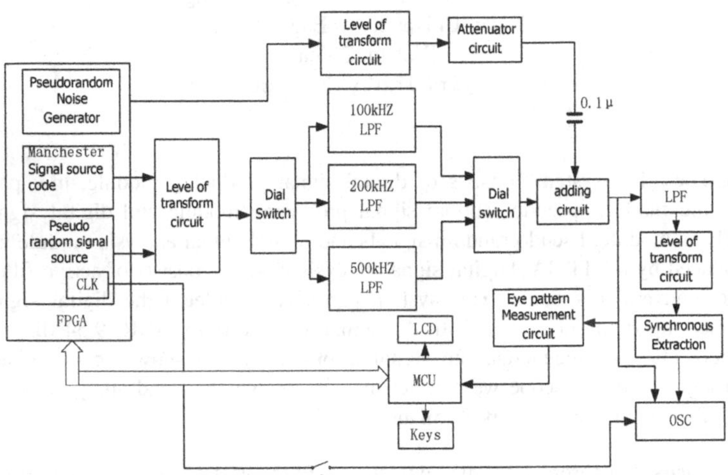

Fig. 1. System Chart

3 Hardware of System

The system developed uses the hardware described previously and it is responsible for making the acquisition, processing and storage of measured data. It consists of the following.

3.1 Low-Pass-Filter

The system contains three low-pass filters .The filter contains capacitances, resistances and operational amplifiers. The cut-off frequencies of three low-pass filters are 100 KHz, 200 KHz and 500 KHz. The frequencies which are 100 KHz and 200 KHz are not beyond the cut-off frequency of NE5532 chip. So the low-pass filters whose cut-off frequency are 100kHz and 200kHz are designed with the NE5532 chip.The system used variable gain amplifier (VGA) to change the gain of the low-pass filters .The VGA is built by TL081.The low-pass filter whose cut-off frequency is 500KHz used AD8039 chip whose cut-off frequency is 50MHz as the amplifier .The low-pass filter can get attenuation that is 40 db per ten times frequency processes .So the system used four orders low-pass filter whose attenuation is theoretically 80 dB per ten times frequency process . The low-pass filter circuit diagram is shown in figure 2.

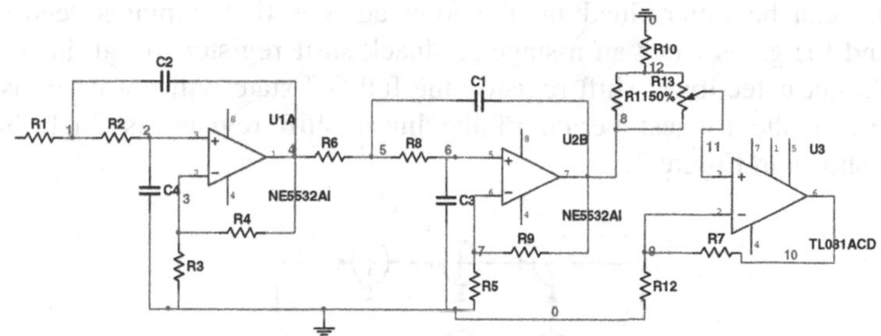

Fig. 2. Low-pass Filter Circuit Diagram

The values of the resistances and capacitances are different from what we calculate, so we change the values of resistances and capacitances in the circuit diagram and test them .The parameters of circuit diagram after adjusting is shown in the Table 1.

Table 1. The Parameters of circuit diagram.

Fc (KHz)	R (KΩ)					
	R1	R2	R3	R4	R5	R6
100	2	2	10	6.2	10	1.67
200	1	1	10	6.2	10	0.91
500	0.47	0.47	10	6.2	10	0.24
Fc (KHz)	R (KΩ)		C (pF)			
	R8	R9	C1	C2	C3	C4
100	1.67	6.2	680	680	680	680
200	0.91	6.2	680	680	680	680
500	0.24	6.2	470	470	470	470

3.2 M Series Digital Signals

The signal source and the simulation signal noise all use the M series produce false random signal, the M series also called maximum length pseudo-random binary sequences, actually it is a special kind of cycle sequence, its value is only 0 or 1 . The characteristic equation of M sequence is expressed as:

$$f(x) = C_0 + C_1 D + \cdots + C_n D_n = \sum_{i=0}^{n} C_i D_i \tag{1}$$

M-sequence can be determined by the loop counter that contains feedback shift registers and the gates . For an n-stage feedback shift register ,it can have up to 2n states, for a linear feedback shift register, the full "0" state will not be transferred to other states, so the longest period of the linear shift register is 2n-1. Schematic diagram is shown in Figure 3.

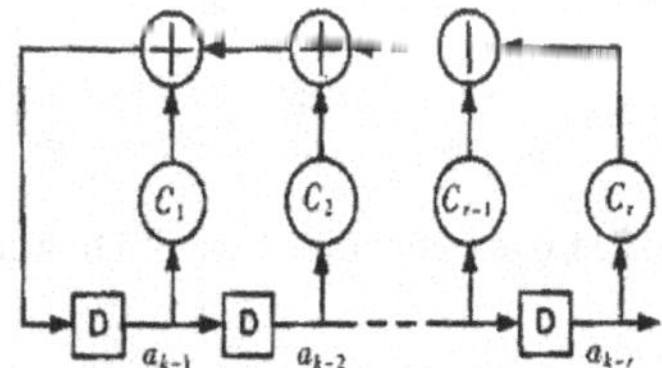

Fig. 3. M-sequence Schematic Diagram

3.3 Manchester Coding

Manchester code used to indicate the edge of 0 or 1, with the rising edge of the system on behalf of 0, with the falling edge of the representation 1. Manchester of the M coding sequence diagram is shown in Figure 4. Clock and its inverted signal were added to the 2-to-1 input, M sequence is used as the selection control signal . when the M-sequence is 1, the output is the clock falling edge, or the output is the clock rising edge.

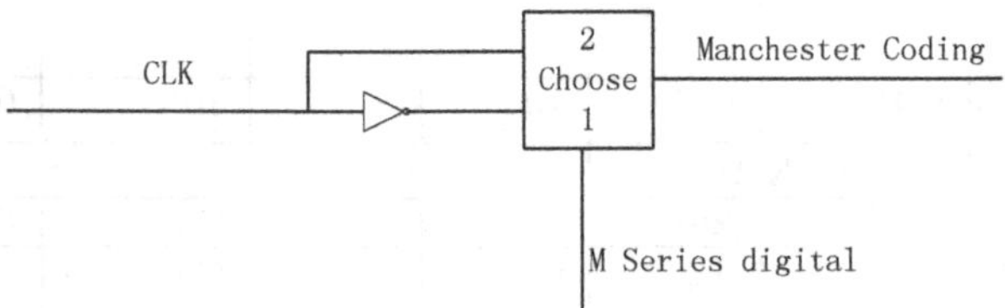

Fig. 4. Manchester sequence diagram

3.4 The Synchronization Signal Extraction

Digital phase-locked loop includes a local oscillator, additional doors, excluding doors, N divider and phase comparator.

Assume that the frequency of input signal is f ,and the output clocks that have two phase in the loop are Nf . And the difference between two phase is π .Two clocks are added after pass the attached gate and additional gate. Then clock enters the N divider. The relationship between the signal from N divider (f_{out}) and the input signal can be divided into the same phase ,lead or lag .

The principles of the output signal that controls the adjusting relationship of phase are shown below. When the phase of the input signal is same with the clock .The frequency of the clock signal which is divided by N is f .When the input signal phase advances, the control reduced doors open .An input pulse counter that is put into the N divider is reduced .The result is that N divider is countered less a pulse than before . The N divider output phase lag 2π / N .When the phase of input signal lags, the control additional doors open. A pulse is added to the N divider ,and the result is that the phase of N divider output is advanced 2π / N. With the constant of external input signal, the loop is ended after several adjustments to establish synchronization[5].

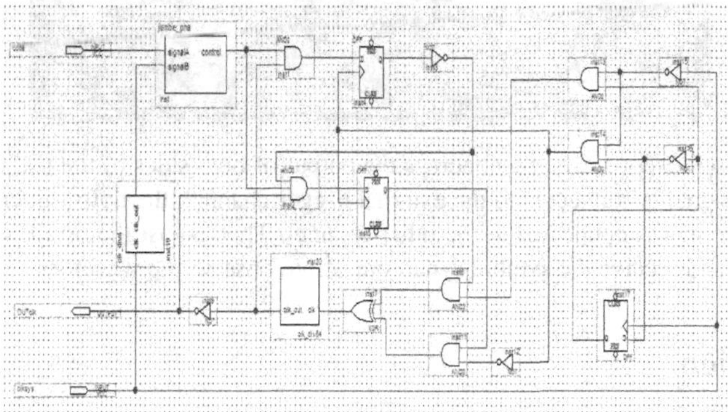

Fig. 5. M-sequence Schematic Diagram

4 Test Methods and Data

4.1 Low-Pass-Filter Circuit Test

Table 2. The test data of low-pass-filter.

Display (kbps)	Output (kHz)	Display (kbps)	Output (kHz)
10	10.00	20	19.99
30	29.99	40	40.04
50	50.01	60	59.98
70	70.00	80	80.03
90	90.03	100	100.01

Using calibrated signal source generates a sine wave add to the signal terminal. Inner 100KHz ~ 500KHz frequency band, with a certain frequency step setting test points, test points at each frequency to 3dB for the step from 0dB to 20dB test seven sets of data. The test data of low-pass-filter circuit diagram is shown in the table 2.

4.2 Eye-Chart Test

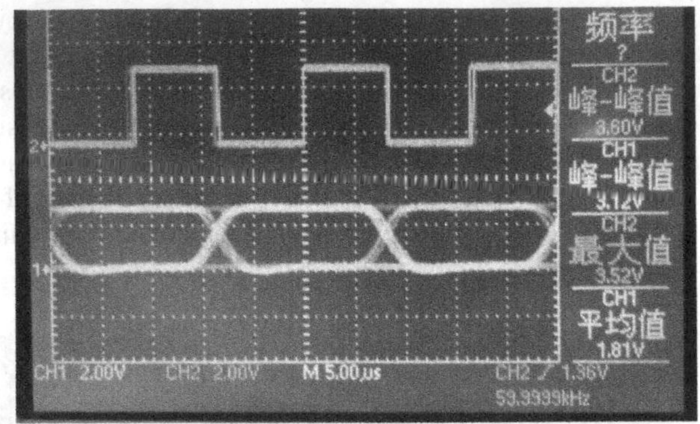

Fig. 6. Use an oscilloscope to connect across the output of both sides of the test circuit. And the extracted clock signal is connected to the external trigger channel; the noise superimposed on the signal output terminal is connected to the other channel. Then adjust the oscilloscope horizontal scanning period and receive the symbol synchronization, observe the waveform.

5 Conclusion

The system's digital signal generator can get the signal whose data rate is from 10 kpbs to 100 kpbs and the 10 kpbs bit rate for the step changing . The absolute error is less than 0.1%. Pseudo-random signal generator can achieve a data rate of 10M pbs signal generation. Low-pass filter can get good signal of the filter whose cutoff frequency are 100kHz,200kHz,500kHz. Digital signal analysis circuits can get the eye-chart when the Signal to Noise Ratio is 0.2 .%.

References

1. Gao, Y.: The design and simulation methods of switched capacitor Filter. Journal of Jilin University, 373–377 (2005)
2. Zhou, R.: The Design and simulation of high-end switched filter, technology of Electronic Measurement, vol. 2, pp. 22–25. Academic, Beijing (2010)
3. Zhang, X.: Design of programmable filter. Technological Development of Enterprise 1-2, 243–246 (2009)
4. Wang, H.: Inspecting FSK signal based on MCU and FPGA. Instrument Technique and Sensor 10, 12–17 (2001)
5. Liao, Y.: Design of programmable filter. Electronic Design Engineering 17, 50–53 (2009)

Using Variable Precision Rough Set Model to Build FP-Tree of Association Rules

SuJuan Qian[1,*] and ZhiQiang Zhang[2]

[1] Zhengzhou Vocational College of Economics and Trade, Zhengzhou, 450006, China
[2] Henan Occupation Technical College, Zhengzhou, 450046, China
zhangzhiqiang70@sina.com

Abstract. The main mission of the variable precision rough set is to solve the problem of non-functional of properties or the classification in uncertain relationship data. FP-TREE of Association rule is one of the important research areas in data mining. Its goal is to discover previously unknown, interesting relationships among attributes from large databases. This paper presents FP-tree of association rules algorithms based on variable precision rough set model in e-commerce. The experiments show the algorithm could flexibly and the experimental results indicate that this method has great promise.

Keywords: association rule, variable precision rough set model, data mining.

1 Introduction

Data Mining: Data mining should firstly determine what is the task or goal about mining, such as data summary, classification, clustering, association rules or sequential patterns discovery etc. After determining the mining task, we must decide what kind of mining algorithms can be used; mining algorithm is the core of data mining. Rough set theory from birth to now only about twenty years, but has been applied to machine learning, pattern recognition, knowledge discovery, decision analysis and process control and other fields. At present, the variable precision rough set model has become an academic hot spot in the field of artificial intelligence research, it has been successfully used in medical diagnosis, engineering fault diagnosis, telecommunications, community planning and global climate research and other fields [1]. The current research of variable precision rough set model is mainly in theory and application two aspects.

High efficient reducing algorithm: To identify all the reduction or optimal reduction of information systems is a NP problem, so now the study focused on the reduction heuristic, parallel algorithm, incremental algorithm of derived rules and so on. But the knowledge acquisition of information systems based on variable precision rough set also has no systematic research, such as classification of the variable precision rough set, attribute reduction, decision-making rules and other aspects of theoretical research and application is not mature enough, which need further

---

* Author Introduce: SuJuan Qian (1980), Female, lecturer, Master, Zhengzhou Vocational College of Economics and Trade, Research area: web mining, data mining.

D. Jin and S. Lin (Eds.): Advances in CSIE, Vol. 1, AISC 168, pp. 623–628.
springerlink.com © Springer-Verlag Berlin Heidelberg 2012

research. Classification and decision-making rules for research: In the rough set data analysis, a group of rough decision-making rules can be formed by rough approximation through decision-making table, which involves consistency rules, but also contains the inconsistency rules, this decision-making rules can be used to accurately classify and forecast the new data.

The expansion model of variable precision rough set: The expansion model established on variable precision rough set model which is available to deal with the mechanism of incomplete information and the noise, such as the variable precision rough set model based on similarity.Association rules is a data mining research is an important research topic. Agrawal, Imieliski and Swami in 1993 for market basket analysis problem first proposed the concept of association rule; its purpose is to find the transaction database, the link between the rules of different commodities.

This article will focus on the non-candidate generation algorithm on the set, the analysis summarizes the results of previous studies, and the paper proposes FP-tree algorithm based on variable precision rough set model for mining maximal frequent item sets and frequent closed item sets mining algorithms.FP-tree of association rules algorithms based on variable precision rough set model is proposed based on integrating of variable precision rough set and data mining. From the algorithm analysis and experimental comparison shows that: For intensive data, Unid_FP-Max algorithm in time and space overhead is less than FP-Max algorithm.

2 The Research of Association Rules Based on FP-Tree Technology

Association rules defined as follows: Let $I = \{i_1, i_2, ..., i_m\}$ is the collection of data items (itemset), which is called the elements of the project (item). Transaction database D is a transaction (transaction) T set, where transaction T is a collection of items, namely: $T \subseteq I$. Corresponding to each transaction has a unique identifier, denoted by TID. Let A be an I in the collection of items, if$A \subseteq T$, then said transaction T contains A.

An association rule is the implication of the form $A \Rightarrow B$style, here$A \subseteq I$, $B \subseteq I$, and $A \cap B = \varnothing$ Rules $A \Rightarrow B$ in the transaction database D, the degree of support (support) that these two items A and B set in the transaction database D, the probability of simultaneously[2]. Denoted as support ($A \Rightarrow B$), that support($A \Rightarrow B$)=P($A \cup B$) Rules $A \Rightarrow B$in the transaction database D, the confidence (confidence) is there in the affairs of the itemset A centralized, key set B also probability, denoted by confidence($A \Rightarrow B$), the confidence($A \Rightarrow B$)=P($B \mid A$).

Association rule mining problem is to find the transaction database D with a user-given minimum support and minimum confidence min_conf min_sup association rules. FP-growth algorithm for mining frequent patterns effectively opened up a new way. Fan et al proposed a modified FP-tree structure and a sub-tree based constrained algorithm for mining frequent item sets [3]. This will be the improved FP-tree is called a one-way FP-tree, because of this new type of tree structure to mining frequent itemsets provides a new idea, it deserves careful study.

Definition 1. Frequent pattern tree (frequent pattern tree) referred to as FP-tree (FP-tree), is one of the following conditions tree.

For the FP-tree contains a node on the item ,α there will be a path from the root to reach α, α is not included in the path where the node part of the path prefix sub-path is called α (prefix α subpart), is called the path of the suffix.

Algorithm 1: FP-tree construction

Input: transaction database D and the minimum support threshold min_sup.

Output: D corresponding to the FP-tree.

Step1: Scanning the transaction database D, containing a D in all the frequent items, F_1, and their respective support. Frequent items of F_1in descending order according to their degree of support to be L;

Step2: Create the root of FP-tree T, the "null" tag. Again scanning the transaction database. For each transaction in D Trans, to elect one of the frequent items sorted according to the order in L. Let the sorted frequent item table $[p \mid P]$, where p is the first frequent item, and P is the remaining frequent items.

Algorithm 2 (FP-growth): frequent pattern growth.

Input: Obtained by the algorithm FP-tree and the minimum support threshold min_sup.

Output: the set of all frequent itemsets.

Methods: call FP-growth (FP-tree, null)

(1) **for** $i = k_{m-1}$ **down to** 1 **do**{

(2) **if** ($ST(k_1, \cdots, k_m)$.*count* [*i*] > =*min_count*) {

(3) *FP*[++ *length*] = *item* (*i*) ;

(4) output *FP* and its support $ST(k_1, \cdots, k_m)$.*count* [*i*]/ *n* ;

(5) build $ST(i)$ based on Unid_FP-tree ;

(6) **if** (there is an non-root node in $ST(i)$)

(7) *mine* ($ST(i)$) ;

(8) *length* - - ;

Theorem 2 for a given transaction database D, minimum support min_sup, algorithm 2 can correctly generate all frequent patterns.

Proof: to be evidence for a given transaction database D and the minimum support threshold, the algorithm output models are frequent, and all the frequent itemsets are generated algorithm correctly.Set $pattern = \{ i_1, i_2, \cdots, i_k \}$is any frequent pattern, its support count is c , i_1 , i_2 , $\cdots$, $i_k$number wereo_1 , o_2 , $\cdots$, o_k. First use, the length of the pattern of induction to prove that when k> 1 时, $ST(o_k, o_{k-1}, \cdots, o_2)$algorithm will be constructed, and $ST(o_k, o_{k-1}, \cdots, o_2)$.

3 Variable Precision Rough Set Model

The reduction of redundancy knowledge is an important part of the rough set theory, due to rough set itself, it is superior to other methods in the knowledge reduction, but the definition of the reduction in the rough set model depends on lower approximation, and the lower approximate calculation is very sensitive to noise data, this results that the attribute reduction greatly influenced by the noise data, further result to many valuable rules can not be extracted, under the system with inconsistent information, the practical application have also encountered difficulties[4]. Therefore, in order to better handle the noise data, increase the necessary redundancy to the attribute reduction process, Ziarko, who proposed a variable precision rough set (Variable Precision Rough Set, VPRS) model, greatly improving the coverage and generalization ability of extraction of rules, to better reflect the data correlation of the data analysis, so as to laid the foundations for obtaining similar decision-making rules.

Set $S=(U,A)$ is an information system, P、$Q \subseteq A$ are respectively the condition attribute set and decision-making attribute set. IND(P), IIND(Q) present the non-distinguish relationship divided by P, Q, the collection of equivalence classes of IND(P) called condition class, represented by U/P, the collection of equivalence classes called decision-making class, represented by U/Q.

Set $\{U, R\}$ is the approximation space, U which is the domain, R is the equivalence relationship on U, $X \subseteq U$ then X about R's lower approximation (Lower Approximation) is defined as 1.

The definition of dependence$\beta$of decision-making attribute set $Q$and condition attribute set P is

$$pos(P,Q,\beta) = \cup_{Y \in U/Q} \underline{IND(P)}_{\beta} Y \tag{1}$$

Similar dependence is the extent of roughs dependence idea, when β=1, it becomes rough dependence. The measure of similar dependence is the evaluation of implementing the classification capability of the object which owns the classification error.

Theorem 2.1: If the condition equivalence classe X is distinguishable on $0.5 < \beta \leq 1.0$, then X is distinguishable on $0.5 < \beta_1 < \beta$.

Theorem 2.2: If the condition equivalence classe X is undistinguishable on $0.5 < \beta \leq 1.0$, then X is undistinguishable on $0.5 < \beta_1 < \beta$

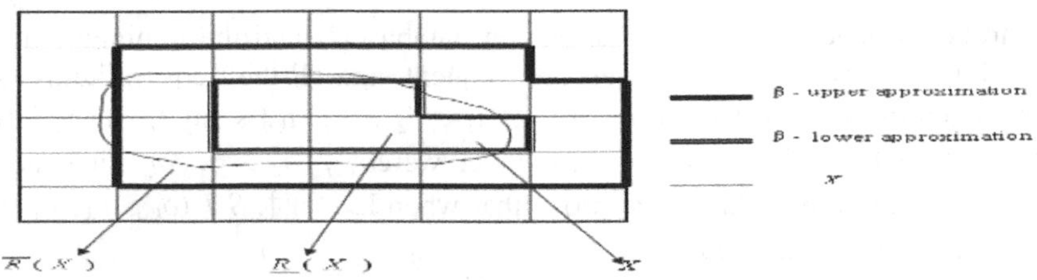

Fig. 1. Sketch map of variable precision rough set.

The upper (lower) distribution consistent set B is the attribute which can maintain every decision-making's the upper (lower) approximation unchanging[5]. It is compatible with the proposition rule generated by C, that is, in the original system and reduction system, the decision-making part of the proposition rules which produced by the same object is same.

4 FP-Tree of Association Rules Algorithms Based on Variable Precision Rough Set Model

FP-Max algorithm is a very effective FP-tree based frequent itemsets mining maximum depth-first algorithm. However, the algorithm is recursive in the mining process requires the generation of a large number of conditional pattern tree, the time and space efficiency is still not high enough.

FP-Max algorithm using divide and conquer idea, after the second pass scanning, the database of all thc frequent information compression is stored in the FP-tree, then then FP-tree is divided into a number of conditional pattern base, and then excavation of these libraries were.

Algorithm 3 (FP-Max): FP-tree based mining maximal frequent itemsets.
Input: the algorithm FP-tree and the minimum support threshold min_sup.
Output: All frequent itemsets MFI.
Method: Call FP-Max (FP-tree, null, MFI).

```
for i=k-1 down to 2 do{
  if i* then{
  if S T (X) .count[i]≥min_count {
    build S T (X∪i) based on S T (X) ;
    if there is an non-root node in S T (X∪i) and S T (X∪i).count[ ]≥min_count
    while (i *){ X=X∪i ;
      i-- ; }
    if S T (X).count[i]≥min_count and I(X∪i) then{
      build S T (X∪i) based on S T (X) ;
      if there is an non-root node in S T (X∪i)
      then call Unid_FP-FCI(S T (X∪i), FCI)
      else I(X∪i) FCI
```

If the minimum support count min_count = 2, scan D, get frequent 1 - itemset, its support count in descending order by to get items - order conversion table. For convenience, in the one-way constraint FP-tree and the sub-tree using serial number on behalf of the project. FP-tree generated by one-way as shown in Figure 1.

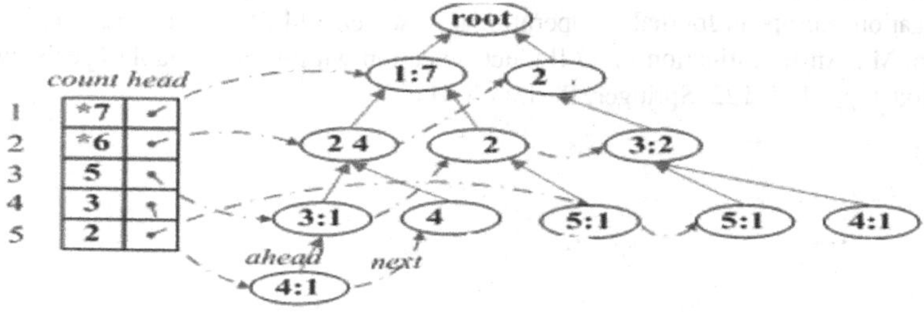

Fig. 2. FP-tree generated by one-way.

Case analyses give a complete calculation process, and proved effective through demonstrating the case. Experimental results show that: the algorithm ensures the compatibility of the proposition rules, and because of the appropriate choice of β value, it significantly increases the credibility of the rules compared with the similar algorithms. This method has certain fault tolerance, can mine a certain degree of consistent decision-making rules from the inconsistent data, to overcome the shortcomings of the standard rough set model is too sensitive to noise data, thereby enhancing the data analysis and robust of processing.

5 Summary

This paper presents FP-tree of association rules algorithms based on variable precision rough set model in e-commerce, analyzes the limitation of the reduction methods based on variable precision rough set theory. Experimental results show that the algorithm ensured the compatibility of the proposition rules, and because of the appropriate given value of β, it significantly increases the credibility of the rule compared with similar algorithms. Design based on one-way FP-tree Unid_FP-FCI algorithm. Preliminary analysis shows that Unid_FP-FCI algorithm is high time and space efficient than CLOSET algorithm. The experiment result shows that the proposed algorithm can improve classification efficiency of the rules than ID3, so it has some practical value.

Acknowledgement. This paper is supported by Education Department of Henan Province, 2011 Natural Science Research Program (2011C520019).

References

1. Mi, J.-S., Wu, W.-Z., Zhang, W.-X.: Approaches to knowledge reduction based on variable precision rough set model. Information Sciences 159, 255–272 (2004)
2. Marcus, A., Maletic, J.I., Lin, K.: Ordinal Association Rules for Error Identification in Data Sets. In: CIKM, pp. 589–591 (2001)
3. Grahne, G., Zhu, J.: Efficiently using prefix-trees in mining frequent itemsets. In: IEEE ICDM 2003 Workshop FIMI 2003 (November 2003)
4. Beynon, M.: Reduces within the variable precision rough sets model: A further investigation. European Journal of Operational Research 134(3), 592–605 (2001)
5. Beynon, M.: An Investigation of -β Reduct Selection within the Variable Precision Rough Sets Model, pp. 114–122. Springer, Berlin (2001)

Design and Implementation of a GIS-Based Management Information System to Protect Suzhou Prime Cropland Preservation

Jiang Ye[1], Cheng Hu[1,*], Gang Chen[1], and Yang Yang[2]

[1] School of Environmental Studies, China University of Geosciences, Wuhan, 430074, China
hu_cheng@cug.edu.cn
[2] College of Information Engineering, China University of Geosciences, Wuhan, 430074, China

Abstract. Managing and protecting prime cropland requires the integration of a large amount of high-quality data from a variety of sources. Due to the limitations of labour-consuming and low-efficiency in cropland protecting processes implementation by traditional approaches, an integrated management information system of crucial importance when conducting a successful protection policy. This paper represents the methodology of design and development of Suzhou prime cultivated land management information system using ArcGIS Developer Kit. A mixed mode of C/S and B/S was used to design system framework. The B/S module (a internet GIS) , mainly for public users, aimed at the land map publishing and general data query, moreover, the C/S part supply the specialized users with the functionality of data management and professional analysis. The two different subsystems separately connected to the same geodatabase through ArcSDE engine. The database was established in a remote server to ensure the consistency of the data and provide the subordinate access through the special data line on the Internet. Some key functions realization were described detailedly including the map publishing using ArcServer, the dynamic management of occupy tocsin, the intelligent land complementing, and the historical backtracking and browse based on the ArcSDE version management technology.

Keywords: ArcServer, ArcEngine, Prime Cropland Preservation, Information System, Suzhou.

1 Introduction

The prime cropland, providing sufficient agricultural products for the increasing population, is vital to the whole national economy development and strategic safety, therefore, prime cultivated land preservation is one of the crucial country policies [1]. Owing to the limitations of labour-consuming and low-efficiency, the traditional implement process of cropland managing model has been challenged. With the rapid development of the computer technology, data communication and information

* Corresponding author.

D. Jin and S. Lin (Eds.): Advances in CSIE, Vol. 1, AISC 168, pp. 629–636.
springerlink.com © Springer-Verlag Berlin Heidelberg 2012

technology in GIS, it has been thought highly of using specialized information system to enhance land management level in recent years [2].

The land resources information system research has been improved evidently in china, especially after 90s. During this period, some application projects were built successfully, however, the research and development of geographic information system in the prime cropland preservation hasn't been attracted enough attention. Some running information systems have no necessary strategies, including the distributed system configuration and ranked operation authority, which usually resulted in the unavailable capacities, such as rapid data inquiry and retrieval, intelligent statistic analysis and dynamic monitoring [3].

Due to the present problems mentioned above, we can introduce ArcGIS technologies into the routine cropland managing tasks, especially the spatial-information management function. It is of help for us to manage, maintain and update the prime cropland space information effectively. ArcSDE can be applied for the mass data management in prime cropland protection; ArcEngine can be used for the design and realization of specialized analysis; ArcServer can support map publishing to users.

Under the complicated specific circumstances of Suzhou prime cropland protection, aiming at the final purpose to realize scientific and automatic prime cropland information management, our research established the Suzhou prime cropland protection management information system to provide the technical support for the dynamic management of principle farmland information.

2 System Framework Design

2.1 System Construction Tasks

1) Integration of the Prime Cropland Data

Successful management of cropland resources is a very complex task; the main challenge is that information related to the cropland is usually accessible in so many ways. Furthermore, the acquisition of cropland data is costly. The numerous means used for collecting data, e.g., geophysical investigations, soil sampling, and irrigation water chemistry analyses, are expensive although necessary for the characterization of cropland quality situation. Hence, the proper recording of these data and ensuring their availability within a unified spatial database become critical for users and maximize the utility of cost-intensive datasets. The cropland data of numerous amount and multi sources were integrated into a group under the guidance of the national data standards. Accessing a spatial database may also assist in the future cultivation planning through the effective use of all preexisting information. Furthermore, databases allow officials and practitioners to have easy access to a large pool of data from which they can draw upon for research or industrial projects.

2) Construction of Prime Cropland Database

As the computational efficiencies of computers and software increase, larger amounts of data may be recorded, treated, and then coupled with GIS software for rapid georeferencing and spatial representation of the data. The principle cropland data was organized into a spatial RDBMS database.

Entity-Relation Diagram analysis method was used to express the relationship between the objects. Furthermore, the data table structure was defined on RDBMS platform according to the national cropland data standards.

3) Development of Prime Cropland Management Information System
The goal of this research is to develop an efficient and intelligent prime cropland management information system. It will be convenient for users to search data, carry out daily business and professional analysis function, so as to improve the utilization degree and use value of prime cropland management information database.

2.2 Data Organization and Management

The system database design is based on the second national land survey technique rules and "TD/T 1019-2009 prime cultivated land database standard".

According to user demands, the involved data was divided into several parts by administrative region, which will be convenient for storage and management. We set up the working area by the county as a unit, each country area of data storage [4].

The system used ArcSDE to connect relational database Oracle10G for storage of mass data. ArcSDE, as the spatial data engine of ArcGIS, is the main users' spatial database access of storage and management in the relational database management system (RDBMS). From the view of spatial data management, ArcSDE is a continuous spatial data model. With this model, we can realize the spatial database management through RDBMS. Blend spatial data in RDBMS, ArcSDE can provide high efficiency service on the spatial and non-spatial data [5].

2.3 System Architecture Design

Based on traditional GIS technology, the prime cropland management information system development usually used B/S mode or C/S mode. These two kinds of model have their advantages and disadvantages.

The subsystem based on C/S mode, for specialized users, was run in Intranet with characteristics of high speed, powerful functionality and tight security, and the other subsystem based on B/S was designed for public users with less functionality in Internet and Browse environment.

In the C/S mode application system, GIS function will be relatively more perfect, but due to restrictions such as security and network factors, data sharing and publishing became difficulty. BS model application system use the Internet Explore to implement GIS function, which is relative convenient. However, in the development of GIS browser to realize senior function is very difficulty [6].

In order to meet the determinate demands of the application system, our system architecture design using C/S and B/S mixed mode, illustrated as the following figure 1.

Among them, the releasing and inquiring functionality of Suzhou city's prime cropland information used B/S mode, which will be the convenient for common users' supervision. While for professional function such as editing maps and statistical analysis, we use C/S mode for the professional management staff in the local area network for internal users.

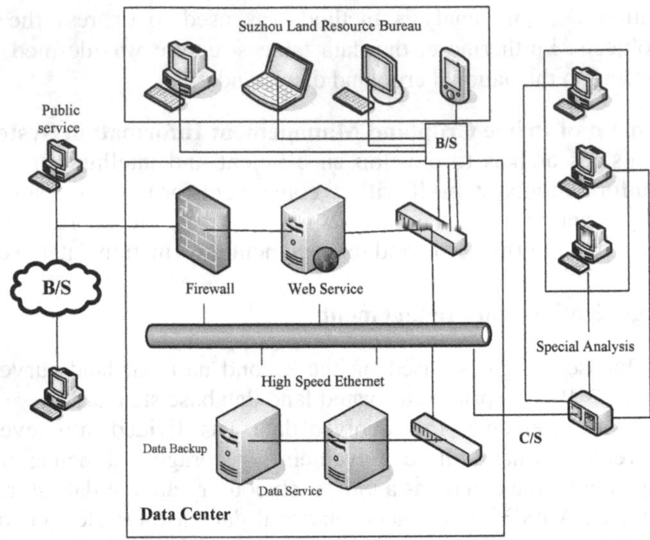

Fig. 1. The Architecture of the System

2.4 System Function Modules Design

According to the user requirement analysis and evaluation, this system function module mainly from two aspects: the prime cropland transaction and the subject analysis. As for the transaction module, there were four basic functions: prime cropland demarcated, prime cropland protection and construction, supervision and management of prime cropland and prime cropland capital management. From the view point of project analysis, system function includes dynamic management function, statistics analysis function, historical data backtracking management and map publishing.

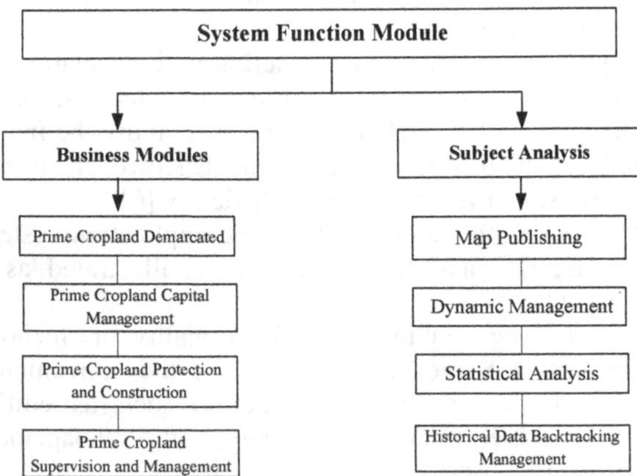

Fig. 2. The System Function Module

3 Key Functionalities Design

3.1 Map Publishing

It is necessary to realize the map issued through the network, for opening the information of the prime cropland preservation and sharing the resources to provide convenient public service. In recent years, the emergence of the Internet has provided convenience for effective sharing of data and technology. Delivering geographic data, maps and various application services through the Internet has been proposed by the modern tendency. B/S module, the subsystem, which is explored based on the Web application and service framework of ArcServer9.3, can publish map and share the data through Internet browser, and provides the public to query the prime cropland information. Meanwhile, the map buffer technique can be used to speed up the user experience [7].

3.2 Dynamic Management

In C/S mode, the subsystem including dynamic inspection, occupy tocsin and intelligent complement farm. The last two are the focal and difficult points for the system realization.

1) Occupy Tocsin

The key of occupy tocsin analysis is to analyze the position, quantity and area of acre which are occupied by construction projects through overlaying the project planning layer and the prime cropland planning spot. According to the results of the analysis, system can make out tocsin to provide decision support for approving project site and building site.

The source of prime cropland planning layer includes three kinds of approaches: (1) loading existed project layer, (2) importing node coordinates to create a temporary layer, and (3) drawing a graphic of polygon directly on the map by mouse.

Through the three ways above, the occupied lands can be positioned and displayed on the system interface, the quantity and detailed information of occupied lands will be shown to the users to analyze comparatively [8].

2) Intelligent Complement Farm

Intelligent complement farm, one of the main system functions, is mainly for city users to search automatically and obtain all the qualified records which are cultivated land complemented from spare land spots mapped by the second national land survey with the exception of prime cropland. At the same time, the users can locate and browse the block information, then export the results in the form of excel, which can provide decision support for the complement farm work [9].

The process of intelligent complement farm is shown as follows. First, filtering the normal farmland spot out of the area used for land as layer A. Secondly, filtering the cultivated land spot from the second national land survey spot layer as layer B. Thirdly, using the layer A and B to realize overlay analysis, filter the overlap part. Finally, filtering the eligible location and area which are set by users from the results of previous step.

Through the four steps above, we can obtain the qualified blocks. The city users also can query the details about each block and send them to own countries in the form of excel that can provide decision support for the complement farm work.

3.3 Statistical Analysis

Statistical analysis function of the system mainly includes prime cropland area and account book statistic etc. Area statistic consists of three types. Users can query through prime cropland preservation area, plot and patch. The system can provide the area of preservation area, cultivated land, agricultural land and prime cropland in the form of excel.

Account book statistic function can provide union query about the attribute information through several layers, then export the statistical results that include county names, town names, preservation area number, preservation plot number, preservation patch number, nature reserves area, preservation plot area, preservation patch area, sheet number, plot number, plot area and so on.

3.4 Historical Backtracking and Browse

The spatiotemporal data management is significant for land and resources management information system, in which the capacity of historical backtracking and browse is a very challenging work. For the multi-user concurrent operation, the most effective method was to introduce the version management into the database engine as to support long transaction processing. The version is the snapshot of the entire database, not the hard copy, which saves the content of database changes only. For the contents that have not changed, the version saves them only once for physical store. In ArcSDE, the system allows multiple versions to generate version for each user. Users can edit the data for a long time in their own version, which will not affect the operation of other users [10].

The process of historical retrospect realization is demonstrated as follows. All the layer data will divide into present-day layers and historical layers, also the present-day and historical database will be established in the data server. One of the databases is for historical data that is added when the version is submitted, the other is for present-day data that is in default version, so the version management and the historical retrospect will complement each other. Table fields of creating time and deleting exist in both kinds of data. On the basis of the relationship between the creating time and deleting time, the system can restore the historical data which is at a certain time in history. The historical data retrospect will be viewed based on the time relationship between the present-day and historical data [11]. The scheme logical process of historical retrospect is shown in the following figure.

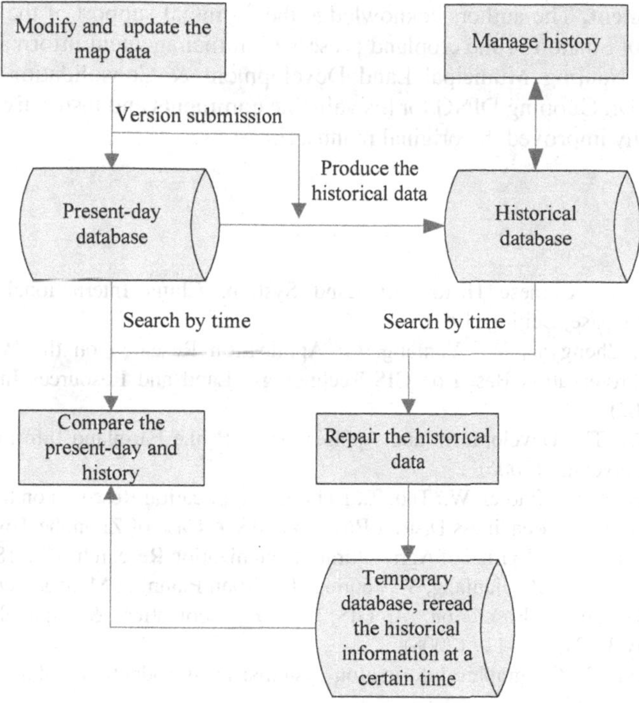

Fig. 3. The Flow Chart of the Historical Backtracking Design

Compared browsing between versions (such as version A and B) can be realized through the following procedures. (1) To connect the version database A. (2) To read the version A layers. (3) To connect the version database B. (4) To read the version B layers. (5) To query the different landmark information between two versions. (6) To load the different version layers up to the AxMapcontrol respectively.

Users can make modification in the new version after the creation based on the current version. Then, the edited version should be submitted to its father version. Also, the historical data can be migrated out of the database when it is too redundancy.

4 Conclusions and Discussions

With the rapid development of land and resources informatization and the management information system, it is inevitable for the system to realize perfect data management function that manage the mass of prime cropland preservation data, which can be used for the problems such as occupancy and complement farm.

The ArcGIS technology mentioned in this paper, including ArcSDE, ArcEngine and ArcServer have been used in the Suzhou prime cropland preservation management information system. But the design and implementation of the technology mentioned above still needs further discussion and perfection, especially for optimized visualization of spatial data and non-georeferenced data under internet circumstance.

Acknowledgment. The authors acknowledge the financial support of the project "The development of Suzhou prime cropland preservation management information system" sponsored by Nanjing Municipal Land Development & Consolidation Center. The authors thank Dr. Guoping DING for his valuable comments and instructive reviews, all of which greatly improved the original manuscript.

References

1. Chuansuo, Y.: Chinese History of Land System. China International Broadcasting Publishing House, Beijing (1990)
2. Xiong, C., Zhengyan, W., Yanfang, L.: Application Research on the Work of Prime Cropland Preservation Based on GIS Technology. Land and Resources Informatization, 37–39 (2002)
3. Hongxia, C.: The Development and Application of Prime Farmland Information System. Xinjiang University (2006)
4. Le, Y., Jianjun, T., Xiaofei, W., Tao, Z., Debi, L.: Application Research on the Designation of Basic Farm land Readiness District Based on GIS-A Case of Zhonghe Town of Xiushan County in Chongqin. Journal of Agricultural Mechanization Research, 178–181 (May 2011)
5. Yongxing, L., Yan, Z., Jianfang, Y., Junting, J.: Urban Planning Management Information System Designing Based on ArcGIS Server. Geomatics & Spatial Information Technology 31(3), 114–117 (2008)
6. Bernhardsen, T.: Geographic information systems: an introduction. John Wiley & Sons, Inc., New York (2002)
7. Xiaosong, L.: The Design and Developm ent of Basic Farm land Geographic Information System. Journal of Chongqing Normal University (Natural Science Edition) 22(2), 68–71 (2005)
8. Wei, W., Xiaoshuai, L., Bin, Z.: Probe of GIS Developing Technology Based on ArcGIS Engine. Science Technology and Engineering 6(2), 176–178 (2006)
9. Qiuying, T., Mingjun, L.: Prime Cropland Preservation Information System Based on ArcEngine. Geospatial Information 5(6), 24–26 (2007)
10. Shilin, Z., Honghua, C., Gangshen, X., Bo, L.: Design and Realization of Edition Management and History Trace Based on ArcSDE. Journal of East China Institute of Technoligy 31(4), 357–360 (2008)
11. Zhenzhou, C., Qingquan, L.: The Design and Implementation of a Version ConArcSDE and ArcGIS Engine. Geomatics & Spatial Information Technology 29(1), 76–78, 97 (2006)

Zigbee Based Dynamic Ranging Approach Applied to Coalmine

ShaoHua Chen, Xun Mao, HaiKuo Zhang, and PengCheng Wang

College of Electrical & Information Engineering, Dalian Jiaotong University, Dalian 116028, P.R. China
chengshineng@163.com

Abstract. A successful new design for coalmine ranging and positioning is presented. According to the investigation on the spot of coalmine, improved Trilateration algorithm is adopted and the RSSI algorithm is used in calculating the signal attenuation during transmission. We raise the new approach with high precision based on the Zigbee technology to achieve dynamic, high precision and real-time ranging. This is of great significance for ensuring road safety in the coal mine.

Keywords: Zigbee, Wireless ranging, Signal attenuation, shadowing, RSSI algorithm.

1 Introduction

Engineering vehicles employed in the coalmine, more often than not, are extremely huge in relative to others. As is shown in Fig.1, taking CAT 797 for example, its length, width, and height is respectively 14.8,9.8,7.7 meters[1]. Due to the blind zone of drivers and surrounding factors like dust, accidents occur frequently along the route of transportation, the approach of crash avoidance and security insurance, namely a way of real-time ranging, is in great need of carrying out.

Fig. 1. Zigbee node distribution schemes

D. Jin and S. Lin (Eds.): Advances in CSIE, Vol. 1, AISC 168, pp. 637–642.
springerlink.com © Springer-Verlag Berlin Heidelberg 2012

Traditional solution used to ranging includes positioning relay, Global Positioning System (GPS) satellite, radar position and supersonic wave [2]. Radar signal attenuates significantly when condition goes bad. GPS is costly and high energy consuming, and its accuracy is not better than 3m [3]. Therefore both the two ways can't meet the demand of coalmine ranging perfectly.

2 Scheme of Ranging

In order to settle the problem above, we present an approach based on Zigbee. The corresponding communication modules are to be set up in Location 1~6 of the truck (showed in Fig.1), and the reverse side of truck can be installed as well. Communication Sensor A is to be installed at the top of the miniature car. Directional antenna can be also installed to make our system collecting information more easily [4]. Thus by calculating the information Sensors collected, the position of all the miniature ones within the communication range can be displayed clearly, and warnings take place when the distance is smaller than the secure minimum radius so as to reduce the occurrence of accidents.

3 The Algorithm

3.1 Trilateration Algorithm

In this section, the Trilateration algorithm [5] is adopted. Fig. 2 is the cutting plane. Module 4, 5, 6 and A are corresponding to the ones in Fig. 1.

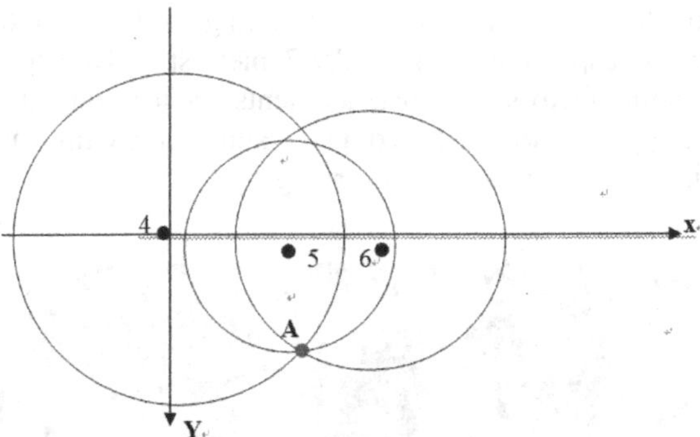

Fig. 2. Vector diagram of Triangle projection range

In this method, establishing the coordinate system in Fig.2, define the coordinates of Point 4, 5, 6 as $(x_4, y_4),(x_5, y_5),(x_6, y_6)$. It stands for the position of truck. Obviously, as long as obtain the horizontal distance Point 4, 5, and 6 to A, namely r_4, r_5, r_6, then coordinate of Point A is able to be carried out by equation(1)[6]

$$\begin{cases}\sqrt{(x-x_4)^2+(y-y_4)^2}=r_4\\ \sqrt{(x-x_5)^2+(y-y_5)^2}=r_5\\ \sqrt{(x-x_6)^2+(y-y_6)^2}=r_6\end{cases} \tag{1}$$

Practically, modules operate in three dimension space. Now we try to convert the visual angle into the horizontal plane, so as to attain the value of r_4,r_5,r_6 combined with the Received Signal Strength Indicator(RSSI) algorithm and the Pythagorean theorem.

3.2 RSSI Algorithm

Signal attenuates regularly in the propagation process. The mechanism of RSSI algorithm is to figure out the distance by comparing the practical signal intensity with the initial signal strength. The calculating formula of RSSI is as follows.

$$p = p_0 + 10n \lg\left[\frac{d}{d_0}\right] + \zeta \tag{2}$$

1. n is path attenuation index, it is decided by the surrounding factors and Signal transmission frequency, usually varied from 2 to 5. The value of n should be chosen according to the actual mine environment.
2. ζ is Shadowing factor which the unit is dB, also called Non line-of-sight influence.It is related with barriers in field environment, such as vehicles, air temperature and so on, nothing to do with the signal propagation distance.

We use the model which contains a number of trucks and small cars to calculate the value of ζ. Suppose there are n communication modules on the truck{a1、a2、... an-1、an}, n>=3 ; m communication modules on the small car {b1、b2、...bm-1、bm}, m>=1. Select a module b_i from small car, and it's sum distance between n modules on the truck is the shortest. We can get a total of n groups a_n to b_i distance and RSSI value (d, p), choose n=j（1<=j<=n）, and take（d_j, p_j）into Equation 2 (d_0, p_0),we can get the value of ζ by the equation(3).

$$\zeta = \sum_{k \neq j}^{n} p - p_j - 10\, n \lg\left[\frac{d}{d_j}\right] \tag{3}$$

Now there is n-1groups data be left over, marked as （d_k, p_k）.

$$\zeta_j = \frac{\sum_{k \neq j}^{n} p_k - p_i - 10\, n \lg\left[\frac{d_k}{d_j}\right]}{n-1} \tag{4}$$

By the weighted average method, the Shielding factor of the average number of modules is:

$$\zeta^{*} = \sum_{j=1}^{n}\left(\frac{p_j \zeta_j}{\sum_{i=1}^{n} p_i}\right) \tag{5}$$

3. p is the practical signal intensity. First, when the distance between the two measurement target changes, RSSI and the distance is not change to maintain a linear relationship. It can be found by experiment when the distance between two targets is short, RSSI values decay faster and when the distance is farther, RSSI values decay slowly.

In addition, multipath propagation is an important factor which has significant impact on the RSSI value. In mining areas, there always have raised dust at any time, accumulation of the coal beside the road and sharp turn. All these are important factors lead to multi-path effects during the signal transmission.

In practical application, we can set a threshold of RSSI in advance. When the measured RSSI value is greater than this critical value, it means the distance between the two vehicles is the "alarm distance" and system will warn our drivers.

To avoid the signal code interference caused by multipath interference, Filtering method has been used at the receive signal. Through Average filtering, we can obtain the average value of RSSI.

$$\overline{RSSI} = \frac{1}{n}\sum_{n=1}^{i} RSSI_i \qquad (6)$$

In this way, by selecting more RSSI values, we may get a result more close to ideal result. Moreover, in order to avoid the impact of jitter value on the average, we can set two thresholds x_{min} and x_{max} according to the corresponding experiment. Only when $x_{min}<RSSI_i<x_{max}$, the value of RSSI will be taken into the equation, which makes result much more accurate.

What has been mentioned above is the standard shadowing modeling program. The calculation of RSSI can be simplified by using the simplified formula (7). Based on the positioning situation between two vehicles, shadowing factor will not be considered under the situation of two-car collision.

$$RSSI = -(10p\lg d + q) \qquad (7)$$

p is the Signal transmission constant, q is the signal intensity in the distance of 1 meter which the unit is dB. d is the distance of Projection Node. The following is the d and q parameter values of some common situation:

Table 1. d and q parameter values in different situations.

environment	p	q (dB)
park	(32.7-36.0)	(2.7-3.4)
bamboo grove	(35.2-38.2)	(4.5-5.4)
sandbeach	(37.5-40.8)	(3.8-4.6)
courtyard	(34.2-39.0)	(2.8-3.8)
Alley	(44.0-48.7)	(2.1-3.0)
grassland	(33.2-36.4)	(3.0-3.9)
rail fence	(34.5-38.2)	(4.6-5.1)

On different occasions, parameters should be measured and take mathematical statistical simulations to determine the optimal parameter values.

4. d_0 is reference distance and p_0 is signal intensity received in the distance of d_0. As the transmission power of the Zigbee module is set, via reference distance d0, p0 can be measured by software in advance [6].

At last, the distances between two modules d_4, d_5, d_6 are able to be obtained.

Considering the distinct conditions of practical environment and hardware, suppose that the value of d_4, d_5, d_6 is respectively 17m, 7m and 12m, and we need to convert them into distances of horizontal plane r_4, r_5, r_6. As Fig. 1 shows, the height of Module 4, 5, 6, that is h_4, h_5, h_6, is respectively 7m, 3m, 5m; h_A, the distance from module installed at the top of the miniature car , is 1.5m. r_4, r_5, r_6 can be calculated through the following equations.

$$r_4 = \sqrt{d_4^2 - (h_4 - h_A)^2}, \quad r_5 = \sqrt{d_5^2 - (h_5 - h_A)^2}, \quad r_6 = \sqrt{d_6^2 - (h_5 - h_A)^2} \tag{8}$$

Put the corresponding data into the Formula 4, 5, 6 to get the value of r_4, r_5, r_6 which is respectively 13.95m, 6.84m, 11.48m. Then incorporate them into (1) to obtain the coordinate of Point A on the horizontal plane.

What needs explaining is that our hardware networking can bring more modules participate in collecting information. In practical condition, it follows that if there are x nodes in the truck, x equation sets are to be listed, and every equation sets has two equations.(showed in (9),and x is the number of nodes installed on the truck ,and those participates are all in horizontal plane)

$$\begin{cases}\sqrt{(x-x_4)^2+(y-y_4)^2}=r_4\\ \sqrt{(x-x_5)^2+(y-y_5)^2}=r_5\end{cases} \begin{cases}\sqrt{(x-x_4)^2+(y-y_4)^2}=r_4\\ \sqrt{(x-x_6)^2+(y-y_6)^2}=r_6\end{cases} \begin{cases}\sqrt{(x-x_5)^2+(y-y_5)^2}=r_5\\ \sqrt{(x-x_6)^2+(y-y_6)^2}=r_6\end{cases} \tag{9}$$

Solve these equation sets, abandon the results which is negative and obtain the approximate result (7.1, 11.2), (8.4, 11), (4.42, 7.33), and then we can find out the average coordinate of A is (6.64, 9.84). Practically, we can choose all modules on the truck to list equation sets, and thus attain the more precise average. At last, by showing the coordinate of A to the driver, accidents can be avoided to a great extent.

Zigbee technology can ensure the information convey distance reach 70m in open area [7]. By using the Trilateration and multiple-node projection ranging method, averaging the data of all modules, the accuracy is within 3m.

4 Conclusion

The method in this paper, which is based on the Trilateration algorithm, bring up a new approach of triangle projection range according to the circumstance of coalmine. It's high precision makes it reliable in dynamic ranging and the analysis in the whole passage has also demonstrated the feasibility of the approach to avoid car accidents in coalmine.

Acknowledgment. This work was supported by the S&T plan projects of Liaoning Provincial Education Department (No. L2010082).

References

1. http://auto.ifeng.com/roll/20091027/135816.shtml
2. Akyildiz, I.F., Su, W., Sankarasubramaniam, Y., et al.. A survey on sensor networks. IEEE Communications Magazine 40, 102–114 (2002)
3. Cullar, D., Estrin, D., Strvastava, M.: Overview of Sensor Network. Computer 37(8), 41–49 (2004)
4. Gungor, V.C., Lambert, F.C.: A Survey on Communication Networks for Electric System Automation. Computer Networks (50), 877–897 (2006)
5. Shen, T.: Study of Localization Algorithm for Wireless Sensor Network. Zhe Jang University (2007)
6. Liu, X.: Research and Application on RSSI Localization Algorithm for Wireless Sensor Network. Northwest University (2009)
7. Pottie, G., Kaiser, W.: Wireless Integrated Network Sensors. Communications of ACM 43(5) (2000)

SD Rats' Fatty Liver Tissue Classification Based on Ultrasound Radiofrequency Signal

XiuQun Xie[1], Yan Luo[2], JieRong Quan[2], Ke Chen[1], and JiangLi Lin[1]

[1] Department of Biomedical Engineering, Sichuan University
linjlscu@163.com
[2] West China Hospital of Sichuan University, Chengdu, China

Abstract. Early diagnostic of mild fatty liver and prediction for the level of fatty liver have important value in clinic. However, it's difficult to diagnose mild fatty liver and grade fatty liver correctly using ultrasonic images. The method based on extracting features from ultrasound radiofrequency (RF) signals was proposed in this paper. Five features were selected: msr(mean/sd), skewness and kurtosis of RF signal envelope, the approximate entropy of signal and the energy ratio of two regional signals. Then back-propagation(BP) network combined with "leave-one-out" cross-validation was employed to classify normal livers and variable degree of fatty livers. The results showed that, the features could distinguish normal livers from pathological livers with accuracy rates of 98%. The accuracy rates of classification were 98.08%, 82.05%, 78.38%, 87.5% for normal liver, mild fatty liver, moderate group and severe group, separately.

Keywords: radiofrequency signal, fatty liver, envelop, network.

1 Introduction

Fatty liver is reversible if diagnosed and treated timely. Especially mild fatty liver can donor for liver patients, which is of important value in medical fields. So far many methods based on ultrasound image have been used to diagnose and grade fatty liver [1,2,3], but the methods always depend on clinic experience and image quality. Meanwhile, part of tissue information was lost during the reconstruction of the image. So the ultrasound RF signal with rich tissue information has potential influence on diagnosing fatty liver. Many researchers have used RF signal in lesion detection, but the papers about grading fatty liver based on RF signal are rarely reported.

Envelope power spectrum of RF signal was used to classify liver tissues(normal, cirrhotic, fat), the accuracy rates was 85.8% to 87.5% discriminating pathological from normal tissue states[4].Meanwhile, the statistical model describing the envelope of the backscattered ultrasonic echoes was applied to ultrasonic tissue characterization of myocardium, coronary atherosclerotic morphology and so on[5-8].Their study showed that ,when the scattering elements were altered, the envelop distribution begined to depart from a Rayleigh scattering process, resulting in an over-all different

D. Jin and S. Lin (Eds.): Advances in CSIE, Vol. 1, AISC 168, pp. 643–647.
springerlink.com © Springer-Verlag Berlin Heidelberg 2012

probability distribution function. The study of Z.D. LIU showed that the ultrasonic RF signal was effective in diagnosing liver fatty degree[9]. This work set out to test whether the new features based on ultrasonic RF signal was sensitive for normal liver and fatty liver steatosis degree comparing with pathological results. Results were presented for classification into four subgroups, and also into two general types, resulting in an improved success rate although still being clinically useful.

2 Method

2.1 Data Acquisition

There were 120 male Wistar rats (Nomal:43,mild:33,moderate:26,severe:13) provided by animal experiment center of Sichuan University, all of which were made definite diagnosis with pathology. Rat livers were scanned using a Siemens Antares system (Siemens Acuson Antares, USA) equipped with the Axius direct ultrasound research interface (URI), which provides 16 bit digitized echo signals at a sampling rate of 40 MHz. Each frame of a scan consisted of 312 scan-lines, and 1296 sample points were collected for each scan-line, corresponding to a scan-depth of approximately 2.5 cm.

Areas of well-defined histology were identified with regions of the ultrasound image and used to select regions-of-interest (ROIs) for subsequent RF signal analysis. Two types of ROIs were selected: One hundred fifty-two ROIs of length 512 sample points and width fifty scan-lines were selected in the following proportions: Normal(52), Mild(39), Moderate(37), Severe(24). An ROI in white region is indicated in Fig. 1a. Three hundred and four ROIs of length 256 sample points and width thirty scan-lines were selected in the following proportions: Normal(104), Mild(78), Moderate(74), Severe(48). Fig.1b present two ROIs with white rectangles in the near and far field, respectively, which were centered at 1cm.

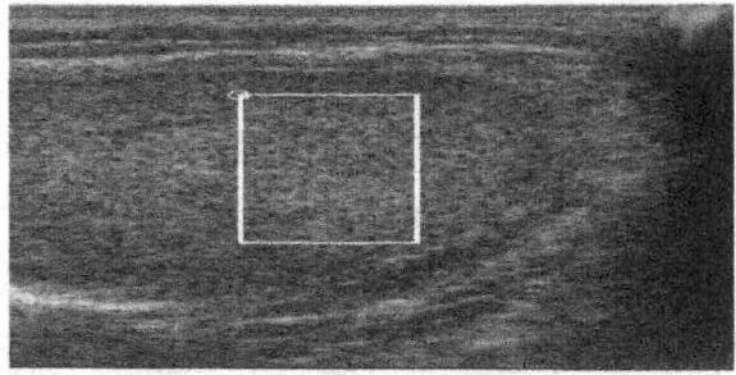

Fig. 1a. ROI of 50 vectors*512 points

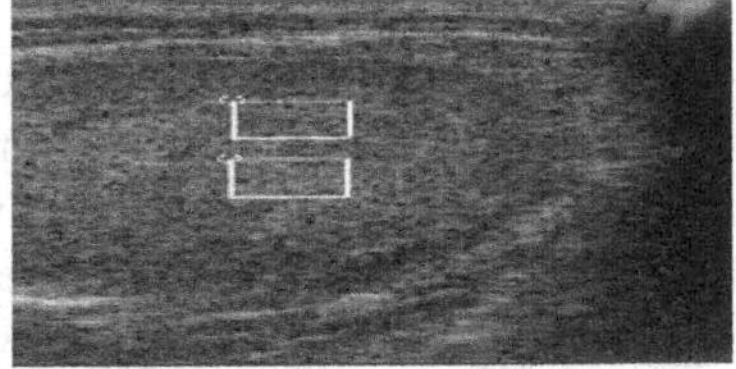

Fig. 1b. ROI of 30 vectors*256 points

2.2 Feature Extracting

1)Parameters of signal envelop: Firstly, the envelop of each signal line of ROI in Fig.1a was obtained by the Hilbert transform, then the statistical parameters of mean/SD ratio[msr], skewness[sk], and kurtosis[ku] were calculated for the envelope, finally these parameters for each vector were averaged for all vectors of the liver ROI in Fig.1a. The three parameters were defined as equation (1) to (3) [5]:

$$msr = u / \sigma . \tag{1}$$

$$sk = \frac{E[(s(t) - u)^3]}{\sigma^3}. \tag{2}$$

$$ku = \frac{E[(s(t) - u)^4]}{\sigma^4}. \tag{3}$$

2)Approximate entropy: Approximate entropy (ApEn) is a "regularity statistic" that quantifies the unpredictability of fluctuations in a time series[10]. To compute ApEn of a time series x(n), n=1,2,...N, first the series of vectors of length m, v(n)=[x(n), x(n+1),...x(n+m-1)]^T is derived from the signal samples x(n). The distance D(i,j) between two vectors v(i) and v(j) is defined as the maximum difference in the scalar components of v(i) and v(j). Then NOD, i.e., the number of vectors j (with j≤N-m+1) such that the distance between the vectors v(j) and the generic vector v(i) (with i≤N-m+1) is lower than r, D(i,j)≤r, is computed. Now define $C_i^m(r)$, the probability to find a vector which differs from v(i) less than the distance r, and $\Phi^m(r)$ as:

$$C_i^m(r) = NOD / (N - m + 1) . \tag{4}$$

$$\Phi^m(r) = (N - m - l)^{-1} \sum \ln C_i^m(r). \tag{5}$$

ApEn is given by:

$$ApEn(m, r) = \Phi^m(r) - \Phi^{m+1}(r) . \tag{6}$$

In this experiment m and r were set to 2 and 0.25*SD(u),respectively, where SD(u) represented the standard deviation of the data of the vector.

3)Energy ratio: The energy of the RF signal was caculated as the sum of each sample's magnitude square, i.e., $E = \sum |x(t)|^2$,where x(t) was RF signal of rat liver. Energy ratio (E_ratio) was given by equation (7):

$$E\_ratio = E_1 / E_2 . \tag{7}$$

Where E_1 was the energy of near field ROI of liver in Fig.1b, and E_2 was the energy of far field ROI of liver in Fig.1b.As well known, the attenuation increased with the level of the fatty liver, which would lead to the change of the value of E_ratio.

2.3 Neural Network Classification

A three-layer BP neural network was designed in this paper, with learning rate of 0.01 and 5 hidden layer nodes to balance the training time and accuracy. A realistic estimate of the classifier error was obtained using "leave-one-out" cross-validation, in which the classification was performed for each sample in turn, based on the data from the data set excluding the samples being classified. That is, 151 samples were trained and 1 sample was tested every time. Finally, we did two types of identification to all samples based on the five features proposed above.

3 Results and Discussion

The means and standard deviations of the five features for each of the four subgroups are shown in Table1.We can find significant difference between normal liver and fatty tissue. Increasing value of msr can be seen progressing from normal through mild up to severe group. This can be attributed to increasing amounts of fat granule in fatty liver, and to the mirror-like elements (such as liver cells, small blood vessels etc) in normal liver. Meanwhile, the signal envelope of fatty liver is more closer to the Rayleigh distribution than normal group. The difference is also reflected in skewness and kurtosis of envelope. Some degree of overlap exists in approximate entropy between mild and moderate liver, as is to be expected due to the continuous nature of fatty liver. The results also show that attenuation of fatty liver is far greater than that of normal liver, which is highly consistent with the clinical findings.

Table 1. Means and standard deviations of five parameters for four tissue types

	msr	sk	ku	ApEn	E_ratio
Normal	1.20±0.15	2.50±0.72	12.72±6.41	0.97±0.09	0.53±0.19
Mild	1.53±0.11	1.63±0.35	7.81±2.16	1.08±0.05	1.00±0.28
Moderate	1.68±0.11	1.20±0.28	5.60±1.60	1.09±0.04	1.23±0.23
Severe	1.69±0.09	1.15±0.25	5.11±1.16	1.10±0.03	1.79±0.44

Following are the results of BP network classification combined with leave-one-out cross-validation. When the liver tissues are grouped to four classes: (1)normal; (2)mild; (3)moderate; and (4)severe. The result is showed in Table 2. Group the livers into two common subgroups: (1)normal group; (2)fatty group. We can find the classification is improved in Table 3, with an overall classification rate of 98%.

Table 2. The result of classification into four classes.

	Normal	Mild	Moderate	Severe	Total	Rate(%)
Normal	51	1	0	0	52	98.08
Mild	2	32	5	0	39	82.05
Moderate	0	4	29	4	37	78.38
Severe	0	0	3	21	24	87.50

Table 3. The result of classification into two classes.

	Normal	Fatty	Total	Rate(%)
Normal	51	1	52	98.08
Fatty	2	98	100	98.00

From the recognition results, the extracted features can be correctly separated normal liver from and fatty liver, especially only two mild fatty livers were identified as normal tissues, which may provide important reference value to clinic diagnosis. Msr is good description of concentrating level of signal envelop. Skewness and kurtosis can quantitatively represent the shape of the distribution. The former refers to the asymmetry of the shape and the latter to the "peakedness" of the distribution. The envelop distribution is more symmetrical and concentrating with the deepening of fatty liver. Severe fatty liver can be characterized by E_ratio which is much larger than the other categories. So these features proposed above may be indicators of aided diagnosis of fatty liver. However, grading of human fatty liver more accurately will be the focus of future study.

Acknowledgments. This study was supported by the Natural Science Foundation of China (30870715 and 30970781).

References

1. Potdukhe, M.R., Karule, P.T.: MLP NN based DSS for analysis of ultrasonic liver image and diagnosis of liver disease. In: Second International Conference on Emerging Trends in Engineering and Technology, ICETET 2009, pp. 67–72 (2009)
2. Thijssen, J.M., Starke, A., Weijers, G., Haudum, A.: Computer-Aided B-Mode Ultrasound Diagnosis of Hepatic Steatosis: A Feasibility Study. IEEE Transactions on Ultrasonics, Ferroelectrics, and Frequency Control 55(6), 1343–1354 (2008)
3. Fukushimal, M., Ogawa, K., Kubota, T., Hisa, N.: Quantitative Tissue Characterization of Diffuse Liver Diseases from Ultrasound Images by Neural Network. In: Nuclear Science Symposium, vol. 2, pp. 1233–1236. IEEE (1997)
4. Lang, M., Ermert, H., Heuser, L.: In vivo study of on-line liver tissue classification based on envelope power spectrum analysis. In: Proceedings of IEEE 1990 Ultrasonics Symposium, pp. 1345–1348 (1990)
5. Komiyama, N., Berry, G.J., Kolz, M.L.: Tissue characterization of atherosclerotic plaques by intravascular ultrasound radiofrequency signal analysis: an in vitro study of human coronary arteries. Am Heart J. 140(4), 565–574 (2000)
6. Wagner, R.F., Insana, M.F., Brown, D.G.: Statistical properties of radio-frequency and envelope-detected signals with applications to medical ultrasound. J. Opt. Soc. Am. 4(5), 910–922 (1987)
7. Clifford, L., Fitzgerald, P.J., James, N.D.: Non-rayleigh first-order statistics of ultrasonic backscatter from normal myocardium. Ultrasound Med. Biol. 19(6), 487–495 (1993)
8. He, P., Greenleaf, J.F.: Application of stochastic analysis to ultrasonic echoes: estimation of attenuation and tissue heterogeneity from peaks of echo envelope. J. Acoust. Soc. Am. 79(2), 526–534 (1986)
9. Liu, Z.D., Lin, J.L.: Quantizing and Grading of Fatty Liver Based on Ultrasonic RF Signals. Journal of Sichuan University (Engineering Science Edition) 43(suppl. 1), 160–164 (2011)
10. Nie, N., Yao, D.Z.: The technology and application of Biomedical signal digital processing, Beijing (2005)

Robust Satellite Attitude Control

Lina Wang, Zhi Li, and Bing Wang

Department of Communication Engineering,
School of Computer and Communication Engineering,
University of Science and Technology Beijing, Beijing 100083, China
{wln_ustb,wangbing1978}@126.com, lizhi_xiaogui@163.com

Abstract. In this paper, the stability of satellite attitude control with the existence of external disturbances is considered. The satellite attitude matrix is represented by quaternion. A robust satellite attitude controller is designed, which introduces the suppression vector into the control law to counter external disturbances. At the same tine stability conditions of the robust attitude controller are presented. Finally, the effectiveness and robustness of the attitude controller is shown through numerical simulation results.

Keywords: attitude control, quaternion, external disturbance, stability.

1 Introduction

The attitude and orbit of a satellite must be controlled so that the satellite's antennas point toward the earth. However, an orbiting satellite is subjected to several gravitational influences. These forces have an effect on a satellite's trajectory, and tend to its attitude and orbit. Therefore, the attitude control system must counter any rotational torque or movement.

A considerable amount of efforts has been devoted to the problem of satellite attitude control over the past several decades. Many control methods have been developed to treat this problem, based mainly on Euler angle, modified Rodrigues parameter vector and quaternion representation [1-6]. This paper focuses on the stability of satellite attitude represented by quaternion. A robust satellite attitude controller is designed with considering the bounded external disturbances.

The rest of paper is organized as follows: Section 2 presents the mathematical model of a satellite control system. The design of attitude controller and stability conditions of the controller are given in Section 3. Section 4 numerical shows the simulation results and analysis. Finally, the conclusions are provided in Section 5.

2 Mathematical Model of Satellite Attitude Control System

The satellite attitude motion can be described by kinematic and dynamic equations.

We use the unit quaternion to represent satellite attitude in order to avoid singularity. Define the unit quaternion as

D. Jin and S. Lin (Eds.): Advances in CSIE, Vol. 1, AISC 168, pp. 649–654.
springerlink.com © Springer-Verlag Berlin Heidelberg 2012

$$\bar{q} = \begin{pmatrix} q \\ q_0 \end{pmatrix} = \begin{pmatrix} \sin(\theta/2)\hat{n} \\ \cos(\theta/2) \end{pmatrix} . \tag{1}$$

where $\hat{n} \in R^3$ is the rotation axis represented by unit vector, θ is the rotation angular, $q \in R^3$ and $q_0 \in R$ are the components of the unit quaternion, which subject to the following constraints:

$$q^T q + q_0^2 = 1 . \tag{2}$$

The kinematic equation represented by the unit quaternion is given by

$$\dot{q} = \frac{1}{2} E(q)\omega = \frac{1}{2}(q_0 I + q^{\times})\omega . \tag{3a}$$

$$\dot{q}_0 = -\frac{1}{2} q^{\mathrm{T}} \omega . \tag{3b}$$

where $\omega = [\omega_1\ \omega_2\ \omega_3]^{\mathrm{T}}$ is the satellite angular velocity vector with respect to the inertial reference frame, expressed in the satellite body-fixed reference frame, I is the 3×3 unit matrix, $q^{\times}$ is the skew symmetric matrix which is defined by

$$q^{\times} = \begin{pmatrix} 0 & -q_3 & q_2 \\ q_3 & 0 & -q_1 \\ -q_2 & q_1 & 0 \end{pmatrix} . \tag{4}$$

The dynamic model of the satellite attitude control system is described by the differential equation in Eq.(5).

$$J\dot{\omega} = -\omega^{\times} J\omega + u + d . \tag{5}$$

where $J = J^{\mathrm{T}} \in R^{3\times 3}$ is the inertia matrix which is a symmetric and positive define matrix, $u \in R^3$ is the vector of control torque, $d = [d_1\ d_2\ d_3]^{\mathrm{T}}$ is the vector of external disturbance which is bounded as $|d_i| \le \delta_i$, where δ_i is a positive constant, for i=1, 2, 3.

3 Controller Design

We consider the control law given in Eq.(6), where k_1 and k_2 are positive constants.

$$u = -k_1\omega - k_2 E^{\mathrm{T}} q . \tag{6}$$

Considering the Lyapunov function candidate

$$V = \frac{1}{2}\omega^{\mathrm{T}} J\omega + k_2 q^{\mathrm{T}} q . \tag{7}$$

Then the time derivate of V is computed according to Eqs.(3), (5) and (6). We have that

$$\dot{V} = \omega^{\mathrm{T}} J\dot{\omega} + 2k_2 q^{\mathrm{T}} \dot{q} = \omega^{\mathrm{T}}(-\omega^{\times} J\omega + u + d) + k_2 q^{\mathrm{T}} E\omega = -k_1\omega^{\mathrm{T}}\omega + \omega^{\mathrm{T}} d \ . \tag{8}$$

When d=0, Eq.(8) can be simplified as $\dot{V} = -k_1\omega^{\mathrm{T}}\omega \le 0$. Since the Lyapunov function candidate V is positive definite and radially unbounded. By LaSalle invariance principle we know that all the trajectories of the closed-loop system which is described by Eqs.(3), (5) and (6) will converge to the maximal invariant set $\psi = \{(\omega, q) : \dot{V} = 0\} = \{(\omega, q) : \omega = 0\}$, which implies that $\dot{\omega} = 0$. Since $\omega = 0$, we have that $\dot{q} = 0$, $\dot{q}_0 = 0$ from Eqs.(3a) and (3b). From Eq.(5), we have that $J\dot{\omega} = -\omega^{\times} J\omega + u$ (d=0) or $u = J\dot{\omega} + \omega^{\times} J\omega = 0$. From Eq.(6), we have that $k_2 E^{\mathrm{T}} q = -u + k_1\omega = 0$. So $q = 0$. Consequently, the maximal invariant set is $\psi = \{(\omega, q) : \omega = 0, q = 0\}$, which is also the stable equilibrium.

When $d \neq 0$, we will present a controller for the attitude control system.

First of all, we introduce a result about Input-to-State Stability [7].

Lemma 1. Let $V : [0, \infty) \times R^n \to R$ be a continuously differentiable function that satisfies the following properties:

$$\alpha_1(\|x\|) \le V(t, x) \le \alpha_2(\|x\|) \ . \tag{9}$$

$$\frac{\partial V}{\partial t} + \frac{\partial V}{\partial x} f(t, x, u) \le -W_3(x) , \ \forall \|x\| \ge \rho(\|u\|) > 0 \ . \tag{10}$$

$\forall (t, x, u) \in [0, \infty) \times R^n \times R^m$, where α_1 and α_2 are class K_∞ functions, ρ is a class K function, and $W_3(x)$ is a continuous positive definite function on R^n. Then, the system is input-to-state stable with $\gamma = \alpha_1^{-1} \circ \alpha_2 \circ \rho$.

Now Eq.(8) can be written as follows

$$\begin{aligned} \dot{V} &= -(k_1 - \theta)\omega^{\mathrm{T}}\omega - \theta\omega^{\mathrm{T}}\omega + \omega^{\mathrm{T}} d \le -(k_1 - \theta)\omega^{\mathrm{T}}\omega - \theta\sum_{i=1}^{3}|\omega_i|^2 + \sum_{i=1}^{3}|\omega_i||d_i| \\ &= -(k_1 - \theta)\omega^{\mathrm{T}}\omega - \sum_{i=1}^{3}|\omega_i|(\theta|\omega_i| - |d_i|) \end{aligned} \ . \tag{11}$$

Therefore, when ω satisfies $|\omega_i| \ge |d_i| / \theta$ for i=1,2,3, we have that $\dot{V} \le -(k_1 - \theta)\omega^{\mathrm{T}}\omega$ where $0 < \theta < k_1$. By Lemma 1, the designed controller can make the closed-loop system achieve input-to-state stable.

Now we will provide an improved controller. In order to counter the effect of external disturbance, we introduce the suppression vector V of external disturbance into the controller law. Let

$$u = -k_1\omega - k_2 E^{\mathrm{T}} q - v \ . \tag{12}$$

where $v = [v_1\ v_2\ v_3]^{\mathrm{T}}$, $v_i = \delta_i \operatorname{sgn}(\omega_i)$ for i=1, 2, 3. $\operatorname{sgn}(x)$ is the symbolic function.

Then we compute the time derivative of Lyapunov function candidate V based on the Eq.(8) and have that

$$\dot{V} = -k_1\omega^{\mathrm{T}}\omega - \sum_{i=1}^{3}\delta_i|\omega| + \sum_{i=1}^{3}|d_i||\omega_i| \ . \tag{13}$$

Therefore, when ω satisfies $|\omega_i| \geq |d_i|/\theta$ for i=1,2,3, $V \leq -\sum_{i=1}^{3}\delta_i|\omega_i|$. According to Lemma 1, we know that the designed controller can make the system which is described by Eqs.(3) and (5) input-to-state stable.

4 Numerical Simulation

In this section, simulations on robust satellite attitude control are conducted in order to demonstrate and verify performance of the designed controller.

The inertial matrix of a satellite is $J = \begin{pmatrix} 15 & 0 & 0 \\ 0 & 20 & 0 \\ 0 & 0 & 10 \end{pmatrix} (\mathrm{kg \cdot m^2})$. k_1=8, k_2=4. And the initial states are $\hat{n} = \begin{pmatrix} 0.5345 \\ 0.2673 \\ 0.8018 \end{pmatrix}$, $\theta = \dfrac{11\pi}{6}$, $\omega(0) = 0$ rad/s. Then the corresponding quaternion representations are $q(0) = \begin{pmatrix} 0.1383 \\ 0.0692 \\ 0.2075 \end{pmatrix}$, $q_0(0) = -0.9659$. At this time, yaw angle, roll angle and pitch angle are 23.56°, 17.21° and 4.98°, respectively. The external disturbance is $d = \begin{pmatrix} 0.01\sin(0.2t) \\ -0.006\sin(0.3t) \\ 0.014\sin(0.4t) \end{pmatrix} (\mathrm{N \cdot m})$.

(1) The control law does not include the suppression vector V of external disturbance, namely $\delta_1 = \delta_2 = \delta_3 = 0$, the simulation results are shown in Figs.1 and 2.

(2) The control law includes the suppression vector V of external disturbance, namely $\delta_1 = 0.01$, $\delta_2 = 0.006$, $\delta_3 = 0.014$, the simulation results are shown in Figs.3 and 4.

(3) If there exist model error and model parameter uncertainty. That is,

$J = \begin{pmatrix} 15+0.5 & 0 & 0 \\ 0 & 20+0.4 & 0 \\ 0 & 0 & 10+0.6 \end{pmatrix} (\mathrm{kg \cdot m^2})$, the simulation result is shown in Fig.5.

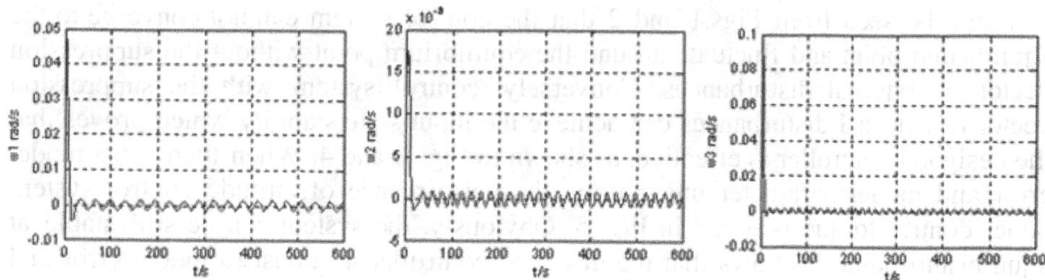

Fig. 1. Angular velocity curve without suppression vector of external disturbance

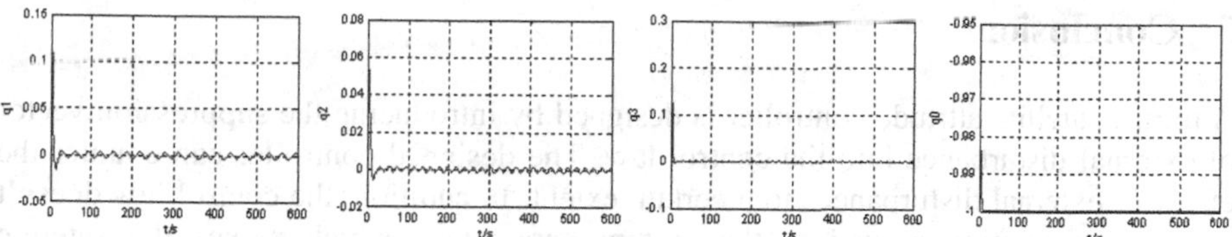

Fig. 2. Quaternion velocity curve without suppression vector of external disturbance

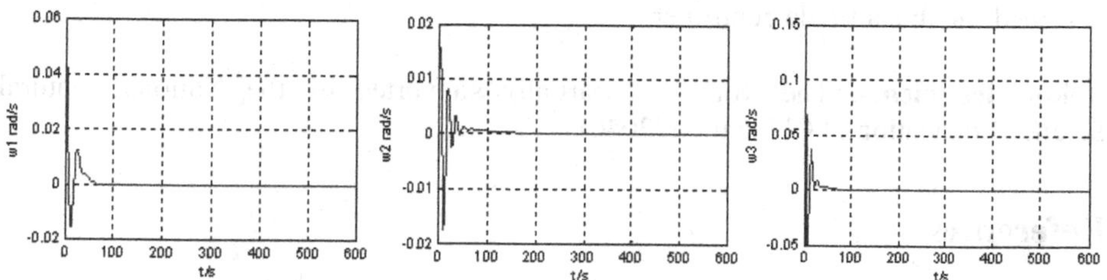

Fig. 3. Angular velocity curve with suppression vector of external disturbance

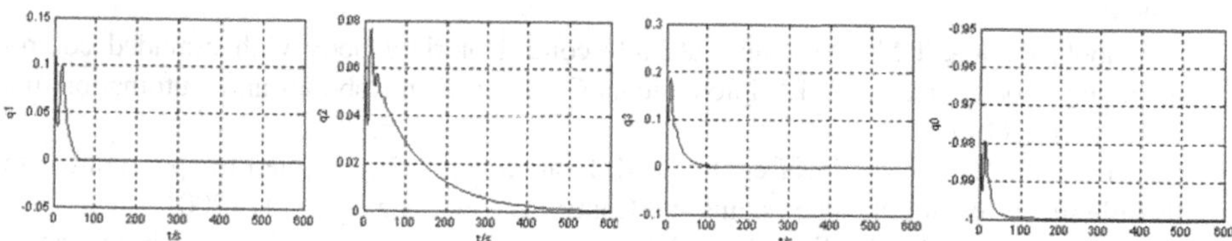

Fig. 4. Quaternion velocity curve with suppression vector of external disturbance

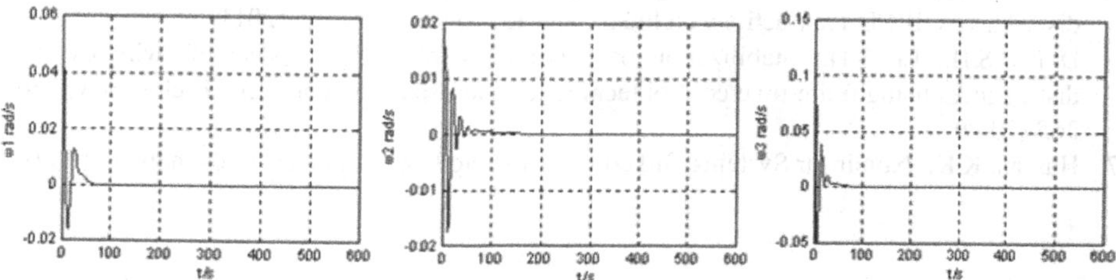

Fig. 5. Angular velocity curve with the existence of model error and model parameter uncertainty

It can be seen from Figs.1 and 2 that the control system can not converge to the equilibrium point and fluctuate around the equilibrium point without the suppression vector of external disturbances. Conversely, control system with the suppression vector of external disturbances can achieve the input state stability which proves that the designed controller is effective, as shown in Figs.3 and 4. When there exist model error and model parameter uncertainty, the performance of attitude control system under control torque is given in Fig. 5. Obviously, the system can be still stable at equilibrium point. It shows that the designed controller is robust to model error and model parameter uncertainty.

5 Conclusions

A robust satellite attitude controller is designed by introducing the suppression vector of external disturbance into the control law. The designed controller can counter the effect of external disturbances to a certain extent. In addition, the control law doesn't contain information related to the system parameters, which makes the satellite attitude control system robust to model error and model parameter uncertainty. Theoretical analysis and simulations verify the effectiveness and robustness of the designed satellite attitude controller.

Acknowledgments. The work was partially supported by the National Natural Science Foundation of China (60872046).

References

1. Joshi, S.M., Kelkar, A.G., Wen, J.T.: Robust attitude stabilization of spacecraft using nonlinear quaternion feedback. IEEE Transactions on Automatic Control, 1800–1803 (1995)
2. Yonmook, P., Jea, T.M.: Nonlinear attitude control of rigid body with bounded control input and velocity-free. In: IEEE International Conference on Robotics and Automation, pp. 1621–1626 (2001)
3. Lin, Y.Y., Lin, G.L.: Nonlinear control with Lyapunov stability applied to spacecraft with flexible structures. Journal of Systems and Control Engineering, 131–141 (2001)
4. Bayat, F., Bolandi, H., Jalali, A.A.: A heuristic design method for attitude stabilization of magnetic actuated satellites. Acta Astronautica, 1813–1825 (2009)
5. Xia, Y.Q., Zhu, Z., Fu, M.Y., Wang, S.: Attitude tracking of rigid spacecraft with bounded disturbances. IEEE Transactions on Industrial Electronics, 647–659 (2011)
6. Dong, S.H., Li, S.H.: Stabilization of the attitude of a rigid spacecraft with external disturbances using finite-time control techniques. Aerospace Science and Technology, 256–265 (2009)
7. Hassan, K.K.: Nonlinear Systems, 3rd edn. Publishing House of electronics industry (2007)

A Coordination Space Based Resource-Centered Dynamic Coordination Approach to Software Systems

Chengyao Wang, Xuefeng Zheng, and Xuyan Tu

School of Computer and Communication Engineering,
University of Science and Technology Beijing, Beijing 100083, China
wangchengyao@ustb.edu.cn

Abstract. The large scale software system often consists of many subsystems or modules. Coordination among subsystems or modules is a key issue. In this paper, a coordination space based resource-centered dynamic coordination model is presented, which includes three kinds of entities: computation entities, coordinators and coordination space. Computation entity contains role, actors and services. Coordination space plays part in the basis of dynamic coordination among computation entities, which mainly contains computation description, resource description, environment description and registry. Coordinator is responsible for coordination among computation entities, which includes communication coordinator, resource coordinator and service coordinator. When new role, actors and services enter to the system, they must login to the registry, and when leaving the system, they must logout to the registry. By this way, description information in coordination space is updated at runtime, so that dynamic coordination is achieved. The implementation approach to dynamic coordination is discussed.

Keywords: Coordination model, resource-centered coordination, coordination space, dynamic coordination.

1 Introduction

Coordination is a fundamental issue for developing a large scale software system. In many disciplines, such as computer science, artificial intelligence, management science, organization theory, and so on, coordination technology is researched. Definitions of coordination are given from different views. Coordination is "the process of managing dependencies among activities" [1]. Coordination is viewed as supporting the activity of managing dependencies and possible conflicts between collaborative agents involved in common and inter-related tasks of a collaborative activity [2]. Malone presented three kinds of basic dependencies: flow, sharing and fit [1]. In reference [3], coordination ideas of centralized and decentralized large scale systems were proposed from the view of Large Systems Cybernetics.

Many coordination models and approaches were presented in computer science, such as Linda [4], LogOp [5], SBC [6], Reo [7][8], ARC [9], ROAD [10]. However most of these models are suitable for the situation that coordinated entities and their behaviors are known at design time, not suitable for dynamic coordination at runtime.

D. Jin and S. Lin (Eds.): Advances in CSIE, Vol. 1, AISC 168, pp. 655–660.
springerlink.com © Springer-Verlag Berlin Heidelberg 2012

The ROAD supports change of roles based on contract and adaptive strategies at runtime, but doesn't support to model interdependency concerned with three or more roles [10]. The complexity of large scale software systems and diversity of their application environment demand software systems to adapt dynamically to change of business environment and user requirements. Existing coordination approaches have no effective mechanisms to achieve dynamic coordination at runtime.

The reminder of this paper is organized as follows. Section 2 presents a coordination space based resource-centered coordination model. In Section 3, the implementation approach to dynamic coordination is proposed. Finally, we conclude our works in Section 4.

2 Coordination Space Based Resource-Centered Coordination Model

We classify coordination into two types: resource-centered coordination and service-centered coordination. The resource-centered coordination is required to be imperative, otherwise system faults may happen. For example, access to critical resources must be exclusive mutually, resource consumers is demanded to wait for data created by the resource producers, clients wait for response of servers, and so on. The service-centered coordination aims at improving system performance, and will not result in system failure even though without coordination. For example, there are two processes which have similar or overlapped functions, if the one is just done, the other can directly use the result to improve efficiency.

Due to limited space, this paper will not discuss service-centered coordination. The resource-centered coordination model includes three parts: computation entities, coordinators and coordination space.

2.1 Computation Entities

In general, the system contains multiple computation entities. The computation entity is consisted of role, actors and services.

Role is the abstract of a group of actors that have the same behaviors. Each computation entity only contains a role. Each actor only belongs to a role.

Actor is a process, and responsible for achieving the computation function of system. Each role may contain several actors. When actor starts, it registers provided services to registry in coordination space. Similarly, actors are needed to logout to registry when leaving the system.

Services are responsible for activities of actors, and different services can run concurrently. Each actor may provide multiple services, and each runtime service corresponds to one activity instance as a working thread.

2.2 Coordinators

The function of coordinators is to achieve coordination among computation entities. Coordinator has three kinds: communication coordinator, resource coordinator and service coordinator.

The communication coordinator is responsible for communication between coordinators and computation entities.

The resource coordinator is responsible for synchronization and mutual exclusion of access to shared resources.

The service coordinator aims to choose proper service for requested activity and schedule services according to service dependencies described in coordination space.

2.3 Coordination Space

Coordination space contains description information of the system for coordination, including computation description, resource description, environment description, registry, and so on.

Computation Description. The computation description contains description information of all roles, actors and services. Role description mainly includes role identifier, name and actors. Actor description mainly includes actor identifier, name, startup path and parameters. Service description mainly contains service handler and its corresponding activity.

Resource Description. The resource description mainly contains resource identifier, name, location, property, number, states, dependency between producers and consumers, and runtime resource object.

Environment Description. The environment description mainly contains system parameters (such as operating system version, timeout, performance, and so on), configuration information (such as startup sequence of role actors, restriction on the number of actors belonging to each role) and runtime context (such as status of service, service availability).

Registry. The registry is aimed to receive login/logout requests from roles/actors, and update description information in coordination space when roles, actors or services change. When new roles, actors and services enter to the system, they must login to the registry firstly. When leaving the system, they must logout to the registry.

In the model presented above, coordination is separated from computation. The coordination behaviors are not hard-wired into the programs when the system is designed, but are independently described in coordination space. Part of description information in coordination space can be updated at runtime so as to adapt to change of requirement and environment.

3 Implementation of Dynamic Coordination

The resource-centered coordination means to coordination shared resources among computation entities. When the shared resource is concerned with multiple services, access to it may be sequential, parallel and exclusive. So properties of resources must be considered.

3.1 Property of Shared Resources

Form the view of divisibility, resources may be divisible or indivisible. If the resource is divisible, it can be properly divided into several separated parts to avoid access conflict. For example, main memory can be divided into multiple areas for processes to access different parts at the same time without conflict. Form the view of reusability, resources may be reusable or non-reusable. If the resource is reusable, it can be used several times. For example, a dada file still can be used by other process after it is read or written by one process. If the resource is non-reusable, it will vanish after being used one time by any process. For instance, the data of pipe will be removed after being read.

So the two properties of shared resources are defined as follows:

Write: indicates the resources are writable, and read only is as default.

Reusable: indicates the resources can be repeatedly used. Otherwise, it is non-reusable.

3.2 Implementation Approach to Coordination

Coordination relationship of resource implies that dependency exists among services access to the same resource. For instances, sharing dependency often implies mutual exclusion of access to the resources, prerequisite dependency usually implies data flow dependency between resource production and consumption. For two services, coordination relationship of resources can be simply classified into the following three types:

Production-production. Two services create the same resource. Because of resource created by multiple producers, coordination is required to avoid repeated generation of the same resource and conflict of resource location.

Production-consumption. One service is responsible for producing resource and the other consumes it only. The shared resource is either arranged in fixed location or transformed from one service to another. Synchronization among services must be achieved.

Consumption-consumption. Two services use the same resource existed. The resource will not be existent after being consumed, if it is not reusable. If resource is reusable, coordination is required among consumers. Otherwise, it is not allowed to use the resource by multiple consumers.

Moreover, the following problems need to be considered:

Access permission. When resource is used, the system must decide whether requestor has right to access to it.

Resource format. In general, for the same resource, different services often provide or require different format. One way is to define uniformed resource format. Another way is to convert to corresponding format before the resource is used.

So different coordination approaches can be used according to different coordination relationships.

Mutual Exclusion among Shared Resources. The following methods can be used:

Limit concurrency. For example, shared resource is copied to multiple addresses, or resource is divided into several different parts for multiple services to use individually.

Use Semaphore provided by Operating Systems. This way is only suitable for endogenous coordination.

Control startup of all services and access to resources by centralized coordinator. This way is able to avoid conflict of access to resources, but it is easy to result in single points of failure.

Autonomous coordination by interaction among services. This way needs high communication costs.

Coordination between Producers and Consumers. There are implicit timing relationship between producers and consumers. Meanwhile, there is mutually excluded sharing relationship among multiple produces or multiple consumers.

From the view of the number of producers and consumers, there are the following four kinds of relationships:

1:1. One producer and one consumer. There is only implicit timing relationship between producer and consumer, and synchronization mechanisms are needed.

1:N. One producer and multiple consumers. Mutual exclusion is needed among consumers, and synchronization is needed between producer and consumers.

N:1. Multiple producers and one consumer. Mutual exclusion is needed among producers, and synchronization is needed between producers and consumer.

M:N. Multiple producers and multiple consumers. Mutual exclusion is needed among producers and among consumers, and synchronization is needed between producers and consumers.

Combined Production. Resources are Created by Multiple Producers Jointly. In real systems, coordination among producers can be achieved by dividing resource into separated multiple parts or introduce an extra service to combine several parts to the whole.

Combined Consumption. Resources are Used by Multiple Consumers Jointly. In real systems, coordination among consumers can be achieved by dividing resource into separated multiple parts or introduce an extra service to remove the whole resource after all combined consumers are complete.

The four coordination states of resource are defined, including *Consumable, Producible, Using* and *Unusable.*

Resource coordinator mainly provides three primitives: initialize(), request() and release(). When resources coordinator starts, the initialize() is invoked to create shared resource objects according to description in coordination space. When service concerned with shared resources is scheduled to run, the request() is invoked by resources coordinator to decide whether the resource is available, if not, the service is appended to the resource waiting queue. When service concerned with shared resources finish, the release() is triggered by resources coordinator to activate a waited service to run or set the resource status to be usable.

4 Conclusion

This paper presents a coordination space based resource-centered dynamic coordination model, which divided software system into computation entities, coordinators and coordination space. Resource-centered coordination relationships are defined. The implementation approach to dynamic coordination is discussed. Dynamic coordination is achieved by means of coordination space. Description information in coordination space is modified automatically when roles, actors and services change at runtime. The approach proposed in this paper can adapt to change of requirement and environment.

Acknowledgments. This work was partly supported by the National Hi-Tech Research and Development Program of China (863 Program) under Grant No. 2009AA062801.

References

1. Malone, T.W., Crowston, K.: The Interdisciplinary Study of Coordination. ACM Computing Surveys 26(1), 87–119 (1994)
2. Coates, G.: Agent Coordination Aided Distributed Computational Engineering Design. Expert Systems with Applications 31, 776–786 (2006)
3. Tu, X.Y., Wang, C., Guo, Y.H.: Large Systems Cybernetics. Beijing University of Posts and Telecommunications Press, Beijing (2005)
4. Carriero, N., Gelernter, D.: LINDA in Context. Communications of the ACM 32, 444–458 (1989)
5. Menezes, R., Omicin, A., Viroli, M.: On the Semantics of Coordination Models for Distributed Systems: The LogOp Case Study. Electronic Notes in Theoretical Computer Science 97, 97–124 (2004)
6. Space Based Computing, http://www.complang.tuwien.ac.at/eva/
7. Arbab, F.: Reo: A Channel-based Coordination Model for Component Composition. Mathematical Structures in Computer Science 14, 329–366 (2004)
8. Clarke, D., Proença, J., Lazovik, A., Arbab, F.: Channel-based Coordination via Constraint Satisfaction. Science of Computer Programming 76(8), 681–710 (2011)
9. Marth, K., Ren, S.P.: The ARC Programming Model-Language Constructs for Coordination. Electronic Notes in Theoretical Computer Science 229, 95–113 (2009)
10. Colman, A., Han, J.: Using Role-based Coordination to Achieve Software Adaptability. Science of Computer Programming 64, 223–245 (2007)

Application and Development of Multimedia Technology

Juan Liu

Jiangxi University of Finance and Economics, Nanchang, Jiangxi, 330032, China
hhyy1992006@126.com

Abstract. This paper introduced the concept of multimedia and multimedia technology.This paper discussed the application of multimedia technology in the education and training, office automation, diagnosis and medical applications and other aspects, and analysed the development trends of multimedia technology .

Keywords: multimedia, multimedia technology, applications, trends.

1 Introduction

"Multimedia" was translated from the English word of "Multimedia", Multimedia is a new word synthesized by the two words of the media and the multiple,media possesses various kinds ,as long as playing a role of pipeline in the process of information transmission. It can be considered a "transfer of informationmedia "which contain usually text, graphics, video images, computer animation, voice ... ... The prefix of Multimedia is result from the word of multiple, and which express the sense of "many".So multiple + media constructing the word of Multimedia means " multiple media ".

Multimedia technology is still in rapid development of a comprehensive electronic information technology.20 years ago,It is multimedia that few people made slide coupled with synchronized sound .Today, with the rapid development of micro-electronics, computer, communications and digital audio-visual technology, multimedia technology is given a new content.It is the text, audio, video, graphics, images, animation and other media information through a computer to digitize the collection, acquisition, compression decompression, editing, storage and other processing, then shown alone or in synthetic form of the integration of technology.Multimedia technology for computer applications to develop a wider area. The current multimedia hardware and software have been able to data, voice and high-resolution images of objects as windows software to do all kinds of treatment.The emergence of multimedia applications in a variety of colorful, original computer technology not only icing on the cake, and complex things become more concrete.Shanghai Municipal Government defines that a clear message is the important constituent of the modern city.By 2007, the initial build the basic framework of "digital city" , in 3 years to 5 years of time to make a multimedia production value of Shanghai reach 100 billion yuan.In Expo 2010 , in urgent need of more ten thousand of professionals working on multimedia presentations people, nearly 200 Expo Hall to be built will involve multimedia presentations and

D. Jin and S. Lin (Eds.): Advances in CSIE, Vol. 1, AISC 168, pp. 661–667.
springerlink.com © Springer-Verlag Berlin Heidelberg 2012

multimedia network communication technology.To promote the industry's rapid development, faster and international standards, the Shanghai municipal government will invest a billion dollars for construction of multimedia technology and equipment, will invest tens of millions of the last stage of the construction of a public open multimedia platform.Meanwhile, under the guidance and support of the Shanghai Multimedia Industry Association, Shanghai has implemented a "multi-media design and application capability assessment" project will train a number of different areas of society to adapt to the requirements of multimedia professional and technical personnel, which will lead to unprecedented procurement and learning climax.To this end the Shanghai Multimedia Industry Association and Shanghai Technology Convention & Exhibition Co. constructed powerful combination, to create a window to pass the government's industrial policy to the multimedia industry of R & D research, sales, agency and other parties , to access to market information, to show promoting new product, to understand the latest status of the domestic and outside peer, to show the latest multimedia technologies and products to professionals and visitors .

2 A Multimedia Technology in Education and Training Applications

The impact of multimedia technology in education is much more than its on other areas.Multimedia technology has changed the traditional teaching methods, teaching materials to make great changes, not only in the text, still images, moving images and voice as well.Application of multimedia technology in teaching is to improve the quality of teaching and effective way of education, in the form of education can also be diverse, and more be interactive learn in a distance,which is much good than traditional classroom teaching methods.

2.1 Effective Learning

This has been acknowledged by many education experts, not only in the interactive teaching, classroom teaching and other aspects of teaching demonstration effect far more than traditional teaching, but also other ways to create a simulation can not achieve authenticity and intimacy.Multimedia teaching not only for students to create a lively learning environment, so that teachers have a high level, high-quality teaching and learning environment.In addition, dangerous and high cost, long cycle simulation of the natural world is also the most adept of multimedia (such as: chemistry and biology, geography and other subjects in the elements and real-world analog). Hypermedia technology is make promote the changes of teaching mode , so that it can be effectively improve the teaching on these aspects of the traditional teacher-centered teaching model to a learner-centered, individual learning to collaborative learning, learning from passive to active exploration.

2.2 Persuasive

Multimedia systems can constitute a strong persuasive programs, such as personal plans, project feasibility studies, reports, the image presentations of enterprises and institutions etc. It makes vivid effect, the user handy for education, especially vocational education, adult education more convincing.Human sensory organs are each to receive information, the best weapon to change the concept of behavior, but they use a variety of multimedia interaction, or both sensory stimulation to produce multiple transmit information.Studies have shown that humans have access to information from the outside 83% visual, 11% through hearing, through visual, auditory information obtained access to information from outside accounted for 94%, so the audio-visual media and computer combination will greatly improving access to effective, lively images, animation, real photos, coupled with voice generated by the simulation results may not be comparable to a single media.

2.3 Best Creative Thinking Training

Provoking the learner's imagination is the most fascinating multi-media where various media combined with the computer can make the human senses and imagination with each other, resulting in an unprecedented space for thinking and creating resources.Multimedia teaching and learning environment are more favourbale to score the capability of heuristic teaching, students exploring , training, creative thinking for teachers.Especially the starting of distance education, multimedia network also shows the training of personnel in education and extraordinary talent, and will effectively promote the development of CAI.

Table 1. The Practice process of making use of multimedia courseware by Authorware

Editing procedures	Detailed operation
Integrated material	First,planning out the entire program schedule in accordance with the requirements of the text script, according to the planning,draw the appropriate icon onto the corresponding position in the flow line from the icon bar, and then edit the icon of the contents and set the relevant parameters
Debugging courseware	Complete Integration is need to test whether the interactive functions are fulfilled by the program check in the program effect.How the multimedia effects,how the program play the system, whether to meet user needs.
Courseware publishing	authorware itself has the function of the publishing process, the program can be packaged intoa variety of practical procedures. The method of operation is as follows: toolbar file menuout ofcloth a package command

3 Multimedia Technology in Office Automation Applications

A few years ago, when leaders or department inspection reports are often read by the competent leadership of the manuscript.In recent years, reports are increasingly using multimedia, multimedia demo prior to report, through the projector for presentations.Because multimedia demo set pictures, sound and animation in one, to report more vivid, richer, more flexible form.As a result of advanced digital imaging and multimedia computer technology to document scanners, fax machines, microfilm documents and communications systems and other modern office equipment, integrated network management together, this will constitute a new office automation system.Office staff no longer have to face a thick document, but easily in front of a computer and a variety of multimedia devices can be office.

4 Multimedia Technology in Other Applications

In recent years, a variety of popular multimedia authoring tool, you can create a variety of electronic publications and a variety of textbooks, reference books, maps, medicine, health, business manuals and game entertainment, and multimedia applications; demonstration systems or information inquiry system; entertainment, video, animation and advertising; dedicated multimedia applications, etc..Currently sold in the domestic market of multimedia applications are: force electronics companies developed "real estate sales advisory system"; Beijing Haida International Computer Software Engineering has introduced the "768 Ventura multimedia TV programs automatically create sound and playsystem "and" Touch the touch screen media query system ", can be used for large-scale real-time sports information inquiry, leading decision-support systems, hotel information inquiry system, guide system, dance hall song settlement system, the store shopping guide system, production of commercial real-time monitoring systems and securitiestrading real-time query system.Currently, multimedia technology, mainly from the following directions.

4.1 Study of Tendency of Multi-media Communication Network Environment

Multimedia environment, research and communication networks, will make stand-alone single point from the distribution of multimedia, collaborative multimedia environment development in the world to establish a global free interactive communication network.Equipment to the network and its applications in research and distribution of online research and information services will be hot.Future multimedia communications will move without time, space, communications and other objects in respect of any constraints and limitations of direction, its goal is "to anyone, at any time, any place with any person, any form of communication.".Human communication through multimedia rapid access to large amounts of information, in turn, means the maximum Youyi creating greater social benefits for the community.

The unique advantage of the media. Make multimedia courseware by using Authorware ,which is divided into the following three steps.

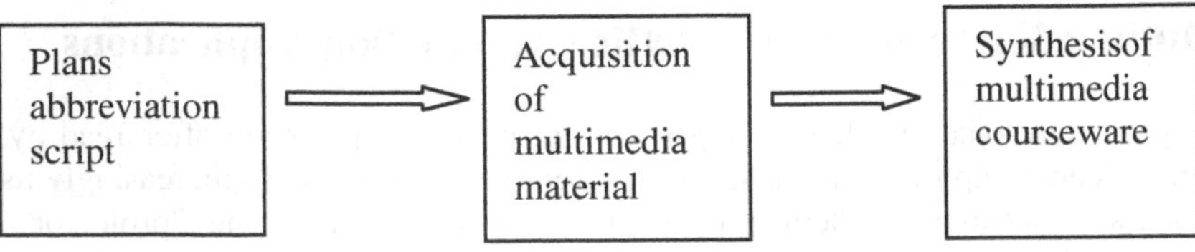

4.2 Multimedia Standard Is Still the Focus of the Study

Study of various types of standard products will help standardize the application .Due to the multimedia industry as the core information security industry breakthrough single limit, covering many industries, and multimedia systems integration characteristics of the standardization of a very high demand, it is necessary to carry out standardization, it is a large-scale multimedia information exchange and industryof the key.

4.3 Multimedia Technology and Peripheral Technology of Virtual Reality Construct

Continues multimedia virtual reality and visualization technology needs complement each other and with the voice, image recognition, intelligent interface technology, to establish high-level virtual reality system.

5 Multimedia Technology Trends

Overall, the positive aspects of multimedia technology development: ① networking trends, technology, communications and broadband network with each other, so that multimedia technology into the research design, business management, office automation, remote education, remote medical care, counseling search, culture and entertainment, automatic monitoring and control and other areas; ② multimedia components, intelligence and embedded technology to improve multimedia performance computer system itself, the development of intelligent home appliances.

5.1 Multimedia Technology Development Trend of Network

Technology innovation and development will, such as servers, routers, switches and other network equipment performance more and more, including the client CPU, memory, graphics card's hardware capabilities, including an unprecedented expansion, it will benefit from unlimited computing andplenty of bandwidth, which allows network applications to change the past, those who passively accept the status of processing information, and a more proactive stance to participate in front of the virtual world.The development of multimedia technology, multimedia computer, the computer will form a better support collaborative work environment, eliminating the barriers of distance, but also to eliminate the barriers of time distance, to provide better information on human services.Interactive, dynamic multimedia technology in a networked environment to create a more vivid 2D and 3D scenes, people can make use of cameras and other equipment, the office and multimedia entertainment tools for collections in the terminal on the calculator, available at any corner of the worldwith counterparts thousands of miles away in real-time video conference to discuss on the market, product design, to enjoy high-quality image display.Generation of user interface (UI) and artificial intelligence (IntelligentAgent) and other network-based, user-friendly, personalized multimedia software application can make different nationalities, different cultural backgrounds and different cultures of the people through "man-machine dialogue", to eliminatethe gap between them, the freedom to communicate and understand.Interactive multimedia technology, the multimedia technology in pattern recognition, holographic images, natural language understanding (speech recognition and synthesis) and the new sensor technology (handwriting input, data gloves, electronic odor synthesizer), etc., based on the use of humana variety of sensory channels and action channels (such as voice, writing, expression, posture, attention, movement and smell, etc.), data gloves and tracking through sign language information extraction specific facial characteristics, synthetic facial movements and expressions, and non-parallelaccurate way to interact with the computer system.

5.2 Multimedia Components, Intelligence and Development Trend of Embedded

Currently, the multimedia computer hardware architecture, video and audio multimedia computer interface software continue to improve, especially the use of the hardware design and software architecture, algorithms, combining programs to multimedia computers to further improve performance, but to meet the multimedia networkenvironmental requirements, the need for further software development and research, the multimedia terminal equipment has a higher and intelligent components, such as multimedia terminals to increase the recognition and text input, and input Chinese speech recognition, natural language understandingand machine translation, identification and understanding of graphics, robotics and computer vision and other visual intelligence.

In the past to consider CPU chip design more computing power, mainly for math and numeric processing, along with multimedia technology and network communication technology, requires a higher CPU chip itself has a comprehensive treatment of its sound, text, graphics, information and communicationfunction, so we can deal with the media information and real-time compression algorithm to achieve CPU chip.From the current development trend of this chip can be divided into two categories: ① based multimedia and communications capabilities.CPU chip integration of the existing computing power, it's designed to be used in multi-media special equipment, home appliances and broadband communications equipment, these devices can replace a large number of ASIC and CPU and other chips.② to general-purpose CPU-based computing, integration of multimedia and communications capabilities, their design goal is compatible with existing computer series, both multimedia and communications features, mainly used in multimedia computers.In recent years, with the development of multimedia technology, TV and PC technology, competition and integration of more and more compelling, traditional TV is mainly used in entertainment, while the PC focuses on access to information.With the development of TV technology, TV viewing watching features, interactive program guide, TV Internet access and other features have emerged.The PC technology programs in the media processing has also been a great breakthrough, enhanced video and audio streaming capabilities, search engines, online TV and other technology appear appropriate, comparisons, send and receive E-Mail, chat and video conferencing terminal function isPC and TV technology, integration points, and digital set-top box technology to adapt to the fusion of TV and PC trend, extending the "information appliance platform" concept, the set of multimedia home shopping, home office, home health care, interactive teaching, interactivegame, video mail and video-on-demand applications for a full, representative of today's embedded multimedia terminal development.Embedded multimedia system can be applied to people living and working in all aspects of industrial control and business management in areas such as intelligent industrial control equipment, POSATM machine, IC cards, etc.; in the family field, such as digital set-top boxes, digital TV, WebTV,Internet refrigerator, air conditioning and other network consumer electronics products, in addition, embedded multimedia systems are still medical electronic devices, multimedia mobile phones, PDAs, car navigation, entertainment, military and other fields have a great prospect.

6 Conclusion

In short, the increasingly wide range of multimedia applications, multimedia technological knowledge is increasingly important.The general trend is a better, more natural interaction, a greater range of information access services for the future of human life to create a more perfect new world of function, space, time and human interaction .

References

1. Huang, C.: The practical application of Multimedia technology. Xuzhou Institute of Technology (2006)
2. The development of geographic information systems and prospects. Annual Report of China's software industry (2000)
3. Wang, H.: Multimedia technology development status and future trends pale. Intelligence Theory and Practice (2001)
4. Zhang, Y.: The global software industry status, trends and challenges. Microsoft China Research (2001)
5. Lu, H.: Video conferencing systems and remote monitoring. Measurement and Control Technology (2000)

Color Image Retrieval Based on Color and Texture Features

Xiuxin Chen[1], Kebin Jia[2], Xiaoqin Lian[1], and Shiang Wei[1]

[1] College of Computer and Information Engineering,
Beijing Technology and Business University, Beijing, China
[2] College of Electronic Information and Control Engineering
Beijing University of Technology, Beijing, China
chenxx19790126.com

Abstract. To solve the problems of sensitivity to little color change, huge computation, low effect that exist in content-based image retrieval methods, a new method is proposed. This retrieval method is based on image color and texture features. First, each part of HSV color space is quantized using non-uniform quantization. Then, to eliminate the quantization error, adding each two neighbor values in the histogram to form a three-dimensional color histogram. A corresponding histogram intersection matching method is proposed at the same time. Moreover, a texture parameter is constructed to remove irrelevant images. Experiments show that the new method is effective to color image retrieval. It is also low in computation and robust to slight color changes.

Keywords: three-dimensional color histogram texture feature, color image retrieval, image retrieval.

1 Introduction

With the rapid development of image capture technology and internet, the number of digital image is getting larger and larger. Traditional keywords-base retrieval method cannot work efficiently anymore. How to get interested images based on their content has been a hot research topic.

Great progress has been achieved in the field of Content Based Image Retrieval(CBIR) in resent years. The color feature [1], texture feature [2], shape feature[3][4] and affine invariant features [5] have all been used in image retrieval. Of all image content features, color and texture are two important features and play an important role in image content.

Color histogram has the advantages of transform invariant, rotate invariant and scale invariant and has been widely used in image retrieval. This paper adds each two neighbor values in the histogram to form a three-dimensional color histogram which makes the method more robust to slight color changes. And also corresponding histogram intersection method is given in this paper. This paper also introduces a texture parameter which can effectively describe the texture feature of the images. This parameter is used to remove the irrelevant images in the retrieval results.

D. Jin and S. Lin (Eds.): Advances in CSIE, Vol. 1, AISC 168, pp. 669–674.
springerlink.com © Springer-Verlag Berlin Heidelberg 2012

2 Three-Dimensional Color Histogram

2.1 Quantization of HSV Color Vectors

In this paper, a (16:4:4) non-uniform quantization method is adopted in which H vector is divided into 16 values and S, V are divided into 4 values separately. The S, V and H values after quantization are shown as (1) and (2).

$$S=\begin{cases}0.075 & s\in[0,0.15)\\ 0.275 & s\in[0.15,0.4)\\ 0.575 & s\in[0.4,0.75)\\ 0.875 & s\in[0.75,1]\end{cases}\qquad V=\begin{cases}0.075 & v\in[0,0.15)\\ 0.275 & v\in[0.15,0.4)\\ 0.575 & v\in[0.4,0.75)\\ 0.875 & v\in[0.75,1]\end{cases}\tag{1}$$

$$H=\begin{cases}0 & h\in(345,360)or(0,15)\\ 20 & h\in(16,25)\\ 35 & h\in(26,45)\\ 50 & h\in(46,55)\\ 68 & h\in(56,80)\\ 94 & h\in(81,108)\\ 124 & h\in(109,140)\\ 153 & h\in(141,165)\\ 178 & h\in(166,190)\\ 205 & h\in(191,220)\\ 238 & h\in(221,255)\\ 265 & h\in(256,275)\\ 283 & h\in(276,290)\\ 303 & h\in(291,315)\\ 323 & h\in(316,330)\\ 338 & h\in(331,345)\end{cases}.\tag{2}$$

Altogether 16*4*4=256 color values are achieved after quantization.

2.2 The Construction of Three-Dimensional Color Histogram

In the image after quantization, its H vector has 16 values and S, V has 4 values separately. So this paper defines an array of 16*4*4 size which is T to calculate the ratio of pixels of each color to the overall pixels. The element in T is defined in (3).

$$T(i,j,k)=\frac{N_{i,j,k}}{M},(1\le i\le 16,1\le j,k\le 4)\tag{3}$$

Where $T(i, j, k)$ means the ratio of pixels whose color value are the ith value in H, the jth value in S and the kth value in V to the overall pixels of the image. $N_{i,j,k}$ is the number of pixels whose color values are values mentioned above. M is the overall pixel number of the image.

The above quantization method is a hard division of color values. Actually, color values are continuous and neighboring values are quite similar. This kind of division will cause problems: similar color values are divided to different values and are considered two different colors. When images are slightly changed in color, the quantization result may be quite different.

To reduce the quantization error, considering the overall number of pixels in two neighboring quantization value remain the same whether the pixels in the division edge are divided to the former value or to the later value, this paper adds the neighboring values in H direction in the three-dimensional color histogram. The result is considered as the final three-dimensional color histogram and is used to achieve image retrieval results.

$$F(i,j,k)=\begin{cases}\dfrac{N_{i,j,k}+N_{i+1,j,k}}{M} & 1\le i\le 15\\[2ex] \dfrac{N_{i,j,k}+N_{1,j,k}}{M} & i=16\end{cases} \tag{4}$$

The size of F is also 16*4*4.

2.3 Color Histogram Matching

Define sample image as Q and the image to be matched is D and their three-dimensional color histograms are F_Q and F_D. Histogram intersection defined in (5) is performed to determine similarity of these two images.

$$S(Q,D)=\frac{\sum_{i=1}^{16}\sum_{j=1}^{4}\sum_{k=1}^{4}\min(F_Q(i,j,k),F_D(i,j,k))}{2M} \tag{5}$$

Where M is the overall number of pixels in the image. *min()* is a function that can give the smallest value. The value range of $S(Q,D)$ is 0 to 1. The more similar the two images are, the bigger the value of S is. For the same images, S is 1.

3 Texture Feature Extraction

To extract the texture feature of an image, the color image is converted to a gray one which is defined as g and the pixel in it is defined $g(x,y)$. The image is divided to 16 blocks according to Fig. 1 and each block is labeled as $B_i(i=1,...,16)$.

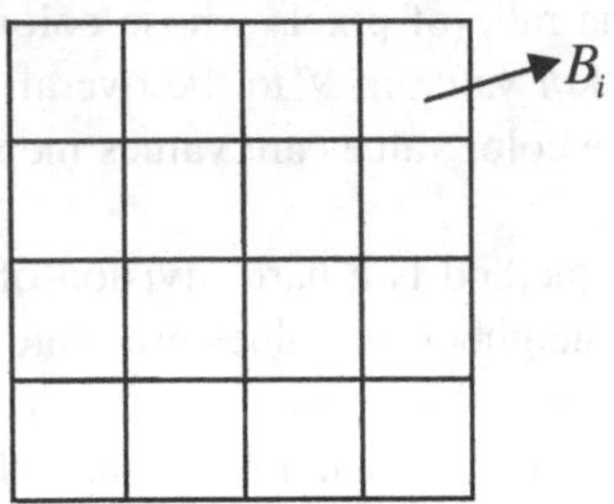

Fig. 1. Division of the image to get texture feature

For images which can not be equally divided, take the maximum part which can be divided by 4 as the image to get 16 blocks. So the width and height of the image to extract texture feature is as (6).

$$\begin{aligned} W &= 4*[W_o/4] \\ H &= 4*[H_o/4] \end{aligned} \tag{6}$$

Where W_o and H_o are the width and height of the original image and [.] means to get integer part of a float number.

The texture feature of an image is computed using (7).

$$T = \sum_{i=1}^{16} \frac{\sqrt{\sum_{(x,y)\in B_i} (g(x,y) - A_i)^2}}{M_i} \tag{7}$$

Where A_i is the average gray value of the block B_i and M_i is the number of pixels in block B_i.

T is used to determine whether the images in the retrieval results achieved through the three-dimensional color histogram based method are similar to example image in texture feature. If the difference of *T* value of a result image and that of the example image is small than a threshold *G*, the result image is considered as a correct result. Otherwise, it will be eliminated from the retrieval result.

4 Experimental Results and Analysis

The testing image set includes three types of Corel image set which are 300 original images. Each one of the original images is slightly changed in color using Photoshop. Altogether 600 images are used in the experiment. The Algorithm is performed using Visual C++ 6.0.

Fig. 2 shows a testing image and 6 most similar images in the retrieval result.

(a) Testing image1

(b) The 6 most similar retrieval result images

Fig. 2. Testing image1 and its 6 most similar images

From Fig. 2 it can be seen that the method introduced in this paper is very effective. The comparing of precision and recall of this method and traditional color histogram method is shown in Table 1.

Table 1. Comparing of this method and other methods

	Traditional color histogram method	Method in this paper
Recall	87%	96%
Precision	81%	94%

From Table 1, it can be concluded that this method is better than traditional color histogram based method. It can be used in color image retrieval.

The experiments are performed under computer of Intel Core 2 Duo CPU, 2.4GHz, 2G RAM and with operating system of Windows XP. The size of images is 384*256. For a image set of 600 images the average time is 2 second which is very fast.

5 Conclusions

Content Based Image Retrieval is a research hot in computer science. It is getting more and more important to practical applications. This paper introduces an effective image retrieval method which is based on the color feature and texture feature of images. Experiments show its effectiveness and robustness to slight color changes. It is small in computation amount and can be used in color image retrieval.

Acknowledgments. This paper is sponsored by the project of Beijing Undergraduate Science Research and Undertaking Action Program (No. 19005114009), Science Research Starting Fund for Young Teachers of Beijing Technology and Business University (2011) and Scientific Innovation Platform: The Research Platform of Signal Detection and Intelligent Signal Processing (No. 201151).

References

1. Pun, C.-M., Wong, C.-F.: Fast and robust color feature extraction for content-based image retrieval. International Journal of Advancements in Computing Technology 3(6), 75–83 (2011)
2. Ning, J., Zhou, H.: Study on image retrieval methods with texture on the basis of compression. Computer Applications and Software 28(6), 254–256 (2011)
3. Kao, C.C., Lai, Y.T., Lin, C.H.: An efficient reflection invariance region-based image retrieval framework. International Journal of Imaging Systems and Technology 20, 155–161 (2010)
4. Yang, X., Yang, X., Tian, X.: Robust approach for 2D shape based image retrieval. Pattern Recognition and Artificial Intelligence 23(5), 738–744 (2010)
5. Mikolajczyk, K., Tuylelaars, T., Schmid, C., et al.: A comparison of affine region detectors. International Journal of Computer Vision 65(1/2), 43–72 (2005)

Automatic Organization of Programming Resources on the Web

Guojin Zhu and Lichao Fu

Dept. of Computer Science and Technology, Donghua University
Shanghai 201620, China
gjzhu.dhu@163.com, dhu.try@gmail.com

Abstract. There are lots of programming problems and their solution reports distributed on the web. These programming resources, which are valuable for programming learners, are not organized for teaching and learning. To address such issue, we propose a method to organize the programming resources automatically on a basis of a predefined hierarchical body of programming knowledge which includes algorithms, data structures, graph theory, number theory, combinatorial mathematics, computational geometry, etc. Our experiment shows that nearly 80 percent of the programming problems with their solution reports could be organized by the proposed method, which finds 830 problems suitable for learners to practice from the first 1000 problems on an online judge, and analyzes 6064 corresponding solutions out of 12174 reports obtained by a search engine.

Keywords: Programming resources, Programming problems, Online judges, Programming solutions, Web resources, Programming knowledge, Programming practice, Automatic organization.

1 Introduction

Programming teaching, which is one of the core curriculums in the basic science education, has attracted increasing attention. More and more universities are participating in ACM International Collegiate Programming Contest (ACM/ICPC) [1], which has became an annual contest for the students to show their innovative capability and problem-solving ability under pressure. Nowadays, there are a number of online judge (OJ) systems available on the web that provides thousands of programming problems gathered from many contests. However, it is difficult to find the problems that are suitable for programming learners.

In the meantime, some contestants usually write down their experiences in participating in ACM/ICPC and upload their problem-solving reports to their own blogs in order to share with other programming enthusiasts. Each of these reports gives a solution to a specific problem, while some reports include other relevant problems that share the same knowledge points with the specific problem. Due to the limited individual ability, however, no one can solve every programming problem in the OJ system. On each of the blogs only limited problems can find their solution reports.

D. Jin and S. Lin (Eds.): Advances in CSIE, Vol. 1, AISC 168, pp. 675–681.
springerlink.com © Springer-Verlag Berlin Heidelberg 2012

In this paper, we propose a method to organize these programming resources automatically into a hierarchical body by identifying programming knowledge in the blogs. To this end, we employ the web technologies [2-4] to find corresponding problem-solving reports for the programming problems. There are two ways to obtain the reports. One is to download them from the official websites of the contests. However, some official websites have been out of date and their links no longer work, while some contest sponsors do not provide any problem-solving report at all. The other is to search the problem-solving reports which are written by excellent contestants who upload them to their own blogs from time to time. We are inclined to the latter.

Usually, it will take learners too much time to understand the ideas or the algorithms in the problem-solving reports. It is desirable that the machine could tell the learners the programming knowledge used in the reports. We achieve it on a basis of hierarchical knowledge representation. The idea of hierarchical knowledge representation is derived from [5-6].

The relationship between problems could be established after identifying the programming knowledge in the reports, leading to organizing the programming resources automatically.

2 Programming Resources

There are some common characteristics among OJ systems. For example, every 100 problems are stored in one volume and their ID numbers are sorted. We can get the problem ID numbers, the titles of the problem statements and their source information by parsing the HTML source codes of pages from OJ systems.

For each problem, we can use a set of keywords (e.g., OJ name, problem ID, problem title) to search its problem-solving reports with a search engine, and parse the HTML source codes of the resulting pages to get the URL information of the reports. Usually, the URLs of problem-solving reports will appear in the first few pages. However, there is much of useless information, such as the information about forums, Wiki, Google Groups, Baidu Wenku, and even the source information of the problem itself. After filtering these useless web pages, the remaining ones can be regarded as candidates that may contain the solutions to the specified problem.

Some of these candidates are summaries of problem classification and/or algorithm classification. Some are too long to be problem-solving reports. After filtering these two kinds of candidates, we extract the body parts from the rest HTML source codes and save them as local files.

For example, a problem named "Genghis Khan the Conqueror" is originally from the ACM/ICPC 2011 Asia Fuzhou Regional Contest. We know that this problem is on the website of the HDU OJ system, whose number is 4126. We search its reports by the keywords "HDU 4126 Genghis Khan the Conqueror" with the Google search engine. By parsing the HTML source code, we can obtain 10 URLs from the first resulting page, but only 5 URLs remain after filtering the useless information and analyzing the content, see Table 1. With these 5 URLs, we can get 5 reports, the first one of which is shown in Fig. 1.

Table 1. Website information by filtering.

No.	Website	Reserved
1	http://attiix.com/2011/11/25/hdu-4126-genghis-khan-the-conqueror/	Yes
2	http://attiix.com/	Yes
3	http://acm.hdu.edu.cn/showproblem.php?pid=4126	No
4	http://acm.hdu.edu.cn/search.php?field=problem&key=Asia	No
5	http://acm.split.hdu.edu.cn/search.php?field=problem&key=asia	No
6	http://www.shuizilong.com/house/archives/3558	Yes
7	http://www.shuizilong.com/house/archives/716	Yes
8	http://www.shuizilong.com/house/archives/3574	Yes
9	http://acmicpc.info/archives/565	No
10	http://en.wikipedia.org/wiki/Genghis_Khan	No

Source @ 2011 Asia Fuzhou Regional Contest

Given a graph (N<=3000, M<=N*N) and some queries(Q<=10000), for each query, specify exactly one edge with a greater value, calculate the corresponding MST.

First we may think of that, the MST won't change much when only one edge is changed. So we can try to find the new MST from the origin MST for each query.

...

Fig. 1. The content of a downloaded report is saved as a local file.

3 Knowledge Identification

There is a lot of information about knowledge points in the report, e.g., Minimum-cost Spanning Tree (MST) in Fig. 1. We can identify the knowledge points on a basis of a multi-tree structured catalogue of programming knowledge.

The root of each tree in the catalogue represents a class of knowledge points, such as sorting, search, graph theory, number theory, computational geometry, and etc. Its child nodes stand for sub-classes. For example, shortest path problems, difference constraints, minimum-cost spanning tree are all belong to graph theory, see Fig. 2. The depth of each tree is not greater than 3. There are 14 classes and about 200 sub-classes in our predefined hierarchical body of programming knowledge, which can be incremented according to the developmental tendency of ACM/ICPC.

Most authors of problem-solving reports prefer the abbreviations for some algorithms and/or data structures, while others do not. We use a group of aliases to stand for the same knowledge point. For example, extended euclid, extended gcd, ext euclid, ext gcd are all the aliases of extended euclid algorithm. Authors may come from different countries around the world. In this work, we only consider the reports in Chinese or in English.

Let T be all the knowledge points in our predefined hierarchical body of programming knowledge. Every knowledge point in T is represented by a group of aliases. Let K_i be the set of knowledge points contained in a report i. The procedure for identifying the knowledge points in a report i is described as follows:

1) Get K_i by matching with the aliases of every knowledge point in T;
2) Expand K_i by adding knowledge points recursively up, until the root node in T.

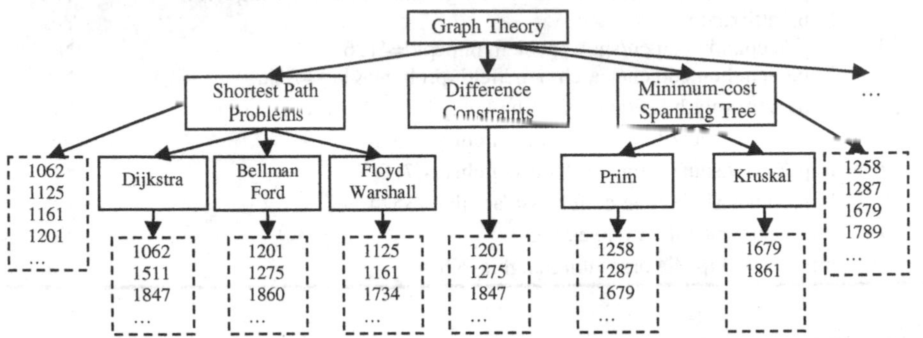

Fig. 2. Graph theory is one of the trees in the predefined hierarchical body of programming knowledge. Each leaf node is the corresponding group of problems.

Some reports contain too many knowledge points. They are not problem-solving reports. They are just web pages that present algorithm classification and/or problem classification. We discard these reports by deleting their knowledge points. The procedure is as follows:

1) Denote the size of K_i by N_i;
2) Remove all of elements in K_i (i.e., $K_i = \varnothing$) if N_i is too large; otherwise, keep K_i unchanged.

After identifying the knowledge points from all the reports of a given problem p, we can determine the mainstream knowledge used to solve the problem p as follows:

1) Let $F = \varnothing$, where F is a list whose elements are knowledge points:
2) For each report of the problem p, insert every element of K_i into the list F, resulting in that F contains every knowledge points from all the reports of the problem p and that a knowledge point may occurs several times in the list F;
3) Let V be the set of all the elements in the list F;
4) For each element j in V, calculate the number of its occurrences in the list F, and denote it as E_j;
5) Calculate the ratio of E_j to M, and denote it as R_j. where M denotes for the largest number of occurrences calculated in the step above;
6) Remove the element j from V if its R_j is too small.

The final resulting elements in V are regarded as the mainstream knowledge for the problem p.

4 Resource Organization

Take the problem named "Genghis Khan the Conqueror" as an example again. Initially, there is only a relationship from its problem-solving reports to the problem (see the dashed arrow labeled *original* in Fig. 3), but no other physical associations exist between the problem, its problem-solving reports and the knowledge points used to solve the problem.

After searching the reports on the web for this problem (indicated by the arrow labeled *search* in Fig. 3) and downloading the contents to local files (see the dashed arrow labeled *download*), we can establish the connection from the problem to its problem-solving reports (see the arrow from the box labeled *problem information* to the box labeled *report files*). With this connection, a programming learner can find the solutions to the problem easily whenever necessary.

Next, we can establish the relationship between the problem and the knowledge points (indicated by the two arrows between them) by identifying the knowledge points in its problem-solving reports (see the dashed arrow labeled *identify*). With this relationship, a programming learner can go forward and back between the problem and the knowledge points freely.

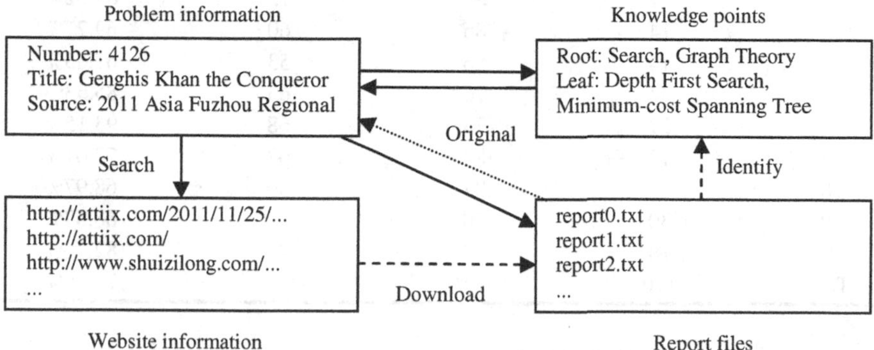

Fig. 3. The relationships are established between a problem, its reports and the knowledge points used to solve the problem.

Finally, we can put the problem into a class of problems that need the same knowledge points to solve. Once a contestant concentrates on some knowledge point in training, we can recommend a group of problems related to this knowledge point. In Fig. 2, each leaf node stands for a group of problems which need the same knowledge point to solve.

5 Experiment and Results

The problems used in this experiment were downloaded from the largest OJ system in China, i.e., Peking University (PKU) OJ. There are over 3000 programming problems which are from various contests over the years in this OJ system.

We selected the first 1000 problems from the first 10 volumes. Our program got rid of 170 very difficult problems (see the bottom of the second column in Table 2). By 'very difficult', we mean that the number of successful submissions is less than 80, and/or the percentage of successful submissions is less than 20%. The rest of 830 problems are regarded as suitable ones (see the bottom of the third column in Table 2). Meanwhile, our program downloaded 12174 corresponding reports by using the function of the Baidu search engine which is the largest Chinese search engine [7]. 6110 reports which are not in the true sense were filtered, and the rest were 6064 solution reports.

As shown in the last 2 columns of Table 2, we found that 662 out of 830 problems (i.e., 79.76%) whose solution reports were found successfully by the process mentioned in the previous section. These 662 problems are called classified problems. This result illustrates that most of the problems can be associated with their reports and knowledge points.

Table 2. Programming problems statistics.

Volume	Difficult	Suitable	Classified	Percentage
1	6	94	82	87.23%
2	11	89	73	82.02%
3	14	86	60	69.77%
4	14	86	53	61.63%
5	24	76	65	85.53%
6	27	73	68	93.15%
7	13	87	67	77.01%
8	13	87	60	68.97%
9	30	70	62	88.57%
10	18	82	72	82.76%
Total	170	830	662	79.76%

Table 3. Root-level knowledge points statistics.

No.	Knowledge point	Number of problems
1	Search algorithm	299
2	Sorting algorithm	26
3	Basic algorithm	173
4	Data structure	109
5	Number theory	36
6	Graph theory	108
7	Network flow	6
8	Dynamic programming	99
9	String processing	19
10	Computational geometry	60
11	Combinatorial mathematics	15
12	Linear algebra	9
13	Probability theory	0
14	Game theory	2
Total		965

After associating each problem with knowledge points identified in its solution reports, it is easy to organize the programming resources. As in Fig. 2, the problems were added to the corresponding leaf nodes of the tree. Table 3 shows the number of problems in each tree. It can be found that some popular knowledge points are used frequently, such as search, data structure and graph theory. But some are used scarcely, such as linear algebra, probability theory and game theory. In addition, the total number (965) is greater than the number of classified problems (662), because it is possible that several knowledge points are associated with the same problem.

6 Conclusion

We have presented a procedure to organize programming problems and their solution reports according to the programming knowledge they share. This procedure includes searching for possible solutions to a given problem, identifying knowledge points in the solution reports, and classifying the programming resources. Our experiment shows that the most programming recourses on the web could be organized automatically on a basis of a hierarchical body of programming knowledge. We believe that this work is helpful for programming teaching and learning.

Acknowledgments. This research is supported by the National Natural Science Foundation of China (NSFC) under Grant No. 60973121.

References

1. ACM/ICPC, `http://cm.baylor.edu/welcome.icpc`
2. Ding, L., Finin, T., Joshi, A., Pan, R., Cost, S.R., Peng, Y., Reddivari, P., Doshi, V., Sachs, J.: Swoogle: a search and metadata engine for the semantic web. In: Proceedings of the 13th ACM Conference on Information and Knowledge Management, New York, USA, pp. 652–659 (2004)
3. Geng, J., Yang, J.: AUTOBIB: Automatic extraction of bibliographic information on the Web. In: Proceedings of the 8th International Database Engineering and Applications Symposium, pp. 193–204 (2004)
4. Lau, K.-N., Lee, K.-H., Ho, Y., Lam, P.-Y.: Mining the web for business intelligence: Homepage analysis in the internet era. Journal of Database Marketing 12(1), 32–54 (2004)
5. Yoon, Y., Lee, C., Lee, G.G.: An effective procedure for constructing a hierarchical text classification system. Journal of the American Society for Information Science and Technology 51(3), 431–442 (2006)
6. Day, M.-Y., Tsai, R.T.-H., Sung, C.-L., Hsieh, C., Lee, C.-W., Wu, S.-H., Wu, K.-P., Ong, C.-S., Hsu, W.-L.: Reference metadata extraction using a hierarchical knowledge representation framework. Proceedings of Decision Support Systems 43(1), 152–167 (2007)
7. Baidu search engine, `http://www.baidu.com`

Mining Source Codes of Programming Learners by Self-Organizing Maps

Guojin Zhu and Chuanfu Deng

School of Computer Science and Technology, Donghua University
Shanghai 200051, China
gjzhu.dhu@163.com, tranfu@qq.com

Abstract. There are lots of source codes on the Internet, which are resources valuable for programming tutoring. In school programming training, for instance, teachers may want to know how many solutions students could use to solve a given problem and which solution is the most popular. For this reason, we propose a method which is based on self-organizing maps to discover the mainstream solution for each problem by mining its corresponding source codes that learners submitted. We believe that the problem that has many solutions is much suitable for programming learners and that the mainstream solution should be recommended. 1510 source codes submitted by 40 students for 60 problems were mined in our experiment. The results show that each problem has 3.26 solutions on average and that 90% problems have their unique mainstream solutions.

Keywords: Data mining, Self-organizing maps, Program understanding, Source code mining, Mainstream solutions.

1 Introduction

Nowadays, there are lots of online-judges that provide thousands of problems for programming learners to solve on a purpose of practice. This results in abundant source codes (submitted by students as solutions to the problems) on the Internet. All these problems and source codes, however, are not organized effectively for programming tutoring [1]. It is often difficult for teachers to find the problems from the Internet that are suitable as examples to tutor their students. We believe that the problem that has many solutions is much suitable for programming learners and that the mainstream solution should be recommended. The purpose here is to find how many solutions a problem may have and which one is the mainstream solution by mining its corresponding source codes that learners submitted.

We use the abstract syntax tree (AST) to represent the program source code. The abstract syntax tree is easy to traverse and operate, and suitable for the representation of expression [2]. We apply the tree edit distance [3-4] algorithm to a simplified version of abstract syntax trees to calculate the similarity between program source codes. We put forward a solution to mine source codes by self-organizing maps (SOM), see Fig.1. The self-organizing maps [5-6] are able to cluster the similar source codes. The source codes that implement the same solution will be gathered

D. Jin and S. Lin (Eds.): Advances in CSIE, Vol. 1, AISC 168, pp. 683–688.
springerlink.com © Springer-Verlag Berlin Heidelberg 2012

together as one cluster [7]. In this way, all source codes of one problem will be divided into several clusters, each representing one solution to the problem and the largest one representing the popular solution.

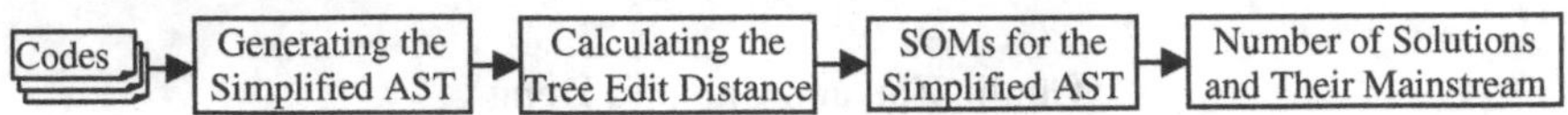

Fig. 1 SOM framework for source codes

2 Abstract Syntax Tree and Tree Edit Distance

An abstract syntax tree (AST) is a tree representation for the abstract syntactic structure of a source code written in a programming language. Each node of the tree denotes a construct occurring in the source code. We convert the source code to an abstract syntax tree and then simplify the abstract syntax tree. We create the abstract syntax tree for each source code by JavaCC [8]. JavaCC is a parser-generator written in Java and allows the generated parsers to produce abstract syntax trees.

Consider a short program named Source Code A written in the C++ programming language as shown in Fig.2. Its abstract syntax tree generated by JavaCC is shown in Fig.3 (a), where some interior nodes with gray background do not contain concrete information. Interior nodes are labeled as non-terminals while the leaf nodes are labeled as terminals by the grammar.

```
#include <iostream>
using namespace std;
int main() {
  return 0;
}
```

```
#include <iostream>
using namespace std;
int main() {
  cout << "hello";
  return 0;
}
```

Fig. 2. The left is Source Code A and the right is Source Code B

The abstract syntax trees created by JavaCC contain more information than we need, e.g., the non-terminals with gray background in Fig.3 (a), which are necessary for parsing, but not for meaning. We only need those nodes that contain concrete information. So we delete the non-terminal tree nodes that do not contain concrete information. Fig.3 (b) is a simplified version of the abstract syntax tree in Fig.3 (a).

The tree edit distance is an algorithm used to compute the difference between two labeled trees. It matches the labeled trees based on simple edit operations of deleting, inserting, and relabeling nodes. These operations are defined as follows. (1) RELABEL, change the label of a tree node. (2) DELETE, delete a non-root tree node v with its parent w, making the children of v become children of w. And (3) INSERT, the complement of delete, insert a tree node v as a child of w, making v the parent of a consecutive subsequence of the children of w. Each of these edit operations has a cost with its value being 1. An edit script S between two trees T_1 and T_2 is a sequence of edit operations turning T_1 into T_2. The cost of S is the sum of the costs of the operations in S. An optimal edit script between T_1 and T_2 is an edit script with the minimum cost between T_1 and T_2. This cost is called tree edit distance.

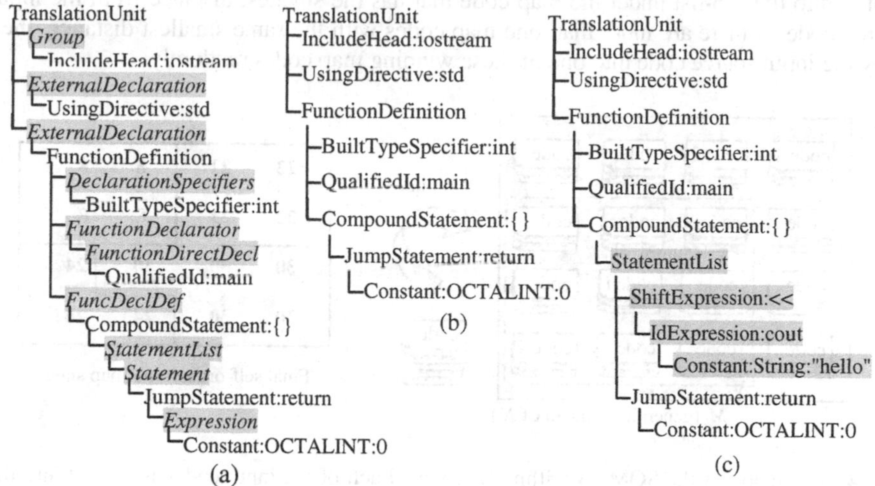

Fig. 3. (a) The abstract syntax tree of Source Code A. The interior nodes with gray background do not contain concrete information. (b) The simplified version of the abstract syntax tree of Source Code A after deleting the interior nodes with gray background. (c) The simplified version of the abstract syntax tree of Source Code B. After deleting 4 nodes with gray background, it will turn into the abstract syntax tree (b). So the tree edit distance between Source Code A and Source Code B is 4.

Fig.3 (c) is another abstract syntax tree which corresponds to the program named Source Code B in Fig.2. This abstract syntax tree will turns into the abstract syntax tree in Fig.3 (b) after deleting its four nodes with gray background. So the tree edit distance between these two trees is 4. This means that the tree edit distance between Source Code A and Source Code B is 4.

It is necessary to simplify abstract syntax trees because the complexity of the tree edit distance algorithm is too high. Computing the tree edit distance will benefit a lot from the simplified version of abstract syntax trees.

3 The Self-Organizing Map

A self-organizing map (SOM) is a type of artificial neural networks that is trained using unsupervised learning to produce a low-dimension (typically two-dimension) [6]. Self-organizing maps are different from other artificial neural networks in the sense that they use a neighborhood function to preserve the topological properties of the input space.

In this work, the input space of the training samples is composed of source codes. The tree edit distance of abstract syntax trees will be used to measure similarities between programming codes. The map has a size k, e.g., $k = 4{\times}4$.

The first step in the algorithm is to initialize some codes in the map, named *map codes*. We select *k* source codes randomly from the input space to initialize the map codes. Then go through all the input source coeds. Each of the input source codes is copied into the sub-list under the map code that has the smallest distance from the input source code. If there are more than one map codes with the same smallest distance, then copy the input source code into one of these winning map codes randomly.

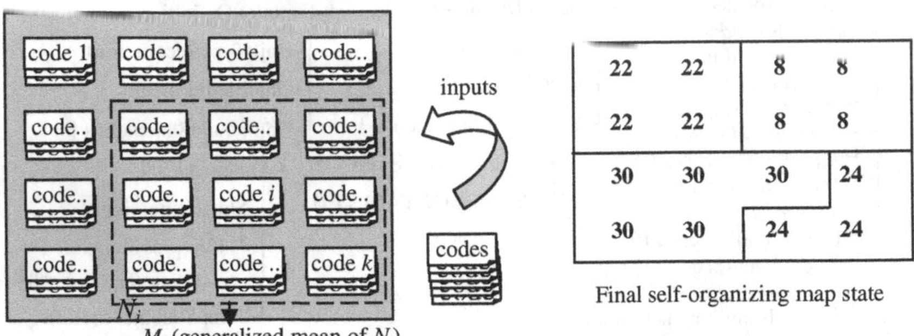

Fig. 4. Illustration of the SOM algorithm for codes. Each of the input codes is copied into the sub-list under the map code that has the smallest tree edit distance from the input code. After the generalized median M_i of the neighborhood set N_i is determined, and the old map code *i* is replaced by M_i. This cycle is repeated until the map codes are not changed any longer. On the right side, for instance, is the final state of the self-organizing map for the problem 1. The map is not changed any longer after repeating SOM cycle 60 times. 34 source codes of the problem 1 were copied into the sub-lists under corresponding map codes. 4 unique map codes with their sub-list sizes were obtained.

We define for each map code a neighborhood set N_i (the set of the map codes located within a certain radius from the map code *i* in the map). Consider the union of all the sub-lists within N_i (shown by the dotted line in Fig.4). The generalized median M_i of the neighborhood set N_i is defined to be identical with the input source codes in the union of the sub-lists in N_i that has the smallest sum of tree edit distances from all the other source codes in N_i. For each N_i, $i = 1, 2, \ldots, k$, the generalized median M_i is defined as the item that satisfies the condition

$$\sum_{code\ (i) \in N_i} d[code\ (i), M_i] = Min\ ,$$

where *Min* denotes the minimum value and the $d[code(i), M_i]$ is the tree edit distance between *code*(*i*) and M_i. The old map codes are replaced by the respective generalized medians in the current operation. This cycle is repeated until the map codes are not changed any longer.

4 Experimental Results

We downloaded 60 problems and gathered 1767 corresponding source codes submitted by 40 students from an online judge. We got rid of 257 source codes that were judged incorrect by the online-judge system, and employed for our experiment the rest of 1510 source codes that were judged correct.

To get 1510 simplified abstract syntax trees corresponding to the 1510 source codes, we create the abstract syntax trees by a tool called JavaCC, and delete the non-terminals which do not contain the double colon (::) and which are not components constructing basic shape of the trees. We use one java implementation of the Zhang Shasha [9] tree edit distance algorithm. The tree format for Zhang Shasha is defined as ordered edges of the form "-". However, the tree format for JavaCC is defined as ordered nodes with dot marks. Thus, we convert the tree format from JavaCC version to Zhang Shasha version.

Fig.5 gives the statistics of all the problems. The results show that the more solutions that a problem has, the less percentage of the mainstream solution. We found that the average number of solutions is 3.26 and that 90% problems have their unique mainstream solutions. The other 10% problems have not the sub-list that is larger than all of others, see the problem numbered 6 (id = 6) in Table 1.

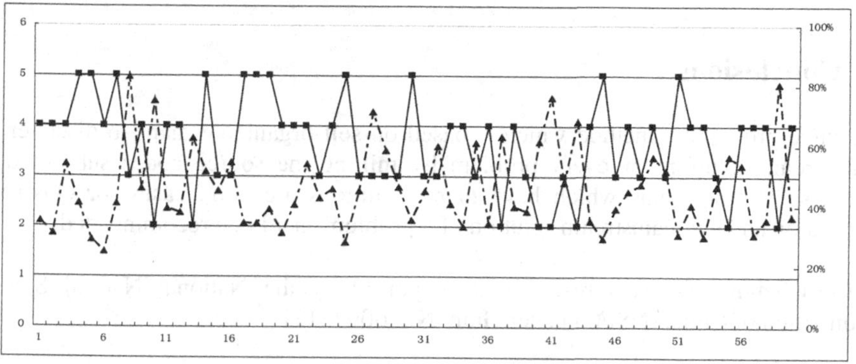

Fig. 5. The horizontal axis represents the 60 problems, the number of problem solutions (solid line) recorded on the left vertical scale, and the percentage of the source codes that use the mainstream solution in all source codes of the problem (dotted line) recorded on the right vertical scale.

Table 1. Problem solving methods.

Id	Num	Solution-1		Solution-2		Solution-3		Solution-4		Solution-5	
4	39	16	41%	10	25%	5	14%	4	10%	4	10%
1	34	12	35%	10	30%	8	23%	4	12%	0	0%
14	33	17	51%	7	21%	3	9%	3	9%	3	9%
16	33	18	54%	12	36%	3	10%	0	0%	0	0%
18	32	11	34%	8	25%	7	21%	5	15%	2	5%
20	32	10	31%	8	25%	8	25%	6	19%	0	0%
21	32	13	40%	9	28%	6	20%	4	12%	0	0%
6	32	8	25%	8	25%	8	25%	8	25%	0	0%
36	32	11	35%	8	25%	8	25%	5	15%	0	0%
52	32	13	40%	9	29%	8	25%	2	6%	0	0%

Table 1 lists the data about problem solving methods of 10 problems. The problem numbered 1 (id = 1) has 34 source codes submitted by students. For convenience, we computed the tree edit distance between each pair of two simplified abstract syntax trees of these 34 source codes using the Zhang Shasha algorithm in advance and stored the 34×34 tree edit distances in a text file for the inputs to the self-organizing map. The first solution (solution-1) contains 12 source codes and the percentage of the 12 source codes in the 34 source codes is 35%.

To mine source codes of the problem numbered 1, we initialized the map size with $k = 4 \times 4$. We found that the map codes were not changed any longer after repeating the self-organizing map cycle 60 times. We got 4 clusters with 4 different areas, see the right side in Fig.4. The source codes which have the same solution were gathered together. We got 4 unique map codes whose id numbers are 30, 22, 8, and 24, respectively. Their sub-list sizes are 12, 10, 8, and 4. It means that there are 4 solutions to the problem 1 and that the mainstream method is the first one having 12 source codes submitted by programming learners. Checking experimental results of the problem 1, we found that the source codes in the same sub-list of a map code implement the same kind of problem-solving method.

5 Conclusions

We have proposed a method which is based on self-organizing maps to discover the mainstream solution for every problem by mining the source codes submitted by learners. The problems which have many solutions are good cases for classroom discussion and the mainstream solution of a problem should be recommended.

Acknowledgments. This research is supported by the National Natural Science Foundation of China (NSFC) under Grant No. 60973121.

References

1. Holmes, R., Murphy, G.C.: Using Structural Context to Recommend Source Code Examples. In: Proceedings of the 27th International Conference on Software Engineering, pp. 117–125 (2005)
2. Koschke, R., Falke, R., Frenzel, P.: Clone detection using abstract syntax suffix trees. In: WCRE 2006, pp. 253–262. IEEE-CS, Washington, DC (2006)
3. Demaine, E.D., Mozes, S., Rossman, B., Weimann, O.: An Optimal Decomposition Algorithm for Tree Edit Distance. In: Arge, L., Cachin, C., Jurdziński, T., Tarlecki, A. (eds.) ICALP 2007. LNCS, vol. 4596, pp. 146–157. Springer, Heidelberg (2007)
4. Bille, P.: A survey on tree edit distance and related problem. Theoretical Computer Science 337, 217–239 (2005)
5. Günter, Bunke: Self-organizing map for clustering in the graph domain. Pattern Recognition Lett. 23, 401–417 (2002)
6. Kohonen, T., Somervuo, P.: How to make large self-organizing maps for nonvectorial data. Neural Networks 15, 945–952 (2002)
7. Zhu, X., Zhu, G.: Self-organizing map for clustering Algorithms in Programming Codes. In: Proceeding of the 3th Business Intelligence and Computational Finance (BIFE 2010), HongKong, China (August 2010)
8. `http://javacc.java.net`
9. `http://web.science.mq.edu.au/~swan/howtos/treedistance/`

Research on New Tabu Search Algorithm for Multi-cargo Loading Problem

Chunyu Ren

School of Information Science and Technology, Heilongjiang University,
150080 Harbin, China
rency2004@163.com

Abstract. This paper studies the loading problems of Multi-category goods with priority. According to the characteristics of model, new tabu search algorithm is used to get the optimization solution. It applies newly improved insertion method to construct initial solution, to improve the feasibility of the solution; centers cubage-weight balance to design dual layered random operation to construct its neighborhood, to boost the efficiency and quality of the searching. Applies auto adaptive tabu length to control the searching capability dynamically; At last, it uses simulated experiments to prove the effectiveness and feasibility of this algorithm, and provides clues for massively solving practical problems.

Keywords: Multi-category Goods, loading problem, improved insertion method, dual layered random operation, auto adaptive tabu length.

1 Introduction

Cargo loading problem is a complex and constraint combinatorial optimization problem, which is NP-hard problem. The main research methods of loading problem include precise algorithm, heuristic algorithm and intelligence optimization algorithm. In the case can be solved, precise algorithm is superior to artificial intelligence algorithm. However, this algorithm can only effectively solve small scale problem [1]. Heuristic algorithm can only offer a local optimal solution of the problem, and is absent of ability of global optimization. Especially when solving the big scale problem, the efficiency is not high [2].

Alvarez-Valdes applied greedy randomized adaptive search algorithm to solving two-dimensional rectangular bar packing problem (2SP, Two-Dimensional Strip Packing Problem) [3]. Burke proposed BF (Best-Fit) algorithm through filling position evaluation [4]. Hifi- proposed hybrid algorithm for two-dimensional packing problems. In this algorithm firstly, get the initial solution of the problem through a constructive way, and then improve the solution quality through the local search or modern heuristic algorithm to achieve satisfied solution [5]. John proposed Hybrid ant colony algorithm which based on Local Search method to solve the container loading problem [6]. Allan proposed hybrid grouping genetic algorithm (HGGA) to solve large-scale packing problem [7]. Bortfeldt proposed partial filling genetic algorithm based on homogenous block idea to solve individual differences problem in cargo

D. Jin and S. Lin (Eds.): Advances in CSIE, Vol. 1, AISC 168, pp. 689–694.
springerlink.com © Springer-Verlag Berlin Heidelberg 2012

container loading [8]. Pu Lei designed heuristic algorithm to solve cargo loading problem with multi-variety [9]. Cao Hongmei designed ant algorithm to solve cargo loading problem with multi-variety [10].

2 Mathematical Model

$$Max \quad Z = \sum_{i\in N}\sum_{j\in K}\lambda g_i x_{ij} + \sum_{i\in N}\sum_{j\in K}(1-\lambda)v_i x_{ij} \tag{1}$$

Constraints:

$$\lambda G\min{}_j \le \sum_{i\in N} g_i x_{ij} \le \eta_1 G_j, \quad j\in K \tag{2}$$

$$(1-\lambda)V\min{}_j \le \sum_{i\in N} v_i x_{ij} \le \eta_2 V_j, \quad j\in K \tag{3}$$

$$\sum_{j\in K} x_{ij} \le 1, \quad i\in N - D_0 \tag{4}$$

In model, all variables can be shown as followings. The maximum loading weight of j tools is G_j. The maximum loading capacity is V_j. The weight of $i(i\in N)$ is g_i. Volume is v_i. $G\min_j$ is the minimum loading weight of j loading tools. $V\min_j$ is the minimum loading capacity of j loading tools. $\eta_2 (0\le\eta_2\le 1)$ is the elasticity coefficient of loading capacity. $D_i (D_i \subset N)$ is the cargo mustering not mixing cargo. $D_0 (D_0 \subset N)$ is the loading cargo muster of priority mandatory installation.

3 Parameter Design for Tabu Search Algorithm

3.1 The Formation of Initial Solution

Step1: Initial data are supposed $n_k = 0$, $V_{sum}^1 = 0$, $G_{sum}^1 = 0$, $S^0 = \Phi$, $K = m$;

Step 2: Calculate $\omega_i = \frac{g_i}{v_i} (i\in N)$, sorting according to non-increasing order. Take this muster as p^1.

Step 3: Suppose $p = p^1$, calculate muster p. Review it from beginning to ending. The order is $p[1], p[N], P[2], P[N-1], \ldots$.

Step 4: if $G_{sum}^1 + g_{p[1]} \le G_j (j\in m)$ and $V_{sum}^1 + v_{p[1]} \le V_j (j\in m)$, $G_{sum}^1 = G_{sum}^1 + 1$ and $V_{sum}^1 = V_{sum}^1 + 1$.

Step5: if $G_{sum}^1 + g_{p[N]} \le G_j(j \in m)$ and $V_{sum}^1 + v_{p[n]} \le V_j(j \in m)$, $G_{sum}^1 = G_{sum}^1 + 1$ and $V_{sum}^1 = V_{sum}^1 + 1$. Otherwise, turn into step 7.

Step 6: Repeat step 4 and 5 to $G_{sum}^1 > G_j(j \in m)$ and $V_{sum}^1 > V_j(j \in m)$.

Step 7: Record current status of G_{sum}^1, V_{sum}^1, S^1, $n_k = n_k + 1$, $K = K - 1$ and $S^2 = S^1 \cup S^0$.

Step8: Suppose $p^1 = p - S^2$;

Step 9: Repeat step 3 to step 8 until to $K = 0$. And turn into $K = 0$.

Step 10: Output the cargo muster $\{S^1, S^2, ..., S^{nk}\}$, total weight of loading cargo G_{sum}, total volume of loading cargo is V_{sum}.

3.2 Inner Neighborhood Operation

Specific procedures as such:

(1) 1-move

$1-move$ is a heuristic algorithm the same as operators (1, 0) and (0, 1), which can effectively improve the quality of solutions and the feasibility of poor solutions.

(2) 2-opt

$k(i)$ signified the neighbor point of the client point i in the route l, and $a(i, j)$ signified to change the direction of the route from i to j. That was in the l route, the client points were: $(0,1,2,...,n,0)$, in it, 0 signified distribution centre. The procedures of the $2-opt$ neighborhood operation were as such:

Step1: $i_1 := 1, i := 0$;

Step2: if $i > n - 2$, end; otherwise, turn to Step3;

Step3: revise $i_2 := k(i_i)$, $j_1 := k(i_2)$, $j := i + 2$;

Step4: if $j > n$, turn to Step8, if not, turn to Step5;

Step5: $j_2 := s(j_1)$, change route l as such (1) $a(i_2, j_1)$, (2) alternately used (i_1, j_1) and (i_2, j_2), substitute (i_1, i_2) and (j_1, j_2);

Step6: If the changed route l_1 is feasible, and better than l, revise l, if not, turn to Step7;

Step7: $j_1 := j_2$, $j := j + 1$, return to Step4;

Step8: $i_1 := i_2$, $i := i + 1$, return to Step2.

3.3 Outer Neighborhood Operation

(1) 2-opt*

That is in the route l, the client points are $(0,1,2,...,n,0)$, in the route k, the client points are $(0,1,2,...,m,0)$, in it, 0 signifies distribution centre.

Step1: Randomly choose n number of client points in the route l, for each client point i, choose client point j nearby the route k, if exist, exchange chains $(i,i+1),(j,j+1)$;

Step2: Conduct $2-opt$ neighborhood operation in the exchanged routes l^1 and k^1, to obtain feasible solution;

Step3: Calculate the exchanged objective function f^1, if $f^1 > f$, turn to Step4; if not, turn to Step5;

Step4: If the current optimal solution does not exist in the tabu list, update tabu list, input the obtained optimal solution into the tabu list, simultaneously remove out the ban-lifted elements; otherwise, turn to Step5;

Step5: $i = i+1$, turn to Step1;

Step6: repeat Step1- 5, till the current optimal solution can not update.

3.4 Adaptive Tabu Length

In order to ensure effectiveness of the tabu list, during the whole process of searching, make $L_{\min}, L_{\max}$ as its variable region $\left[a\sqrt{N}, b\sqrt{N}\right]$, in it $0<a<b$. So the tabu length L's variable scope is the formula as the following:

$$L = \lambda L_{\min} + (1-\lambda)L_{\max} \tag{5}$$

In the formula, $L_{\min}$ and $L_{\max}$ are the upper and lower bound of tabu length L's dynamic change respectively, N refers to the number of clients, the weighing coefficient is $0 \le \lambda \le 1$.

4 Experimental Calculation and Result Analysis

The data of study is from reference [8, 9], utilize TBJ10 type container to load bulk cargos of 42 freight invoices. The maximum loading weight of TBJ10 type container is $G = 10t$. The maximum loading capacity is $V = 16.81m^3$. The cargo of No.1 freight invoice must be loaded firstly. And nature and package of cargo don't occur conflicting. Confirm the scheme of maximum loading rate.

4.1 Solution of New Tabu Search Algorithm

This algorithm adopts the following parameters as part. The maximum iterative times are max_ $iter$ =300, tabu length is $\alpha = 2$, $\beta = 3$, $\lambda = 0.5$, and candidate solution amount is 20. The concrete route can be seen in table 1.

Table 1. Optimal results by new tabu search algorithm

No.	1 4 7 21 34 39
Loading utilization /%	95.94
Capacity utilization /%	99.88

4.2 Solutions by Genetic Algorithm

Reference [9] is adopted genetic algorithm to get the solution. The main parameters are as followings. Group scale is $N = 80$. The maximum iterative times is $K = 300$, crossover operator is $p_c = 0.95$, mutation operator is $p_m = 0.01$, initial temperature is T_0 =250, temperature coefficient $\delta = 0.89$, and randomly get the solutions for 30 times. The optimal loading weight and capacity utilization rate is 83.80% and 91.13%. The concrete loading schemes can be shown in Table 2.

Table 2. Optimal results on solving by GA

No.	1 9 18 23 28 33 40
Loading utilization /%	83.80
Capacity utilization /%	91.13

4.3 Solutions by Ant Colony Algorithm

Reference [10] is adopted by ant colony algorithm to get the solution through multi-times solving. The concrete parameters are as followings. $\alpha = 1$, $\beta = 5$, $\rho = 0.9$, $\tau_0 = 0.5$, $\eta_1 = \eta_2 = 1$, $m = 30$, $n_c = 200$. The optimal loading weight and capacity utilization rate are 81.19% and 98.75%. The concrete scheme can be shown in Table 3.

Table 3. Optimal results on solving by CA

No.	1 3 7 18 29 36
Loading utilization /%	81.19
Capacity utilization /%	98.75

4.4 Analysis on Three Algorithms

Compared the optimal scheme of reference [9, 10], experiments proved that this algorithm can achieve not only better calculating results, but also better calculation efficiency and quicker convergence rate. That can be shown in Table 4.

Table 4. Comparison among GA, CA and Algorithm of this study

	GA	**CA**	**NTS**
Loading utilization /%	83.80	81.19	95.94
Capacity utilization /%	91.13	98.75	99.88

5 Conclusions

In general, the proposed algorithm has strong searching ability and high solving high quality. Therefore, it is more practical significance and value so as to reduce operating cost and improve economic benefit. At the same time, the proposed algorithm doesn't set opposition between utilization rate of loading weight and capacity. On the contrary, the study designs this problem on the whole and fully and balanced utilizes the weight and capacity of loading tools.

References

1. Chen, C.S., Lee, S.M., Shen, Q.S.: An analytical model for the container loading problem. European Journal of Operational Research 80, 68–76 (1995)
2. Shachnai, H., Tamir, T.: Polynomial time approximation schemes for class-constrained packing problems. Journal of Scheduling 4, 312–338 (2001)
3. Alvarez-Valdes, R., Parreno, F., Tamarit, J.M.: Reactive GRASP for the strip packing problem. Computers & Operations Research 35, 1065–1083 (2008)
4. Burke, E.K., Kendall, G., IPhitwell, G.: A New Placement Heuristic for the Orthogonal Stock-Cutting Problem. Operations Research 52, 655–671 (2004)
5. Hifi, M., Hallah, R.M.: A hybrid algorithm for the two-dimensional layout problem: the cases of regular and irregular shapes. International Transactions in Operational Research 10, 1–22 (2003)
6. Levine, J., Ducatelle, F.: Ant Colony Optimization and Local Search for Bin Packing and Cutting Stock Problems. Journal of the Operational Research Society, Special Issue on Local Search 55, 705–716 (2004)
7. Tucker, A., Crampton, J., Swift, S.: RGFGA: An Efficient Representation and Crossover for Grouping Genetic Algorithms. Evolutionary Computation 13, 477–499 (2005)
8. Bortfeldt, A.: Eine Heuristic fuel Multiple Container loading problem. OR Spektrum 22, 239–262 (2000)
9. Bu, L., Yin, C.-Z., Pu, Y.: Genetic algorithm for optimal arrangement of general piece goods. Journal of Traffic and Transportation Engineering 34, 84–87 (2004)
10. Cao, H., Gao, L., Zhang, T.: Ant algorithm for optimal loading of multi category goods. Computer and Communications 26, 11–14 (2008)

New Genetic Algorithm for Capacitated Vehicle Routing Problem

Chunyu Ren and Shiwei Li

School of Information Science and Technology, Heilongjiang University, 150080 Harbin, China
rency2004@163.com

Abstract. In logistic transport industry, individual demands and the diversity requirements are matters in transport operation; this paper focused on the vehicle routing problem. New genetic algorithm is used to get the optimization solution in the paper. First of all, use natural number coding to simplify the problem; apply insertion method to improve the feasibility of the solution. Secondly, use individual amount control choice strategy to guard the diversity of group; this approach also improved route crossover operation to avoid destroying goods. Finally, the good performance of improved algorithm can be proved by experiment calculation and examples.

Keywords: CVRP, individual amount control, improved route crossover operation, natural number coding, new genetic algorithm.

1 Introduction

Vehicle Routing Problem is a typical NP problem. The research method of VRP mainly includes precise algorithm, heuristic algorithm and intelligent optimized algorithm. Fisher put forward three subscribed vehicle flow equation to solve VRP problems with capacity constraints [1]. Later, Laporte applied two subscribed vehicle flow equation to solve VRP problems with both capacity and distance constraints [2]. Ergun proposed simple moving rules, which based on the method he constitutes large-scale neighboring scope search algorithm that meets the moving rules [3]. Derigs applied climbing heuristic algorithm to solve VRP [4].

The intelligent algorithm applied more widely. Bell modified the ant colony algorithm, which is used to solve Travelling Salesman Problems and this algorithm conduct multi-path search method and proposed the ant colony algorithm applicable to solve VRP problems [5]. Yuvraj applied large-scale ant colony algorithm to solve VRP problems with backhauls. This algorithm constructs new rules and local search strategy with multi routes [6]. Ali Haghania applied genetic algorithm to study dynamic VRP problem with the features of travelling time changed timely and different types of vehicles with different capacities [7]. Doris used Fuzzy Clustering and GA to construct another hybrid GA to solve the VRP problem featured by dynamic delivery and pickup through many types of vehicles [8]. Bent applied two phase algorithm to solve VRP problem, at the first phase, used simulated annealing algorithm to optimize the number of vehicles; used taboo search algorithm to

D. Jin and S. Lin (Eds.): Advances in CSIE, Vol. 1, AISC 168, pp. 695–700.
springerlink.com © Springer-Verlag Berlin Heidelberg 2012

optimize the sequence of the customers so as to make the shortest distance [9]. Ganesh firstly used clustering algorithm and the nearest neighbor search algorithm to construct initial solution, then used genetic algorithm to solve it [10].

2 Mathematical Model

$$Z = Min\sum_{i\in S}\sum_{j\in S}\sum_{k\in V} X_{ijk} d_{ij} \tag{1}$$

Constraints:

$$\sum_{i\in H}\sum_{j\in S} q_i X_{ijk} \le W_k, \quad k\in V \tag{2}$$

$$\sum_{i\in S} X_{ijk} = Y_{ik}, \quad j\in S, \quad k\in V \tag{3}$$

$$\sum_{j\in S} X_{ijk} = Y_{ik}, \quad i\in S, \quad k\in V \tag{4}$$

$$\sum_{i\in S}\sum_{j\in S} x_{ijk} \le |m|-1, \quad \forall m \subseteq \{2,3...,n\}, \quad k\in V \tag{5}$$

In the formula: $G\{g_r|r=1,...,R\}$. A series of aggregations of distribution centre in the place R ; $H\{h_i|i=R+1,..R+N\}$ is a series of clients' aggregations in the place N ; $S\{G\}\cup\{H\}$ is the combination of all distribution centres and clients. $V\{v_k|k=1,..K\}$ is travel vehicle k 's aggregation; q_i is the demand amount of client $i(i\in H)$; W_k is travel vehicle k 's loading capacity; d_{ij} is the linear distance from client i to client j .

3 Parameter Design for New Genetic Algorithm

3.1 The Formation of Initial Solution

Given h_k as the total number of client nodes served by vehicle k , aggregation $R_k = \{y_{ik}|0\le i\le h_k\}$ to correspond the client nodes served by the number k vehicle. The procedures as such:

Step1: Order vehicles' initial remaining load capacity: $w_k^1 = w_k$, $k=0$, $h_k = 0$, $R_k = \Phi$;

Step2: The demand amount corresponding to the i client node in a route q_i , order $k=1$;

Step3: if $q_i \le w_k^1$, then order $w_k^1 = Min\{(w_k^1 - q_i), w_k\}$, if not turn to Step6;

Step4: if $w_k^1 - q_i \le w_k$, and $D_{i-1} + D_i \le D_k$; then $R_k = R_k \bigcup \{i\}$, $h_k = h_k + 1$ if not turn to Step6;

Step5: if $k > K$, then $k = K$, otherwise, $k = k$;

Step6: $k = k + 1$, turn to Step3;

Step7: $i = i + 1$, turn to Step2;

Step8: repeat Step2-7, K recorded the total used vehicles, R_k and recorded a group of feasible routes.

3.2 Selecting Operator

The paper adopts fitness distribution according to different proportion. Obviously, if the individual fitness is bigger, then the selected probability is bigger. Select two father individual randomly every time to continue the next operation.

$$P_i = f_i / \sum_{i=1}^{n} f_i \tag{6}$$

According to model theorem of genetic algorithm, sample amount of model H in t generation is the following formula.

$$m(H,t) \equiv m(H,0).(1+c)^t \tag{7}$$

Here, $m(H,0)$ is the sample amount of model in initial population. Algorithm restrains to the partial optimal solution when sample number of model H reach to certain quantity.

Step one: t-1 generation group forming after t-1 genetic operation can have selecting operation to create group $p(t)$ according to proportion fitness.

Step two: Calculate every individual number in group $p(t)$.

Step three: Have the following operation to group $p(t)$ to create group $p^1(t)$. If some individual number exceeds the marginal value ε of t individual, delete this individual so as to control individual number in the extent ε .otherwise, copy all individuals.

Step four: If the number of group $p^1(t)$ is less than group scale N, then randomly operate N- $p^1(t)$ new individuals. And new individual can take part in following cross and mutation operation.

3.3 Crossover Operator

The improved route crossover operation is as followings.

Step1: Randomly select two chromosomes in father generation. According to conventional mode, get two groups of feasible vehicle routes after having feasible research on routes.

Step2: Generate mode H to every group of vehicle route, which mode is a bunch of randomly generating binary system number, its size is the amount of every groups of vehicles in length, that is, every binary system location is corresponding to one vehicle route.

Step3: In one father generation $P_i(i=1,2)$, to “1” route in mode H, directly copy it to its filial generation S. To “0” route in mode H, form into one new order according to its successive sequence in another father generation $P_j(j=1,2, j \neq i)$.

Step4: To every requirement dot in 1, successively insert into the place of first satisfying restriction condition in their respective filial generation in term of their successive order in sequence.

Step5: If one requirement dot cannot insert into any existing routes, it will generate one new route which requirement dot inserted to guarantee the principle of all requirement dot are satisfied.

Step6: Repeat them in turn until all requirement dots in generating sequence are inserted into one route. After finishing this operation, generate two new generations s_1, s_2 .

3.4 Mutation Operator

The mutation strategy of this study is adopted 2- commutation mutation strategy, namely, randomly selecting mutated individual chromosome according to some mutation probability and two gene locations in this chromosome, exchanging gene in two places and form into new gene clusters. If it is continuously appeared with zero code in gene clusters, exchange zero code and non-zero code in random place.

4 Experimental Calculation and Result Analysis

To test new genetic algorithm performance, its large-scale vehicle routing problem can give better results. In this paper, Christofides and Elion's E-n33-k4 VRP test problems is to verify. The test question is: in a distribution system, there is a logistics center, 32 customer needs point of the task, vehicle capacity of 8000. The number of vehicles is 4. Other data can be found in other literature. It is required that rationally arrange distribution vehicle so as to the shortest delivery mileage.

4.1 Solution of New Genetic Algorithm

The main parameters: population size of 30, the maximum number of iterations is 200; crossover 0.80, mutation operator is 0.01. Randomly solve ten times.

Here, the corresponding optimal total length of 837.672. The concrete route can be seen in table 1 and figure 1.

Table 1. Optimal results by new genetic algorithm

No.	Running route	Distance
1	0-4-7-9-8-32-11-12-2-0	167.719
2	0-3-5-6-10-18-19-21-20-22-23-24-25-17-13-0	266.662
3	0-1-15-26-27-16-28-29-0	246.859
4	0-30-14-31-0	156.432
Total vehicles	4	
Total distances	837.672	

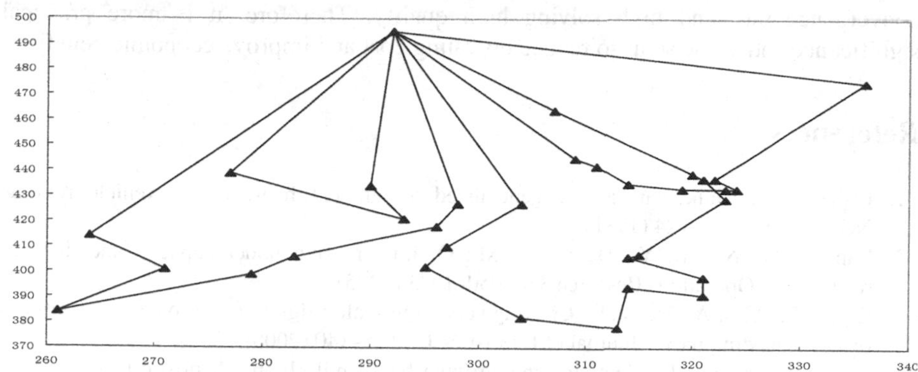

Fig. 1. Optimization route of CVRP using new genetic algorithm

4.2 Analysis on Three Algorithms

An example of E-n33-k4, using different algorithms(Simple Genetic Algorithm, SGA and Immune Genetic Algorithm, IGA) for 10 times, calculated results and comparison of statistical data such as Table2. The population size M is 200, the evolution generation is 500.

Table 2. Comparison of Optimal Calculation Results for Ten Times Using SGA and IGA

	1	2	3	4	5	6	7	8	9	10
SGA	940	912	916	898	893	940	928	889	911	912
IGA	888	874	882	852	867	916	862	873	862	905

Compared the optimal scheme of reference [11], experiments proved that this algorithm has a strong search capability, high computational efficiency and high quality on algorithm solving.

Table 3. Comparison Three Algorithms

Algorithm	Optimal Results
Simple Genetic Algorithm(SGA)	889
Immune Genetic Algorithm(IGA)	852
This Algorithm	837.672

5 Conclusions

In general, the proposed new genetic algorithm has strong searching ability, rapid convergence rate and high solving high quality. Therefore, it is more practical significance and value so as to reduce operating cost and improve economic benefit.

References

1. Fisher, M.L., Jaikumar, R.: A generalized assignment heuristic for vehicle routing. Networks 11, 109–124 (1981)
2. Laporte, G., Nobert, Y., Desrocher, M.: Optimal routing under capacity and distance restrictions. Operations Research 33, 1050–1073 (1985)
3. Ergun, J., Orlin, A., Steele, F.: Creating very large scale neighborhoods out of smaller ones by compounding move. Journal of Heuristics 12, 115–140 (2006)
4. Derigs, U., Kaiser, R.: Applying the attribute based hill climber heuristic to the vehicle routing problem. European Journal of Operational Research 177, 719–732 (2007)
5. Bell, J.E., McMullen, P.R.: Ant colony optimization techniques for the vehicle routing problem. In: Advanced Engineering Informatics, vol. 18, pp. 41–48 (2004)
6. Gajpal, Y., Abad, P.L.: Multi-ant colony system (MACS) for a vehicle routing problem with backhauls. European Journal of Operational Research 196, 102–117 (2009)
7. Haghania, A., Jung, S.: A dynamic vehicle routing problem with time dependent travel times. Computers & Operations Research 32, 2959–2986 (2005)
8. Seza, D., Comes, C.E., Nueza, A.: Hybrid adaptive predictive control for the multi-vehicle dynamic pick up and delivery problem based on genetic algorithms and fuzzy clustering. Computers and Operations Research 35, 3412–3438 (2008)
9. Bent, R., van Hentenryck, P.: A two-stage hybrid local search for the vehicle routing problem with time windows. Transportation Science 38, 515–530 (2004)
10. Ganesh, K., Narendran, T.T.: A cluster and search heuristic to solve the vehicle routing problem with delivery and pick up. European Journal of Operational Research 17, 699–717 (2007)
11. Gang, S., Yuan, W.: Research of improved immune clonal algorithms and its applications. In: The International Conference on Computational Intelligence for Measurement Systems and Applications, vol. 1, pp. 11–13 (2009)

Study on 3D Reconstruction of Imaging Lidar

Wei Shao, HongSheng Lin, GuoEn Wang, and YongTao Tang

Department of Information Technology, Naval Petty Officer Academy, Bengbu 233012, China

Abstract. The imaging lidar acquires feature points by using laser pulse to scan the object. And the data can be used to reconstruct the object in the computer. First, 3D coordinates of the feature points can be acquired by the scan equation of the lidar. Then, in order to smooth the 3D data, a improved Laplacian method is proposed. Last, the methods put forward in this paper are validated with experiment, and the results show that the methods are effective.

Keywords: Imaging lidar, 3D reconstruction, Laplacian, Taubin filter.

1 Introduction

The imaging lidar, which integrates laser technology and radar technology, can be used to scan object by laser pulse. And then angle, distance and gray intensity of the feature point of the object can be acquired. Using the data, we can reconstruct the object in the computer by displaying in distance image and gray image[1,5]. We get distance information of the feature points by using the imaging lidar to scan the object, while 3D coordinates of the feature points are acquired by the scan equation of the lidar. In the process of the data acquisition, some noise data have turned up because of the noise in the receiver circuit and error made by the scan mirror. So the data acquired is to be filtered and thus curved surface reconstructed become smooth[2].

2 Experimental Device

The laser unit used in this project is made by Dr. ChunSheng HU who is from College of Optoelectronic Science and Engineering, National University of Defense Technology[3]. Figure 1 shows structure of the experimental unit.

The imaging lidar composes of transmitter, receiver, scanner and processor. Transmitter beams a laser to the object under the control of the processor. Receiver gets the laser echo reflected from the object, and then changes it to electric signal and amplifies it. Besides controling transmitter to transmit laser, processor measures the time interval of emitting and receiving. Base on the time interval, we can acquire a distance image. Another function of the receiver is getting a gray image from output of the receiver. And the processor stores and displays the distance image and gray image. The scanner makes the field of view of the transmitter and receiver synchronous.

D. Jin and S. Lin (Eds.): Advances in CSIE, Vol. 1, AISC 168, pp. 701–707.
springerlink.com © Springer-Verlag Berlin Heidelberg 2012

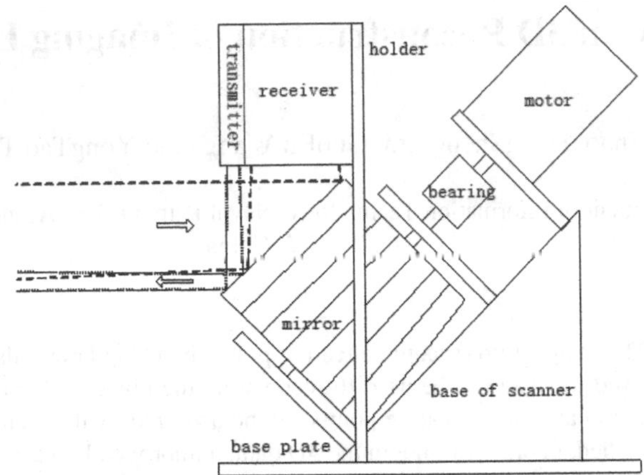

Fig. 1. Structure of imaging lidar.

The procession of the 3D reconstruction is showed in Figure 2. Two main issues are to be considered in the 3D reconstruction process. First how to get 3D coordinates from distance data of the feature points. And second how to filter the data that contain noise.

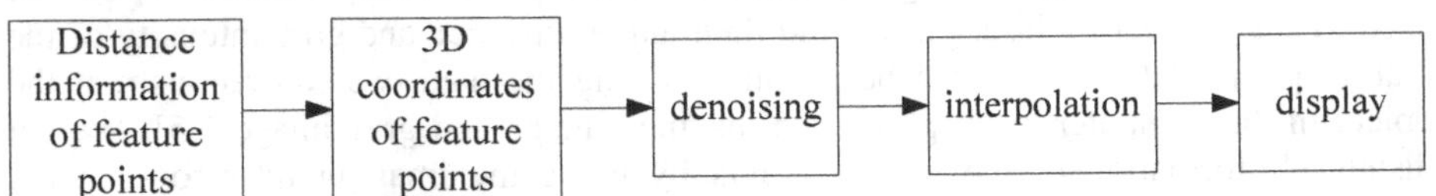

Fig. 2. Basic procession of 3D reconstruction.

3 Scan Equation

We build a lidar coordinates system as showing in figure 3. We place axis of the scanner in the YOZ plane. And it crosses axis OY with point A and crosses axis OZ with point B. The angle between axis of the scanner and the XOY plane $\angle OAB = \alpha = 45^\circ$. $\overrightarrow{CD}$ is the unit normal vector of mirror N of the scanner(the value of N is 1～16). And it is located in the plane YOZ. $\overrightarrow{CE}$ is the unit normal vector of mirror i of the scanner when the scanner rotates a angle of ϕ. The angle between axis of the scanner and $\overrightarrow{CD}$、$\overrightarrow{CE}$ is γ. That is $\angle ACD = \angle ACE = \gamma$.

It is assumed that the scanner rotate a angle of β as showing direction in figure 3($\beta \in (-\frac{\pi}{N}, \frac{\pi}{N})$). The unit vector of axis of the scanner $\overrightarrow{BA}$ is $(0, \cos\alpha, -\sin\alpha)$. The unit vector of $\overrightarrow{CD}$ is $(0, \cos(\gamma-\alpha), \sin(\gamma-\alpha))$. Assumed that coordinates of the unit vector of $\overrightarrow{CE}$ is (x_0, y_0, z_0), we can get equations (1) from figure 3.

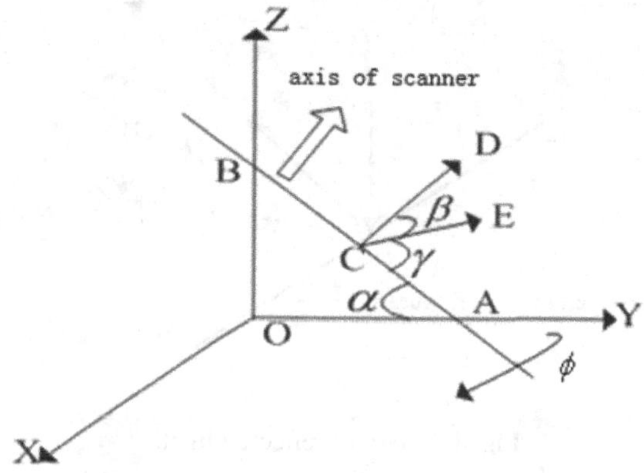

Fig. 3. Scanning schematic plot of imaging lidar.

$$\begin{cases} \overrightarrow{CD} \bullet \overrightarrow{CE} = \left|\overrightarrow{CD}\right|\left|\overrightarrow{CE}\right| \cos\beta \\ \overrightarrow{BA} \bullet \overrightarrow{CE} = \left|\overrightarrow{BA}\right|\left|\overrightarrow{CE}\right| \cos\gamma \\ \left|\overrightarrow{CE}\right| = 1 \end{cases} . \tag{1}$$

Replace $\overrightarrow{CD}$, $\overrightarrow{CE}$, $\overrightarrow{BA}$ with expression of vector , we can get coordinates of $\overrightarrow{CE}$ showing as below.

$$\begin{cases} x_0 = \dfrac{(2\cos\beta\cos^2\gamma - \cos^2\beta - \cos 2\gamma)^{\frac{1}{2}}}{\sin\gamma} (\beta \geq 0) \\ x_0 = -\dfrac{(2\cos\beta\cos^2\gamma - \cos^2\beta - \cos 2\gamma)^{\frac{1}{2}}}{\sin\gamma} (\beta < 0) \end{cases} . \tag{2}$$

$$y_0 = \frac{\sin\alpha\cos\beta + \sin(\gamma\text{-}\alpha)\cos\gamma}{\sin\gamma} . \tag{3}$$

$$z_0 = \frac{\cos\alpha\cos\beta\text{-}\cos(\gamma\text{-}\alpha)\cos\gamma}{\sin\gamma} . \tag{4}$$

After we get coordinates of unit normal vector of reflector and direction of incident light, we can get coordinates of reflected light. We make mirror of num N as research object, and given a incident light enters mirror of num N in the direction of $\overrightarrow{ZO}$ as showing in figure 3. Figure 4 shows the schematic plot.

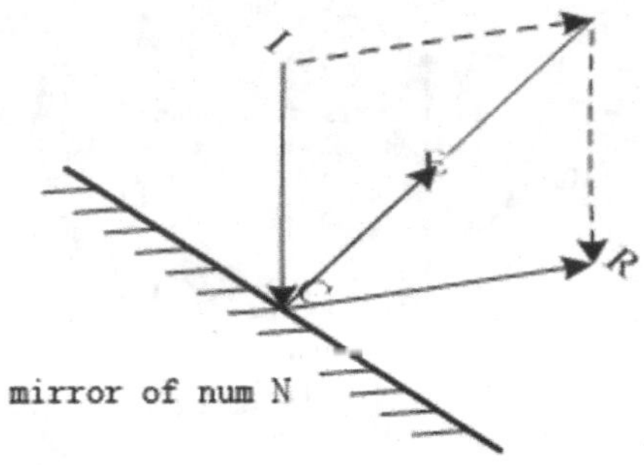

Fig. 4. Vector of reflected light.

Assumed that coordinates of the unit vector of reflected light $\overrightarrow{CR}$ is (x_r, y_r, z_r), coordinates of the unit vector of incident light $\overrightarrow{IC}$ is $(0, 0, -1)$, then we can get a equation from law of reflection as below.

$$\overrightarrow{CR} = -2\left|\overrightarrow{IC} \bullet \overrightarrow{CE}\right| \overrightarrow{CE} + \overrightarrow{IC} . \tag{5}$$

We can get coordinates of reflected light $\overrightarrow{CR}$.

$$\begin{aligned} x_r &= 2z_0x_0 \\ y_r &= 2z_0y_0 \\ z_r &= 2z_0^{\ 2} - 1 \end{aligned} \quad . \tag{6}$$

Replace x_0 、y_0、z_0 with expression (2), (3), (4), We can get coordinates of reflected light $\overrightarrow{CR}$ showing as below.

$$\begin{cases} x_r = 2\dfrac{(\cos\alpha\cos\beta - \cos(\gamma-\alpha)\cos\gamma)(2\cos\beta\cos^2\gamma - \cos^2\beta - \cos 2\gamma)^{\frac{1}{2}}}{\sin^2\gamma} (\beta \geq 0) \\ x_r = -2\dfrac{(\cos\alpha\cos\beta\text{-}\cos(\gamma\text{-}\alpha)\cos\gamma)(2\cos\beta\cos^2\gamma - \cos^2\beta - \cos 2\gamma)^{\frac{1}{2}}}{\sin^2\gamma} (\beta < 0) \end{cases} \tag{7}$$

.

$$y_r = \frac{\cos^2\beta\sin 2\alpha - \cos^2\gamma\sin(2\gamma - 2\alpha) + 2\cos\beta\cos\gamma\sin(\gamma - 2\alpha)}{\sin^2\gamma} . \tag{8}$$

$$z_r = 2\left(\frac{\cos\alpha\cos\beta - \cos(\gamma\text{-}\alpha)\cos\gamma}{\sin\gamma}\right)^2 - 1 . \tag{9}$$

Given coordinates of feature point as (x, y, z), we can get a parameter equation of the feature points base on the unit vector of reflected light (x_r, y_r, z_r) and distance information R.

$$\begin{aligned} x &= x_r R \\ y &= y_r R \, . \\ z &= z_r R \end{aligned} \tag{10}$$

4 Fairing

In the process of data acquisition, some noise data have turned up because of the noise in the receiver circuit and error made by the scan mirror. And these data will greatly affect 3D reconstruction. So a denoising method is to be applied for the data acquired.

There is many fairing methods for 3D data. One of the typical method is Laplacian fairing method. Laplacian method adjusts vertex for fairing to its linear combination of first order neighbourhood. Because of its linear time and space complexity, Laplacian method becomes one of the widely used fairing methods. Another kind of fairing method bases on optimization algorithms. It minimizes some energy function to fairing 3D data. And thus it can minimize deformation of some geometrical parameter in theory. But for 3D model of large scale, this method calls for lots of calculating cost.

In this paper, we use improved Laplacian method to filter the lidar data.

4.1 Laplacian Method

Laplacian fairing method is a common and simple method[6]. It uses laplacian operator for every vertex, which is showing below.

$$\nabla^2 = \frac{\partial^2}{\partial x^2} + \frac{\partial^2}{\partial y^2} + \frac{\partial^2}{\partial z^2} \, . \tag{11}$$

Assumed that some vertex is $p_i = (x_i, y_i, z_i)$, then fairing process can be regarded as a diffusing process.

$$\frac{\partial p_i}{\partial t} = \lambda L(p_i) \, . \tag{12}$$

This method acts on every vertex, and move it to geometric gravity center of neighborhood.

$$L(p_i) = p_i + \lambda\left(\frac{\sum_j w_j q_j}{\sum_j w_j} - p_i\right), j = 1, 2 \ldots, k \, . \tag{13}$$

Among them, q_i is regarded as one neighbor of p_i. λ is a small positive number.

The second order laplacian operator is showed below.

$$L^2(p_i) = \frac{\sum_j w_j L(q_j)}{\sum_j w_j} - L(p_i)\,. \tag{14}$$

4.2 Improved Laplacian Method

Taubin proposed a low-pass filter algorithm for fair surface[4]. With this algorithm, fairing very large surface becomes affordable. By combining this algorithm with Laplacian method we obtain a very effective fairing technique.

$$T(p_i) = (1-\mu L)(1+\lambda L)(p_i) = p_i - (\mu - \lambda)L(p_i) - \mu\lambda L^2(p_i)\,. \tag{15}$$

Where $\mu > \lambda > 0$.

5 Results

In this section we present a result to demonstrate the effectiveness and efficiency of the system. Combining all the methods presented in the paper, we carry out a series of experiments based on Visual C + + integrated development environment and OpenGL. Figure 5 shows the lidar image acquired by scanning the back of a person, and visible image of the same scene. Figure 6 shows the lidar image that after filtering.

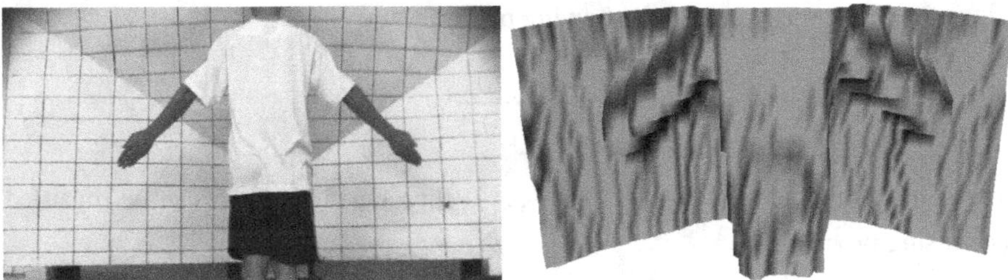

Fig. 5. Lidar image acquired by scanning the back of a person(right), and visible image of the same scene(left).

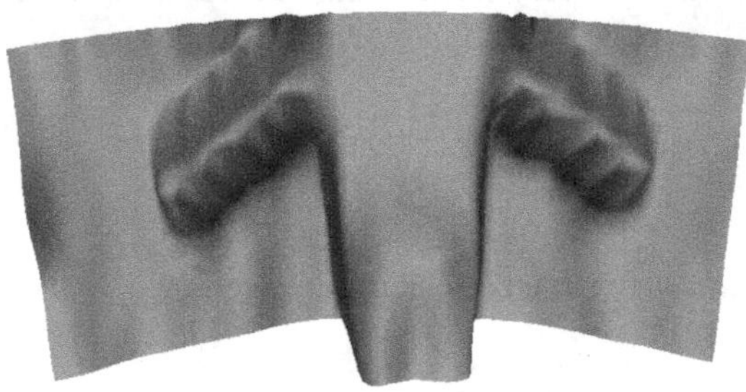

Fig. 6. Lidar image after filtering.

6 Conclusion

Through the result of the experiment, we can see that the image after filtering is very different from that of no filtering. It shows that the method used in this paper is effective. As far as software and reconstruction tasks are concerned, many issues have not been yet considered in this paper. For example, the fairing method we adapted is a iteration procedure, so the num of iterations and scale factor is very important for the method. In terms of future work, we plan to study on the characteristics of lidar signal, and thus can choose better parameters.

References

1. Sequeira, V., Ng, K., Wolfart, E., Goncalves, J.G.M., Hogg, D.: Automated Reconstruction of 3D Models from Real Environments. ISPRS Journal of Photogrammetry and Remote Sensing (Elsevier) 54, 1–22 (1999)
2. Taubin, G.: Curve and surface smoothingwithout shrinkage. Technical Report RC-19536, IBM Research (1994)
3. Hu, C.-S.: Investigation into the High-speed Pulsed Laser Diode 3D-imaging Ladar. Ph.D. thesis, National University of Defense Technology (2005)
4. Taubin, G.: A signal processing approach to fair surface design. In: ACM SIGGRAPH Conference Proceedings, pp. 351–358 (1995)
5. Stamos, I., Allen, P.K.: 3-D Model Construction Using Range and Image Data. In: IEEE International Conference on Computer Vision and Pattern Recognition, vol. I, pp. 531–536 (2000)
6. Zhu, X.-X.: The Modeling Technics of Free Curve and Surface. Science Press, Beijing (2000)

The Organization and Design of A Self-Repairing SoC

Wei Zhang and Limin Liu

Institute of Embedded Systems IT School, Huzhou University
Huzhou, Zhejiang, 313000, China
zhangwei@hutc.zj.cn, liulimin@ieee.org

Abstract. SoC, System on a Chip, is the advanced hardware of embedded systems. A self-repairing SoC, SRSoC is ideal solution for some embedded applications. In this paper, the organization and design of a SRSoC are discussed. The organization is based on multiple MCU cores. The design is a cooperation implement of hardware and software.

Keywords: SoC, embedded systems, self-repairing design, hardware and software.

1 Introduction

Embedded systems are the most popular solutions of IT application systems. Actually an embedded solution is to be designed with cooperation of hardware and software. For the hardware, SoC, System on a Chip, is ideal. It is to integrate some application system on one semiconductor chip [1]. When a system is integrated on a chip, but not one or more PCBs, the system should be with smaller size, higher reliability, faster operation and lower cost for some applications. Therefore, SoC, is the evolution and advanced organization of embedded systems. In fact, hardware to replace some software is a useful feature of SoC [2]. For some specific applications, the replacement would obtain better operating efficiency.

There are two main fields, design and organization, of SoC research. The design is normally concerned to IC design and microelectronics. The organization research may be to build SoC in various architectures. For the organization, some functions or parameters of SoC can be modified. The SoC is called reconfigurable SoC, RSoC [3-4]. When the SoC can update or modify some functions itself, the SoC will be a self-repairing SoC, or SRSoC.

2 Organization of A SRSoC

As above description, SRSoC is a kind of SoC. It can update or modify its functions or organization. If the repairing process is “dynamically”, the SRSoC is more satisfied. Here, the “dynamically” means the updating with no stopping, no resetting and bumpless transferring for the operation.

There are different organizations of SRSoC for various applications. The SoC with delta MPU cores is a kind of SRSoC. An organization of SRSoC is indicated in Fig.1. MPU0 is a monitor core; MPU1 and MPU2 are OP, operation, and BK, backup, cores.

D. Jin and S. Lin (Eds.): Advances in CSIE, Vol. 1, AISC 168, pp. 709–712.
springerlink.com © Springer-Verlag Berlin Heidelberg 2012

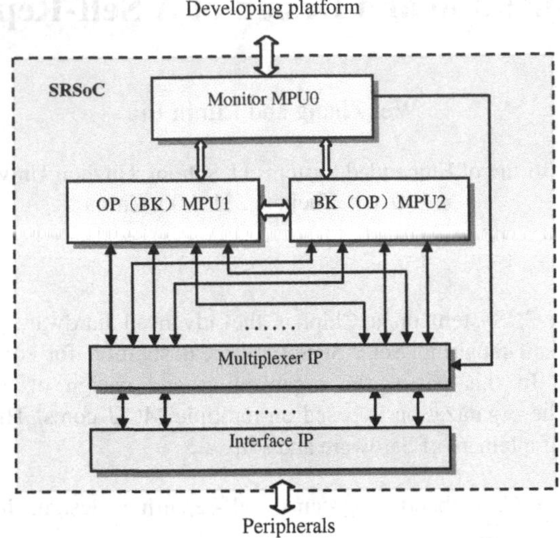

Fig. 1. An indication of SRSoC with multi MPU cores

From Fig.1, a SRSoC with multi-MPU cores or delta MPU cores. The updating ability of the SRSoC is more powerful. It can do the highest level updating for a SoC. In this updating, except outline, all contents of SoC, such as MPU, interfaces and instructions, can be modified.

In a normal operation, just one MPU may be in work. When users ask to update or modify the SRSoC functions, MPU0 can instruct backup core to be as the updating core, and reconfigure or reset up the MPU core. Then the new core replaces the operation core in an application.

The monitor can manage and control the updating. It also guarantees to make a smooth transferring or bumpless switching.

The multiplexer IP connect to output lines both of MPU1 and MPU2, and choose one of them as system output. In fact, it is a switch circuit for output. The multiplexer is controlled by monitor MPU0.

The output of every unit of multiplexer is managed by monitor core. When the control signal from the monitor is 1, for instance, the MPU1's output is as output of the multiplexer. In contrast, the signal from monitor is 0, then, the output of multiplexer is to connect MPU2's output.

Since there is not any difference for application object from the ex-behavior, such as signals on the output and parameters for management, of the SRSoC, the updating and transferring are smooth and satisfied.

Therefore, the SoC with delta MPU cores is a dynamic self-repairing SoC, DSRSoC, an advanced SRSoC.

3 Cooperative Design of the SRSoC

Functions of SoC can be designed in hardware or software. The advantage of hardware for an application is in higher speed operation, higher reliability, smaller memory size, less programming hours and better security. The disadvantage may be high cost and less flexible for some utilities. For a normal case, the advantage of software includes more flexible and lower cost, and main disadvantage is less efficient operation and more memory size [5].

A SoC application is a special design for some object. Its requirement about flexibility is not so high. Since it is built on a PLD chip, its hardware cost is more fixable in some level. Therefore, the disadvantage of hardware is not very obvious for a SoC. For most SoC applications, more hardware functions mean more operation efficiency, and the cost is similar, the flexibility is acceptable. If possible, it is better for SoC hardware to replace software in most functions. However, for a SoC system, the application software is essential yet. Some hardware of PLD can be instead of low level software, but not all. Therefore, cooperative design of HW and SW is a better choice for most SoCs.

Since SRSoC is a new concept, and its design and application are modified frequently. Its design is normally based on some PLD devices, such as FPGA, with VHDL language [6-7]. Actually, a SRSoC solution is complex. Therefore, the SRSoC modeling should be behavior modeling or hybrid modeling, but not structural modeling in VHDL development. It is because that the structural modeling is depended on lower-level components. If without suitable components, the result of structural modeling is not satisfied.

When a SoC is built by FPGA, the updating or reconfigurable procedures are easier. It is because of FPGA is a kind of PLD device to be easily modified. New products of FPGA can be made with about 180,000 LE (Logic Element) on a chip. That means more than one hundred MPU core, each to take 400-1700 LE, may embed into one FPGA chip. And, to build multi MPU SoC is possible. On the other hand, new FPGA products supply enough resource for SRSoC yet.

The frame of a SRSoC based on VHDL is as follows.

```
Library
Entity SRSoC_swt is
Port( );
End SRSoC_swt;
Architecture struct of SRSoC_swt t is
         Signal
         Constant
Begin
Res: process;
Clk :clk_div;
SWITCH: process;
  Begin
    If init then
      Initialization;
    Elsif sg then
      CheckSignal;
```

```
      Elsif mtst then
         Match State;
      Elsif switch then
        Switch MPU;
    End if;
  End process;
  Us: ustimer;
  Ms: mstimer;
  End struct;
```

There are some analysis and decision algorithms in memory of a SoC. They are software to be run in monitor MPU. According to the algorithms, monitor MPU determines reconfiguration of backup MPU and switches it into operation.

Therefore, a solution of the SRSoC with delta multi-core must be a cooperation of software and hardware designs.

4 Conclusions

A SRSoC is composed of three MPU, monitor, operating and backup MPU. When the SoC is required updated or repaired, the monitor controls backup MPU to be modified. The design is based on FPGA with VHDL. It is a typical co-design of hardware and software. For most cases of SRSoC applications, cooperative design of hardware and software is the more efficient way.

Acknowledgments. This research was supported in part by the National Natural Science Foundation of China under grant 60872057, by Zhejiang Provincial Natural Science Foundation of China under grants R1090244, Y1101237, Y1110944 and Y1100095. We are grateful to NSFC, ZJNSF and Huzhou University.

References

1. Saleh, R., Wilton, S., et al.: System-on-chip: reuse and integration. Proceedings of the IEEE 94, 1050–1069 (2006)
2. Sifakis, J.: Embedded systems design - Scientific challenges and work directions. In: Proceedings of DATE 2009, p. 2. IEEE Press, Nice (2009)
3. Liu, L., Luo, X.: The Reconfigurable IP Modules and Design. In: Proc. of EMEIT 2011, Harbin, China, pp. 1324–1327 (August 2011)
4. Ostua, E., Viejo, J., et al.: Digital Data Processing Peripheral Design for an Embedded Application based on the Microblaze Soft Core. In: Proc. 4th South. Conf. Programmable Logic, San Carlos de Bariloche, Argentina, pp. 197–200 (March 2008)
5. Liu, L.: A Hardware and Software Cooperative Design of SoC IP. In: Proc. of CCIE 2010, Wuhan, China, pp. 77–80 (June 2010)
6. Kilts, S.: Advanced FPGA Design: Architecture, Implementation, and Optimization. Wiley, New Jersey (2007)
7. Dimond, R.G., Mencer, O., Luk, W.: Combining Instruction Coding and Scheduling to Optimize Energy in System-on-FPGA. In: Proceedings of 14th Annual IEEE Symposium on Field-Programmable Custom Computing Machines, Napa, CA, USA, pp. 175–184 (April 2006)

An Immune Control for Multivariable Industrial Systems

Wei Zhang and Limin Liu

Institute of Embedded Systems IT School, Huzhou University
Huzhou, Zhejiang, 313000, China
zhangwei@hutc.zj.cn, liulimin@ieee.org

Abstract. The traditional control for some multivariable industrial processes is not available. An intelligent control is a better for those cases. An immune control, similar to expert system and fuzzy control, is discussed in this paper. For some complex industrial systems, the technology, to study operations of human experts, based on artificial intelligent and fuzzy system is the more useful and efficient.

Keywords: MIMO systems, intelligent control, fuzzy systems, control simplication.

1 Introduction

Some complex multivariable system is hardly to be controlled in traditional method. An intelligent control is the batter solution for most industrial multivariable process[1-2]. Fuzzy control, neural network and knowledge-based system are three main branches of intelligent control system[3-4]. In resent years, more application systems are developed in different fields, such as industrial processes, electromechanical systems, instruments, automobiles, communications, internet and neural network techniques[5-7].

The immune control is an efficient solution with intelligent regulation. It can study and take the actions of some good operators. Actually it is an expert control system. It is available for multivariable process control.

A typical multivariable industrial process is pulverizing system. It makes raw coal into powder that has dryness and fineness requested and is blown into boilers for firing.

There three acting variables in this process, FC: amount of feed coal, HA: hot air, RA: recycle air; and three controlled variables, P: entry subatmospheric pressure, ΔP: pressure difference (for load), T: outlet temperature.

The procedure of pulverizing coal is that raw coal from bucket through feeder is mixed with hot and recycle air, then enters ball mill to be grounded, is made powder to output into separator. In separator, the certified powder is conveyed to the bank, others return the mill. If the ball mill run in the rated output, the operation is most efficient and obtains better benefit.

D. Jin and S. Lin (Eds.): Advances in CSIE, Vol. 1, AISC 168, pp. 713–716.
springerlink.com © Springer-Verlag Berlin Heidelberg 2012

2 Traditional Control Design

The traditional control systems for ball mill are designed as three single loop systems which are:

a. Pressure Difference
   to be controlled Feed Coal;
b. Temperature
   to be controlled Hot Air;
c. Subatmospheric Pressure
   to be controlled Recycle Air.

With closed coupling among the three loops, the scheme is not successful. The automatic control systems cannot be used.

A decoupling method with diagonal matrix is feasible in theory. The transfer functions are given as follows.

$G_{11}(s)$=T(s)/$m_1(s)$ =0.94/(80s+1)^3
$G_{21}(s)$=P(s)/$m_1(s)$ =1.6/(8s+1)
$G_{31}(s)$=ΔP(s)/$m_1(s)$ = -0.02
$G_{12}(s)$=T(s)/$m_2(s)$ =0.17/[(60s+1)^3(45s+1)]
$G_{22}(s)$=P(s)/$m_2(s)$ =0.54/(11s+1)
$G_{32}(s)$=ΔP(s)/$m_2(s)$ =0.44/[(11s+1)(8s+1]
$G_{13}(s)$=T(s)/$m_3(s)$ =1/770s(80s+1)
$G_{23}(s)$=P(s)/$m_3(s)$ =0.1e^{-240s}/(250s+1)
$G_{33}(s)$=ΔP(s)/$m_3(s)$ =1/[1425s(80s+1)]

where T(s) represents Temperature, P(s) for Subatmospheric Pressure and ΔP(s) for Pressure Difference; $m_1(s)$, $m_2(s)$ and $m_3(s)$ separately stand for Hot Air, Recycle Air and Feed Coal.

When $$\mathbf{C(s)}=\begin{bmatrix} T(s) \\ P(s) \\ \Delta P(s) \end{bmatrix} \quad \mathbf{M(s)}=\begin{bmatrix} m_1(s) \\ m_2(s) \\ m_3(s) \end{bmatrix}$$

$$\mathbf{G(s)}=\begin{bmatrix} G_{11}(s) & G_{12}(s) & G_{31}(s) \\ G_{21}(s) & G_{22}(s) & G_{32}(s) \\ G_{31}(s) & G_{32}(s) & G_{33}(s) \end{bmatrix}$$

Then **C(s)=G(s)M(s)**

If to get $$\mathbf{G^*(s)}=\begin{bmatrix} G_{11}(s) & 0 & 0 \\ 0 & G_{22}(s) & 0 \\ 0 & 0 & G_{33}(s) \end{bmatrix}$$

And **G(s)D(s)= G*(s)**

We can obtain the decoupling matrix **D(s)** from

$$\mathbf{D(s)= G^{-1}(s)\, G^{*}(s)}$$

where $\mathbf{G^{-1}(s)}$ is an inverse matrix on **G(s)**.

To solve a ninth order matrix is not easy. Even if the solution can be found in theory, it is not available for the practical applications. At least two problems have to make the control through the mathematical methods hardly. First, a lot of steel balls filled in the pulverizing equipment will affect the dynamic characteristic of the system when the numbers of balls and wear-and-tear are varied. On the other hand, the humidity and quality of raw coal also influence the process control.

3 Intelligent Immune Control Solution

In fact, operators do not make so complicated calculation and still manage the system better. This shows that the system can be controlled by non-mathematics way. Based on an imitated control, intelligent fuzzy controller for the ball mill can be designed. The core of control algorithm is fuzzy rule sets.

In this case, deviation of variables is quantized into nine fuzzy levels as:

PV	Positive Very large
PL	Positive Large
PM	Positive Medial
PS	Positive Small
ZO	ZerO
NS	Negative Small
NM	Negative Medial
NL	Negative Large
NV	Negative Very large

The rules suited to F rule inference can be set. They are expressed as follows.

Rule 1	IF	Pressure Difference is PL
	THEN	Feed Coal adjusts in NL
	AND	Hot Air adjusts in NS
	AND	Recycle Air adjusts in NS
Rule 2	IF	Pressure Difference is NL
	THEN	Feed Coal adjusts in PL
	AND	Hot Air adjusts in PS
	AND	Recycle Air adjusts in PS
Rule 3	IF	Pressure Difference is PS
	THEN	Feed Coal adjusts in NS
Rule 4	IF	Pressure Difference is NS
	THEN	Feed Coal adjusts in PS

…

Although there are a lot of rules from excellent operators, not all of them are available for an application. Since sampling period of controller is not longer than 50 ms, it is almost impossible that the controlled variables of the system occur PL or NL in the twinkling. Hence, rules dealt with NS or PS, such as Rule 3 and Rule 4, are generally inquired by the inferring engine within the controller.

In practical cases, Hot Air and Recycle Air are not often adjusted. The main controlled variable is Pressure Difference. Correspondingly, acting variable Feed Coal is an important role in the control system. According to the experience of the operators, the control of Feed Coal depends on not only Pressure Difference but also the status of the coal and mill. Congestion in the entry of a ball mill is a frequent trouble in operation. It can be described by a rule as:

IF Pressure Difference is PL or PVL
AND Temperature is PL or NL
AND Subatmospheric Pressure is NL
THEN congestion warning .
Other rules for warning or advising have to be designed.

The implementation of the controller is not difficult. Its simulation is built. The result of the simulation shows that the control scheme is feasible.

4 Conclusions

The intelligent control is suitable for some practical industrial applications. The immune control adapts expert system and fuzzy control technology. Its solution is available for most multivariable industrial systems.

Acknowledgments. This research was supported in part by the National Natural Science Foundation of China under grant 60872057, by Zhejiang Provincial Natural Science Foundation of China under grants R1090244, Y1101237, Y1110944 and Y1100095. We are grateful to NSFC, ZJNSF and Huzhou University.

References

1. Haber, R.E., Alique, J.R.: Fuzzy Logic-Based Torque Control System for Milling Process Optimization. IEEE Transactions on Systems, Man, and Cybernetics 37, 941–950 (2007)
2. Sofianos, N.A., Kosmidou, O.I.: Guaranteed cost LMI-based fuzzy controller design for discrete-time nonlinear systems with polytopic uncertainties. In: Proc. of MED 2010, Marrakech, Morocco, pp. 1383–1388 (2010)
3. Chang, Y.: Intelligent Robust Tracking Control for a Class of Uncertain Strict-Feedback Nonlinear Systems. IEEE Transactions on Systems, Man, and Cybernetics 39, 142–155 (2009)
4. Ling, P., Wang, C., Lee, T.: Time-Optimal Control of T-S Fuzzy Models via Lie Algebra. IEEE Trans. on Fuzzy Systems 17, 737–749 (2009)
5. Szabat, K.: Robust control of electrical drives using adaptive control structures — a comparison. In: Proc. of ICIT 2008, Chendu, China, pp. 1–6 (April 2008)
6. Liu, L.: A Control Based on Rule Updating for Non-Linear Systems. In: Proc. of 2011 CCDC, Mianyang, China, pp. 3094–3097 (May 2011)
7. Hua, C., Wang, Q., Guan, X.: Robust Adaptive Controller Design for Nonlinear Time-Delay Systems via T–S Fuzzy Approach. IEEE Trans. on Fuzzy Systems 17, 901–910 (2009)

Design and Development of Sports E-commerce Based on the Sports Industry Information

Yin Ji

Shandong Jiaotong University, Jinan Shandong, China

Abstract. It is crucial to carry out sports industry technology with the development of economy globalization. E-commerce is important way to materialize the sports industry information. This paper intends to construct the system of sports e-commerce from the prospect of goal of design, the function as well as data design and technology. It also discusses the importance of e-commerce building and suggests some strategies related.

Keywords: sports industry, informationalization, sports e-commerce.

1 Introduction

1.1 The Sports Industry Informatization

Sports industry refers to a gathering of sports activities or branches that support the society with mutual competitive sports products or services, while informationalization stands for a historical process where we can make the best of information and technology, facilitate information communication and knowledge sharing. Sports industry informationalization is to increase the quality of industry economy through the use of information technology.

1.2 Brief Introduction on Sports E-commerce System

e-commerce means to realize the commercial exchange and administration management with the help of computer and internet. In a narrow sense, it is some business or on-line business which has been operated through the net and technology including varieties of services for government, enterprises, finance organization and so on.

Sports electronic business refers to the physical production and a series of electronic trading activities such as : production management, sport products marketing network, electronic payment, logistics management and customer management. The platform is based on information technology and Internet system as the basis, completion of the sports products from the production, supply, sell a series of management process. Electronic commerce is in the development of sports industry advanced components.

D. Jin and S. Lin (Eds.): Advances in CSIE, Vol. 1, AISC 168, pp. 717–722.
springerlink.com © Springer-Verlag Berlin Heidelberg 2012

2 The Sports Electronic Business System Analysis

2.1 Objectives and Tasks

The system goal is clear. Sports product electronic commerce technology platform research and application development of sports products, accelerate the information circulation, promote the sports product commercialization, promote the development of sports information. The sports to the internationalization of. The market development direction. The task of the system are the following: (1) physical information sharing, accelerate the sports information circulation, maintain physical stability. (2) increase the sports product sales channels, change traditional sale pattern, increase business income. Let both sides of supply and demand the maximum possible direct transactions, reduce transaction costs, achieve a win-win situation of production and consumption. (3) in favor of setting up sports corporate image, increase market competitiveness of sports products.

2.2 System Module

The system of the service object is a sports enthusiast. Fitness, sports enterprise, national sports management department. According to the system construction goal, tasks and service object, the system is divided into a development module. (1) Sports products supply and demand information management system. (2) The sports products market and price management information system. (3) Sports products online trading system. (4) Sports products online auction system. (5) Sports products e-commerce platform management system.

2.3 Architecture

Software size determines the size of the software itself structure. The prevalence of three layer architecture. MVC framework, namely : the presentation layer, business logic layer, control layer. The presentation layer by JSP is to achieve, by SERVLET to realize control layer, business logic layer to realize by JAVABEAN.

3 The Analysis on Sports E-business System Function

The system is mainly to serve with the demand and supply of information and inquiries, price, trade show, online auction of four major functions. At the same time provide sports products e-commerce platform management capabilities. Details are as follows: management system, the main function of the completion of the is, rights management, user management, market order management and administrator management, member of statistical information, delete invalid information, decision support information, visit volume statistics work. Online transaction function, this part is the key part, relates to the system function and sports products division, modular design. Total design product display order, product management, order management, information statistics and four functional modules. The market and price information management system function, market and price information management system display market pricing, real-time quotes and information management three modules. Market quotation function module is divided into: the market price query, query, price of variety of market situation reviews features such as information management function module is divided into: market management, category management, market quotation management. Instant quote management four functions. These price information for the transaction of reliability. Online auction system function. The Department to teach members to extract land, senior, gold member can auction items. Ordinary members can only shoot buy products. The system includes the product to add, delete, modify, search. Given the price of the product and product introduction. The product is an attempt to auction system function, increase trade interest. Supply and demand information management system function, supply and demand information management system is the electronic commerce system is the most important function of. Having the function of intuitive, easy to use, high rate of use of the advantages of. Today, the sports are widely used and the most typical supply and demand information release platform is the sports department information center developed the "one-stop" supply and demand information release system. With the information, publishing, management and other functions, these functions are not technically difficult to achieve.

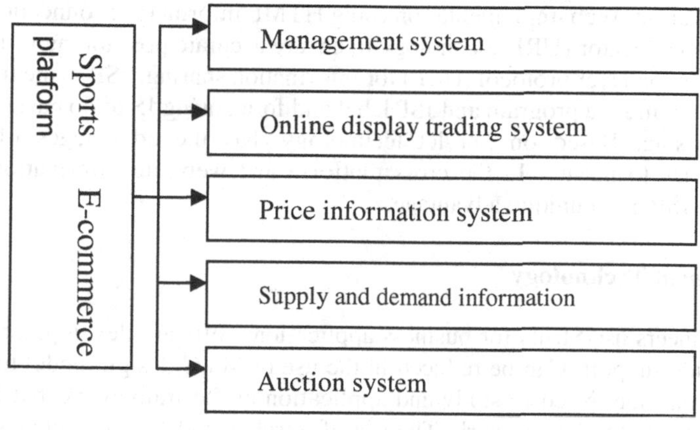

Fig. 1. System Module

4 Sports E-business System Design Data

The domain model driven approach and the top-down approach to database design, first analyzes the system business, according to the definition of object functions. Object to meet the packaging characteristics, ensure and responsibilities of related data items are defined in a data item within the object, can describe the responsibilities. No description of deficiency. And an object and only a duty, an object to be responsible for two or more than two duties, should be split.

For all tables with primary and foreign keys build an index, to build up the combination property index to improve the search efficiency. Based on the domain model for mapping database tables. You should reference database design second paradigm: a table all non-key attributes are dependent on the entire keyword. Keywords can be an attribute, can also be a plurality of a set of properties, it shall ensure the key can ensure that only sexual. The domain model in every object has a duty, the data in the object not exist transitive dependence. This kind of train of thought of the database table structure design from the beginning that meet the third paradigm: a table should meet the second paradigm. And the attribute does not exist between transitive dependence. Design of the table to good use, mainly reflected in the query is associated more than one table and also the need to use complex techniques of SQL. Design of the table as possible to reduce the data redundancy, to ensure the accuracy of the number, effective control redundancy helps to improve database performance. According to the above requirements, the system uses the ORACLE database.

5 Sports E-business System Core Technology

5.1 WEB Technology

Web to provide people with information and information service, resource sharing and information sharing. Web implementation using HTML information connection; with a uniform resource locator (URI) technology to ensure accurate positioning of the cyber source. Application layer protocol (HrrP) for information sharing. JSP in the traditional HTML page to join Java program and JSP label and form, using JS to do basic webpage logical processing. Based on Servlet technology, has the advantages of Servlet technology, easy to master. In the cross platform and web site information system development, JSP has a unique advantage.

5.2 Component Technology

Software engineers use Struts for business applications software development of each layer to provide support. Can be reduced in the use of MVC design model to develop web application time. Need to study and application of the framework, but it will be completed some of the heavy work. The use of Servlets and JSP advantages to build scalable applications, Struts is a good choice. The system uses struts to build the business layer.

Hibernate is open-source object-relational mapping framework, it undertook to JDBC lightweight package, Java programmers to use objects to manipulate database programming thinking, realize the database object oriented programming. Hibernate can be applied in the use of JDBC occasions, also can be in Servlet / JSP Web applications to use, it provides a data processing model.

Spring is committed to J2EE of the coating solution. Not only focus on a single layer scheme. Is a set of technical system. It can be said that the Spring is enterprise application development. A station type" option, well throughout the presentation layer, business layer and the persistence layer. The technique is to some extent a platform technology, is the core of reverse control and dependency injection. Spring technology for the optimization of system development has important applications in.

Flex technology has opened a new era of webpage graphic design, graphic is always the most intuitive display mode, the multimedia technology has the best on the function and infection, web page technique to the direction of development. Flex introduced object oriented programming script ActionScript3.0, and establish a similar to the Java Swing class library and phase, component (component), which can produce dramatic impact page display effect, in this system can make the rich application.

Ajax technology asynchronous corresponding alleviated on certain level the development of network bandwidth. Based on the application of data, the needs of the user data table. Can be independent of the actual webpage server to obtain and can be dynamically written webpage, to slow Web application experience of desktop applications like colouring. This will increase the WEB response speed, more rapid and user interaction.

5.3 MVC Technology

MVC mode is designed to realize the dynamic programming, the program can be revised and extended to simplify procedures, a portion of the repeating utilization become possible. Software system based on the basic part of their separation and also gives the basic part of its function. Professionals can use its expertise in packet using the JSP processing page display.

6 Sports E-business System Implementation

6.1 Hardware Implementation

Hardware platform using the LINUX operating system. The software uses the TOMCAT WEB server, database server using SUN v890 as hardware platform, operating system using SOLARIS, database software using ORACLE. Stability of the hardware system, efficient database software, large capacity database space, advanced development platform, to build a fully functional sharing platform.

6.2 Plan Implementation

According to the investigation and analysis of various types of information resources, according to the content and structure of different, take different resource integration scheme. And then design the different tables and database, are treated separately. In the project implementation should pay attention to the following points: the first is the bottom-up unidirectional information resource sharing of T second is down from one-way information resources sharing. The third category is the system each module resource problem independently, using a separate database.

6.3 Systems Integration Implementation

System integration technology to realize the system integration services and the related technology, with emphasis on the technology about system technology. System integration technology is to realize the information collection, transmission, exchange, storage, processing and use of the integration as the goal, the corresponding development of the (physical)" integrated" system interconnect technology, to achieve" integration" of the software integration and data integration technology. The former settlement information transmission and exchange. The latter resolved information collection, storage, processing and utilization.

References

1. Jiang, X.P.: Electronic Commerce and Network Marketing. Tsinghua University press, Beijing (1998)
2. Zhang, Z.: E-commerce on Improving International Competitiveness of Sports Goods Industry. Market Modernization 8, 139–140 (2007)
3. Dong, L.: China Sports E-business Present Situation Investigation and Countermeasure Research. Journal of Shandong Sports Institute 1, 32–34 (2005)
4. Lu, J.: The network economy and the construction of club culture. Journal of Nanjing Sport Institute 1, 6–9 (2005)
5. Li, X.: Informatization and Economic development. China Development Press, Beijing (2000)

Author Index